# 14th International Conference on the Physics of Highly Charged Ions 2008



# (HCI 2008)

Journal of Physics: Conference Series Volume 163

## Tokyo, Japan
## 1-5 September 2008

**Editors:**

Toshiyuki Azuma
Nobuyuki Nakamura
Chikashi Yamada

ISBN: 978-1-61738-371-7

**Printed from e-media with permission by:**

Curran Associates, Inc.
57 Morehouse Lane
Red Hook, NY 12571
www.proceedings.com

**Some format issues inherent in the e-media version may also appear in this print version.**



Printed by Curran Associates, Inc. (2010)



Institute of Physics
Dirac House, Temple Back
Bristol BS1 6BE UK

Phone: 44 1 17 929 7481
Fax: 44 1 17 920 0979

techtracking@iop.org

# TABLE OF CONTENTS

## REVIEWS AND PROGRESS REPORTS



## FUNDAMENTAL ASPECTS, STRUCTURE AND SPECTROSCOPY

## COLLISIONS WITH ELECTRONS, IONS, ATOMS AND MOLECULES

## INTERACTIONS WITH CLUSTERS, SURFACES AND SOLIDS



## INTERACTIONS WITH PHOTONS, PLASMAS AND STRONG FIELD PROCESSES

## PRODUCTION, EXPERIMENTAL DEVELOPMENTS AND APPLICATIONS

# PREFACE

This volume contains the Proceedings of the 14th International Conference on the Physics of Highly Charged Ions (HCI2008), held at the University of Electro-Communications, Chofu, Tokyo, Japan from 1–5 September 2008. This series of conferences began in Stockholm, Sweden in 1982 and has since been held every other year; in Oxford, UK (1984), Groningen, the Netherlands (1986), Grenoble, France (1988), Giessen, Germany (1990), Manhattan, Kansas, USA (1992), Vienna, Austria (1994), Omiya, Japan (1996), Bensheim, Germany (1998), Berkeley, USA (2000), Caen, France (2002), Vilnius, Lithuania (2004) and Belfast, UK (2006).

Highly charged ions (HCI), which are defined as highly ionized (i.e. positively charged atomic) ions here, mainly exist in hot plasmas such as the solar corona and fusion plasmas. It is true that its importance in plasma physics has driven researchers to the spectroscopic studies of HCIs, but the spectroscopy of few-electron ions is not only important for plasmas but also interesting for fundamental atomic physics. Electrons moving fast near a heavy nucleus give a suitable system to test the fundamental atomic theory involving relativistic and quantum electro-dynamic effects in a strong field. Also, the huge potential energy of a HCI induces drastic reaction in the interaction with matter. This unique property of HCIs, coupled with the recent development of efficient ion sources, is opening the possibility to utilize them in new technologies in the field such as nano-fabrication, surface analysis, medical physics, and so on. Hence, this conference is recognized as a valuable gathering place for established practitioners and also for newcomers; we exchange information, we are introduced to the subject itself, and to unexpected interfaces with other fields.

On 31 August, the day before the opening of HCI2008, we welcomed the delegates at the university's restaurant—and we were greeted with an unusually heavy summer shower! The conference then opened on the morning of 1 September. In the evening we invited the foreign delegates to the Tokyo-EBIT laboratory and entertained them with sushi and YEBISU beer at the lab (YEBISU is a popular brand of beer in Japan and also the nickname of the Tokyo-EBIT project, Young Electron Beam Ion Source Unit). Cans of YEBISU beer also helped us to make the poster sessions lively, on the evenings of 2 and 4 September. The best poster of each session was selected by the vote of International Advisory Board, and two young scientists were awarded the prizes. On the afternoon of 3 September, we took an excursion to Lake Yamanaka near Mount Fuji for a conference outing. The conference dinner was then held at a hotel alongside Lake Yamanaka from where we really enjoyed the view of Mount Fuji. In the summer it is quite rare to have clear weather, but a *teru-teru bozu* (a Japanese paper doll with a wish for good weather) placed on the top page of the conference WEB site brought us clear skies for the dinner.

About 200 scientists from 22 countries registered for the HCI2008. The number of invited talks was 18 and the contributed papers 206. These proceedings contain 9 invited talks and 107 contributed articles. Following the previous conference, they are grouped into the five categories of: (1) Fundamental aspects, structure and spectroscopy; (2) Collisions with electrons, ions, atoms and molecules; (3) Interactions with clusters, surfaces and solids; (4) Interactions with photons, plasmas and strong field processes, and (5) Production, experimental developments and applications.

We are happy that the conference was completely successful, and this success owes much to the sponsors. We acknowledge the support from the Japan Society for the Promotion of Science (JSPS), The International Union of Pure and Applied Physics (IUPAP), Matsuo Foundation, Iwatani Naoji Foundation, CASIO Science Promotion Foundation, and The Society for Atomic Collision Research.

The next International Conference on the Physics of Highly Charged Ions will be held in Shanghai under the organization of Professor Y Zou of Fudan University and her colleagues. We wish the organizers of HCI2010 all possible success and are looking forward to meeting you all in Shanghai.

**Toshiyuki Azuma, Nobuyuki Nakamura and Chikashi Yamada**
Editors

14th International Conference on the Physics of Highly Charged Ions (HCI 2008) IOP Publishing
Journal of Physics: Conference Series **163** (2009) 012001 doi:10.1088/1742-6596/163/1/012001

# Astrophysical Relevance of Storage-Ring Electron-Ion Recombination Experiments

**Stefan Schippers**
Institut für Atom- und Molekülphysik, Justus-Liebig-Universität, 35392 Giessen, Germany
E-mail: `Stefan.Schippers@iamp.physik.uni-giessen.de`

**Abstract.** The relevance of storage-ring electron-ion recombination experiments for astrophysics is outlined. In particular, the role of low-energy dielectronic-recombination resonances is discussed. A bibliographic compilation of electron-ion recombination measurements with cosmically abundant ions is provided.

## 1. Introduction

Heavy-ion storage rings equipped with electron coolers serve as an excellent experimental environment for electron-ion collision studies. Some recent studies of dielectronic recombination (DR) focussed on high-resolution spectroscopy of highly-charged ions. Highlights of this research are the measurement of the hyperfine induced decay rate of the $1s^2\,2s\,2p\ ^3P_0$ state in berylliumlike $Ti^{18+}$ [1] utilizing DR at the storage ring TSR of the Heidelberg Max-Planck-Institute for Nuclear Physics, the observation of the isotope shift in DR of three-electron $Nd^{57+}$ [2] using different isotopes of this ion at GSI's storage ring ESR and the observation of the hyperfine splitting of $Sc^{18+}$ low-energy DR resonances at the TSR high-resolution electron target [3]. The latter experiment resulted in the determination of the $Sc^{18+}(2s_{1/2} - 2p_{3/2})$ energy splitting with an uncertainty of only 4.6 ppm which is less than 1% of the few-body effects on radiative corrections [4]. Since these exciting developments have already been reviewed recently [5], the present review focusses on the relevance of storage-ring electron-ion experiments for astrophysics.

Storage-ring experiments provide particularly valuable information on DR in low-temperature plasmas such as photoionized plasmas that occur, e.g., in active galactic nuclei (AGN) in the vicinity of super-massive black holes [6]. In such plasmas highly charged ions exist at relatively low temperatures. For many ions, the DR rate coefficients, that determine the charge balance in these plasmas, depend sensitively on the low-energy DR resonance structure at relative electron-ion energies below $\lesssim 3$ eV. In the following, the influence of low-energy DR resonances on electron-ion recombination rate coefficients in low-density plasmas and recent efforts of building a recombination data-base for astrophysical modeling of photoionized plasmas are briefly discussed. Finally, a compilation of experimental results for the astrophysically most abundant ions is presented.

## 2. Low-energy dielectronic recombination resonances

Figure 1a shows the measured $Mg^{8+}$ recombination rate coefficient [7] at electron-ion collision energies below 200 meV. The strongest dielectronic recombination (DR) resonance occurs at 21 meV. The inset shows that the rate coefficient in plasma changes by more than an order of magnitude when the low-energy resonances are shifted by as little as $\pm 100$ meV. Usually, for ions with more than three electrons, the theoretical uncertainties for DR resonance energies are much larger. Correspondingly, the uncertainties of calculated recombination rate coefficients in a plasma can be rather large, especially at low temperatures where photoionized plasmas exist. As an example figure 1b compares the experimentally derived $Mg^{8+}$ rate coefficient in a plasma with two recent state-of-the-art calculations. The deviations of the theoretical from the experimental results are as large as a factor of 2 and, additionally, there is considerable discrepancy between both theoretical curves.

This situation can be regarded as typical. Even state-of-the-art atomic codes are not always capable of providing reliable low-temperature DR plasma rate coefficients for ions. The difficulty in calculating sufficiently precise DR resonance energies is rooted in the many-body nature of the problem. Only in special cases, where one electron is outside a closed shell, relativistic many-body perturbation theory (RMBPT) yields reliable low-energy DR resonance parameters as has been demonstrated, e.g., for Li-like fluorine [8], sodium [9], and scandium [3; 4; 10] as well as for Na-like silicon [11]. However, RMBPT cannot easily be extended to more complex ionic systems and, therefore, cannot satisfy the vast astrophysical atomic data needs.

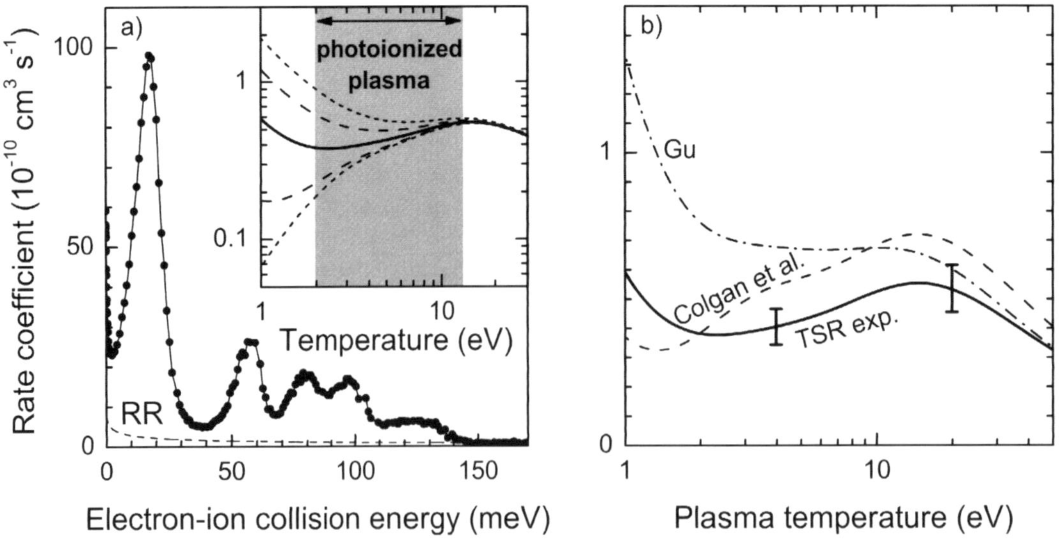


**Figure 1.** a) Experimental merged-beams rate coefficients for the recombination of electrons with Be-like $Mg^{8+}$ ions at low electron-ion collision energies measured at the Heidelberg heavy-ion storage ring TSR [7]. The theoretically calculated contribution by non-resonant radiative recombination (RR) is shown as a dashed line. The inset shows the effect of hypothetical resonance shifts by $\pm 50$ meV (dashed curves) and by $\pm 100$ meV (dotted curves) on the $Mg^{8+}$ recombination-rate coefficient in a plasma. The temperature range where $Mg^{8+}$ occurs in a photoionized plasma is highlighted. b) The experimental $Mg^{8+}$ recombination-rate coefficient in a plasma [7] at low temperatures compared with recent state-of-the-art theoretical results by Colgan et al. [12] (dashed curve) and Gu [13] (dash-dotted curve).

## 3. Dielectronic recombination data for photoionized plasmas

Cosmic atomic plasmas can be divided into broad classes, collisionally ionized plasmas (CP) and photoionized plasmas (PP) [6]. Historically, most theoretical recombination data were calculated for CP where highly charged ions exist only at rather large temperatures, e.g., in the solar corona. At these temperatures, recombination rate coefficients are largely insensitive to low-energy DR resonances. Consequently, the theoretical uncertainties are much smaller at higher than at lower plasma temperatures which are typical for PP. Until recently, theoretical recombination rate coefficients were mainly calculated for the CP temperature ranges (see e.g. [14]). If these data are used for the astrophysical modeling of PP, inconsistencies arise. This has been noted, e.g., in the astrophysical modeling of x-ray spectra from AGN by Netzer [15] and Kraemer et al. [16]. On the basis of astrophysical modeling these authors have suspected that the DR rate coefficients for iron ions with an open M-shell ($Fe^{1+}$–$Fe^{15+}$) from the widely used compilation of Arnaud & Raymond [17] are much too low in the PP temperature range.

These findings motivated storage-ring recombination measurements with iron M-shell ions. Figure 2 shows results, that were obtained for Al-like $Fe^{13+}$ ions. The low-energy merged-beams recombination rate coefficient (figure 2a) is dominated by strong DR resonances that decisively determine the low-temperature recombination rate coefficient in a PP (dashed line in figure 2b). As suspected, the experimentally derived PP rate coefficient is considerably higher — by up to orders of magnitude — than the early theoretical result from the compilation of Arnaud & Raymond [17] (figure 2b). Such discrepancies have also been found for other iron M-shell ions

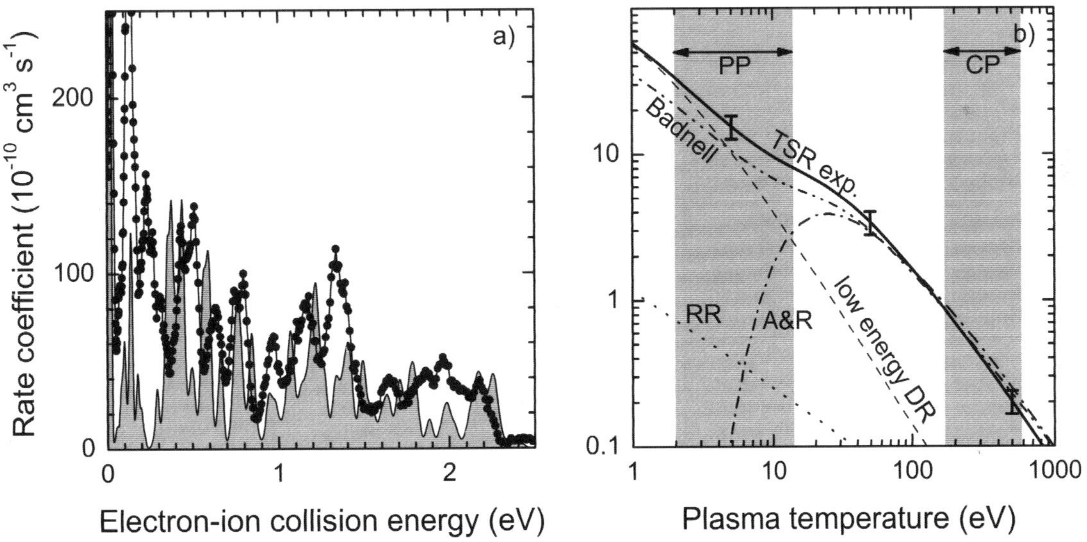


**Figure 2.** a) Experimental (filled circles) [18] and theoretical (shaded curve) [19] merged-beams rate coefficient for the recombination of electrons with Al-like $Fe^{13+}$ ions at low electron-ion collision energies. b) $Fe^{13+}$ recombination-rate coefficient in a plasma. The experimental result from the TSR (full curve) is compared with the theoretical rate coefficient from the widely used compilation of Arnaud & Raymond [17] (dash dotted curve) and the recent state-of-the-art calculation of Badnell [19] (dash-dot dotted curve). The contribution of RR to the latter is shown as a dotted curve. The dashed curve is the contribution of the low-energy DR resonances shown in panel a) to the total plasma rate coefficient. The temperature ranges where $Fe^{13+}$ exists in photoionized plasmas (PP) and collisionally ionized plasmas (CP) are highlighted.

(see references in table 1).

It is clear, that the large discrepancies between the experimental and the early theoretical rate coefficients are due to a simplified theoretical treatment that was geared towards CP and more or less disregarded low-energy DR in order to keep the calculations tractable. Modern computers allow more sophisticated approaches, and recent theoretical work has aimed at providing a more reliable recombination data-base by using state-of-the-art atomic codes [20]. Badnell and coworkers [21] have calculated DR rate coefficients for finite-density plasmas. Results have been published for the isoelectronic sequences from H-like to Mg-like [22] (and references therein). Complementary RR rate coefficients are also available [23]. Independently, Gu calculated DR and RR rate coefficients for selected ions of astrophysical interest [13; 24; 25]. Figure 1b shows the $Mg^{8+}$ DR results from these two data sets. Although the new theoretical work removes the striking low-temperature disagreement that was found between experimental and early theoretical results, uncertainties remain as discussed above.

For the recombination of $Fe^{13+}$ a detailed comparison between the experimental and new theoretical results was presented by Badnell [19]. Experimental and calculated low-energy merged-beams recombination rate coefficients are shown in figure 2a. Although calculated and measured DR resonance structures are approximately of the same height there are many differences in the details. Fortuitously, these differences largely average out in the plasma recombination rate coefficient. Still, the theoretical plasma rate coefficient is up to 50% smaller than the experimental one in the PP temperature range (figure 2b).

In conclusion, storage ring experiments are presently the only reliable source for low-temperature DR data. Moreover, they provide valuable benchmarks for the further development of the theoretical methods.

## 4. Bibliography of experimental results from storage rings

For the interpretation and understanding of astronomical observations numerical models of astrophysical plasmas are used [26]. As an input these models require atomic data especially for the 15 most abundant elements [27]. Their relative abundances in the solar photosphere [28] are depicted in figure 3. Table 1 lists those astrophysically relevant ions for which experimental results on electron-ion recombination have been obtained from storage-ring measurements. The corresponding references are given. More comprehensive but now somewhat outdated compilations which additionally contain results from early single-pass merged-beams experiments as well as results for other ions have been published previously [29–31].

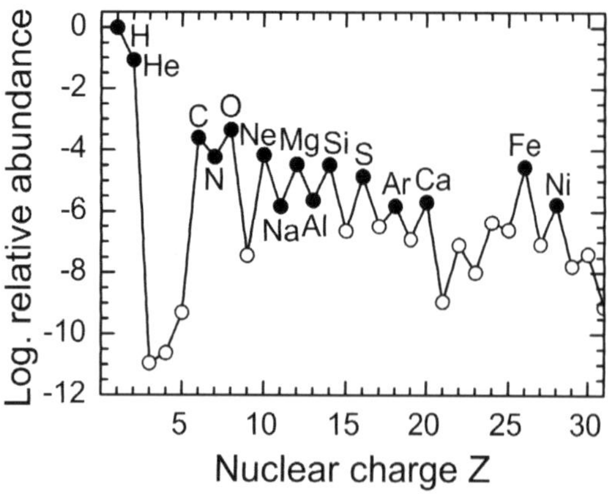


**Figure 3.** Abundances of the elements in the solar photosphere relative to hydrogen [28]. The astrophysically most relevant elements are marked by solid circles. Elements with nuclear charge $Z > 30$ have relative abundances below $10^{-8}$. Iron is the heaviest element that is produced by nuclear fusion in stars. Iron ions play a prominent role in x-ray astronomy [32] due to their large relative abundance and their high nuclear charge.

**Table 1.** Bibliographic compilation of storage-ring electron-ion recombination experiments with astrophysically relevant ions. For each ion the primary charge state before recombination is listed.

| ion | reference | ion | reference | ion | reference |
|---|---|---|---|---|---|
| H 1+ | [33; 34] | Ne 6+ | [35] | Fe 7+ | [36; 37] |
| He 1+ | [38–41] | 7+ | [42–44] | 8+ | [36; 37] |
| 2+ | [34] | Ne 10+ | [34; 45] | 9+ | [46] |
| C 2+ | [47] | Na 8+ | [9] | 10+ | [46] |
| 3+ | [48; 49] | Mg 8+ | [7] | 13+ | [18] |
| 4+ | [50; 51] | Si 3+ | [11; 52; 53] | 14+ | [54] |
| 5+ | [55] | 11+ | [56; 57] | 15+ | [58] |
| 6+ | [59] | 14+ | [34] | 16+ | [60] |
| N 3+ | [47] | S 5+ | [61] | 17+ | [62; 63] |
| 4+ | [44; 64] | 15+ | [65] | 18+ | [63; 66] |
| 7+ | [34] | Ar 7+ | [61] | 19+ | [67] |
| O 4+ | [47] | 13+ | [45; 68] | 20+ | [69] |
| 5+ | [70; 71] | 15+ | [72] | 21+ | [69] |
| 7+ | [73] | | | 22+ | [74] |
| | | | | Ni 17+ | [75; 76] |
| | | | | 25+ | [77] |

**Acknowledgments**
The author is grateful to Michael Lestinsky, Alfred Müller, Daniel Wolf Savin, Eike Schmidt and Andreas Wolf for long-standing fruitful collaboration and likewise to all his other collaborators who are too many to list them all. This research was supported in part by the German federal research-funding agency DFG under contract no. Schi 378/5.

**References**

[1] Schippers S, Schmidt E W, Bernhardt D, Yu D, Müller A, Lestinsky M, Orlov D A, Grieser M, Repnow R and Wolf A 2007 *Phys. Rev. Lett.* **98** 033001

[2] Brandau C, Kozhuharov C, Harman Z, Müller A, Schippers S, Kozhedub Y S, Bernhardt D, Böhm S, Jacobi J, Schmidt E W, Mokler P H, Bosch F, Kluge H J, Stöhlker T, Beckert K, Beller P, Nolden F, Steck M, Gumberidze A, Reuschl R, Spillmann U, Currell F, Tupitsyn I I, Shabaev V M, Jentschura U D, Keitel C H, Wolf A and Stachura Z 2008 *Phys. Rev. Lett.* **100** 073201

[3] Wolf A, Buhr H, Grieser M, von Hahn R, Lestinsky M, Lindroth E, Orlov D A, Schippers S and Schneider I F 2006 *Hyperfine Interact.* **172** 111

[4] Lestinsky M, Lindroth E, Orlov D A, Schmidt E W, Schippers S, Böhm S, Brandau C, Sprenger F, Terekhov A S, Müller A and Wolf A 2008 *Phys. Rev. Lett.* **100** 033001

[5] Schippers S 2008 *Nucl. Instrum. Methods* B in print (*Preprint* `arXiv:0805.4381 [physics.atom-ph]`)

[6] Savin D W 2007 *J. Phys. Conf. Ser.* **88** 012071

[7] Schippers S, Schnell M, Brandau C, Kieslich S, Müller A and Wolf A 2004 *Astron. Astrophys.* **421** 1185

[8] Tokman M, Eklöw N, Glans P, Lindroth E, Schuch R, Gwinner G, Schwalm D, Wolf A, Hoffknecht A, Müller A and Schippers S 2002 *Phys. Rev.* A **66** 012703

[9] Nikolić D, Lindroth E, Kieslich S, Brandau C, Schippers S, Shi W, Müller A, Gwinner G, Schnell M and Wolf A 2004 *Phys. Rev.* A **70** 062723

[10] Kieslich S, Schippers S, Shi W, Müller A, Gwinner G, Schnell M, Wolf A, Lindroth E and Tokman M 2004 *Phys. Rev.* A **70** 042714

[11] Orban I, Lindroth E, Glans P and Schuch R 2007 *J. Phys.* B **40** 1063

[12] Colgan J, Pindzola M S, Whiteford A D and Badnell N R 2003 *Astron. Astrophys.* **412** 597–601

[13] Gu M F 2003 *Astrophys. J.* **590** 1131

[14] Mazzotta P, Mazzitelli G, Colafrancesco S and Vittorio N 1998 *Astron. Astrophys.* **133** 403

[15] Netzer H 2004 *Astrophys. J.* **604** 551

[16] Kraemer S B, Ferland G J and Gabel J R 2004 *Astrophys. J.* **604** 556

[17] Arnaud M and Raymond J 1992 *Astrophys. J.* **398** 394

[18] Schmidt E W, Schippers S, Müller A, Lestinsky M, Sprenger F, Grieser M, Repnow R, Wolf A, Brandau C, Lukić D, Schnell M and Savin D W 2006 *Astrophys. J.* **641** L157

[19] Badnell N R 2006 *J. Phys.* B **39** 4825

[20] Badnell N R 2007 *J. Phys. Conf. Ser.* **88** 012070

[21] Badnell N R, O'Mullane M G, Summers H P, Altun Z, Bautista M A, Colgan J, Gorczyca T W, Mitnik D M, Pindzola M S and Zatsarinny O 2003 *Astron. Astrophys.* **406** 1151; `http://amdpp.phys.strath.ac.uk/tamoc/DATA/`

[22] Altun Z, Yumak A, Yavuz I, Badnell N R, Loch S D and Pindzola M S 2007 *Astron. Astrophys.* **474** 1051

[23] Badnell N R 2006 *Astrophys. J. Suppl. Ser.* **167** 334

[24] Gu M F 2004 *Astrophys. J. Suppl. Ser.* **153** 389

[25] Gu M F 2003 *Astrophys. J.* **589** 1085

[26] Ferland G J 2003 *Annu. Rev. Astron. Astrophys.* **41** 517

[27] Kallman T R and Palmeri P 2007 *Rev. Mod. Phys.* **79** 79

[28] Asplund M, Grevesse N and Sauval A J 2006 *Nucl. Phys.* A **777** 1–4

[29] Müller A 1995 *Nucl. Fusion Suppl.* **6** 59

[30] Schippers S, Bartsch T, Brandau C, Gwinner G, Linkemann J, Müller A, Saghiri A A and Wolf A 1999 *Phys. Scr.* **T80** 314

[31] Schippers S 2002 *Habilitation Thesis* (Justus-Liebig-Universität Giessen, Germany) appendix A `http://geb.uni-giessen.de/geb/volltexte/2002/909/`

[32] Kahn S M, Behar E, Kinkhabwala A and Savin D W 2002 *Philos. Trans. R. Soc. London Ser.* A **360** 1923

[33] Biedermann C, Gao H, Schuch R, Zong W, Asp S, DeWitt D R and Kuiper H 1995 *J. Phys.* B **28** 505

[34] Gao H, Schuch R, Zong W, Justiniano E, DeWitt D R, Lebius H and Spies W 1997 *J. Phys.* B **30** L499

[35] Orban I, Böhm S, Loch S and Schuch R 2008 *Astron. Astrophys.* **489** 829

[36] Schmidt E W, Schippers S, Brandau C, Bernhardt D, Yu D, Müller A, Lestinsky M, Sprenger F, Hoffmann J, Orlov D A, Grieser M, Repnow R, Wolf A, Lukić D, Schnell M and Savin D W 2007 *J. Phys. Conf. Ser.* **58** 223

[37] Schmidt E W, Schippers S, Bernhardt D, Müller A, Hoffmann J, Lestinsky M, Orlov D A, Wolf A, Lukić D V, Savin D W and Badnell N R 2008 *Astron. Astrophys.* in print

[38] Tanabe T, Tomizawa M, Chida K, Watanabe T, Watanabe S, Yoshizawa M, Muto H, Noda K, Kanazawa M, Ando A and Noda A 1992 *Phys. Rev.* A **45** 276

[39] Haar R R, Tanis J A, Plano V L, Zaharakis K E, Graham W G, Mowat J R, Ellison T, Jacobs W W and Rinckel T 1993 *Phys. Rev.* A **47** R3472

[40] DeWitt D R, Schuch R, Quinteros T, Gao H, Zong W, Danared H, Pajek M and Badnell N R 1994 *Phys. Rev.* A **50** 1257

[41] DeWitt D R, Lindroth E, Schuch R, Gao H, Quinteros T and Zong W 1995 *J. Phys.* B **28** L147

[42] Zong W, Schuch R, Gao H, DeWitt D R and Badnell N R 1998 *J. Phys.* B **31** 3729

[43] Böhm S, Schippers S, Shi W, Müller A, Djurić N, Dunn G H, Zong W, Jelenković B, Danared H, Eklöw N, Glans P and Schuch R 2001 *Phys. Rev.* A **64** 032707

[44] Böhm S, Müller A, Schippers S, Shi W, Fogle M, Glans P, Schuch R and Danared H 2005 *Astron. Astrophys.* **437** 1151

[45] Gao H, DeWitt D R, Schuch R, Zong W, Asp S and Pajek M 1995 *Phys. Rev. Lett.* **75** 4381

[46] Lestinsky M et al. 2008 in preparation

[47] Fogle M, Badnell N R, Glans P, Loch S D, Madzunkov S, Abdel-Naby S A, Pindzola M S and Schuch R 2005 *Astron. Astrophys.* **442** 757

[48] Mannervik S, deWitt D, Engström L, Lidberg J, Lindroth E, Schuch R and Zong W 1998 *Phys. Rev. Lett.* **81** 313

[49] Schippers S, Müller A, Gwinner G, Linkemann J, Saghiri A A and Wolf A 2001 *Astrophys. J.* **555** 1027

[50] Kilgus G, Habs D, Schwalm D, Wolf A, Schuch R and Badnell N R 1993 *Phys. Rev.* A **47** 4859

[51] Mannervik S, Asp S, Broström L, DeWitt D R, Lidberg J, Schuch R and Chung K T 1997 *Phys. Rev.* A **55** 1810

[52] Orban I, Glans P, Altun Z, Lindroth E, Källberg A and Schuch R 2006 *Astron. Astrophys.* **459** 291

[53] Schmidt E W, Bernhardt D, Müller A, Schippers S, Fritzsche S, Hoffmann J, Jaroshevich A S, Krantz C, Lestinsky M, Orlov D A, Wolf A, Lukić D and Savin D W 2007 *Phys. Rev.* A **76** 032717

[54] Lukić D, Savin D W, Schnell M, Brandau C, Schmidt E W, Böhm S, Müller A, Schippers S, Lestinsky M, Sprenger F, Wolf A, Altun Z and Badnell N 2007 *Astrophys. J.* **664** 1244

[55] Wolf A, Berger J, Bock M, Habs D, Hochadel B, Kilgus G, Neureither G, Schramm U, Schwalm D, Szmola E, Müller A, Wagner M and Schuch R 1991 *Z. Phys.* D **21** S69

[56] Kenntner J, Linkemann J, Badnell N R, Broude C, Habs D, Hofmann G, Müller A, Pindzola M S, Salzborn E, Schwalm D and Wolf A 1995 *Nucl. Instrum. Methods* B **98** 142

[57] Bartsch T, Müller A, Spies W, Linkemann J, Danared H, DeWitt D R, Gao H, Zong W, Schuch R, Wolf A, Dunn G H, Pindzola M S and Griffin D C 1997 *Phys. Rev. Lett.* **79** 2233

[58] Linkemann J, Kenntner J, Müller A, Wolf A, Habs D, Schwalm D, Spies W, Uwira O, Frank A, Liedtke A, Hofmann G, Salzborn E, Badnell N R and Pindzola M S 1995 *Nucl. Instrum. Methods* B **98** 154

[59] Gwinner G, Hoffknecht A, Bartsch T, Beutelspacher M, Eklöw N, Glans P, Grieser M, Krohn S, Lindroth E, Müller A, Saghiri A A, Schippers S, Schramm U, Schwalm D, Tokman M, Wissler G and Wolf A 2000 *Phys. Rev. Lett.* **84** 4822

[60] Schmidt E W, Bernhardt D, Hoffmann J, Lestinsky M, Lukić D V, Müller A, Orlov D A, Savin D W, Schippers S, Wolf A and Yu D 2008 *J. Phys. Conf. Ser.* these proceedings

[61] Orban I, Böhm S, Trotsenko S and Schuch R 2008 unpublished

[62] Savin D W, Bartsch T, Chen M H, Kahn S M, Liedahl D A, Linkemann J, Müller A, Schippers S, Schmitt M, Schwalm D and Wolf A 1997 *Astrophys. J.* **489** L115

[63] Savin D W, Kahn S M, Linkemann J, Saghiri A A, Schmitt M, Grieser M, Repnow R, Schwalm D, Wolf A, Bartsch T, Brandau C, Hoffknecht A, Müller A, Schippers S, Chen M H and Badnell N R 1999 *Astrophys. J. Suppl. Ser.* **123** 687

[64] Glans P, Lindroth E, Badnell N R, Eklöw N, Zong W, Justiniano E and Schuch R 2001 *Phys. Rev.* A **64** 043609

[65] Wolf A 1992 Report MPI H - V15 -1992 Max-Planck-Institut für Kernphysik Heidelberg

[66] Savin D W, Kahn S M, Linkemann J, Saghiri A A, Schmitt M, Grieser M, Repnow R, Schwalm D, Wolf A, Bartsch T, Müller A, Schippers S, Chen M H, Badnell N R, Gorczyca T W and Zatsarinny O 2002 *Astrophys. J.* **576** 1098

[67] Savin D W, Behar E, Kahn S M, Gwinner G, Saghiri A A, Schmitt M, Grieser M, RRepnow, Schwalm D, Wolf A, Bartsch T, Müller A, Schippers S, Badnell N R, Chen M H and Gorczyca T W 2002 *Astrophys. J. Suppl. Ser.* **138** 337

[68] DeWitt D R, Schuch R, Gao H, Zong W, Asp S, Biedermann C, Chen M H and Badnell N R 1996 *Phys. Rev.* A **53** 2327–2336

[69] Savin D W, Kahn S M, Gwinner G, Grieser M, Repnow R, Saathoff G, Schwalm D, Wolf A, Müller A, Schippers S, Závodsky P A, Chen M H, Gorczyca T W, Zatsarinny O and Gu M F 2003 *Astrophys. J. Suppl. Ser.* **147** 421

[70] Böhm S, Schippers S, Shi W, Müller A, Eklöw N, Schuch R, Danared H, Badnell N R, Mitnik D and Griffin D C 2002 *Phys. Rev.* A **65** 052728

[71] Böhm S, Müller A, Schippers S, Shi W, Eklöw N, Schuch R, Danared H and Badnell N R 2003 *Astron. Astrophys.* **405** 1157

[72] Zong W, Schuch R, Lindroth E, Gao H, DeWitt D R, Asp S and Danared H 1997 *Phys. Rev.* A **56** 386

[73] Kilgus G, Berger J, Blatt P, Grieser M, Habs D, Hochadel B, Jaeschke E, Krämer D, Neumann R, Neureither G, Ott W, Schwalm D, Steck M, Stokstad R, Szmola E, Wolf A, Schuch R, Müller A and Wagner M 1990 *Phys. Rev. Lett.* **64** 737

[74] Savin D W, Gwinner G, Grieser M, Repnow R, Schnell M, Schwalm D, Wolf A, Zhou S G, Kieslich S, Müller A, Schippers S, Colgan J, Loch S D, Chen M H and Gu M F 2006 *Astrophys. J.* **642** 1275

[75] Fogle M, Eklöw N, Lindroth E, Mohamed T, Schuch R and Tokman M 2003 *J. Phys.* B **36** 2563

[76] Fogle M, Badnell N R, Eklöw N, Mohamed T and Schuch R 2003 *Astron. Astrophys.* **409** 781

[77] Schippers S, Bartsch T, Brandau C, Müller A, Gwinner G, Wissler G, Beutelspacher M, Grieser M, Wolf A and Phaneuf R A 2000 *Phys. Rev.* A **62** 022708

Journal of Physics: Conference Series **163** (2009) 012002 doi:10.1088/1742-6596/163/1/012002

# Spectroscopy of Highly Charged Ions in Solar and Astrophysical Plasmas

**Tetsuya Watanabe** [1, 2]

[1] National Astronomical Observatory, Japan

2-21-1 Osawa, Mitaka, Tokyo 181-8588 Japan

[2] Graduate School for Advanced Studies

Hayama, Miura-gun, Kanagawa Pref., 240-0193 Japan

E-mail: watanabe@uvlab.mtk.nao.ac.jp

**Abstract**. Spectroscopic observation of EUV emission lines in the transition-region and corona provide unique information on physical conditions in these outer atmospheres of the Sun. The EUV Imaging Spectrometer (the EIS) on board the *Hinode* mission is capable of observing, for the first time in Solar EUV observations, spectra and monochromatic images of possibly non-ionization-equilibrium plasmas in the solar transition-region and corona at two-wavelength bands of 170 - 210Å and 250 - 290Å, with typical time-resolutions of 1 - 10 seconds. Dynamic plasma acceleration and heating are found to take place in the solar atmospheres, and they are confined in tiny structures. A time-dependent collisional-radiative model for the iron plasmas is developed to diagnose temperatures and densities in the outer atmospheres of the Sun; no systematic models yet exist for iron ions at the ionization stages of L- and M-shells, which are very important for coronal plasma diagnostics. Adopting the best available theoretical calculations of atomic parameters of these iron ions, as well as generating the experimental data by a compact electron beam ion trap (EBIT), is essential to one of the aims of our research that the mechanism of coronal heating is explored via accurate diagnostics information obtained by the EIS instrument.

## 1. Introduction

Matter in the Universe is not distributed uniformly; some is very hot, highly ionized, may not be in thermal equilibrium, or even not thermalized. Spectroscopy of lines originating from highly charged ions is therefore of crucial importance to understand the characteristics of these hot plasmas. One of the fastest ways to understand the characteristics of astrophysical plasmas is to understand those of solar plasma, because it is believed that most of the phenomena happening elsewhere in the Universe also take place on the Sun. In addition, the Sun serves as "a laboratory in space," which shows interesting phenomena of plasma physics on a gigantic scale. Spectroscopic observation of EUV emission lines in the transition-region and corona provide unique information on the physical conditions in these outer atmospheres of the Sun.

In this paper, a progress report of our research project is presented by briefly introducing new understanding of observations in the laboratory, the universe, especially in the Sun, and, of theoretical predictions. The newly-constructed Sun-observing satellite, *Hinode* presented very dynamic nature of magnetic fields and plasmas in the solar outer atmospheres.

An analysis tool is developed to handle these plasmas in non-equilibrium states with improved atomic datasets, which are obtained by evaluating the theoretical calculations and by laboratory measurements. The ultimate goal is that the time-dependent collisional-radiative model will be applied to solar coronal plasmas to understand the mechanism of coronal heating.

## 2. Progress of EUV spectroscopy for Solar and Astrophysical Plasmas

### 2.1. Non-solar observations

Since the launch of EUVE [1] in 1990, a new era of EUV and X-ray spectroscopy started in astrophysics with the availability of high resolution spectrographs. At the beginning of the 21st century, X-ray satellites such as Chandra [2] and XMM-Newton [3] enabled grating spectroscopy in X-ray regions, though it was a major loss of scientific capability on board *Suzaku*, that the micro-calorimeter failed before the start of its observation. Emission lines from hot plasmas in stars, especially in galaxies, and/or clusters of galaxies, are for the first time revealed by these highly resolved spectra in EUV and X-ray wavelengths.

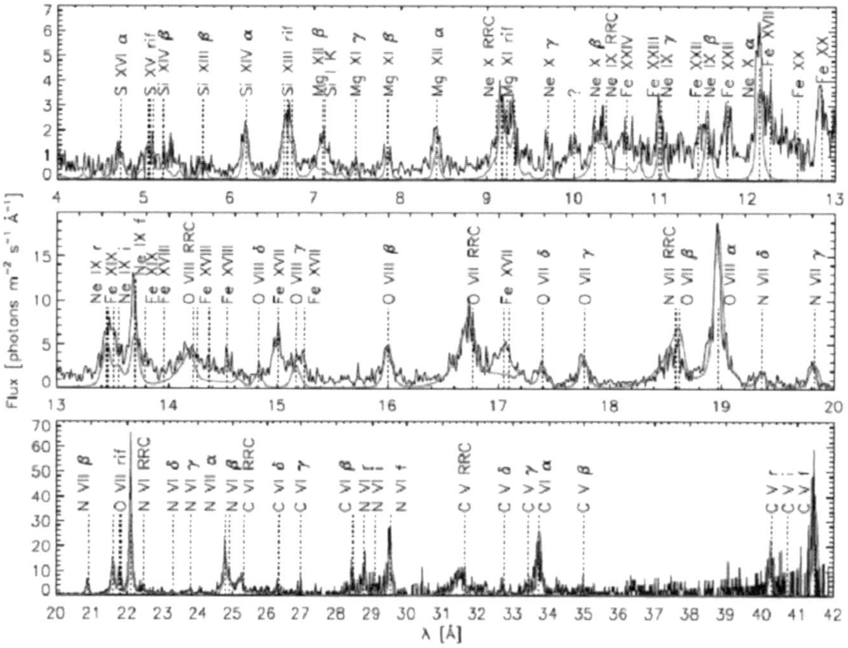


Fig 1 Spectrum of the first order of the primary region of NGC1068, together with the model: The model does not yet include Fe L-shell transitions [4].

Fig. 1 shows a spectrum of NGC1068 taken by Chandra in its primary region [4]. Similar spectra are also taken by RGS on board XMM-Newton. Highly charged ions of various elements are clearly seen. Diagnostics such as that of using the density-sensitive line ratios in the He-like triplets of N, O, and Ne are possible. In this example, however, these forbidden lines appear significantly stronger than predicted. This is likely due to inner shell ionization in the corresponding Li-like ions, which enhance only the forbidden line of the triplet.

### 2.2. Solar observation; *Hinode* Mission and EUV Imaging Spectrometer on board *Hinode*

*Hinode* was successfully launched from Uchinoura Space Centre (USC)/JAXA at 18:36:00 on 22-Sep-2006 (UT); "*Hinode*" is the sunrise in Japanese. Its science objectives are to understand the basic questions of solar physics; a) the coronal heating mechanism, b) the origin of strong magnetic fields, and c) the trigger mechanism of solar flares along with other questions.

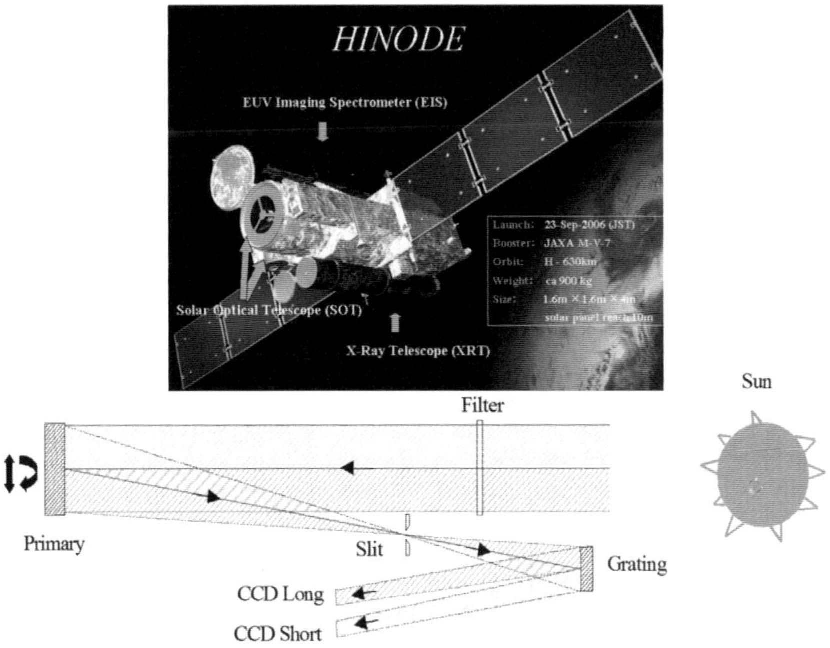


Fig 2 The 22nd Japanese Scientific Satellite, "*Hinode*," and the optical layout of EIS

Table 1. EIS Parameters

| | |
|---|---|
| Wavelength Bands | 170 - 190Å & 250 - 290Å |
| Peak Effective Areas | $0.30cm^2$ & $0.11\ cm^2$ |
| Primary Mirror | 15cm diameter, two Mo/Si multilayer coating |
| Grating | Toroidal & laminar, 4200 grooves·$mm^{-1}$, two Mo/Si multilayer coating |
| CCD Cameras | Two back-thinned E2V CCDs, 2048x1024x13.5μm pix. |
| Plate Scales | 13.53μm/arcsec at CCD; 9.40μm/arcsec at slit |
| Spatial Resolution (pix) | 2 arcsec (1 arcsec) |
| Field of View | 6 arcmin x 8.5 arcmin |
| Raster | 1 arcsec in 0.7sec, min. step 0.123 arcsec |
| Slit/Slot Widths | 1, 2, 40, & 266 arcsec |
| Spectral Resolution | 47mÅ (FWHM) at 185Å; 1pix = 22mÅ, approx. 25km $s^{-1}$/pix |
| Temperature Coverage | log T = 4.7 – 7.3 K |
| CCD Frame Read Time | 0.8 sec |
| Line Observation | Simultaneous observation of up to 25 lines |

It was successfully put into a sun-synchronous polar orbit at an altitude of 680 km above sea level, *Hinode* started initial scientific observations in December 2006. A guest-investigation-type program was announced in the Call for Proposals. This program encourages the world-wide scientific involvement in the *Hinode* observations. All the data taken by the telescopes, as well as those for house-keeping purposes, have been immediately open to the public and available since 27-May-2007 with instruction for data users.

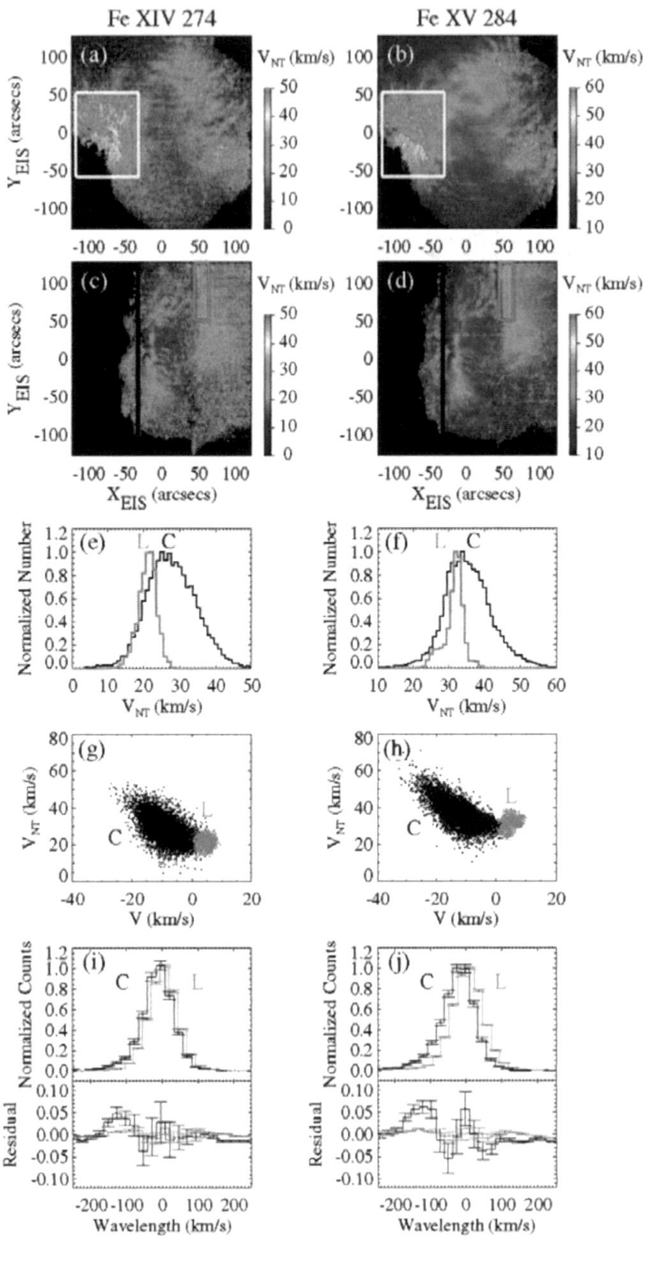


Fig 3 Left: Fe XIV 274 line and Right: Fe XV 284 line observations. From the top to bottom, non-thermal velocity maps at disk centre, those at the west limb, histograms of non-thermal velocity, scatter plots between Doppler and non-thermal velocities, and normalized line profiles with their residual from a single Gaussian fitting at a foot point of a coronal loop are shown. C and L in panels (e) to (j) indicate an observation at disk centre and limb, within a white or red box in VNT maps, respectively.

Three telescopes are accommodated on board the *Hinode* mission; Solar optical telescope (SOT), X-ray telescope (XRT), and EUV imaging spectrometer (EIS). The SOT instrument consists of the Optical Telescope Assembly (OTA), and the Focal Plane Package (FPP). It is a Gregorian telescope with an effective aperture of 50 cm that is aiming at diffraction-limited performance of the OTA with the tip-tilt movable mirror (TCM) system, namely, 0.2 – 0.3 arc seconds at the wavelengths 380 - 670 nm. The X-ray telescope (XRT) is a grazing incidence mirror telescope accommodating a 2k × 2k CCD as a detector. The spatial resolution is improved to 1 arcsec, about three times better than that of *Yohkoh*/SXT, and its measurable plasma temperature range is increased to 1 – 20 million K.

The major improvement of the EIS telescope throughput was achieved by applying multi-layer coatings to the primary mirror and the toroidal grating (groove density = 4200gr/mm) used in the EUV wavelength range. The two wavelength bands observed with the EIS instrument are 170 – 210Å and 250 – 290 Å and include the emission lines of He II and Fe VIII – Fe XXIV, lines formed in the temperature range $4.7 < \log T < 7.2$ [5].

The parameters of the instrument are summarized in table 1. The optical layout is shown in Fig.2. The plate scale is such that one CCD detector pixel corresponds to 1 arcsec in the spatial direction along the slit and the spectral resolution is 0.00223Å/pixel. Four kinds of slit/slots, slits of 1 and 2 arcsec(s) and slots of 40 and 266 arcsecs, are accommodated and can be interchanged in a housing near the focal plane of the primary mirror. A total of 25 emission line windows can be selected on two CCDs, dedicated to the shorter and longer wavelength bands, respectively. The small-scale velocity structure and non-equilibrium nature of solar coronal plasmas can be observed for the first time in these EUV wavelengths with this capable instrument.

### 2.3. New discoveries by the *Hinode* Mission

More than 40 papers describing the first results of *Hinode* were published in the special issue of Publication of Astronomical Society in November 2007 [6]. A Science special issue accommodates 9 papers [7]. New discoveries and findings during the first year observations of *Hinode* can be summarized as follows.

New discoveries that *Hinode* made so far during the first year of operation are: 1) Strong magnetic fields (B> 1 kG) were observed in the polar regions, where the magnetic fields have previously been thought to be weak and vertical; 2) Horizontal strong magnetic fields, though transient by nature, were observed in the quiet sun and in the polar region as well. They are numerous; 3) Convective collapse, one of the mechanisms to make up strong magnetic fields in the photosphere, could be observed in the diffraction limited magnetograph images; 4) Alfvén waves were seen to propagate throughout the photospheres and the chromospheres; 5) Jets and/or supersonic flows were also present everywhere in the quiet sun and also in the coronal holes.

First detections and detailed analyses are continuing for the following science topics: 1) Collapsing processes in sun spots; cancellation and escape of magnetic fluxes from sunspots will be studied in detail with continuous magnetograph observations; 2) Flows associated with flares will be studied. Inflows/outflows to/from the magnetic reconnection sites, chromospheric evaporations and down flows above flare arcades will be closely investigated with all three instruments; 3) Turbulence or turbulent flows at the foot points of coronal loops seem universal in most active regions. This will provide clues to allow identification of the mechanism of coronal heating; 4) Temperature structures along with height will be investigated through XRT and EIS by calibrating their diagnostic capabilities; 5) Polar jets found in the polar coronal holes will be studied in detail, to understand their contribution to coronal heating, polar plumes, and the fast and slow solar winds; 6) Oscillations are seen everywhere in the photosphere. Umbral oscillations in particular will reveal a method for energy transport in strong magnetic fields; 7) Numerous interesting phenomena, such as flows at the roots of the fast solar wind, coronal hole formation, white light flares and others will be further discussed in detail.

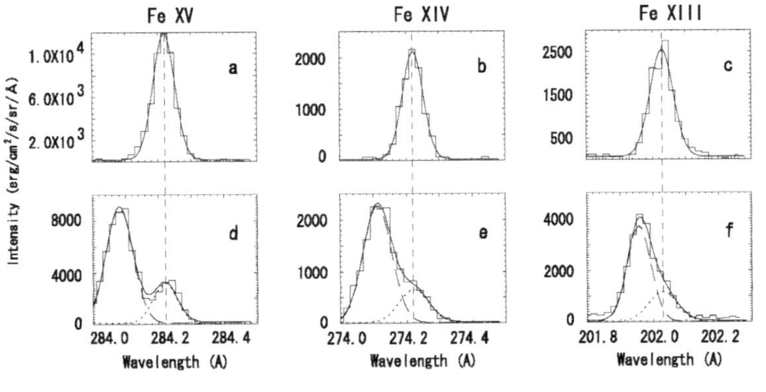


Fig 4 Sample of spectra taken from corona high-temperature emission lines (Fe XV, Fe XIV, Fe XIII): The vertical axes show the intensity, and the horizontal axes show the wavelength. The vertical dashed lines show the line center of the emission lines, which are emitted from stationary ions. The upper panels show reference emission lines, which seem to be emitted from a stationary plasma, to determine the line center, (a)–(c); the lower panels show the emission lines of the plage region, which exhibit strong blueshifts.

### 2.4. EIS discoveries

The EIS instrument on board the *Hinode* mission is unique, as it is capable of taking solar EUV emission line spectra at its high spectral resolution, for the first time, to analyze their profiles. The dynamics and fine structure of plasma motions in the transition region and corona are thus revealed by this instrument.

Fig. 3 shows the structure of non-thermal line broadenings in an active region [8]. Coronal lines always show additional line width to that expected from thermal Doppler motions. In the disk centre observation subsonic up flow motions of tens of kms$^{-1}$ and enhanced non-thermal velocities have been found near the foot points of the active-region loops assuming a single Gaussian approximation for the emission-line profiles. When the same part of the active region is observed near the limb, both up flows and enhanced non-thermal velocities essentially decrease. There is a strong correlation between Doppler velocity and non-thermal velocity. Significant deviations from a single Gaussian profile are found in the blue wings. These suggest that there are unresolved high-speed up flows that might be closely related to coronal heating mechanisms.

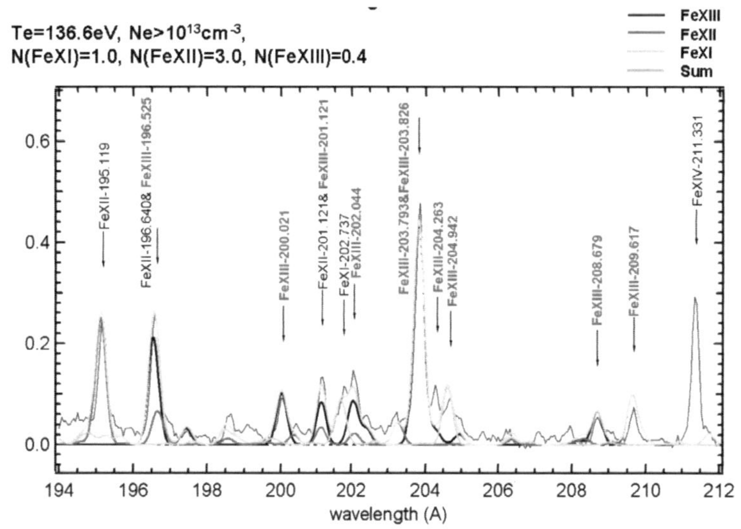


Fig 5 LHD spectra in the wavelengths of 194 -212 Å overlaid by theoretical predictions of the model.

Temperature-dependent plasma motions (up flows) are observed in association with a transient coronal hole produced by a big flare (X3.2) that occurred on December 13 2006 [9]. Multi-wavelength spectral observations allow us to determine velocities from the Doppler shifts at different temperatures. Strong up flows as well as stationary plasma have been observed in the Fe XV line 284.2 Å (log T (K) = 6.3) in this region. The strong up flows reach almost 150kms,$^{-1}$ estimated by a two-component Gaussian fitting in Fig. 4. On the other hand, at a lower corona/transition-region temperature (He II 256.3Å, log T (K) = 4.7), very weak up flows, almost stationary, have been observed. We find that these up flow velocities clearly depend on the temperature with the hottest line, Fe XV, showing the fastest up flow velocity, and the second-highest line, Fe XIV, showing the second-highest up flow velocity (130kms$^{-1}$). All velocities are below the speed of sound. The trend of the up flow dependence on temperature dramatically changes at 1 MK.

## 3. Development of time-dependent collisional-radiative model for iron

In order to understand the dynamic solar plasmas introduced in the previous sections, a time-dependent collisonal-radiative model for iron has been developed to analyze the data taken by the EIS instrument, and to diagnose temperatures and densities of those plasmas in the outer atmospheres of the Sun. No systematic models yet exist for iron ions of L- and M-shells, which are very important for coronal plasma diagnostics.

Atomic data for these ions were calculated by the HULLAC code [10]. Electron collision rates on Fe XIII are adopted from Aggarwal and Keenan [11], which is the latest calculation by R-matrix method.

For developments of the analysis tool with high precision, selections of atomic data (wavelength (level energy), radiative transition probabilities, collisional excitation cross sections etc.) and constructions of the collisional-radiative model for spectral line analysis are important. Atomic data for iron ions of Fe X to Fe XV were surveyed and evaluated, and most recommended data were determined. Parameters for analytical fitting functions were obtained and provided [12].

EUV spectra in the wavelengths of 170 – 190 Å were taken at the LHD, by injecting iron TESPEL (tracer-encapsulated solid pellets). Data taken by this experiment is analyzed. Calculations of collisional excitation cross sections of Fe XIII lines were compared with LHD spectra with cross sections of proton collisions taken into account, and Watanabe et al. [13] compared the density-sensitive Fe XIII line ratios seen in EIS, and checked the consistency of theoretical calculations of line intensity.

## 4. Compact electron beam ion trap for spectroscopy of moderate

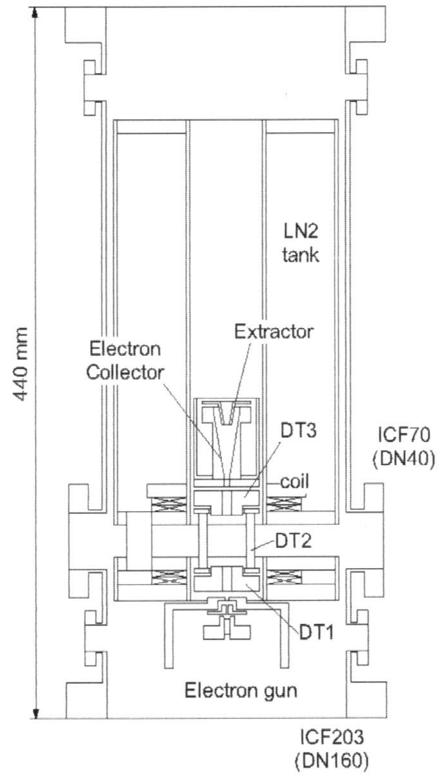


Fig 6 Schematic drawing of the compact electron beam ion trap

During the course of our development for the model atom and ions, it was found that data accommodated in the iron model come almost entirely from theoretical calculations, and no systematic experimental data exist. Therefore, a compact electron beam ion trap (EBIT) has been fabricated [14]. EBITs have been designed to be operated with rather high electron beam energy (~ 10keV or more). With such a high-energy electron beam, light and moderate elements are easily ionized to few electron or bare ions. On the other hand, spectroscopic studies of moderately charged ions which still retain many electrons are more important in solar coronal plasma diagnostics.

A new compact EBIT was developed for spectroscopic studies of moderate charge state ions, especially iron ions with charge states of Fe VI to Fe XVII relevant to the diagnostics of the solar corona. The electron energy range of the present EBIT is 100–1000eV, so that the size of the device has been much reduced from that of ordinary EBITs. In addition, a superconducting wire with a high critical temperature is used for the central coil so that it can be operated at liquid nitrogen temperatures.

Spectra of iron ions in moderate charged states will be measured to give experimental line intensity ratios used for tool to diagnose solar coronal plasmas.

## 5. Summary

Progress of high-resolution spectroscopy in astrophysics was briefly reported, especially for solar plasma diagnostics. The EIS instrument on board the *Hinode* mission is a suitable instrument to understand the mechanism of coronal heating. A time-dependent model iron atom/ions was constructed to simulate these plasmas, using best available data sets, both from theoretical calculations and laboratory experiments.

**Acknowledgement**
*Hinode* is a Japanese mission developed and launched by ISAS/JAXA, collaborating with NAOJ as domestic partner, NASA and STFC (UK) as international partners. Scientific operation of the *Hinode* mission is conducted by the *Hinode* science team organized at ISAS/JAXA. This team mainly consists of scientists from institutes in the partner countries. Support for the post-launch operation is provided by JAXA and NAOJ (Japan), STFC (U.K.), NASA (U.S.A.), ESA, and NSC (Norway).

This work is partly carried out at the NAOJ Hinode Science Centre, which is supported by the Grant-in-Aid for Creative Scientific Research: "The Basic Study of Space Weather Prediction" from MEXT, Japan (Head Investigator: K. Shibata), generous donations from Sun Microsystems, and NAOJ/NINS internal funding. It is also supported by NINS inter-institute collaborative program for Creation of New Research Area (Head Investigator: T. Watanabe), and by NIFS/NINS under the project of Formation of International Network for Scientific Collaborations (Head Investigator: H. Yamada).

**References**

[1] S. Bowyer and R.F. Malina, in "Extreme Ultraviolet Astronomy," ed. R.F. Malina and S. Bowyer, New York: Pergamon Press, pp. 397 – 408 (1991).

[2] Weisskopf, M. C., Brinkman, B., Canizares, C., Garmire, G., Murray, S., Van Speybroeck, L. P.: Publ. Astron. Soc. Pacific, 114, 1 – 24 (2002). See also: http://cxc.harvard.edu/

[3] J. W. den Herder, A. C. Brinkman, S. M. Kahn, G. Branduardi-Raymont, K. Thomsen, H. Aarts, M. Audard, J. V. Bixler, A. J. denBoggende, J. Cottam, T. Decker, L. Dubbeldam, C. Erd, H. Goulooze, M. Guedel, P. Guttridge, C. J. Hailey, K. AlJanabi, J. S. Kaastra, P. A. J. deKorte, B. J. vanLeeuwen, C. Mauche, A. J. McCalden, R. Mewe, A. Naber, F. B. Paerels, J. R. Peterson, A. P. Rasmussen, K. Rees, I. Sakelliou, M. Sako, J. Spodek, M. Stern, T. Tamura, J. Tandy, C. P. deVries, S. Welch, and A. Zehnder: Astron. Astrophys., 365, L7 (2001).

[4] A. C. Brinkman, J.S. Kaastra, R. L. J. van der Meer, A. Kinkhabwala, E. Behar, S. M. Kahn, F. B. S. Paerels, and M. Sako: Astron. Astrophys., 396,761 (2002).

[5] J. L. Culhane, L. K. Harra, A. M. James, K. Al-Janabi, L. J. Bradley, R. A. Chaudry, K. Rees, J. A. Tandy, P. Thomas, M. C. R. Whillock, B. Winter, G. A. Doschek, C. M. Korendyke, C. M. Brown, S. Myers, J. Mariska, J. Seely, J. Lang, B. J. Kent, B. M. Shaughnessy, P. R. Young, G. M. Simnet, C. M. Castelli, S. Mahmoud, H. Mapson-Menard, B. J. Probyn, R. J. Thomas, J. Davila, K. Dere, D. Windt, R. Hagood, R. Moye, H. Hara, T. Watanabe, K. Matsuzaki, T. Kosugi, V. Hansteen and Ø. Wikstol: Solar Phys., 243, 19 - 61 (2007).

[6] Publication of Astronomical Society of Japan Special Issue (2007) vol. 59-S3 contains 43 papers of the initial results obtained by the *Hinode* mission.

[7] Science vol. 318 (2007 December 7 issue, pp. 1574 – 1599) contains 9 scientific papers based on *Hinode* observation.

[8] H. Hara, T. Watanabe, L. K. Harra, J. L. Culhane, P. R. Young, J. T. Mariska, and G. A. Doschek: Astrophys. J. in press (2008).

[9] S. Imada, H. Hara, T. Watanabe, S. Kamio, A. Asai, K. Matsuzaki, L. K. Harra, J. T. Mariska: Publ. Astron. Soc. Japan, 59 (S3), S793 – S799.

[10] Bar-Shalom, A., Klapisch, M., Oreg, J. 2001, J. Quant. Spectr. Rad. Transf., 71, 179.

[11] K. M. Aggarwal, F. P. Keenan, 2005, Astrophys. J., 429, 1117.

[12] N. Yamamoto, T. Kato, H. Funaba, K. Sato, N. Tamura, S. Sudo, P. Beiersdorfer, and J. K. Lepson: Astrophys. J. in press (2008).

[13] T. Watanabe, H. Hara, N. Yamamoto, D. Kato, H. A. Sakaue, I. Murakami, T. Kato, N. Nakamura, and P. R. Young: submitted to Astrophys. J. (2008).

[14] N. Nakamura, H. Kikuchi, H. A. Sakaue, and T. Watanabe: Rev. Sci. Instr., 79, 063104 (2008).

# Emission and absorption in laser produced plasmas: processes and applications

**G. O'Sullivan[1], A. Cummings[1], C. Z. Dong[2], P. Dunne[1], P. Hayden[1], O. Morris[1], E. Sokell[1], F. O'Reilly[1], M. G. Su[2] and J. White[1]**

1. School of Physics, University College Dublin, Belfield, Dublin 4, Ireland.

2. College of Physics and Electronic Engineering, Northwest Normal Ubniversity, Lanzhou, 730070, China

gerry.osullivan@ucd.ie

**Abstract**. Laser produced plasmas have been used for many years as intense sources of extreme ultraviolet (EUV) and soft x-ray radiation. Depending on the choice and composition of target the EUV spectra can be dominated by line, unresolved transition array (UTA) or continuum emission. Line and UTA emission have found application in various proposed EUV sources for lithography, which is based on the availability of mirrors with high reflectivity in a 2% bandwidth at a wavelength of 13.5 nm. The results of recent experimental measurements of absolute in-band and out of band intensity, ion distribution and debris are presented. It was found that because of opacity effects, the conversion efficiency is sensitive to ion density and laser wavelength. Various schemes to improve the conversion efficiency are discussed as are the results of recent plasma modelling calculations.
In addition, laser produced plasmas of some high Z elements emit intense line free continua over extensive energy ranges. Some recent results on inner shell photoabsorption spectra of Sn, I and Au ions obtained using these continua are presented also. .

## 1. Introduction

In this paper a brief overview of laser produced plasma properties is given in section 1. Section 2 deals with applications to extreme ultraviolet lithography (EUVL), while section 3 will look at applications of laser produced plasmas to study ionic photoabsorption.

### *1.1. Laser produced plasmas*

When a high power pulsed laser is incident on a solid target a hot transient plasma is produced whose average electron temperature, $T_e$, is essentially determined by the laser pulse power density $\Phi$ and wavelength $\lambda$, according to $T_e(eV) \approx bA^{1/5}(\lambda^2\Phi)^{3/5}$ while the average charge $\approx 0.67\ (AT_e)^{1/3}$, where A is the atomic number. Absorption of the laser pulse proceeds in an interaction region typically ~100 μm deep above the target surface at the so-called critical density $n_{ec}$ where the plasma frequency matches the laser frequency. Since $n_{ec} \sim 10^{21}/\lambda^2$ cm$^{-3}$, ($\lambda$ in μm) the electron density is thus strongly influenced by laser wavelength, $10^{19}$ cm$^{-3}$ for $CO_2$ irradiation, $10^{21}$ cm$^{-3}$ in the case of Nd:YAG. The corresponding maximum ion densities are then typically an order of magnitude lower in each case. Behind the critical density layer, in the denser region, a shock wave propagates into the target and to conserve momentum, the plasma expands rapidly in the opposite direction usually along the incident beam direction; typical expansion velocities for illumination by a Q-switched laser delivering a few hundred mJ in approximately 10 ns are of the order of $10^6$ cm/s. The ions are accelerated after the lighter more rapidly moving electrons and thus the plasma consists of a hot core with the most highly charged ions emitted strongly in the direction of the incident laser beam and successively lower stages in cones of increased apex angle about this direction. Hence opacity issues are important and absorption by lower ion stages can dramatically reduce the emitted EUV flux. Because of their lower initial densities, $CO_2$ generated plasmas are optically thinner than those from Nd:YAG or visible/UV lasers which is an important consideration in designing sources of line or continuum radiation. Moreover, it has been shown that the EUV and x-ray emission from the plasma core in turn can produce a secondary cold plasma surrounding the main plume and absorption in this plasma can influence radiation transport close to the target surface [1].

1.2 *Radiation from Laser Produced Plasmas*
The EUV spectrum from a laser produced plasma consists of lines (bound-bound transitions), and because of the high densities encountered, lines from high n states are usually not seen, recombination radiation (bound-free transitions) which scales as $\zeta^4$ where $\zeta$ is the average ionic charge and bremsstrahlung (free-free). For an optically thin plasma: $P_{lines}$:$P_{recomb}$:$P_{brem}$ ~ 100:10:1. In some cases the lines group together to form an unresolved array (UTA) [2] where the linewidth is typically greater than the spacing between individual line spacing. An example of such an EUV spectrum is shown below.

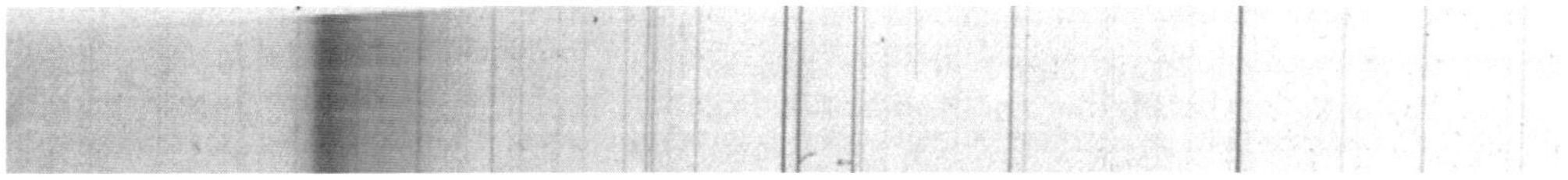

**Figure 1.** EUV spectrum from a target containing Ce in a plastic substrate.

## 2 Extreme UV lithography

2.1 *Background*
Conventionally semiconductor devices are fabricated using a lithographic process, the most recent phase of which is referred to as deep-ultraviolet lithography (DUVL). It is well known that the shorter the wavelength of light used in lithography the smaller the feature size that can be fabricated. The next stage in the semiconductor lithography roadmap is Extreme-UV Lithography (EUVL) which utilises a molybdenum silicon (Mo/Si) multi-layer mirror system to collect and focus EUV radiation in narrow band around a wavelength of 13.5nm. The short wavelength will result in a higher density of transistors and therefore faster processors. High temperature plasmas, e.g., Gas Discharge Plasmas (GDP) or Laser Produced Plasmas (LPP) are the sources of choice as they can be tuned to emit at this 13.5nm wavelength. Already it is well known that the optimum electron temperature in such plasmas should be in the 30-40 eV range, the optimum density is around $10^{18}$ ions $cm^{-3}$ and the optimum size should be hundreds of microns [3-5]. For a single pulse typically $10^{18}$ atoms/ions are ablated and can be an important source of contamination in an EUV tool.

Initially, the need to reduce particulate emission led to the choice of xenon, which as an inert gas should provide a debris-free source. There is a line group in the spectrum of xenon which has been shown by a number of researchers to arise from $4d^8$-$4d^7 5p$ transitions in Xe XI by comparison with atomic structure calculations [6]. The population of this ion stage needs to be optimised to attain maximum intensity. Note that, depending on the type of source and hence ion density, optimisation is not always necessarily the same as maximising a given ion population because of opacity effects. Considerable work has been expended on exploring the feasibility of using laser produced plasmas of xenon clusters, produced by supersonic jets or gas puffs from nozzles and solid xenon targets [7]. The optimum in-band efficiency obtainable from laser produced plasmas appears to be of the order of 1%; therefore xenon is not the optimum choice if higher conversion is required. In studies on Xe spectra the bulk of the resonance emission was always found to lie in the 10–12-nm region and it was shown by Fahy *et al.*[8] that potentially 5 times as much flux could be obtained within a 2% bandwidth at 11 nm compared to 13.5 nm. The transitions responsible arise from $4p^64d^n - 4p^54d^{n+1} + 4d^{n-1}4f$ lines which essentially overlap in adjacent ion stages and merge to form a UTA [9].

The same unresolved transition array as that in xenon at 11 nm is also observed in adjacent elements. From work, ongoing at UCD since the late 1970s, tin was already identified as potentially the strongest emitter at 13.5 nm since the UTA is centred near that wavelength. It was further shown that if the tin concentration was reduced to approximately 10%, the peak brightness actually increased due to the reduction in opacity effects. Furthermore, if the remaining 90% of the constituents were low-Z materials the radiation emitted was concentrated in a band 1–2-nm wide centred near 13.4 nm and the ordinary recombination continuum from the plasma was suppressed in comparison with that from a pure tin target. In fact if pure targets are used the recombination continuum is greatly enhanced as is the contribution from satellite emission which has been shown to be particularly important at higher ion densities [10].

The current status of EUVL is that the initial prototype or α-tools have been supplied by ASML, Nikon and Canon. These tools, originally based on xenon plasmas, will all run with tin discharge plasmas because of their superior conversion efficiency (CE). However, although the α-tools are based on discharge produced plasmas (DPP), it is not certain that DPP sources will be able to deliver the powers, >115 W into 2π sr at intermediate focus (IF), that will be required for a manufacturing tool. Consequently, laser-produced plasma (LPP) sources, which are more readily scalable and permit an increase in collection efficiency by up to a factor of three, since they can use normal rather than grazing incidence collection remain of significant interest to EUVL source developers.

One drawback however, is that normal incidence systems are vulnerable to sputtering and erosion by fast ions. However these can be deflected by electromagnetic fields while plasma curtains have been shown to be quite effective in removing neutrals. To operate at high repetition rates, the use of mass limited droplet systems has been pioneered by Richardson *et al.*[11]. In this approach the number of tin ions is $\sim 10^{17}$ and all are ionised and emit a number of times during the laser pulse. 100 kHz injection has been successfully demonstrated for such droplet based systems while laser operation has been achieved at frequencies up to 30 kHz. Moreover, CE values of 4.5% have been obtained using $CO_2$ laser based systems, compared with typical values of 2.5% for DPP sources [12].

2.2 *Experimental Results from Nd:YAG Plasmas*

2.2.1 *Conversion Efficiencies:* Using a Jenoptic 0.25 m Spectrograph with an absolutely calibrated CCD detector, spectra were recorded at a range of laser pulse power densities from both pure tin and glass targets containing 5% tin by number. This concentration had been shown to be optimum in the

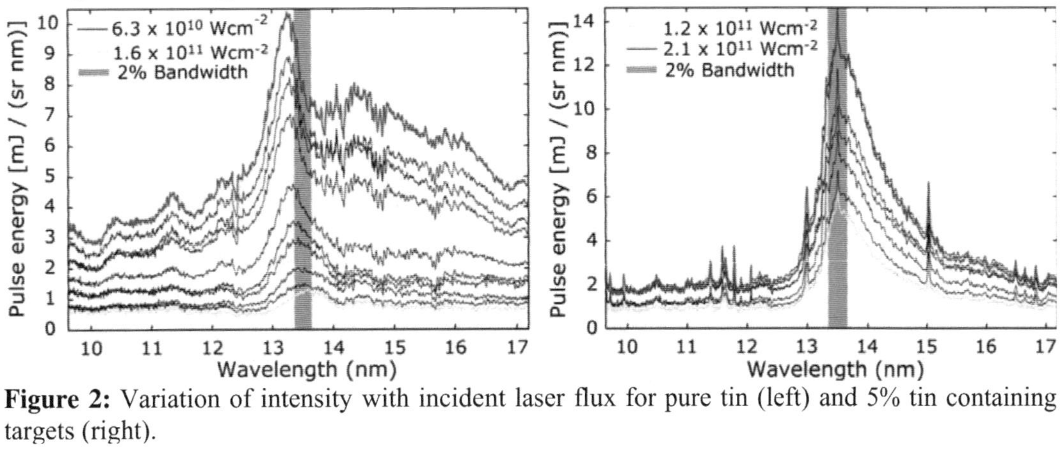


**Figure 2:** Variation of intensity with incident laser flux for pure tin (left) and 5% tin containing targets (right).

trade-off between emission which in an optically thin regime scales linearly with number density and absorption which scales exponentially. The results are shown in Figure 2 and translate to maximum in-band CEs of 2.3% into 2π sr for $\Phi = 1.6 \times 10^{11}$ $Wcm^{-2}$ for pure tin and 2.9% into 2π sr for $\Phi = 2 \times 10^{11}$ $Wcm^{-2}$ for the target containing 5% tin [13].

2.2.2. *Isotropy of emission:* The above data were recorded from slab targets irradiated at normal incidence and viewed at 45% to the target normal. Isotropy of emission was assumed. However in the introduction it was noted that the ion distribution varies with angle and consequently one would expect both emission and absorption to be sensitive functions of angle. Moreover, the plasma core is typically 220-300 μm in diameter but only ~ 100 μ deep so geometrical effects are important also. In a detailed study it was found that maximum emission was obtained when the plasma was viewed close to normal incidence and the intensity dropped by a factor of 2 in going from 20° to 90°. The average integrated flux was essentially the same as that recorded at 45 °. Thus care is needed in performing CE measurements and comparisons and the disagreement among a number of published CEs.

2.2.3. *Effect of a prepulse:* Previous work with laser produced plasmas has shown that increased UV and x-ray flux may be obtained if a prepulse is used to produce a seed plasma that is then rapidly heated by the main pulse [14]. In an experiment to explore the effects of a prepulse on inband emission from a tin plasma, a 170 ps , 500 mJ, 1.06 μm pulse was divided into differing pre-/ main pulse fractions and the delay between them was varied using an optical delay line. The results are shown in Figure 3 where it is seen that at an interpulse delay of 5.5ns, for a pre-/main pulse ration of 1/9, an increase in intensity of 80% was achieved.

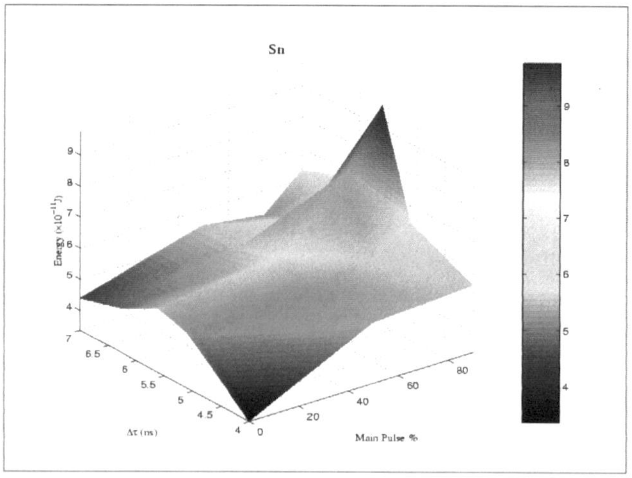


**Figure 3.** Effects of using prepulses: EUV intensity as a function of interpulse delay and fraction.

In a subsequent experiment the effects of heating a plasma produced by a 14 ns pulse with a 170 ns one was also investigated. In this case a 75% increase of in-band CE was obtained using a pure tin target pointing to the possibility of obtaining an overall efficiency of ~ 4%.

2.3. *Theoretical modelling of LPPs for EUVL applications:*
Considerable effort has been expended on modeling the emission from LPPs in order to identify the optimum set of operating parameters to optimize the in-band CE. These have included both steady state and time dependent approaches.

2.3.1. *Steady state plasmas*. In this approach ion fractions as a function of electron temperature were calculated assuming a collisional radiative equilibrium (CRE) [15]. The emission spectra from $4p^6 4d^n - 4p^5 4d^{n+1} + 4d^{n-1}4f + 4d^{n-1}5p$ transitions were calculated for each ion stage and then weighted with the ion fraction appropriate to a particular electron temperature and summed to give a spectrum for that temperature. The results, with the line distributions fitted to Gaussians and assuming optical transparency are shown in Figure 4. From this figure it is seen that the optimum electron temperature is in the 30-40 eV range.

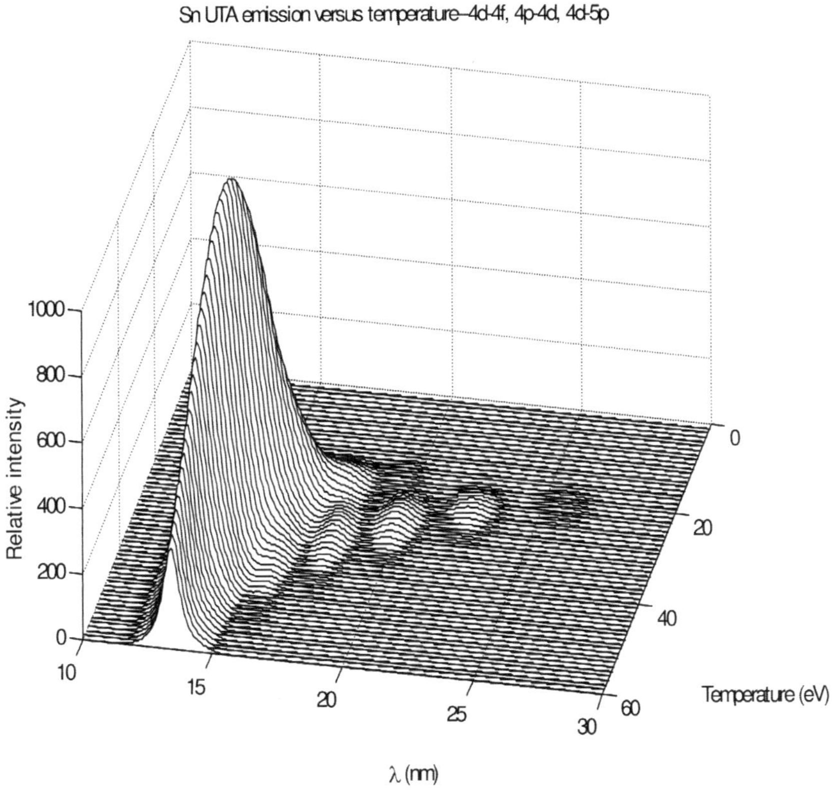


**Figure 4:** Calculated emission from an optically thin Sn plasma as a function of electron temperature.

2.3.2. *Time dependent modeling*: For a more realistic description, one has to allow for plasma expansion and optical opacity effects and perform a full time dependent calculation. A number of different results using 1- and 2-D models to describe the plasma expansion have been obtained and all point to a CE of about 3% being attainable with Nd:YAG plasmas, which is limited by absorption in the outer layers. If the opacity could be reduced and the most intense emission zones imaged directly, then a CE of 5-6% should be achievable [3, 4]. One way to reduce the amount of self absorption in the plume could be to use a modified laser pulse profile. Indeed a recent calculation using the Z* code has shown that if a flat-top rather than Gaussian profile were used the CE is enhanced by almost 100% [5].

An alternative approach would be to use a lower density plasma and this scenario is realized in a $CO_2$ plasma. Here, because of the lower cutoff density, electron and ion densities are only 1% of those of Nd:YAG plasmas and the emission approximates better to that shown in Figure 2. for 5% Sn targets. CEs of 4-4.5% [16,17] have been measured with such plasmas and the best laser source now available comprises mass limited droplet injection, with Nd:YAG pumping to vapourise and ionize it that then expands to a size where the average electron density matches the $CO_2$ pulse which then illuminates it [17]. The disadvantage is that currently $CO_2$ lasers have pulse lengths in the 40 - 50 ns range and the plasma expands to a lower suboptimal density within this time frame so only the early part of the pulse is fully utilised. For bulk targets, preliminary Z* calculations show that the EUV output remains essentially flat regardless of the length of the laser pulse for pulses in excess of 10 ns duration, while for significantly shorter pulses the EUV CE is reduced, presumably because of the importance of reflection at the beginning of the ablation process.

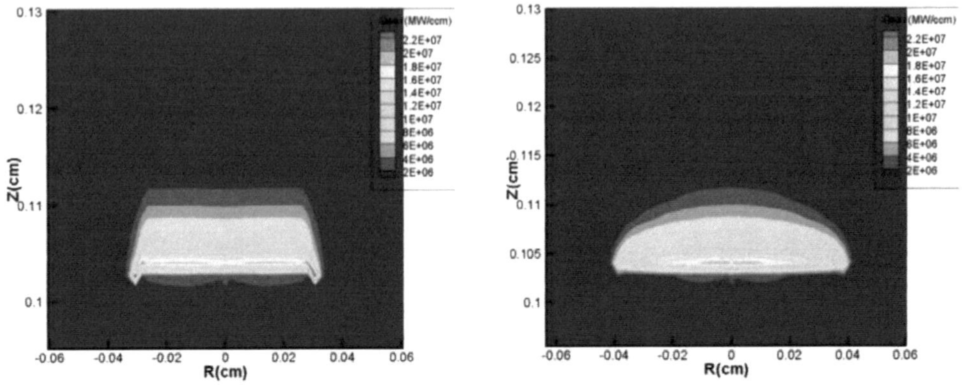


**Figure 5.** EUV emission from slab targets produced by Nd:YAG pulses with different spatial profiles: Flat-top (left) and Gaussian (right). Note that for observation from the sides, the distance to the peak emission region is shorter for the flat-top pulse. This leads to an enhancement of EUV CE by almost a factor of 2.

## 3. Photoabsorption studies.

The EUV spectra of the elements from Z = 62 to 74 (i.e. samarium to tungsten) produced by focusing the output pulse of a Q-switched laser onto a solid target at power densities in the range $10^{11}$ to $10^{12}$ $Wcm^{-2}$ contain extensive regions of line free continua [17]. In particular the spectrum of samarium is line free from 3 –200 nm, while with increasing Z the long wavelength limit decreases, until at tungsten the continuum is line free from 3-13 nm although the line density is very low below 20 nm. Since the intensity of the recombination radiation scales with $<\zeta>^4$ where $\zeta$ is the ionic charge, its importance is expected to increase with Z since the average ionic charge increases as the ionisation potentials are on average lower for a given $\zeta$ with increasing Z and the ions spend more time in the interaction region because of their increased mass. However such behaviour cannot explain the absence of lines over extensive energy ranges in the heavier lanthanides since the extent of the continuum actually decreases with Z. Furthermore past Z = 74 lines are again present throughout the XUV in plasmas of all elements investigated up to and including uranium. It was shown that, because of the effects of 4*f* orbital collapse, 4*f* electrons are present in the ground configurations of all elements in the range $62 \leq Z \leq 74$ in the ion stages produced in a typical Q-switched laser plasma where the emission is in general dominated by species with charge states ranging from 4+ to 16+ [17]. At the lower Z end of this group of elements the 4*f* binding energy in the neutral atom is lower than either the 5s or 5p but, because of the tendency of levels to regroup according to principal quantum numbers with increasing ionisation, the 4*f* binding energy is greater than either the 5s or 5p at the fifteenth or sixteenth ion stage [18]. At intermediate ion stages there is a near degeneracy first of 4*f* and 5p and later of 4*f* and 5s energies and as a result the ground configuration in most ion stages contains an open 4*f* subshell. These near degeneracies produce a vast array of low lying excited states which greatly increases the number of possible transitions. It was also shown that the statistics of the levels follow the same behaviour as compound nuclear states and may form a quantum chaotic system [19]. The lines which appear in the spectra of higher Z elements of this group only do so because the 4*f* subshell is filled or contains at most one vacancy, and ground configurations of many stages involve only open 5s or 5p subshells which give rise to relatively simple spectra. From these considerations it was deduced that the transitions blended to form weak bands of quasi-continua which blended with and enhanced the recombination continuum. These plasmas were shown to provide a stable, compact and bright alternative source of XUV radiation to the synchrotron in experiments where a knowledge of absolute XUV flux or polarisation was not required. In this mode they have been used for applications that include photoabsorption studies using the dual laser plasma (DLP) technique.

The basis of the DLP technique is that a Q-switched laser (1 Hz, 0.8J in 45 ns) may be focused using a cylindrical or spherical lens onto the element of interest at a relatively low power density so that the maximum ion stage obtained does not radiate itself in the EUV while a second laser pulse is focused tightly onto a W or Sm target to yield a backlighting continuum. By varying the interplasma time delay or probing different plasma regions, the absorption of particular ion stages may be differentiated and optimized [20].

A typical result of an experiment on iodine photoabsorption at two different time delays is shown below [21] where they are compared with calculations using the Cowan code [22]. In obtaining the theoretical spectra, cross-sections were obtained for each term of the open 4d subshell calculations which were then weighted and summed assuming a Boltzmann distribution for a particular plasma

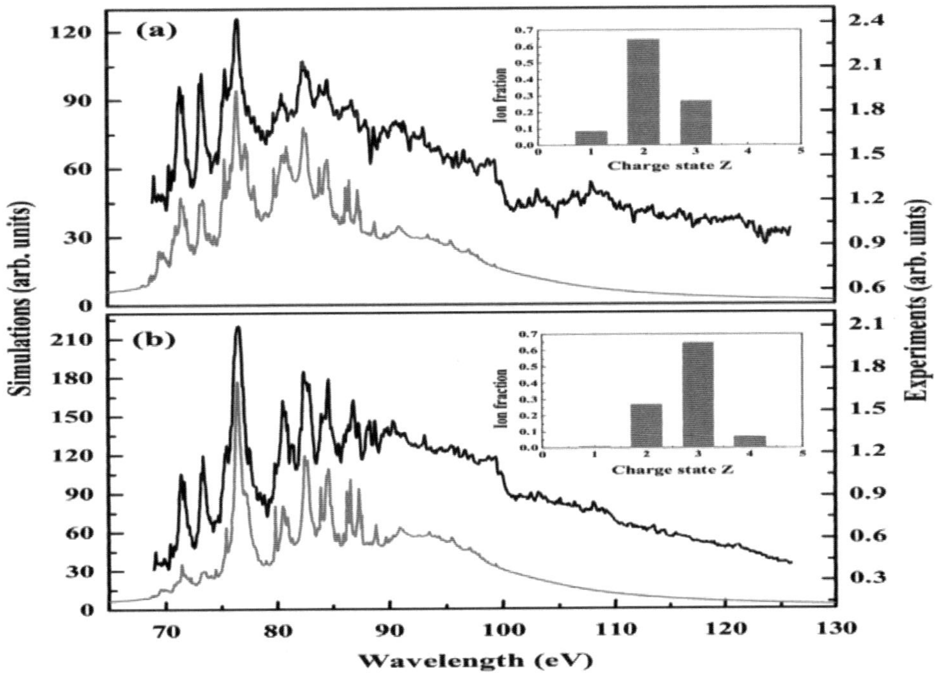


**Figure 6:** 4d photoabsorption in iodine plasmas recorded at two different time delays (black) compared with calculations. The differences arising from having different ion populations is clearly evident.

temperature as described elsewhere [23]. This process was repeated for each ion stage and the contribution of each to the absorption profile then summed again assuming the same constant temperature. For all but closed shell species it is difficult to isolate a spectrum free of excited state absorption. The situation is particularly complex for open 4d and 4f ions and it was recently found to be impossible to observe a simple spectrum in higher ions of the Xe isolectronic sequence again reflecting the effects of 4f and 5p level crossing.

**Acknowledgement:**

The support of Science Foundation Ireland and the China-Ireland Collaboration Fund is gratefully acknowledged.

## References

1 Filevitch J *et al. Phys. Rev.* E 2003 **67** 056409
2 Bauche J, Bauche-Arnoult C and Klapisch M 1987 *Adv. At. Mol. Phys.* **23** 131
3 Nishihara K, Sasaki A, Sunahara A and Nishikawa T 2006 *EUV Sources for Lithography*, (SPIE press, ed. V Bakshi, 339
4 White *et al.* 2007 *J. Appl. Phys.* **101** 043301
5 White J. 2008 *et al. Appl. Phys. Lett.* **93** 151501
6 S S Churilov, Y N Joshi, J Reader and R R Kildiyarova 2004 *Phys. Scr.* **70** 126
7 Shields H *et al.* 2002 *Proc SPIE* **4688** 94
8 Fahy K *et al.* 2004 *J Phys. D: Appl. Phys.* **37** 3225
9 O'Sullivan G and Faulkner R 1994 *Opt. Eng.* **33** 3978
10 Sasaki A J. *et al.* 2008 *Phys. Conf. Series* **112** 042062
11 George S A *et al.* 2007 *Opt. Express* **15** 16348
12 *Proceedings of 6th EUVL Symposium* ed V. Bakshi 2007 Sapporo Japan
13 Hayden P *et al.* 2006 *J. Appl. Phys.* **99** 093302
14 Kodama K and Mochizuki T 1987 *Opt. Lett.* **12** 990
15 Colombant D F and Tonon G F 1973 *J. Appl. Phys.* **44** 3524
16 Ueno Y et al. 2007 *Appl. Phys. Lett.* **91** 231501
17 Fujioka S *et al.* 2008 *Appl. Phys. Lett.* (in Press)
18 Carroll P K and O'Sullivan G 1982 *Phys. Rev.* A **25** 275
19 O'Sullivan G *et al.* 1999 *J. Phys. B:Atom. Molec. Opt. Phys.* **32** 1893
20 Costello J T *et al.* 1991 *Phys. Scr.* **T34** 77
21 Su M G *et al.* 2008 *Phys. Rev.* A **78** 043401
22 Cowan R D 1981 *"The Theory of Atomic Structure and Spectra"* University of California Press
23 D'Arc R, Costello J T, McGuinness C and O'Sullivan G 1999 *J. Phys. B: Atom. Molec. Opt. Phys.* **32** 4859

# Non-statistical population of $1s2s2p\,^4P_J$ states by charge transfer into multiply charged ions

**T.J.M. Zouros**[1], **Béla Sulik**[2], **L. Gulyás**[2], **K. Tőkési**[2]
[1]Department of Physics, University of Crete, P.O. Box 2208, 71003 Heraklion, Crete, Greece
[2]Institute of Nuclear Research of the Hungarian Academy of Sciences (ATOMKI), H-4001 Debrecen, Hungary

E-mail: tzouros@physics.uoc.gr

**Abstract.** The experimentally observed enhancement of $(1s2s\,^3S)n\ell\,^4P$ to $(1s2s\,^3S)n\ell\,^2P$ populations in collisions of He-like ions with $H_2$ and He targets is discussed and explained by a state-selective cascade feeding mechanism. Continuum Distorted Wave (CDW) and Classical Trajectory Monte Carlo (CTMC) calculations show that large populations of higher lying $(1s2s\,^3S)n\ell\,^{2,4}L$ Rydberg states for $\ell$=$L$=0, $n$-1 and $n = 2-8$ can be formed in 0.25-2 MeV/u collisions by direct capture to metastable $F^{7+}(1s2s\,^3S)$ ions. These states subsequently decay through a cascade sequence of radiative or Auger transitions dictated by angular momentum and spin selection rules. All radiative (E1) and Auger transitions for $n \leq 5$ were evaluated using the Cowan package for Hartree-Fock calculations. The $(1s2s\,^3S)n\ell\,^2L$ doublets were found to Auger decay rapidly to the $1s^2$ level allowing for negligible cascade feeding of lower lying doublets including the $1s2s2p\,^2P_{\pm}$. Alternatively, Auger decay of the $(1s2s\,^3S)n\ell\,^4L$ quartets is blocked by spin selection rules. Decay can proceed, however, through dipole transitions, but only to other *quartets*, including the lowest lying $1s2s2p\,^4P$ state, which in this way acts as an "excited ground state" eventually collecting most cascade flux. This selective cascade feeding mechanism thus leads to the non-statistical enhancement of the $1s2s2p\,^4P$ quartet population over those of the $1s2s2p\,^2P_{\pm}$ doublet populations providing an alternative explanation to the recently suggested novel mechanism of dynamical Pauli exchange.

## 1. Introduction

Cascade feeding of atomic levels has been intensely investigated from the early days of beam-foil spectroscopy [1, 2, 3, 4, 5] to the more recent investigation of wake field effects in fast ions emerging from thin foils [6], the production of hollow atoms by charge transfer in highly charged ion-surface interactions [7], K x-ray emission due to charge exchange between solar wind heavy ions and cometary gases [8], as well as fusion plasma research [9] and references therein. While the process itself is well understood, a detailed cascade feeding investigation is experimentally impractical and theoretically very tedious since a great many transitions are involved. Thus, few complete studies exist and these have been mostly performed at low collision energies. At higher collision energies in the MeV/u range electron capture is known to primarily populate the K- or L-shells and cascade feeding is usually considered to be unimportant.

More relevant to this work dealing with the production of $1s2s2p\,^4P$ states in ion-atom collisions, early Auger and x-ray beam-foil investigations of fluorine ions [1, 2] suggested the existence of long-lived Li-like Rydberg $(1snln'l'\,^4L)$ *quartet* states that should strongly feed lower lying levels. Stolterfoht *et al.* [4] suggested that the observed strong population enhancement of the neon $1s2s2p\,^4P$ state above statistics, observed in 200 MeV $Xe^{31+}$ collisions with neon, could be due to cascade feeding from such higher-lying quartet states produced in the collision. About a decade later, Bliman *et al.* [10] found that higher lying $(1s2s\,^3S)n\ell\,^{2,4}L$ (with $L = \ell$) states were produced by direct $n\ell$ capture in slow 3.8 keV/u collisions of $O^{6+}$ with He. Anthony *et al.* [11] and Lee *et al.* [12, 13] determined the production cross sections $\sigma(^4P)$ and $\sigma(^2P_\pm)$ for populating the $1s2s2p\,^{2,4}P$ states in mixed initial state $(1s^2/1s2s\,^3S)$ He-like $C^{4+}$ and $F^{7+}$ collisions with $H_2$ and He using high resolution (zero-degree) Auger electron projectile spectroscopy [14, 15], demonstrating the importance of both electron capture and transfer excitation (TE) mechanisms over the collision energy range of 0.25 – 2 MeV/u. However, all these suggestions of cascade feeding were never backed up by any type of detailed analysis.

More recently, cascade feeding of the $1s2s2p\,^4P$ state from higher lying $1s2sn\ell\,^4L$ states ($n > 2$) formed by transfer-loss (TL) mechanisms in collisions of ground-state Li-like $O^{5+}(1s^22s)$ and $F^{6+}(1s^22s)$ ions with He and $H_2$ was shown [16] to largely account for the existing more than a factor of 10 disagreement between experimentally measured TL cross sections and independent particle model TL calculations with $n = 2$ [17, 18]. In fact, CDW calculations showed $n\ell$ transfer into the $1s2s\,^3S$ levels of $O^{6+}$ and $F^{7+}$ for $n = 3-7$ to be as important as capture to just $n = 2$, even for relatively fast ~1 MeV/u collisions [16]. Thus, it has become clear that cascade feeding can also be important even in faster collisions and needs to be carefully considered where relevant.

In 2004 Tanis *et al.* [19] reported on the ratio $R{=}\sigma(^4P)/[\sigma(^2P_-) + \sigma(^2P_+)]$, where $\sigma(X)$ with $X = 1s2s2p\,^{2,4}P$ are the production cross sections of the corresponding $F^{6+}$ states in collisions of $F^{7+}$ ions with He and Ne. Auger projectile electron spectroscopy measurements at 1.1 MeV/u found $R$=2.9 for He and $R$=2.5 for Ne. Spin statistics, for purely $2p$ *single* electron capture ($\sigma = \sigma_{2p}$) to the $1s2s\,^3S$ long-lived beam component, give $R$=2 [20, 21]. This seemingly anomalous enhancement of $R$ was interpreted [19] as evidence for a new dynamical $2p$ capture mechanism named *Pauli exchange* involving projectile and target electrons with the same spin alignment. In 2008, Tanis *et al.* [22], using a technique for producing practically pure ground state He-like beams [23, 24, 25], experimentally subtracted the ground state component from the mixed beam spectrum to obtain $1s2l2l'$ intensities due solely to the $C^{4+}(1s2s\,^3S)$ component simplifying the theoretical interpretation. However, as in their 2004 article, the suggested Pauli exchange mechanism still remained non-amenable to calculation. Nonetheless, the cascade mechanism was also acknowledged and supporting theoretical calculations suggested that about half of the observed enhancement could indeed be attributed to cascade effects.

In this report, we give a brief account of our cascade feeding analysis resulting in the Auger electron emission from the $F^{6+}(1s2s2p)\,^4P$ and $^2P_\pm$ states produced in 0.25-2 MeV/u collisions of $F^{7+}$ ions with He and $H_2$ gas targets as already also presented in Ref. [26] with some additional information. Both our CDW [27] and CTMC [28] results indicate that a substantial number of $(1s2s^3S)n\ell\,^{4,2}L$ Rydberg levels with $\ell = L$ and $n = 2-8$ are populated by single $n\ell$ electron transfer to the metastable $1s2s\,^3S$ ion beam component. An in-depth radiative cascade feeding analysis based on Hartree-Fock calculations of all radiative (E1) and Auger transitions for $n \leq 5$ is presented. A cascade mechanism preferentially feeding the $1s2s2p\,^4P$ becomes evident. Above the collision energy of 0.7 MeV/u, this satisfactorily accounts for the observed experimental results of Ref. [19] as well as the older measurements of Lee *et al.* [13, 12] for $F^{7+}$ collisions on both He and $H_2$. Below 0.7 MeV/u, however, large still unexplained discrepancies persist.

## 2. Theoretical results

In Fig. 1, CDW total single capture cross sections $\sigma_n$ are shown as a function of the principal quantum number $n$. The effective ion charge seen by each shell was computed using Slater screening [29] known to improve capture results [27]. Additional CTMC calculations performed at 0.25, 0.5 and 1.1 MeV/u (not shown) were found in good overall agreement with the CDW results for the $n$-distributions [Fig. 1 (right)] and $nl$ distributions (not shown). Since the CTMC

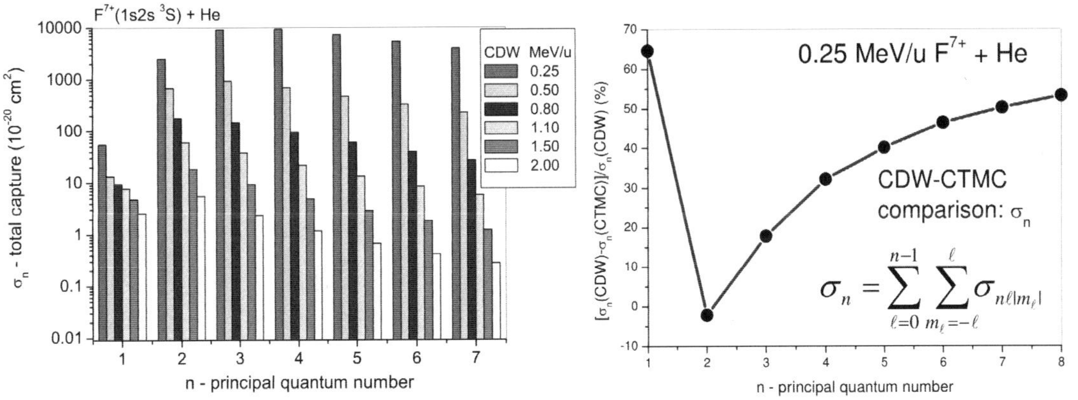


**Figure 1.** (Color on line) Calculated single capture cross section $\sigma_n$ for $F^{7+}(1s2s\,^3S)$+He collisions leading to the production of $1s2snl$ levels. (Left) CDW results for 0.25–1 MeV/u collisions. For each $n$ results have been summed over all available orbital momenta $\ell$ and total spin $s = 3/2, 1/2$ values. (Right) Relative difference between CDW and CTMC at 0.25 MeV/u. Overall agreement between the two calculations for $n > 1$ is seen to be within ∼50%.

is a *non*-perturbative method such agreement provides additional support as to the accuracy of our CDW capture results.

The well-known Cowan Hartree-Fock package [30, 31] was used to calculate all relevant $F^{6+}(1s2sn\ell\,^{2,4}L_J)$ Li-like energy levels, dipole and Auger transition rates for principal quantum number $2 \leq n \leq 5$ and $\ell = 0,\, n-1$ with more accurate rates for $n = 2, 3$ transitions taken from Refs. [32, 33, 34]. In Fig. 2 of Ref. [26], the energy level scheme with transition rates was shown and will not be repeated here. The underlying selective cascade feeding mechanism becomes instantly apparent: the $(1s2s\,^3S)n\ell\,^2L$ doublets are found to Auger decay strongly to the K-shell (thick blue transition lines), while the $(1s2s\,^3S)n\ell\,^4L$ quartets cannot as they are blocked by spin selection rules. Radiative E1 transitions, however, are readily allowed, but only to lower lying *quartets*. Thus, radiative branching ratios, $\mathrm{BR}_{rad}$, for transitions between quartets are much larger than for corresponding transitions between doublets, effectively resulting in the strong cascade feeding of the lowest lying $1s2s2p\,^4P$ states only [26].

Separate quartet and doublet cascade transition matrices were constructed as negligible cross feeding was assumed due to spin conservation. A detailed time-dependent cascade feeding analysis [35] was performed which used the computed capture cross sections $q(n\ell m_\ell)$ to provide the initial $t = 0$ $(1s2s\,^3S)n\ell\,^{2S+1}L_J$ population cross sections $\sigma(nLSJ)$ according to spin and angular momentum coupling statistics requiring $\sigma(nLSJ) = \frac{(2J+1)}{6(2L+1)}\sum_{m_\ell=-\ell}^{l}\sigma_{n\ell|m_\ell|}$. This is seen to give the correct 2:1 ratio of $1s2s2p\,^4P$ to $1s2s2p\,^2P$ configuration populations [21], i.e. $\sum_J\sigma(n=2, L=1, S=3/2, J)/\sum_J\sigma(n=2, L=1, S=1/2, J) = 2$. We note that a mixed $(1s2s\,^3S, 1s^2)$ component He-like beam also allows for the population of the $1s2s2p\,^2P_\pm$ levels by *non-negligible* Transfer-Excitation (TE) from the $F^{7+}(1s^2)$ ground state [13, 12] included here.

In Fig. 2 (left) our results are compared to the 0° Auger electron emission single differential cross sections (SDCS) measurements of Lee *et al.* [13, 12] based on the metastable beam fractions determined by Teresawa *et al.* [36]. Good overall agreement is observed between theory and experiment above about 0.7 MeV/u. The Auger Resonant Transfer Excitation (RTEA) and Non-resonant Transfer Excitation (NTEA) contribution assessment is also included. For the $H_2$ target, the distinct hump in the doublet SDCS due to RTEA [13, 12] is very nicely reproduced by our calculations [37, 15]. The energy dependence of the data is well reproduced even on an absolute scale when cascade contributions are included, while theory clearly underestimates the

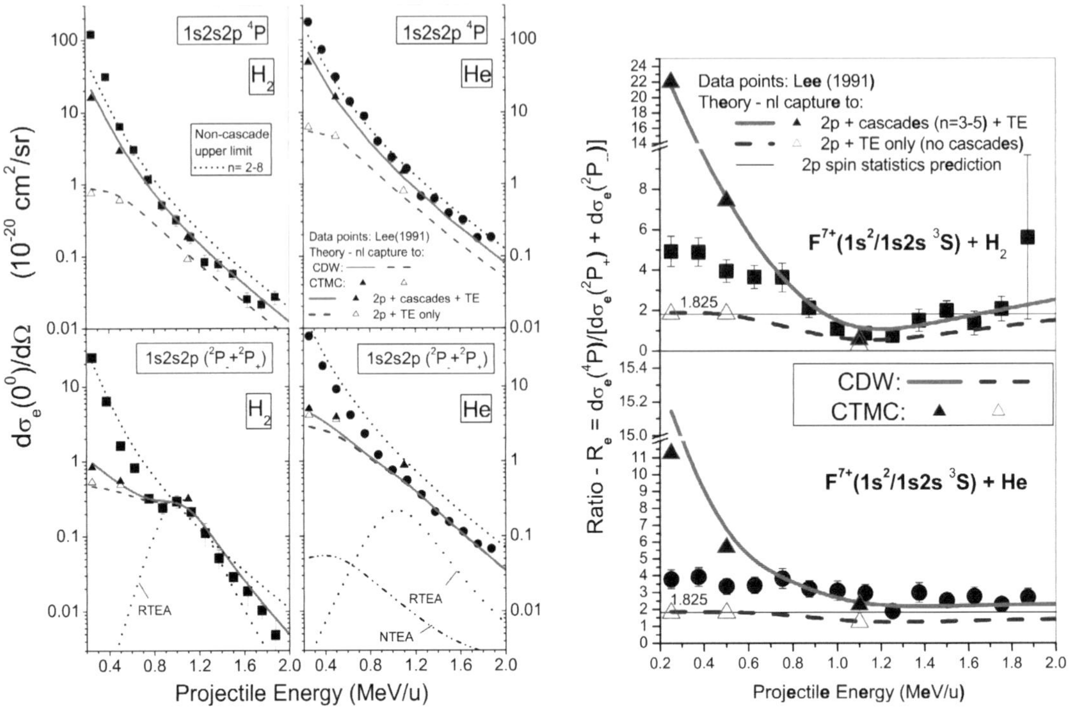


**Figure 2.** (Color on line) Absolute 0° Auger emission SDCS for the $1s2s2p\,^4P$ and the sum of $1s2s2p\,^2P_{\pm}$ levels for $H_2$ and He (left) and their ratio $R_e$ (right). Data points from the measurements of Lee *et al* [13, 12]. Lines and triangles: Theoretical CDW and CTMC results, respectively. Solid red lines and filled triangles include cascade feeding from $(1s2s\,^3S)n\ell$ with $n = 3-5$, while dashed blue lines and open triangles do not ($n = 2$ only). Transfer-Excitation (TE) Auger contributions (both RTEA and NTEA) to the $^2P_{\pm}$ from the $1s^2$ are also shown. TE contributions to the $^4P$ are assumed negligible [12]. An upper limit estimate to capture into the sum of $n = 2-8$ levels is also shown. (Right) Ratio $R_e$ of experimental and theoretical Auger SDCS shown on left. Expected theoretical spin statistics ratio assuming isotropy is 1.825 [26].

experiment without the cascade contributions. Furthermore, the cascade contributions are seen to be needed only in the case of the $1s2s2p\,^4P$ results consistent with our model. With decreasing collision energy below 0.7 MeV/u, cascade feeding becomes increasingly more intense and even the $1s2s2p\,^2P_{\pm}$ levels start to receive some cascade contributions (see also Fig. 3 of Ref. [26]). Our cascade analysis for $n \leq 5$ seriously underestimates both quartet and doublet production at low collision energies for reasons not yet understood. The disagreement is worse for doublets from collisions with $H_2$ targets, seemingly ruling out non-resonant transfer excitation (NTE) as a possible cause, since this is known to be much weaker for $H_2$ than for He [37].

To estimate additional possible contribution from higher lying levels with $n > 5$, an upper limit CDW estimate on the Auger SDCS is also provided in Fig. 2. It ideally assumes *all* capture to $n = 2-8$ levels ends up directly into the $1s2s2p\,^4P$ or $^2P_{\pm}$ states. In reality, cascade mechanisms will transfer this population to the lower-lying levels with decreasing efficiency as the number of steps needed is increased (cascade order [5]). Below 0.7 MeV/u, the measured SDCS for the $^4P$ are seen to lie above this limit, while for the $^2P_{\pm}$, they lie mostly below. Consequently, the theoretically available direct capture population seems to be clearly insufficient in the case of the $^4P$ levels. In the case of the $^2P_{\pm}$ levels this is not as clear. Either higher capture populations are needed under the present inefficient cascade transfer to the $^2P_{\pm}$, or a much more efficient cascade transfer mechanism with the present calculated capture populations. Clearly, at low energy more $^2P$ intensity is missing than $^4P$ as is evident from the overestimation of their theoretically calculated ratio $R_e$ (right).

To further investigate this puzzling state of affairs we suggest the following possibilities: (a) "cross feeding" between the quartet and doublet populations, considered here to be negligible to first-order, could transfer available capture flux from doublets to quartets this way resolving both puzzles, (b) low energy capture could be more intense than what has been calculated, (c) calculated low energy $\ell$-distributions could be wrong particularly for the highest $\ell$-levels (the Yrast levels) known to contribute most to cascade feeding [5], (d) de-alignment effects at 0° observation [38, 39] need to be carefully considered given possible redistribution of magnetic sub-state populations due to cascade feeding [40, 41]. When included just for 2p capture they were found to affect Auger yields within a factor of 2, particularly for He. Best overall agreement, however, was found assuming isotropy as shown here, (e) A new mechanism further enhancing capture cross sections at the lowest collision energies might be needed. However, such a mechanism would need to enhance both the $^4P$ and $^2P_{\pm}$ population levels and not just the $^4P$ as proposed by Tanis *et al.* since both seem to be in need of additional capture flux as already seen in Fig. 2. Furthermore, as shown in the recent $C^{4+}$ results on Pauli exchange [22], the experimental ratio $R$ increases from 6 to 8 as the collision energy increases from 0.5 to 1 MeV/u. This is particularly puzzling given that most exchange interactions usually decrease in strength as the required interaction time is also correspondingly decreased [17], (f) the measured cross sections of Lee *et al.* might be too large at the lowest collision energies if computed with the wrong metastable fraction since this was not measured under the same conditions, but taken from the older measurements of Teresawa *et al.* [36].

The ratio of quartet to doublet SDCS is plotted in Fig. 2 (right). $R_e$ with both cascade and TE corrections (red line) is clearly in better agreement with the data than without the cascade corrections (blue line). For $H_2$ agreement is excellent, even reproducing the minimum around 1.1 MeV/u due to RTE, while for He it is also as good. However, below 0.7 MeV/u, the effect of the observed disagreement in the absolute values of the computed SDCS is clearly evident.

## 3. Summary and Conclusions

In conclusion, we have shown through extensive CDW and supporting CTMC single capture calculations that $1s2sn\ell$ levels of both quartets and doublets can be expected to be strongly populated for $n \lesssim 8$ in collisions $F^{7+}(1s2s\,^3S)$ ions with $H_2$ and He targets. These states subsequently decay through a cascade sequence of radiative or Auger transitions dictated by angular momentum and spin selection rules. A detailed time-dependent cascade feeding calculation including all radiative (E1) and Auger transitions for $n \leq 5$ shows the $(1s2s\,^3S)n\ell\,^2L$ doublets to Auger decay rapidly to the $1s^2$ level allowing for negligible cascade feeding of lower lying doublets including the $1s2s2p\,^2P_{\pm}$. Alternatively, Auger decay of the $(1s2s\,^3S)n\ell\,^4L$ quartets is blocked by spin selection rules, with decay proceeding through intense dipole transitions, but only to other *quartets*, including the lowest lying $1s2s2p\,^4P$ state. This results in the strong enhancement of the $1s2s2p\,^4P$ level populations relative to the similarly configured $1s2s2p\,^2P_{\pm}$ doublets. Our calculations are in excellent agreement with experimental results when cascade contributions are included over the energy range of 0.7-2 MeV/u. However, at collision energies below 0.7 MeV/u, where cascade feeding is most intense, strong disagreement still persists. This is particularly puzzling given the good agreement above 0.7 MeV/u. Some suggestions for further investigation are proposed. Clearly more theoretical work is needed, particularly at the lower collision energies, to arrive at a more satisfactory explanation. Additionally, it would also be interesting to experimentally investigate the isoelectronic behavior of $R$.

## Acknowledgments

We would like to thank Alexander E. Kramida (Atomic Physics Division, NIST) for his generous help in providing us with a working version of the Cowan code for the PC [31], setting up the Li-like energy level input and sharing his extremely useful Perl scripts used to extract all information from the Cowan output. We also thank Tom Gorczyca and John Tanis (Western Michigan University), Tom Kirchner (Clausthal University of Technology) and Dominique Vernhet (INSP) for useful clarifications. This work was supported by the Greek-Hungarian S & T cooperation grants GR-11/03 and OTKA grants K73703 and IN64319.

## References

[1] I. A. Sellin *et al.*, Phys. Rev. Lett. **27**, 1108 (1971).
[2] P. Richard *et al.*, Phys. Rev. Lett. **30**, 888 (1973).
[3] L. J. Curtis, in *Beam Foil spectroscopy*, edited by S. Bashkin (Springer–Verlag, Berlin, 1976), p. 63.
[4] N. Stolterfoht, D. Schneider, R. Mann, and F. Folkmann, J. Phys. B **10**, L281 (1977).
[5] S. M. Younger and W. L. Wiese, Phys. Rev. A **17**, 1944 (1978).
[6] E Lamour, B Gervais, J P Rozet and D Vernhet, Phys. Rev. A **73**, 042715 (2006).
[7] N. Stolterfoht *et al.*, Phys. Rev. A **52**, 445 (1995).
[8] P. Beiersdorfer, J. H. Scofield, and A. L. Osterheld, Phys. Rev. Lett. **90**, 235003 (2003).
[9] P. Beiersdorfer *et al.*, Phys. Rev. Lett. **85**, 5090 (2000).
[10] S. Bliman *et al.*, J. Phys. B **25**, 2065 (1992).
[11] J. M. Anthony *et al.*, J. de Phys. (Paris) **48**, C9 301 (1987).
[12] D. H. Lee *et al.*, Phys. Rev. A **44**, 1636 (1991).
[13] D. H. Lee *et al.*, Nucl. Instrum. Methods Phys. Res. B **56/57**, 99 (1991).
[14] N. Stolterfoht, Phys. Rep. **146**, 315 (1987).
[15] T. J. M. Zouros and D. H. Lee, in *Accelerator-based atomic physics techniques and applications*, edited by S. M. Shafroth and J. C. Austin (AIP Press, Woodbury, NY, 1997), pp. 426–79.
[16] T. J. M. Zouros, B. Sulik, L. Gulyás, and A. Orbán, J. Phys. B **39**, L45 (2006).
[17] T. J. M. Zouros, D. H. Lee, and P. Richard, Phys. Rev. Lett. **62**, 2261 (1989).
[18] B. Sulik, T. J. M. Zouros, A. Orbán, and L. Gulyás, J. Electron Spectrosc. and Relat. Phenom. **114-116**, 191 (2001).
[19] J. A. Tanis *et al.*, Phys. Rev. Lett. **92**, 133201 (2004); Phys. Rev. Lett. **96**, 019901 (2006).
[20] E. P. Benis *et al.*, Phys. Rev. A **69**, 052718 (2004).
[21] E. P. Benis *et al.*, Phys. Rev. A **73**, 029901 (2006).
[22] D. Strohschein *et al.*, Phys. Rev. A **77**, 022706 (2008).
[23] M. Zamkov *et al.*, Phys. Rev. A **64**, 052702 (2001).
[24] M. Zamkov, E. P. Benis, P. Richard, and T. J. M. Zouros, Phys. Rev. A **65**, 062706 (2002).
[25] E. P. Benis, M. Zamkov, P. Richard, and T. J. M. Zouros, Nucl. Instrum. Methods Phys. Res. B **205**, 517 (2003).
[26] T. J. M. Zouros, B. Sulik, L. Gulyas, and K. Tokesi, Phys. Rev. A **77**, 050701 (2008).
[27] L. Gulyás, P. D. Fainstein, and T. Shirai, Phys. Rev. A **65**, 052720 (2002).
[28] K. Tökési and A. Kövér, J. Phys. B **33**, 3067 (2000).
[29] J. C. Slater, Phys. Rev. **36**, 57 (1930).
[30] R. D. Cowan, *The theory of atomic structure and spectra* (University of California Press, Berkeley, CA, 1981).
[31] A. E. Kramida, *Cowan's code package for calculation of atomic properties, adapted for PC*, Institute for Spectroscopy RAS, Troitsk, Russia, *http://das101.isan.troitsk.ru/cowan.htm.*
[32] M. H. Chen, B. Crasemann, and H. Mark, Phys. Rev. A **27**, 544 (1983).
[33] M. H. Chen, Atomic Data and Nuclear Data Tables **34**, 301 (1986).
[34] B. F. Davis and K. T. Chung, Phys. Rev. A **36**, 1948 (1987); Phys. Rev. A **39**, 3942 (1989); Phys. Rev. A **41**, 5844 (1990).
[35] L. J. Curtis, American J. Phys. **123**, 1123 (1968).
[36] M. Terasawa *et al.*, Phys. Rev. A **27**, 2868 (1983).
[37] T. J. M. Zouros, C. P. Bhalla, D. H. Lee, and P. Richard, Phys. Rev. A **42**, 678 (1990).
[38] B. Cleff and W. Mehlhorn, J. Phys. B **7**, 593 (1974).
[39] N. Stolterfoht *et al.*, Phys. Rev. A **48**, 2986 (1993).
[40] N. R. Badnell, M. S. Pindzola, and D. C. Griffin, Phys. Rev. A **41**, 2422 (1990).
[41] N. R. Badnell, Phys. Rev. A **44**, 1554 (1991).

Journal of Physics: Conference Series **163** (2009) 012005 doi:10.1088/1742-6596/163/1/012005

# Fragmentation and desorption in low-energy highly charged ion collisions with molecules and surfaces

**K Motohashi**

Department of Applied Physics, Tokyo University of Agriculture and Technology, 2-24-16 Koganei-shi, Tokyo 184-8588, Japan

motohasi@cc.tuat.ac.jp

**Abstract**. In order to study secondary-ion emission in low-energy highly charged ion collisions with molecules and surfaces, we performed coincidence measurements of secondary ions and scattered ions, scattered neutral atoms or secondary electrons. Fragmentation and desorption processes induced by electron captures were successfully extracted by observing the scattered ions/atoms with small scattering angles. Momentum imaging of the secondary ions offers a new analysis method when combined with translational energy spectroscopy or energy-gain spectroscopy of scattered ions. This technique was successful in clarifying the reaction pathways of the electronic transitions of molecules and following the dissociation processes in collisions between $Ar^{q+}$ ($q$ = 3–12) and $CF_4$ and $N_2$ molecules. We also successfully performed secondary ion mass spectroscopy of the topmost layers of the surfaces in glancing collisions between $Ar^{8+}$ and both GaN (0001) and ($000\bar{1}$) surfaces.

## 1. Introduction

The electron capture cross-section increases almost linearly with an increase in the charge state of incident ions [1]. In addition to a large cross-section that varies with the charge state of ions, the electron capture processes in collisions with highly charged ions (HCIs) have the important characteristic of *state selectivity* due to their resonant feature called the reaction window [2, 3]. The width of the reaction window is proportional to the collision velocity taking into account the uncertainty, with the barrier height formed in the HCI and target material. The probability of electron capture, especially in low-energy collisions, increases rapidly around a collision energy that matches the binding energy of the electron. On the other hand, fragmentation of molecules and desorption of surface atoms are often induced after electron capture, because the valence electrons corresponding to chemical bonds are captured preferentially by the HCI. In the case of soft collisions, e.g. less correspondence of an interaction with small impact parameters, fragmentation, desorption or sputtering with pure electronic transitions should occur. Soft, state-selective and efficient secondary-particle emission is expected.

To investigate the electronic transitions in HCI–molecule and HCI–surface interactions followed by secondary-particle emissions, time-of-flight mass spectroscopy of secondary ions was performed in coincidence with scattered ions and atoms.

## 2. Experimental setup and procedures

### *2.1. Ion–molecule collisions—Coincidence measurements between momentum-imaging spectroscopy of fragment ions and translational-energy spectroscopy of scattered ions*

Figure 1 shows a schematic illustration of the experimental setup. This setup [4, 5] makes it possible to perform momentum-imaging spectroscopy (MIS) of secondary ions [6] and translational energy spectroscopy (TES) of scattered ions [7] simultaneously. We observed processes with large impact parameters and small recoil energies by selecting small scattering angles $\theta_f$ ($\theta_f \leq 0.5°$) and small energy gain $Q$ with respect to the incident energy $E_0$ ($Q/E_0 \leq 1 \times 10^{-2}$) [4]. Limited processes without nuclear momentum transfer were observed selectively under these experimental conditions.

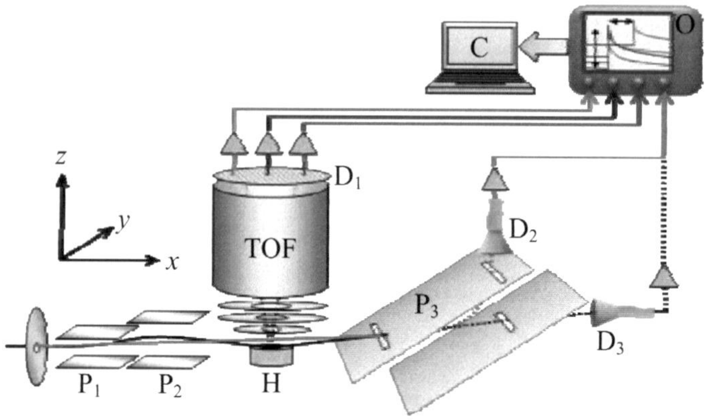


**Figure 1.** A schematic illustration of the experimental setup $P_1$–$P_3$, Parallel plates; H, Holder; $D_1$–$D_3$, Detectors; O, Oscilloscope; C, Computer; Red or black line, HCI trajectory in collisions with molecules or surfaces (A nozzle was replaced with the holder in ion-molecule collision experiments.)

The incident energy of $Ar^{q+}$ ($q$ = 3–12) was maintained at 8.00 keV. The ions were extracted from an electron beam ion source. The energy gain of the scattered ions was analysed using a 45° parallel plate ($P_3$ in Figure 1). The energy resolution of the analyser ($\Delta E/E$) was $2 \times 10^{-4}$. Target molecules were introduced from an effusive nozzle. Note that the nozzle was replaced with the sample holder (H), although it is not illustrated in Figure 1. The red curve in Figure 1 represents a typical HCI trajectory. The flight time $t$ and the position signals $x$ and $y$ of fragment ions hitting the two-dimensional position-sensitive detector ($D_1$) were recorded using a digital oscilloscope triggered by a scattered ion signal of the $D_2$ detector. A three-dimensional (3D) momentum vector of a fragment ion just after electron transfer collision was reconstructed from signals consisting of ($t$, $x$, $y$). The background pressure of the collision chamber was maintained below $2 \times 10^{-6}$ Pa.

### *2.2. Ion–surface collisions–Coincidence detection between scattered atoms/ions and secondary ions in glancing collisions*

A pair of parallel plates ($P_1$ and $P_2$ in Figure 1) was set in front of the target surface to realize glancing collisions without rotating the surface [10–13]. This setup allows us to continue extracting secondary ions along the surface normal direction, even during glancing collisions between the HCIs and surfaces. This is important for the MIS of secondary ions. The angle of incidence $\theta_i$ was controlled by the bias voltage applied to the deflectors. The angle $\theta_i$ is almost independent of the charge state $q$ under a constant acceleration voltage of the HCIs and a constant bias voltage applied to the deflectors. The initial kinetic energy along the surface normal direction, $E_0\sin^2\theta_i$, is very small (≤3.0 eV) when $E_0 = 2.50 \times q$ keV ($q \leq 14$) and $\theta_i \approx 0.5°$. Kinetic sputtering processes are quite suppressed under this experimental condition. Thus, potential sputtering processes due to electron capture were predominantly observed.

In earlier experiments, coincidence detection was performed between secondary ions and scattered atoms that were completely neutralized at the surface [8–11]. The 45° parallel plates ($P_3$) were used to eliminate all scattered ions. The flight time and position signals of secondary ions were also recorded using a digital oscilloscope triggered by a scattered atom signal of the $D_3$ detector.

A novel time-of-flight secondary ion mass spectrometer (TOF-SIMS) and a low-energy scattered ion spectrometer (LEIS) have been developed recently [12]. The kinetic energy and charge state of the scattered ions can be analysed using a Wien filter and the parallel plates, respectively. Coincidence detection between scattered and secondary ions makes it possible to separate potential sputtering from kinetic sputtering.

## 3. Results and discussion

### *3.1. Ion-molecule collisions*

Figure 2 shows a typical energy gain spectrum of scattered ions ($Ar^{10+}$) in $Ar^{11+}$ + $CF_4$ collisions. The most intense peak labelled S corresponds to a single electron transfer (SET), whereas the three peaks labelled D, T and Q correspond to double (DET), triple (TET) and quadruple electron transfer (QET) processes, respectively, followed by one, two and three Auger electrons. The multiple electron transfer process followed by autoionization is called transfer ionization (TI) [13]. It was found [4, 5] that the $q$-dependence of the energy-gain peaks of each electron transfer process can be fitted with a curve calculated by the classical over-barrier model (COB) [2]. Electron transfer processes can be specified by analysing the energy-gain spectrum.

In the next step, we performed coincidence measurements between the momenta of secondary ions and the energy gains of the scattered ions. Figure 3 shows 3D-momentum images of $CF_3^+$ ions and the kinetic energy released (KER) distribution in collisions of $Ar^{8+}$ + $CF_4$. Each $x$ and $z$ axis in the figure is parallel to the $Ar^{8+}$ beam and TOF axis, respectively. (The coordinates are shown in Figure 1.) The momentum images of $CF_3^+$ ions were taken in coincidence with those of $Ar^{7+}$ ions having an energy gain of $Q$ = 15.4 eV (SET). The entire reaction process, including electron capture and fragmentation, was clarified by the energy balance distributed to the final products (or $Ar^{7+*}$, F and $CF_3^+$). The peak KER is 0.28 ± 0.01 eV. This value is in good agreement with those measured by fragment ion-photon coincidence experiments [16, 17]. Thus, the processes were assigned as follows:

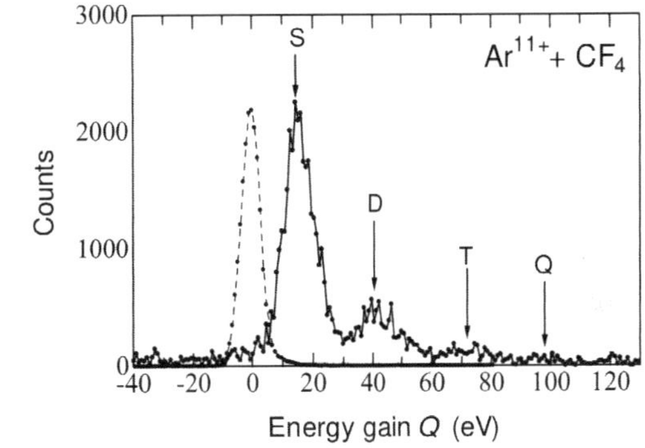


**Figure 2.** Energy gain spectrum of $Ar^{10+}$ in collisions of $Ar^{11+}$ + $CF_4$
The broken curve is the energy spectrum of the primary $Ar^{11+}$ beam. The arrows S, D, T and Q indicate the peak energies of single-, double-, triple- and quadruple-electron transfer, respectively, of the TI processes.

$$Ar^{8+}\,(2p^6\,{}^1S_0) + CF_4\,(\tilde{X}\ {}^1A_1)$$

$$\rightarrow Ar^{7+*}\,(2p^65d\ {}^2D_J) + CF_4^{+*}\,(\tilde{C}\ {}^2T_2) + 15.4\ \text{eV}$$

$$\quad \Rsh\ CF_4^{+*}\,(\tilde{A}\ {}^2T_2) + h\nu$$

$$\qquad \Rsh\ F + CF_3^+ \qquad (1)$$

Some structures seem to appear in the momentum images (a) and (b) of Figure 3. However, further investigation requires more precise measurements. In the case of DET processes, triple coincidence data, e.g. data for a scattered ion, the first and the second arrived fragment ion, were observed, because doubly charged molecular ions produced just after the DET often dissociate into two ionic fragments. Figure 4(a) shows the TOF correlation plot of $F^+$ and $CF_3^+$ ions produced in the DET of $Ar^{8+}$ + $CF_4$ collisions. The horizontal axis is the TOF of the first arrived $F^+$ ion, and the vertical

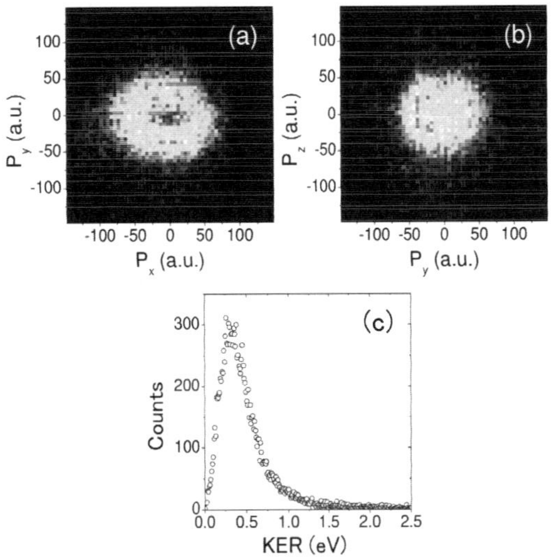


**Figure 3.** 3D-momentum images of $CF_3^+$ ions and the kinetic energy released distribution (KERD) in the SET process of $Ar^{8+}$ + $CF_4$

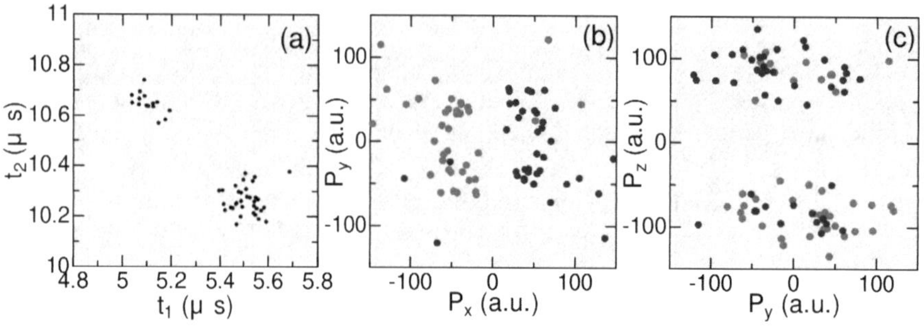


**Figure 4.** (a) TOF correlation plot, (b) $P_x$–$P_y$ and (c) $P_y$–$P_z$ momentum images of $F^+$ (red) and $CF_3^+$ (blue) ions produced in the DET of $Ar^{8+}$+$CF_4$

axis is that of the second arrived $CF_3^+$ ion [4]. One island around $t_1 \sim 5.1$ μs and $t_2 \sim 10.7$ μs marks ion-pair formation of forwards emitted $F^+$ and backwards emitted $CF_3^+$ with respect to the detector, whereas another island around $t_1 \sim 5.5$ μs and $t_2 \sim 10.3$ μs marks that of backwards emitted $F^+$ and forwards emitted $CF_3^+$. Figure 4(b) and (c) show the momentum images of $F^+$ and $CF_3^+$ ions from the correlation data. Note that each momentum vector was plotted in the centre-of-mass frame of the two fragment ions. The momenta of $F^+$ and $CF_3^+$ ions were clearly separated along the $x$-axis, or $Ar^{8+}$ beam axis, as shown in Figure 4(b), although there is almost no anisotropy in $P_y$–$P_z$, as shown in Figure 4(c). This directly suggests that the $CF_4$ molecule prefers to orient parallel to the HCI-beam direction heading in a fluorine atom in a specific DET-TI process [4], as described below.

$$Ar^{8+}(2p^6\,{}^1S_0) + N_2(X^1\Sigma_g^+) \rightarrow Ar^{6+*}(2p^6 4\ell n\ell) + N_2^{2+*}({}^1\Pi_u,\ {}^1\Delta_g \text{ or } {}^3\Delta_u)$$
$$\rightarrow Ar^{7+}(2p^6 3s\,{}^2S_{1/2}) + e^- + N^+ + N^+ \qquad (2)$$
$$Ar^{8+}(2p^6\,{}^1S_0) + CF_4(\tilde{X}\ {}^1A_1) \rightarrow Ar^{6+*}(2p^6 4f5\ell) + CF_4^{2+*}(1t_1^{-1}4t_2^{-1}\,{}^3T_1)$$
$$\rightarrow Ar^{7+}(2p^6 3s\,{}^2S_{1/2}) + e^- + F^+ + CF_3^+ \qquad (3)$$

In the $Ar^{8+}$ + $N_2$ collision experiment, we made sure that there was no significant anisotropy in the $P_x$–$P_y$ momentum images of forwards emitted and backwards emitted $N^+$ ions produced in the same DET process [5]. The physics of this phenomenon are not yet clear, but we can assume that the DET processes in collisions of HCI with $CF_4$ molecules have large cross-sections when an HCI approaches an F atom along the F–C bond. One reason for making this assumption is that according to an evaluation by a simple model calculation based on the COB [2, 16], double electron capture is found to be possible even in the straight-line trajectory with a large impact parameter. However, a more in-depth examination is required.

### *3.2. Ion-surface collisions*

We performed coincidence measurements of scattered neutral atoms and secondary ions in the first step. These experiments were performed by applying a high voltage (+3.0–4.0 kV) to the 45° parallel plate to eliminate all scattered ions, as described in 2.2.

Figure 5 shows the momentum image of protons emitted from a GaN(0001) surface interacting with $Ar^{8+}$ at $\theta \sim 0.5°$. This is the first 3D visualization of desorption induced by slow HCIs having kinetic energy of less than 2.0 eV in the surface normal direction [8]. To measure the charge-state dependence of the proton desorption yield in glancing-angle collisions, we conducted HCI-collision experiments using a polycrystalline aluminium surface. The steady state surface coverage of hydrocarbon molecules was maintained at ~1 during the experiment, because desorption induced by HCI collisions and adsorption of hydrocarbon molecules were balanced under an HCI flux of $1 \times 10^6$ ions $cm^{-2}s^{-1}$ and a residual gas pressure of $1 \times 10^{-7}$ Pa at room temperature [10]. The proton emission intensity normalized by the number of incident ions is shown in Figure 6. Note that the intensity is not equal to the sputtering yield, because the detection efficiencies of $Ar^{q+}$ ions, Ar atoms and protons have not been calibrated. It is clear that our experimental data saturates more remarkably than the data that did not include glancing collisions [17]. However, such saturation enhancement was not observed in GaN(0001) or 6H-SiC(0001) surfaces, even in glancing collisions [8]. According to a simple model calculation that includes image acceleration [10, 11], the saturation was clearly enhanced, as shown in Figure 6. Therefore, we conclude that shortening of the interaction time above the surface due to image acceleration of incident ions is one reason for the remarkable saturation of proton emission intensity in glancing collisions with an aluminium surface.

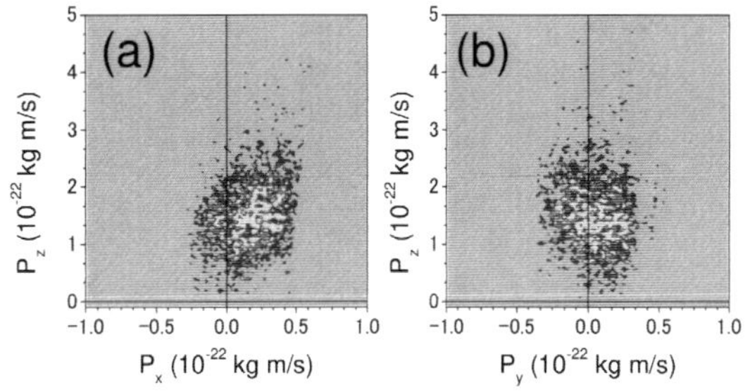


**Figure 5.** 3D-momentum images [(a) $P_x$–$P_z$ and (b) $P_y$–$P_z$] of protons desorbed from a GaN (0001) surface interacting with $Ar^{8+}$ at a glancing angle [10]

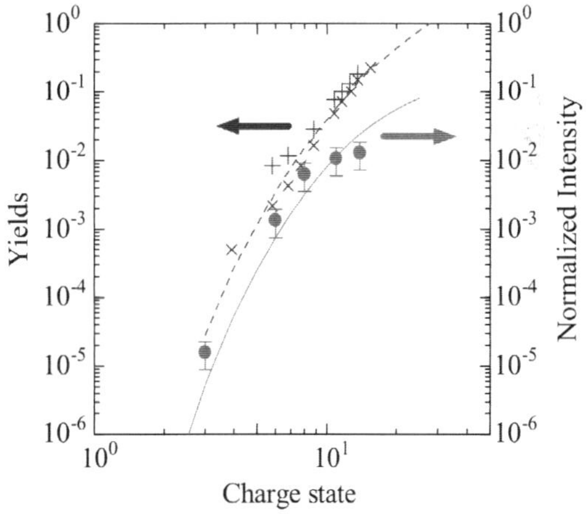


**Figure 6.** Number of protons detected per incident ion (right axis) and sputtering yields (left axis) as a function of the charge state of $Ar^{q+}$ ($q$ = 3–14). •, present experiment (Al) [12]; + and ×, untreated surfaces [19]; solid curve, present calculation [12, 13]; broken curve, model calculation [19]

Do substrate atoms emit as secondary ions in glancing collisions? The answer may be yes for semiconducting surfaces, but no for conducting and insulating surfaces. No secondary ions of substrate atoms were found on Al, Cu, Ag, Au and $Al_2O_3$ surfaces in coincidence experiments between scattered atoms and secondary ions, whereas $Ga^+$, $Si^+$, $N^+$ and $SiC^+$ were found on GaN and SiC surfaces [8]. These results suggest that the capture of conducting electrons by HCIs induces no secondary ion emission in metal surfaces [18], whereas positive hole charges prevent HCIs from approaching the surface of insulators, i.e. the 'trampoline effect' [19].

Here, I would like to show an interesting result of GaN(0001) and $(000\bar{1})$ surfaces. Wurtzite GaN has two lattice polarities of (0001) (Ga face; +c) and $(000\bar{1})$ (N face; −c) [20]. A gallium atom on the topmost layer combines with three nitrogen atoms of the bulk GaN in the (0001) surface, whereas a nitrogen atom on the topmost layer combines with three gallium atoms in the $(000\bar{1})$ surface. TOF-SIMS spectra measured on GaN surfaces with two different lattice polarities are shown in Figure 7. The intensity of $Ga^+$ ions on the (0001) surface was 4.7 times larger than that on the $(000\bar{1})$ surface, whereas the intensity of $N^+$ ions on the $(000\bar{1})$ surface was 1.7 times larger than that on the (0001) surface [9, 11]. This means that the topmost atoms on the Ga or N face were preferentially sputtered. The smaller enhancement factor in the case of $N^+$ may be due to the small probability of ionization by electron capture. This is direct evidence of *selective sputtering of the topmost substrate atoms*, or *SIMS with atomic-depth resolution*. This will be a key technique for surface analysis in the near future. More 'soft' collisions of HCIs with higher charge state are required for higher selectivity or resolution.

We are developing a novel TOF-SIMS spectrometer combined with a LEIS analyser in order to realize atomic layer–by-layer chemical analysis. We have succeeded in obtaining some preliminary data on proton desorption [12]. Proton emission may be connected in plural kinetic energies of scattered ions in $Ar^{6+}$–GaN(0001) collisions ($\theta \sim 12°$) [21].

## 4. Summary

To investigate the secondary-ion emissions induced by slow, highly charged ions, time-of-flight mass spectroscopy of secondary ions was performed in coincidence with scattered ions and atoms. This allowed the successful clarification of the dissociation pathways and electronic states of the molecular ions produced just after electron transfer. Selective sputtering of the topmost substrate atoms was observed in GaN single crystals in glancing angle collisions.

It was demonstrated that state- or atomic-layer-specific observation in fragmentation or desorption was possible by observing limited processes with small scattering angle. This fact suggests that both molecules and topmost atoms on surfaces may be excited state-selectively in electron capture processes with HCIs.

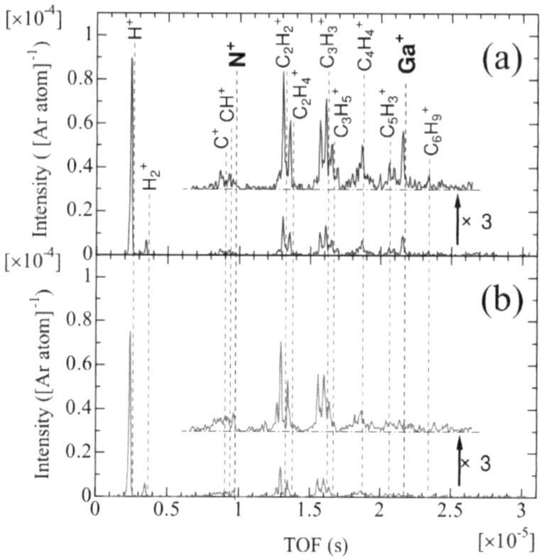


**Figure 7.** TOF-SIMS spectra of (a) GaN (0001) and (b) (000$\bar{1}$) surfaces interacting with $Ar^{8+}$ at a glancing angle

## 5. Acknowledgements

This work was partly supported by a Grant-in-Aid for Scientific Research (B) (15340128 and 19360016). The author would like to thank Prof. Seiji Tsurubuchi for fruitful discussion.

## References

[1] Kimura M, Nakamura N, Watanabe H, Yamada I, Danjo A, Hosaka K, Matsumoto A, Ohtani S, Sakaue H A, Sakurai M, Tawara H and Yoshino M, 1995 J. Phys. B: At. Mol. Opt. Phys. **28** L643
[2] Niehaus A, 1986 J. Phys. B: At. Mol. Opt. Phys. **19** 2925
[3] Burgdörfer J, Lerner P and Meyer F W, 1991 Phys. Rev. A **44** 5674
[4] Motohashi K and Tsurubuchi S, 2003 J. Phys. B: At. Mol. Opt. Phys. **36** 1811
[5] Tsurubuchi S and Motohashi K, 2003 J. Phys. B: At. Mol. Opt. Phys. **36** 3847
[6] Eppink A T J B and Parker D H, 1997 Rev. Sci. Instrum. **68** 3477
[7] Yaltkaya S, Kamber E Y and Ferguson S M, 1993 Phys. Rev. A **48** 382
[8] Motohashi K, Tsurubuchi S and Koukitu A, 2005 Nucl. Instrum. Meth. Phys. Res. B **232** 254
[9] Motohashi K, Hosoya K, Imano M, Tsurubuchi S and Koukitu A, 2007 Surf. Sci. **601** 5304
[10] Motohashi K and Tsurubuchi S, 2007 Nucl. Instrum. Meth. Phys. Res. B **264** 15
[11] Motohashi K, Tsurubuchi S and Koukitu A, 2007 J. Phys. Conf. Ser. **80** 012015
[12] Motohashi K e-J. Surf. Sci. Nano Tech. (submitted)
[13] Ohtani S, Kaneko Y, Kimura M, Kobayashi N, Iwai T, Matsumoto A, Okuno K, Takagi S, Tawara H and Tsurubuchi S, 1982 J. Phys. B: At. Mol. Opt. Phys. **15** L533
[14] Müller U, Lange M, Haas W and Brenn R, 1994 J. Chem. Phys. **100** 5550
[15] Furuya K, Koto E and Ogawa T, 2001 J. Phys. B: At. Mol. Opt. Phys. **34** 1405
[16] Motohashi K and Tsurubuchi S, Conference program with Abstracts of 23rd International Conference on Photonic, Electronic and Atomic Collisions (ICPEAC) 2003 We161.pdf
[17] Burgdörfer J and Yamazaki Y, 1996 Phys. Rev. A **54** 4140
[18] Hayderer G, Cernusca S, Hoffmann V, Niemann D, Stolterfoht N, Schmid M, Varga P, Winter HP and Aumayr F, 2001 Nucl. Instrum. Meth. Phys. Res. B **182** 143
[19] Wirtz L, Reinhold C O, Lemell C and Burgdörfer J, 2003 Phys. Rev. A **67** 012903
[20] Koukitu A, Mayumi M and Kumagai Y, 2002 J. Cryst. Growth **246** 230
[21] Motohashi K, Flores M and Kanai Y J. Phys. Conf. Ser. (submitted)

14th International Conference on the Physics of Highly Charged Ions (HCI 2008) IOP Publishing
Journal of Physics: Conference Series **163** (2009) 012006 doi:10.1088/1742-6596/163/1/012006

# Progress of the spectroscopy research platform at the Shanghai electron beam ion trap

**Roger Hutton, Ke Yao, Jun Xiao, Yang Yang, Di Lu, Yang Shen, Yunqing Fu, Xuemei Zhang and Yaming Zou**

The Key Lab of Applied Ion Beam Physics, Fudan University, Educational Ministry, China

*Shanghai EBIT Lab., Modern Physics Institute, Fudan University, Shanghai 200433, China*

zouym@fudan.edu.cn

**Abstract**. In this report we will focus on spectrometer development, spectroscopic studies and a few other recent developments at the Shanghai Electron Beam Ion Trap, EBIT laboratory. Currently the Shanghai EBIT has three spectrometers covering totally the wavelength region of 1 to 10000 Å. Two of these instruments are home made. A flat crystal spectrometer covers the wavelength range of around 1 – 20 Å while a flat field instrument covers the range of around 20 – 400 Å. The 3rd instrument is a commercial McPherson 225 normal incidence spectrometer. All spectrometers employ CCD cameras for photon detection. The Shanghai EBIT is also equipped with high purity Germanium detectors for, amongst other things, dielectronic recombination studies and time evolution studies of ion distributions. To back up these experimental studies computer codes have been developed for calculation of charge state balances etc. Parallel to the experimental program we have also developed experience at running a number of atomic structure codes (MCHF, MCDF, FAC) for various systems, e.g. the M3 decay of the $3d^94s\ ^3D_3$ for Ni-like ions.

## 1. Introduction

The Shanghai Electron Beam Ion Trap, EBIT, project was initiated in the year 2002 and there are a number of papers describing the history and progress of this instrument, see [1] and references therein. Basically the Shanghai EBIT belongs to the class of so-called Super EBITs, i.e. the maximum design electron beam energy is 200 keV. At the time of writing this paper, an energy of 135 keV has been achieved with an electron beam current of 150 milliamps. In this paper we will describe the suite of available and under-development spectrometers for spectroscopic studies of highly charged ions. Our design aim is to cover the wavelength region from around 1 – 10000 Å using three instruments: a flat crystal spectrometer for the region of 1 – 24 Å, a flat field aberration corrected instrument for the 20 – 400 Å region and finally a 1 meter normal incidence spectrometer for the 400 – 10000 Å wavelength region. Each of these separate wavelength regions will be described in the following sections.

## 2. The 1 – 24 Å wavelength region.

For the wavelength region below around 10Å there are no real choices other than crystal elements if wavelength dispersive instruments are to be used. There are other options for this wavelength region but these are better classified as energy dispersive, such as micro-calorimeters. Micro-calorimeters are already employed at the Livermore and NIST EBIT laboratories, see [2] and [3] respectively. There are various ways the crystal can be mounted to form an X-ray crystal spectrometer. We have chosen the flat crystal geometry. Flat crystal spectrometers are also in operation at the Tokyo [4] and Heidelberg [5] EBITs. There are a number of reasons why this geometry was chosen, one being the fact that the alignment is not so critical, which can be important in an EBIT when the exact location of the electron beam is difficult to define. The spectrometer and its connection to the EBIT are shown schematically in figure 1.

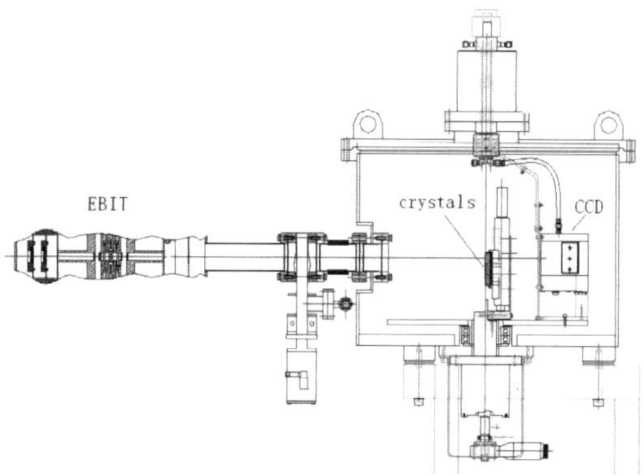


**Figure 1**: Schematic side-view diagram of the flat crystal spectrometer at the Shanghai EBIT. The diffraction plane is oriented perpendicular to the vertical electron beam. The crystals and the CCD are mounted inside the vacuum chamber and a fully automatic control system is used to rotate the crystals and the CCD.

The design of the flat crystal spectrometer allows for 3 crystals to be positioned in the vacuum chamber at any one time. The photon detector, which is an Andor DX436 CCD is positioned inside the vacuum chamber and in this way Bragg angles from 15 to 75 degrees are possible. The CCD chip has 2048x2048 13.5 micron pixels. All movements, i.e. changing the crystal and turning either the crystal or CCD can be done remotely via computer control. The spectrometer has been tested of-line using a home made X-ray generator and good resolution was observed. Recently it has been used to study M-shell X-ray spectra of highly charged Au ions, see section and [6]. In the Au work a resolution, $\lambda/d\lambda$, of around 3000 using crystals of LiF(200) (2d=0.4027nm) and $SiO_2$(1010) (2d=0.8512nm) was obtained.

### 2.1. The 20 – 400 Å wavelength region.

This spectral region is traditionally studied using a grazing incidence spectrometer. The problem in using such instruments is that not only is the incidence angle high, the diffracted angle is also high. If the instrument is to use a multi-channel detector, such as a CCD, it must be positioned tangential to the Rowland circle and the incoming photons will have an angle high with respect to the CCD chip normal, hence the chip will act as a mirror and reflect a percentage of the photons. This percentage will depend strongly on the incidence angle, i.e. on the photon wavelength. One solution to this reflectivity problem is to use the detector oriented so that its center is perpendicular to the incoming photons, see [7] where a CCD was adapted for Beam-Foil spectroscopy. Unfortunately only the center of the detector will be at the focus of the spectrometer. A more elegant solution to this problem is afforded by using so-called flat-field gratings. The properties of such gratings were first elucidated in the early 1980's by Harada and Kita et al., [8, 9]. It was found that if the line spacing was allowed to vary across the surface of the grating a solution to the imaging equations could be found where the image plane was, to a very good approximation, flat. This is of course idea for a multi-channel detector such as a CCD. Many spectrometers have been designed using such gratings but we will limit our discussion to those used at EBITs. So far flat-field spectrometers are used at the EBITs in Livermore [10] NIST [11] and good results have been obtained. The current flat-field spectrometer at the Shanghai EBIT relies on standard Hitachi gratings. Due to the fixed focal length of these gratings and the geometry of the Shanghai EBIT a technique of refocusing the spectrometer for a distant object will be required [12].

### 2.2. The 400 – 10000 Å wavelength region.

It is customary to cover the wavelength region over about 2000 Å with some sort of normal incidence spectrometer, normal incidence implying that the incidence and diffraction angles are kept close to the grating normal. This lower wavelength cut off comes from the fact that the spectrometer chamber can be de-coupled from any light source vacuum chamber by quartz a lens/window, hence it is a property of the coupling optics and not of the spectrometer geometry. Again there are a number of geometries that can be chosen, based on either plane or curved diffraction gratings. It is quite possible to use the normal incidence geometry down to wavelengths of around 185 Å [13] but to do this efficiently, the fewer reflections the better as the reflectivity of all materials drops as a function of wavelength. Hence a concave grating working in the Rowland circle geometry would appear as the best choice as only one reflection is needed. The more usual geometry for spectroscopy in the VUV is that of grazing incidence as the reflectivity of most materials becomes good at high angles to the optic normal as discussed in the previous section. The reason why normal incidence optics can be considered, even in the wavelength region where materials have low reflectivities, is because larger f-numbers, i.e. light collection angles, can be used. Grazing incidence leads to high aberrations, in particular astigmatism, leading to a limitation on the size of the grating for decent imaging [14]. Hence we have chosen to use a classic 1 meter normal incidence spectrometer from the McPherson company, the model 225 Nova instrument. A similar, in Geometry, 1 meter normal incidence instrument (Acton) has been previously implemented at the Livermore EBIT [15]. In [15] it was possible to position the spectrometer close enough to the EBIT that the electron beam could act as the spectrometer entrance slit. Diameters of electron beams in EBITs are on the order of 50 – 100 μm and hence equivalent to typical slit widths. Due to the construction of the Shanghai EBIT this is not possible and hence we use a focusing mirror to image the electron beam onto the entrance slit of the spectrometer. Currently this imaging is done by a spherical mirror operating at an incidence angle of 67.5 degrees. This will of course lead to some aberrations of the image. The mirror was designed to give maximum light throughput at the entrance slit. However, due to the weak nature of EBIT light sources this mirror is in the process of being replaced by a toroid. The mirror chamber acts as a differential pumping stage, isolating the very good vacuum of the EBIT from the $10^{-7}$ Torr vacuum of the spectrometer. The arrangement of the spectrometer and mirror chamber is shown in figure 2.

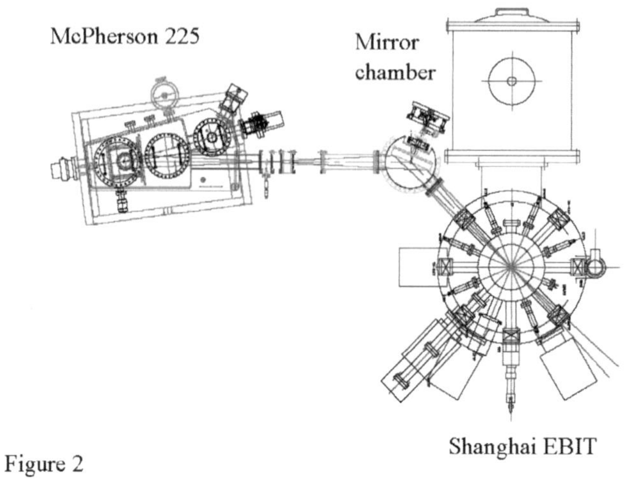


**Figure 2:** Implementation of a normal incidence spectrometer at the Shanghai EBIT.

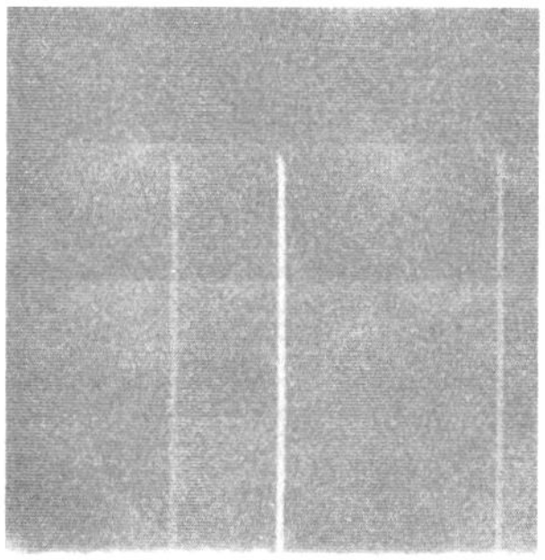

**Figure 3:** Spectral lines from metastable levels of relatively low charge states of krypton taken using the McPherson spectrometer centred at 500 Å. The slits were open at 4mm, hence the broad lines found under these conditions. The EBIT was operated at 1.3 keV and a current of only 5 milliamps. The lines are possibly 4p-4d transitions in Kr VII – Kr X although analysis is ongoing.

2.3. Other Instruments

The Shanghai EBIT has two high purity Germanium, hpg, detectors; (i) a standard detector which is positioned outside the EBIT, i.e. around 60 cm from the electron beam and (ii) an Iglet detector which is installed inside one of the viewing arms of the EBIT, and hence can get closer to the electron beam. These hpg detectors can be used for (a) monitoring the charge states and elements inside the EBIT and (b) measuring di-electronic recombination, DR, cross sections. The DR cross-sections are measured by recording the arrival of X-rays as a function of the electron beam energy. Similar techniques are used at all EBIT laboratories and more details of the procedure used in Shanghai can be found in [16]. One extra feature concerning DR measurements at the Shanghai EBIT is the ability to measure the electron beam energy to quite high precision using a high voltage divider [17]. We have also developed a slit-imaging system which can be used for measuring the width of the electron beam. This device and its use is described in [18].

## 3. Some results.

As mentioned above the flat crystal spectrometer has been used to study M-shell transitions in highly ionized gold. These experiments were done using electron beam energies between 5-20 keV, with electron beam currents of 20 to 80 mA. The gold atoms were introduced into the EBIT from a Metal Vapor Vacuum Arc (MEVVA) ion source. Argon was used as a partial calibration for this experiment and He and Li-like Ar lines are shown in figure 3. From these lines we can deduce that the spectrometer resolution is around 3000. The spectra are under analysis and a sample spectrum is shown in figure 4. More details of this experiment will be available shortly [6].

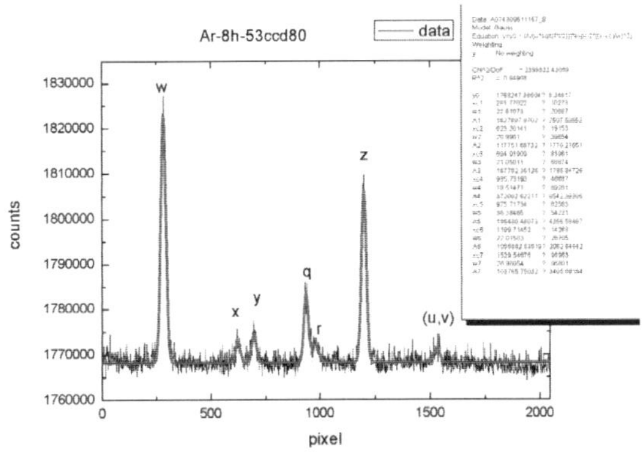


**Figure 4:** This shows He and Li-like Argon lines and indicates a spectrometer resolution of around 3000.

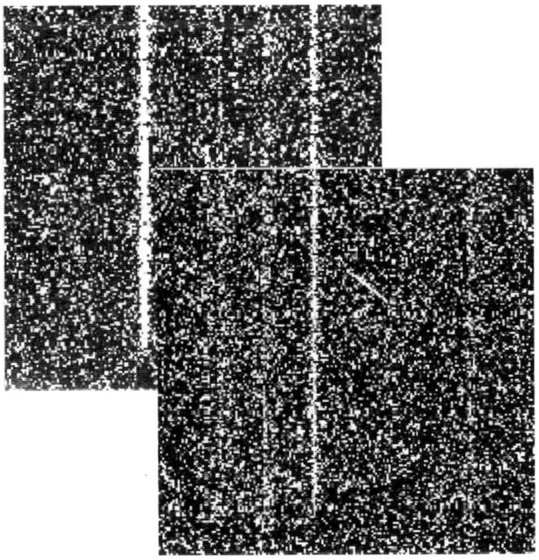

**Figure 5:** part of the M X-ray spectrum of Au taken using the crystal spectrometer. The electron beam energy was 18 keV, 68 milliamps, and the crystal (Si(111)) operated at a Brag angle of 39.5 degrees for the left most spectrum. The brighter line on the left hand side of the CDD centre in the top left spectrum is the Z line of He-like Ar at 3104 eV. Livermore EBIT [19] sees lines at 3012 and 3132 eV and predicts other lines at 3100 and 3136 eV using the HULLAC codes for charge states Ni-Kr like, so we may assume our charge state balance is different as we see an Au line at 3064 eV. Further analysis is ongoing.

## 4. Future Developments.

As mentioned above the spherical focusing mirror for the normal incidence spectrometer is due to be replaced by a toroidal mirror. This will increase the light collection properties by a factor of about 10. A flat field spectrometer based on customized gratings has been designed and is currently under construction. This instrument has some interesting properties and uses 3 gratings to cover the wavelength region of 20 – 400 Å. It also has one grating to give a survey spectrum.

## 5. Other activities.

In anticipation of future experiments we have initiated an active program of investigating the effects of the hyperfine interaction on the lifetimes of atomic energy levels. This program originated in a statement made in a paper from 1991, "The hyperfine interaction has been proposed to quench metastable beam fractions from an ECR" [20] and updated now with a study of the decay channels available to the $3d^94s\ ^3D_3$ level in Ni-like ions. In particular Ni-like Xe was studied as there was a substantial discrepancy between theory and experimental results of Träbert et al [21]. This discrepancy was explained by hyperfine effects in [22,23] and confirmed by more refined experiments in [24]. The $^3D_3$ level was previously considered to decay only through an M3 decay. However, for nuclei with a non-zero spin the hyperfine interaction will mix levels of different j quantum numbers, i.e. j is no longer a good quantum number and the f quantum number becomes the important quantity. F is the vector sum of the nuclear spin and the atomic angular momentum, i.e. F = I + J. Hence the 3D3 will no longer decay by purely an M3 decay. The $^3D_3$ level will mix with the $^{3,1}D_2$ levels where the mixing with the $^3D$ dominated due to the closeness in energy. As the $^3D_2$ has a much faster E2 decay the effect of this albeit small mixing will be very noticeable on the lifetime of the $^3D_3$ level. In fact each f sub-level of the $^3D_3$ will have a different lifetime. Lifetimes now become f-dependent and as an example the two f–sub levels of the $^3D_3$ for the 129 isotope of Ni-like Xe have lifetimes differing by a factor of 7 [22,23]. Since then other Ni-like ions have been studied [25] and other iso-electronic sequences are under investigation. Results for Ne-like ions are presented in these proceedings [26]. Many cases of reduced lifetimes for forbidden transitions were found. In other cases one can expect redistribution of spectral intensity, as found for Ga II in [27]. Cleary by introducing changes in level lifetimes we can also expect a change in branching fractions, which means a redistribution of line intensities. In the case of Ga II it was not possible to simulate astrophysical spectra where Ga II was observed until the hyperfine interaction was included in the calculations of the atomic properties. Similar effects can be expected in other systems when ever there is a change in an atomic level lifetime. In more recent experiments, the effects of the hyperfine interaction have been studied using heavy ion storage ring techniques [28]. It remains to be seen what impact f-dependent lifetimes will have on the field of, for example, astrophysical plasma diagnostics, its "predecessor" j-dependent lifetimes certainly has a well documented significance to both atomic physics and plasma diagnostics. An interesting effect along similar lines was shown by Beiersdorfer et al when they observed magnetic sensitive lines [29]. In that paper they reported on the single photon decay of the $2p^53s\ ^3P_0$ level in Ne-like Ar induced by the magnetic field in their light source, namely the Livermore EBIT. The intensity off this line, with respect to lines from the decay of the other $2p^53s$ levels, increased as a function of the magnetic field. These effects are showing that a very small mixing, induced by say the hyperfine interaction or the Zeeman effect, may have very little effect on the energy of an atomic level, but depending on the detailed atomic structure, can have a very big influence on the lifetimes of forbidden transitions. For example the magnetic sensitive line mentioned above had a zero single photon decay rate in the absence of the magnetic field ($2p^6\ ^1S_0 - 2p^53s\ ^3P_0$). The field induced decay rates are similar in size to the M2 decay of the 2p53s 3P2 level. More studies along these lines are anticipated. At first though one may imagine these small effects leading to shorter lifetimes as only affecting a very specific area of a spectrum, however deeper thought reveals that changing lifetimes of metastable levels can have consequences throughout a spectrum. Quenching of metastable levels means that population mechanisms can be drastically changes and hence these effects need fuller study.

**Acknowledgements.**
This work was supported by the National Natural Science Foundation of China under Grant No. 10434050 and also partly supported as a Shanghai Leading Academic Discipline Project under Project No. B107.

## References

[1] M. He *et al.*, J. Phys. Conf. Series **58** (2007) 419
[2] see the contribution by F. Porter in this volume.
[3] E. Silver *et al.*, Astrophys. J. **541** (2000) 495.
[4] N. Nakamura, Rev. Sci. Instrum. **71** (2000) 4065
[5] J. Braun *et al.*, Rev. Sci. Instrum. **76** (2005) 073105
[6] J. Xiao *et al.* work in progress and to be published.
[7] R. Hutton *et al.*, Physica Scripta **T80** (1999) 532
[8] T. Harada and T. Kita, Appl. Opt. **19** (1980) 3987
[9] T. Kita, T. Harada, N. Nakano, and H. Kuroda, Appl. Opt. **22** (1983) 512
[10] P. Beiersdorfer *et al.*, Rev. Sci. Instrum. **75** (2004) 3724
[11] B. Blagojević *et al.*, Rev. Sci. Instrum. **76** (2005) 083102
[12] X.H. Xie et al., Chinese Physics Letters **13** (2004) 1826
[13] A. Redfors and U. Litzen, J.O.S.A. **B6** (1989) 1447
[14] J.E. Mack, J.R. Stehn and B. Edlén JOSA **22** (1932) 245.
[15] S.B. Utter *et al.*,Rev. Sci. Instrum. **70** (1999) 288
[16] W.D. Chen *et al.*, Physics of Plasmas accepted (2008)
[17] W.D. Chen *et al.*, Submitted to Rev. Sci. Instrum. (2008)
[18] Y. Yang *et al.*, Submitted to Nucl. Instrum. and Meths A (2008)
[19] M. May *et al.*, Phys. Rev. **E68** (2003) 03642.
[20] R. Hutton *et al.*, Phys. Rev. **A44** (1991) 243.
[21] E. Träbert *et al.*, Phys. Rev. **A73** (2006) 022508
[22] K. Yao *et al.*, Phys. Rev. Lett. **97** (2006) 183001
[23] K. Yao *et al.*, Phys. Rev. Lett. **98** (2007) 269903
[24] E. Träbert, P. Beiersdorfer, and G. V. Brown, Phys. Rev. Letter **98** (2007) 263001
[25] M. Andersson *et al.*, Phys. Rev. **A77** (2008) 042509
[26] see the contribution by M. Andersson in this volume.
[27] M. Andersson, P. Jönsson and H.J. Sabel, J. Phys. **B39** (2006) 4239
[28] S. Schippers *et al.*, Phys. Rev. Letts **98** (2007) 33001.
[29] P. Beiersdorfer, J.H. Scofield and A.L. Osterheld, Phys. Rev. Lett. **90** (2003) 235003

# Two-loop QED contributions tests with mid-$Z$ He-like ions

**K Kubiček[1], H Bruhns[2], J Braun[1], J R Crespo López-Urrutia[1] and J Ullrich[1]**

[1]Max-Planck-Institut für Kernphysik, 69117 Heidelberg, Germany
[2]Present address: Columbia Astrophysics Laboratory, Columbia Universtity, NY, NY 10027-6601, USA

E-mail: katharina.kubicek@mpi-hd.mpg.de

**Abstract.** We report about high-precision wavelength determination of H-like and He-like ions at the Heidelberg Electron Beam Ion Trap (HD-EBIT). The experiment was carried out with a novel flat crystal (Si-111) spectrometer applying a reference technique without collimation. The result for the transition energy of the $1s2p\ ^1P_1 \rightarrow 1s^2\ ^1S_0$ resonance line in He-like $S^{14+}$ agrees with theoretical predictions with an experimental accuracy five times higher than earlier experiments. The same line in $Ar^{16+}$ was measured with a relative uncertainty of $\delta\lambda/\lambda = 2 \times 10^{-6}$, a factor of 2.5 more accurate than any X-ray wavelength in highly charged ions ever reported, and for He-like ions probes QED two-electron and two-photon radiative corrections. Beside relative wavelength measurements absolute ones were carried out using the Bond method. The results point at the possibility of establishing absolute Lyman-$\alpha_1$ transition X-ray wavelength standards in the future.

## 1. Introduction

Any experiment intended to test QED and the small effects it predicts have to be very precise. In 2005 T.W. Hänsch was awarded the Nobel Prize for his determination of the frequency of the $1s - 2s$ transition in hydrogen with a relative accuracy of 2 parts in $10^{14}$ [1, 2]. Current theory can predict nearly all decimal figures of the experimental result correctly, but is affected by limitations in the knowledge of the internal structure of the proton, which presently introduces the largest uncertainty in the calculations.

Highly charged ions (HCIs) offer important advantages for studying QED effects scaling with the nuclear charge $Z$ to fourth power, thus increasing with higher charge state of an ion and becoming easier to measure. Moreover, in such systems the electric field strength experienced by the 1s electron can go up to $10^{16}$ V/cm ($U^{91+}$). This strength of the interactions between electrons and the nucleus cannot be described anymore in the perturbative approach which has been very successful in the case of the hydrogen atom. In addition, due to the high binding energy of their electrons for an accurate description of those ions relativistic effects become essential.

Further theoretical challenges arise in systems involving more than one bound electron. Paramount examples for relativistic systems strongly affected by QED are highly charged ions belonging to the He isoelectronic sequence. In order to predict the energies of electronic states there, one has to solve a many-body problem, which can be treated only approximately. There are various theoretical approaches to that end. For He-like ions, accurate correlated two-electron non-relativistic wave functions perturbatively corrected for QED and relativistic effects are applied by the unified method (UM) of Drake [3]. The all-order (AO) method of Plante [4] has a different ansatz, namely employing a fully relativistic two-body calculation accounting for QED and electronic interaction by introducing the Breit approximation. While AO calculations use the QED corrections derived from the UM, a recent calculation by Artemyev *et al.* [5] generates a new, *ab initio* QED prediction within an all-order approach here referred to as bound-state QED (BSQED). Notice that the UM should describe systems at low $Z$ more accurately since in this regime correlation effects are more relevant than relativistic ones, while the AO and BSQED methods are to be expected better at high $Z$, where relativistic effects become dominant. For medium $Z$ ions, $Z \approx 26$, both relativistic and correlation effects are of comparable size and therefore accurate description of all relativistic, correlation and QED effects is necessary.

From an experimental point of view, the strongest transitions in HCI appear in the X-ray region. The most accurate transition wavelengths in that region ($K_\alpha$ lines) are known with an accuracy of $\sim 1$ ppm [6] only, ten orders of magnitude lower than in the visible range. For He-like systems several measurements of the $1s2p\,^1P_1 \rightarrow 1s^2\,^1S_0$ ( "$w$" [7]) resonance transition ranging from S ($Z = 16$) [8] to Kr ($Z = 36$) [9] have been carried out. Deslattes *et al.* [10] reached an accuracy of ($\Delta\lambda/\lambda = 12\,\text{ppm}$) for $Ar^{16+}$ ($Z = 18$), the hitherto highest for such a measurement. To distinguish among the various theoretical approaches, the uncertainty has to be reduced by a factor of 5 compared to this experiment. The current experiment at the Heidelberg Electron Beam Ion Trap (HD-EBIT) contributes to alleviate this situation.

## 2. The experiment

The classical crystal spectrometer is still widely used to study X-ray transitions at high resolution, since there are currently no appropriate laser sources available in the X-ray region. Since X-ray spectrometers using curved crystals have a complicated focussing geometry and are difficult to describe, a flat crystal spectrometer type with rotatable crystal (Si-111) was chosen for in our experiment.

Transition energy determination implies measuring the Bragg angles which becomes a quite challenging task if an accuracy at the low ppm is aimed, since an offset of the angular scale cannot be avoided. In contrast, angular differences can be determined very precisely. Thus, for high precision measurements X-ray reference lines are usually required, normally $K_\alpha$ transitions of copper, molybdenum and tungsten [11], limiting the accuracy which can be reached to the precision to which their energies are known, $\sim 1$ ppm. These transitions as reference lines, however, have serious limitations at higher levels of accuracy. Since to generate such lines neutral atoms embedded in the bulk have to be excited, multi-electron effects caused by all bound electrons, i.e. so called satellite transitions, lead to asymmetric lineshapes. These asymmetries cannot be predicted by theory at the required level of accuracy. Therefore it has been proposed [12, 13] to use other lines not affected by these problems as reference lines. Suitable transitions are the $Ly_\alpha$ transitions (used in our experiment), since they can be determined reliably by theory.

In our experiment we use the HD-EBIT for the production of HCI. The advantage of our method in comparison to others such as the measurements applying the recoil ion technique [29, 10], deceleration and electron capture method [14] or using tokamak plasmas [15] is the spectral purity, i.e. satellite-free lines (Fig. 1) and the absence of Doppler shift. Thus using an EBIT for the production of the ions studied later on it should be possible to establish Lyman-$\alpha$ wavelengths as high-precision, calculable atomic X-ray standards. This standard is requested compulsory with regard to future laser resonance fluorescence studies at synchrotron facilities and free-electron lasers [16] as well as for the development of X-ray microcalorimeters [17].

Normally, performing high precision measurements on transition wavelengths in the X-ray

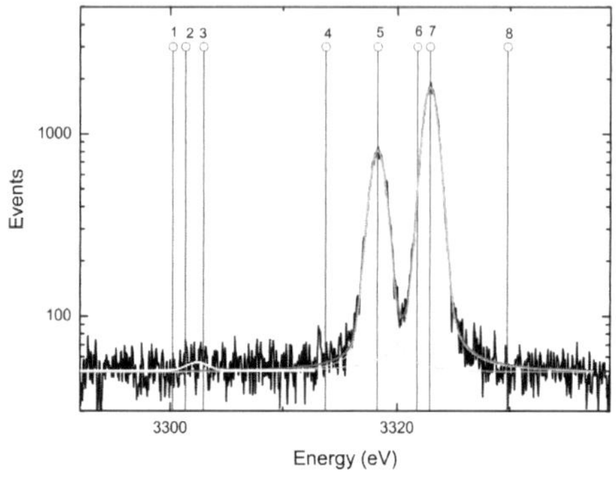


**Figure 1.** X-ray spectrum of the Lyman-$\alpha_1$ (7) and Lyman-$\alpha_2$ (5) lines in $Ar^{17+}$ (HD-EBIT) fitted with a Voigt profile shown together with predicted [18] satellites (1-4, 6, 8) on a logarithmic scale. One can see that satellite "6" could have an influence on the peak position of the Lyman-$\alpha_1$ line which can be calculated by fitting a Gaussian to satellite "2" since the excited state is the same for both satellites. The transition probability of satellite "2" (E1 transition) is $\sim$ 80 times larger than the one of satellite "6" (M1 transition). With the branching ratio mentioned above, the intensity of satellite "6" is $<$ 0.03% of the intensity of the Lyman-$\alpha_1$ transition, i.e. insignificant. Notice also that the resolution is better than FWHM $<$ 1.8 eV at 3323 eV.

region the incoming direction of the X-rays is fixed by collimation with slits which leads to a decreasing X-ray flux, not desirable in the view of the low intensity (generated by $\sim$ 1 million of trapped ions) emitted by an EBIT. To avoid these problems, in our experiment a reference method avoiding collimation, the "a/b-method" [19, 20], developed in our group by J. Braun and H. Bruhns was applied.

### *2.1. The a/b-method using light fiducials*

The central idea of this new measuring is to monitor the incoming direction of the X-rays without significant loss of X-ray flux using two laser beams originating from a virtual source overlapping with the trapped ions as angle and direction references ("fiducials"). These laser beams are reflected by the same crystal onto a CCD detector according to the usual optical reflection law. The ratio of the distances from the X-ray position on the detector to the light fiducial positions is a function of the incoming direction of the X-rays and depends on the Bragg angle. A central advantage of the a/b-method is that this ratio, as a result of similar triangles' theorem, does not depend on the exact locations of the crystal or the camera, as long as the points from which the X-rays and the light fiducials originate coincide, thus generating triangles with a common vertex. To align the virtual source of the laser fiducials, two retractable lenses collect visible light emitted by trapped B-like $Ar^{13+}$ ions which is imaged onto the CCD detector. The virtual source is then adjusted to overlap with the imaged light. A detailed description of the a/b-method can be found in [19].

### *2.2. Combination of the a/b-method with the Bond method*

To perform absolute wavelength measurements we combined the a/b-method with the Bond method [21]. The rotatable flat crystal reflects X-rays towards either one of two symmetrically arranged, windowless, cryogenically cooled CCD detectors. This angle is determined by the a/b-method, and still contains an offset. In a second step, a rotation by $180° - 2\Theta_B$ where $\Theta_B$ is the Bragg angle of the line of interest leads to a reflection of the X-rays towards the other CCD detector. This second angle containing the same offset is also determined. In this way the exact Bragg angle $\Theta_B$ can be derived from the measured angular difference in the crystal angle.

*2.3. Experimental parameters*

For the production and trapping (trapping region: cylindrical volume of 100 $\mu$m diameter and 4 cm length) of $Ar^{16+}$, $Cl^{16+}$, $S^{14+}$ needed for the measurements in our experiments the electron beam current was set to about $\sim$ 400 mA at an electron beam energy of $E_e = 9.5$ keV. In order to keep the temperature-stabilized laboratory undisturbed for several days during our measurements, the EBIT and the spectrometer were computer controlled and worked without operator intervention during the runs.

## 3. Results and conclusions

For alignment of the crystal, the angle $\phi$ between the Si (111) plane and the crystal surface was determined in a separated setup to be $\phi = 0.0585° \pm 0.002°$ [28]. The crystal is mounted on a holder which allows to rotate it around the surface normal axis. This way the orientation of $\phi$ can be adjusted perpendicular to the scattering plane thus not having any influence on the measured difference angle. The orientation of the Si (111) plane with respect to the crystal shape needed for this purpose was known with a precision of $\pm 2°$. After performing this alignment procedure the remaining error contribution caused by this process was calculated to be 0.5 ppm [20]. Smaller errors arise from the change of the lattice spacing due to temperature fluctuations, adjustment of the visible light assembly (see 2.1), alignment of the spectrometer with respect to the trap center and stability of the light fiducials.

Besides counting statistics, the remaining leading error contribution to our *absolute* measurements is caused by the fact that the EBIT trapping volume is a line source emitting X-rays which generate a series of circular segments on the CCD cameras after Bragg reflection. Therefore a projection of the X-ray spectra on the energy dispersive axis results in intrinsically asymmetric line profiles leading to a calculable shift of the line centroid to about 40 ppm higher energies [20]. A direct observation of the line bending is not possible since symmetric line broadening effects, e.g. thermal broadening effects, lead to a modification of the line shape. If the X-ray emission center lies outside the scattering plane this shift is partially compensated, in our case by $\sim$ 15 ppm [28], since both the effective crystal lattice spacing and the line curvature shift are modified, and a final uncertainty of the *absolute* wavelength measurements of $\sim$ 20 ppm cannot be eliminated. Performing *relative* wavelength measurements the final uncertainty can be significantly reduced by putting these systematic contributions leading to a shift of the line centroid in a correction angle $\Delta\Theta_{corr}$ and adding it to the absolutely measured Bragg angle $\Theta_a$ of the line of interest. $\Delta\Theta_{corr}$ is obtained as the difference between the predicted $Ar^{17+}$ Lyman-$\alpha_1$ Bragg angle $\Theta_t$ derived from the theoretical wavelength $\lambda_t$ [22] and the measured absolute Lyman-$\alpha_1$ Bragg angle. A detailed description of this procedure can be found in [28]. A scaling from $\Theta_t$ to $\Theta_a$ was necessary since the correction angle $\Delta\Theta_{corr}$ depends on the Bragg angle. The scaling factor was derived from a ray-tracing simulation [20] of the X-ray trajectories.

Our results listed with $1\sigma$ error bars are shown in Table 1. The results were converted to energy using the conversion factor $hc = 1.239841875(31) \times 10^{-6}$ eV m [11]. Figure 2 shows

**Table 1.** Present experimental results and theoretical values.

| Transition | $E_{theo}$ (eV) | $\mathbf{E_{exp}}$ (eV) | **Error** (ppm) |
|---|---|---|---|
| $Cl^{16+}$ Ly-$\alpha_1$ | 2962.352 [22] | **2962.344(30)** [28] | **10.0** |
| $S^{14+}$ $w$ | 2460.629 [5] | **2460.641(32)** | **13.0** |
| $Ar^{16+}$ $w$ | 3139.582 [5] | **3139.583(6)** [28] | **1.9** |

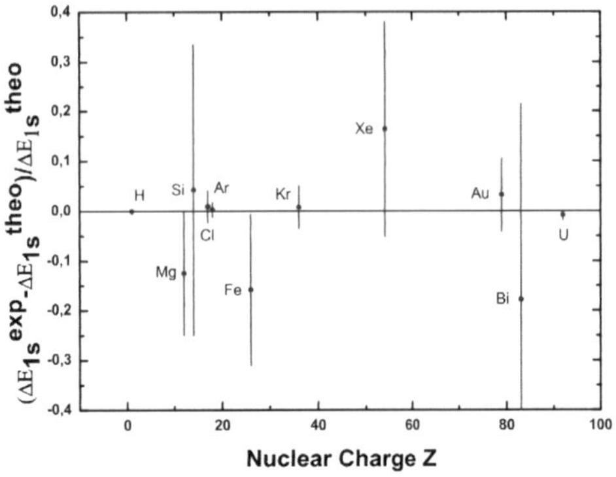


**Figure 2.** Comparision of theoretical predicted [22] and experimental measured ground state Lamb shifts for H-like ions. H:(Z=1) [2], Mg:(Z=12) [26], Si:(Z=14) [27], Cl:(Z=17) [28], Ar:(Z=18) [29], Fe:(Z=26) [30], Kr:(Z=36) [31], Xe:(Z=54) [32], Au:(Z=79) [33], Bi:(Z=83) [34], U:(Z=92) [35]. Notice that the only absolute wavelength measurements displayed here are the one on magnesium [26] and silicon [27].

the relative deviation of the experimetally measured $\Delta E_{1s}^{exp}$ and theoretically predicted $\Delta E_{1s}^{theo}$ Lamb shift [22] in hydrogenlike ions as a function of the nuclear charge $Z$. Within the error bars all experimental results agree with the theoretical calculations, especially good for hydrogen and uranium while unfortunately for the mid-$Z$ region experimental precision is much lower (as well as the accuracy reached for our results of the chlorine Lyman-$\alpha_1$ transition energy [28]), not allowing to establish transitions in hydrogenlike ions as X-ray standard lines. For this purpose theoretical predictions still have to be tested on a higher level of accuracy. Consequently, currently huge efforts are made to align the spectrometer in such a way that the line source center lies inside the scattering plane in order to minimize the above mentioned uncertainty of our *absolute* wavelength measurements. Nevertheless, as we compare our 2007 result for the chlorine Lyman-$\alpha_1$ transition energy with former results [23, 24, 25], all results reported earlier had an uncertainty bigger than 100 meV. Our result has an error bar of only 30 meV [28], enabling to test the vacuum polarization contribution to the $^1S_0$ state of 68 meV [23]. Our recent result for the $S^{14+}$ $w$ transition, $\lambda_w = 5038.695(74)$ mÅ, or $E_w = 2460.641(32)$ eV, has an error of only 12.8 ppm, a factor of $\sim 3$ better than the former most precise measurement of this transition energy [8]. This result is in excellent agreement with all theoretical predictions [3, 4, 5] within its $\sim 13$ ppm uncertainty. However, the error bar has to be reduced in ongoing experiments to be able to distinguish between the different theoretical calculations on a level of (finally) 1 ppm. We would like to point out our 2007 result [28] for the $Ar^{16+}$ $w$ transition, $\lambda_w = 3949.066(8)$ mÅ, or $E_w = 3139.583(6)$ eV, which has an uncertainty of only 2 ppm [28]. This is 6 times more accurate than the most precise measurement in any He-like ion, namely in $Ar^{16+}$ recoil ions [10] and until now a factor 2.5 better than any X-ray wavelength in HCIs ever reported [29]. To calibrate our spectra we used the theoretical $Ar^{17+}$ Lyman-$\alpha_1$ wavelength instead of the experimental one obtained [29] because the uncertainty of that experimental value, 5 ppm, exceeds the theoretical one by far. While the measurement of Deslattes *et al.* [10] gave an indication that the w transition energy could be smaller than the predicted theoretical values, our result is in excellent agreement with state-of-the-art BSQED calculations by Artemyev *et al.* [5] and the results of the AO method [4] within its $1\sigma$ error estimate. The UM prediction [3] lies just inside our 2 ppm error bar. This could be due to an incomplete description of relativistic effects at $Z = 18$ in UM [28]. Table 2 lists theoretical one- and two-electron QED contributions to the $1s2p\ ^1P_1$ and $1s^2\ ^1S_0$ states in the $Ar^{16+}$ ion (from [5]). As seen there, our result probes, at a level of 6% of their total, all two-electron QED contributions to the $^1S_0$ ground state energy, i. e., screened self-energy, screened vacuum polarization and two-photon exchange diagrams. For example, the change of the ground state energy due to vacuum polarization effects (7.2 meV) is

**Table 2.** Theoretical one- and two-electron QED contributions (in eV) to the $1s2p\ ^1P_1$ and $1s^2\ ^1S_0$ states in $Ar^{16+}$ (BSQED taken from [5]). Total 1-el.: sum of one-electron contributions; Scr. SE: screened self-energy; Scr. VP: screened vacuum polarization; 2-ph. exch.: exchange of two virtual photons; total 2-el.: sum of two-electron contributions.

| State | **Total 1-el.** | Scr. SE | Scr. VP | 2-ph. exch. | **Total 2-el.** |
|---|---|---|---|---|---|
| $^1S_0$ | **-1.1310** | 0.1116 | -0.0072 | -0.0091 | **0.0953** |
| $^1P_1$ | **-0.0062** | 0.0031 | -0.0001 | 0.0001 | **0.0031** |

1.2 times larger than our present experimental accuracy. Our experimental value is also sensitive to the total one-electron QED correction of the $^1P_0$ excited state, but the uncertainty is still too low to probe the total 2-electron contribution (3 meV). Thus, although our results now agree excellently with the predictions of BSQED [5] on the present level of accuracy, and are in good agreement with [3] and [4] at $Z = 18$, further tests require extending measurements over a broader range of elements.

## References

[1] Niering M *et al* 2000 *Phys. Rev. Lett.* **84** 5496–99
[2] Fischer M *et al* 2004 *Phys. Rev. Lett.* **92** 230802
[3] Drake G W F 1988 *Phys. Rev. Lett.* **74** 4791
[4] Plante D R, Johnson W R and Sapirstein J 2004 *Phys. Rev.* A **49** 3519–30
[5] Artemyev A N, Shabaev V M, Yerokhin V A, Plunien G and Soff G 2005 *Phys. Rev.* A **71** 062104
[6] Hölzer G, Fritsch M, Deutsch M, Härtwig J and Förster E 1997 *Phys. Rev.* A **56** 4554
[7] Gabriel A H 1972 *Mon. Not. R. Astron. Soc.* **160** 99
[8] Schleinkofer L, Bell F, Betz H D, Trollmann G and Rothermel J 1982 *Phys. Scr.* **25** 917
[9] Widmann K, Beiersdorfer P and Decaux V 1996 *Phys. Rev.* A **53**, 2200–05
[10] Deslattes R D, Beyer H F and Folkmann F 1984 *J. Phys.* B **17**, L689–94
[11] Mohr P J *et al* 2007 http://physics.nist.gov/constants
[12] Flowers J L, Klein H A, Knight D J E and Margolis H S 1984 *NPL Report CBTLM* **11**
[13] Anagnostopoulos D F, Gotta D, Indelicato P and Simonis L M 2003 *Phys. Rev. Lett.* **91** 240801
[14] Beyer H F, Indelicato P, Finlayson K D, Liesen D and Deslattes R D 1991 *Phys. Rev.* A **43** 223–7
[15] Marmar E S, Rice J E, Källne E, Källne J and LaVilla R E 1986 *Phys. Rev.* A **33** 774–7
[16] Epp S W *et al* 2007 *Phys. Rev. Lett.* **98** 183001
[17] Porter F S *et al* 2004 *Rev. Sci. Inst.* **75**, 3772
[18] Vainshtein L A 1978 *At. Data Nucl. Data Tables* **21** 1
[19] Braun J, Bruhns H, Trinczek M, Crespo López-Urrutia J R and Ullrich J 2005 *Rev. Sci. Instrum.* **76** 073105
[20] Braun J 2006, Ph.D. Thesis, Max-Planck-Institut für Kernphysik, Heidelberg
[21] Bond W L 1960 *Acta Cryst.* **13** 814
[22] Johnson W R and Soff G 1985 *At. Data Nucl. Data Tables* **33** 405–46
[23] Källne E, Källne J, Richard P and Stöckli M 1984 *J. Phys.* B **17** L115–20
[24] Richard P *et al* 1984 *Phys. Rev.* A **29**, 2939–42
[25] Deslattes R D, Schuch R and Justiniano E 1985 *Phys. Rev.* A **32**, 1911–13
[26] Hölzer G *et al* 1998 *Phys. Rev.* A **57** 945–48
[27] Tschischgale J *et al* 2002 *Can. J. Phys.* **80**, 867–74
[28] Bruhns H, Braun J, Kubiček K, Crespo López-Urrutia J R and Ullrich J 2007 *Phys. Rev. Lett.* **99** 113001
[29] Beyer H F, Deslattes R D, Folkmann F and LaVilla R E 1985 *J. Phys.* B **18** 207–15
[30] Briand J P *et al* 1983 *Phys. Rev.* A **28** 1413–17
[31] Tavernier M, Briand J P, Indelicato P, Liesen D and Richard P 1985 *J. Phys. B: At. Mol. Phys.* **18** L327–30
[32] Briand J P *et al* 1989 *Europhys. Lett.* **9** 225
[33] Beyer H F *et al* 1994 *Phys. Lett.* A **184** 435–9
[34] Stöhlker T *et al* 1992 *Phys. Lett.* A **168** 285–90
[35] Gumberidze A *et al* 2004 *Phys. Rev. Lett.* **92** 203004

# Theoretical progress in studying the characteristic x–ray emission from heavy few–electron ions

**Andrey Surzhykov[1,2], Stephan Fritzsche[2,3], Nikolai M. Kabachnik[2,4] and Thomas Stöhlker[1,2]**

[1]Physikalisches Institut, Universität Heidelberg, D–69120 Heidelberg, Germany

[2]Gesellschaft für Schwerionenforschung (GSI), D–64291 Darmstadt, Germany

[3]Frankfurt Institute for Advanced Studies, D–60438 Frankfurt am Main, Germany

[4]Institute of Nuclear Physiscs, Moscow State University, Moscow 119991, Russia

E-mail: surz@physi.uni-heidelberg.de

**Abstract.** Recent theoretical progress in the study of the x-ray characteristic emission from highly-charged, few-electron ions is reviewed. These investigations show that the bound–state radiative transitions in high–Z ions provide a unique tool for better understanding the interplay between the structural and dynamical properties of heavy ions. In order to illustrate such an interplay, detailed calculations are presented for the $K\alpha_1$ decay of the helium–like uranium ions $U^{90+}$ following radiative electron capture, Coulomb excitation and dielectronic recombination processes.

## 1. Introduction

Different processes that are known to occur in relativistic collisions of heavy, few-electron ions with atomic or electronic targets lead to the production of excited ionic states. In the high-$Z$ domain, the de-excitation of these states then usually results in the emission of one (or several) x-ray photons until the ground state is reached. The investigation of this characteristic x–ray emission has recently attracted much theoretical and experimental interest since it may improve our understanding of the electronic structure and dynamical behaviour of heavy atomic systems [1, 2]. A series of measurements have been carried out, especially at the GSI storage ring in Darmstadt, in order to analyze the angular–resolved radiation from the bound–state transitions in helium–like uranium $U^{90+}$ ions [3, 4, 5]. Being performed differential in photon emission angle, these experiments appeared to be very sensitive to electron–electron and electron–photon interactions in the strong–field domain.

It is well known, however, that the angle–differential properties of the characteristic radiation usually depend on both, the structure of the ions and the way how the excited ionic states were formed originally [6, 7]. In order to interpret correctly the GSI experimental data on the x–ray emission from heavy projectiles, following relativistic ion–atom collisions, one needs to be able to describe the *interplay* between the structure and dynamics in course of the "excitation–and—decay" processes. For high–$Z$, few–electron ions, moreover, such a description should account for the relativistic and many–body effects.

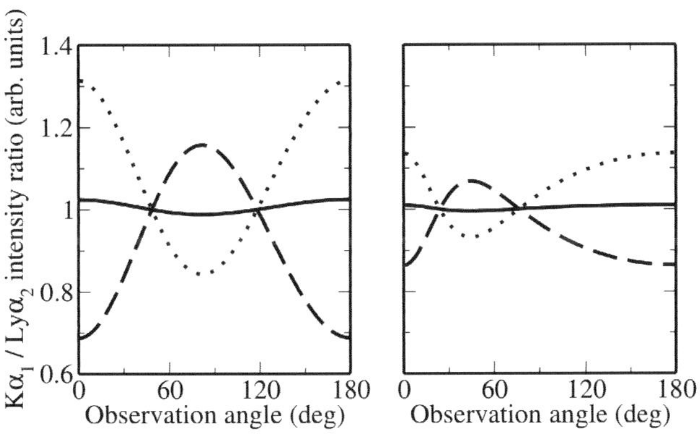


**Figure 1.** Angular distribution of the $K\alpha_1$ radiation (solid line) following the REC into (initially) hydrogen–like $U^{91+}$ ions at the projectile energies $T_p$ = 10 MeV/u (left panel) and $T_p$ = 400 MeV/u (right panel). In addition, the individual contributions from the $^1P_1 \to {}^1S_0$ (dashed line) and $^3P_2 \to {}^1S_0$ (dotted line) transitions are shown. All the data are normalized with respect to the intensity of the isotropic Lyman–$\alpha_2$ radiation following the capture into bare $U^{92+}$ ions, cf. Ref.[4].

In this contribution, we shall summarize the results from our recent theoretical studies on the production of excited ionic states and their subsequent radiative decay. Emphasis in this short review will be placed on the angular properties of characteristic $K\alpha_1$ emission from helium–like uranium $U^{90+}$ ions. In Section 2, we first discuss the emission which follows the radiative electron capture (REC) into the $1s\,2p_{3/2}$ excited ionic states. In particular, we argue that the isotropy of the $K\alpha_1$ ($1s\,2p_{3/2}\,{}^{1,3}P_{1,2} \to 1s^2\,{}^1S_0$) line, reported in Refs. [4, 5], results from a mutual compensation of the strongly anisotropic $^1P_1 \to {}^1S_0$ electric dipole ($E1$) and $^3P_2 \to {}^1S_0$ magnetic quadrupole ($M2$) fine–structure components. In contrast to REC, no compensation arises for the radiative decay following Coulomb excitation since the latter process leads to an almost exclusive population of the $1s\,2p_{3/2}\,{}^1P_1$ fine–structure level. Results of our multi–configuration Dirac–Fock (MCDF) calculations for the magnetic sublevel population of this state and, hence, anisotropy of the subsequent $1s\,2p_{3/2}\,{}^1P_1 \to 1s^2\,{}^1S_0$ radiation are displayed in Section 3. Beside the radiative electron capture and Coulomb excitation, in Section 4 we also discuss the angular properties of the $K\alpha_1$ hypersatellite (HS) photons following the $K-LL$ dielectronic recombination (DR) of initially hydrogen–like $U^{91+}$ ions. In this Section, emphasis is placed especially on the effects of higher (non–dipole) contributions to the electron–photon interaction in the bound–state transitions. Finally, a brief summary of our theoretical results is given in Section 5.

## 2. $K\alpha_1$ decay following radiative electron capture: The fine–structure effects

During the last two decades, a number of experiments have been performed at the GSI storage ring in order to investigate various *charge transfer* processes in relativistic ion–atom collisions. In the radiative electron capture, for example, a free (or quasi–free) electron is captured into a bound state of the projectile ion, accompanied by the simultaneous emission of a recombination photon. If such a capture proceeds into some excited state, the residual ion will decay towards some lower-lying level under the emission of characteristic radiation. Angular properties of this characteristic emission have been the subject of intense experimental and theoretical studies for many years [3]. For example, several measurements have been performed to explore the electron capture into the $1s_{1/2}2p_{3/2}\,{}^{1,3}P_{1,2}$ states of (initially) hydrogen–like $U^{91+}$ uranium ions and its subsequent $K\alpha_1$ ($^{1,3}P_{1,2} \to {}^1S_0$) decay [4]. This bound–bound transition gave rise to an almost *isotropic* angular distribution of the characteristic radiation, quite in contrast to what is expected from a "one–electron" model for which the Lyman–$\alpha_1$ ($2p_{3/2} \to 1s_{1/2}$) decay following REC into (initially) bare ions exhibits a strong angular dependence [8].

In order to understand the qualitatively different behaviour of the Lyman-$\alpha_1$ and $K\alpha_1$ angular–dependent emission, a detailed theoretical analysis has been carried out within the framework of the density matrix theory. In this framework, the population of the excited ionic

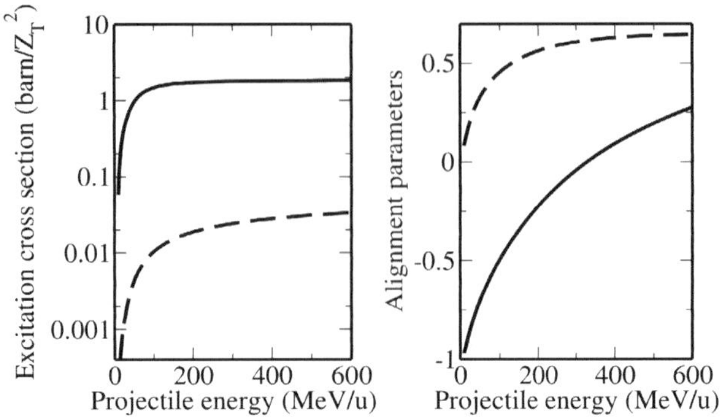


**Figure 2.** Total excitation cross sections (left panel) and alignment parameters (right panel) of the $1s\,2p_{3/2}\,{}^1P_1$ (solid line) and $1s\,2p_{3/2}\,{}^3P_2$ (dashed line) states of helium–like uranium $U^{90+}$.

states and their subsequent decay are treated as two steps, well separated in time. In the first step, the radiative electron capture usually leads to an alignment of the resulting (excited) ion which is described by means of so–called alignment parameters $\mathcal{A}_k$. The properties of the decay photons, which are emitted in the second step, are then closely related to these alignment parameters. That is, the angular distribution of the characteristic radiation is given in general by [7, 9]:

$$W(\theta) \sim 1 + \sum_k \beta_k \, \mathrm{P}_k(\cos\theta) \, , \tag{1}$$

where the *anisotropy* parameters $\beta_k$ are uniquely defined by the alignment $\mathcal{A}_k$ and by the total angular momenta of the initial and final ionic states. For example, the anisotropy parameters of the $1s\,2p_{3/2}\,{}^1P_1 \to 1s^2\,{}^1S_0$ electric dipole (E1) and $1s\,2p_{3/2}\,{}^3P_2 \to 1s^2\,{}^1S_0$ magnetic quadrupole (M2) transitions read as:

$$\beta_2\left({}^1P_1 \to {}^1S_0\right) = \frac{1}{\sqrt{2}}\mathcal{A}_2(J=1) \, , \tag{2}$$

and

$$\beta_2\left({}^3P_2 \to {}^1S_0\right) = -\sqrt{\frac{5}{14}}\,\mathcal{A}_2(J=2) \, , \tag{3}$$

respectively. Apart from the second–rank parameter $\beta_2$, the anisotropy of the ${}^3P_2 \to {}^1S_0$ radiation is also described by the forth–rank parameter $\beta_4$. For the REC into fast, high–$Z$ projectiles, however, this parameter appears to be very small and, therefore, its contribution to the angular distribution of the decay photons M2 can be neglected.

As seen from Eqs. (1)–(3), the angular dependence of the subsequent photon emission is described by the Legendre polynomials $P_2(\cos\theta)$, weighted by the alignment parameters $\mathcal{A}_2$ and some geometrical factors, that arise with different signs for the decay of the ${}^1P_1$ and ${}^3P_2$ states. Therefore, as long as the ${}^1P_1 \to {}^1S_0$ and ${}^3P_2 \to {}^1S_0$ transitions are *not resolved* experimentally, their – measured – superposition (K$\alpha_1$ line) may exhibit an (almost) isotropic behaviour even if the individual components of this line are strongly anisotropic. By performing a more detailed analysis we have found that the angular dependence of the $K\alpha_1$ radiation of helium–like ions follows the typical $\sim 1 + \beta_2 P_2(\cos\theta)$ shape, but with an overall anisotropy parameter [10]:

$$\beta_2^{\rm eff}(K\alpha_1) \;=\; N_{J=1}\,\mathcal{A}_2(J=1)\frac{1}{\sqrt{2}} - N_{J=2}\,\mathcal{A}_2(J=2)\sqrt{\frac{5}{14}} \, , \tag{4}$$

which depends on both, the alignment parameters $\mathcal{A}_2(J=1)$ and $\mathcal{A}_2(J=2)$ and the weight factors $N_{J=1}$ and $N_{J=2}$, respectively. These weights describe the contribution of the individual ${}^1P_1 \to {}^1S_0$ and ${}^3P_2 \to {}^1S_0$ fine–structure transitions to the overall $K\alpha_1$ and are given by the relative population of the ${}^{1,3}P_{1,3}$ levels.

As seen from Eq. (4), any analysis of the angular distribution of the $K\alpha_1$ radiation can be traced back to the computation of the alignment and the relative population of the ${}^{1,3}P_{1,3}$ excited ionic states. We have performed these calculations within the Multiconfiguration Dirac–Fock

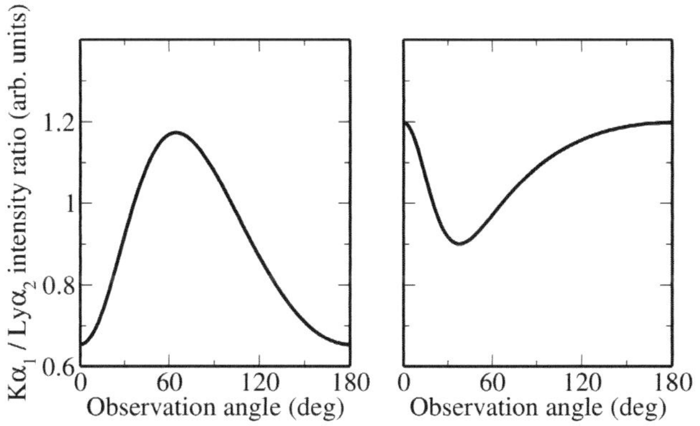


**Figure 3.** Angular distribution of the K$\alpha_1$ radiation following the $K \rightarrow L_3$ excitation of helium–like uranium ions $U^{90+}$ with energies $T_p$ = 100 MeV/u (left panel) and $T_p$ = 600 MeV/u (right panel). All the data are normalized with respect to the intensity of the isotropic Lyman–$\alpha_2$.

(MCDF) approach for the projectile energies in the range $10 \le T_p \le 600$ MeV/u. For all these energies, the $\beta_2^{\rm eff}$ was found to be less than 0.025, in line with an isotropic angular distribution. In Fig. 1, for example, the angular distribution of the K$\alpha_1$ decay (solid line) is displayed for the collision energies $T_p$ = 10 and 400 MeV/u. As seen from this figure, a (nearly) isotropic behavior of the $K\alpha_1$ radiation is observed even though the individual $^1P_1 \rightarrow {}^1S_0$ (dashed line) and $^3P_2 \rightarrow {}^1S_0$ (dotted line) components of this line are strongly anisotropic.

## 3. K$\alpha_1$ decay following Coulomb excitation: Analysis of the population dynamics

Beside the radiative electron capture, the *Coulomb excitation* of few–electron ions in collision with atoms and molecules also leads to the formation of excited ionic states. Again, these excited states are very likely aligned and associated with an anisotropic angular distribution and non–zero polarization of the subsequent decay radiation. In the past, the angle– and polarization–resolved analysis of such a characteristic radiation has been of ongoing interest owing to its importance for the diagnostics of high-temperature laboratory and astrophysical plasmas [11]. In particular, a number of experiments have been performed at the GSI storage ring in order to investigate the population of the $1s\,2p_{3/2}$ ionic states following Coulomb excitation of helium–like $U^{90+}$ projectiles in collision with $N_2$ target molecules [4]. In contrast to the REC, this excitation process was found to result in a strongly *anisotropic* K$\alpha_1$ radiation, and, hence, has initiated a more detailed analysis in order to account for the collision dynamics of the helium–like projectiles.

In Ref. [12], the Coulomb excitation of few–electron heavy ions in relativistic collisions with low–$Z$ target atoms (or protons) has been recently explored by us within the framework of the first–order perturbation theory and combined with multiconfiguration Dirac–Fock (MCDF) computations. Special emphasis was placed on the effective parameter $\beta_2^{\rm eff}(K\alpha_1)$ which reflects the anisotropy of the K$\alpha_1$ decay radiation. While the general expression (4) for this parameter does not depend on the particular excitation process, its numerical values are different if evaluated for the radiative electron capture and Coulomb excitation. For the latter process, our MCDF calculations clearly indicate that the $K \rightarrow L_3$ excitation of helium–like uranium ions results in an almost selective population of the $1s\,2p_{3/2}\,{}^1P_1$ ionic state as it was observed earlier in Ref. [13]. The total cross section for this singlet state is at least 50 times larger than the population cross section for the $1s\,2p_{3/2}\,{}^3P_2$ level [cf. left panel of Fig. 2]. Such a selective population leads to the fact that only $1s\,2p_{3/2}\,{}^1P_1$ state contributes to the characteristic K$\alpha_1$ radiative whose anisotropy parameter:

$$\beta_2^{\rm eff}(K\alpha_1) \quad \approx \quad \mathcal{A}_2(J=1)\frac{1}{\sqrt{2}}, \tag{5}$$

is easily obtained from Eq. (4) by taking $N_{J=1} = 1$ and $N_{J=2} = 0$.

As seen from Eq. (5), the anisotropy of the K$\alpha_1$ radiation following Coulomb excitation of the helium–like projectiles is directly related to the alignment of the $1s\,2p_{3/2}\,{}^1P_1$ state. Therefore,

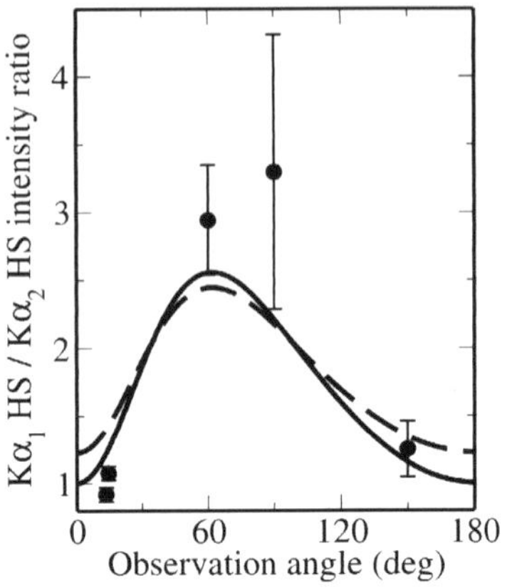


**Figure 4.** Intensity ratio of the $K\alpha_1/K\alpha_2$ HS lines for helium–like uranium following the $K-LL$ DR. Calculations are performed within the E1 approximation (dashed line) as well as by accounting of all allowed multipole channels (solid line) and are compared with experimental data (filled circles) by Ma and co–workers [5].

a series of MCDF computations have been carried out to analyze this alignment for the K–shell excitation of uranium $U^{90+}$ projectiles with energies in the range $10 \le T_p \le 600$ MeV/u. The results of these calculations are shown in the right panel of Fig. 2 and indicate that the parameter $\mathcal{A}_2(^1P_1)$ is large and negative for low–energy collisions with $T_p \le 200$ MeV/u. This refers to a predominant population of the $M_J = 0$ magnetic substate and therefore to an alignment, i.e. an electron density that is directed perpendicular to the beam. In fact, such a result is consistent with the classical picture in which the transferred orbital angular momentum is found perpendicular to the collision direction. For the higher collision energies, however, the classical picture is no longer valid due to the increasing role of relativistic and magnetic effects. These effects result in an ion alignment along the beam direction and subsequently in large positive values of the parameter $\mathcal{A}_2(^1P_1)$ [cf. right panel of Fig. 2].

The strong energy dependence of the alignment parameter $\mathcal{A}_2(^1P_1)$ results in qualitatively different emission patterns of the $K\alpha_1$ radiation following Coulomb excitation in (relatively) slow and fast ions–atom (ion–electron) collisions. For example, as seen from Fig. 3, while the large positive parameter $\mathcal{A}_2(^1P_1) = 0.28$ for $T_p = 600$ MeV/u clearly favors a *co–linear* photon emission along the beam direction, a remarkable $K\alpha_1$ emission *perpendicular* to this direction can be observed for the decay of $1s\,2p_{3/2}\,{}^1P_1$ state produced with alignment $\mathcal{A}_2(^1P_1)$ = -0.49 at the energy $T_p = 100$ MeV/u.

## 4. $K\alpha_1$ decay following dielectronic recombination: Multipole–mixing effects

Until now we have discussed the angular properties of the $K\alpha_1$ photons following the radiative electron capture and Coulomb excitation of heavy, few–electron ions. During recent years, moreover, special interest has been paid also to the radiative decay of the doubly–excited resonances produced in course of the $K-LL$ *dielectronic* recombination of (finally) helium–like $U^{90+}$ ions. In the experiments by Ma and coworkers [5], for example, the angular distributions of the $K\alpha_{1,2}$ hypersatellite and satellite lines have been measured and compared with the theoretical predictions [14] based on Dirac's relativistic theory. Although a reasonable agreement between experiment and theory was found for most groups of lines, a rather remarkable discrepancy remained for the angular distribution of the $K\alpha_1$ hypersatellite lines that arise from the decay of the $L_{1/2}L_{3/2}$ resonances. For this line group, the theoretically predicted anisotropy of the $K\alpha_1$ HS underestimates the observations by more than a factor of two [5, 14], a discrepancy that has not been resolved until today.

As it was already argued in Ref. [5], the observed stronger anisotropy of the $K\alpha_1$ hypersatellite line might be caused by the higher (non–dipole) channels in the bound–bound transitions. Since these—non–dipole effects—have not been taken into account in the previous case studies, we have recently re–considered the radiative decay of the doubly–excited ionic states to include a full multipole decomposition of the electron–photon interaction [15]. Special emphasis has been placed on the four levels $2s_{1/2}\,2p_{3/2}\ :J = 1,2$ and $2p_{1/2}\,2p_{3/2}\ :J = 1,2$ which belong to the $L_{1/2}L_{3/2}$ resonant group and on their subsequent radiative decay. In particular, we have studied the *non* electric–dipole effects on the fine–structure characteristic transitions in helium–like uranium $U^{90+}$ ions. We have shown that the interference between the allowed multipole components of the electron–photon interaction can easily decrease or enhance the anisotropy of individual fine–structure transitions by more than a factor of six [15].

A direct comparison of the calculated and measured angular distributions for *individual* $K\alpha_1$ fine–structure HS lines is presently impossible because of several restrictions that are inherent in existing relativistic collision experiments. Apart from the uncertainty in the central energy and energy profile of the ion beam, there are difficulties associated with the Compton profile of the target electrons as well as with the relatively low resolution of the x–ray detectors. Therefore, in order to understand the experimental data on the $K\alpha_1$ HS decay of four unresolved $L_{1/2}L_{3/2}$ resonances, the various fine–structure transitions belonging to the $K\alpha_1$ line have to be summed *incoherently* by using the calculated branching ratios and the population cross sections for the doubly–excited $2s_{1/2}2p_{3/2}$ and $2p_{1/2}2p_{3/2}$ states. Resulting angular distribution is presented in Fig. 4 together with the experimental data by Ma and co–workers [5]. As seen from this figure, the multipole–mixing effects result in about a 30 % enhancement of the anisotropy of the $K\alpha_1$ line and, hence, in a better agreement with the experimental results if compared to the theoretical predictions obtained within the E1 approximation only.

## 5. Conclusion

In conclusion, the recent theoretical progress in studying the characteristic emission from few–electron, heavy ions has been reviewed. Special attention was paid to the angular distribution of the $K\alpha_1$ x–rays following radiative as well as dielectronic recombination of (initially) hydrogen–like $U^{91+}$ and Coulomb excitation of helium $U^{90+}$ uranium projectiles. Together with the experimental data, our calculations demonstrate that the angle-resolved x-ray spectroscopy provides a sensitive tool for studying the relativistic and many-body effects on the structure and dynamics of high-$Z$, few–electron ions.

## Acknowledgments

A. S. acknowledges the support from the Helmholtz Gemeinschaft (Nachwuchsgruppe VH–NG–421). S. F. acknowledges the support by DFG (FR 1251/13–1) and GSI under the project No. KS–FRT. N.M.K. acknowledges the hospitality and financial support from GSI.

## References

[1] Eichler J and Meyerhof W 1995 *Relativistic atomic collisions* (San Diego: Academic Press)
[2] Marrs R E, Beiersdorfer P and Schneider D 1994 *Physics Today* **47** 27
[3] Fritzsche S, Indelicato P and Stöhlker Th, *J. Phys. B* **38** S707
[4] Gumberidze A *et al.* 2003 *Hyperfine Inter.* **146/147** 133
[5] Ma X, Mokler P H, Bosch F, Gumberidze A, Kozhuharov C, Liesen D, Sierpowski D, Stachura Z, Stöhlker Th and Warczak A 2003 *Phys. Rev. A* **68** 042712
[6] Percival I C and Seaton M J 1958 *Phil. Trans. Roy. Soc. London* **113** 251
[7] Balashov V V, Grum–Grzhimailo A N and Kabachnik N M 2000 *Polarization and Correlation Phenomena in Atomic Collisions* (New York: Kluwer Academic Plenum Publishers)
[8] Surzhykov A, Fritzsche S, Gumberidze A and Stöhlker Th 2002 *Phys. Rev. Lett.* **88** 153001
[9] Surzhykov A, Jentschura U D, Stöhlker Th and Fritzsche S 2006 *Phys. Rev. A* **73** 032716
[10] Surzhykov A, Jentschura U D, Stöhlker Th and Fritzsche S 2006 *Phys. Rev. A* **74** 052710
[11] Reed K J and Chen M H 1993 *Phys. Rev. A* **48** 3644
[12] Surzhykov A, Jentschura U D, Stöhlker Th, Gumberidze A and Fritzsche S 2008 *Phys. Rev. A* **77** 042722
[13] Stöhlker Th *et al.* 1998 *Phys. Rev. A* **57** 845
[14] Zakowicz S, Harman Z, Grün N and Scheid W 2003 *Phys. Rev. A* **68**, 042711
[15] Fritzsche S, Kabachnik N M and Surzhykov A 2008 *Phys. Rev. A* accepted

# Momentum distribution of highly charged ions formed by strong laser fields

**S. Yoshida[1], K. I. Dimitriou[1,2,3], J. Burgdörfer[1], H. Shimada[4,5,6], H. Oyama[4] and Y. Yamazaki[4,5]**

[1]Institute for Theoretical Physics, Vienna University of Technology, Vienna, Austria
[2]Physics Department, National Technical University, and Theoretical and Physical Chemistry Institute, National Hellenic Research Foundation, Athens, Greece
[3] Department of Physical Sciences and Applications, Hellenic Army Academy, Vari, Greece
[4]Atomic Physics Laboratory, RIKEN, Saitama, Japan
[5]Graduate School of Arts and Sciences, University of Tokyo, Tokyo, Japan

E-mail: shuhei@concord.itp.tuwien.ac.at

**Abstract.** Momentum distributions of highly charged ions produced by a strong laser pulse with duration of $\sim$ 200 fs and maximum intensity of $\sim$50-70 PW/cm$^2$ are studied. Due to the spatial variation of the laser intensity within the beam focus, different charge states are produced during a single laser shot. The measured widths of the momentum distribution of the product ions exhibit a linear dependence on the ionization potential of the ions for various rare gas atoms. Such a linear dependence is analyzed theoretically using the single active electron model.

## 1. Introduction

Due to technological advances high-power lasers have become available which are strong enough to multiply ionize atoms and produce highly charged ions [1, 2]. The dynamics of atoms in such strong lasers can be studied by measuring not only the ion yield but also the momentum distribution of the product ions [3, 4, 5]. This provides deeper insights into the ionization dynamics. The measured distributions of the multiply charged ion momentum parallel to the polarization axis of the electric laser field is observed to be sensitive to whether the ionization is sequential or non-sequential [3, 6]. For weak laser intensities non-sequential ionization involving strong electron-electron correlations (rescattering) [7, 8] dominates leading to two distinct peaks with a dip at $p = 0$ in the momentum distribution. On the other hand, strong lasers ionize electrons from an atom sequentially and the resulting momentum distribution is Gaussian-like with a peak at $p = 0$ [4, 9].

In this paper we explore the regime with a very intense laser field by which up to 8-fold charged ions can be produced via sequential ionization. The driving field considered is a linearly polarized laser field approximated by

$$F(t) = F_{\max} f(t) \cos(\omega t + \phi) \quad (-\tau < t < \tau)\,, \tag{1}$$

[6] Present address: Department of Photo-Molecular Science, Institute for Molecular Science, Oakazaki, Japan

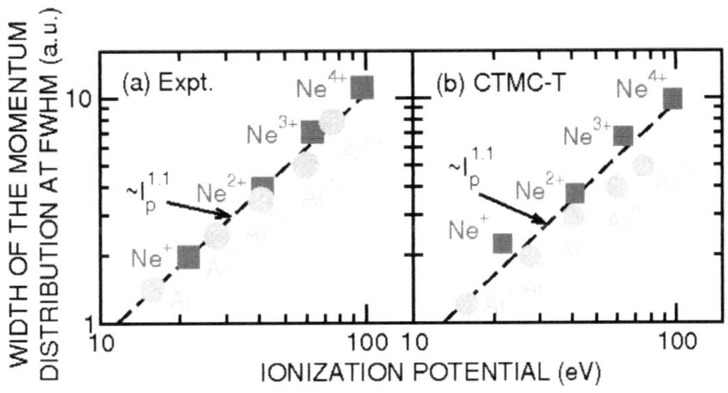


**Figure 1.** Measured (a) and calculated (b) width (FWHM) of the ion momentum distribution as a function of ionization potential $I_p$. The atoms are driven by the laser field (50-70PW/cm$^2$, 775nm, 200fs). The results are fitted to a power law $\propto I_p^{1.1}$ indicated by the dashed line.

where $F_{\rm max}$ is a peak field strength, $\omega$ a laser frequency, $\phi$ the carrier envelope phase, and $f(t) = \cos^2(\pi t/(2\tau))$ the approximate laser envelope. (For long pulses of about 200 fs (FWHM), the carrier envelope phase $\phi$ is unlikely to be important and neglected ($\phi = 0$) in the following.) Due to the focusing of the laser, $F_{\rm max}$ depends on the position of atoms relative to the focal point of the laser. The closer the atoms are to the focal point, the larger the maximum strength $F_{\rm max}$ is. As a consequence, the spatial spread of the atomic beam in the interaction region results in a broad charge state distribution within a single laser shot. The momentum distribution of the produced ion $A^{q+}$ is thus that integrated over a range of laser intensities : from the ionization threshold of $A^{q+}$ to the intensity for which the ion yield of the next charge state becomes dominant. The main contributions result from the laser intensity near the upper boundary for which the yield of $A^{q+}$ is largest. In such a strong field sequential ionization is the dominant process and the momentum distribution is expected to be Gaussian. The width of the distribution is observed to display a nearly linear behavior as a function of the ionization potential of each ion (Ne and Ar, see Fig. 1a). Such a dependence is also observed for other rare gas target atoms [5].

The theoretical investigation of multiple ionization is not an easy task. As the number of electrons involved increases, the ab-initio quantum calculation of the many-body dynamics becomes beyond the reach of currently available computing power. When each electron is ionized sequentially, however, we can simplify the model drastically: the ionization of each electron is analyzed by the single active electron model. In the next section, we briefly review the method to simulate sequential ionization processes which reproduces the observed behavior in the measured width of the momentum distributions [9]. In Sec. 3, we introduce an even simpler model to understand the origin of the nearly linear scaling observed in the measured results. We show that for very strong laser fields with a duration of 200 fs the electron is ionized (detached) even before the laser field reaches its maximum amplitude. While the electron will experience the maximum amplitude of the laser after ionization, its final momentum is determined by the temporal laser amplitude at the time of ionization rather than the maximum laser amplitude $F_{\rm max}$. This temporal laser amplitude features the linear dependence on $q$ which eventually leads to the linear scaling of the momentum distribution width.

## 2. Simulation of sequential ionization by the single active electron model

We briefly review the quasi-classical method used to simulate the momentum distribution of highly charged ions by a strong laser field (see details in [9]). In sequential ionization processes, the detachment of each electron is simulated independently within a single active electron model using the classical Monte Carlo method including tunneling (CTMC-T) [10, 9]. We consider the production of $A^{q+}$ ion by the laser field with a given peak field strength $F_{\max}$ [Eq. (1)]. The detachment of the $i^{\text{th}}$ electron is simulated by the Hamiltonian

$$H_i = \frac{p^2}{2} + V_i(r) + zF(t) \tag{2}$$

where $V_i(r)$ is the model potential [11] representing the core potential of the $A^{i+}$ ion screened by bound electrons. The interactions between the electron to be detached and $i-1$ electrons which are detached earlier than the $i^{\text{th}}$ electron are neglected. (Atomic units are used throughout unless otherwise stated.) The equations of motion are solved numerically for an ensemble of trajectories (microcanonical ensemble with $H_i = I_p(i)$) representing the ground state wavefunction of the $i^{\text{th}}$ electron with the ionization potential $I_p(i)$. Each trajectory is allowed to tunnel through the potential barrier formed by the core potential and the laser field, with the tunneling probability estimated semiclassically, so that the tunneling ionization can be properly represented without solving the Schrödinger equation. The final momentum distribution of the electron can be determined by statistically sampling the final momenta of trajectories of the ensemble. Due to momentum conservation, this electron momentum distribution is considered to be identical to the ion momentum distribution. When the simulated momentum distributions are compared with the measured data, they are integrated over a range of laser peak field strength $F_{\max}$ weighted by the distribution of the laser field strength within the interaction region ($\propto 1/F_{\max}$) and the ionization probability.

For the momentum distribution of the $A^{q+}$ ion, the $q^{\text{th}}$ detached electron has the largest momentum and provides the main contribution to the overall width of the ion momentum distribution. Considering only the main contribution from the $q^{\text{th}}$ electron we could well reproduce the measured width $\Delta p$ for different rare gas atoms. The estimated width depends almost linearly on the ionization potential $I_p(q)$ (Fig. 1b).

## 3. Analysis using the strong field approximation

When we consider the production of the $A^{q+}$ ion by the laser with a given peak field $F_{\max}$, the peak field has to be large enough to detach $q$ electrons but should not be too strong to detach the $(q+1)^{\text{th}}$ electron. Figure 2 displays the ionization probability of $Ne^{q+}$ ions (Eq. (2) with $i = q$) as a function of $F_{\max}$. In this calculation, the hydrogenic approximation for the core

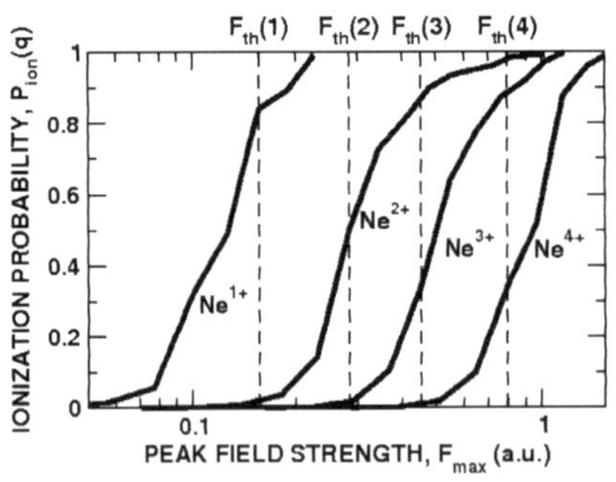


**Figure 2.** Ionization probability (solid lines) of Ne ions as a function of peak field strength $F_{\max}$ of the laser pulse (775nm, 200fs). The vertical dashed lines are the classical ionization thresholds $F_{\text{th}}(q)$.

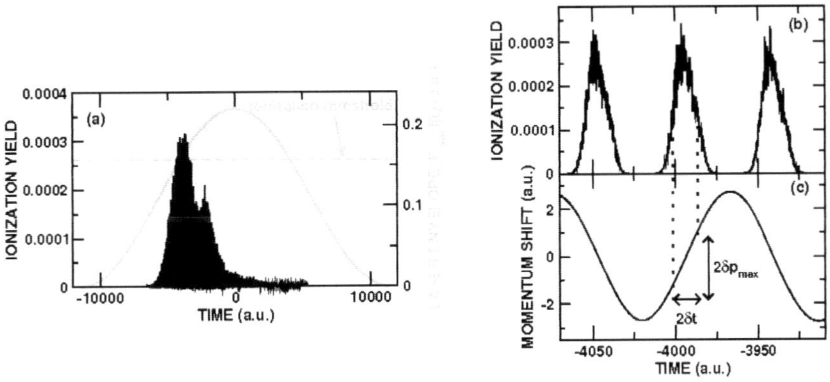


**Figure 3.** (a) Ionization yield of $Ne^{1+}$ ion as a function of time. Ne atoms are subject to the laser (50-70PW/cm$^2$, 775nm, 200fs). The laser envelope (solid light blue/gray line) and the ionization threshold $F_{\rm th} = I_p^2/4$ (dashed line) are also plotted. (b) A zoom-up of the ionization yield around the time for which the yield reaches maximum. (c) The momentum shift (Eq. (4)) in the same time interval as in (b). The vertical dotted lines indicate the times $t_{\rm tnn} \pm \delta t$ at which the ionization yield reaches the half of the maximum value within a single peak. The corresponding momentum shift $\delta p_{\rm max}$ is the width of the final momentum distribution (see the main text for the details.)

potential $V(r) = -q/r$ was used for simplicity. Near the classical ionization threshold

$$F_{\rm th}(q) = I_p(q)^2/(4q) \tag{3}$$

the yield of $Ne^{q+}$ ion saturates while the next charge state starts appearing around its threshold $F_{\rm th}(q+1)$. Thus, the momentum distribution is integrated over a range $F_{\rm th}(q) < F_{\rm max} < F_{\rm th}(q+1)$. While the peak laser amplitude exceeds the ionization threshold $F_{\rm th}$, the potential barrier at the saddle point ($x = y = 0$) is lowered below the binding energy $-I_p$ of the $q^{\rm th}$ electron. However, the potential energy ($E_z = -I_p - (p_x^2 + p_y^2)/2$) of the electron can be below the barrier preventing the ionization over the barrier. Thus it is important to include the tunneling effect into the ionization dynamics even when the peak laser amplitude is above the threshold value. For long ($\sim$ 200fs) laser pulses, the electron will be ionized mostly even before the laser reaches its maximum amplitude. The tunneling rates are small but the electron is exposed to weak fields for long times due to the slow ramping of the laser causing the tunneling probability to be non-negligible. Figure 3a displays the $Ne^{1+}$ ion yield by the laser field with a peak field strength of $F = 1.4F_{\rm th}(q = 1)$. The electron is ionized before the laser field reaches the maximum. When the laser amplitude (solid light blue/gray line) is near the threshold (dashed line), the ionization takes place with the largest probability.

To determine the final momentum of the ionized electron the tunneling time $t_{\rm tnn}$ at which electron tunnels through the potential barrier plays an important role. After ionization the laser amplitude increases towards its maximum value $F_{\rm max}$ and the core potential becomes less important. In the strong field approximation, the dynamics of the tunneled electron follows the equation of motion $dp_z/dt \simeq -F(t)$ decoupled from the other degrees of freedom. After the interaction with the laser thus the electron momentum becomes $p_z(t > \tau) = p_z(t = t_{\rm tnn}) + \delta p$ with

$$\delta p = -\int_{t_{\rm tnn}}^{\tau} F(t)dt \simeq \frac{F_{\rm max}}{\omega} f(t_{\rm tnn}) \sin(\omega t_{\rm tnn}) \tag{4}$$

($\tau$ : the time at which the laser pulse ends, Eq. (1)) [12]. As the tunneling probability becomes maximum at $\omega t_{\rm tnn} \simeq \pm\pi/2$ within a half optical cycle ($t = -4000$a.u. in Fig. 3b), the momentum

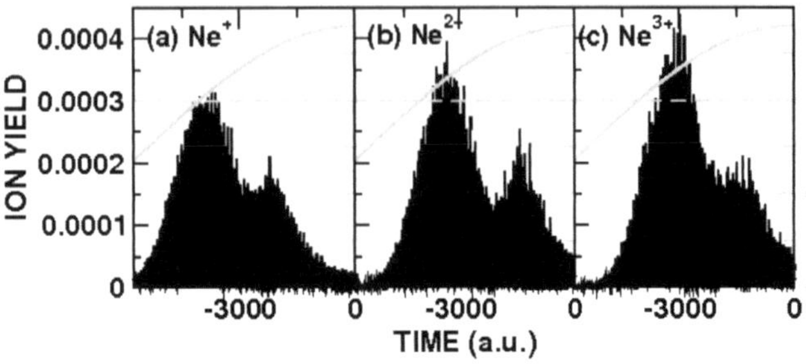


**Figure 4.** Ionization yield of $Ne^{q+}$ ion ($q = 1, 2, 3$) as a function of time. Ne atoms are subject to the laser (50-70PW/cm$^2$, 775nm, 200fs). The laser envelope (solid light blue/gray line) and the ionization threshold $F_{\rm th} = I_p^2/4$ (dashed line) are also scaled by the threshold value $F_{\rm th} = I_p(q)^2/(4q)$

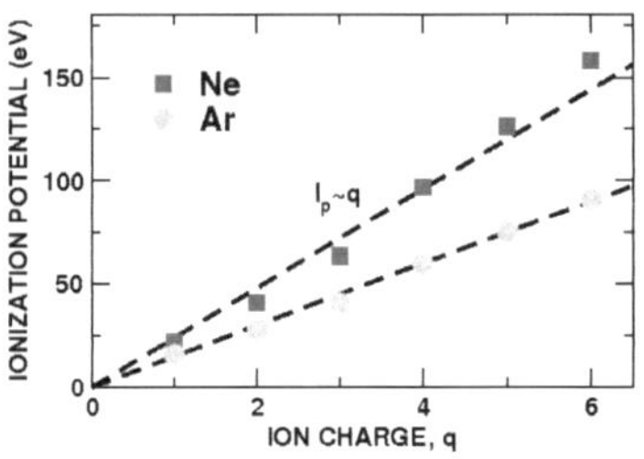


**Figure 5.** $q^{\rm th}$ ionization potential $I_p(q)$ for Ne and Ar plotted as a function of $q$. The ionization potential is fitted for each atom by a linear scaling $I_p(q) \propto q$ indicated as the dashed line.

shift becomes nearly zero $\sin(\omega t_{\rm tnn}) \sim 0$ (Fig. 3c). Accordingly, as the ionization yield is non-negligible for a finite duration $\pi/(2\omega) - \delta t < t_{\rm tnn} < \pi/(2\omega) + \delta t$ (Fig. 3b), the electron momentum is shifted at maximum $\delta p_{\rm max} = F_{\rm max} f(t_{\rm tnn})\delta t + O(\delta t^3)$ (Fig. 3c). Considering that $p_z(t_{\rm tnn}) \simeq 0$ at tunneling the width of the momentum distribution for the ionized electron is given by $2\delta p_{\rm max}$ and the width is simply determined by the temporal laser amplitude at tunneling, $F_{\rm tnn} = F_{\rm max} f(t_{\rm tnn})$. The ionization yield is non-negligible over many optical periods. For the quantitative evaluation of the momentum distribution width, therefore, the average of momentum shift $\delta p_{\rm max}$ over those optical periods has to be calculated. However, this averaging broadens the distribution similarly for all ions with different $I_p(q)$ without changing the $I_p$-dependence of the distribution width $\delta p_{\rm max}$. Thus the scaling behavior of the momentum distribution width can be analyzed without this averaging taken into account. As seen in Fig. 3a the electron is ionized with the largest probability when the laser amplitude is around the threshold value $F_{\rm tnn} \simeq F_{\rm th}(q)$. The temporal laser amplitude $F_{\rm tnn}$ at the largest tunneling probability can be more precisely determined as the field strength for which the static ionization rate (of the order of $10^{-2}$ a.u. for Ne with $F = F_{\rm th}$ with only a small dependence on $q$) multiplied by an optical cycle (of the order of $10^2$ a.u. for $\omega = 0.059$ a.u.) becomes on the order of one. Indeed, for Ne, the laser amplitude $F_{\rm tnn}$ at the maximum tunneling probability slightly increases with increasing $q$ ($F_{\rm tnn} \simeq F_{\rm th}$ for $q = 1$, $1.08F_{\rm th}$ for $q = 2$, and $1.15F_{\rm th}$ for $q = 3$ (Fig. 4)). Thus an additional $q$ dependence of the temporal laser amplitude at tunneling can be fitted as $F_{\rm tnn} \simeq q^{0.1}F_{\rm th}(q)$. Thus the width of the final momentum distribution can be estimated as $2q^{0.1}F_{\rm th}(q)\delta t$. We note that in the ionization yields (Fig. 4a-c) the 2nd peak about 2000.a.u after the 1st peak is observed. In the CTMC-T, the trajectories oriented along the laser polarization have larger ionization rates than those oriented perpendicular to the laser polarization. The latter is responsible for the 2nd maximum in the ionization yield.

The interval $\delta t$ depends only on the laser frequency, i.e., $\delta t \sim \omega^{-1}$ and is rather independent of the ionization potential $I_p(q)$ or the charge $q$. According to Eq. (3), therefore, the width $\delta p_{\rm max}$ of the momentum distribution is proportional to $I_p^2$. This scaling appears to contradict the linear behavior observed in Fig. 1. One key factor to resolve this contradiction is the $q$-dependence of the ionization potential. Figure 5 shows an approximately linear dependence of the ionization potential $I_p(q)$ for Ne and Ar, i.e., $I_p(q) \propto q$. This linear dependence is well known, in particular, for the systems, such as $C_{60}$, involving many equivalent electrons [13]. For Ne and Ar, the same applies when $q$ is small. For large values of $q$, slight deviations from the linear scaling are expected (see the onset in Fig. 5 ). For the observed ions in Fig. 1 ($q \leq 5$), it is safe to assume that the ionization potential is given by $I_p(q) = qI_p(1)$. The width of the momentum thus becomes proportional to $q^{1.1}$, explaining the linear dependence on the ionization potential with the same spices of atom (Ne or Ar) seen in Fig. 1.

Within this simplified model there is a small target-atom-dependence of the momentum width $\delta p$ through the 1st ionization potential, i.e, $\delta p \propto I_p(1)^2 q^{1.1}$. This target atom dependence is hardly seen in the measured momentum width. One known problem of the CTMC-T method is that the tunneling probabilities for hydrogenic atoms are underestimated compared to those predicted by the corresponding quantum simulations [10]. Moreover, the tunneling probabilities for rare gas atoms involving multiple electrons may cause additional uncertainties in the estimation of ionization probabilities. More accurate evaluation of the ionization probabilities may modify the momentum distribution width by a different constant factor for each target atom and narrow the small difference between the measured and calculated results. On the other hand, it is interesting to measure the width for the target atoms with relatively large 1st ionization potential, such as He, which would reveal more prominently the target atom dependence if any.

## 4. Conclusion

We showed that the width of the ion momentum distribution driven by an intense (50-70 PW/cm$^2$) and long (200 fs) laser pulse can be well represented by the single active electron model assuming that sequential ionization dominates the ionization process. The width is determined through the dynamics after the electron is detached from the core ion. In the strong field regime, the effect of the core potential does not play an important role and the temporal laser amplitude at the time when the electron is detached determines the final width of the momentum distribution. The obtained width explains the nearly linear behavior of the distribution width as a function of ionization potential.

## Acknowledgments

This research is supported by the FWF-SFB-F016 (Austria) and EU-HITRAP Project number HPRI-CT-2001-50067. JB also acknowledges support from the RIKEN Eminent Scientist program.

## References

[1] Yamakawa K, Akahane Y, Fukuda Y, Aoyama M, Inoue N, Ueda H and Utsumi T 2004 *Phys. Rev. Lett.* **92** 123001
[2] Palaniyappan S, DiChiara A, Chowdhury E, Falkowski A, Ongadi G, Huskins E L and Walker B C 2005 *Phys. Rev. Lett.* **94** 243003
[3] Rudenko A, Zrost K, Schröter C D, de Jesus V L B, Feuerstein B, Moshammer R and Ullrich J 2004 *J. Phys. B* **37** L407
[4] Shimada H, Nakai Y, Oyama H, Ando K, Kambara T, Hatakeyama A and Yamazaki Y 2005 *Nuc. Inst. and Meth. B* **235** 221
[5] Shimada H 2007 *Multiple ionization of atoms by an intense laser field: momentum spectroscopic approach* Ph.D. thesis University of Tokyo
[6] Rudenko A, Ergler T, Zrost K, Feuerstein B, de Jesus V L B, Schröter C D, Moshammer R and Ullrich J 2008 *J. Phys. B* **41** 081006
[7] Ho P J, Panfili R, Haan S L and Eberly J H 2005 *Phys. Rev. Lett.* **94** 093002
[8] Chen J, Liu J, Fu L B and Zheng W M 2000 *Phys. Rev. A* **63** R11404
[9] Dimitriou K I, Yoshida S, Burgdörfer J, Shimada H, Oyama H and Yamazaki Y 2007 *Phys. Rev. A* **75** 013418
[10] Cohen J S 2001 *Phys. Rev. A* **64** 043412
[11] Szydlik P P and Green A E S 1974 *Phys. Rev. A* **9** 1885
[12] Feuerstein B, Moshammer R and Ullrich J 2000 *J. Phys. B* **33** L823
[13] Seifert G, Vietze K and Schmidt R 1996 *J. Phys. B* **29** 5183

# Microcalorimeter observations of $L$-shell spectra of Ne- through Fe-like Au ions in an EBIT

**E. Träbert[1,2], S. B. Hansen[1], G.V. Brown[1], P. Beiersdorfer[1], K. Widmann[1], H. K. Chung[1]**

1 High Temperature and Astrophysics Division, Lawrence Livermore National Laboratory, Livermore, CA 94550-9234, U.S.A.
2 Astronomisches Institut, Ruhr-Universität Bochum, D-44780 Bochum, Germany

E-mail: traebert@astro.rub.de

**Abstract.**
L-shell x-ray emission of Au has been observed of the plasma that flows from reduced-size hohlraums under high-power laser irradiation. In order to provide tools for determining the temperature, charge state distribution and mean charge of such a plasma, we have measured the $L$-shell emission from highly charged gold ions in the SuperEBIT electron beam ion trap under bombardment by electrons at energies from 10 to 18 keV. The emission was recorded with an x-ray microcalorimeter, featuring an instrumental line width of 10 eV in the region of primary interest. Lines from ironlike $Au^{53+}$ through neonlike $Au^{69+}$ ions were identified. We find that the strong $3d_{5/2} => 2p_{3/2}$ emission features are well separated for at least the highest 13 charge states and can be used to diagnose the charge state distribution.

## 1. Introduction

In the original concept of inertial confinement fusion (ICF) experiments, laser light was supposed to depose energy into the outer shell of a fuel pellet which then would undergo compression by shock waves. To improve symmetry, an intermediate stage has been added in which the pellet is placed inside a gold-coated cylindrical hohlraum of centimeter size. Many beams of laser light enter the hohlraum through the openings at the ends of the cylinder and strike the inner wall. A plasma is generated that produces x rays; these x rays provide a more uniform energy flux than the original lasers that then heats the fuel pellet. Experiments of this type are going on at various laboratories.

It would be interesting to find out how to deposit the available laser energy in the smallest possible space enclosed by walls of high-$Z$ material. For this purpose, reduced-size (half a millimeter) hohlraums with only one opening have been constructed. In these 'hot' hohlraums, the plasma can fill the volume during the laser pulse, and then spills out. In fact, the effluent plasma interacts with the incoming laser light and limits further access to the hohlraum [1, 2]. In order to learn about this interaction, it is necessary to characterize the effluent plasma by density, temperature, charge state distribution, and mean charge of the high-$Z$ material.

In gold plasmas with temperatures of tens of keV, the dominant charge state is likely to be neonlike $Au^{69+}$. The $3d_{5/2} \rightarrow 2p_{3/2}$ transitions of Ne-like ions and the neighboring ions of lower charge have been seen to form a well-spaced "picket-fence" pattern [3, 4], similar to observations on M-shell transitions of gold and tungsten [5, 6, 7]. The shape of this pattern depends on the abundance of the ions of each charge state and thus on the ionization balance which, in turn, depends on the electron temperature. This radiation pattern, therefore, can serve as a means for determining both the average ionization state of the plasma and its electron temperature.

## 2. Experiment

We have recorded the $3d_{5/2} \rightarrow 2p_{3/2}$ emission pattern of highly charged gold ions in an electron beam ion trap under well controlled laboratory conditions. The adjustable beam energy of the device allows us to step through the charge states, enabling unambiguous identification of the strong $L$-shell lines and line blends, in this experiment of $Au^{53+}$ through $Au^{69+}$ (Fe- through Ne-like ions), as well as the determination of charge state distributions.

The experiment was performed at the SuperEBIT electron beam ion trap [8] at the Lawrence Livermore National Laboratory. The device has been optimized for spectroscopic studies of highly charged ions [9], including the study of spectra needed for diagnosing high energy density plasmas [10]. About every two seconds, a laser injection system supplied Au to the ion trap, where it was quickly ionized by the electron beam. Every 19 s the content of the trap was dumped to halt the accumulation of possible contaminants, such as barium and tungsten, and then the trap cycle was repeated. X-ray emission from the ions in the trap was detected by means of a cryogenic microcalorimeter built by Goddard Space Flight Center [11]; out of the array of 32 small HgTe absorbers ("pixels"), 12 were read out at the same time. In order to record x rays with energies as high as 18 keV, the microcalorimeter was operated at a (slightly elevated) temperature of T=65 mK. The individual pixel pulse height was calibrated against n=1-2 lines of He- and H-like ions of Ar, Ni, and Ge, and then all pixel spectra were summed. For single x-ray transitions at 8 to 10 keV, a line width (FWHM) of about 10 eV was found, which corresponds to a resolving power of about 900. Spectral features measured with larger line widths (15 to 18 eV) reveal the presence of line blends. The quantum efficiency of the HgTe absorbers in the detector varies significantly over the range of the present observations. Corrections for filter and Be window transmission and for detection efficiency have been applied.

The experiment primarily aimed to measure the 2-3 transitions of Au emitted in the x-ray energy range from 9 to 12 keV. The highest charge state of interest was $Au^{69+}$ (Ne-like); the ionization potential (IP) of 18 keV of this ion was chosen as the highest electron beam energy used in the measurements. The ionization potentials of the next lower charge state ions are 8.4 keV and consecutively lower. However, the IP refers to the valence electron (n=3 for those ions below the Ne-like ion), whereas the x-ray transitions of present interest relate to inner-shell excitation.

Figure 1 shows sections of emission spectra obtained at electron beam energies from 10 to 18 keV. Evidently the Au emission varies with the electron beam energy: as $E_{beam}$ decreases, the intensity of major emission features shifts to lower energies. The picket-fence structure of the $3d_{5/2} \rightarrow 2p_{3/2}$ emission features in our spectra extends to ironlike $Au^{53+}$, with a picket-fence spacing of about 40 eV. The highest-energy peak of each line group belongs to the Ne-like ion. Line blending becomes significant for charge states below about vanadiumlike $Au^{56+}$, and for those ions full modeling is needed to predict the emission. However, in higher-temperature plasmas, where the average ion charge $\langle Z \rangle$ is well above 56+, the $3d_{5/2} \rightarrow 2p_{3/2}$ features provide a reasonably good indication of the charge balance even in the absence of detailed modeling.

## 3. Atomic structure and collisional-radiative calculations

We have used the Flexible Atomic Code (FAC) of Gu [12] to generate fine-structure atomic levels and rate data for Au ions from H-like ($Au^{78+}$) to Zn-like ($Au^{49+}$). The atomic structure and rate data were then used in the collisional-radiative and spectral synthesis code SCRAM [13] to calculate level populations, charge state distributions, and emission spectra. At the low densities of the electron-beam ion trap, where electronic state populations are overwhelmingly concentrated in ground state configurations and line emission tends to be limited by direct collisional excitation and ionization, a set of "coronal" configurations is deemed adequate to describe the major transitions with reasonable accuracy.

At higher densities, multi-step processes lead to significant population in multiply excited configurations, opening up new channels for population transfer and giving rise to strong satellite emission features. In order to cope with that complexity (in a reasonable approximation), we use a hybrid atomic structure scheme in which the fine-structure "coronal" configurations are supplemented with relativistic configuration averaged levels ($nlj$ terms), taking single and double excitations, CI corrections and more into account. This hybrid atomic structure approach [13] with CI correction preserves much of the accuracy of fine structure modeling and permits us to model complex electronic systems with reasonable completeness. For calculations of $\langle Z \rangle$, we further supplement the hybrid model with screened hydrogenic superconfigurations up to $n = 10$ employing the simplest approximation options described in [14]. $\langle Z \rangle$ calculations by the hybrid model and FLYCHK [14] indicate that the emission patterns observed from electron beams of 10 to 18 keV (see Figure 1) can be associated with a Maxwellian $T_e$ of 5 to 15 keV. The first dozen charge states below the Ne-like ion can be associated with individual peaks in our spectra, but the complexity increases with the number of electrons. Most of the lower charge states contribute to several peaks at a time.

At temperatures above 10 keV or densities below about $10^{19}\mathrm{cm}^{-3}$, our calculations are fairly insensitive to $n_e$. At lower temperatures and at densities near or above $10^{12}\mathrm{cm}^{-3}$, the calculated mean charge state $\langle Z \rangle$ is sensitive to $n_e$ due to increasing populations of multiply excited configurations. With increasing densities, the major transition features are broadened and can be shifted to lower energies by as much as 10 eV as the populations of multiply excited states with weakly bound spectator electrons increase. This is way beyond the density range in EBIT. However, the density effect is clearly present in some beam-foil spectra [16] which could not be interpreted properly at the time: The beam-foil x-ray lines associated with Ne-like ions are in good agreement with atomic structure calculations, but those of Na- and Mg-like ions are shifted to lower energies, because they depend on satellite line contributions excited at solid state density in the exciter foil.

## 4. Acknowledgements

This work was performed under the auspices of the U.S. Department of Energy by Lawrence Livermore National Laboratory under Contract DE-AC52-07NA27344. SH was supported in part by LLNL LDRD 05-erd-068. ET acknowledges travel support from the German Research Association (DFG).

## References

[1] Hinkel D E, Schneider M B, Baldis H A, *et al* 2005 *Phys. Plasmas* **12** 056305
[2] Hinkel D E, Schneider M B,Young B K, *et al* 2006 *Phys. Rev. Lett.* **96** 195001
[3] Beiersdorfer P, Bitter M, von Goeler S, *et al* 1986 *Phys. Rev.* A **34** 1297
[4] Beiersdorfer P, von Goeler S, Bitter M, *et al* 1988 *Phys. Rev.* A **37** 4153
[5] May M J, Fournier K B, Beiersdorfer P, *et al* 2003 *Phys. Rev.* E **68** 036402
[6] May M J, Beiersdorfer P, Jordan N, *et al* 2005 *Nucl. Instrum. Meth.* B **235** 231
[7] Neill P, Harris C, Safronova A S, *et al.* 2004 *Can. J. Phys.* **82** 931
[8] Knapp D A, Marrs R E, Elliott S R, *et al* 1993 *Nucl. Instrum. Meth.* A **334** 305
[9] Beiersdorfer P 2008 *Can. J. Phys.* **86** 1
[10] Schneider M B, Mancini R, Widmann K *et al* 2008 *Can. J. Phys.* **86** 259
[11] Porter F S, Beiersdorfer P, Boyce K R, *et al* 2008 *Can. J. Phys.* **86** 231
[12] Gu M F 2008 *Can. J. Phys.* **86** 675
[13] Hansen S B, Bauche J, Bauche-Arnoult C, and Gu M F 2007 *High Energy Density Physics* **3** 109
[14] Chung H K, Chen M H, Morgan W L, *et al* 2005 *High Energy Density Physics* **1** 3
[15] Brown G V, Hansen S B, Träbert E, *et al* 2008 *Phys. Rev.* E **77** 066406

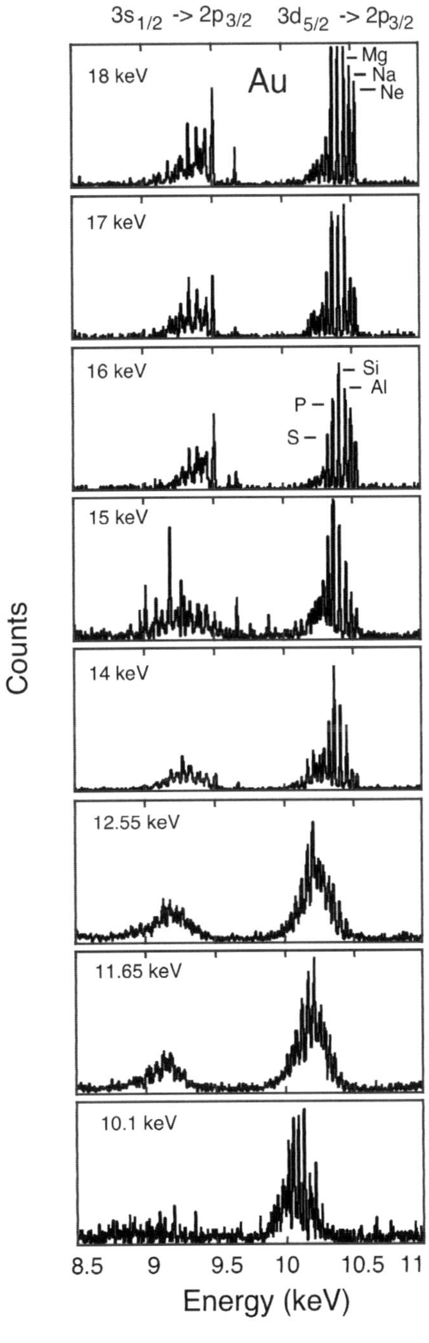


**Figure 1.** Range of the $n = 2 - 3$ emission of Au ions at eight electron beam energies. The microcalorimeter covered a spectral range from 0.5 to 18 keV. The transition arrays are identified on top. The picket-fence pattern of the lines from Ne- through S-like ions appears most clearly above 13 keV electron beam energy.

[16] Chandler G A, Chen M H, Dietrich D D, *et al* 1989 *Phys. Rev.* A **39** 565

Journal of Physics: Conference Series **163** (2009) 012011 doi:10.1088/1742-6596/163/1/012011

# Quasirelativistic *ab initio* study of Gallium like Molybdenum and Tungsten

**O Rancova, P Bogdanovich and R Karpuškienė**
Institute of Theoretical Physics and Astronomy of VU, A. Goštauto st. 12, 01108 Vilnius, Lithuania
pavlas@itpa.lt

**Abstract**. Possibilities of the new quasirelativistic approach designed for *ab initio* calculations of spectral characteristics of highly charged ions and heavy atoms are briefly described and illustrated with an example of a study of spectral characteristics of gallium-like Molybdenum and Tungsten.

## 1. Brief description of the method

Within the quasirelativistic approach the main relativistic effects are taken into account already in obtaining the radial orbitals. Nevertheless the orbitals remain one component and independent on $j$. Some peculiarities of the well-known quasirelativistic approximation by R. D. Cowan, described in the monograph [1] and references therein, caused a demand of creating this different approach.

The general form of the quasirelativistic Hartree-Fock (QRHF) equations used by us is described in [2]. Taking into account the corrections for $p$ electrons presented in paper [3] brings about the quasirelativistic equations of the following form:

$$\left\{\frac{d^2}{dr^2}-\frac{l(l+1)}{r^2}-V(nl\,|\,r)-\varepsilon_{nl}\right\}P(nl\,|\,r)-X(nl\,|\,r)+\frac{\alpha^2}{4}\frac{D(nl\,|\,r)P(nl\,|\,r)}{1-\frac{\alpha^2}{4}\left(\varepsilon_{nl}+V(nl\,|\,r)\right)}$$
$$+\frac{\alpha^2}{4}\left(\left(\varepsilon_{nl}+V(nl\,|\,r)\right)^2P(nl\,|\,r)+\left(\varepsilon_{nl}+V(nl\,|\,r)\right)X(nl\,|\,r)\right)=0. \qquad (1)$$

Here $X(nl|r)$ denotes the exchange part of the potential, and $V(nl|r)$ - the direct part of it including an interaction of an electron with a nucleus $U(r)$ and with other electrons $V(nl\,|\,r)=U(r)+Y(nl\,|\,r)$. The contact interaction with a nucleus is denoted $D(nl|r)$ and the last term describes the mass dependence on the velocity.

The main distinctions between the approach under investigation and the well-known computer code by R. D. Cowan [1] are the following:

- No statistical potentials are used. There are only conventional self-consistent field direct $V(nl|r)$ and exchange $X(nl|r)$ potentials in QRHF.
- The finite size of the nucleus is taken into account in the determination of the potential $U(r)$ while solving the quasirelativistic equations [4].
- The mass-velocity term splits into two parts – the direct potential and the exchange one.
- The contact interaction term contains only the nucleus potential $U(r)$ derivative in the numerator. No two-electron potentials are included in the numerator.

- Only the direct part of the potential $V(nl|r)$ is included into the denominator of the contact interaction with a nucleus term.
- The contact interaction with a nucleus is taken into account not only for $s$-electrons, but also for $p$-electrons with some additional corrections made [3]:

$$D(nl\,|\,r)=\left(\delta(l,0)+\frac{1}{3}\delta(l,1)\right)\frac{dU(r)}{dr}\left(\frac{d}{dr}-\frac{1}{r}\left(\alpha^2Z^2\delta(l,1)\left(-\frac{37}{30}-\frac{5}{9n}+\frac{2}{3n^2}\right)+1\right)\right). \quad (2)$$

- The electron contact interaction may be taken into account when calculating the potential functions forming $V(nl|r)$ and $X(nl|r)$ [2].

Exploitable radial orbitals (RO) are already quasirelativistic; therefore the energy operator matrix must be calculated with account of the relativistic corrections as it is done within Breit-Pauli (BP) approach. The integral describing the one-electron part of the energy operator independent on the total momentum of an electron $j$ can be written down from the equation (1). The one-electron part of the spin-orbit interaction integral is described as follows [3]:

$$\eta_{QR}(nl)=\frac{\alpha^2}{8}\int_0^\infty\frac{P(nl\,|\,r)\frac{1}{r}\frac{dU(r)}{dr}P(nl\,|\,r)}{1-\frac{\alpha^2}{4}\left(\varepsilon_{nlj}+V(r)\right)}dr. \quad (3)$$

Remaining relativistic interactions of the BP approach are two-electron ones. Taking account of these interactions in the case of quasirelativistic ROs is not different from the case of the conventional non-relativistic functions.

The transformed radial orbitals with the variable parameters (TRO) are used in this work to describe the virtual excitations in the configuration interaction (CI) approach. This method for the non-relativistic functions was developed in [5] and successfully applied to take into account the correlation effects in many calculations. TRO is obtained from the RO of the investigated configuration $K_0$ using the following transforming function:

$$P_{TRO}(nl\,|\,r)=N\left(f(k,m,A,B)P(n_0l_0\,|\,r)-\sum_{n'<n}P(n'l\,|\,r)\int_0^\infty P(n'l\,|\,r)f(k,m,A,B)P(n_0l_0\,|\,r)dr\right). \quad (4)$$

Here $N$ is a normalization factor. The first term in the parentheses represents the transformation itself and the second one ensures the orthogonality of the radial orbitals. In the present work the exponential transformation is used:

$$f(k,m,A,B)=r^k\left(1+Ar^m\right)\exp(-Br);\quad k=l_0-l,l_0-l+2,...,l_0-l+6;\quad m=0,2,4,6;\quad A,B>0. \quad (5)$$

The parameters $k$ and $m$ can only be integers with a step equal to 2. This is because the quasirelativistic RO in the vicinity of the origin expand in powers of $r$ only of one parity depending on the value of $l$ [4]. The criterion in determining the optimal values of integer parameters $k$ and $m$ and real parameters $A$ and $B$ is the maximum of the averaged energy correction, expressed in the second order of perturbation theory. More detailed description of the method used one can find in [6].

## 2. Results and discussion

The spectral characteristics of highly charged ions were obtained employing the quasirelativistic RO for the calculation of the multiconfiguration energy operator matrix within the BP approach (QR+CI). The calculations executed within the described approach were reduplicated by analogical calculations based on the usual non-relativistic radial orbitals (HF+CI). All procedures of TRO obtaining, admixed configurations and their term selection, and others were the same in both calculations. The procedures have been described elsewhere for HF+CI case (see for example [7]). It allows one to evaluate the advantages and new possibilities in using the quasirelativistic RO against the conventional non-relativistic RO when one performs the CI calculation using transformed radial orbitals in the same way. The solution of Hartree-Fock equations and calculations of the angular parts of matrix elements were performed by the codes from the known program complex [8]. All other calculations were completed using our own computer programs.

The energy spectra, the probabilities of the electric dipole transitions and the radiation lifetimes were calculated for nine configurations: $4s^24p$, $4s4p^2$, $4p^3$, $4s^24d$, 4s4p4d, $4p^24d$, $4s^24f$, 4s4p4f and $4s4d^2$ of $Mo^{+11}$ and $W^{+43}$ ions. A part of the results for $Mo^{+11}$ with experimental data available is presented in Table 1. As seen from the table, the deviations from the experimental energy values are several times smaller within QR+CI approach than those of HF+CI. The mean square deviations presented at the end of the table confirm this fact. It is necessary to mention that the level characteristics presented in the first column of the table correspond to the experimental data from [9]. At the same time in the case of two pairs of close lying levels displayed in bold both theoretical calculations reveal opposite positions of the mentioned terms when the maximum expansion coefficients of the multiterm function are used for classification. The comparison of the lifetimes presented in two last columns shows, that in this case switching from the non-relativistic RO to the quasirelativistic ones does not have significant influence on the values.

**Table 1.** Spectral characteristics of $Mo^{+11}$.

| | | Enegy levels ($cm^{-1}$) | | | Lifetimes (ns) | |
|---|---|---|---|---|---|---|
| Term | J | Exp. [9] | HF+CI | QR+CI | HF+CI | QR+CI |
| $4s^2\,4p^1\,^2P$ | 0.5 | 0 | 0 | 0 | - | - |
| $4s^2\,4p^1\,^2P$ | 1.5 | 28464 | 25100 | 26700 | - | - |
| $4s^1\,4p^2\,(^3P)\,^4P$ | 0.5 | 197794 | 194400 | 197300 | 5.87 | 5.33 |
| $4s^1\,4p^2\,(^3P)\,^4P$ | 1.5 | 211589 | 206100 | 209900 | 32.80 | 30.50 |
| $4s^1\,4p^2\,(^3P)\,^4P$ | 2.5 | 224236 | 217700 | 222000 | 6.56 | 5.99 |
| $4s^1\,4p^2\,(^1D)\,^2D$ | 1.5 | 262318 | 258500 | 262900 | 0.1930 | 0.1930 |
| $4s^1\,4p^2\,(^1D)\,^2D$ | 2.5 | 269162 | 263400 | 268400 | 0.3050 | 0.3150 |
| $\mathbf{4s^1\,4p^2\,(^1S)\,^2S}$ | 0.5 | 297054 | 295700 | 299900 | 0.0351 | 0.0346 |
| $\mathbf{4s^1\,4p^2\,(^3P)\,^2P}$ | 0.5 | 325519 | 321200 | 326900 | 0.0329 | 0.0332 |
| $4s^1\,4p^2\,(^3P)\,^2P$ | 1.5 | 332036 | 327100 | 333000 | 0.0221 | 0.0220 |
| $4s^2\,4d^1\,^2D$ | 1.5 | 424004 | 425600 | 426800 | 0.0126 | 0.0126 |
| $4s^2\,4d^1\,^2D$ | 2.5 | 428285 | 429700 | 430900 | 0.0143 | 0.0146 |
| $\mathbf{4p^3\,^2D}$ | 1.5 | 493598 | 488100 | 494500 | 0.0374 | 0.0388 |
| $\mathbf{4p^3\,^4S}$ | 1.5 | 507733 | 499700 | 507000 | 0.0464 | 0.0451 |
| $4p^3\,^2D$ | 2.5 | 510263 | 502700 | 510000 | 0.1080 | 0.1100 |
| $4p^3\,^2P$ | 0.5 | 546449 | 540700 | 548400 | 0.0460 | 0.0464 |
| $4p^3\,^2P$ | 1.5 | 559781 | 550900 | 559500 | 0.0461 | 0.0466 |
| Mean square deviation | | | **5500** | **1600** | | |

In Table 2 the characteristics of E1 transitions from the excited states to the levels of $^2P$ term of the ground configuration $4s^24p$ of $W^{+43}$ ion are presented. The data for E2 and M1 transitions within the ground configuration are shown in the last row. They are compared with the results of the calculations introduced in paper [10] performed within Dirac-Fock approach using GRASP (GR) code and with the wavelength measurement results in Tokamak (Tok) plasma from the same paper. As seen from the table the application of HF+CI approach leads to rather significant inaccuracies of wavelengths. It is quite explicable as the usual BP approach is not intended to perform calculations for ions charged so highly. Application of the quasirelativistic RO in analogous calculations increases essentially the accuracy of the results. Though in this case the obtained QR+CI wavelengths are inferior to the pure relativistic results, the deviations are quite acceptable. The oscillator strengths (gf) obtained within two approaches are in rather good agreement. At the same time the emission transition probabilities (A) differ essentially. It is due to the strong dependence of the probabilities on the accuracy of the transition energy calculations, which are described much better within QR+CI approach.

**Table 2.** Characteristics of the transitions to the levels of the ground configuration $4s^24p$ of $W^{+43}$ ion.

| | | λ (Å) | | | | gf | | A ($s^{-1}$) | |
|---|---|---|---|---|---|---|---|---|---|
| $J_{final}$ | $LSJ_{initial}$ | HF+CI | QR+CI | GR[10] | Tok.[10] | HF+CI | QR+CI | HF+CI | QR+CI |
| 0.5 | $4d\ ^2D_{1.5}$ | 51.62 | 47.12 | 47.60 | 47.91 | 1.55 | 1.72 | $9.71\times10^{11}$ | $1.29\times10^{12}$ |
| 0.5 | $4s4p^2\ ^2P_{0.5}$ | 67.96 | 59.30 | 60.20 | 60.63 | 0.758 | 0.764 | $5.48\times10^{11}$ | $7.24\times10^{11}$ |
| 0.5 | $4s4p^2\ ^2D_{1.5}$ | 68.84 | 59.98 | 61.19 | 61.14 | 0.816 | 0.801 | $2.87\times10^{11}$ | $3.71\times10^{11}$ |
| 0.5 | $4s4p^2\ ^4P_{0.5}$ | 133.22 | 125.76 | 128.99 | 128.24 | 0.151 | 0.157 | $2.84\times10^{10}$ | $3.30\times10^{10}$ |
| 1.5 | $4s4p^2\ ^2D_{2.5}$ | 140.80 | 133.51 | 136.82 | 135.34 | 0.268 | 0.254 | $1.50\times10^{10}$ | $1.59\times10^{10}$ |
| 0.5 | $4p\ ^2P_{1.5}$ E2 | 155.00 | 122.38 | 125.87 | 126.39 | $1.38\times10^{-6}$ | $2.46\times10^{-6}$ | $9.55\times10^{4}$ | $2.74\times10^{5}$ |
| | M1 | | | | | $3.48\times10^{-5}$ | $4.41\times10^{-5}$ | $2.41\times10^{6}$ | $4.90\times10^{6}$ |

The presented examples reveal that application of the quasirelativistic RO within CI approach allows one to increase significantly the accuracy of the obtained results and essentially broadens the framework of the BP approximation.

**Acknowledgment**
This work, partially supported by the European Communities under the contract of Association between EURATOM/LEI FU06-2006-00443, was carried out within the framework of the European Fusion Development Agreement. The views and opinions expressed herein do not necessarily reflect those of the European Commission.

This work was also partially supported by Joint Taiwan-Baltic Research Project.

A part of the calculations was performed using computational resources of the European Commission project RI026715 BalticGrid and LitGrid project.

**References**
[1] Cowan R D 1981 *The theory of atomic structure and spectra* (Berkeley and Los Angeles: University of California Press).
[2] Bogdanovich P and Rancova O 2006 *Phys. Rev.* A **74** 052501
[3] Bogdanovich P and Rancova O 2007 *Phys. Rev.* A **76** 012507
[4] Bogdanovich P and Rancova O 2002 *Lithuanian J. Phys.* **42** 257;
Bogdanovich P, Jonauskas V and Rancova O 2005 *Nucl. Instr. Meth.* B **235** 145
[5] Bogdanovich P and Karpuškienė R 1999 *Lithuanian J. Phys.* **39** 193
[6] Bogdanovich P and Rancova O 2008 Phys. Scr. **78** 045301
[7] Bogdanovich P 2005 *Nucl. Instr. Meth.* B **235** 92
[8] Froese Fischer C 2000 *Comput. Phys. Commun.* **128** 635
[9] Ralchenko Yu, Kramida A E, Reader J, and NIST ASD Team 2008 *NIST Atomic Spectra Database* (version 3.1.5)
[10] Pütterch T *et al.* 2005 *J. Phys.* B **38** 3071

Journal of Physics: Conference Series **163** (2009) 012012 doi:10.1088/1742-6596/163/1/012012

# The hyperfine quenching of polarized two-electron ions in an external magnetic field

**A Bondarevskaya[1], A Prozorov[1], L Labzowsky[1,2], G Plunien[3], D Liesen[4] and F Bosch[4]**

[1] Physics Faculty, St. Petersburg State University 198504, Uljanovskaya 3, Petrodvorets, St. Petersburg, Russia

[2] St. Petersburg Nuclear Physics Institute 188300, Gatchina, St.Petersburg, Russia

[3] Institut für Theoretische Physik, Technische Universität 01062, Dresden, Germany

[4] GSI Helmholtzzentrum für Schwerionenforschung Gmb H 64291, Planckstr. 1, Darmstadt, Germany

E-mail: nastya.spb@list.ru

**Abstract.** The hyperfine quenching (HFQ) mechanism of metastable states in polarized He-like heavy ions is considered. The lifetime dependence of these states on the ion polarization in an external magnetic field is established. This dependence is presented for the $2^3P_0$ state of the europium ($Z = 63$) ion and is proposed as a method for the measurement of the ion polarization in the experiments for the search of parity violating effects.

## 1. Introduction

The hyperfine quenching (HFQ) in the isoelectronic sequence of $He$-like highly charged ions (HCI) was successfully used for the determination of the transition probabilities and the fine structure intervals in [1]-[5]. The hyperfine quenching mechanism consists of mixing of the metastable levels like $2^1S_0, 2^3P_0$ etc with levels $2^3S_1, 2^3P_1$ by the hyperfine interaction. Then the transitions from $2^1S_0, 2^3P_0$ to the ground state become open. In [6] it was proposed to use the HFQ for observing the parity nonconservation (PNC) effect in He-like Eu. This effect consists of the mixing between $2^1S_0$ and $2^3P_0$ states by the PNC (weak) interaction. The choice of Eu (Z=63) is determined by the near-crossing of the metastable $2^3P_0$ level with the opposite parity $2^1S_0$ level and by the inequality $\Gamma(2^3P_0) > \Gamma(2^1S_0)$ where $\Gamma$ denotes the total width. The latter inequality allows to avoid a huge background effect in the proposed PNC experiment. The experiment assumes the employment of a polarized ion beam. The possible beam polarization method was described in [7]. In the present paper we discuss a method of measuring the beam polarization via the same HFQ mechanism in an external magnetic field. We will show that the HFQ probability in an external magnetic field depends on the beam polarization and this dependence can be measured by standard experimental HFQ techniques [1]-[5]. In the case of the $2^1S_0$, $2^3P_0$ states with zero total electron angular momentum the ion polarization exists as nuclear polarization. We should add that polarized ion beams can be used not only for PNC experiments but also for many other purposes, e.g. for testing the time-reversibility [8]. The values of the energies of the $2^3P_0$, $2^3S_1$ and $2^1S_0$ states are taken from the most accurate recent calculations [9].

## 2. The HFQ width of the metastable level in an external magnetic field.

The decay rate for the HFQ transition $FIJ \to F'IJ' + \gamma$ with the admixture of the $FIJ''$ state is given

$$W_0^{HFQ}(FIJ \to F'IJ') = \frac{1}{\Delta E^2}\frac{1}{2F+1}\sum_{M_F M'_F M} \times$$

$$|\langle FM_F IJ|\hat{H}_{hf}|FM_F IJ''\rangle\langle FM_F IJ''|V^{E(M)}_{wjM}|F'M'_F IJ'\rangle|^2 \tag{1}$$

Here I,J are the nuclear and the total electron angular momenta of the ion, $FM_F$ is the total angular momentum and its projection, $|FM_F IJ\rangle$ is the wave function for the hyperfine structure level described by the quantum numbers $FM_F IJ$. In Eq.(1) the wave functions $|FM_F IJ\rangle$, $|F'M'_F IJ'\rangle$ correspond to the initial, final state and $|FM_F IJ''\rangle$ corresponds to the state, which is admixed to $|FM_F IJ\rangle$ by the hyperfine interaction. In case of the HFQ decay the $2^3P_0$ state, $|FM_F IJ\rangle \equiv |FM_F 2^3P_0\rangle$, $|F'M'_F IJ'\rangle \equiv |F'M'_F 1^1S_0\rangle$ and $|FM_F IJ''\rangle \equiv |FM_F 2^3P_1\rangle$. The hyperfine magnetic-dipole interaction operator in Eq.(1) is denoted as $\hat{H}_{hf}$. The notation $V^{E(M)}_{wjM}$ is employed for the photon emission operator [10].

An expression for the $W_H^{HFQ}$ is given

$$W_H^{HFQ}(FIJ \to F'IJ') = \frac{1}{\Delta E^2}\frac{1}{2F+1}\sum_{M_F M'_F M} \times \tag{2}$$

$$2Re < FM_F IJ \mid \hat{H}_{hf} \mid FM_F IJ'' >^* < FM_F IJ \mid \vec{\mu}\vec{H} \mid FM_F IJ'' >$$

$$|< FM_F IJ'' \mid V^{E(M)}_{\omega jM} \mid F'M'_F IJ' >|^2$$

where $\vec{\mu}\vec{H}$ is the interaction of the magnetic moment of an ion with an external magnetic field.

## 3. The HFQ transition for the polarized ions in the presence of an external magnetic field

In case of the polarized ions, according to [7] we have

$$\sum_{M_F} n_{FM_F} M_F = F\lambda_F \tag{3}$$

where $n_{FM_F}$ are the occupation numbers for the Zeeman sublevels $FM_F$ and $\lambda_F$ is the "degree of polarization" also introduced in [7].

In the quasiclassical description the degree of polarization is defined by

$$\lambda_F = \lambda(\vec{\xi}\vec{h}) \tag{4}$$

where $\lambda$ is the quasiclassical degree of polarization, $\vec{\xi}$ is the unit quasiclassical polarization vector and $\vec{h} = \frac{\vec{H}}{|\vec{H}|}$.

Since the field-independent contribution $W_0^{HFQ}$ is polarization-independent we obtain the final expression for the polarized ions in the presence of an external magnetic field in the form

$$W_0^{HFQ} + W_H^{HFQ} = W_0^{HFQ}[1 + a_H(F, JJ'', I; n_1j_1l_1, n_2j_2l_2)\lambda(\vec{\xi}\vec{h})] \tag{5}$$

where $a_H$ is some coefficient, depending on the hyperfine and Zeeman matrix elements. The indices $n_1j_1l_1$ and $n_2j_2l_2$ are the standard one-electron quantum numbers for the two-electron configuration (in our case $2^1S_0$).

## 4. The HFQ transition $2^1S_0 \to 1^1S_0$ for the polarized ions in the presence of an external magnetic field

Evaluation of the constant $a$ for the case of $2^1S_0 \to 1^1S_0$ transition in the magnetic field $H = 1T$ gives

$$a_H = 1.1 \times 10^{-4} \tag{6}$$

The smallness of this coefficient is connected with the smallness of the Zeeman matrix elements compared to the $hfs$ matrix element in HCI.

There are methods of the ion beam polarization measurement where the polarization effect is larger than in Eq.(6) [11], [12]. However, our proposal has some advantage since it can be performed within the same PNC experiment. The polarization can be observed from the difference between two signals with the opposite directions of the magnetic field

$$dW^{HFQ} = 2\lambda a_H W_0^{HFQ} \tag{7}$$

where $dW^{HFQ}$ is the signal difference. Another possibility is switching off the magnetic field. The signal then is

$$dW^{HFQ} = \lambda a_H W_0^{HFQ} \tag{8}$$

For $2^1S_0 - 1^1S_0$ transition the two-photon decay is dominant. However it does not change the value of the effect (Eq.(6)) but influences only the statistics of the experiment. We assume that this statistics is enough for the observation of the polarization effect.

## 5. Summary

We have proposed a method for measuring heavy ion beam polarization in the HFQ transitions in $He$-like HCI. The method consists of the observation of the asymmetry of the HFQ photon emission with respect to the external magnetic field. The same HFQ transitions were discussed earlier in the context of the possible observation of the PNC effects. The accuracy necessary for the measurement of the degree of polarization $\lambda$ does not exceed the accuracy achieved in the modern HFQ transition measurements.

*Acknowledgements*

The work of A.B, A.P. and L.L was partly supposed by the RFBR grant $\aleph^0 = 08 - 02 - 00026$. They are also grateful to GSI for the hospitality during their visits.

## 6. References

[1] Marrus R, Simionovici A, Indelicato P, Dietrich D D, Charles P, Briand J P, Finlayson K, Bosch F, Liesen D and Parente F 1989 *Phys. Rev. Lett.* **63** 502
[2] Dunford R W, Liu C J, Last J, Berrah-Memsur N, Vondrasek R, Churah D A and Curtis L J 1991 *Phys. Rev.* A **44** 764
[3] Indelicato P, Birkett B B, Briand J P, Charles P, Dietrich D D, Marrus R and Simionovici A 1992 *Phys. Rev. Lett.* **68** 1307
[4] Birkett B B, Briand J P, Charles P, Dietrich D D, Finlayson K, Idelicato P, Liesen D, Marrus R and Simionovici A 1993 *Phys. Rev. Lett.* A **47** R2454
[5] Simionovici A, Brikett B B, Briand J P, Charles P, Dietrich D D, Finlayson K, Indelicato P, Liesen D and Marrus R 1993 *Phys. Rev.* A **48** 1695
[6] Labzowsky L N, Nefiodov A V, Plunien G, Soff G, Marrus R and Liesen D 2001 *Phys. Rev.* A **63** 054105
[7] Prozorov A, Labzowsky L, Liesen D and Bosch F 2003 *Phys. Lett.* B **574** 180
[8] Khriplovich I B 1998 *Phys. Lett.* B **444** 98; 2000 *Hyperfine Interact.* **127** 365
[9] Artemyev A N, Shabaev V M, Yerokhin V A, Plunien G and Soff G 2005 *Phys. Rev.* A **71** 062103
[10] Labzowsky L, Klimchitskaya G and Dmitriev Yu 1993 *Relativistic Effects in the Spectra of Atomic System*, IOP Publishing, Bristol and Philadelphia

[11] Surzhykov A, Fritzsche S, *Stöhlker* Th and Tashenov S 2005 *Phys. Rev. Lett.* **94** 203202
[12] Tashenov S, *Stöhlker* Th, Banas D, Beckert K, Beller P, Beyer H F, Bosch F, Fritzsche S, Gumberidze A, Hagmann S, Kozhuharov C, Krinds T, Liesen D, Nolden F, Protic D, Sierpowski D, Spillmann U, Steck M and Surzhykov A 2006 *Phys. Rev. Lett.* **97** 223202

Journal of Physics: Conference Series **163** (2009) 012013 doi:10.1088/1742-6596/163/1/012013

# Hyperfine dependent lifetimes in Neon like ions

**M. Andersson[1,2], R. Hutton[1,2] and Y. Zou[1,2]**

[1]Shanghai EBIT Lab, Institute of Modern Physics, Fudan University, Shanghai, P.R. China
[2]The Key Lab of Applied Ion Beam Physics, Ministry of Education, P.R. China

E-mail: martinan@fudan.edu.cn

**Abstract.** As part of the investigation of what could be considered as the third generation of forbidden transitions, an investigation of the hyperfine quenching of the two metastable levels, $2p^53s\ ^3P_2$ and $2p^53s\ ^3P_0$, in the Neon like iso-electronic sequence has been performed. It is shown that the lifetime of the first excited level, $^3P_2$, is sensitive to hyperfine quenching all along the sequence $Z = 13 - 79$, and accurate predictions of this lifetime can only be obtained by including hyperfine quenching into the theoretical model. $2p^53s\ ^3P_0$ is the third excited level in all Neon like ions up to $Z \approx 51$. An investigation of the hyperfine quenching of this state has been performed for ions up to $Z = 49$, and it is shown that the lifetime of this state is very sensitive to hyperfine quenching in the beginning of the iso-electronic sequence, but the sensitivity decreases going to the higher end.

## 1. Introduction

As part of our ongoing investigation of what could be considered as the third generation of forbidden transitions [1, 2, 3, 4], an investigation of the influence of hyperfine quenching on two metastable levels in the Neon like iso-electronic sequence, have been performed. The first excited level in this sequence, $2p^53s\ ^3P_2$, can only decay to the ground state through a magnetic quadrupole (M2) transition. In the presence of a nuclear spin, the hyperfine interaction introduce a mixing with $2p^53s\ ^3P_1$ and $^1P_1$ respectively, opening up a hyperfine induced electric dipole (hpf-E1) transition channel to the ground state. Calculations of the hyperfine quenching of the $^3P_2$ hyperfine levels were performed for all stable isotopes with a nuclear spin in the range $Z = 13 - 79$, and it was investigated how this influenced the lifetime of these states.

For all ions up to $Z \approx 51$, $2p^53s\ ^3P_0$ is the third excited level. Going higher in the sequence, the $2p^53d\ ^3P_0$ level comes plunging down and becomes the lowest excited $J = 0$ level [5]. The $2p^53s\ ^3P_0$ level can only decay through a magnetic dipole (M1) and an electric quadrupole (E2) transition withing the triplet. The hyperfine interaction opens an additional hpf-E1 transition channel to the ground state and calculations of this quenching of the $^3P_0$ hyperfine level were performed for all stable isotopes with a nuclear spin up to $Z = 49$, and it was investigated how this influenced the lifetime of this state and the spectrum of the ion.

To our knowledge, besides Helium like systems, not much work have been performed concerning hyperfine quenching of states which can decay not only through a two-photon transition. Besides [1, 2, 3, 4], [6] investigated the hyperfine quenching of the lowest lying $^3P_2$ state in neutral Mg, Ca, Sr and Yb in connection with the design of an ultra precise atomic clock and [7] measured the hyperfine dependent lifetimes of the $3d^94s\ ^3D_3$ state in Ni-like $^{129}Xe$ confirming our earlier predictions in [1, 2].

## 2. Method

In the presence of a nuclear spin, there is a non-central interaction between the electrons and the nuclear electromagnetic moments. This interaction splits each fine structure level into multiple closely spaced hyperfine levels which is defined by the quantum number $F$. The hyperfine interaction can also introduce a mixing between states with different $J$ quantum numbers and potentially opening up new transition channels. In the Ne-like system, the hyperfine interaction between $2p^5 3s\ ^3P_0$ ($^3P_2$) and $2p^5 3s\ ^3P_1$ and $^1P_1$ respectively opens up hpf-E1 transition channels to the ground state. To calculate the rate of these transitions, first order perturbation calculations were performed.

The $^3P_0$ can only mix with these two states through magnetic dipole hyperfine interaction. Using off-diagonal hyperfine interaction constants, $A$ (see for example [8]), the rate of the hyperfine induced transition to the ground state can be calculated using

$$A_{hpf-E1}(^1S_0,^3P_0) = \frac{(2\pi)^3}{9\hbar\lambda^3} I(I+1) \left| \frac{A(^1P_1,^3P_0)}{E_{^3P_0} - E_{^1P_1}} \langle ^1S_0 || \mathbf{D}^{(1)} || ^1P_1 \rangle + \frac{A(^3P_1,^3P_0)}{E_{^3P_0} - E_{^3P_1}} \langle ^1S_0 || \mathbf{D}^{(1)} || ^3P_1 \rangle \right|^2 \tag{1}$$

where $\langle ^1S_0 || \mathbf{D}^{(1)} || ^1P_1 \rangle$ and $\langle ^1S_0 || \mathbf{D}^{(1)} || ^3P_1 \rangle$ are the reduced transition matrix elements of the transitions to the ground state from $2p^5 3s\ ^1P_1$ and $^3P_1$ respectively.

The corresponding expression for the $^3P_2$ hyperfine induced transition rate are more complicated since this quenching can occur both due to the nuclear dipole and electric quadrupole hyperfine interaction. Using off-diagonal $A$ and $B$ hyperfine constants [8], the rate can be calculated using

$$\begin{aligned} A_{hpf-E1}(^1S_0,^3P_2) = \frac{(2\pi)^3}{9\hbar\lambda^3} \left[(I+F+1)^2 - 4\right] \left[4 - (I-F)^2\right] \Bigg| \frac{1}{E_{^3P_2} - E_{^1P_1}} \Bigg[ \frac{A(^1P_1,^3P_2)}{2} \\ + \frac{\sqrt{3}[F(F+1) - I(I+1) - 3]}{4I(2I-1)} B(^1P_1,^3P_2) \Bigg] \langle ^1S_0 || \mathbf{D}^{(1)} || ^1P_1 \rangle + \frac{1}{E_{^3P_2} - E_{^3P_1}} \\ \times \left[ \frac{A(^3P_1,^3P_2)}{2} + \frac{\sqrt{3}[F(F+1) - I(I+1) - 3]}{4I(2I-1)} B(^3P_1,^3P_2) \right] \langle ^1S_0 || \mathbf{D}^{(1)} || ^3P_1 \rangle \Bigg|^2 \end{aligned} \tag{2}$$

Using off-diagonal hyperfine interaction constants were convenient for our project since all $F$ dependence are gathered in a simple algebraic expression. Also by knowing the constants for one isotope of an ion, they can easily be rescaled to other isotopes.

## 3. Result

In figure 1 we present two plots for an overview of the results. The $2p^5 3s\ ^3P_0$ hyperfine level have three different transition channels. It can decay through M1 transitions to the $^3P_1$ hyperfine levels, through E2 transitions to the $^3P_2$ hyperfine levels and through a hpf-E1 transition to the ground state. The E2 transition rate is though very small compared to the M1 rate so we ignore the former and treat the $^3P_0$ hyperfine level as having two transition channels. In the left of the plots in figure 1 we have plotted the branching ration, $Q_A$, between the hpf-E1 and the M1 transition, defined as

$$Q_A = A_{hpf-E1}/A_{M1} \tag{3}$$

Besides telling us the predicted ratio between the two lines in a spectrum, $Q_A$ times 100 is the percent with which the hyperfine quenching is shortening the lifetime of the $^3P_0$ hyperfine level. From this plot it is found that this hyperfine level is very sensitive to hyperfine quenching and the transition channel to the ground state is in many cases the dominant one in the beginning of the iso-electronic sequence. Going to the higher end, the general trend is that the $M1$ transition

**Figure 1.** Branching ratio between the hpf-E1 $2p^6\ ^1S_0 - 2p^53s\ ^3P_0$ and the M1 $2p^53s\ ^3P_1 - 2p^53s\ ^3P_0$ M1 transitions in the left plot. In the right plot the branching ratio between the hpf-E1 and the M2 transition channels $2p^6\ ^1S_0 - 2p^53s\ ^3P_2$.

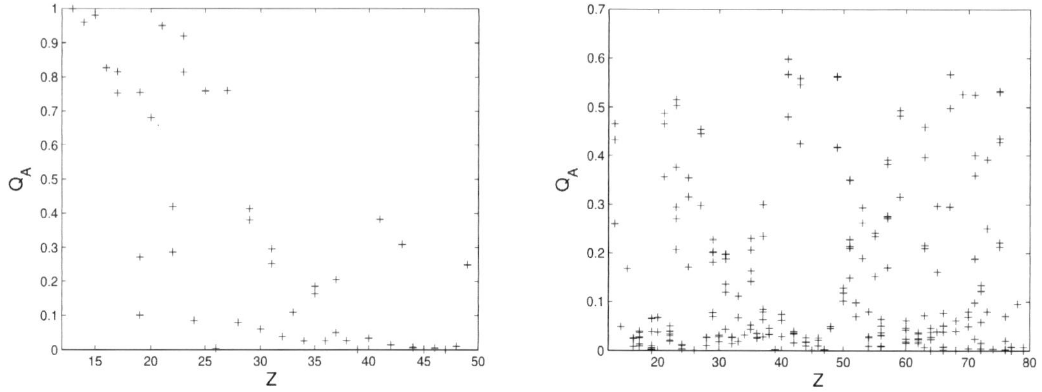


channel becomes heavily dominant. This can be understood from that the system becomes more and more pair coupled so the energy gap between the $^3P_0$ and the $^3P_1$ hyperfine levels increase quickly. Since the M1 transition rate depends on this energy gap to the power of three, it increases very rapidly. Even so, if the nuclear magnetic dipole moment is large, the hyperfine quenching could still have a rather large impact also in the end of the sequence. For example is it shortening the lifetime of the hyperfine level in $In^{39+}$ by about 25%.

Comparing the spectra from two different isotopes of the same ion, one with nuclear spin and one without, could show some dramatic differences. Since the hyperfine interaction opens up a new transition to the ground state, this additional line should be seen in the former spectrum but being absent in the latter. The hyperfine quenching would also decrease the intensity of the $2p^53s\ ^3P_1 - 2p^53s\ ^3P_0$ transition, so also in this part of the spectra should differences be expected. This M1 transition attracted some interest in the 80:s, and it was studied for different ions in the solar corona to experimentally determine the fine structure splitting in Ne-like ions. It was found though that these lines many times were surprisingly weak and in some cases could not even be found, see for example [9]. To our knowledge, this was never understood, but we believe that we have found parts of the explanation.

In the right plot of figure 1 we present the result for the $2p^53s\ ^3P_2$ hyperfine levels along the iso-electronic sequence. In the absence of a nuclear spin this state can only decay through a M2 transition to the ground state. In the presence of nuclear electromagnetic moments, the hyperfine interaction opens an additional hpf-E1 transition channel to the ground state. Depending on the nuclear spin $I$, the isotope has two to five different hyperfine levels and one or two of these have no hyperfine quenching. To present the importance of the hyperfine quenching, we have plotted the $F$-dependent branching ratio, $Q_A$, between the hpf-E1 and the M2 transition in figure 1. Just as above is the $Q_A$ value a variable telling us how much the lifetime of the hyperfine level is shortened due to the hyperfine quenching. It is found that there is no such dramatic effects for the $^3P_2$ hyperfine levels as there were for the $^3P_0$ hyperfine level. On the other hand does the importance of the hyperfine quenching not seam to decrease going along the sequence, but the sensitivity tends to be constant. Whether the hyperfine quenching has a large or a small impact on the lifetime is entirely determined by the size of the nuclear electromagnetic moments and throughout the sequence there are examples of hyperfine levels which have their lifetime halved due to the hyperfine quenching.

To obtain a better understanding of the importance of the two different types of hyperfine

**Figure 2.** The ratio between the $A_{hpf-E1}(2p^6\ ^1S_0, 2p^53s\ ^3P_2)$ including both types of hyperfine interaction and only including magnetic dipole hyperfine interaction in the plot to the left. To the right, the corresponding relative lifetimes.

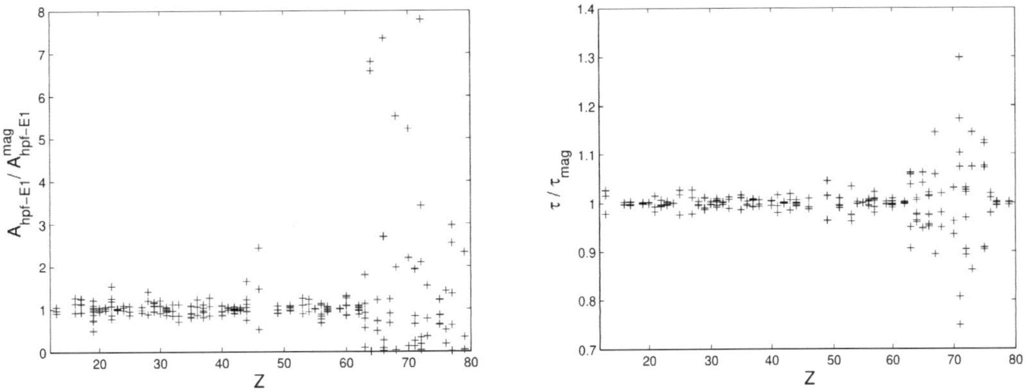


interaction, we also investigated the contribution from the electric quadrupole hyperfine interaction. In figure 2 we present two different plots. In the left one we have plotted the hyperfine induced transition rate, $A_{hpf-E1}$, including both types of hyperfine interaction, relative the transition rate $A^{mag}_{hpf-E1}$ where only the magnetic dipole hyperfine interaction was included. From this plot it is found that the mixing in equation 2 due to the two types of interactions could have both constructive and destructive interference. It is also found that even for isotopes where the nuclear electric quadrupole moment is small compared to the nuclear magnetic dipole moment, the electric quadrupole hyperfine interaction still have a rather large impact on the hyperfine induced transition rate. For some isotopes, the electric quadrupole hyperfine interaction is the most dominant one. For example have we left out two of the results for $^{153}$Eu from our plot since the relative transition rates were about 17 and 18.

In the right plot of figure 2 we have plotted the lifetime of the $2p^53s\ ^3P_2$ hyperfine levels relative the lifetime if only the magnetic dipole hyperfine interaction was included. Except for a few isotopes with large nuclear electric quadrupole moment, the inclusion of this interaction only changed the lifetime with a few percent. This is of course directly related to the relative size of the hpf-E1 and the M2 transition rates.

## Acknowledgment
This work was supported by the National Natural Science Foundation of China under Grant No. 10434050 and China Postdoctoral Science Foundation. This work was also partly supported by Shanghai Leading Academic Discipline Project, Project No. B107 and Kungliga Fysiografiska Sällskapet i Lund.

## References
[1] Yao K, Andersson M, Brage T, Hutton R, Jönsson P and Zou Y 2006 *Phys. Rev.* Lett. **97** 183001
[2] Yao K, Andersson M, Brage T, Hutton R, Jönsson P and Zou Y 2007 *Phys. Rev.* Lett. **98** 269903
[3] Andersson M, Yao K, Hutton R, Zou Y, Chen C Y and Brage T 2008 *Phys. Rev.* A **77** 042509
[4] Andersson M, Liu Y, Chen C Y, Hutton R, Zou Y, and Brage T 2008 *Phys. Rev.* A (Submitted)
[5] Dong C, Xie L, Wan J and Jiang J 2005 *Chinese Physics* **14** 1008
[6] Porsev S G and Derevianko A 2004 *Phys. Rev.* A **69** 042506
[7] Träbert E, Beiersdorfer P and Brown G V 2007 *Phys. Rev. Letter* **98** 263001
[8] Jönsson P, Parpia F A and Froese Fischer C 1996 *Comput. Phys. Commun.* **96** 301
[9] Kastner S O 1983 *Astrophys J.* **275** 922

Journal of Physics: Conference Series **163** (2009) 012014 doi:10.1088/1742-6596/163/1/012014

# QED calculation of interelectron interaction corrections for transition probabilities in two-electron ions

**Oleg Yu. Andreev**

V. A. Fock Institute of Physics, St. Petersburg State University, Ulyanovskaya 1, 198504, Petrodvorets, St. Petersburg, Russia

E-mail: olyuan@gmail.com

**Leonti N. Labzowsky**

V. A. Fock Institute of Physics, St. Petersburg State University, Ulyanovskaya 1, 198504, Petrodvorets, St. Petersburg, Russia
Petersburg Nuclear Physics Institute, 188300, Gatchina, St. Petersburg, Russia

**Günter Plunien**

Institut für Theoretische Physik, Technische Universität Dresden, Mommsenstraße 13, D-01062, Dresden, Germany

**Abstract.** Accurate QED calculations of the transition probabilities between the quasidegenerate two-electron configurations $(1s2p)\,^1P_1, ^3P_1$ and the ground state $(1s1s)\,^1S_0$ ($E1$ transitions) are presented for highly charged ions with nuclear charge numbers $Z = 10-92$. A special technique has been developed to master the bad convergence of the perturbation expansion for the interelectron interaction in the low $Z$ region.

## 1. Introduction

Highly charged ions (HCI), in particular, two-electron HCI, considered in the present work are under intensive experimental and theoretical investigation during the last decades. HCI allows for tests of QED in strong electric fields. Moreover, HCI can be used for the investigation of fundamental problems beyond QED: for probing of the Standard Model [1] and for the search of the variation of fundamental constants [2, 3]. In many cases the two-electron ions are preferable for performing the corresponding experiments (see, *e.g.*, [1, 2]).

For intermediate $Z$ values a variety of fully relativistic methods which partly include the QED effects were developed. One of the most successful methods was developed by Johnson and coauthors (see [4] for the calculation of the energy levels and [5] for the calculation of transition probabilities). It is based on a combination of the Configuration Interaction (CI) method and the Relativistic Many Body Perturbation Theory (RMBPT) approach. However, compared to the exact QED approach, RMBPT calculations are incomplete due to the approximate treatment of retardation effects and the neglect of the negative-energy contributions, including the neglect crossed-photon Feynman graphs.

On the other side, the interelectron interaction was partly included in CI-RMBPT calculations up to an infinite order. This is especially important for the evaluation of closely-lying (quasidegenerate) levels. The latter situation often occurs for low-lying excited levels in He-like HCI. Rigorous QED calculations of the energy levels were performed in several works: in [6] within the Two-Time Green's Function method, in [7] within the Covariant Evolution Operator method and in [8] within Line Profile Approach (LPA). All approaches mentioned, merge partly the relativistic many-body theory and QED and allow for the inclusion of QED corrections in a consistent way.

A detailed investigation of the transition probabilities for two-electron ions has been presented in Ref. [5], where the transitions probabilities for low-lying two-electron configurations have been calculated for ions within the entire range of nuclear charge numbers $Z$. The first exact QED evaluation of transition probabilities for two-electron configurations has been presented in [9] involving nondegenerate states with emission of magnetic photons. QED calculations of the transition probabilities for quasidegenerate states up to now were missing. In this paper, we perform such calculations for the first time. This rigorous approach is applied for the evaluation of transition probabilities, in particular, for the quasidegenerate states. It can now be applied for the evaluation of transition probabilities in the intermediate $Z$ region. Accordingly, accurate QED calculations for various properties (ionization and capture cross-sections, dielectronic recombination cross-sections, *etc.*) of HCI in the intermediate $Z$ region become available. It also makes actual a calculation of the radiative corrections to the transition probabilities in two-electron ions. Up to now these corrections were obtained only for one-electron ions.

## 2. Line profile approach

Within the framework of the Line Profile Approach (LPA) the properties of an ion are derived here from the line profile associated with the process of elastic scattering, where an ion in the ground state absorbs and emits a photon. The resonances of this scattering process correspond to the energy levels. In order to describe an energy level by two parameters (energy and width) we parameterize the line profile by a Lorentz contour, which is determined by the position of the resonance (*i.e.*, energy) and its width. For details concerning the LPA and its application to the evaluation of energies we refer to [8, 10]. The application of the LPA to the evaluation of transition probabilities is presented in [11, 12].

Within the framework of the LPA the energy levels are derived from the matrix $V$, which is evaluated with employment of the QED perturbation theory order by order

$$V(\omega) \;=\; V^{(0)} + \Delta V^{(1)} + \Delta V^{(2)}(\omega) + \dots \,. \tag{1}$$

The matrix $V(\omega)$ depends on $\omega$, which has a physical meaning of the scattered photon frequency. The matrix $V^{(0)}$ contains the corresponding Dirac energies. The matrix $V^{(1)}$ includes one-photon exchange corrections. The matrix $V^{(2)}(\omega)$ includes two-photon exchange corrections ("box" and "cross" graphs). In this work we consider only the interelectron interaction corrections.

To evaluate eigenvector of the matrix $V$ corresponding to a reference state $n_g$ we compose a set of configurations $g$ including the reference state $n_g$ and the corresponding mixing configurations. Then, it is convenient to write the matrix $V$ in block form

$$V \;=\; \begin{bmatrix} V_{11} & V_{12} \\ V_{21} & V_{22} \end{bmatrix}, \tag{2}$$

where the matrix $V_{11}$ is built on the configurations from the set $g$. The matrix $V_{11}$ can be diagonalized numerically

$$V_{11}^{\mathrm{diag}} \;=\; B^t V_{11} B \,, \tag{3}$$

where $B$ is an orthogonal matrix.

The eigenvector of the matrix $V$ can be evaluated with use of the perturbation theory

$$\Phi_{n_g} = \sum_{k_g \in g} B_{k_g n_g} \Psi^{(0)}_{k_g} + \sum_{\substack{k \notin g \\ l_g \in g}} (\Delta V_{21})_{k l_g} \frac{B_{l_g n_g}}{E_{n_g} - E^{(0)}_k} \Psi^{(0)}_k + \dots , \qquad (4)$$

where $E_{n_g}$ are the eigenvalues of the matrix $V_{11}$ and $E^{(0)}_k$ are sums of the Dirac energies. The functions $\Psi^{(0)}$ are two-electron functions written in the $j$–$j$ coupling scheme.

The amplitude of the decay process from the initial state $I$ to the final state $F$ with emission of photon $\omega_0$ can be written as $U_{I \to F} = (\Xi(\omega_0))_{\Phi_F \Phi_I}$, where $\Phi_I$ and $\Phi_F$ are the eigenvectors corresponding to the $I$ and $F$ states, respectively. The operator $\Xi(\omega_0)$ is evaluated with employment of the QED perturbation theory (see [11, 12]).

## 3. Numerical results and discussion

In Tables 1,2 we present numerical results for $E1$-transition probabilities between $(1s2p)\,^3P_1$, $^1P_1$ and $(1s1s)\,^1S_0$ two-electron configurations. The calculation of the transition probabilities are performed within the framework of QED utilizing the "velocity" and "length" gauges for the emitted photons, $W_V$ and $W_L$, respectively. The one- and two-photon exchange corrections to the matrix $V$ are taken from [8]. In the columns $W_L^{1\text{ph}}$ we give transition probabilities calculated with only one photon exchange correction taken into account, *i.e.*, we recalculate $W_L$ where we set $\Delta V^{(2)} \equiv 0$ in Eq. (1). The columns $W_L^{\text{RMBPT}}$ display our results of the recalculation of the $W_L$ values within RMBPT. In the last column we give the results of the RMBPT calculations by of Johnson *et al* [5]. In [5] for $(1s2p)\,^3P_1$, $^1P_1$ configurations exclusively "length" gauge was employed. The values are given in units $s^{-1}$. Digits in square brackets denote the power of 10.

The diagonalization of the matrix $V_{11}(\omega)$ (Eq. (3)) (see [11, 12]) means that we partly take into account the photon exchange corrections in all orders. This violates the gauge invariance and, accordingly, explains the deviation between $W_L$ and $W_V$. This deviation also helps us to estimate the contribution of the higher order terms in the expansions Eqs. (1), (4). This contribution is larger for the small $Z$ values, where the convergence of the perturbation theory in the interelectron interaction is poorer.

The difference between the data in the columns $W_L$ and $W_L^{\text{RMBPT}}$ determines the nonradiative QED corrections. For small $Z$ values the considered configurations are strongly mixed. The transition probabilities for $^3P_1$ levels are very sensitive to the mixing matrix $B$ what explains the large value of the QED corrections for small $Z$ values. Note, that the transition probability for $^3P_1$ level is by several orders of magnitude smaller, than for $^1P_1$ level. This means that the relative correction to the decay of $^3P_1$ level due to the change of the matrix $B$ is essentially larger than the correction to the decay of the $^1P_1$ level. For large $Z$ values the mixing of the configurations is small and the QED corrections mainly appear as QED corrections to the matrix $(\Delta V_{21})_{kl_g}$ in $\Phi_{n_g}$ (see Eq. (4)).

The perturbation theory applied to the construction of the matrix $V$ (Eq. (1)) and the one applied to the diagonalization of $V$ (Eq. (4)) are different. The comparison of the columns $W_L$ and $W_L^{1\text{ph}}$ reveals the importance of the two photon corrections to the matrix $V$ for convergence of the series of the perturbation theory. If the set $g$ is enlarged by excited configurations, the values $W_L^{1\text{ph}}$ would approach the values $W_L$. However, in order to achieve good agreement for small $Z$, would need to include up to 4000 significant configurations in the set $g$. The matrix $V$ is well investigated for the purpose of calculating the energies of the configurations [6, 8], so the evaluation of the mixing matrix $B$ in higher orders is a more efficient technique, than the enlargement of the set $g$. The presented technique for the calculation of the transition probabilities is a rigorous QED procedure and allows for an improvement of the accuracy of the calculation by taking into account the QED corrections of higher orders.

**Table 1.** $E1$-transition probabilities ($s^{-1}$) between $(1s2p)\,^3P_1$ and $(1s1s)\,^1S_0$ configurations.

| $Z$ | $W_L^{1ph}$ | $W_L^{RMBPT}$ | $W_L$ | $W_V$ | Ref. [5] |
|---|---|---|---|---|---|
| 10 | 3.096[ 9] | 5.211[ 9] | 5.350[ 9] | 5.094[ 9] | 5.356[ 9] |
| 18 | 1.391[12] | 1.793[12] | 1.799[12] | 1.777[12] | 1.799[12] |
| 26 | 3.925[13] | 4.482[13] | 4.421[13] | 4.396[13] | 4.421[13] |
| 30 | 1.160[14] | 1.258[14] | 1.251[14] | 1.246[14] | 1.251[14] |
| 40 | 6.846[14] | 7.047[14] | 7.013[14] | 6.996[14] | 7.011[14] |
| 50 | 2.104[15] | 2.132[15] | 2.123[15] | 2.120[15] | 2.120[15] |
| 60 | 4.838[15] | 4.874[15] | 4.855[15] | 4.850[15] | 4.845[15] |
| 70 | 9.472[15] | 9.523[15] | 9.497[15] | 9.489[15] | 9.460[15] |
| 80 | 1.672[16] | 1.680[16] | 1.674[16] | 1.673[16] | 1.668[16] |
| 92 | 3.007[16] | 3.020[16] | 3.008[16] | 3.006[16] | 2.994[16] |

**Table 2.** $E1$-transition probabilities ($s^{-1}$) between $(1s2p)\,^1P_1$ and $(1s1s)\,^1S_0$ configurations.

| $Z$ | $W_L^{1ph}$ | $W_L^{RMBPT}$ | $W_L$ | $W_V$ | Ref. [5] |
|---|---|---|---|---|---|
| 10 | 8.607[12] | 8.538[12] | 8.538[12] | 8.103[12] | 8.851[12] |
| 18 | 1.069[14] | 1.061[14] | 1.061[14] | 1.047[14] | 1.070[14] |
| 26 | 4.611[14] | 4.551[14] | 4.553[14] | 4.527[14] | 4.566[14] |
| 30 | 7.857[14] | 7.747[14] | 7.754[14] | 7.720[14] | 7.763[14] |
| 40 | 2.234[15] | 2.211[15] | 2.215[15] | 2.210[15] | 2.212[15] |
| 50 | 5.096[15] | 5.064[15] | 5.074[15] | 5.065[15] | 5.057[15] |
| 60 | 1.013[16] | 1.009[16] | 1.011[16] | 1.010[16] | 1.006[16] |
| 70 | 1.819[16] | 1.813[16] | 1.816[16] | 1.814[16] | 1.805[16] |
| 80 | 3.010[16] | 3.002[16] | 3.008[16] | 3.006[16] | 2.986[16] |
| 92 | 5.046[16] | 5.034[16] | 5.045[16] | 5.041[16] | 5.001[16] |

## 4. Acknowledgments

This work was supported by DFG, the INTAS-GSI grant 06-1000012-8881 and the RFBR grant 05-02-17483. O.Yu.A. and G.P. acknowledge financial support from DFG.

## 5. References

[1] Labzowsky L N, Nefiodov A V, Plunien G, Soff G, Marrus R and Liesen D 2001 *Phys. Rev. A* **63** 054105
[2] Andreev O Y, Labzowsky L N, Plunien G and Soff G 2005 *Phys. Rev. Lett.* **94** 243002
[3] Schiller S 2007 *Phys. Rev. Lett.* **98** 180801
[4] Plante D R, Johnson W R and Sapirstein J 1994 *Phys. Rev. A* **49** 3519
[5] Johnson W R, Plante D R and Sapirstein J 1995 *Adv. At. Mol. Opt. Phys.* **35** 255
[6] Artemyev A N, Shabaev V M, Yerokhin V A, Plunien G and Soff G 2005 *Phys. Rev. A* **71** 062104
[7] Lindgren I, Salomonson S and Åsén B 2004 *Phys. Rep.* **389** 161
[8] Andreev O Y, Labzowsky L N, Plunien G and Soff G 2004 *Phys. Rev. A* **69** 062505
[9] Indelicato P, Shabaev V M and Volotka A V 2004 *Phys. Rev. A* **69** 062506
[10] Andreev O Y, Labzowsky L N, Plunien G and Soff G 2001 *Phys. Rev. A* **64** 042513
[11] Andreev O Y, Labzowsky L N, Plunien G and Solovyev D A 2008 *Phys. Rep.* **455** 135
[12] Andreev O Y, Labzowsky L N and Plunien G *arXiv:0808.3868v1 [physics.atom-ph]*

# The $^1P^o$ Resonant States of LiII Between the N=2 and 3 Thresholds

**Ming-Keh Chen**
Department of Physics, National Chung-Hsing University, Taichung, Taiwan 40227
E-mail: mkc1@phys.nchu.edu.tw

**Abstract.** We studied the lowest five doubly excited $^1P^o$ resonant states of $Li^+$ between the N=2 and 3 thresholds by the saddle-point complex rotation method with B-spline functions. The doubly excited states are grouped in Rydberg series labeled by the quantum numbers K, T, A. Our results of energies and widths are in good agreement with other theoretical and experimental results.

## 1. Introduction

Doubly excited states of two-electron atom and ions have attracted physicists to study experimentally and theoretically. The doubly excited $^1P^o$ resonant states of $Li^+$ have been observed firstly by Dual Laser Plasma technique [1]. They were also studied experimentally by beam-foil techniques[2, 3], and using the ejected electron spectroscopy method[4]. After high-resolution monochromators at synchrotron-radiation facilities have been developed, high-resolution measurements on $Li^+$ ion were performed at the Advanced Light Source [5] and Super-ACO [6]. Diehl *et al* [6] have measured the doubly excited states of $Li^+$ below n=2 threshold. More recently, absolute cross-section measurements for the double photoexcitation of $Li^+$ ion have been performed by using the photo-ion merged-beam endstation. Accurate measurements on several members of the $^1P^o$ $Li^+$ ($_2(0,1)^+{}_n$ , $_3(1,1)^+{}_n$ ) ion have been made [7, 8, 9]. On the theoretical side, some resonances have been studied by the complex rotation method [10], the diagonalization approximation[11], the pseudo-potential-Feshbach method [12], the saddle-point complex-rotation method[13],the variation method with combining Hylleraas and incomplete hydrogenic wave functions [15], the density-functional method [14], the R-matrix method [7, 8, 9] , the saddle-point complex rotation method [16], and the SCUNC method [17].

## 2. Theory

In the present work, we studied doubly excited $^1P^o$ states of $Li^+$ between the n=2 and 3 thresholds by saddle-point complex rotation method with B-spline functions[18]. In a Configuration Interaction scheme [19, 20, 21, 22], we constructed the wave functions in terms of B splines of order k and total number N, defined between two end points, $r_{min}$=0 and $r_{max}$= R, and built vacancies into the wave functions. With an exponential sequence, we have the trial wave function for a two-electron system

$$\Psi = A(1 - P(\vec{r_1})(1 - P(\vec{r_2})) \sum_{i,j,l_1,l_2} C_{i,j,l_1,l_2}\Phi_{i,j}(\vec{r_1}, \vec{r_2})Y^{LM}_{l_1,l_2}\chi(1,2) \quad (1)$$

with

$$\Phi_{i,j}(\vec{r_1},\vec{r_2}) = \frac{B_{i,k}(r_1)}{r_1}\frac{B_{j,k}(r_2)}{r_2}, \tag{2}$$

$$Y^{LM}_{l_1,l_2} = \sum_{m_1,m_2} < l_1 l_2 m_1 m_2 | LM > Y_{l_1,m_1}(\hat{r_1}) Y_{l_2,m_2}(\hat{r_2}), \tag{3}$$

and

$$i \geq j - jm, \tag{4}$$

where L is the total angular momentum and the numbers i and j are positive integers, which are not larger than N, and jm is a specific selected integer. A is the antisymmetrization operator, $\chi(1,2)$ is the spin wave function, and $P(\vec{r})$ is a projection operator. For the present,1s,2s and 2p orbitals are the vacancy orbitals.

$$P(\vec{r}) = P_{1s}(\vec{r}) + P_{2s}(\vec{r}) + P_{2p}(\vec{r}) \tag{5}$$

$$P_{nl}(\vec{r}) = |\phi_{nl} >< \phi_{nl}| \tag{6}$$

We assume $\phi_{nl}$ to be nl hydrogenic orbital with effective nuclear charge, q. The saddle-point variation is carried out by first minimizing the energy with respect to $C_{i,j,l_1,l_2}$ and the set of B-spline basis functions, and then maximizing the energy with respect to the effective nuclear charge, q, to obtain the saddle-point energy and wave function. After the saddle-point variation is carried out, we calculate the resonance energy and width by a complex-rotation method. The trial wave functions are composed of the saddle-point wave functions( eq.1) and the open-channel components, We choose the open-channel components to be:

$$\Psi_{open} = A \sum_i \psi_i(\vec{r_1}) \sum_{kc=L}^{KO} C_{kc} u_{kc}(\vec{r_2}) Y^{L,M}_{1,L} \chi(1,2), \tag{7}$$

with

$$u_{kc}(\vec{r_i}) = r_i^{kc} e^{-\beta r_i}, \tag{8}$$

and

$$Y^{LM}_{1,L} = \sum_{m_1,m_2} < 1 L m_1 m_2 | LM > Y_{1,m_1}(\hat{r_1}) Y_{L,m_2}(\hat{r_2}), \tag{9}$$

where $\psi_i$ is the 1s,2s and 2p radial wave function of hydrogenlike ions. The non-negative integer, KO, is chosen to be large enough to ensure the accuracy of the resonance energy and width in the calculation by the complex-rotation method. In the complex-rotation calculation, each radial coordinate $r_i$ in $u_{kc}(\vec{r_i})$ of the open-channel components, $\Psi_{open}$ takes the form $r_i e^{i\theta}$.

## 3. Results

In the present, we studied the lowest five doubly excited $^1P^o$resonant states of $Li^+$ between the N=2 and 3 thresholds. The values of R for end points are chosen to be 400 a.u. so that the saddle-point energies of the lowest five doubly excited $^1P^o$resonant states of $Li^+$ between the N=2 and 3 thresholds are converged to within an uncertainty of $10^{-8}$ a.u. We included 8-15 partial waves in calculating $^1P^o$ states of $Li^+$ for the saddle-point energies are converged to within an uncertainty of $10^{-7}$ a.u.. By examining convergence behaviors, it is believed that the accuracy of the present results is comparable with other theoretical results. We also calculated the expectation value of the angle between the two electrons, $< \theta_{12} >$ ,the average values of r for the inner electron and outer electron, $< r_< >$ and $< r_> >$. The results are shown in Table

Journal of Physics: Conference Series **163** (2009) 012015 doi:10.1088/1742-6596/163/1/012015

**Table 1.** The expectation values of $\cos\theta_{12}$, $r_<$, $r_>$, r and $r^2$.(in a.u.)

| n | $<\cos\theta_{12}>$ | $<r_<>$ | $<r_>>$ | $<r>$ | $<r^2>$ | $<\cos(\theta_{12})>$ | $<r_<>$ | $<r_>>$ | $<r>$ | $<r^2>$ |
|---|---|---|---|---|---|---|---|---|---|---|
| | | $_3(1,1)_n^+$ | | | | | $_3(-1,1)_n^+$ | | | |
| 3 | -.223 | 3.51 | 5.62 | 4.57 | 24.1 | .262 | 3.49 | 5.94 | 4.72 | 26.3 |
| 4 | -.267 | 4.10 | 9.45 | 6.77 | 59.6 | | | | | |
| | | $_3(2,0)_n^-$ | | | | | $_3(0,0)_n^-$ | | | |
| 4 | -.589 | 3.88 | 9.08 | 6.48 | 52.5 | .0064 | 3.65 | 9.19 | 6.42 | 53.2 |

**Table 2.** Non-relativistic energies and widths for doubly excited $^1$P$^o$ resonances of Li$^+$ between the n= 2 and 3 thresholds( in a.u.).A(-m) means A$\times$ $10^{-m}$.a: the present results, b: theory[8], and c: experiment.[8].

| | n | -E | width | -E | width | -E | width | -E | width |
|---|---|---|---|---|---|---|---|---|---|
| | | $_3(2,0)_n^-$ | | $_3(1,1)_n^+$ | | $_3(0,0)_n^-$ | | $_3(-1,1)_n^+$ | |
| a | 3 | | | 0.828710 | 9.81(-3) | | | 0.732843 | 2.46(-3) |
| [10] | | | | 0.8297 | 1.0(-2) | | | 0.7328 | 2.5(-3) |
| [12] | | | | 0.8288 | 1.1(-2) | | | 0.7320 | 2.5(-3) |
| [13] | | | | 0.828755 | 1.040(-2) | | | 0.732690 | 2.484(-3) |
| [15] | | | | 0.8359 | | | | | |
| b | | | | 0.8275 | 1.036(-2) | | | | |
| c | | | | 0.8241±0.0026 | | | | | |
| a | 4 | 0.6871264 | 4.58(-5) | 0.659366 | 5.26(-3) | 0.655662 | 6.91(-5) | | |
| [12] | | 0.6872 | 6.6(-5) | 0.6598 | 6.6(-3) | 0.6555 | 8.8(-5) | | |
| [13] | | 0.6871259 | 4.563(-5) | 0.659379 | 5.278(-3) | 0.655667 | 6.957(-5) | | |
| b | | | | 0.6581 | 5.207(-3) | | | | |
| c | | | | 0.6566±0.002 | | | | | |

1. From $<\cos\theta_{12}>$, $<r_<>$ and with Lin's results[23, 24], The doubly excited states are grouped in Rydberg series labeled by the quantum numbers K, T, A. The average value of K $\approx$ $<r_< \cos\theta_{12}>$. For the lowest five $^1$P$^o$ resonant states, there are two members in the $_3(1,1)_n^+$ series, one member in the $_3(2,0)_n^-$, $_3(-1,1)_n^+$ and $_3(0,0)_n^-$ series. The numbers, n in $_3(K,T)_n^A$ are assigned according to the effective principal quantum numbers $n^*$, which are calculated from the equation $E_n$=-$(Z-1)^2/2\ n^{*2}$. $E_n$ is measured from the 3l states of Li$^{+2}$.

The resonant energies and widths for Li$^+$ are shown and compared with other theoretical[8, 10, 12, 13, 15] and experimental[8] results in Table 2.

Our results for energies agree with other theoretical results[13] within first 4-5 digits, with the theoretical results of pseudo-potential-Feshbach Method[12] within first 3-4 digits, but with the experimental and theoretical results[8] within first 2 digits. The agreement amongst the various theoretical results is better for $_3(2,0)_n^-$ and $_3(0,0)_n^-$ series than that for $_3(1,1)_n^+$ and $_3(-1,1)_n^+$ series. It may be due to the stronger electron-electron correlation in the $_3(1,1)_n^+$ and $_3(-1,1)_n^+$ Li$^+$. For the widths, our results agree well with other theoretical results[10, 12, 13, 8] except the theoretical results[12] for $_3(2,0)_4^-$, $_3(1,1)_4^+$ and $_3(0,0)_4^-$ states. Scully *et al*[8] did not reported the experimental results of widths for these resonant states. The agreement amongst the various theoretical and experimental results is not as good as that for the $^1$P$^o$ states of Li$^+$

below n=2 threshold[16]. Therefore, we hope our present work will stimulate further theoretical and experimental results to understand these stats more.

## Acknowledgments

This work is supported by National Science Council Grant No. NSC95-2112-M-005-006-MY2.

[1] Carroll P K and Kennedy E T 1977 *Phys. Rev. Lett.* **38** 1068
[2] Bruch R, Paul G, Andra J and Lipsky L 1975 *Phys. Rev.* A **12** 1808
[3] Ziem P, Bruch R and Stolterfoht N 1975 *J. Phys.* B **8** L480
[4] Rødbro M, Bruch R and Bisgaard P 1979 *J. Phys.* B **12** 2413
[5] Diehl S *et al* 1997 *J. Phys. B* **30** L595
[6] Diehl S *et al* 1999 *J. Phys.* B **32** 4193
[7] Scully S W J *et al* 2005 *J. Phys.* B **38** 1967
[8] Scully S W J *et al* 2006 *J. Phys.* B **39** 3957
[9] Scully S W J *et al* 2007 *J. Phys. Conf. Ser.* **58** 387
[10] Ho Y K 1979 *J. Phys. B* **12** 387
[11] Wague A 1987 Z. Phys. D **6** 337
[12] Bachau H *et al* 1991 *At. Data Nucl. Data Tables* **48** 167
[13] Chung K T 1998 *At. Data Nucl. Data Tables* **69** 101
[14] Roy A K, Singh R and Deb B M 1997 *J. Phys.* B **30** 4763
[15] Biaye M *et al* 2005 *Phys. Scr.* **72** 373
[16] Chen M-K 2007 *J. Phys. Conf. Ser.* **58** 125
[17] Sakho I *et al* 2008 *Eur. Phys. J.* D **47** 37
[18] deBoor C 1978 *A Practical Guide to Splines* (New York: Springer)
[19] Chen M-K 2005 *Chinese J. of Phys.* **43** 329
[20] Chen M-K 2003 *Eur. Phys. J.* D **21** 13
[21] Chen M-K 1999 *Phys. Rev.* A **60** 2565
[22] Chen M-K 1997 *Phys. Rev.* A **56** 4537
[23] Lin C-D 1982 *Phys. Rev.* A **25** 1535
[24] Lin C-D 1984 *Phys. Rev.* A **29** 1019

# Observation and modelling of hollow multicharged ion x-ray spectra radiated by laser produced plasma

**A.Ya. Faenov[1,2], T.A. Pikuz[1,2], I.Yu. Skobelev[2], Y. Fukuda[1], J. Colgan[3], J. Abdallah Jr.[3]**

[1]Kansai Photon Science Institute (KPSI), Japan Atomic Energy Agency (JAEA), Kizugawa-shi, Kyoto, Japan
[2]Joint Institute for High Temperatures, Russian Academy of Sciences, Moscow, Russia
[3]Theoretical Division, Los Alamos National Laboratory, Los Alamos, New Mexico, USA

E-mail: faenov.anatoly@jaea.go.jp

**Abstract**. The role of the highly charged hollow ions in X-Ray emission plasma spectra is investigated for 2 cases: 1) plasma obtained under irradiation of Ar clusters by ultrashort laser pulses and 2) Mg-plasma heated by a short-wavelength long (nanosecond) laser pulse. Experimental measurements are presented. Calculations in support of these measurements have been performed using a detailed atomic kinetics model with the ion distributions found from solution of the time-dependent rate equations.

## 1. Introduction

Plasmas of multicharged ions (Z =10 – 20) with electron temperatures 10 – 100 eV and electron densities $10^{22} – 10^{23}$ cm$^{-3}$ are produced usually by interaction of high-contrast femtosecond laser pulses with solids or clusters. The same plasma parameters are also realized when nanosecond short-wavelength laser pulses (KrF or XeCl or 3$^{d}$, 4$^{th}$ harmonics of Nd:glass lasers) interact with solids. Highly charged ions in a plasma with such parameters are already weakly- coupled and it is very interesting to investigate their radiation properties. It has been shown in [1-7] that the X-Ray emission spectra of such plasmas contain some exotic spectral lines caused by radiative transitions in so called “hollow ions”, that are the highly charged ions with an empty inner K-shell (KK-hollow ions) or with vacancies in K- and L-shells (KL-hollow ions).

In the present work the role of highly charged hollow ions in X-Ray emission plasma spectra is investigated for 2 cases: 1) plasma obtained under irradiation of Ar clusters by ultrashort laser pulses and 2) Mg-plasma heated by a short-wavelength long (nanosecond) laser pulse.

## 2. The role of hollow ions in the spectrum of an ultrashort-pulse-laser-driven Ar cluster target

The Ar cluster experiments were performed with the JAEA (Kyoto, Japan) Ti: sapphire laser system based on the technique of chirped pulse amplification, which was designed to generate 20 fs pulses and is capable of producing a focused intensity up to $10^{20}$ W/cm$^2$ [8]. In this study the amplified pulses were compressed to 30 fs by a vacuum pulse compressor yielding maximum pulse energy of 360 mJ. Two types of experiments were performed. In the first set of experiments, only one double Pockels cell was used after the regenerative amplifier, which allowed us to reduce the laser prepulse by up to $5\times10^{-4}$ compared with the main laser pulse. In addition, experiments were performed where after the regenerative amplifier; the laser pulse is passed through two double Pockels cells, which allows the possibility of reducing the laser prepulse even further. The final contrast ratio between the prepulse and main laser pulse, which precedes it by 10 ns, was better than $4.6\times10^{-6}$. The experiments were carried out with various laser energies (49 - 115 mJ), and in a wide range of laser pulse durations from 30 up to 1000 fs, which corresponds to laser intensities of about $6\times10^{16}$-$2\times10^{18}$ W/cm$^2$.

A cluster-gas target was produced by expanding 60-bar Ar gas into a vacuum using a pulsed valve connected with a special nozzle consisting of three truncated cones with different apex angles. The nozzle has the capability to produce Ar clusters with an average diameter of around 1.5 μm. To minimize the tendency of the laser prepulse to destroy the clusters, we purposely used in our experiments very big clusters, since the rate of cluster decay is primarily determined by the number of atoms in the cluster [9,10]. In such cases, the use of micron-sized clusters can significantly reduce the cluster's sensitivity to the laser prepulse, preclude low-density preplasma formation, and thus guarantee direct interaction between the high-density cluster and the main fs pulse.

The spatially resolved X-ray spectra were measured using focusing spectrometers with spatial resolution [11]. This spectrometer recorded He-like spectra of Ar including the associated satellite lines and neutral $K_\alpha$.

The effects of hollow atoms on the observed emission spectra was found to be difficult to isolate, due to the close proximity of strong lines from transitions within the F- and O-like ion stages of Ar, as well as the nearby $K_\alpha$ lines. Detailed theoretical calculations made using the ATOMIC plasma kinetics code [12], which include configurations of KL-hollow ions, have allowed exploration of these emission lines.

It is found (see for more details [13]) that the closest agreement with the experimental measurements comes from a calculation where the electron temperature is 50 eV, with 3% of these electrons at 5 keV, and an atom density of $10^{22}$/cm$^3$ (see Fig. 1). Transitions in the hollow ions give significant (about 50%) contribution only to the strong peak at 4.17 Å. The smaller peak is formed by transitions in F- and O-like ions.

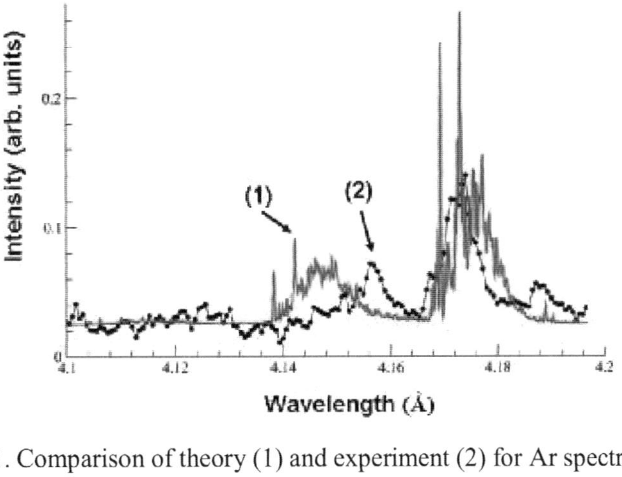


Fig. 1. Comparison of theory (1) and experiment (2) for Ar spectra.

## 3. Mixed Unresolved Transition Array (MUTA) calculations of a Mg hollow ion spectrum

Previous experiments [6] carried out using a low intensity XeCl excimer laser facility with an energy of 2 J and a pulse duration of 12 ns (laser intensity on the target is 4 x $10^{12}$ W/cm$^2$) using FSSR spectrometers [11], have allowed us to measure low intensity X-ray spectra simultaneously with high spectral and spatial resolution (see Fig. 2). These spectra (especially near the target) show unusual quasi-continuous line emission produced by the laser irradiation of Mg targets. This emission occurs at wavelengths just shorter than the He-like resonance line at 9.17 Å. Note that the observed unusual quasi-continuous line emission is lying between the resonance $Ly_\alpha$ line of the H-like ion and resonance lines $He_\alpha$ from the He-like ion, and that they are strongest near the target surface. Typical laser produced plasmas emit satellite lines at wavelengths just longer than the $He_\alpha$ line. These types of transitions have been studied in detail and relative intensities of spectral lines can often be predicted by collisional-radiative models for a prescribed plasma density and temperature. The satellite lines that are commonly observed in these plasmas are generally due to $1s2pnl$-$1s^2nl$ transitions in the Li-like ion. Lines occurring at wavelengths just between the $Ly_\alpha$ and $He_\alpha$ lines belonging to transitions from autoionizing states in He-like ions are rarely observed.

It is clear that the spectra observed are caused by very non-equilibrium conditions because of their unusual characteristics. Local thermodynamic equilibrium (LTE) and typical collisional-radiative modeling would not normally predict such intense spectral structures relative to the resonance line in the wavelength region in question. The first requirement is to identify the emitting species. An initial guess for line emission in this region could be n = 3 to n = 1 transitions in moderately ionized magnesium. Another possibility is n = 2 to n = 1 transitions in KK-hollow ions. Previous theoretical analysis [6] confirmed that such complicated structure in the spectral range 8.5-9.1 Å could belong to spectra from such transitions. However, those calculations were incomplete, and questions remain about the mechanisms of hollow atom formation in such a laser-produced plasma and additional modeling is necessary to address such issues.

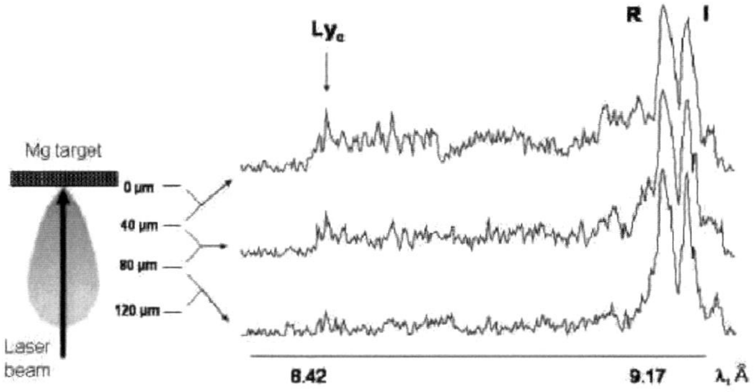


Fig. 2. Traces of spatially resolved X-ray emission spectra of Mg plasma in the spectral range of 8.4 - 9.3 Å heated by a low intensity XeCl nanosecond laser. The strongest spectral lines belong to resonance $Ly_a$ line of Mg XII and the resonance (R) and intercombination (I) lines of Mg XI as marked.

The solution of the rate equations for the Mg plasma was carried out using the Los Alamos code ATOMIC, which has been described previously [12]. The MUTA capability allows us to obtain ionic populations at the configuration-average level while retaining the ability to provide a spectral description comparable in accuracy to a detailed (fine-structure) approach, where all lines are included explicitly in the calculation. This development was crucial since it enabled us to include kinetics for configurations where two or more electrons were promoted from the K and L shells within a calculation that could be completed in a timely manner (compared with a full fine-structure kinetics calculation). Since the plasma was formed from a very long laser pulse, the steady-state rate equations were solved for various electron temperatures and densities. A small fraction of the electron density (typically 0.1% or less) was at a temperature of 2 keV to simulate a 'hot electron' tail.

In this paper we have used the latest developments in atomic kinetics modeling [14], coupled with the use of a comprehensive configuration dataset, to model complex spectra created from the interaction of a nanosecond XeCl laser on a Mg target. The experimental spectral lines which arise from 1s-3l transitions, 1s-4l transitions, and transitions involving hollow ions are, on the whole, quite well reproduced by the calculations (see Fig. 3). It should be emphasized that the traditional kinetic model without KK- and KL-hollow ion transitions gives no spectral lines in the region 8.55 – 9.15 Å. Inclusion of a small fraction of hot electrons is necessary to reproduce the experimental spectra, although we must also make the assumption of quite large temperature and density gradients within the bulk electron distribution. We find a reasonable reproduction of the experimental spectra by assuming small parts of the plasma are relatively cool and dense, and larger parts of the plasma are hot and less dense (see for details [15]). Even at the high densities that we have assumed in our plasma, the line broadening of the satellite lines on which this study is focused is fairly small. We look forward to applying similar calculations to other experimental studies of hollow ion spectral lines, since the investigation of these exotic configurations is increasingly becoming accessible using modern laser technology and X-ray spectrometers with high spectral and spatial resolution.

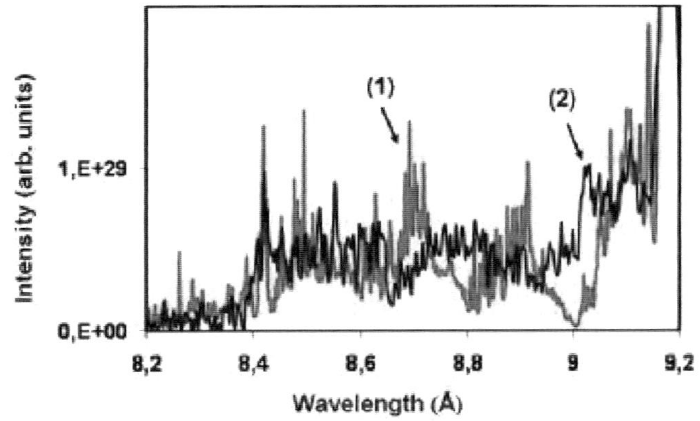


Fig. 3. (1) - Mg spectra from mixed ATOMIC calculation. (2) - Mg spectra from the experimental measurements of [6]

## 4. Acknowledgments

This work was partly supported under the auspices of the US Department of Energy through the Los Alamos National Laboratory, by the Japan Ministry of Education, Science, Sports and Culture, Grant-in-Aid for Kiban A No. 20244065, Kibam B No. 21360364, by the RFBR (Projects No. 09-02-92382 -MNKS_a, by ISTC grant 3504 and by the RAS Presidium Program of basic researches No. 12.

## References

[1] McPherson A, Thompson B.D, Borisov A.B. et al. 1994 Nature **370**, 631
[2] Gauthier J.-C, Geindre J.-P, Audebert P. et al., 1995 Phys. Rev. E. **52**, 2963
[3] Faenov A.Ya., Abdallah J. Jr., Clark R.E.H. et al. 1997 Proceedings of SPIE-97, **3157**,10
[4] Urnov A.M., Dubau J., Faenov A.Ya et al., 1998 JETP Letters **67**, 489
[5] Faenov A.Ya., Magunov A.I., Pikuz T.A. et al., 1999 Physica Scripta **T80**, 536
[6] Abdallah J. Jr., Skobelev I.Yu., Faenov A.Ya. et al., 2000 Quantum Electronics **30**, 694
[7] Rosmej F.B., Faenov A. Ya., Pikuz T.A. et al. 1999 J. Phys. B. **32**, L107
[8] Akahane Y., Fukuda J., Aoyama M. et al., 2006 Rev. Sci. Instrum. **77**, 023102
[9] Boldarev A.S., Gasilov V.A., Faenov A.Ya. et al. 2006 Rev. Sci. Instrum. ,**77**, 083112
[10] Faenov A.Y., Skobelev I.Y., Magunov A.I. et al., 2001 Proc. SPIE **4504**, 121
[11] Faenov A.Ya,. Pikuz S. A, Erko A. I. et al., 1994 Physica Scripta **50**, 333
[12] Magee N.H, Abdallah J.Jr., Colgan, et al., 2004 11th APS Topical Conference on Atomic Processes in Plasmas, (Eds: J. S. Cohen, S. Mazevet, and D. P. Kilcrease) AIP Conference Proceedings, Melville, New York, p. l68
[13] Colgan J., Abdallah J, Jr., Faenov A.Ya. et al., 2008 Laser and Particle Beam, **26**, 83
[14] Mazevet S. and Abdallah J. Jr., 2006 J. Phys. B **39**, 3419
[15] Colgan J, Abdallah J, Jr., Faenov A.Ya, Pikuz T.A and Skobelev I.Yu. 2008 Physica Scripta,**78**, 015302

# A tale of two lines: Searching for the $5s - 5p$ resonance lines in Pm-like ion spectra

**E. Träbert[1,2], M. J. Vilkas[3] and Y. Ishikawa[3]**

1 Astronomisches Institut, Ruhr-Universität Bochum, D-44780 Bochum, Germany
2 High Temperature and Astrophysics Division, Lawrence Livermore National Laboratory, Livermore, CA 94550-9234, U.S.A.
3 Department of Chemistry, University of Puerto Rico, P. O. Box 23346, San Juan, Puerto Rico 00931-3346, U.S.A.

E-mail: traebert@astro.rub.de

**Abstract.**
Highly charged ions in the promethium sequence have been suggested to show spectral features resembling the alkali sequence ions. Guided by calculations, the $5s - 5p$ resonance lines have been sought in a variety of experiments. In the light of the most extensive calculations of Pm-like ions yet, applying relativistic multi-reference Møller-Plesset second-order perturbation theory, the experimental evidence is reviewed and the line identification problem assessed.

## 1. Introduction

In 1980, Curtis and Ellis [1] suggested on the basis of Hartree-Fock calculations that Pm-like ions (61 electrons) with a nuclear charge $Z > 73$ have alkali structure with a ground configuration $4f^{14}5s$. Dirac-Fock calculations by Theodosiou and Raftopoulos [2] added important relativistic corrections and then predicted the turning point at which $4f^{14}5s\ ^2S_{1/2}$ rather than $4f^{13}5s^2\ ^2F^o_{7/2}$ becomes the ground state to lie near $Z = 77$ instead of $Z = 74$. Since then, the prospective $5s - 5p$ resonance transitions have been searched for in the spectra of tokamaks, laser-produced plasmas, vacuum sparks, foil-excited ion beams, ion-atom collisions, and electron beam ion traps. A number of identifications of these lines in Pm-like ions of W ($Z = 74$) through U ($Z = 92$) have been proposed, but at wavelengths that are neither in agreement with theory nor consistent within the data set. We have undertaken calculations of the term energies in Pm-like ions using second-order single-reference Møller-Plesset (MP2) and multi-reference Møller-Plesset (MR-MP) theory, both as a demonstration of the capabilities of our calculational approach and in order to compare with the experimental observations. Our computational procedures have been described elsewhere [3, 4, 5, 6].

One of our results is that the $4f^{14}5s\ ^2S_{1/2}$ level is the ground state from Pt ($Z = 78$) onwards. In W, an element of high interest for the ITER fusion reactor experiment [7, 8], the $5s$ level is only the $24^{th}$ lowest level - far above the true ground state $4f^{13}5s^2\ ^2F_{7/2}$. Similarly, the $4f^{14}5p$ states are actually not the lowest odd-parity levels in the ions up to $Pb^{21+}$, the levels being affected by numerous crossings with $4f^{12}5s^25p$ levels. The admixture of those states is so significant that it requires an extensive multi-reference treatment of the correlation effects in the theoretical approximations. This finding immediately suggests that the lines of primary interest might not be standing out as the resonance in the true alkali sequence spectra do, but be accompanied by many others of comparable, if not higher, intensity.

## 2. Review of experiments

When wide humps were observed in the EUV spectra of fusion plasma devices like ORMAK at Oak Ridge [9], a search started for signature lines of Pm-like ions that would in turn reveal the charge state distribution and thus the temperature of such plasmas. A beam-foil experiment (with fast Au ($Z = 79$) ions) at Brookhaven [10] remained inconclusive. A sequence of follow-up experiments at Bochum [11, 12] found signature lines in neighbouring lower charge states, and clusters of partly resolved lines that were believed to arise from the many core-excited states of Pm-like ions. Among the weak lines nearby are candidates for the $5s - 5p$ transitions, at wavelengths just short of various predictions.

Hutton *et al.* [13] have reported on identifications of two $5s-5p$ lines in the Pm-like spectrum of W ($Z = 74$) as observed in the Berlin electron beam ion trap (EBIT). Such EBITs are low-density devices, and the core excited level decays that swamp the beam-foil spectra are not expected to show at all. The charge state distribution in an EBIT can be varied by adjusting the electron beam energy. Based on such a variation, a number of lines were given charge state identifications or even line labels; however, without any known reference line in any of the tungsten spectra, such a result remains doubtful. The wavelengths of the lines identified with the $5s - 5p$ transitions differ from prediction by clearly different amounts, which signals consistency problems. Quite possibly no more than one of the two lines is from Pm-like W, if any.

Hutton *et al.* [13] have reported one line ($5s - 5p_{1/2}$) from a beam-foil experiment on Pb ($Z = 82$) done at RIKEN (Japan), with a wavelength a little shorter than predicted, just as observed with the candidate lines in the Bochum beam-foil spectrum. The spectrum is shown in a recent paper by Wu and Hutton [14].

Uranium in a (low density) tokamak plasma [15] is the heaviest element of which a line has been assigned to the $5s - 5p_{3/2}$ transition in a Pm-like ion so far, at a wavelength slightly longer than predicted. Unfortunately, in a tokamak there is no reliable way of actually determining the charge state of a highly charged ion, and hence the proximity of the selected line to the calculated wavelength of a specific ion transition may be fortuitous. The authors of ref. [15] note that, according to their collisional-radiative model calculations, the Pm-like ion resonance lines have to be accompanied by stronger lines from the $4d^{13}5s^2$ -$5s5p$ array and other, similar satellites that were not identified in the experiment.

## 3. Discussion

In short, although identifications of $5s - 5p$ lines of Pm-like ions have been suggested for W, Au, Pb, and U, the evidence is meagre. To our present judgment, none of the experimental investigations has unequivocally demonstrated seeing the $5s-5p$ "alkali-metal like ion resonance lines" yet. Evidently, the set of suggested identifications presented by Hutton *et al.* [13] is not intrinsically consistent and therefore hardly convincing; of the six lines used by Hutton *et al.* for a comparison to theory, five are known to be wrongly associated with $4f^{14}5s - 5p$ transitions or at least strongly suspected to be. Apparently there are only three line identifications that are not yet discounted; all three are from beam-foil spectra: two lines of Au [12] and one of Pb [13]. The three experimental wavelengths are about 1% shorter than our calculations (see [6]) predict. That is a poorer agreement than we have achieved in our work on fewer-electron ions. However, the three experimental wavelengths are not particularly accurate: the 1 Å measurement uncertainty corresponds to about 0.3 to 0.5%. There certainly is room for improvement on the experimental side.

In the present context, the best spectra (with isolated lines and the least clutter by unresolved core-excited state contributions) are those obtained at the Berlin EBIT [13]. However, our calculations confirm the much earlier calculations by Theodosiou and Raftopoulos [2] in that the spectrum of Pm-like ions of W has practically no alkali-like character, and the $5s-5p$ transitions cannot stand out at this low ion charge. The question arises as to what actually has been seen at the Berlin EBIT. Simulated Pm-like W spectra, modeled once for high density (based on our MR-MP calculations) and once for low density (EBIT) conditions (based on a collisional-radiative model using the FAC code [16]) are shown in figure 1. The low-density model spectrum resembles the Berlin observations, but holds almost as many lines of ions of a single charge state as the EBIT spectra that surely represent a charge state distribution of several charge states of tungsten (in addition to lines from He). Calculations of the level structure and spectra of $W^{7+}$ through $W^{13+}$ [17] indicate a number of possible line identifications. It seems likely that among the lines seen are some of the Pm-like ion, but the $5s-5p$ transitions are not prominent among them.

For a meaningful analysis of the W spectra obtained at the Berlin EBIT, a more complete set of observations would be helpful. Moreover, the calculations have to evolve towards matching the configurations and levels treated in accurate calculations (like our MR-MP) with those of collisional-radiative models based on FAC. However, many levels cannot be identified easily, because the total angular momentum of some of the electrons is no longer a good quantum number. The FAC code has proven to be robust and versatile. The accuracy of the FAC level energies, however, is not sufficient for spectroscopic applications within a given atomic shell. Considering that hundreds of levels have to be included in the model output, the substitution of accurate MR-MP level energies into FAC will have to be automated.

Concerning the experimental pursuit of alkali-ness in Pm-like spectra, the most viable option is provided by electron beam ion traps. This device offers an adjustable charge balance similar to the beam-foil interaction process, but EBITs work better for more highly charged ions. Studies in search for the $5s-5p$ transitions outside a closed-shell core in alkali-metal like ions would best start with U and then continue towards lighter elements. Pm-like $U^{31+}$ with its ionization potential of only about 1.1 keV is accessible at all present-day EBITs.

## 4. Acknowledgements

We thank C Biedermann (Berlin) and R Hutton (now at Shanghai) for access to unpublished material, and M F Gu (Livermore) for providing synthetic spectra on request. This research was supported in part by the US-Israel Binational Science Foundation. ET acknowledges travel support from the German Research Association (DFG). Some of this work was performed under the auspices of the US Department of Energy by Lawrence Livermore National Laboratory under Contract DE-AC52-07NA27344.

## References

[1] Curtis L J and Ellis D G 1980 *Phys. Rev. Lett.* **45** 2099
[2] Theodosiou C E and Raftopoulos V 1983 *Phys. Rev.* A **28** 1186 (1983).
[3] Vilkas M J and Ishikawa Y 2003 *Phys. Rev.* A **68** 012503 (2003).
[4] Vilkas M J and Ishikawa Y 2005 *Phys. Rev.* A **72** 032512
[5] Vilkas M J, Ishikawa Y and Träbert E 2006 *J. Phys. B: At. Mol. Opt. Phys.* **39** 2195
[6] Vilkas M J, Ishikawa Y and Träbert E 2008 *Phys. Rev.* A
[7] Peacock N J, OMullane M G, Barnsley R, and Tarbutt M (2008) *Can. J. Phys.* **86** 277
[8] Skinner C H 2008 *Can. J. Phys.* **86** 285
[9] Sugar J and Kaufman V 1980 *Phys. Rev.* A **21** 2096
[10] Johnson B M, Jones K W, Kruse T H, Curtis L J, and Ellis D G 1982 *Nucl. Instrum. Meth.* **202** 53
[11] Träbert E and Heckmann P H 1986 *Z. Phys.* D **1** 381
[12] Kaufman V, Träbert E, Heckmann P H, Möller G, Lodwig V, and Blanke J H 1990 *Phys. Scr.* **42** 705

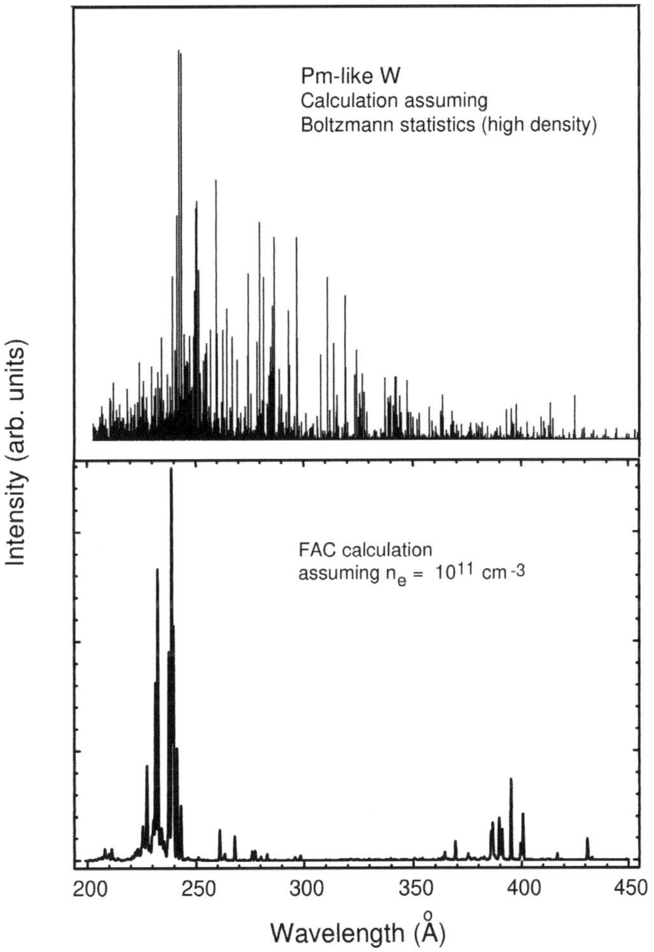


**Figure 1.** Simulated spectrum of the Pm-like ion $W^{13+}$ in the range of the Berlin EBIT spectra shown by Hutton *et al.* [13, 14]. The upper simulation is based on the present MR-MP calculation and assumes high electron density (Boltzmann level population). The lower spectrum (courtesy of M F Gu (Livermore)) is based on FAC code calculations, for a low-density plasma ($n_e = 10^{11}$ cm$^{-3}$) as is typical for an EBIT.

[13] Hutton R, Zou Y, Reyna Almandos J, Biedermann C, Radtke R, Greier A., and Neu R 2003 *Nucl. Instrum. Meth. Phys. Res.* B **205** 114
[14] Wu S and Hutton R 2008 *Can. J. Phys.* **86** 125
[15] Fournier K B, Finkenthal M, Lippmann S, Holmes C P, Moos H W, Goldstein W H, and Osterheld A L 1994 *Phys. Rev.* A **50** 3727
[16] Gu M F 2008 *Can. J. Phys.* **86** 675
[17] Vilkas M J, Ishikawa Y and Träbert E (work in progress)

Journal of Physics: Conference Series **163** (2009) 012018 doi:10.1088/1742-6596/163/1/012018

# Spectroscopy of multiply charged titanium ions in high-density magnetic fusion plasmas

**Joel H T Clementson**[1,2], **Peter Beiersdorfer**[1], **and Reginald D Wood**[1]

[1]Lawrence Livermore National Laboratory, Livermore, California 94550, USA

[2]Atomic Physics Division, Lund Institute of Technology, SE-221 00 Lund, Sweden

E-mail: clementson@llnl.gov

**Abstract.** The M-shell line emission from multiply charged titanium ions has been investigated at the Sustained Spheromak Physics Experiment in Livermore. Titanium was introduced into the relatively low-temperature, high-density magnetically confined spheromak plasmas using a titanium gettering system. The measurements were done using a high-resolution grazing-incidence spectrometer with a 1200 lines/mm grating and a Photometrics charged-coupled device camera. Spectral lines from the transition array $3s^2 3p^k$ - $3s^2 3p^{k-1} 3d$ in argon-like $Ti^{4+}$, chlorine-like $Ti^{5+}$, and sulfur-like $Ti^{6+}$ have been observed in the 240 - 370 Å interval.

## 1. Introduction

Titanium is a transition element of interest in magnetic confinement fusion (MCF) research. Since 1974 titanium has been used in fusion experiments to reduce impurity concentrations and control hydrogen recycling [1]. As radiation losses have large effects on the energy balance in fusion plasmas, it is of utter importance to reduce ion impurities to be able to heat plasmas to thermonuclear temperatures. Titanium is deposited on plasma-facing surfaces by means of the so-called titanium gettering method. In this process titanium is evaporated inside the MCF vacuum vessel and deposited on the interior surfaces. This way the titanium will not only reduce the impurity concentrations by binding to various gas molecules present in the device, but also by burying impurities on the walls [1].

Ti has been spectroscopically investigated in many MCF experiments. Especially x-ray spectra of the K-shell emission from highly charged titanium ions have been observed in several tokamaks [2, 3, 4, 5, 6, 7], but also EUV measurements have been performed, see e.g. Ref. [8] for observations of strong Ti lines on the Princeton Large Torus (PLT) tokamak. More recently Ti lines have unexpectedly appeared in EUV spectra from the National Spherical Torus Experiment (NSTX), i.e. from plasmas where no Ti gettering is used [9]. The study of titanium spectra in fusion plasmas is also of general interest to better understand the spectra of multiply charged ions from mid-Z elements in both laboratory and astrophysical environments. When the 3p subshell gets partially stripped, starting with chlorine-like ions, the spectra get relatively complex due to the numerous coupling combinations of angular momenta.

Here we have investigated the titanium 3p - 3d M-shell EUV emission at the Sustained Spheromak Physics Experiment (SSPX) [10, 11]. The SSPX spheromak research facility, in operation from 1999 to 2007 at the Lawrence Livermore National Laboratory in California, was a project among innovative confinement concepts in magnetic fusion research, aimed at investigating the feasibility of spheromaks as potential fusion reactors [12]. Typical electron densities of SSPX plasmas were a few times $10^{14}$ cm$^{-3}$, and were thereby denser than most tokamak core plasmas. This made SSPX a good test bed for the study of spectra not readily attainable in other devices. The objective of our investigations was to determine the radiative losses of the spheromak by identifying the impurity ions and their respective charge states present in the plasmas.

## 2. Measurements

The SSPX spheromak produced toroidal hydrogen plasmas confined in a copper flux conserver coated with tungsten to minimize sputtering of wall material. On the plasma-facing components a 10 nm layer of titanium was deposited by titanium gettering [12]. The titanium deposition effectively helped to reduce the amount of impurity ions, but, as a side effect, also introduced titanium ions into the spheromak plasmas. For our measurements we utilized this as a convenient injection method.

The instrument used for the titanium spectroscopy was a flat-field grating instrument as described in Ref. [13]. Employing a 1200 lines/mm grating, the high-resolution EUV spectrometer achieved a line width of around 0.3 Å full width at half maximum. The spectrometer had a tangential line of sight of the spheromak magnetic axis through the midplane of the toroidal plasmas. The spectra were time-integrated and recorded with a Photometrics charged-coupled device camera.

The spectra were wavelength calibrated using first and second order lines from oxygen ions present in SSPX [14, 15] and lines from the Lyman series in helium, which was puffed into the device to provide reference lines. By analyzing many SSPX shots we could recognize which lines belonged to Ti and which originated from other plasma impurities such as C, N, O, and W.

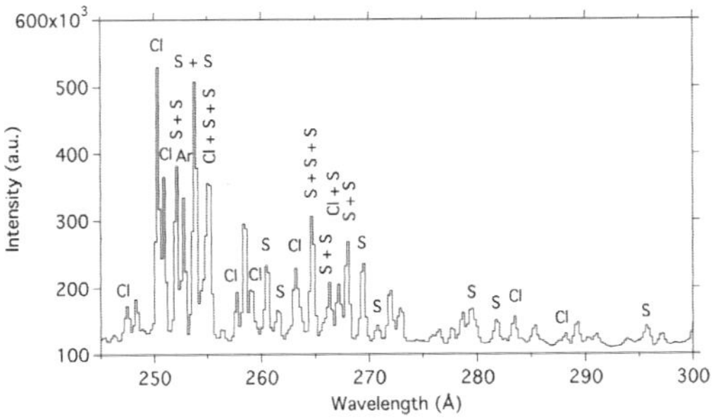


**Figure 1.** Spectrum from SSPX shot # 18633 showing strong titanium lines labeled according to charge state. Also shown are O V at 248.5 Å, and O IV at 272.1 and 279.8 Å.

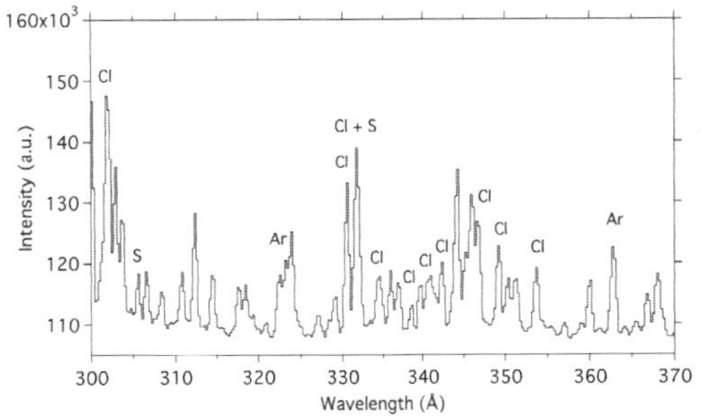


**Figure 2.** SSPX shot # 18633. Titanium lines, He II at 303.8 Å, C IV at 312.4 Å, and second order lines of O VI at 300.2, 346.0, 368.0 Å and O V at 303.0, 340.4, and 344.4 Å.

## 3. Results

By studying subsequent discharges after titanium gettering it was clear the strong line emission observed in the 240 - 280 Å interval originated from M-shell titanium ions. The lines were very strong in the first shots after the titanium deposition and got weaker as the Ti wore off the walls during discharges and got pumped out from the vacuum vessel. We have identified strong lines from four times ionized argon-like through six times ionized sulfur-like titanium. The strongest lines are of the $3s^23p^k$ - $3s^23p^{k-1}3d$ transition array, which is listed in Table 1. We have used line lists from Kelly [16] and the NIST Atomic Spectra Database [17] to identify the observed features. The listed titanium data are taken from measurements by Svensson and Ekberg 1968 [18] and Svensson 1976 [19].

| Ion | Lower level | Upper level | λ (Å) | Ion | Lower level | Upper level | λ (Å) |
|---|---|---|---|---|---|---|---|
| $Ti^{4+}$ | $3s^23p^6\ ^1S_0$ | $3s^23p^53d\ ^1P_1$ | 253.0 | $Ti^{6+}$ | $3s^23p^4\ ^3P_1$ | $3s^23p^3(^2P)3d\ ^1D_2$ | 248.0 |
| | $^1S_0$ | $^3D_1$ | 323.4 | | $^3P_2$ | $^3D_1$ | 250.9 |
| | $^1S_0$ | $^3P_1$ | 363.1 | | $^3P_2$ | $^3D_2$ | 252.2 |
| | | | | | $^1D_2$ | $^1F_3$ | 252.3 |
| $Ti^{5+}$ | $3s^23p^5\ ^2P_{3/2}$ | $3s^23p^4(^1D)3d\ ^2D_{3/2}$ | 247.5 | | $^1S_0$ | $^1P_1$ | 252.6 |
| | $^2P_{3/2}$ | $^2D_{5/2}$ | 250.5 | | $^3P_1$ | $^3D_1$ | 253.8 |
| | $^2P_{1/2}$ | $^2D_{3/2}$ | 251.1 | | $^3P_2$ | $^3D_3$ | 254.0 |
| | $^2P_{3/2}$ | $^2P_{3/2}$ | 255.4 | | $^3P_0$ | $^3D_1$ | 254.7 |
| | $^2P_{1/2}$ | $^2P_{1/2}$ | 257.9 | | $^3P_1$ | $^3D_2$ | 255.1 |
| | $^2P_{1/2}$ | $^2P_{3/2}$ | 259.2 | | $^1D_2$ | $^1D_2$ | 260.7 |
| | $^2P_{3/2}$ | $^2S_{1/2}$ | 263.2 | | $^3P_2$ | $3s^23p^3(^2D)3d\ ^1P_1$ | 261.9 |
| | $^2P_{1/2}$ | $^2S_{1/2}$ | 267.3 | | $^3P_2$ | $3s^23p^3(^2P)3d\ ^3P_1$ | 263.9 |
| | $^2P_{3/2}$ | $3s^23p^4(^1S)3d\ ^2D_{5/2}$ | 282.2 | | $^3P_2$ | $^3P_2$ | 264.8 |
| | $^2P_{3/2}$ | $^2D_{3/2}$ | 283.6 | | $^3P_1$ | $3s^23p^3(^2D)3d\ ^1P_1$ | 265.0 |
| | $^2P_{1/2}$ | $^2D_{3/2}$ | 288.4 | | $^3P_1$ | $3s^23p^3(^2P)3d\ ^3P_0$ | 265.1 |
| | $^2P_{3/2}$ | $3s^23p^4(^1D)3d\ ^2F_{5/2}$ | 301.9 | | $^3P_0$ | $3s^23p^3(^2D)3d\ ^1P_1$ | 266.0 |
| | $^2P_{3/2}$ | $3s^23p^4(^3P)3d\ ^2D_{5/2}$ | 330.7 | | $^3P_2$ | $^3S_1$ | 266.5 |
| | $^2P_{3/2}$ | $^4P_{1/2}$ | 331.8 | | $^3P_1$ | $3s^23p^3(^2P)3d\ ^3P_1$ | 267.1 |
| | $^2P_{3/2}$ | $^2D_{3/2}$ | 334.5 | | $^3P_1$ | $^3P_2$ | 268.0 |
| | $^2P_{1/2}$ | $^4P_{1/2}$ | 338.3 | | $^3P_0$ | $^3P_1$ | 268.1 |
| | $^2P_{1/2}$ | $^2D_{3/2}$ | 341.1 | | $^1D_2$ | $^3D_2$ | 268.5 |
| | $^2P_{3/2}$ | $^2P_{3/2}$ | 342.6 | | $^3P_1$ | $3s^23p^3(^2D)3d\ ^3S_1$ | 269.8 |
| | $^2P_{3/2}$ | $^2P_{1/2}$ | 346.7 | | $^3P_0$ | $^3S_1$ | 270.7 |
| | $^2P_{1/2}$ | $^2P_{3/2}$ | 349.6 | | $^1D_2$ | $^1P_1$ | 279.5 |
| | $^2P_{1/2}$ | $^2P_{1/2}$ | 353.9 | | $^1D_2$ | $3s^23p^3(^2P)3d\ ^3P_1$ | 281.9 |
| | | | | | $^1D_2$ | $^3P_2$ | 282.9 |
| | | | | | $^1D_2$ | $3s^23p^3(^2D)3d\ ^1F_3$ | 296.1 |
| | | | | | $^1S_0$ | $^1P_1$ | 305.7 |
| | | | | | $^1D_2$ | $^1D_2$ | 332.1 |

**Table 1.** $3s^23p^{k-1}3d$ resonance transition arrays in $Ti^{4+}$ through $Ti^{6+}$. All the data are taken from the NIST Atomic Spectra Database [17].

Shown in Figure 1 is a spectrum from SSPX shot # 18633 displaying the strong titanium lines in the 240 - 300 Å interval. In this discharge the charge balance seems to be dominated by the chlorine- and sulfur-like ions, indicating an electron temperature of around 50 - 70 eV. In the spectrum there are additional lines (some that are blended with those from titanium), mainly from oxygen. The $Ti^{6+}$ line at 254.0 Å has previously been observed in the PLT tokamak [8].

The longer wavelengths are shown in Figure 2. The Ti lines are here much weaker and, again, lines from other plasma impurities interfere with the titanium spectra. In this interval several strong lines from O V and O VI appear in second order. The $Ti^{5+}$ line at 338.3 Å has not previously been observed and we use the Ritz wavelength from the NIST database for our identification.

Most of the 3p - 3d transitions listed in Table 1 are observed in SSPX spectra. The identification of a few lines are somewhat uncertain due to line blends. Due to blends and overlapping lines it is a difficult task to determine the intensity of the titanium lines and the charge state distribution. This is especially true for blends with oxygen lines, which always exist in the spectra. We are planning to further investigate the spectra from higher charge states of titanium present in SSPX as well as to perform collisional radiative modeling of the spheromak plasmas. This will help us better understand the M-shell emission from transition elements in high-density magnetic fusion plasmas.

## Acknowledgments

This work was performed under the auspices of the U.S. Department of Energy by Lawrence Livermore National Laboratory under contracts W-7405-ENG-48 and DE-AC52-07NA-27344. The authors would like to thank Dr. Harry McLean for supporting the project, as well as Ed Magee for technical assistance.

## References

[1] Braams C M and Stott P E 2002 *Nuclear Fusion - Half a Century of Magnetic Confinement Fusion Research* (Institute of Physics Publishing)
[2] Bely-Dubau F, Faucher P, Steenman-Clark L, Bitter M, von Goeler S, Hill K W, Camhy-Val C and Dubau J 1982 *Phys. Rev. A* **26** 3459–3469
[3] Bitter M, Hill K W, Zarnstorff M, von Goeler S, Hulse R, Johnson L C, Sauthoff N R, Sesnic S and Young K M 1985 *Phys. Rev. A* **32** 3011–3029
[4] Kato T, Morita S and Masai K 1987 *Phys. Rev. A* **36** 795–803
[5] Pacella D and Leigheb Mattioli M 1998 *Phys. Scr.* **57** 265–271
[6] Lee P, Lieber A J and Wojtowicz S S 1985 *Phys. Rev. A* **31** 3996–3998
[7] Lee P, Lieber A J and Chase R P 1985 *Phys. Rev. Lett.* **55** 386–389
[8] Davé J H, Feldman U, Seely J F, Wouters A, Suckewer S, Hinnov E and Schwob J L 1987 *J. Opt. Soc. Am. B* **4** 635–643
[9] Beiersdorfer P, Lepson J K, Bitter M, Hill K W and Roquemore L 2008 *Rev. Sci. Instrum.* **79**
[10] Hooper E B, Pearlstein L D and Bulmer R H 1999 *Nucl. Fusion* **39**
[11] Hudson B, Wood R D, McLean H S, Hooper E B, Hill D N, Jayakumar J, Moller J, Montez D, Romero-Talámas C A, Casper T A, Johnson III J A, LoDestro L L, Mezonlin E and Pearlstein L D 2008 *Phys. Plasmas* **15** 056112
[12] Wood R D, Hill D N, Hooper E B, Buchenauer D, McLean H, Wang Z, Woodruff S and Wurden G 2001 *J. Nucl. Mater.* **290-293** 513–517
[13] Clementson J, Beiersdorfer P and Magee E W 2008 *Rev. Sci. Instrum.* **79**
[14] Clementson J, Beiersdorfer P, Gu M F, McLean H S and Wood R D 2008 *J. Phys. Conf. Ser.* In press
[15] Wilcox P G, Safronova A S, Kantsyrev V L, Safronova U I, Williamson K M, Yilmaz M F, Clementson J, Beiersdorfer P and Struve K W 2008 *Rev. Sci. Instrum.* **79**
[16] Kelly R L 1987 *J. Phys. Chem. Ref. Data* **16**
[17] Ralchenko Y, Kramida A E, Reader J and NIST ASD Team (2008) NIST ASD Atomic Spectra Database (version 3.1.5) [online at http://physics.nist.gov/asd3, august 25 2008]
[18] Svensson L Å and Ekberg J O 1967 *Ark. Fys.* **37** 65–84
[19] Svensson L Å 1976 *Phys. Scr.* **13** 235–239

# EUV spectra from highly charged tin ions observed in low density plasmas in LHD

**C Suzuki[1], T Kato[1], K Sato[1], N Tamura[1], D Kato[1], S Sudo[1], N Yamamoto[2], H Tanuma[3], H Ohashi[3], S Suda[3], G O'Sullivan[4] and A Sasaki[5]**

[1] National Institute for Fusion Science, 322-6 Oroshi-cho, Toki 509-5292, Japan
[2] Institute of Laser Engineering, Osaka University, 2-6 Yamadaoka, Suita 565-0871, Japan
[3] Tokyo Metropolitan University, 1-1 Minami-Osawa, Hachioji 192-0397, Japan
[4] University College Dublin, Belfield, Dublin 4, Ireland
[5] Advanced Photon Research Center, Japan Atomic Energy Agency, 8-1 Umemi-dai, Kizugawa 619-0215, Japan

E-mail: csuzuki@lhd.nifs.ac.jp

**Abstract.** In order to meet the increasing needs for experimental databases of tin (Sn) extreme ultraviolet (EUV) spectra relevant to the next generation EUV lithography, we have measured EUV spectra from highly charged tin ions in low density plasmas produced in the Large Helical Device (LHD) at the National Institute for Fusion Science. A small amount of tin was introduced into a background high temperature and low density hydrogen plasma by injecting a pellet. The EUV spectra were monitored by a grazing incidence spectrometer whose wavelength resolution is about 0.01 nm. Two different types of spectral features were found in the LHD depending on the discharge condition. The well known dense spectral structure around 13.5 nm is measured when the plasma is rapidly cooled and approaching radiative collapse after the pellet injection, while the sparse spectrum with several unidentified discrete lines from 13.8–14.6 nm is observed if the plasma is cooled more slowly and higher temperature is maintained for a while. According to comparison with other charge-separated experimental data, the dominant charge states in the former case are $Sn^{11+}$–$Sn^{14+}$. The latter case may be explained by considering the spectral lines from charge states higher than $Sn^{19+}$.

## 1. Introduction

Recently the needs for experimental databases of extreme ultraviolet (EUV) radiation from highly charged tin (Sn) ions have been increasing in the context of the development of light sources around 13.5 nm for the next generation semiconductor lithography process [1]. Though laser and/or discharge produced high density ($>10^{23}$ m$^{-3}$) tin plasmas are promising candidates for these light sources, EUV spectra of such plasmas tend to be complex due to the effects of line broadening, optical thickness and satellite lines. Furthermore, spatial and temporal scales of changes in plasma parameters are extremely steep and rapid in these plasmas, which makes it difficult to benchmark theoretical models with experimental results.

On the other hand, low density plasmas generated in magnetically confined devices utilized for fusion research are more suitable for the benchmarking because they can easily generate highly charged ions of intentionally injected impurity materials (e.g. tin) under optically thin conditions. In addition, they have relatively mild temperature and density gradients which are easily controlled and can be measured by reliable diagnostic tools.

In fact, we have already studied xenon (Xe) EUV spectra from low density plasmas generated in the Compact Helical System (CHS) and the Large Helical Device (LHD) which are fusion-oriented torus devices at the National Institute for Fusion Science [2, 3]. Since the electron temperature easily exceeds 1000 eV in the core region, LHD plasmas can be utilized as a light source from highly charged ions by injecting impurity materials of interest. In this study, we have observed EUV spectra from highly charged tin ions in LHD plasmas. Solid tin is introduced by a tracer encapsulated solid pellet (TESPEL) [4] injected into a background hydrogen plasma. The measured spectral features are discussed based on comparisons with other experiments and theoretical models.

## 2. Experimental details

The LHD is one of the largest devices in the field of magnetically confined fusion research which generates large volume (30 $m^3$) plasmas with the shape of a helical torus whose average major and minor radii are 3.9 m and 0.6 m, respectively. The plasmas are confined in nested magnetic surfaces formed by unclosed field lines under a magnetic field of 2.75 T at the plasma center. The typical plasma density is much lower ($\leq 10^{20}$ $m^{-3}$) than that of laser or discharge produced high density plasmas. Spatial profiles of electron density and temperature are measured by an existing laser Thomson scattering diagnostic system [5]. In this study, a small amount ($\simeq$0.1 % of bulk ion) of tin was introduced into the high temperature ($\simeq$1000 eV) and low density ($\simeq 10^{19}$ $m^{-3}$) hydrogen plasma by injecting the TESPEL.

The EUV spectra were monitored by a grazing incidence spectrometer SOXMOS [6] whose groove density and focal length are 600 $mm^{-1}$ and 1 m, respectively. The spectral range was fixed to 11.0–15.2 nm in this study. The integration time of the detector of the SOXMOS spectrometer was typically 200 ms and was adjusted according to the brightness. The overall spectral resolution is about 0.01 nm. The viewline of the spectrometer is slightly tilted against the equatorial plane within the horizontally elongated plasma cross section because the core temperature tends to be too high to observe the EUV spectra.

The wavelength of the spectrometer was carefully calibrated by observing iron lines as an intrinsic impurity material in the plasma without tin injection. The spectral lines from iron ions whose wavelengths are already known were observed under the same spectrometer setup. Then these wavelengths were fitted to the dispersion relation of the SOXMOS spectrometer determined by the geometrical arrangement. Consequently, we could determine the absolute wavelength with an accuracy of ±0.02 nm over the whole spectral region.

## 3. Results and discussion

We have observed more than ten EUV spectra in various timings in several discharges under different conditions. Consequently, the measured spectral feature can be categorized into two types depending on the discharge condition. The examples of the two types of discharge condition are displayed in Fig. 1 by drawing the time sequences of heating, stored energy ($W_p$) and total radiation power ($P_{rad}$). The discharges were initiated by electron cyclotron heating (ECH) followed by three neutral beam injection (NBI) heatings. When the pellet was injected at 1.3 s, the total radiation power rapidly increased and the stored energy began to decrease simultaneously due to radiative cooling. The cooling process is apparently faster in case (a) than in case (b) as observed in the waveforms of the stored energy. This difference results in the two types of spectral feature shown in the top row of Fig. 2, where the measured EUV spectra (normalized) around 13.5 nm integrated during the shaded area in Fig. 1 (a) and (b) are drawn. The plasma approached radiative collapse and the total radiated power finally began to increase at the end of the integration time in the case (a), while the radiated power remained

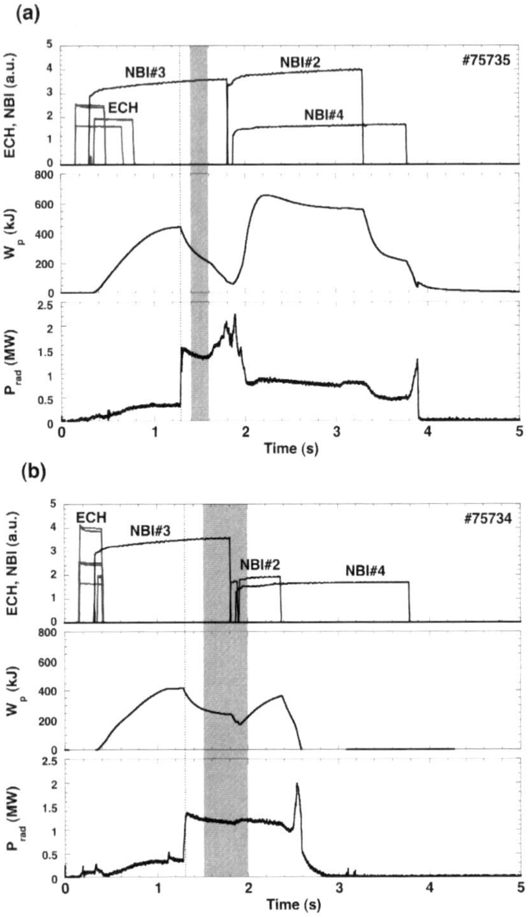


**Figure 1.** Time sequences of heating, stored energy ($W_p$) and total radiation power ($P_{rad}$) in the two different LHD discharges with the tin pellet injection. The cases of (a) fast and (b) slow radiative cooling are displayed. The shaded area denotes the integration time of the detector corresponding to the spectra in the top row of Fig. 2.

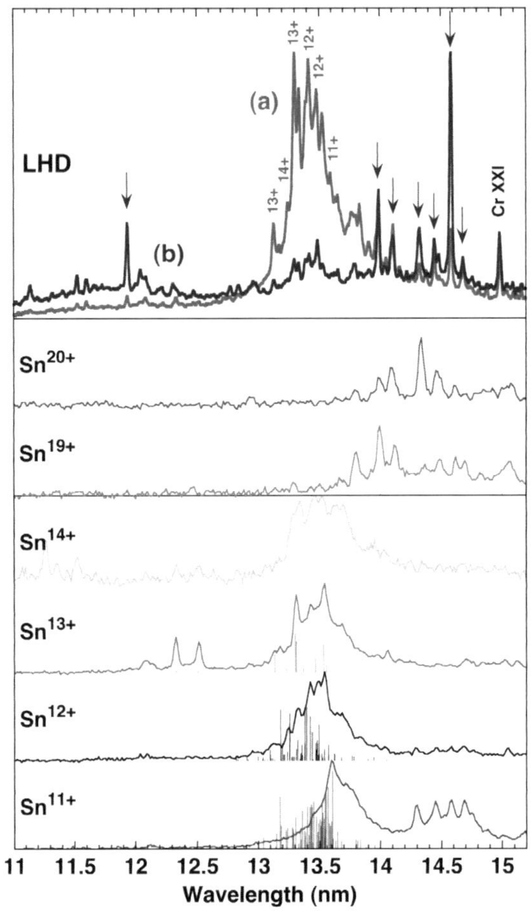


**Figure 2.** The measured EUV spectra for the cases (a) and (b) in Fig. 1, and comparisons with other charge-separated experimental data for $Sn^{11+}$–$Sn^{14+}$, $Sn^{19+}$ and $Sn^{20+}$. Spectra and vertical bars in the middle and bottom rows are the experimental data from charge exchange collisions [9] and vacuum spark discharge [10], respectively.

stable during the integration time in the case (b). The line-averaged electron density was about $3\times10^{19}$ $m^{-3}$ in both cases. The feature of spectrum (a) is characterized by the well known dense structure around 13.5 nm, while the more sparse spectrum with several discrete lines at 11.94 nm and in 13.9–14.7 nm (indicated by arrows) was observed in (b). The line at 14.98 nm arises from an intrinsic chromium impurity in the plasma.

According to theoretical calculations of tin ion energy levels, the dense spectral structure in 13.5 nm region arises from the $4d$-$4f$ unresolved transition array (UTA) of moderately charged open $4d$ subshell tin ions [7]. Recently charge-separated experimental spectra of tin ions have been reported by charge exchange collisions between charge-selected tin ion beams (up to $Sn^{20+}$) and rare gas targets [8, 9], which are shown in the middle and bottom rows of Fig. 2 for $Sn^{11+}$–$Sn^{14+}$, $Sn^{19+}$ and $Sn^{20+}$. Additional charge-separated spectral data obtained by fitting data from Cowan code calculations to the experimental spectra in a vacuum spark discharge have recently been published for $Sn^{8+}$–$Sn^{14+}$ [10]. The statistically weighted Einstein A coefficients ($gA$ values) listed in this reference are also plotted in Fig. 2 by vertical bars. As a result of the comparisons, the dominant emitters in case (a) are found to be $Sn^{11+}$–$Sn^{14+}$ and several prominent peaks are identified as indicated in the top row of Fig. 2. These charge states would become dominant when the electron temperature is around 50 eV assuming coronal ionization equilibrium. However, the actual electron temperature near the point of the pellet ablation is much higher (300–400 eV), which implies that the plasma lies far from the ionization equilibrium due to the effects of radial transport of tin ions.

As for the sharp discrete lines in 13.9–14.7 nm that are relatively more intense in case (b), $4d$-$5p$ or $4d$-$4f$ lines of charge states lower than $Sn^{11+}$ would appear in this wavelength range. However, the dominant charge states in case (b) are expected to be higher than that in case (a) considering the difference in the speed of radiative cooling. Actually spectral features similar to these lines can be found in the charge exchange data for $Sn^{19+}$ and $Sn^{20+}$ as shown in the middle row of Fig. 2. Hence it can be speculated that charge states higher than $Sn^{19+}$ could be observed in this case. The lines at 11.94 nm and 14.58 nm are not found in the other experimental data. The brightest line at 14.58 nm may arise from $Sn^{21+}$ because a $4p$-$4d$ spectral line appears nearby at 14.62 nm in a Hullac code calculation [11]. Since the experimental data of charge exchange collisions are unavailable for charge states higher than $Sn^{21+}$ at present, we intend to make comparison with the other theoretical calculations (e.g. Cowan code) to identify them.

**Acknowledgments**
The authors acknowledge the LHD experimental group for their assistance. This work was partly supported by the MEXT under the leading project for EUV lithography source development.

## References

[1] Nishihara K *et al.* 2008 *Phys. Plasmas* **15** 056708
[2] Suzuki C *et al.* 2005 *J. Plasma Fusion Res.* **81** 480
[3] Kato T *et al.* 2008 *J. Phys.* B **41** 035703
[4] Sudo S 1993 *J. Plasma Fusion Res.* **69** 1349
[5] Narihara K, Yamada I, Hayashi H and Yamauchi K 2001 *Rev. Sci. Instrum.* **72** 1122
[6] Schwob J L, Wouters A W and Suckewer S 1987 *Rev. Sci. Instrum.* **58** 1601
[7] White J, Hayden P, Dunne P, Cummings A, Murphy N, Sheridan P and O'Sullivan G 2005 *J. Appl. Phys.* **98** 113301
[8] Tanuma H, Ohashi H, Fujioka S, Nishimura H, Sasaki A and Nishihara K 2007 *J. Phys.: Conf. Series.* **58** 231–234
[9] Ohashi H, Tanuma H, Fujioka S, Nishimura H, Sasaki A and Nishihara K 2007 *J. Phys.: Conf. Series.* **58** 235–238
[10] Churilov S S and Ryabtsev A N 2006 *Phys. Scr.* **73** 614
[11] Sasaki A 2008 private communication

Journal of Physics: Conference Series **163** (2009) 012020 doi:10.1088/1742-6596/163/1/012020

# EUV spectroscopy of highly charged iron ions with a low energy compact EBIT

**H A Sakaue[1], D Kato[1], N Nakamura[2], E Watanabe[2], N Yamamoto[3], C Chen[4] and T Watanabe[5]**

[1]National Institute for Fusion Science,Toki, Gifu 509-5292, Japan
[2]Institute for Laser Science, The University of Electro-Communications, Chofu, Tokyo 182-8585, Japan
[3]Institute of Laser Engineering, Osaka University, Suita, Osaka 565-0871, Japan
[4]Institute of Modern Physics, a Fudan University, Shanghai 200433, P. R. China
[5]National Astronomical Observatory of Japan, Mitaka, Tokyo, 181-8588, Japan

E-mail: sakaue@nifs.ac.jp

**Abstract.** We measured extreme ultraviolet (EUV) spectra of highly charged iron ions (Fe IX - Fe XV) in the range of wavelength 130-300 Å by a new compact electron beam ion trap (EBIT) which was developed for low electron energy operation. The electron energy of this compact EBIT is controllable from 100 eV to few keV, therefore it is appropriate for spectroscopic investigation of the moderately charged ions. A slitless flat-field EUV spectrometer developed exclusively for this compact EBIT is mounted with it and high sensitivity and high resolution spectroscopic analysis of highly charged iron ions become possible. The electron energy dependence of highly charged iron ion spectra were measured in the range of electron energy of 200-1350 eV. We could get the spectral information of wide range of charge state from Fe IX to Fe XV, and moreover, various charge state ions could be selectively produced with a narrow charge state distribution by adjusting the electron beam energy.

## 1. Introduction

The research of highly charged ions (HCIs) with an electron beam ion trap (EBIT) [1] advanced widely in the field of atomic and molecular physics. Although it can be used as an ion source similar to an electron beam ion source [2], it is a powerful tool especially for spectroscopic studies of trapped HCIs. In particular, the spectroscopic investigation of HCIs which are moderate charge state ions attract attention not only in atomic physics but also in some research fields recently. For example, to develop an extreme ultraviolet (EUV) light source for the next-generation lithography, the atomic data of highly charged tin ions with charge states around 10 are strongly needed [3, 4]. Another example is the diagnostics of nuclear fusion and astrophysical plasmas. The spectra of highly charged iron ions with charge states around 10 are observed for diagnostics of the solar corona by the launched satellite Hinode recently [5, 6, 7, 8]. The atomic data of impurity highly charged iron ions are needed for the spectroscopic diagnostics of high temperature and density controlled plasma like a Large Helical Device (LHD) [9, 10]. For International Tokamak Experimental Reactor (ITER), which is the next-generation fusion device, the spectroscopic data of tungsten ions with moderate charge states around 20 are needed [11] because tungsten is one of the candidate of diverter materials.

We developed the compact EBIT [12] for low energy, which is named "CoBIT", and an EUV spectrometer for exclusive use. We carried out the EUV spectroscopy of such a moderate charge state ions. In this paper, we present the observation EUV spectra (130-300 Å) of highly charged iron ions in CoBIT. The electron energy range is from 200 to 1350 eV and the observed charge states of iron ions are from 8+ to 14+. The detail of the new device and electron energy dependence of highly charged iron ion spectra are discribed in this paper.

## 2. Experiments

The experimental apparatus consists of the CoBIT, which is highly charged ion source, and the EUV spectrometer. A schematic drawing of the CoBIT and the EUV spectrometer are shown in Figure 1. The EBIT mainly consists of an electron gun, an ion trap (drift tube), an electron collector, a superconducting coil, and liquid nitrogen tank. We used the same electron gun as the Tokyo-EBIT [13, 14, 15] which had a good performance. The perviance of the gun is about 0.4 μperviance, so that an electron current of 10 mA can be obtained with an anode voltage of 1 kV. At the CoBIT, it is possible to decelerate the electron beam between the anode and drift tube. The electron beam current is about 20mA in electron energy 400 eV by such a decelerating operation. The high critical temperature superconducting Helmholtz coil which can be used at the liquid nitrogen temperature is mounted around the drift tube. The electron beam emitted from the electron gun is accelerated (or decelerated) toward the drift tube and compressed by the magnetic field produced by the superconducting Helmholtz coil. The maximum central magnetic field produced by the present coil is about 0.2 T with a maximum current of 50 A. This magnetic field 0.2 T is enough to produce the HCI of moderate charge state. The trap consists of three drift tubes (DTs). DTs are generally cylindrical electrodes, however, for high accessibility to the trap center, we employed DT2 which consists of six poles surrounding the electron beam. For example, by surrounding the trap by a parabola, ellipse or Wolter mirror, the visible, EUV and x-ray spectroscopy can be performed with high efficiency. After passing through the drift tube, the electron beam is collected by the electron collector which has a cooling system. All the electrodes in this CoBIT are fixed in the liquid nitrogen tank with ceramic insulators (Shapal M-soft) which have high thermal conductivity to keep them at low temperature. The electron collector, at which 10 W is consumed at the maximum, is also cooled through this thermal conductivity and so the present EBIT is compact. The operating electron current is 10-20 mA at electron energy 200-2000 eV, and very stable operation is possible during a long exposure time (several hours). When we operate the CoBIT, the background pressure in it is very important parameter. It is necessary to keep the pressure of the CoBIT less than $10^{-8}$ Pa during the measurement, however, the pressure of spectrometer which connected to the CoBIT is about $10^{-7}$ Pa. Therefore, the quadrate pipe (6W×26H×60L mm) was installed between the CoBIT and spectrometer for differential pumping.

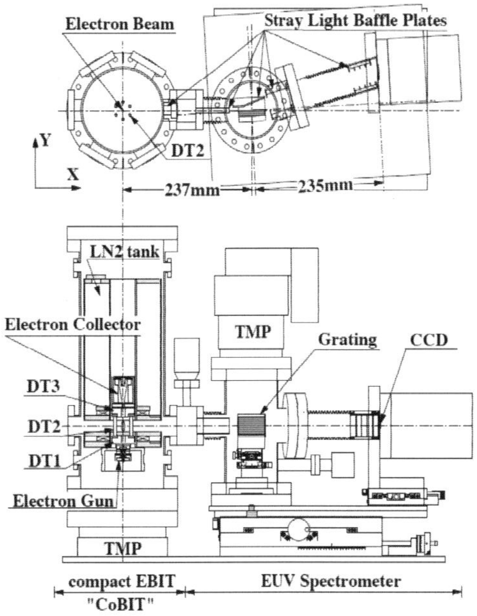


**Figure 1 Cross sectional view of compact EBIT and EUV spectrometer**

We developed an slitless EUV spectrometer which was exclusive use of CoBIT. In general, it is necessary to install a entrance slit in front of the diffraction grating, but we designed the spectrometer of slitless type because an EBIT represents a line source with a width of about 0.1 mm. Therefore, high efficient spectroscopic measurement became possible. On the other hand, we paid attention enough for a stray light countermeasure because of the slitless spectrometer. The designed spectrometer is flat-field grazing-incidence (87deg.) spectrometer with a 1200 grooves/mm laminar-type replica diffraction grating (30-002, Shimadzu Corporation). The diffraction grating is installed in the chamber center precisely and can be moved to the X,Y axis directions at minimum resolution 0.01 mm and be adjusted it to the design position. The spectrometer is mounted with the rotary table whose the axis of rotation agrees with the diffraction grating central axis. Then the rotary table is adjustable with precision of turning angle +/- 0.02 deg. to set a spectrometer to angle-of-incidence 87 deg.. In addition, the diffraction grating holder has an adjustment mechanism which adjust the blast angle by αβ-axis goniometers. The position and angle adjustments of spectrometer are possible while measuring a spectrum by these adjustment mechanisms and we can perform optimization smoothly. The detector is the back-illuminated CCD camera (Princeton Instruments PIXIS-XO: 400B). This camera has 1340×400 pixels whereby the dimension of each pixel is 20× 20 μm$^2$. The CCD camera installed in 235 mm from the diffraction grating position is movable on the focusing plate and fixed on the position of the measuring wavelength area. The wavelength range by this grating is from 50 to 300 Å. The high resolution EUV spectroscopy is possible at wavelength range from 10 to 60 Å by only changing the another diffraction grating (30-003:2400 grooves/mm, grazing-incidence 88.65 deg.).

## 3. Results and Discussion

EUV spectra of highly charged iron ion are shown in Figure 2. The electron energy dependence of highly charged iron ion spectra were measured in the energy range 200-1350 eV and the wavelength range 130-300 Å. The broad emission lines around 190-200 Å are 2nd order lines from an impurity HCI of Ba evaporated from the cathode. It is called UTA (Unresolved Transition Array), and the impurity HCIs of O and N were observed. At the electron energy range 600-1350 eV, the emission lines were not observed at all in the measurement wavelength range of 130-300 Å. The strong emission line of Fe XV (284.16 Å) were observed at 500 eV. As electron energy decreases from 500 to 300 eV, the dominant emission lines transfer from Fe XV to Fe IX successively. The emission lines of highly charged iron ion were almost not confirmed in this wavelength region at 200 eV. In spectrum of each electron energy, only emission lines from two or three kinds of charge states of iron HCI are observed. We could measure the EUV spectra with a narrow charge state distribution of HCIs at each electron energy in this way. This narrow charge state distribution of HCIs is very sensitive to the pressure in ion source, therefore it is

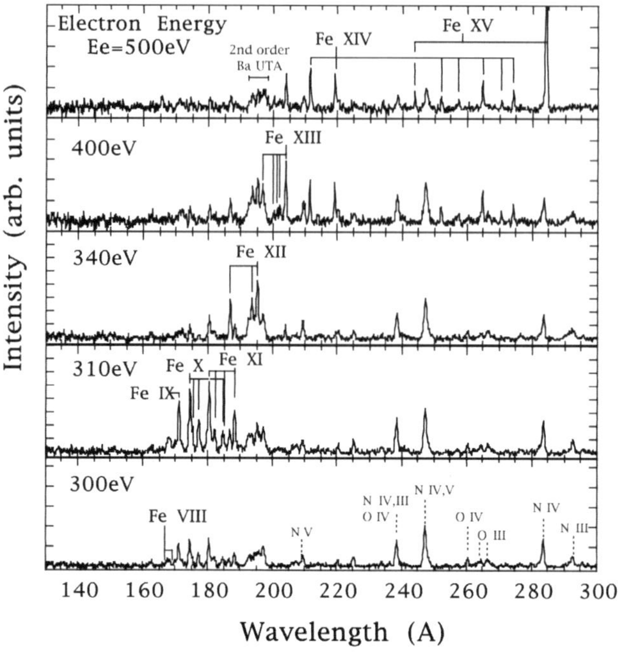


Figure 2 EUV spectra of highly chargd iron ions obtained with the compact EBIT. For all the spectra, the exporsure time were about 20 minutes.

Table 1 configurations of observed lines.

| Ion | lower | Upper | Wavelength(Å) [16] | |
|---|---|---|---|---|
| Fe XV | $3s\,3p\ ^1P_1$ | $3s\,3d\ ^1D_2$ | 243.794 | |
| | $3s^2\ ^1S_0$ | $3s\,3p\ ^1P_1$ | 284.164 | |
| Fe XIV | $3s^23p\ ^2P_{1/2}$ | $3s^23d\ ^2D_{3/2}$ | 211.331 | |
| | $3s^23p\ ^2P_{3/2}$ | $3s^23d\ ^2D_{5/2}$ | 219.136 | |
| | $3s^23p\ ^2P_{1/2}$ | $3s\,3p^2\ ^2P_{3/2}$ | 252.188 | |
| | $3s^23p\ ^2P_{1/2}$ | $3s\,3p^2\ ^2P_{1/2}$ | 257.377 | |
| | $3s^23p\ ^2P_{3/2}$ | $3s\,3p^2\ ^2P_{3/2}$ | 264.785 | |
| | $3s^23p\ ^2P_{3/2}$ | $3s\,3p^2\ ^2P_{1/2}$ | 270.511 | |
| | $3s^23p\ ^2P_{1/2}$ | $3s\,3p^2\ ^2S_{1/2}$ | 274.203 | |
| Fe XIII | $3s^23p^2\ ^1D_2$ | $3s^23p\,3d\ ^1F_3$ | 196.525 | |
| | $3s^23p^2\ ^3P_1$ | $3s^23p\,3d\ ^3D_2$ | 200.021 | |
| | $3s^23p^2\ ^3P_1$ | $3s^23p\,3d\ ^3D_1$ | 201.121 | |
| | $3s^23p^2\ ^3P_0$ | $3s^23p\,3d\ ^3P_1$ | 202.044 | |
| | $3s^23p^2\ ^3P_2$ | $3s^23p\,3d\ ^3D_3$ | 203.826 | |
| Fe XII | $3s^23p^3\ ^2D_{3/2}$ | $3s^23p^2(^3P)3d\ ^2F_{5/2}$ | 186.856 | *1 |
| | $3s^23p^3\ ^2D_{5/2}$ | $3s^23p^2(^3P)3d\ ^2F_{7/2}$ | 186.880 | *1 |
| | $3s^23p^3\ ^4S_{3/2}$ | $3s^23p^2(^3P)3d\ ^4P_{3/2}$ | 193.509 | |
| | $3s^23p^3\ ^2D_{3/2}$ | $3s^23p^2(^1D)3d\ ^2D_{3/2}$ | 195.119 | *2 |
| | $3s^23p^3\ ^4S_{3/2}$ | $3s^23p^2(^3P)3d\ ^4P_{5/2}$ | 195.119 | *2 |
| Fe XI | $3s^23p^4\ ^3P_2$ | $3s^23p^3(^4S)3d\ ^3D_3$ | 180.407 | *3 |
| | $3s^23p^4\ ^3P_1$ | $3s^23p^3(^4S)3d\ ^3D_2$ | 182.173 | |
| | $3s^23p^4\ ^1D_2$ | $3s^23p^3(^2D)3d\ ^1D_2$ | 184.800 | |
| | $3s^23p^4\ ^3P_2$ | $3s^23p^3(^2D)3d\ ^3P_2$ | 188.219 | |
| Fe X | $3s^23p^5\ ^2P_2$ | $3s^23p^4(^3P)3d\ ^2D_3$ | 174.534 | |
| | $3s^23p^5\ ^2P_1$ | $3s^23p^4(^3P)3d\ ^2D_2$ | 175.266 | |
| | $3s^23p^5\ ^2P_2$ | $3s^23p^4(^3P)3d\ ^2P_2$ | 177.243 | |
| | $3s^23p^5\ ^2P_1$ | $3s^23p^4(^3P)3d\ ^2P_1$ | 180.450 | *3 |
| | $3s^23p^5\ ^2P_2$ | $3s^23p^4(^1D)3d\ ^2S_1$ | 184.542 | |
| Fe IX | $3s^23p^6\ ^1S_0$ | $3s^23p^53d\ ^1P_1$ | 171.075 | |

*1,*2,*3 blend

necessary to keep it the low pressure stably even when iron atom is injected. The configurations of ions and wavelength of observed lines are listed in Table 1. Almost all observed lines are transitions to the ground states of each highly charged iron ion. The intensity of emission lines which are transition to the upper level of fine structure *J* in ground states is depend on the electron density strongly. These emission line are studied especially for diagnostics of nuclear fusion and astrophysical plasmas. Moreover, the spectra obtained with CoBIT are useful for testing atomic codes in plasma diagnostics since an EBIT is a simple plasma source consisting of trapped HCIs and a quasi-monoenergetic electron beam whose the energy and current are controllable.

In summary, for the HCIs spectroscopic studies of moderate charge states, we developed the compact EBIT and an EUV spectrometer for exclusive use of it. Then high efficiency and high resolution EUV spectroscopic measurement become possible. Moreover, there is an advantage of this apparatus at HCIs spectroscopy, because various charge state ions can be selectively produced with a narrow charge state distribution by adjusting the electron beam energy in this CoBIT. In addition, the study of electron density dependence of the highly charged ion spectrum is possible by changing an electron beam current. This CoBIT will become the powerful and versatile tool for studying HCIs of moderate charge states.

**Acknowledgement**
This work was partly supported by the JSPS-CAS Core-University Program in the field of "Plasma and Nuclear Fusion" and performed under the Research Cooperation program in the National Institute of Natural Sciences (NINS).

**References**

[1] R. E. Marrs, M. A. Levine, D. A. Knapp, and J. R. Henderson, Phys. Rev.Lett. **60**, 1715 1988 .
[2] E. D. Donets and V. P. Ovsyanniko, Sov. Phys. JETP **53**, 466 1981 .
[3] G. O'Sullivan, A. Cummings, P. Dunne, K. Fahy, P. Hayden, L. McKinney,N. M. E. Sokell, and J. White, AIP Conf. Proc. **771**, 108 2005 .
[4] H. Ohashi, H. Tanuma, S. Fujioka, H. Nishimura, A. Sasaki, and K. Nishimura,J. Phys.: Conf. Ser. **58**, 235 2007 .
[5] T. Watanabe, T. Kato, I. Murakami, and M. Yamamoto, AIP Conf. Proc.**901**, 215 2007 .
[6] T. Watanabe and the EIS Team: Emission Line Imaging Spectroscopy for Diagnosing of Solar Outer Atmospheres, 2007, Plasma and Fusion Res., Volume 2, S1011-1 - S1011-6.
[7] T. Watanabe, T. Kato, I. Murakami, N. Yamamoto: 2007, Solar and LHD Plasma Diagnostics, AIP conf. ser. 901, Proc. 5th International Conference on Atomic and Molecular Data and Their Applications (ICAMDATA), eds. E. Roueff, pp. 215 - 220.
[8] T. Watanabe, H. Hara, N. Yamamoto, D. Kato, H. A. Sakaue, I. Murakami, T. Kato, N. Nakamura, and P. R. Young: 2008, FeXIII Density Diagnostics in the EIS Observing Wavelengths, submitted to ApJ.
[9] I. Murakami et al, J. Phys. B**39**, 2917 (2006)
[10] T. Watanabe, "Solar and LHD Plasma Diagnostics," 2007, Atomic andMolecular Data and Their Applications, AIP Conf. Proc., 901, pp.215 – 220.
[11] R. Radtke, C. Biedermann, P. Mandelbaum, and J. L. Schwob, J. Phys.:Conf. Ser. **58**, 113 2007 .
[12] N. Nakamura, H. Kikuchi, H. A. Sakaue and T. Watanabe, Rev. Sci. Instrum. 79, 063104 (2008)
[13] N. Nakamura, A. Endo, Y. Nakai, Y. Kanai, K. Komaki, and Y. Yamazaki, Rev. Sci. Instrum. **75**, 3034 2004 .
[14] F. J. Currell, J. Aiken, K. F. Dunn, X. Lu, R. W. McCullough, E. J. Sokell, B. E. ORourke, V. Krastev, and H. Watanabe, Nucl. Instrum. Methods **B205** (2003)230-233
[15] M. Sakurai, F. Nakajima, T. Fukumoto, N. Nakamura, S. Ohtani, and S.Mashiko, J. Phys.: Conf. Ser. **2**, 52 2004 .
[16] NIST online Atomic Database ; http://physics.nist.gov/PhysRefData/ASD/

# Measurement of the $K\beta_2/K\beta_1$ ratio in heliumlike krypton

**P Beiersdorfer[1], M F Gu[1], G V Brown[1], H Chen[1], R Kelley[2], C A Kilbourne[2], F S Porter[2], A J Smith[3], D B Thorn[1]**

[1]Lawrence Livermore National Laboratory, Livermore, CA 94550, USA
[2]NASA Goddard Space Flight Center, Greenbelt, MD 20770, USA
[3]Department of Physics, Morehouse College, Atlanta, GA 30314, USA

E-mail: beiersdorfer@llnl.gov

**Abstract.**
We report the measurement of the $K\beta_2/K\beta_1$ ratio of He-like krypton using the SuperEBIT electron beam ion trap at the Lawrence Livermore National Laboratory. The energy of these lines are about 15 keV, which is twice as high as the energy of such lines measured before. A comparison with theoretical predictions shows poor agreement, confirming the trend uncovered earlier where the measured result is considerably larger than predicted.

## 1. Introduction

In a previous measurement of the intensity ratio of the $1s3p\ ^1P_3 \rightarrow 1s^2\ ^1S_0$ intercombination line, commonly labeled $K\beta_2$, to that of the $1s3p\ ^1P_3 \rightarrow 1s^2\ ^1S_0$ resonance line, commonly labeled $K\beta_1$, it was shown that theoretical predictions underestimated the measured values [1]. This study was carried out for ions with intermediate atomic number $12 \leq Z \leq 26$. The discrepancy with theory was largest for the line ratios from the ions with lowest atomic number, i.e., for magnesium, and diminished for the ratios from the ions with highest atomic number under study, i.e., iron. The measurements were performed on the EBIT-II electron beam ion trap [2] at the Lawrence Livermore National Laboratory. Most of these measurements were carried near the threshold energy for excitation of the $K\beta$ transitions.

Among several reasons for the discrepancy, the effects of polarization of the emitted X-ray lines were mentioned as a possible cause, as the unidirectional beam in an electron beam ion trap creates linearly polarized light. This effect was amplified because the x rays were detected with crystal spectrometers, which have different reflectivities for the two polarization components.

In this paper, we report the measurement of the $K\beta_2/K\beta_1$ ratio of He-like krypton using an x-ray microcalorimeter [3]. This instrument is insensitive to the polarization of the incident photons. The present measurement extends the previous study to an ion with considerably higher atomic number. In fact, the x-ray energy of the $K\beta$ lines of krypton is about twice that of the corresponding lines in iron.

## 2. Experiment

The measurement was performed on the SuperEBIT electron beam ion trap at the Lawrence Livermore National Laboratory. Like in the earlier measurements [1], the photons were detected in the plane perpendicular to the electron beam. Although the microcalorimeter is an energy dispersive device whose quantum efficiency is independent of the polarization degree of the x-ray photons, the linear polarization of the photons given off by the trapped ions affects the measured ratio through the anisotropy factor. However, such a dependence on the degree of polarization is much weaker than the strong effect on a measurement made with a crystal spectrometer.

The measurements were carried out with an electron beam current of about 160 mA and at a beam energy of 86 keV. The spectrum, which was measured with the microcalorimeter over a period of 28 hours and collected during five run days, is shown in Fig. 1. Data from 20 pixels were added together.

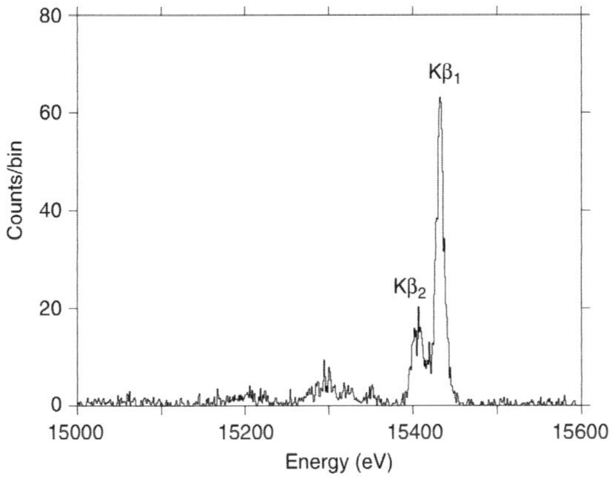


**Figure 1.** X-ray emission of the $n$=3 → $n$=1 transitions of heliumlike krypton measured with the microcalorimeter on SuperEBIT.

The measured ratio is $I_{K\beta_2}/I_{K\beta_2} = 0.37 \pm 0.06$. The uncertainty is about thrice that solely due to counting statistics. The reason is the that the $K\beta_2$ peak is somewhat ill formed, which introduces an amount of arbitrariness in fitting the lines and thus an uncertainty that goes beyond statistical uncertainty. The less than perfect line shape is probably caused by the low count rate of the experiment, which makes proper adding of individual pixels difficult.

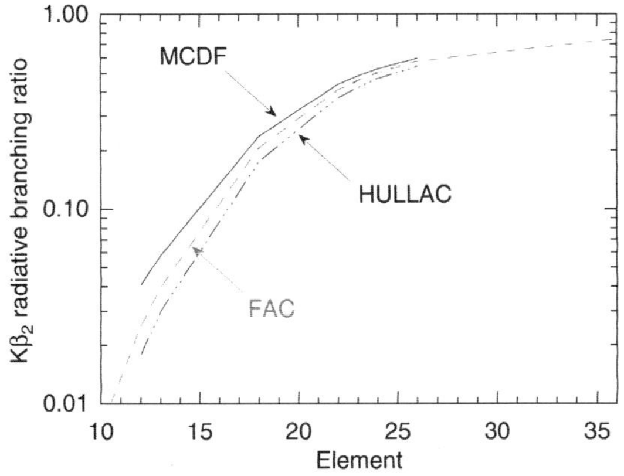


**Figure 2.** Comparison of the radiative branching ratios of the $1s3p\ ^3P_1$ level for producing the $K\beta_2$ transition calculated by our FAC calculations (shown as a red dashed line) with values from MCDF and HULLAC calculations reported in [1].

## 3. Comparison with theory

We have carried out calculations of the atomic data needed for predicting the $I_{K\beta_2}/I_{K\beta_2}$ ratio using the Flexible Atomic Code (FAC) [4]. All levels up to $n = 6$ were included in the collisional radiative model. The calculations include the radiative branching ratio to the $1s^2\ ^1S_0$ level from the $1s2p\ ^3P_1$ level. The results for heliumlike ions between oxygen and krypton are shown in

Fig. 2, where they are compared to values calculated before [1] using two different codes. The agreement with the earlier calculations is excellent.

We have also calculated the radiative branching ratio of the decay of the $1s3p\ ^1P_1$ level. That ratio remains nearly constant from oxygen ($\beta_r$=0.95) to krypton ($\beta_r$=0.93).

We have also calculated the linear x-ray polarization of the krypton $K\beta_1$ and $K\beta_2$ lines as a function of the electron-impact energy. The results are shown in Fig. 3. At the beam energy of our experiment the polarization is $P = 0.270$ for $K\beta_1$ and $P = 0.258$ for $K\beta_2$. In other words, there is no difference in the respective values of polarization, and no adjustments need to be made to account for differences in polarization and angular anisotropies.

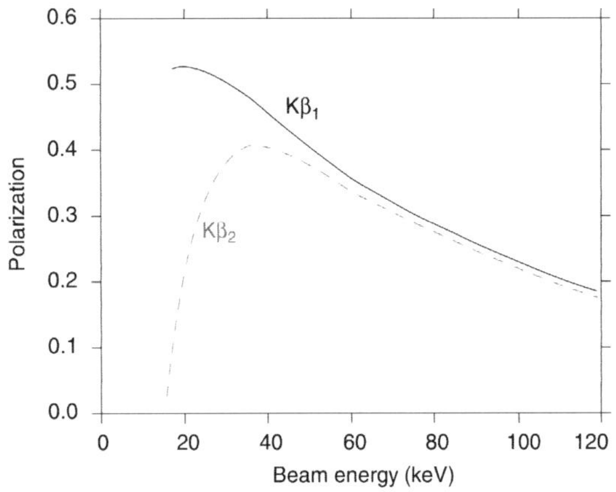


**Figure 3.** Calculated values of the linear polarization of $K\beta_1$ (solid line) and $K\beta_2$ (dashed line) as a function of electron beam energy.

We used FAC to calculate $I_{K\beta_2}/I_{K\beta_2}$ as a function of beam energy, as shown in Fig. 4. The ratio varies near threshold and approaches a constant value at high energy. It assumes a value of 0.22 at 86 keV. This values is about 40% less than measured, as indicated in Fig. 4.

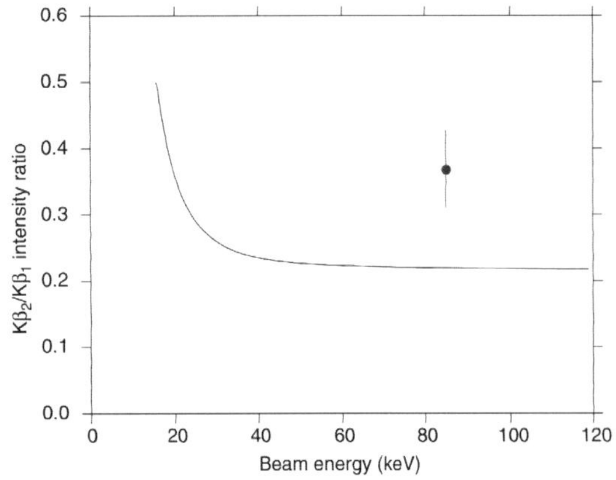


**Figure 4.** Comparison of calculated and measured values for the intensity ratio of $K\beta_1$ and $K\beta_2$.

## 4. Discussion

The measured intensity ratio is significantly larger than predicted. This has been found before in the study of heliumlike ions of magnesium through iron [1]. Our measurement, however, was carried not close to the threshold energy for excitation, as was done in the earlier study, but it was carried out at very high electron energy. in fact, the measurement was carried at a beam energy close to six threshold units. Unfortunately, this means that $Kr^{35+}$ ions were present in the trap during the measurement. Both radiative electron capture from the interaction with the electron beam and charge exchange with neutral background ions may add to the population of the $1s3p\ ^3P_1$ level and thus enhance the $K\beta_2$ line preferrentially compared to the $K\beta_1$ line. Modeling of this enhancement is beyond the scope of the present study, as there is no definitive diagnostic that can quantify the amount of charge exchange occuring in the trap.

Future measurements will need to study the line ratio near threshold, i.e., close to 15,500 eV where no $Kr^{35+}$ ions are produced. Such a measurement was not possible with our setup, as the count rate in the $K\beta$ lines near a beam energy of 15.5 keV would have been too weak to allow for a meaningful addition of the spectra recorded by individual pixels of the microcalorimeter. However, we have now developed new techniques to sum very weak signals [5, 6]. Moreover, the utilization of our EBIT-I facility, where we can achieve a higher beam current at the required beam energy than on SuperEBIT, will increase the amount of signal collected in the future.

## Acknowledgments

This work was performed under the auspices of the U S Department of Energy by Lawrence Livermore National Laboratory under Contract DE-AC52-07NA27344.

## References

[1] Smith A J, Beiersdorfer P, Reed K J, Osterheld A L, Decaux V, Widmann K and Chen M H 2000 *Phys. Rev. A* **62** 012704
[2] Beiersdorfer P 2008 *Can. J. Phys.* **86** 1
[3] Porter F S, Brown G V, Boyce K R, Kelley R L, Kilbourne C A, Beiersdorfer P, Chen H, Terracol S, Kahn S M and Szymkowiak A E 2004 *Rev. Sci. Instrum.* **75** 3772
[4] Gu M F 2008 *Can. J. Phys.* **86** 675
[5] Thorn D B, Brown G V, Clementson J H T, Chen H, Chen M H, Beiersdorfer P, Boyce K R, Kilbourne C A, Porter F S and Kelley R L 2008 *Can. J. Phys.* **86** 241
[6] Thorn D B, Gu M F, Brown G V, Beiersdorfer P, Porter F S, Kilbourne C A and Kelley R L 2008 *Rev. Sci. Instrum.* (in press)

Journal of Physics: Conference Series **163** (2009) 012022 doi:10.1088/1742-6596/163/1/012022

# Survey of the K-shell emission from heliumlike ions with an X-ray microcalorimeter

**P Beiersdorfer[1], G V Brown[1], J H T Clementson[1], M Frankel[1], M F Gu[1], S M Kahn[2], R Kelley[3], C A Kilbourne[3], F S Porter[3], D Thorn[1], E Träbert[1]**

[1]Lawrence Livermore National Laboratory, Livermore, CA 94550, USA
[2]Department of Physics, Stanford University, Stanford, CA 94305, USA
[3]NASA Goddard Space Flight Center, Greenbelt, MD 20770, USA

E-mail: beiersdorfer@llnl.gov

**Abstract.**
The Electron Beam Ion Trap Microcalorimeter Spectrometer (ECS) routinely surveys the K-shell x-ray spectra that fall into the energy range between 200 eV and 14,000 eV. The spectra serve as *in situ* energy references and include the K-shell emission from ions between boron at the low-energy end and krypton at the high end. Example spectra are presented of the $n = 2 \rightarrow n = 1$ emission from heliumlike noble gas ions neon, argon, and krypton, from the heliumlike transition metals iron and nickel, as well as the heliumlike ions of boron, silicon, sulfur, and germanium.

## 1. Introduction

The K-shell emission from highly charged heliumlike ions has played an important role in the spectroscopy of high-temperature plasmas. In magnetic fusion plasmas, for example, the emission from heliumlike argon, titanium, iron, and nickel is routinely observed to determine the central ion temperature [1-3], while the K-shell emission from boron, carbon, nitrogen, and oxygen is measured to study the concentration of indigenous impurities [4-6]. The emission from heliumlike ions of carbon, oxygen, neon, silicon, sulfur, argon, calcium, iron, and nickel is very prominent in the solar corona [7-10] and astrophysical plasmas [12, 13]. The K-shell emission of highly charged ions also dominates the x-ray spectra from comets [14].

The K-shell emission of hydrogenlike and heliumlike ions from an electron beam ion trap (EBIT) has been studied soon after the first such machine, EBIT-I, was put into operation [15]. Early on, the spectra have been studied with solid-state detectors or high-resolution crystal spectrometers, and many of their properties have been determined during these measurements [16-21].

Because the $n = 2 \rightarrow n = 1$ K-shell emission energies are rather well known both theoretically and experimentally, the lines have been used as wavelength references to measure the transition energies of lines from more complex systems. The K-shell emission lines of hydrogenlike and heliumlike ions were first used on tokamaks to measure the wavelengths of L-shell transitions of high-Z ions [22, 23]. The same technique was then applied to measuring the transition energies of neonlike and nickellike ions on the EBIT-I, EBIT-II, and SuperEBIT devices at Livermore

Journal of Physics: Conference Series **163** (2009) 012022 doi:10.1088/1742-6596/163/1/012022

[24-28]. Continuing this tradition, we now use the K-shell emission of highly charged ions as calibration references to determine the energy scale of the microcalorimeter spectrometer used on SuperEBIT. The energy scale is only close to linear in the first half of its working range so that calibration points are especially needed at the higher energies.

In the following we present typical measurements of the K-shell emission of heliumlike ions measured on SuperEBIT with the EBIT calorimeter spectrometer (ECS).

## 2. Experiment and sample spectra

The ECS instrument [29] was implemented on SuperEBIT in 2007. It represents a major upgrade of the earlier x-ray microcalorimeter [30], which was in use up to that time. The ECS consists of a $6 \times 6$ array of x-ray detector elements (pixels), where each detector functions independently. Half of the array utilizes HgTe absorbers that are 8 $\mu$m thick. These pixels are used for analyzing the x-ray emission up to roughly 15 keV. The other half utilizes HgTe absorbers that are 100 $\mu$m thick. These are used to detect x rays with energies up to roughly 60 keV. Here we only present results obtained with the low-energy pixels.

Spectra of the emission from heliumlike germanium (injected by laser blow-off) and krypton (injected as a gas) are shown in Fig. 1. The energies of the lines from these ions provide calibration at the high-end of the low-energy pixels of the ECS. Spectra of the emission from heliumlike iron and nickel are shown in Fig. 2. The resolution of the ECS is about 4.5 eV at 6 keV, which is over twice the resolution achieved in 2000 [31]. There is a $< 2$ eV variation in this number; the value, however, can be larger, if the instrument is operated higher than the typical 50 mK working temperature in order to expand its working range to higher x-ray energies.

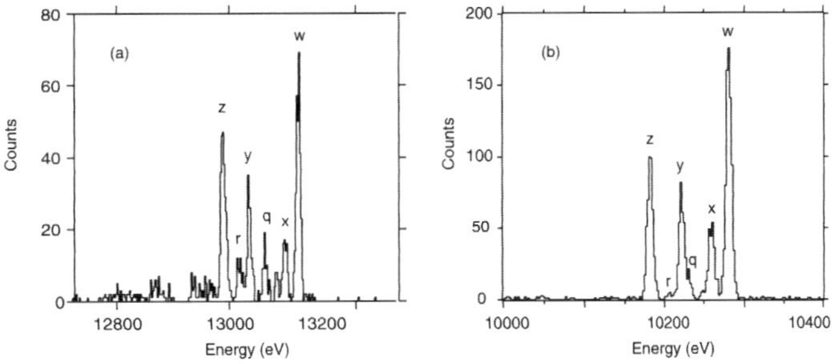


**Figure 1.** X-ray emission of heliumlike (a) krypton and (b) germanium measured with the ECS instrument on SuperEBIT.

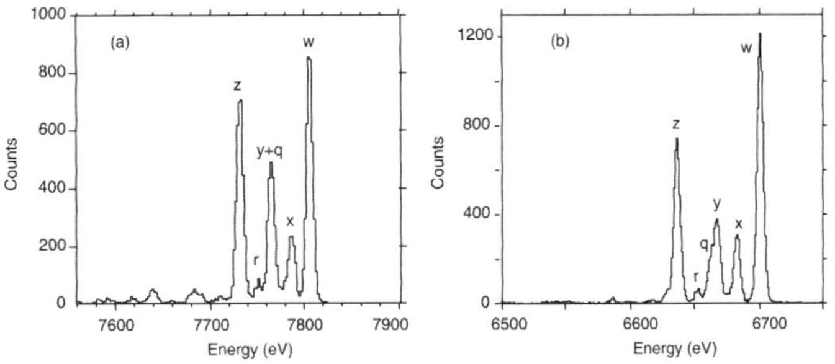


**Figure 2.** X-ray emission of heliumlike (a) nickel and (b) iron measured with the ECS instrument on SuperEBIT.

The spectra show the four heliumlike lines commonly labeled $w$, $x$, $y$, and $z$. These correspond to the transitions $1s2p\ ^1P_1 \rightarrow 1s^2\ ^1S_0$, $1s2p\ ^3P_2 \rightarrow 1s^2\ ^1S_0$, $1s2p\ ^3P_1 \rightarrow 1s^2\ ^1S_0$, and

$1s2p\ ^3S_1 \rightarrow 1s^2\ ^1S_0$, respectively. We have also labeled two of the prominent lithiumlike innershell satellites $q$ and $r$, which correspond to the transitions $1s2s2p\ ^2P_{3/2} \rightarrow 1s^22s\ ^2S_{1/2}$ and $1s2s2p\ ^2P_{1/2} \rightarrow 1s^22s\ ^2S_{1/2}$, respectively.

In Fig. 3 we present sample spectra of the emission from heliumlike oxygen, fluorine, neon, silicon, sulfur, and argon. Oxygen, silicon, and argon are routinely found in SuperEBIT, although their sources are not yet clear - this is especially true for silicon.

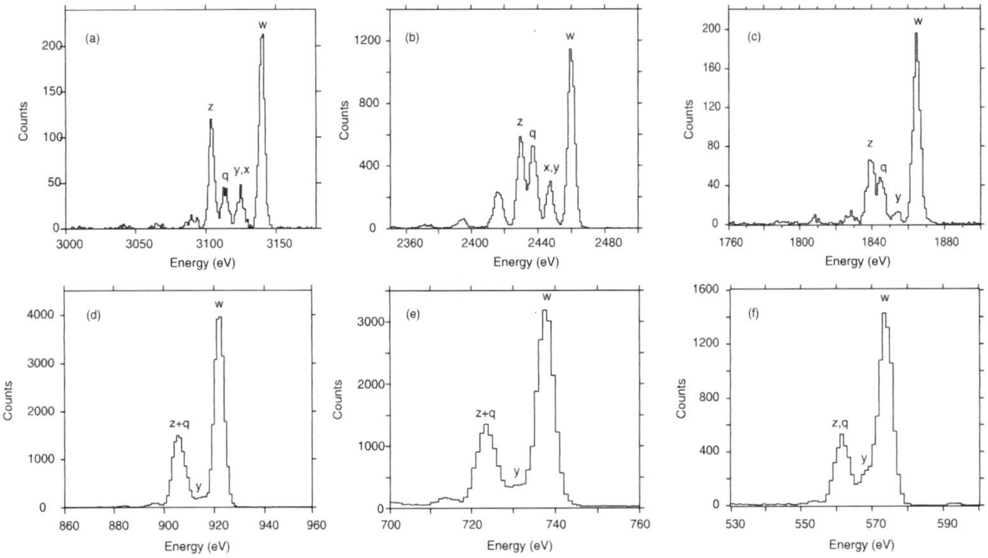


**Figure 3.** X-ray emission of heliumlike (a) argon, (b) sulfur, (c) silicon, (d) neon, (e) fluorine, and (f) oxygen measured with the ECS instrument on SuperEBIT.

The ECS employs ultrathin foils as thermal shields. The combined thickness of these foils is about 2182 Å of polyimide and 1434 Å of aluminum. This means that the ECS is sensitive to photons below the carbon edge. A sample spectrum of the K-shell emission of boron is shown in Fig. 4. The energy of the $1s2p\ ^1P_1 \rightarrow 1s^2\ ^1S_0$ emission of heliumlike boron, i.e. $B^{3+}$, is at 200 eV. The spectrum was obtained by injecting trimethyl borate into SuperEBIT via its gas injection system.

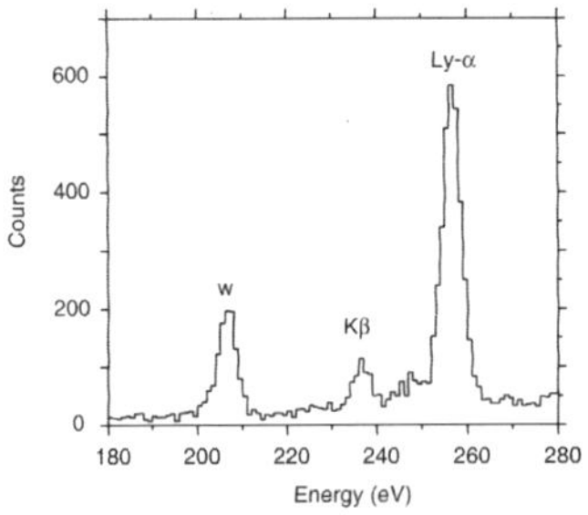


**Figure 4.** X-ray emission of heliumlike boron (line $w$ and its $n$=3 $\rightarrow$ $n$=1 counterpart labeled K$\beta$) and hydrogenlike boron (labeled Ly$\alpha$) measured with the ECS instrument on SuperEBIT.

## 3. Discussion

All spectra were formed by direct electron-impact excitation. The ionization balance varies somewhat from spectrum to spectrum, as seen in the relative amplitudes of lines $q$ and $w$. It is interesting to see that the position of $q$ moves closer to $w$ for ions with higher atomic number $Z$. Also the intercombination lines get stronger as $Z$ increases. Line $x$ is basically absent for ions with $Z$ as high as neon, but it becomes very prominent in argon and above. Line $z$ is strong, even in low-$Z$ spectra. This is in part because it is excited by innershell ionization, i.e., by the process $1s^22s + e \rightarrow 1s2s + 2e$.

Although krypton is the element with the highest $Z$ considered here, we point out that the emission from heliumlike ions as high as praseodymium has now been measured with a microcalorimeter on SuperEBIT [32].

**Acknowledgments**
This work was performed under the auspices of the U S Department of Energy by Lawrence Livermore National Laboratory under Contract DE-AC52-07NA27344 and was supported in part by NASA's Astronomy and Physics Research and Analysis Program via grants to LLNL, Stanford, and GSFC.

**References**

[1] Bitter M, von Goeler S, Horton R, Goldman M, Hill K, Sauthoff N R and Stodiek W 1979 *Phys. Rev. Lett.* **42** 304
[2] Källne E, Källne J, Marmar E S and Rice J E 1985 *Phys. Scr.* **31** 551
[3] Bitter M. *et al* 2003 *Phys. Rev. Lett.* **91** 265001
[4] Beiersdorfer P, Bitter M, Roquemore L, Lepson J K and Gu M-F 2006 *Rev. Sci. Instrum.* **77** 10F306
[5] Chowdhuri M B S M and Goto M 2008 *Appl. Optics* **47** 135
[6] Stratton B C, Bitter M, Hill K W, Hillis D L and Hogan J T 2008 *Fusion Sci. Techn.* **53** 431
[7] Doschek G A, Feldman U, Landecker P B and McKenzie D L 1981 *Astrophys. J.* **249** 372
[8] Faucher P, Loulergue M, Steenman-Clark L and Volonté S 1983 *Astron. Astrophys.* **118** 147
[9] Doschek G A and Cowan R D 1984 *Astrophys. J. (Suppl.)* **56** S67
[10] Sylwester J, Sylwester B and Phillips K J H 2008 *Astrophys. J. (Lett.)* **681** L117
[11] Sylwester J, Sylwester B, Landi E, Phillips K J H and Kuznetsov V D 2008 *Adv. Space Res.* **42** 838
[12] Ness J-U, Mewe R, Schmitt J H M M, Raassen A J J, Porquet D, Kaastra J S, van der Meer R L J, Burwitz V and Predehl P 2001 *Astron. Astrophys.* **367** 282
[13] Mewe R, J J Raassen, Kaastra J S, van der Meer R L J and Brinkman A C 2003 *Adv. Space Res.* **32** 2059
[14] Krasnopolsky V A, Greenwood J and Stancil P C 2004 *Space. Sci. Rev.* **113** 271
[15] Beiersdorfer P 2008 *Can. J. Phys.* **86** 1
[16] Knapp D A, Marrs R E, Levine M A, Bennett C L, Chen M H, Henderson J R, Schneider M B and Scofield J H 1989 *Phys. Rev. Lett.* **62** 2104
[17] Beiersdorfer P, Chen M H, Marrs R E, Schneider M B and Walling R S 1991 *Phys. Rev. A* **44** 396
[18] Chantrenne S, Beiersdorfer P, Cauble R and Schneider M B 1992 *Phys. Rev. Lett.* **69** 265
[19] Wong K, Beiersdorfer P, Chen M H, Marrs R E, Reed K J, Scofield J H, Vogel D A and Zasadzinski R 1993 *Phys. Rev. A* **48** 2850
[20] Widmann K, Beiersdorfer P, Decaux V and Bitter M 1996 *Phys. Rev. A* **53** 2200
[21] Beiersdorfer P *et al* 1996 *Phys. Rev. A* **53** 3974
[22] Beiersdorfer P, Bitter M, von Goeler S, Cohen S, Hill K W, Timberlake J, Walling R S, Chen M H, Hagelstein P L and Scofield J H 1986 *Phys. Rev. A* **34** 1297
[23] Beiersdorfer P *et al* 1988 *Phys. Rev. A* **37** 4153
[24] Beiersdorfer P, Chen M H, Marrs R E and Levine M A 1990 *Phys. Rev. A* **41** 3453
[25] Beiersdorfer P 1991 *Nucl. Instrum. Methods* **B56/57** 1144
[26] Elliott S, Beiersdorfer P and Nilsen J. 1994 *Phys. Scr.* **49** 556
[27] Beiersdorfer P, Elliott S R and Nilsen J 1994 *Phys. Rev. A* **49** 3123
[28] Nilsen J *et al* 1994 *Phys. Rev. A* **50** 2143
[29] Porter F S *et al* 2008 *Can. J. Phys.* **86** 231
[30] Porter F S, Brown G V, Boyce K R, Kelley R L, Kilbourne C A, Beiersdorfer P, Chen H, Terracol S, Kahn S M and Szymkowiak A E 2004 *Rev. Sci. Instrum.* **75** 3772
[31] Porter F S *et al* 2000 *Proc. SPIE* **4140** 407
[32] Thorn D B, Brown G V, Clementson J H T, Chen H, Chen M H, Beiersdorfer P, Boyce K R, Kilbourne C A, Porter F S and Kelley R L 2008 *Can. J. Phys.* **86** 241

# ANALYSIS OF EUV SPECTRA FROM HIGHLY CHARGED IRON IONS WITH A COMPACT EBIT

**N. Yamamoto[1], H. A. Sakaue[2], D. Kato[2], I. Murakami[2], T. Kato[2], N. Nakamura[3], E. Watanabe[3] H. Nishimura[1] and T. Watanabe[4]**

[1] Institute of Laser Engineering, Osaka University, Suita, Osaka 565-0871, Japan
[2] National Institute for Fusion Science, Toki, Gifu 509-5292, Japan
[3] Institute for Laser Science, The University of Electro-Communications, Chofu, Tokyo 182-8585, Japan
[4] National Astronomical Observatory of Japan, Mitaka, Tokyo, 181-8588, Japan

yamamoto-no@ile.osaka-u.ac.jp

**Abstract**. For benchmark tests of the spectral analysis tool for solar plasma, applying the theoretical plasma radiation model to the laboratory plasma is important. In this paper, a compact-EBIT (Electron Beam Ion Trap) was adopted as the laboratory plasma, and extreme UV spectra of Fe ions were measured by the flat-field grazing-incidence spectrometer with no slit. In order to analyze the spectra from plasma generated by mono-energetic electron beam, our model was revised in term of an electron velocity distribution. Consequently, the measured spectra by the compact-EBIT were analyzed by our revised model.

## 1. Introduction

Solar observations have been progressed by HINODE satellite that was launched on 2006 September 23 JST [1]. HINODE satellite has a high precise spectrometer EIS (EUV Imaging Spectrometer). EIS measure EUV spectra in the range of 170-210Å and 250-290Å. In these wavelength ranges, many M-shell Fe ion lines and some L-shell ion lines are measured. These ions are found in the electron temperature of several ten eV to few hundred eV (few keV in flare). M-shell Fe ion lines are often applied for density diagnostics in the region between transition and corona of the Sun. Therefore, the plasma diagnostic tool for line intensity analysis is important and the tool should be evaluated by applying to some laboratory plasmas where the parameters are under control.

In the previous study, the benchmark tests by CHIANTI code [2] and our original model including the most recent atomic data have been carried out by applying to EIS's spectra from solar plasma [3]. Our model was also applied to LHD (large helical device) plasma at NIFS (National Institute for Fusion Science) as the benchmark test [4]. In this study, spectra measured by an EBIT (electron beam ion trap) [5] are adopted as a benchmark test. The generated plasma by EBIT has a simpler plasma parameter profile than magnetic confined plasma. However, an electron energy of the Tokyo-EBIT is much too high to produce M-shell iron ions. Therefore we build a new compact-EBIT [6]. In this paper, we present the analysis of the first data obtained with the compact-EBIT.

## 2. Experiments

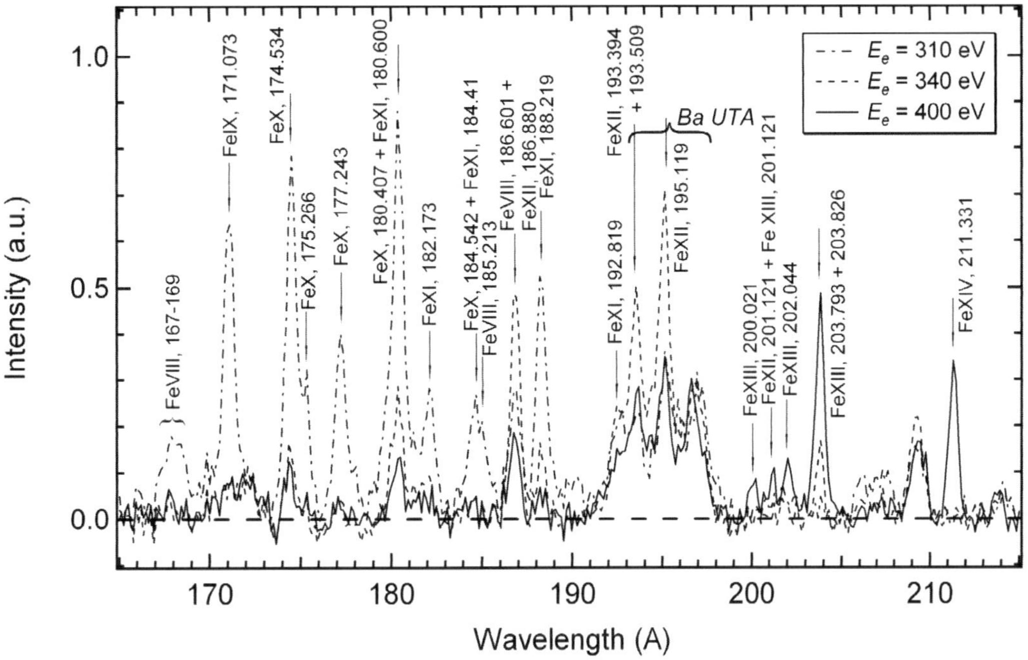


**Figure 1.** The measured spectra by the compact-EBIT for different electron energy with line identifications. Solid, dashed and dot-dashed lines are spectra by electron beam energy $E_e$ of 400 eV, 340 eV and 310 eV, respectively.

The compact-EBIT was developed in order to generate plasmas at lower electron energies than in the Tokyo-EBIT. The experimental setup and conditions are found in ref. [7]. In these experiments, Fe was injected by using ferrocene ($Fe(C_5H_5)_2$)). The flat-field grazing-incidence spectrometer with no slit consists of 1200 grooves/mm laminar-type replica diffraction grating (30-002, Shimazu Corporation) and the back-illuminated CCD camera (Princeton Instruments PIXIS-XO:400B). EUV spectra in the range of 130-300Å can be measured by the spectrometer.

Figure 1 shows the measured spectra at electron energy of 310 eV, 340 eV and 400 eV. The electron current of the cathode in the compact-EBIT is 20mA. With decreasing electron energy, spectral lines of lower ionized ions become strong. Spectral resolution of the present spectra was FWHM ~ 0.83 Å by Gaussian function fitting of some lines. Since the spectrometer has no slit, the resolution depends on plasma size of the compact-EBIT. Spectral lines were identified by using the line database of NIST [8] and CHIANTI. In the measured spectra by electron energy of 310 eV, lines of Fe IX-XI are strong. Broad lines of 192-198Å belong to a UTA (Unresolved Transition Array) of Ba ions from the cathode.

## 3. Collisional-radiative model

### 3.1. Electron velocity distribution

We had constructed a collisional-radiative model for Fe ion lines analysis [4]. In this study, we revised the model about electron velocity distribution for spectra analysis of the EBIT. An electron velocity in the compact-EBIT has a vector of the beam direction and electron energy is monochromatic, i.e. it has no temperature. It means that electron velocity distribution to calculate transition rates is not Maxwellian distribution. As electron velocity distribution $f$, delta function,

$$f(v) = \delta(v - v_e) \quad (1)$$

was adopted, where $v_e$ is electron velocity at electron beam energy of $E_e$. The rate coefficient $C_e$ on the velocity average of a cross section $\sigma$ is $C_e = \langle v\sigma \rangle = v_e \sigma(v_e)$ and the formula becomes very simple.

### 3.2. Atomic data

Our model includes the following number of energy levels with fine structure: 1073 for Fe VIII, 1084 for Fe IX, 1189 for FeX, 1287 for Fe XI, 917 for Fe XII, 332 for Fe XIII and 243 for Fe XIV ($3 \le n \le 5$ & $0 \le l \le 4$). Level energies of lower excited states ($n \le 3$) were corrected by NIST database. Atomic data were calculated by HULLAC code [9]. Since the electron velocity distribution is a delta function, cross sections at beam energy are important. The effects of resonance in the excitation cross section at threshold energy range do not affect excitation rates. Therefore we did not use cross sections by the recent relative R-matrix calculation. On the other hand, for radiative transition rates the data of the recent relative R-matrix calculation (ref. [10] for Fe X, ref. [11] for Fe XI and ref. [12] for Fe XIII) were used. In our model, atomic processes include excitation / de-excitation / ionization by electron impact, radiative transition, radiative recombination, three-body recombination.

### 3.3. Rate equation and quasi-steady state solutions

The spectral line intensity is calculated by using the population density $N_i$ of an excited states $i$ under certain conditions. The population density is determined by including various atomic processes. The population densities are derived as follows,

$$\frac{dN_i}{dt} = -N_i \sum_j W_{ij} + \sum_j W_{ji} N_j = 0 \quad (2)$$

where $W_{ij}$ is total transition rate from an $i$-state to a $j$-state. The population density is able to re-write as

$$N_i = r_i^{ion} N_e N_Z + r_i^{rec} N_e N_{Z+1} \quad (3)$$

where $r_i^{ion}$ and $r_i^{rec}$ are population density coefficients of ionization and recombination components, $N_z$ and $N_{z+1}$ are ion densities of ionization degree of $+Z$ and $+(Z+1)$, respectively. By eq. (2) and (3), differential equations of $N_i$ can be changed to differential equations of $r_i$ and ion densities $N_Z$ become arbitrary parameters. In this paper, population densities of ionization components only are calculated

## 4. Results and Discussions

Figure 2 shows comparisons of the measured spectra with the calculated spectra. In addition, ion abundances for Fe VIII-XIV based on the calculated spectra are also shown. In the measured spectra with the electron energy of 310 eV, lines of Fe XIII-XIV do not appear. In the range of 165-190Å, line spectral structure can be re-produced by our model. Electron density was estimated by a line ratio, 175.27 / 177.24Å of Fe X. The sum of the ion abundances of Fe X and Fe XI were about 70%. The measured spectra with electron energy of 400 eV have weak lines of Fe VIII-XI. Ion abundance of Fe XII was determined by line of 186.88Å because of UTA of Ba in range of 192-198Å. Electron density was estimated to be 1-10 $\times$ $10^{10}$ $cm^{-3}$ by line intensities of 186.88Å / 195.12Å (with UTA of Ba) of Fe XII and 195.53 (with UTA of Ba) / 203.8Å of Fe XIII. The ion abundance of Fe XIV was the largest, because of electron energy of 400 eV corresponds to ionization energy of FeXIV. Although line intensities of Fe XIII in the measured spectra by electron energy of 400 eV are strong, the ion abundance of Fe XIII was <10%.

## 5. Summary

In this paper, we identified spectral line of the measured spectra by the compact-EBIT in detail and the measured spectra were fitted by our model for a mono-energetic electron beam. Consequently, electron density and the ion abundance were estimated.

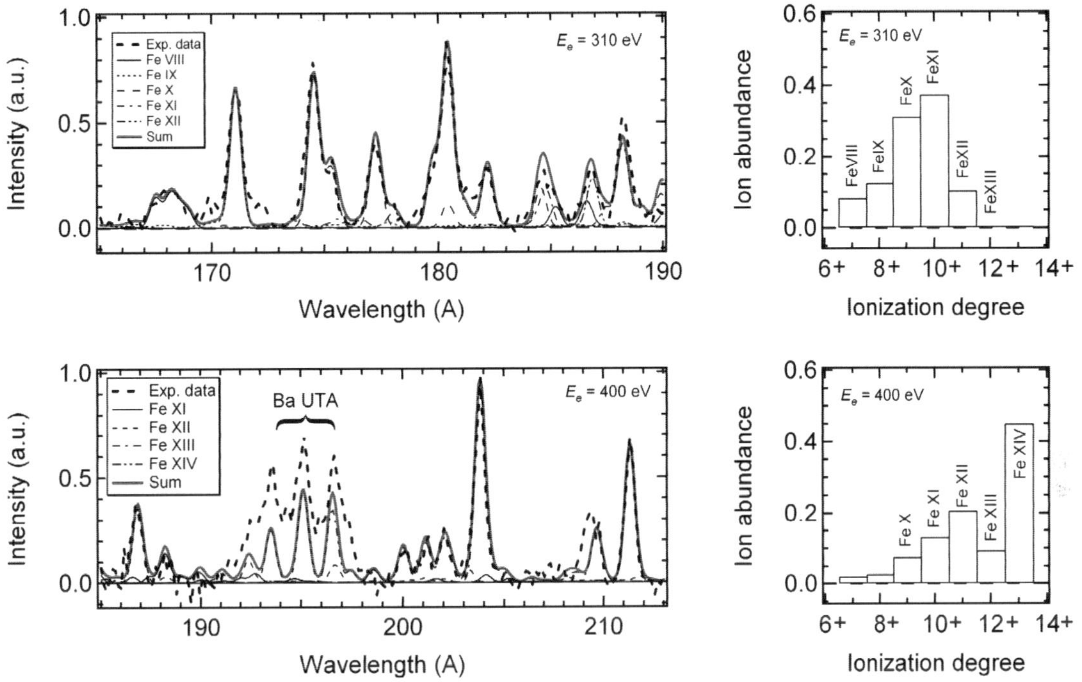


**Figure 2.** Comparison with the measured spectra and the calculated spectra at electron beam energy of 310 eV (top panel) and 400 eV (bottom panel). In the calculated spectra, electron densities are $1 \times 10^{11}$ cm$^{-3}$ for 310 eV and $5 \times 10^{10}$ cm$^{-3}$ for 400 eV (corresponding to electron current of ~20 mA). Left figures show comparisons with spectra and right figures show the ion abundance of FeVIII-XIV in calculation by our model.

**Acknowledgements**
This work was supported by the Solar and Heliospheric Physics Program of the National Aeronautics and Space Administration under grant NNX07AH986. Travel support was provided by National Science Foundation grant INT-0300708.

**References**

[1] T. Watanabe, et al., Publ. Astron. Soc. Japan **59** S669 (2007).
[2] G. Del Zanna, et al.,CDS Software Note, N. 50 (1997).
[3] T. Watanabe et al., ApJ., to be published.
[4] N. Yamamoto et al., ApJ. **689**, 646 (2008).
[5] R. E. Marrs, M. A. Levine, D. A. Knapp, and J. R. Henderson, Phys Rev. Lett. **60**, 1715 (1988).
[6] N. Nakamura, H. Kikuchi, H. A. Sakaue, and T. Watanabe, Rev. Sci. Instrum. **79**, 063104 (2008).
[7] H. A. Sakaue et al. this conference series, submitted.
[8] T. Shirai, J. Sugar, A. Musgrove and W. L. Wiese, J. Phys. Chem. Ref. Data, Monograph **8** (2000). (NIST online Atomic database ; http://www.physics .nist.gov/PhysRefData/ASD/).
[9] A. Bar-Shalom, M. Klapisch, J. Oreg, J. Quant. Spectr. Rad. Transf. **71**, 179 (2001).
[10] K. M. Aggarwal and F. P. Keenan, A&A **427**, 763 (2004).
[11] K. M. Aggarwal and F. P. Keenan, Mon. Not. R. Astron. Soc. **338**, 412 (2003).
[12] K. M. Aggarwal and F. P. Keenan, A&A **429**, 1117 (2005).

Journal of Physics: Conference Series **163** (2009) 012024 doi:10.1088/1742-6596/163/1/012024

# Energy levels, oscillator strengths, and effective collision strengths for transitions in Fe XIV

**K M Aggarwal**[1], **T Kato**[2] **and F P Keenan**[1]

[1]Astrophysics Research Centre, School of Mathematics and Physics, Queen's University Belfast, Belfast BT7 1NN, Northern Ireland, UK

[2] National Institute for Fusion Science, Oroshi-cho, Toki, Gifu, 509-5292 Japan

E-mail: K.Aggarwal@qub.ac.uk

**Abstract.** In this paper we report atomic data, namely energy levels, radiative rates, lifetimes, collision strengths, and excitation rate coefficients, for transitions among 162 levels of the ($1s^22s^22p^6$) $3s^23p$, $3s3p^2$, $3s^23d$, $3p^3$, 3s3p3d, $3p^23d$, $3s3d^2$, $3p3d^2$, $3d^3$, 3s3p4s, and $3s^24\ell$ configurations of Fe XIV. For the calculations of energy levels and radiative rates both the GRASP and FAC codes have been adopted, whereas for the collision strengths only FAC is employed.

## 1. Introduction

Emission lines of Fe XIV have been widely observed in a variety of solar and other astrophysical plasmas, and over a wide wavelength region varying from optical to extreme ultra-violet. The strongest observed forbidden transition is from the coronal green line ($3s^23p$) $^2P^o_{1/2}$ - $^2P^o_{3/2}$ with the wavelength 5303 Å. Many of the line pairs are density and/or temperature sensitive, and hence provide useful information about physical conditions of the plasmas. However, to reliably analyse observations, atomic data are required for many parameters, including energy levels, radiative rates (A- values), and excitation rates. Since experimental data are not available, except for some energy levels, theoretical results are required.

Considering the importance of Fe XIV many calculations have been performed in the past, particularly for the A- values, such as by Huang [1], Storey et al. [2], Safronova et al. [3], Gupta and Msezane [4], Froese-Fischer et al. [5], Dong et al. [6], and Wei et al. [7]. However, most of these calculations are confined to transitions among the lowest 40 levels of the $n$ = 3 configurations, although Gupta and Msezane and Wei et al. have also included some levels of the $n$ = 4 configurations. Similarly, in a more recent paper Tayal [8] has reported energy levels, radiative rates, and effective collision strengths for *some* transitions among the lowest 59 levels of the ($1s^22s^22p^6$) $3s^23p$, $3s3p^2$, $3s^23d$, $3p^3$, 3s3p3d, and $3p^23d$ configurations of Fe XIV. Furthermore, most of these workers have reported A- values for the electric dipole (E1) transitions alone, whereas in the modelling and diagnostics of plasmas corresponding results are also required for other types of transitions, namely electric quadrupole (E2), magnetic dipole (M1), and magnetic quadrupole (M2). Similarly, for the modelling of plasmas atomic data for *all* transitions are preferable. Therefore, the *aim* of the present work is to extend and improve upon the earlier available calculations, apart from reporting a *complete* set of results for *all* transitions, which can be confidently applied in plasma modelling.

## 2. Calculations

For our calculations we have adopted the GRASP (general-purpose relativistic atomic structure package) code, originally developed by Grant et al. [9] but revised by Dr. P. H. Norrington. It is a fully relativistic code, based on the $jj$ coupling scheme. Further relativistic corrections arising from the Breit interaction and QED effects have also been included. Additionally, we have used the option of *extended average level* (EAL), in which a weighted (proportional to $2j+1$) trace of the Hamiltonian matrix is minimized. This produces a compromise set of orbitals describing closely lying states with moderate accuracy. A series of calculations have been performed with increasing amount of CI (configuration interaction) included with up to $n = 5$ orbitals, which generate up to 332 levels. However, the presentation of the results is confined to the lowest 162 levels of the $3s^23p$, $3s3p^2$, $3s^23d$, $3p^3$, 3s3p3d, $3p^23d$, $3s3d^2$, $3p3d^2$, $3d^3$, 3s3p4s, and $3s^24\ell$ configurations alone. Among these the lowest 135 levels were included by Tayal [8], who adopted the multi-configuration Hartree-Fock (MCHF) code of Froese-Fischer [10] for calculating energy levels and A- values, and the $R$-matrix code of Berrington et al. [11] for calculating collision strengths ($\Omega$) and subsequently the effective collision strengths ($\Upsilon$).

Additionally, in order to make a rigorous accuracy assessment of our results, we have also performed calculations from the *Flexible Atomic Code* (FAC) of Gu [12], available from the website `http://kipac-tree.stanford.edu/fac`. This is a fully relativistic code and provides a variety of atomic parameters. Furthermore, the code yields results comparable to those obtained from other atomic structure codes such as CIV3 and GRASP - see, for example, Aggarwal et al. [13] and references therein, and scattering codes such as $R$-matrix - see, for example, Aggarwal et al. [14] and references therein. Furthermore, FAC is based on the well-known and widely-used *distorted-wave* (DW) method and yields results for collision strengths ($\Omega$) comparable with those from the $R$-matrix method, particularly for the allowed transitions, as demonstrated in several of our earlier papers on a variety of ions, including many of iron, i.e. Fe XI - Fe XXVI.

As from GRASP, with FAC also we have performed a series of calculations with increasing amount of CI with up to $n = 5$ orbitals, generating a maximum of 3146 levels. However, for the calculations of $\Omega$ and $\Upsilon$ values, only 162 levels of the $3s^23p$, $3s3p^2$, $3s^23d$, $3p^3$, 3s3p3d, $3p^23d$, $3s3d^2$, $3p3d^2$, $3d^3$, 3s3p4s, and $3s^24\ell$ configurations have been considered. These calculations will provide a direct comparison with the most recent results of Tayal [8].

## 3. Energy levels, radiative rates, and lifetimes

Our energy levels from GRASP and FAC agree with each other in both magnitude and orderings, and the agreement with the experimental values compiled by NIST (`http://physics.nist.gov/PhysRefData`) is also highly satisfactory, generally within 1%. Similarly, the agreement with the energy levels of Tayal [8] is generally satisfactory, but our energy for the $3s^23p$ $^2P^o_{3/2}$ level is particularly closer to the experimental result, whereas Tayal's value is lower by 6%. Similarly, the two sets of A- values from GRASP and FAC agree within 20% for a majority of strong transitions ($f \geq 0.01$), although differences are larger for the weaker ones. Tayal has reported A- values for only some of the transitions among the lowest 59 levels alone, and there is no discrepancy with his results for the strong transitions. A complete set of A- values among 162 levels and for all four types of transitions can be obtained on request from KMA (K.Aggarwal@qub.ac.uk). Furthermore, detailed results for transitions among the lowest 332 levels, along with lifetimes, are reported in a separate paper by Aggarwal et al. [15].

## 4. Collision strengths and effective collision strengths

We have calculated values of $\Omega$ (which is related to the better known quantity collision cross section $\sigma$ as $\Omega = {k_i}^2\omega_i\sigma_{ij}(\pi {a_0}^2)$, where ${k_i}^2$ is the electron energy and $\omega_i$ is the statistical weight of the lower level $i$) for all the 13 041 transitions among the above listed 162 levels of Fe XIV over a wide energy range up to $\sim$ 160 Ryd. This energy range is similar to that of Tayal [8], who calculated values of $\Omega$ up to E = 150 Ryd. No comparisons can be made with the calculations of Tayal, because he has not reported values of $\Omega$. However, comparisons can be made for $\Upsilon$ values, which are obtained after integrating the $\Omega$ values over a *Maxwellian* distribution of electron velocities. A comparison of our values of $\Upsilon$ with those of Tayal in the common temperature range of $3\times10^5 \leq T_e \leq 10^7$ K is generally satisfactory for the *allowed* transitions, as shown in Fig. 1 for three resonance (allowed) transitions, namely 1-8 ($3s^23p\ ^2P^o_{1/2}$ - $3s3p^2\ ^2S_{1/2}$), 1-9 ($3s^23p\ ^2P^o_{1/2}$ - $3s3p^2\ ^2P_{1/2}$), and 1-10 ($3s^23p\ ^2P^o_{1/2}$ - $3s3p^2\ ^2P_{3/2}$). Some anomaly between the two independent calculations, particularly towards the lower end of the temperature range, is perhaps due to the inclusion of resonances by Tayal. Since resonances for allowed transitions are often not prominent, their effect on the calculations of $\Upsilon$ is confined to very low temperatures alone. Similarly the slightly anomalous behaviour of the $\Upsilon$ values of Tayal towards the higher temperatures for these transitions is perhaps due to the limited range of partial waves with angular momentum $J \leq 25$ included by him. This range of partial waves is insufficient for the convergence of $\Omega$ for a majority of the allowed and some forbidden transitions, particularly those among the excited levels. Since he has accounted for the contribution of higher neglected partial waves through a top-up, there is no major discrepancy between the two calculations for a majority of the *common* allowed transitions, although the $\Upsilon$ values differ by $\sim$ 10% over the entire temperature range for some of the transitions, such as 1-8. A part of the differences are due to the corresponding differences in the f- values. As an example, our f- value for the 1-8 transition is 0.199 whereas that of Tayal is 0.191, i.e. lower by 4%.

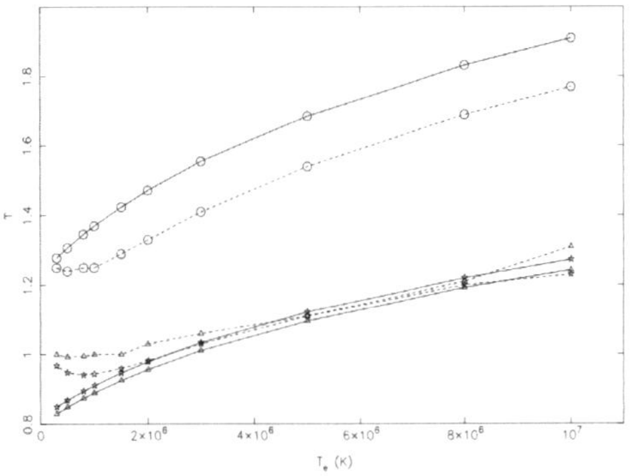

**Figure 1.** Comparison of effective collision strengths between the present work (continuous curves) and of Tayal [8] (broken curves) for the 1-8 (circles: $3s^23p\ ^2P^o_{1/2}$ - $3s3p^2\ ^2S_{1/2}$), 1-9 (triangles: $3s^23p\ ^2P^o_{1/2}$ - $3s3p^2\ ^2P_{1/2}$), and 1-10 (stars: $3s^23p\ ^2P^o_{1/2}$ - $3s3p^2\ ^2P_{3/2}$) *resonance* transitions of Fe XIV.

In Fig. 2 we compare the $\Upsilon$ values for three transitions among the excited levels, namely 2-10 ($3s^23p\ ^2P^o_{3/2}$ - $3s3p^2\ ^2P_{3/2}$), 2-12 ($3s^23p\ ^2P^o_{3/2}$ - $3s^23d\ ^2D_{5/2}$), and 5-27 ($3s3p^2\ ^4P_{5/2}$ - $3s3p3d\ ^4D^o_{5/2}$). For these transitions also there is no (major) discrepancy with the corresponding results of Tayal [8], particularly at higher temperatures. However, differences between the two calculations are significant for a majority of the *forbidden* and some *intercombination* transitions, because Tayal has resolved resonances in the thresholds region whereas we have not. Resonances often enhance the values of $\Upsilon$, by up to an order of magnitude (or even more), depending upon the transition and/or the temperature. Therefore, for the transitions for which Tayal has reported results ($\sim$ 5%), his values of $\Upsilon$ may be preferable, but only for the forbidden transitions. Since his results for both A and $\Upsilon$ values are restricted to a subset of the total 13 041 transitions, these are of limited applications. Therefore, under the circumstances, a complete set of our results for *all* transitions can be confidently applied for the modelling of plasmas. The entire set of data for all transitions for A and $\Upsilon$ values over a wide range of temperatures are reported in a separate paper [15].

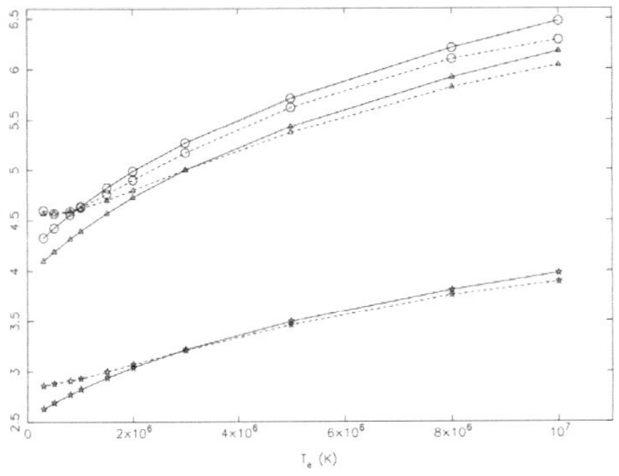

**Figure 2.** Comparison of effective collision strengths between the present work (continuous curves) and of Tayal [8] (broken curves) for the 2-10 (circles: $3s^23p$ $^2P^o_{3/2}$ - $3s3p^2$ $^2P_{3/2}$), 2-12 (triangles: $3s^23p$ $^2P^o_{3/2}$ - $3s^23d$ $^2D_{5/2}$), and 5-27 (stars: $3s3p^2$ $^4P_{5/2}$ - $3s3p3d$ $^4D^o_{5/2}$) transitions among excited levels of Fe XIV.

## 5. Conclusions

In this paper results for energy levels, radiative rates, and effective collision strengths for transitions in Fe XIV have been discussed. For calculating energy levels and A- values, both GRASP and FAC codes have been used whereas for the $\Upsilon$ values FAC alone is adopted. Energy levels and A- values agree in general with the already available data in the literature, but the present results cover a wider range of levels and hence the transitions. Similarly, for $\Upsilon$ values there is a general agreement for many allowed transitions with the most recent calculations of Tayal [8], but differences are notable for the forbidden transitions, because resonances have not been resolved in the present work. Detailed results in numerical form for all atomic parameters can be obtained on request from KMA (K.Aggarwal@qub.ac.uk). Furthermore, a larger calculation involving 332 lowest levels of Fe XIV has also been recently reported by Aggarwal et al. [15]. This paper contains detailed comparisons as well as a complete set of data for all atomic parameters.

## Acknowledgments

This work has been financed by the EPSRC and STFC of UK. FPK is grateful to AWE Aldermaston for the award of a William Penney Fellowship and KMA would like to acknowledge the hospitality of NIFS enjoyed during a visit in November 2007.

## References

[1] Huang K N 1989 *At. Data Nucl. Data Tables* **34** 1
[2] Storey P J, Mason H E and Young P R 2000 *Astron. Astrophys. Supl.* **141** 285
[3] Safronova U I, Sataka M, Albritton J R , Johnson W R and Safronova M S 2002 *Phys. Rev.* **A 65** 022507
[4] Gupta G P and Msezane A Z 2005 *At. Data Nucl. Data Tables* **89** 1
[5] Froese-Fischer C and Tachiev G 2006 *At. Data Nucl. Data Tables* **92** 607
[6] Dong C Z, Kato T, Fritsche S and Koike F 2006 *Month. Not. R. Astron. Soc.* **369** 1735
[7] Wei H L, Zhang H, Ma C W, Zhang J Y and Cheng X L 2008 *Phys. Scr.* **77** 035301
[8] Tayal S S 2008 *Astrophys. J. Supl.* **178** 359
[9] Grant I P, McKenzie B J, Norrington P H, Mayers D F and Pyper N C 1980 *Comput. Phys. Commun.* **21** 207
[10] Froese-Fischer C 1991 *Comput. Phys. Commun.* **64** 369
[11] Berrington K A, Eissner W B and Norrington P H 1995 *Comput. Phys. Commun.* **92** 290
[12] Gu M F 2003 *Astrophys. J.* **582** 1241
[13] Aggarwal K M, Tayal V, Gupta G P and Keenan F P 2007 *At. Data Nucl. Data Tables* **93** 615
[14] Aggarwal K M, Hamada K, Igarashi A, Jonauskas V, Keenan F P and Nakazaki S 2008 *Astron. Astrophys.* **484** 879
[15] Aggarwal K M, Kato T, Keenan F P and Murakami I 2009 *Astron. Astrophys.* submitted

# Study of inter sub-shell and inter shell electron correlations in 4d open-shell heavy atomic ions

**Fumihiro Koike**

Physics Laboratory, School of Medicine, Kitasato University 1-15-1 Kitasato,Sagamihara 228-8555 Japan

E-mail: koikef@kitasato-u.ac.jp

**Abstract.** The effect of correlations between $4p$, $4d$, and $4f$ has been studied extensively. The characteristic spectral structures of $4p-4d$ and $4d-4f$ optical transitions, due to the unique structures of $N=4$ open sub-shells in heavy atomic ions, have been studied theoretically. To gain an insight of this effect, a series of careful MCDF calculations for $4d^q$ ($q=0$ to 10) atomic ions with atomic numbers $Z=48$ to 56 has been carried out. The difference of orbital energy differences between $4p$ and $4d$ orbitals and $4d$ and $4f$ orbitals coincidently falls within the range of a few % for almost all the atomic ions investigated.. The $4p^6 4d4f$ and $4p^5 4d^3$ configurations may mix strongly, and the optical $4p-4d$ and $4d-4f$ transitions may take place coherently, providing us with quite a peculiar EUV emission spectrum. The effect of spectral narrowing and shift is expected to be quite common to the atomic specieds with the atomic numbers in the range $Z=48$ to 56.

## 1. Introduction

The extreme ultra-violet light (EUVL) emissions of 4d open-shell atomic ions are of interest in relation to the semiconductor technologies and also to the study of highly charged heavy atomic ions in plasmas. Extensive efforts have been made in recent years for understanding the emission spectra from the plasmas [1–6]. Those spectral structures are of interest also from the view point of academic atomic spectroscopy, whereas they are obtained mainly on the purpose of EUV light source development.

During the last years, a series of sophisticated charge transfer experiments have been carried out for xenon ions [7,8] and for tin ions [8,9] by Tanuma and his co-workers. In these experiments, a beam of charge selected atomic xenon or tin ions has been introduced into the gas chamber and the EUV emissions from the excited atomic ions that were created by collisional electron capture from neutral gas atoms have been observed. Because the gas pressure has been kept as low as to maintain a single collision condition, their individual spectral data are of the ions with a unique charge state. This gives a great advantage for comparison of the data with theory.

To give a theoretical counterpart of these experimental data, we performed a set of accurate calculations of electronic states and optical processes based on a Multi-Configuration Dirac-Fock (MCDF) approximation. In these computations, the General purpose Relativistic Atomic Structure Program 92 (GRASP92) [10] and the Relativistic Atomic Transition and Ionization Properties (RATIP) [11] computer codes are used. An advantage of these programs is that we can treat the two electron non-local exchange integrals as they are [6]. In this context, we can properly evaluate the electron correlations through configuration interactions including the excited orbitals in CSF's in a sophisticated way.

To gain, especially, an insight of the role of the intra $N$-shell optical transitions, which are relevant for the 13.5 nm EUV emissions from $4d$-open shell Sn ions, we calculated the $4p-4d$ and $4d-4f$ transitions for ions with atomic numbers $Z$ ranging 48 to 56. We have pointed out the important role of the configuration interactions between the $4p$ sub-vallence hole configurations and $4f$ intra-shell excited configurations. It has been pointed out that the modifications of the optical emission spectral structures due to those type of configuration interactions are common to those ionic species.

We describe the procedure and the result of the present theoretical caiculations in the next section. In section 3, we discuss the physical aspect of the electron correlations in the relevant system. And, finally, in section 4, we give a short summary of the present study.

## 2. The Calculation of $4p-4d$ and $4d-4f$ Transition Arrays of $Z = 46-56$ Ions

One of the best candidates for 13.5 nm region EUV light source are considered to be of the intra $N$ shell ($n = 4$ shell) transitions of tin (Sn) multiplly charged atomic ions. It is normally indispensable to take into account the electron correlations if we are to evaluate the the transition energies within the accuracy of a few electron volts, because the correlation energy of the atomic valence electrons falls in this range. To gain an insight of the effects that are pointed out by

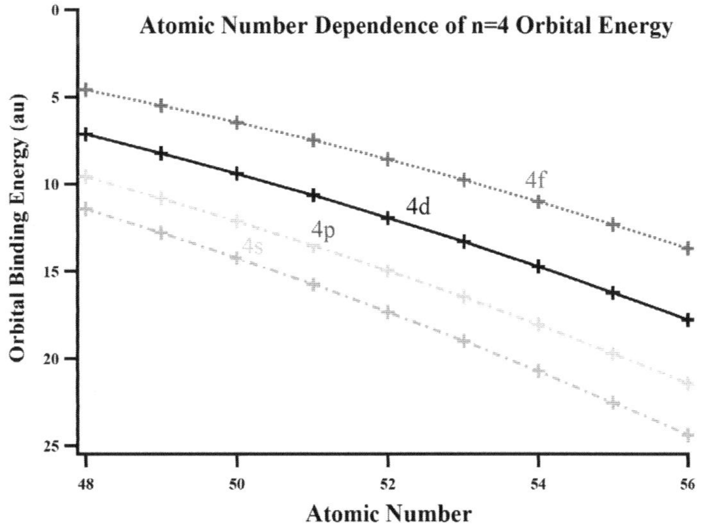


**Figure 1.** Atomic number $Z$ dependence of the energies of $N = 4$ atomic orbitals. The calculated range of the atomic number is $Z = 48$ to $Z = 56$. Dotted curve: $4f$ orbital, Solid curve: $4d$ orbital, Dot-dashed curve in upper entry: $4p$ orbital, and Dot-dashed curve in lower entry: $4s$ orbital.

O'Sullivan and Faukner [12], we have carried out careful MCDF calculations for $4d^q$ ($q = 0$ to 10) atomic ions with atomic number $Z = 48$ to 56. Although, they argued only the Sn atomic ions, those effects should be present also in other atomic species. As an example, we show ,in Fig.2, the single electron atomic orbital energies for Sr-like atomic ions of $Z = 48$ to 56. To obtain the single electron atomic orbitals and their energies, we have made their MCDF optimizations including the basis wavefunctions up to $6p$, $5d$, and $5f$. In this figure,

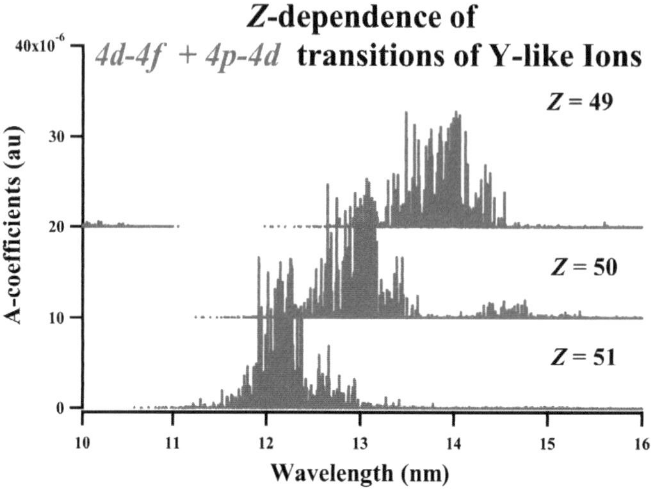


**Figure 2.** Atomic number $Z$ dependence of Einstein's A-factors for $4d-4f+4p-4d$ transitions of Y-like ions. The interference effects of $4d-4f$ and $4p-4d$ transitions are taken into account by configuration interaction calculations.

we can find that the differences of orbital energies between $4p$ and $4d$ orbitals, and $4d$ and $4f$ orbitals coincide within the range of a few %. The $4p^6 4d4f$ and $4p^5 4d^3$ configurations may mix strongly, and the optical $4p-4d$ and $4f-4d$ transitions may take place coherently, providing us with quite a peculiar EUV emission spectrum. The result of the orbital energy calculation stands for the argumentation made by O'Sullivan and FauknerOSullivan1994. Also, we find, from this figure, that we may expect that the so called the effect of spectral narrowing and shift is quite common to the atomic specieds with the atomic numbers in the range $Z=48$ to 56. As an example of the effect of spectral narrowing and shift, we illustrate, in Fig.3 the calculated Einstein's $A$-factor distributions of Y-like ions of atomic species with $Z=49, 50, \mathrm{and} 51$. In this figure, we can also observe the change of the spectral shape in the atomic number $Z$. Due to the interference between the $4p^6 4d^4 - 4p^6 4d^3 4f$ and $4p^6 4d^4 - 4p^5 4d^5$ transitions, we can observe a strong enhancement of the EUV emissions in 13.5 nm region of $Z=50$, which provides us with the narrowing and shift in appearance of the emission spectra. Due to the term splitting both in the ground and the excited states, the spread of the $A$ factor distribution would have been a couple of ten eV without the interference between the $4p-4d$ and $4d-4f$ transitions. Although the figures are not illustrated in the present paper, we have observed quite a similar interference effect in transitions for wide range of atomic numbers, say, for the range $Z=48$ to 56, and also for wide range of occupation numbers of $N$ shells.

## 3. Discussion

The precise and acculate theoretical data of the atomic structures and dynamics that includes the information about the relativistic effect and the electron-electron correlation effects provide us with a good base for precision analysis of highly charged atomic ions in a plasma.. Owing to the general nature of many electron atoms or atomic ions, the inclusion of the effect of configuration interactions among virtually excited states is indispensable to obtain physically realistic atomic state wavefunctions. Especially, it is important to take into account the modification of the atomic state wavefunctions due to the interactions among various many-electron configurations, which can be evaluated by only an SCF iterations in MCDF calculations. We can point out also that the spectral shift and narrowing should be the effects which are quite common to the $4d$ open shell atomic ion species with moderate atomic numbers. From the view point of atomic multi-configuration Dirac-Fock calculations, which can provide us with plausible shapes for the single electron orbitals by the procedure of non-linear numerical orbital SCF optimization, these effects are due to the coincident agreement of the orbital energy differences among the $N = 4$ sub-shells. This coincident agreement itself is the outcome of electron correlations, and a multi-configuration SCF can account for such correlations in the single electron orbital wavefunctions.

## Acknowledgments
This work is partly supported by the Leading Project for advanced semiconductor technology of MEXT. This work was partly supported by the JSPS-CAS Core-University Program in the field of Plasma and Nuclear Fusion.

## References
[1] Gillaspy J D 2006 *EUV Sources for Lithography Edited by V. Bakshi, SPIE Press Bellingham, Washinton, USA* 47
[2] Tolstikhina Yu et al 2006 *EUV Sources for Lithography Edited by V. Bakshi, SPIE Press Bellingham, Washinton, USA* 113
[3] G. O'Sullivan G et al 2006 *EUV Sources for Lithography Edited by V. Bakshi, SPIE Press Bellingham, Washinton, USA* 149
[4] Saloman E B 2004 *J. Phys. Chem. Ref. Data* **33** 921
[5] Sasaki A et al 2006 *Proc. of SPIE* **6151** 61513W.
[6] Koike F, Fritzsche S 2007 *J. Phys. Conf. Ser.*bf 58 157.
[7] Tanuma H et al 2005 *Nucl. Instr. Meth.* **B235** 331
[8] Tanuma H et al 2007 *J. Phys. Conf. Ser.* **58** 231
[9] Ohashi H et al 2007 *J. Phys. Conf. Ser.* **58** 235
[10] Parpia F A, Fischer C F and Grant I P 1996 *Commpt. Phys. Commun.* **94** 249
[11] Fritzsche S 2001 *J. Elec. Spec. Rel. Phenom.* **114-116** 1155
[12] O'Sullivan G, Faukner R 1994 *Opt. Eng.* **33**, 3978.

# Doppler-tuned Bragg Spectroscopy of Excited Levels in He-Like Uranium: a discussion of the uncertainty contributions

**M. Trassinelli[1], A. Kumar[2,3], H.F. Beyer[2], P. Indelicato[4], R. Märtin[2,5], R. Reuschl[2,5] and Th. Stöhlker[2,6]**

[1]Institut des Nanosciences de Paris, CNRS UMR7588 and Université Pierre et Marie Curie-Paris 6, Campus Boucicaut, 140 rue de Lourmel, F-75015 Paris, France
[2]Gesellschaft für Schwerionenforschung, Planckstr. 1, D-64291 Darmstadt, Germany
[3]Nuclear Physics Division, Bhabha Atomic Research Centre, Mumbai - 400 085, India
[4]Laboratoire Kastler Brossel, École Normale Supérieure; CNRS UMR8552; Université Pierre et Marie Curie-Paris6; Case 74, 4 Place Jussieu, F-75005 Paris, France
[5]Institut für Kernphysik, Universität Frankfurt, D-60325 Frankfurt, Germany
[6]Physikalisches Institut, Heidelberg, Philosophenweg 12, D-69120 Heidelberg, Germany

E-mail: `martino.trassinelli@insp.jussieu.fr`

**Abstract.** We present the uncertainty discussion of a recent experiment performed at the GSI storage ring ESR for the accurate energy measurement of the He-like uranium $1s2p\,^3P_2 \to 1s2s\,^3S_1$ intra-shell transition. For this propose we used a Johann-type Bragg spectrometer that enables to obtain a relative energy measurement between the He-like uranium transition, about 4.51 keV, and a calibration x-ray source. As reference, we used the $K\alpha$ fluorescence lines of zinc and the Li-like uranium $1s^2 2p\,^2P_{3/2} \to 1s^2 2s\,^2S_{1/2}$ intra-shell transition from fast ions stored in the ESR. A comparison of the two different references, i.e., stationary and moving x-ray source, and a discussion of the experimental uncertainties is presented.

## 1. Introduction

We present the uncertainty discussion of a recent experiment performed at the GSI (Darmstadt, Germany) for the accurate energy measurement of the He-like uranium $1s2p\,^3P_2 \to 1s2s\,^3S_1$ intra-shell transition. This measurement allows, for the first time, to test two-photons Quantum Electrodynamics in He-like heavy ions. In this article we describe the techniques adopted in the measurement where x rays emitted from fast ions in a storage ring have been detected with a Bragg spectrometer. In particular, we study the contribution of the different uncertainties related to the relativistic velocity of the ions when a stationary or moving calibration x-ray source is considered. Additional details of such an experiment can be found in Ref. [1].

## 2. Description of the set-up

The experiment was performed at the GSI experimental storage ring ESR [2] in August 2007. Here, a H-like uranium beam with up to $10^8$ ions was stored, cooled, and decelerated to an energy of 43.57 MeV/u. Excited He-like ions were formed by electron capture during the interaction of the ion beam with a supersonic nitrogen gas-jet target. At the selected velocity, electrons

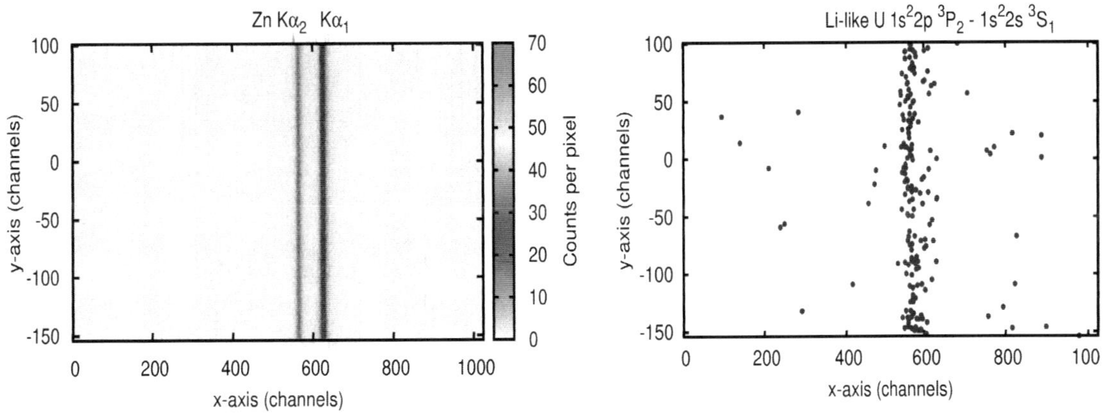


**Figure 1.** Reflection of the zinc $K\alpha$ (left) and Li-like uranium intra-shell $1s^22p\,^2P_{3/2} \rightarrow 1s^22s\,^2S_{1/2}$ transition (right) on the Bragg spectrometer CCD. The transition energy increases with the increasing of x-position. The slightly negative slope of the line is due to the relativistic velocity of the Li-like ions.

are primarily captured into shells with principal quantum number of $n \leq 20$, which efficiently populate the $n = 2\ ^3P_2$ state via cascade feeding. This state decays to the $n = 2\ ^3S_1$ state via an electric dipole (E1) intra-shell transition (branching ration 30%) with the emission of photons of an energy close to 4.51 keV detected by a Bragg spectrometer.

The crystal spectrometer [3] was mounted in the Johann geometry in a fixed angle configuration allowing for the detection of x rays with a Bragg angle $\Theta$ around 46.0°. The spectrometer was equipped with a Ge(220) crystal cylindrically bent, with a radius of curvature $R = 800$ mm, and a newly fitted x-ray CCD camera (Andor DO420) as position sensitive detector. The imaging properties of the curved crystal were used to resolve spectral lines from fast x-ray sources nearly as well as for stationary sources [4]. For this purpose, it was necessary to place the Rowland-circle plane of the spectrometer perpendicular to the ion–beam direction. For a minimizing the systematic effects due to the ion velocity and alignment uncertainties, the observation angle $\theta = 90°$ was chosen.

The value of the ion velocity $v$ was selected such that the photon energy, $E$ in the ion frame, was Doppler-shifted to the value $E_{\text{lab}} = 4.3$ keV in the laboratory frame, where $E_{\text{lab}} = E/[\gamma(1 - \beta\cos\theta)]$, with $\beta = v/c$ ($c$ is speed of light) and $\gamma = 1/\sqrt{1-\beta^2}$. This value of $E_{\text{lab}}$ was chosen to have the He-like uranium spectral line position on the CCD close to the position of the 8.6 keV $K\alpha_{1,2}$ lines of zinc, which were observed in second order diffraction. The zinc lines were used for calibration and they were produced by a commercial x-ray tube and a removable zinc plate between the target chamber and the crystal. An image of the zinc $K\alpha$ lines from the Bragg spectrometer is presented in Fig. 1 (left side).

As an alternative method to the measurement of the He-like uranium intra-shell transition, we used a calibration line originating from fast ions, rather than one from the stationary source. For this purpose the $1s^22p\,^2P_{3/2} \rightarrow 1s^22s\,^2S_{1/2}$ transition in Li-like U at $4459.37 \pm 0.21$ eV [5, 6] was chosen. At the ESR, the Li-like ions were obtained by electron capture into He-like uranium ions. To match the energy of the He-like transition, an energy of 32.63 MeV/u was used to Doppler-shift the Li-like transition. An image of the Li-like uranium transition in the Bragg spectrometer is presented in Fig. 1 (right side).

Starting from the Bragg's law in differential form, $\Delta E \approx -E\,\Delta\Theta/\tan\Theta$, one obtains an approximate dispersion formula that is valid for small Bragg angle differences $\Delta\Theta$. Taking into account the relativistic Doppler effect, the measured value of the He-like U transition is given by

$$E = E_0 \frac{n}{n_0} \frac{\left(1 - \delta(E_0)/\sin^2\Theta_0\right)}{\left(1 - \delta(E)/\sin^2\Theta\right)} \frac{\gamma(1-\beta\cos\theta)}{\gamma_0(1-\beta_0\cos\theta)} \left(1 + \frac{\Delta x}{\tan\Theta_0\ D}\right), \quad (1)$$

where $n_0$ and $n$ are the diffraction order of the He-like U and reference lines, respectively, $\Theta_0$ and $\Theta = \Theta_0 + \Delta\Theta$ the correspondent Bragg angle, $\Delta x$ is the position difference of the spectral lines on the CCD along the dispersion direction and $D$ is the crystal–CCD distance. $\delta(E)$ is the deviation of the index of refraction $n_r(E) = 1 + \delta(E)$ of the crystal material from the unity, which depends on the energy $E$ of the reflected x ray ($\delta \sim 10^{-5} - 10^{-6}$ typically).

In the case $n = n_0$, the corrections due to the refraction index and other energy dependent corrections for curved crystals [7, 8] are negligible. In the case of a stationary calibration source, $\gamma_0 = 1$ and $\beta_0 = 0$.

The measured value of the He-like uranium transition energy and additional information can be found in Ref. [1]. In the following section we present in detail the analysis of the systematic uncertainties.

## 3. Evaluation of the experimental uncertainties

One of the main sources of uncertainty in the present experiment is the low amount of collected data (see Fig. 1, right). This limits the accuracy of the He- and Li-like uranium line position, i.e., the accuracy of $\Delta x$, which is proportional to the energy uncertainty, $(\delta E)_{\mathrm{stat}} \propto \delta(\Delta x)$ (see Eq. (1)). In our specific experiment, characterized by the parameters listed in Table 1, numerical values of $(\delta E)_{\mathrm{stat}}$ are presented in Table 2. Due to the high statistics in the stationary calibration source measurement, the Zn $K\alpha$ spectrum, $(\delta E)_{\mathrm{stat}}$ is $\sim \sqrt{2}$ smaller than when the moving Li-like ion emission is used as a reference.

In the case of systematic uncertainties, three major sources dominate: the accuracy of the reference energy $E_0$, of the ion velocity and of the observation angle $\theta$. Similarly to the statistical uncertainty, the contribution of $\delta E_0$ is much smaller when Zn lines are used for calibration instead of the Li-like U transition (see Table 2). This is due to the high accuracy of the zinc $K\alpha$ transition energy, which in the case of $K\alpha_1$ is $8638.906 \pm 0.073$ eV [9], compared to the Li-like U transition accuracy of 0.21 eV [5, 6].

If on one hand, Doppler tuning of the photon energy in the laboratory frame produce two important systematic uncertainty contributions, due to the ions velocity and the observation angle; on the other hand, it allows for detecting the different spectral lines in the same narrow spatial region of the CCD detector, i.e., $\Delta x/D \ll 1$. This results in a drastic reduction of other systematic effects such as the influence of uncertainty of the crystal–CCD distance $D$, the accuracy of the CCD pixel size [10], the accuracy of the inter-plane distance of the crystal and effects from the optical aberrations in the Johann geometry set-up. Systematic uncertainties related to the relativistic velocity of the ions are treated in detail in the following subsections.

### *3.1. Ion velocity uncertainty*

The ion velocity in the storage ring is imposed by the velocity of the electrons in the electron cooler [2]. This is related to the cooler voltage $V$ by the simple relations

$$\gamma = 1 + \frac{eV}{m_e c^2}, \qquad \beta = \frac{\sqrt{2\frac{eV}{m_e c^2} + \left(\frac{eV}{m_e c^2}\right)^2}}{1 + \frac{eV}{m_e c^2}} \simeq \sqrt{2\frac{eV}{m_e c^2}}, \quad (2)$$

where $m_e$ and $e$ are the mass and charge of the electron, respectively. The factor $eV/(m_e c^2)$ is in general very small, of the order of $4 \times 10^{-2}$ in our specific case. The cooler voltage uncertainty $\delta V$ propagates to the energy uncertainty via the parameters $\gamma$ and $\beta$ in Eq. (1). More precisely,

$(\delta\gamma)_V = e/(m_e c^2)\,\delta V$ and $(\delta\beta)_V \simeq [e/(2m_e c^2 V)]^{1/2}\,\delta V$. We note that $\delta\gamma/\delta\beta = \mathcal{O}(\sqrt{eV/(m_e c^2)})$. For this reason an observation angle of 90°, where the effect of $\delta\beta$ is minimal, was chosen. In the following formulas we will consider only the case $\theta = 90°$.

The uncertainty of the cooler voltage has two principal sources: the accuracy of the absolute value and its linearity. The relation between the real voltage value $V_{\rm real}$ and the set value $V$ can be written as $V = a + b\,V_{\rm real}$, where the single uncertainties on the factors $a$ and $b$ has to be considered for our analysis. With this notation, for an observation angle of $\theta = 90°$, the uncertainty propagation to the energy value is, for the case of a stationary calibration source,

$$\frac{(\delta E)_a}{E} = \frac{1}{\gamma}\left|\frac{\partial\gamma}{\partial a}\right|\delta a = \frac{e}{m_e c^2}\frac{1}{\gamma}\,\delta a, \qquad \frac{(\delta E)_b}{E} = \frac{1}{\gamma}\left|\frac{\partial\gamma}{\partial b}\right|\delta b = \frac{eV}{m_e c^2}\frac{1}{\gamma}\,\delta b, \tag{3}$$

and

$$\frac{(\delta E)_a}{E} = \frac{\gamma_0}{\gamma}\left|\frac{\partial}{\partial a}\left(\frac{\gamma}{\gamma_0}\right)\right|\delta a = \frac{e^2}{m_e^2 c^4}\frac{|V-V_0|}{\gamma\,\gamma_0}\,\delta a,$$
$$\frac{(\delta E)_b}{E} = \frac{\gamma_0}{\gamma}\left|\frac{\partial}{\partial b}\left(\frac{\gamma}{\gamma_0}\right)\right|\delta b = \frac{e}{m_e c^2}\frac{|V-V_0|}{\gamma\,\gamma_0}\,\delta b, \tag{4}$$

in the case of a moving calibration source, where $V_0$ and $\gamma_0$ are the corresponding parameters and where the approximation $V \approx V_{\rm real}$ has been applied. A reduction of the uncertainties is obtained when the moving calibration source is used. The uncertainty $(\delta E)_a$ due to the offset error of $V$ is drastically decreased, by a factor $e|V-V_0|/(m_e c^2\gamma_0) \ll 1$, here as the uncertainty due to the linearity is also reduced, but only by a factor $|V-V_0|/(V\gamma_0)$. Numerical values for our experiments are given in Table 1 and 2.

### 3.2. Observation angle uncertainty

If the effect of the uncertainty due to the ion velocity is minimal when $\theta = 90°$, the effect of the uncertainty $\delta\theta$ of the observation angle itself is maximal. In the case $\theta = 90°$, starting from Eq. (1), for a stationary calibration source we have

$$\frac{(\delta E)_\theta}{E} = \frac{1}{1-\beta\cos\theta}\left|\frac{d}{d\theta}(1-\beta\cos\theta)\right|_{\theta=90°}\delta\theta = \beta\delta\theta, \tag{5}$$

and

$$\frac{(\delta E)_\theta}{E} = \frac{1-\beta_0\cos\theta}{1-\beta\cos\theta}\left|\frac{d}{d\theta}\left(\frac{1-\beta\cos\theta}{1-\beta_0\cos\theta}\right)\right|_{\theta=90°}\delta\theta = |\beta-\beta_0|\delta\theta. \tag{6}$$

**Table 1.** Principal parameters of the ion beams for the He- and Li-like transition measurement (see text).

| | He-like U | Li-like U |
|---|---|---|
| $\beta$ | 0.295578 | 0.257944 |
| $\gamma$ | 1.046771 | 1.035026 |
| $V$(Volt) | 23900 | 17898 |
| $\theta$ | $90.00° \pm 0.38°$ | |
| $a$(Volt) | $0 \pm 10$ | |
| $b$ | $1. \pm 2\times10^{-4}$ | |

**Table 2.** Different uncertainty contributions (in eV) when Li-like uranium and zinc transitions are used as the reference.

| | Li-like U | Zn $K\alpha$ |
|---|---|---|
| $(\delta E)_{\rm stat}$ | 0.43 | 0.30 |
| $(\delta E)_{E_0}$ | 0.21 | 0.04 |
| $(\delta E)_a$ | $1\times10^{-3}$ | 0.08 |
| $(\delta E)_b$ | 0.01 | 0.04 |
| $(\delta E)_\theta$ | 0.11 | 0.88 |
| $(\delta E)_{\rm TOT}$ | 0.50 | 0.93 |

for a moving calibration source. Again, analogously to the calculation of the preceding subsection, a reduction of a factor $|\beta - \beta_0|/\beta$ in the uncertainty is obtained when x rays from fast ions are used for the calibration. In our experimental set-up, the value of $\delta\theta$ is principally due to the accuracy of the position of the gas-jet target with respect to the main axis of the spectrometer ($\pm 0.5$ mm). Numerical values for our experiments are in Table 1 and 2. A direct evaluation of deviation $\Delta\theta$ from 90° can be obtained via Eq. (6) with the measurement of the Li-like uranium energy $4460.12 \pm 0.31$ eV (statistical uncertainty only) using the zinc $K\alpha$ lines as reference, and its comparison with the literature value $4459.37 \pm 0.21$ eV [5, 6]. We estimated $\Delta\theta = -0.37° \pm 0.18°$ in agreement with the expected deviation (see Table 1).

## 4. Conclusions

We present the uncertainty discussion of an accurate energy measurement on He-like uranium intra-shell transition obtained via Bragg spectroscopy of Doppler-tuned x rays emitted from fast ions. We have evaluated and compared the systematic uncertainties when a stationary or moving calibration source is used. In particular in our experiment, the use of the x-ray emission of fast Li-like uranium ions as reference enables to reduce systematics uncertainties by a factor of about 4.

## References

[1] Trassinelli M, Kumar A, Beyer H, Indelicato P, Märtin R, Reuschl R, Kozhedub Y, Brandau C, Bräuning H, Geyer S, Gumberidze A, Hess S, Jagodzinski P, Kozhuharov C, Trotsenko S, Weber G and Stöhlker T 2008 *submitted to Phys. Rev. Lett.*

[2] Franzke B 1987 *Nucl. Instrum. Meth. Phys. Res. B* **24-25** 18–25

[3] Beyer H F, Indelicato P, Finlayson K D, Liesen D and Deslattes R D 1991 *Phys. Rev. A* **43** 223

[4] Beyer H F and Liesen D 1988 *Nucl. Instrum. Meth. Phys. Res. A* **272** 895–905

[5] Beiersdorfer P, Knapp D, Marrs R E, Elliott S R and Chen M H 1993 *Phys. Rev. Lett.* **71** 3939

[6] Beiersdorfer P 1995 *Nucl. Instrum. Meth. Phys. Res. B* **99** 114–116

[7] Cembali F, Fabbri R, Servidori M, Zani A, Basile G, Cavagnero G, Bergamin A and Zosi G 1992 *J. Appl. Crystallogr.* **3** 424–31

[8] Chukhovskii F N, Hölzer G, Wehrhan O and Förster E 1996 *J. Appl. Crystallogr.* **29** 438–445

[9] Deslattes R D, Kessler Jr E G, Indelicato P, de Billy L, Lindroth E and Anton J 2003 *Rev. Mod. Phys.* **75** 35–99

[10] Indelicato P, Le Bigot E O, Trassinelli M, Gotta D, Hennebach M, Nelms N, David C and Simons L M 2006 *Rev. Sci. Inst.* **77** 043107 (pages 10)

Journal of Physics: Conference Series **163** (2009) 012027 doi:10.1088/1742-6596/163/1/012027

# Two-photon decay in highly charged heavy ions: Spectral shape of the $2E1$ $(2\,^1S_0 \rightarrow 1\,^1S_0)$ in He-like tin

**A. Kumar[1,2†], S. Trotsenko[1,3], A.V. Volotka[4], D. Banaś[5], H.F. Beyer[1], H. Bräuning[1], S. Fritzsche[1,6], A. Gumberidze[1,7], S. Hagmann[1,3], S. Hess[1,3], C. Kozhuharov[1], G. Plunien[4], R. Reuschl[1,3], U. Spillmann[1,3], M. Trassinelli[1,8], G. Weber[1,6] and Th. Stöhlker[1,6]**

[1] Gesellschaft für Schwerionenforschung, 64291 Darmstadt, Germany
[2] Nuclear Physics Division, Bhabha Atomic Research Centre, Mumbai - 400 085, India
[3] Institut für Kernphysik, Universität Frankfurt, Germany
[4] Institut für Theoretische Physik, TU Dresden, Germany
[5] Institute of Physics, Jan Kochanowski University, Świętokrzyska 15, 25-406 Kielce, Poland
[6] Physikalisches Institut, Ruprecht-Karls-Universität Heidelberg, Germany
[7] Laboratoire Kastler Brossel, Université Pierre et Marie Curie, Paris, France
[8] Institut des Nanosciences de Paris, CNRS and Université Pierre et Marie Curie-Paris 6, F-75015 Paris, France

E-mail: †a.kumar@gsi.de

**Abstract.** We have measured the two-photon $2E1$ transition from the singlet $1s2s\,^1S_0$ to the $1s^2\,^1S_0$ ground state of helium-like tin. The experiment was performed by using initially Li-like tin ions in collision with a nitrogen gas target at the ESR storage ring at the energy of 300 MeV/u. The two-photon emission arises from the decay of the exclusively formed $1s2s$ states following the ionization of one K-shell electron from the Li-like ions. The obtained spectral shape of the $1s2s\,^1S_0 \rightarrow 1s^2\,^1S_0$ $(2E1)$ transition was compared with the one calculated within a relativistic approach. The spectral shape is displayed and its sensitivity with regard to details in the theoretical description of the $2E1$ decay is discussed.

## 1. Introduction

For helium-like ions, the single-photon emission from the excited $1s2s\,^1S_0$ to their $1s^2\,^1S_0$ ground state is strictly forbidden owing to the conservation of angular momentum selection rule and, therefore, the upper $1s2s\,^1S_0$ state decays predominantly via the simultaneous emission of two photons [1, 2]. In this decay, of course, the sum of the energies of both photons is equal to the total energy of the $1s2s\,^1S_0 \rightarrow 1s^2\,^1S_0$ transition, and the $E1E1$ decay probability clearly dominates over all the other allowed combinations of multipoles, e.g., M1M1, E1M2..., which are known to contribute less than 0.5% to the total two-photon decay rates [3, 4]. In contrast to a single-photon emission, the $2E1$ photons form a continuous energy distribution from zero up to the maximum $1s2s\,^1S_0 \rightarrow 1s^2\,^1S_0$ transition energy. Moreover, the two-photon decay is sensitive to the electronic structure of the ions as a whole since, within the framework of perturbation theory, all intermediate bound and continuum states have to be taken into account in order to obtain the transition probabilities.

Several lifetime measurements have been performed in the past for the $1s2s\ {}^1S_0$ state, and especially for low- and medium-$Z$ ions along the helium isoelectronic sequence [5]. In these case studies, the lifetime of the $1s2s\ {}^1S_0$ state was found to be in good agreement with a number of nonrelativistic and relativistic calculations. However, these lifetime measurements could test only the total transition probability as obtained from the integration over all the energies in the photon distribution. Furthermore, such measurements are difficult to be perform also for high-$Z$ ions, even with the presently available techniques, owing to the rather short lifetime of the $1s2s\ {}^1S_0$ state. On the other hand, measurements of the $2E1$ energy distribution are possible today, even for the heaviest (helium-like) ions.

For helium-like ions, the nonrelativistic calculations predict an increase in the full width at half maximum (FWHM) for the spectral distribution as the nuclear charge $Z$ increases [6], whereas the relativistic calculations [3, 4] only show an increase of the FWHM up to about $Z$=20, and followed by a monotonic decrease for all larger $Z$. This dependence of the two-photon energy distribution is caused by the interplay between the electron-electron correlation and relativistic effects. Therefore, a measurement of the spectral shape of the $1s2s\ {}^1S_0 \rightarrow 1s^2\ {}^1S_0$ two-photon transition along the helium isoelectronic sequence probes uniquely our understanding of the electronic structure of the simplest multi-electron systems. Therefore, a great deal of effort have been spent in the past for an accurate determination of the spectral distribution for the two-photon decay in heliumlike ions [7, 8, 9, 10]. Most of the studies were carried out by using the conventional photon-photon coincidence technique. But although these measurements verified all the general trends in the two-photon emission, they were not (yet) able to test the predictions from the relativistic theory due to the lack of accuracy, the low statistics as well as a number of further experimental difficulties.

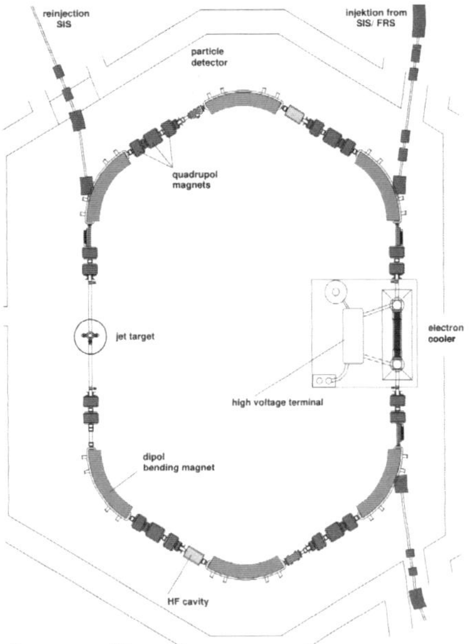


**Figure 1.** Schematic view of the ESR storage ring at the GSI Darmstadt, showing the electron cooler, the gas-jet target, particle detector and several other components [11].

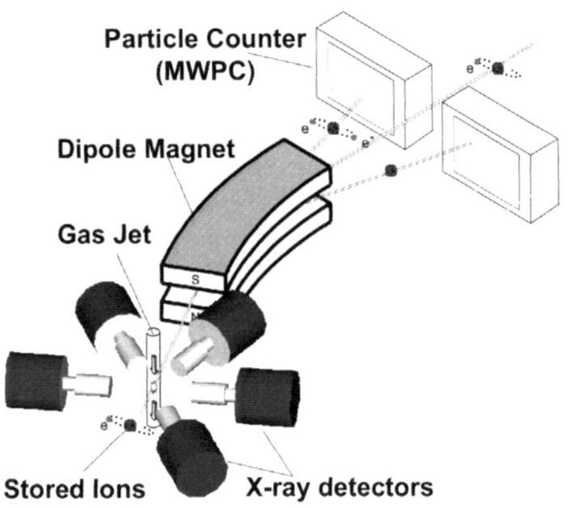


**Figure 2.** Typical experimental arrangement at the gas-jet target of the storage ring ESR [11]. Near to the target, the ion/gas-jet collision zone is surrounding by an array of solid state detectors.

In this contribution, we report about *preliminary* experimental results on the spectral shape of the $1s2s\ ^1S_0 \rightarrow 1s^2\ ^1S_0$ transition of helium-like tin, measured by using an alternative method for the selective K-shell ionization of initially Li-like ions. Importantly, the spectra recorded by this technique are the cleanest spectra obtained so far, and the systematic accuracy along with the counting statistics have been improved substantially when compared with previous studies. Our *preliminary* results show that the experimental spectral confirm the prediction of fully relativistic calculations and might allow us to discriminate the results against the results from the nonrelativistic theory.

## 2. Experimental details

The experiment was performed at the storage ring ESR with 300 MeV/u Li-like tin ions colliding with a nitrogen gas-jet target [11] under single collision conditions. A schematic view of the ESR storage ring and the experimental setup at the target area is depicted in Fig. 1 and Fig. 2, respectively. The produced x-rays were detected by a germanium detector at an observation angle of 35° with respect to the ion-beam direction. Since the x-ray source is moving, the Doppler broadening in the x-rays was reduced by a 4-mm copper-lead slit, placed in front of the detector. After the collision of the initially Li-like projectiles with the $N_2$ molecules, the produced He-like ions were detected by a dipole magnet and registered by means of a multi-wire particle detector. The energy and efficiency calibration of the detector was performed by using calibrated radioisotopes. For further details of the experimental setup we refer the reader to Ref. [12].

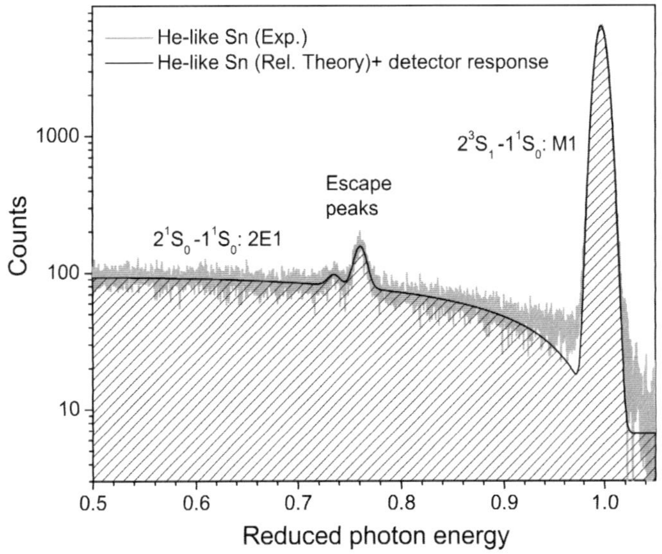


**Figure 3.**
Upper half of the spectral distribution of the $1s2s\ ^1S_0 \rightarrow 1s^2\ ^1S_0$ $2E1$ transition together with the intense peak of the $1s2s\ ^3S_1 \rightarrow 1s^2\ ^1S_0$ $M1$ transition of the projectile x-ray spectra (gray line) for $Sn^{47+} \longrightarrow N_2$ collisions measured in coincidence with the $Sn^{48+}$ ions. *Preliminary* comparison of the measured $2E1$ spectral distribution of He-like Sn with the calculated distribution (and convoluted with the simulated detector response function, the shadowed area) [3, 4]. The reduced photo energy on the x-axis denotes the fraction of total transition energy that is carried away by either one of the photons.

## 3. Results and discussion

The x-ray spectra were recorded in coincidence with the helium-like $Sn^{48+}$ ions. The measured spectra show a broad continuum that results from the two-photon decay of the $1s2s\ ^1S_0 \rightarrow 1s^2\ ^1S_0$ transition and the rather intense and monoenergetic line from the $1s2s\ ^3S_1 \rightarrow 1s^2\ ^1S_0\ M1$ transition, cf. Fig. 3. The absence of any additional x-ray line from the decay of the $1s2p$ or some higher levels indicates that the present collision system leads to a very selective K-shell ionization of the lithium-like tin ions and to the formation of the corresponding $1s2s\ ^{1,3}S$ states of helium-like tin. Such a high level of selectivity of the K-shell ionization has recently been reported especially for relativistic collisions of lithium-like U ions (cf. [12]).

For the further analysis of the observed x-rays, let us choose the upper half of the $2E1$ spectral distribution (cf. Fig. 3) because: (1) this energy distribution symmetric with regard to an equal energy sharing [3, 4, 6] and, hence, either of the halves can be utilized to provide the $Z$-dependent information; (2) higher-energy half is clean and independent of the detector efficiency. The response function of the detector was simulated for the $M1$ transition and compared to the measured $M1$ line intensity. The detector simulations were found to be in good agreement with the measured position and intensity of the K-escape peaks. Furthermore, the theoretical $2E1$ spectral distribution was convoluted with the detector response function for comparison with the measured spectral shape. The intensity of the theoretical two-photon decay continuum was normalized to the measured spectra at the reduced photon energy of 0.50, i.e., for half of the energy of the $1s2s\ ^1S_0 \rightarrow 1s^2\ ^1S_0\ 2E1$ transition (Fig. 3). In order to check the sensitivity of the energy distribution with regard to changes in the nuclear charge $Z$, the measured spectral shape for tin has been compared with relativistic calculations [3, 4] for different helium-like ions (convoluted with the simulated detector response function) and by using the MINUIT minimization code. Our *preliminary* results show a good agreement with the relativistic calculations only if a proper nuclear charge were chosen, thus confirming the sensitivity on $Z$.

In summary, we performed the first measurement of the spectral shape of the two-photon decay of the $1s2s\ ^1S_0$ state of helium-like tin. The measurements have been carried out by using a novel experimental approach in which initially lithium-like projectiles were utilized to collide with a gaseous target and to form the desired $1s2s\ ^1S_0$ of the helium-like ions. The *preliminary* results are in good agreement with fully relativistic calculations.

## Acknowledgments

The support by the DAAD (A.K.) and the Alexander von Humboldt Foundation (M.T.) is gratefully acknowledged. D.B. is supported by Polish Ministry of Education and Science under Grant No. 1P03B00629. A.V.V. acknowledges useful discussions with V. M. Shabaev. The support by I3 EURONS (EC contract no. 506065) is gratefully acknowledged.

## References

[1] Göppert-Mayer M 1931 *Ann. Phys (Leipzig)* **9** 273
[2] Breit G and Teller E 1940 *Astrophys. J.* **91** 215
[3] Derevianko A and Johnson W R 1997 *Phys. Rev. A* **56** 1288
[4] Volotka A, (unpublished).
[5] Mokler P H and Dunford R W 2004 *Physica Scripta* **69** C1 and references therein.
[6] Drake G W F 1986 *Phys. Rev. A* **34** 2871
[7] Mokler P H *et al* 1990 *Phys. Rev. Lett.* **65** 3108
[8] Schäffer H W *et al* 1999 *Phys. Rev. A* **59** 245
[9] Ali R *et al* 1997 *Phys. Rev. A* **55** 994
[10] Schäffer H W *et al* 1999 *Phys. Lett. A* **260** 489
[11] Eichler J and Stöhlker Th 2007 *Physics Reports* **429** 1
[12] Rzadkiewicz J *et al* 2006 *Phys. Rev. A* **74** 012511

14th International Conference on the Physics of Highly Charged Ions (HCI 2008) IOP Publishing
Journal of Physics: Conference Series **163** (2009) 012028 doi:10.1088/1742-6596/163/1/012028

# Experimental rate coefficient for dielectronic recombination of neonlike iron forming sodiumlike iron

**E. W. Schmidt[1], D. Bernhardt[1], J. Hoffmann[2], M. Lestinsky[2,3], D. V. Lukić[3,4], A. Müller[1], D. A. Orlov[2], D. W. Savin[3], S. Schippers[1] and A. Wolf[2]**

[1]Institut für Atom- und Molekülphysik, Justus-Liebig-Universität Giessen, Germany
[2]Max-Planck-Institut für Kernphysik, Heidelberg, Germany
[3]Columbia Astrophysics Laboratory, Columbia University, New York, NY, 10027, USA

E-mail: Eike.W.Schmidt@iamp.physik.uni-giessen.de

**Abstract.** The rate coefficient for dielectronic recombination (DR) of Ne-like $Fe^{16+}$ forming Na-like $Fe^{15+}$ was measured employing the merged electron-ion beams technique at the heavy-ion storage-ring TSR of the Max-Planck-Institut für Kernphysik in Heidelberg, Germany. In the electron-ion collision energy range of 240–840 eV the merged-beams recombination rate coefficient is dominated by DR associated with $2s^2\,2p^6\ {}^1S_0 \rightarrow 2s^2\,2p^5\,3d\ {}^1P_1$ core excitation. The experimental $Fe^{16+}$ DR plasma rate coefficient is derived from the measured merged-beams rate coefficient. It is in good agreement with recent theoretical results.

## 1. Introduction

In our ongoing effort of providing reliable DR rate coefficients for astrophysical applications we have measured the dielectronic recombination (DR) rate coefficient of Ne-like $Fe^{16+}$ forming Na-like $Fe^{15+}$ employing the merged electron-ion beams technique at a heavy-ion storage-ring. The present measurement bridges the gap between our previous recombination experiments for $Fe^{13+}$–$Fe^{15+}$ and $Fe^{17+}$–$Fe^{22+}$ ions (see [1] and references therein). Moreover, it is the first merged-beams measurement for an ion of the Ne-like isoelectronic series. Therefore, this measurement represents an important benchmark for theoretical calculations.

Because of the closed-shell structure of the Ne-like $Fe^{16+}(1s^2\,2s^2\,2p^6)$ parent ion DR can only proceed via core excitations where the principal quantum number $N$ of the "active" core electron changes by $\Delta N \geq 1$. In open L-shell ions, in contrast, also $\Delta N = 0$ core-excitations are also possible which lead to DR resonances at rather low energies. Conversely, $\Delta N = 1$ resonances appear usually at higher energies because of the higher excitation energies involved. For $Fe^{16+}$, DR proceeds dominantly via the $2p^6 \rightarrow 2p^5\,3d\ {}^1P_1$ core excitation [2]. The corresponding series of $2p^5\,3d\ ({}^1P_1)\ nl$ DR resonances ranges from about 300 eV to the series limit at 825.7347 eV [3]. Accordingly, the electron-ion collision energy range 240–840 eV was investigated in the present experiment.

[4] Permanent address: Institute of Physics, Pregrevica 118, 11080 Belgrade, Serbia

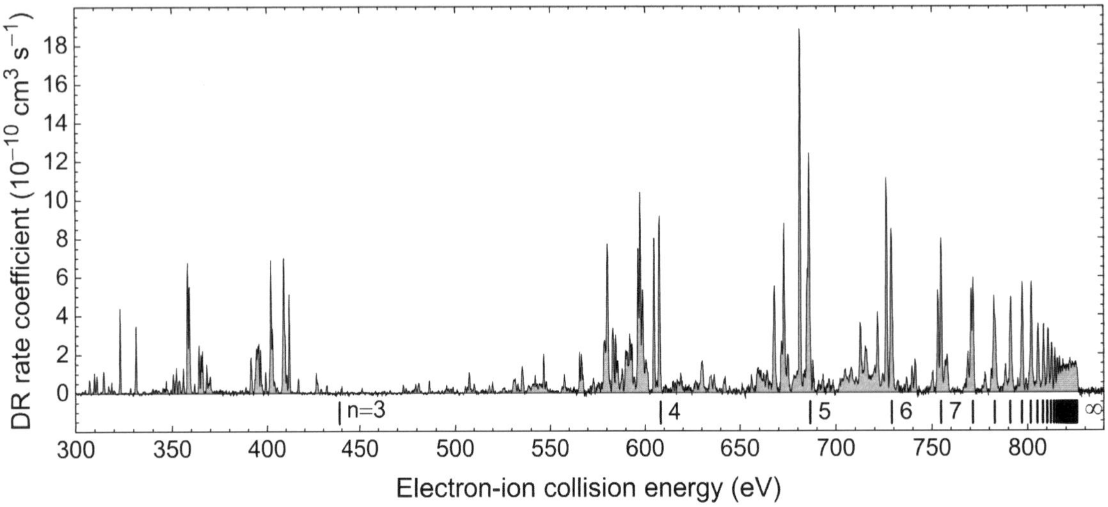


**Figure 1.** Experimentally derived merged-beams dielectronic recombination (DR) rate coefficient of $Fe^{16+}$ forming $Fe^{15+}$. The vertical bars below the spectrum mark the resonance positions of the $2s^2\,2p^5\,3d\ ^1P_1\,nl$ Rydberg series calculated in a hydrogenic approximation by using the Bohr formula. The $2s^2\,2p^6\ ^1S_0 \to 2s^2\,2p^5\,3d\ ^1P_1$ excitation energy was taken from [3].

## 2. Experiment

The experiment was performed at the heavy-ion storage-ring TSR of the Max-Planck-Institut für Kernphysik (MPI-K) in Heidelberg. Here only a brief account on the experimental and data-analysis procedures for electron-ion recombination measurements can be given. Details can be found in previous publications (e.g. [4] and references therein). The $^{56}Fe^{16+}$ ion beam was provided by the MPI-K accelerator facility at an energy of about 5 MeV/u. Ion bunches were continuously injected into the TSR and cooled permanently by the TSR electron cooler in one of the straight sections of the storage-ring. This continuous injection and cooling scheme enhances the duty cycle as compared to the usual measurement procedure, where the ion current in the storage-ring is accumulated and cooled before the recombination measurement starts.

In another straight section of the TSR, the ion beam was overlapped with the electron beam of the TSR electron target. The electron-ion collision energy $E_{\rm rel}$ was varied by changing the laboratory energy of the target electron beam. The longitudinal electron beam temperature was $kT_{\|} = (6.7 \pm 0.3) \times 10^{-5}$ eV as determined from the DR resonance line-shapes in the 800–815 eV energy range. Correspondingly, the experimental energy spread $\Delta E \approx [16(\ln 2)\,E_{\rm rel} kT_{\|}]^{1/2}$ varied from 0.47 eV to 0.63 eV across the experimental collision-energy range.

Recombined ions were separated from the primary beam in the first ring dipole-magnet downstream of the electron target and were detected by a single-particle detector with nearly 100% detection efficiency. Finally, the experimental merged-beams DR rate coefficient (figure 1) was derived by subtracting from the measured recombination spectrum all contributions due to radiative recombination (RR) and electron capture from the residual gas. The uncertainty of the merged-beams recombination rate coefficient is ±20% at a 67% confidence level over the entire energy range. It should be noted, that in the present experiment the use of the continuous injection scheme leads to a larger uncertainty of the ion current measurement and, consequently, of the merged-beams recombination rate coefficient than usual. This uncertainty stems mostly from uncertainties in the ion current measurement. Moreover, the subtraction of the continuous recombination background leads to an additional uncertainty of the merged

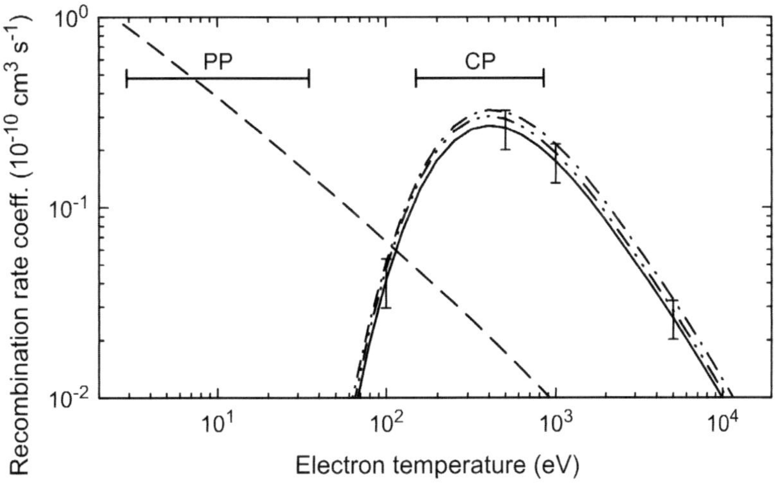


**Figure 2.** Plasma recombination rate coefficients of $Fe^{16+}$ forming $Fe^{15+}$. The solid line represents the experimentally derived DR plasma rate coefficient. The error bars give the systematic errors. Also shown are the recent theoretically calculated results of [5] (dash-dotted curve) and [2] (dash-dot-dotted curve). The dashed curve is a theoretically calculated radiative-recombination (RR) plasma rate coefficient [6]. The temperature ranges where $Fe^{16+}$ is expected to form in photoionized plasmas (PP) and collisionally ionized plasmas (CP) are marked.

beams rate coefficient of $\pm 2 \cdot 10^{-12}$ cm$^3$ s$^{-1}$ in the 240–840 eV energy range.

## 3. Results and Discussion

The resulting merged-beams DR rate coefficient is shown in Figure 1. In accord with the theoretical expectation [2], the spectrum is dominated by $2s^2\,2p^5\,3d\ (^1P_1)\ nl$ DR resonances associated with the $2s^2\,2p^6\,^1S_0 \to 2s^2\,2p^5\,3d\ ^1P_1$ core excitation. The corresponding Rydberg series limit at about 825 eV is clearly visible. No DR resonances have been detected at electron-ion collision energies below 300 eV.

In case of $\Delta N = 0$ DR field ionization of loosely bound high-$n$ Rydberg states in the charge analyzing dipole magnet can have drastic effects on the recombination spectrum [7] since recombination to states with a Rydberg quantum number $n$ exceeding the field ionization cutoff $n_F$ is essentially not detected. For $\Delta N = 1$ DR the DR resonance strength drops more rapidly with increasing $n$ than for $\Delta N = 0$ DR [8; 9]. In the present experiment resonances with $n > n_F = 53$ would contribute only negligibly to the DR rate coefficient close to the $2s^2\,2p^5\,3d\ (^1P_1)\ nl$ series limit.

The DR plasma rate coefficient as a function of electron temperature was derived by convoluting the merged-beams DR rate coefficient with a Maxwellian electron energy distribution. Experimental field-ionization effects were disregarded. By extrapolation of the measured $2s^2\,2p^5\,3d\ (^1P_1)\ nl$ DR resonance strengths to $n > n_F$ we estimated that this introduces only a negligible additional uncertainty of the plasma rate coefficient of at most 0.6% at temperatures higher than 90 eV. The absolute uncertainty of the merged-beams rate coefficient background leads to a relative uncertainty of the plasma rate coefficient of below 10% above 90 eV electron temperature. Adding these uncertainties in quadrature to the 20% relative uncertainty of the merged beams rate coefficient, yields a total uncertainty of the experimental DR plasma rate coefficient of 22% at a 67% confidence level, in the 90–10000 eV temperature range.

The temperature ranges where $Fe^{16+}$ ions form in photoionized plasmas (PP) and collisionally ionized plasmas (CP) can be inferred from ionization balance calculations [10; 11] suggesting that $Fe^{16+}$ reaches more than 10% of its peak abundance in a PP (CP) in the 3–35 eV (150–850 eV) temperature range. In figure 2 the present experimental DR plasma rate coefficient is compared with recent theoretical results [2; 5]. A comparison with earlier theoretical plasma rate coefficients for DR and RR can be found in [2]. As an additional reference a recently calculated RR plasma rate coefficient [6] is also given. $Fe^{16+}$ RR dominates over DR in the PP temperature range, whereas DR dominates in the CP zone.

Within the experimental error margin both recent theoretical DR results agree with the experimental result. In the CP temperature range the theoretical result of [2] ([5]) is up to 14% (23%) larger than the experimental DR plasma rate coefficient. The unusually good agreement between the experimental and theoretical $Fe^{16+}$ DR rate coefficients is due to the fact that only high energy $\Delta N = 1$ DR resonances contribute significantly to the plasma rate coefficient. At high energies the theoretical uncertainties of DR resonance positions are not as critical as for low-energy $\Delta N = 0$ DR where uncertainties in DR resonance positions of 0.1 eV can result in order of magnitude uncertainties of the low-temperature plasma rate coefficient [1].

## Acknowledgments
The excellent support of the MPI-K accelerator and storage-ring crews is gratefully acknowledged. This work was supported in part by the German federal research-funding agency DFG under contract no. Schi 378/5. M.L., D.V.L., and D.W.S. were supported in part by the NASA Astronomy and Physics Research and Analysis program and the NASA Solar and Heliospheric Physics program.

## References
[1] Schippers S 2008 *J. Phys. Conf. Ser.* these proceedings (*Preprint* arXiv:0808.3366 v1[astro-ph])

[2] Zatsarinny O, Gorczyca T W, Korista K, Badnell N R and Savin D W 2004 *Astron. Astrophys.* **426** 699–705

[3] Ralchenko Y, Jou F C, Kelleher D E, Kramida A E, Musgrove A, Reader J, Wiese W L and Olsen K 2007 *NIST Atomic Spectra Database (version 3.1.1)* (Gaithersburg, MD: National Institute of Standards and Technology) `http://physics.nist.gov/asd3`

[4] Schmidt E W *et al.* 2007 *Phys. Rev.* A **76** 032717

[5] Gu M F 2003 *Astrophys. J.* **590** 1131–1140

[6] Badnell N R 2006 *Astrophys. J. Suppl. Ser.* **167** 334–342

[7] Schippers S, Müller A, Gwinner G, Linkemann J, Saghiri A A and Wolf A 2001 *Astrophys. J.* **555** 1027–1037

[8] Savin D W *et al.* 2002 *Astrophys. J.* **576** 1098–1107

[9] Schippers S, Schnell M, Brandau C, Kieslich S, Müller A and Wolf A 2004 *Astron. Astrophys.* **421** 1185–1191

[10] Kallman T R, Palmeri P, Bautista M A, Mendoza C and Krolik J H 2004 *Astrophys. J. Suppl. Ser.* **155** 675–701

[11] Bryans P, Badnell N R, Gorczyca T W, Laming J M, Mitthumsiri W and Savin D W 2006 *Astrophys. J. Suppl. Ser.* **167** 343–356

# Evaluation of direct ionization cross sections for $C_{60}$ by electron interaction

**Neeraj Kumar and Satyendra Pal**
Department of Physics, M.M.H. College, Ghaziabad-201 001 (UP) INDIA
s.pal@india.com

**Abstract**. Partial ionization cross sections corresponding to the production of cations $C_{60}{}^{q+}$($q$=1-3) in the electron impact ionization of $C_{60}$ have been calculated using a modified Jain-Khare semiempirical formalism. The cross sections calculated in the energy range varying from ionization thresholds to 1000 eV revealed a reasonable good agreement with the available experimental and the theoretical data.

## 1. Introduction

The study of $C_{60}$ has attracted an enormous amount of interest in recent years, both from the experimental and the theoretical viewpoints because of its characteristic electronic structure reflecting its unique geometry. The exceptional stability, very high symmetry and several other molecular properties may provide the basis for important applications of this novel, natural state of carbon [1-2]. From this point of view of studying the electron ionization of this molecule, it offers a good opportunity for observing a very large molecule thus providing interesting tests of our understanding of ionization.

In the present work, we have extended and generalized the modified Jain-Khare semiempirical formalism for the evaluation of the partial ionization cross sections corresponding to the formation of the cations in the electron impact ionization of molecules [3-4] to the electron impact ionization of $C_{60}$. In the electron impact ionization of $C_{60}$, we have considered the $C_{60}{}^{q+}$ ($q$=1-3) ions produced in the direct single ionization processes for which the reliable theoretical and the experimental data have been reported in the literature. The recent experimental studies on the total single and partial ionization cross sections for $C_{60}$ include those of Baba et al. [5], Vostrikov et al. [6], Märk and co-workers [7-10] and Itoh et al. [11]. From a theoretical standpoint, calculations for $C_{60}$ are particularly challenging. The rigorous quantum mechanical approach for the calculations for molecules is limited to the application of simple molecules. Recently, Deutsch et al. [12] extended their DM formalism and the semiempirical concept based on the modified additive rule (MAR) [13] for the calculations for the ionic, fragment and total cross sections for fullerenes like $C_{60}$ and $C_{70}$. The present calculations for the ionization cross sections leading to the formation of singly, doubly, and triply charged ions in direct ionization of $C_{60}$ reveal a reasonable good agreement with the established experimental [8-11] and the theoretical data [12-13] in the energy range varying from the ionization threshold to 1000 eV.

## 2. Theoretical

The present calculations are carried out using the modified semi-empirical formalism developed by Khare and co-workers (see the discussion in Refs. 3-4). In brief, the ionization cross sections corresponding to the formation of the $i^{th}$ type of ion in the ionization of a molecule by the incident electron of energy $E$ is given by

$$Q_i(E)=\frac{4\pi a_0^2 R}{E}\left[\frac{E}{(E-I_i)\left(1+\frac{I_i}{E}\right)}\left(M_i^2(E)-\frac{R}{E}S_i(E)\right)\ln[1+C_i(E-I_i)]+\frac{R(E-I_i)}{E}S_i\int_0^{(E-I_i)}\frac{1}{(\varepsilon^3+\varepsilon_0^3)}\left(\varepsilon-\frac{\varepsilon^2}{(E-\varepsilon)}+\frac{\varepsilon^3}{(E-\varepsilon)^2}\right)d\varepsilon\right] \quad (1)$$

$M_i^2(E)$ and $S_i(E)$ are the integrals over the oscillator strengths, defined elsewhere [3]. $W (= \varepsilon + I_i)$, $I_i$, $a_0$, $\varepsilon_0$, $C_i$, $S_i$ and $R$ are energy loss suffered by the incident electron, the ionization threshold for the production of $i^{th}$ type of ion, the Bohr radius, energy parameter, collision parameter, number of ionisable electrons and Rydberg constant, respectively.

The present formulation requires the dipole oscillator strengths $df_i/dW$, as a major input, that is directly proportional to the photoionization cross section. In the present study, we have used the photoionization cross section data for the different cations in the energy range from ionization threshold to 25 eV from the measurement of Jaensch and Kamke [14]. In the energy range 26 to 130eV, the same were taken from the established data of Reinkoster et al. [15]. Above 130 to 280 eV, we have employed the total valance-shell photoionization cross section data measured by Berkowitz [16] and above 280 eV the same was extrapolated by Thomas-Reiche-Kuhn sum rule [16]. In their measurements, they estimated the total experimental uncertainty by ±20%. The total photo- absorption cross sections have been distributed into ionic fragments using the constant ionization efficiency above dipole breakdown limit. The collision parameter $C_i$ and energy parameter $\varepsilon_0$ were evaluated as for other polyatomic molecules [3-4]. The vertical onsets or the ionization potentials corresponding to the various cations are given elsewhere [14-16]. In the present evaluations of cross sections, the estimated error is expected as same as for the measurement of the photoionization cross sections.

## 2. Results and Discussion

Figure 1 shows the comparison of our partial ionization cross section profiles with the experimental [8-11] and the theoretical data [13]. It is interesting to note that the cross sections corresponding to singly charged ion is much higher than the doubly charged ions. Similarly the cross sections for triply charged ions are much smaller in comparison to singly and doubly charged ions. Although this is in contrast to the situation known for ordinary larger polyatomic molecules, where the parent ion is usually almost nonexistent, with the exception for the case of some hydrocarbons, it is nevertheless in accordance with our studies of ionization and fragmentation of the polyatomic molecules [4]. The reason for the larger cross section lies, on the one hand, in the fact that the production of doubly and triply charged ions is energetically more favourable. In addition to sheer size of $C_{60}$ and the fact that the secondary electrons from an initial single ionization process may be ejected into the empty centre of the case and subsequently interact anew with the electron cloud of the $C_{60}$ enhances the chance for the occurrence of inelastic-multiple electron collisions within the quantum system, thereby increasing the possibility for the production of highly charged states.

Formation of $C_{60}^+$ is particularly interesting case where the experimental and theoretical data provided by Mark and co-workers revealed a very unusual shape of cross sections. In the energy range from the ionization threshold to 200 eV, our calculated results are in good agreement with the corrected absolute experimental data of Matt et al. [8] (after Foltin et al. [9]), Tarnoviski et al. [10] and the theoretical data [13]. Above 200 eV, a slightly deviation has been noticed. On the other hand our results show a satisfactory agreement with the experimental data of Itoh et al. [11] available in the asymptotic region (400 - 1000 eV). For the sake of brevity, the experimental data of Baba et al. [5] and Vostrikov et al [6] are not shown in the comparison with our results for $C_{60}^+$. These data-sets exhibit much larger cross sections and quite different cross section shapes, strongly suggest that these data suffered by the unreliable vapour pressure data used for calibration, and in addition by discrimination effects leading to artefacts in the measured cross section curves.

The results of our calculations of the cross sections for the formation of doubly charged ion following electron impact on $C_{60}$ compare quite well with the experimental data sets [9-11]. However, our results in the energy range 100-300 eV, underestimate to the theoretical data of Deutsch et al. [13] but generally lie within the described uncertainties. In low energy region i.e. from the ionization threshold to 100 eV, our results corresponding to the formation of triply charged ions are in good agreement with the experimental and theoretical data sets [9,11,13]. In intermediate and the asymptotic regions, our results are just double to the calculations of Deutsch et al. [13] as well as the experimental data [9,11]. However, a qualitative agreement of the evaluated cross sections with these experimental and the theoretical data has been noticed. In case of the quantitatively dominating molecular ions $C_{60}^{+}$ and $C_{60}^{2+}$, our results are in satisfactory agreement with the recent measurements [9-11] and the theoretical calculations [13] within certain error bars (±20%). In case of triply charged ion, a large disagreement has been noticed. Nevertheless, considering that this is a comparison with the absolute data where error bars of these sets of data are easily in the 5% to 20% regime (in particular taking into account that the calculations are depending on the accuracy of the experimental input parameters) the agreement is reasonably good.

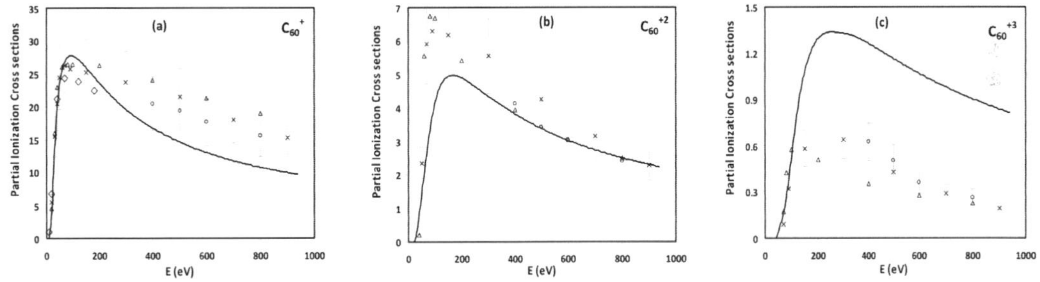


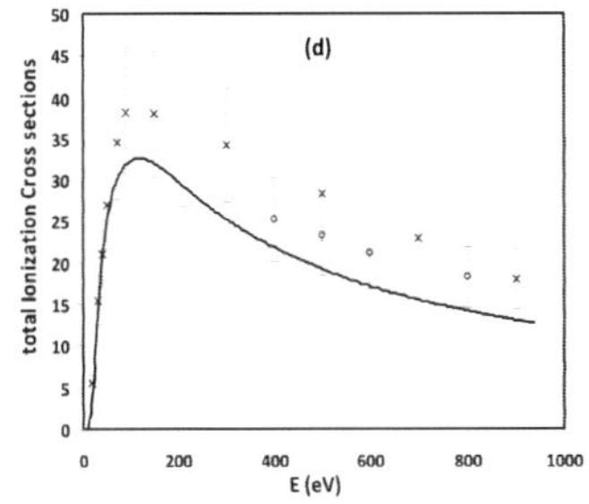


**Fig.1:** Partial and total ionization cross section profiles (in the units of $10^{-16}$ $cm^2$) for the electron impact ionization on $C_{60}$ (designated with solid lines) in comparison with the experimental data designated by: X– Foltin et al. [9], ◊- Tarnovski et al. [10] and O- Itoh et al. [11] and the theoretical data designated by: Δ- Deutsch et al. [13].

On the other hand, the partial cross sections when added upto a total ionization cross section is in good agreement with the experimental data [11]. Our results underestimate the total cross sections measured by Foltin et al. [9] but lie within the experimental uncertainties. Note that the DM formalism calculations [12] and additive calculations [13] for total ionization cross sections (not shown) are much higher than our calculations and other data-sets [9,11]. Conceptually, the DM calculation relies on the additive rule, which builds the total ionization cross section by adding up contributions from the different $C_{60}$ orbitals. This approach might not be applicable to a case-like structure such as $C_{60}$. It could be argued, for instance, that for the incoming projectile only about half of the 60 carbon atoms that constitute the $C_{60}$ case are contributing to the cross sectional area as seen by the projectile (see for instance Ref. 9 for detailed discussion). As already discussed above, there is a poor agreement between our results and the two earlier experimental data sets [5-6] both in terms of the cross section values and in terms of cross section shapes. However, the Jain-Khare formalism has limitations in applications to the cross section calculations and it depends on the accuracy of required input data. The present calculations for the partial and the total ionization cross sections satisfy the necessary consistency checks to access their consistency and reliability. The consistency checks derived from the fact that the total electron impact ionization cross section is equal to the sum of the partial ionization cross sections. This condition is used in the summation method for calibration purposes. In low energy limits, close to the onset of ionization, the shapes of the partial and total electron impact ionization cross section curves are governed by the certain threshold law [17]. The precise shape of the cross section in this region is especially important in determination of respective ionization thresholds, to compare with those derived by other means. The most surprising result in the present study is that the relative magnitude of the partial ionization cross sections is quite different from those of ordinary polyatomic molecules [4].

In conclusion, the modified Jain-Khare semiempirical formulation for the calculations for the partial ionization cross sections of molecules by electron impact has been extended and generalized to the calculations of the cross sections for the single and multiple ionization of fullerene $C_{60}$. Comparison of the calculated partial ionization cross sections for the production of singly and doubly charged ions following the electron impact on $C_{60}$, with the experimental and the theoretical data show a satisfactory agreement (within the composite error bars) in the complete energy range covered in the calculations. For triply charged ion, a considerable disagreement with the established experimental and theoretical data, except low energy region, has been noticed. The present extended approach also provides the framework to develop predictive capabilities that allow us to evaluate the partial differential and partial integral cross sections for fullerenes.

**Acknowledgements**
NK is thankful to the University Grants Commission, New Delhi for awarding project associateship (JRF) with Grant No. F.30-10(2004)(SR).

**References**
[1] "*The Chemical Physics of Fullerenes 10 (and 5) Years Later*" ed Andrson W 1996 (Kluwer Academic: Dordrecht)
[2] Campbell E E B and Rohmund F 2000 *Rep. Prog. Phys.* **63** 1061
[3] Pal S and Prakash S 1998 *Rapid Commun. Mass Spectrom.* **12** 297
[4] Pal S 2008 *Phys. Scr.* **77** 055304 and references therein
[5] Baba M S, Narasimhan T S L, Balasubramanian R and Mathews C K 1992 *Int. J. Mass Spectrom. Ion Proc.* **114** R1
[6] Vostrikov A A, Dubov D Y and Agarkov A A 1995 *Tech. Phys. Lett.* **21** 715
[7] Dünser B, Lezius M, Scheier P, Deutsch H and Märk T D 1995 *Phys. Rev. Lett.* **74** 3364
[8] Matt S, Dünser B, Lezius M, Deutsch H, Becker K, Stamatovic A, Scheier P and Märk T D 1996 *J. Chem. Phys.* **105** 1880
[9] Foltin V, Foltin M, Matt S, Scheier P, Becker K, Deutsch H and Märk T D 1998 *Chem. Phys. Lett.* **289** 181 and references therein
[10] Tarnovski V, Kurunczi P, Matt S, Märk T D, Deutsch H and Becker K 1998 *J. Phys.* B **31** 3043
[11] Itoh A, Tsuchida H, Miyabe K, Majima T, Imanishi N 1999 *J. Phys.* B**32** 277
[12] Deutsch H, Becker K, Pittner J, Bonacic- Koutecky V, Matt S and Märk T D 1996 *J. Phys.* B**29** 5175
[13] Deutsch H, Scheier P, Becker K and Märk T D 2003 *Int. J. Mass Spectrom.* **223-224** 1 &253
[14] Jaensch R and Kamke W 2000 *Mol. Mater.* **13** 143
[15] Reinkoster A, Korica S, Prumper G, Viefhaus J, Godehusen K, Schwarzkopf O, Mast M and Becker U 2004 *J.Phys.* B**37** 2135
[16] Berkowitz J 1999 *J. Chem. Phys.* **111** 1446
[17] *"Secondary Electron Spectra by Charged Particles Interactions"* International Commission on Radiation Units and Measurements 1996 ICRU Report No.5

# Determination of angular cross sections for electron dissociative ionization of the $CCl_4$ molecule

**Satyendra Pal, Anshu and Neeraj Kumar**
Department of Physics, M.M.H. College, Ghaziabad- 201 001 (UP) India

**E-mail:** s.pal@india.com

**Abstract.** Secondary electron energy and angle dependent differential cross sections for the product ions $CCl_3^+$, $CCl_2^+$, $CCl^+$, $C^+$, $Cl^+$, $Cl_2^+$ and $CCl_3^{++}$ resulting from the dissociative ionization of $CCl_4$ by electron collision have been calculated at fixed incident electron energy of 100 eV. The Jain-Khare semiempirical formulation that requires the oscillator strength data as input has been employed. The corresponding partial and total integral ionization cross sections have also been evaluated in the energy range varying from the ionization threshold up to 1000 eV. The partial and the total ionization cross sections are compared with the previously reported experimental and theoretical results leading to a reasonably good agreement.

## 1. Introduction

The present work- one in a series from our research plan dealing with the qualitative and quantitative knowledge of the electron impact ionization of free gaseous species [1-2] -is devoted to the carbon tetrachloride molecule ($CCl_4$). The $CCl_4$ molecule has attained a widespread interest in recent decades due to its applications in the modeling and optimization of industrial plasmas. It is used as a reactive etching gas for silicon wafers in semiconductor industry [3]. It is also an undesired hazardous pollutant with a large residence life time in the upper atmosphere and is a markedly efficient contributor to the greenhouse effect. Hence one object of this paper is to provide the reliable partial differential and the integral ionization cross sections to those who are working on the applications of $CCl_4$ in modeling and in understanding the effects in the atmosphere.

This paper reports the spectrum of the single differential cross sections with the energy of secondary electron(s) produced in the ionization of $CCl_4$ molecule by electron collision at a fixed incident electron energy of 100 eV, employing a semiempirical formalism based on the Jain and Khare approach (see for instance Refs.1-2). At this incident electron energy, the doubly differential cross sections as a function of angle and secondary electron energy have also been calculated. To the best of our knowledge, no experimental and/or theoretical data are available for comparison of our calculations for differential cross sections. Thus, the corresponding derived integral cross sections in terms of the partial ionization cross sections leading to the formation of various cations viz. $CCl_3^+$, $CCl_2^+$, $CCl^+$, $C^+$, $Cl^+$, $Cl_2^+$ and $CCl_3^{++}$ through the dissociation of $CCl_4$ by electron collision and their sum (total ionization cross section) show an acceptable agreement with the available experimental [4-7] and theoretical data [8-9].

## 2. Theoretical

The present calculations are carried out using the modified semi-empirical formalism developed by Khare and co-workers [1-2]. In brief, the single differential cross sections in the complete solid angle Ω as a function of secondary electron energy ε corresponding to the production of the $i^{th}$ type of ion in the ionization of a molecule by incident electron of energy $E$ is given by

$$Q_i(E,W)=\frac{a_0^2R^2}{E}\left[(1-\frac{\varepsilon}{(E-I_i)})\frac{1}{W}\frac{df_i(W,0)}{dW}\ln[1+C_i(E-I_i)]+\frac{1}{E}S\frac{(E-I_i)}{(\varepsilon_0^3+\varepsilon_i^3)}\left(\varepsilon-\frac{\varepsilon^2}{(E-\varepsilon)}+\frac{\varepsilon^3}{(E-\varepsilon)^2}\right)\right]\int d\Omega \quad (1)$$

Where $W$ (= $\varepsilon + I_i$ ) is defined as energy loss suffered by the incident electron. $I_i$, $a_0$, $\varepsilon_0$, $C_i$, $S$ and $R$ are the ionization threshold for the production of $i^{th}$ type of ion, Bohr radius, energy parameter, collision parameter, number of ionisable electrons and Rydberg constant, respectively.

In the present formulation, the dipole oscillator strengths $df_i/dW$ are the key parameters. The oscillator strength is directly proportional to the photoionization cross section [11]. In the present study, we have used the partial photoionization cross section data in the energy range from ionization threshold to 80 eV provided by Burton et al. [10] using dipole (e,e+ion) spectroscopy with resolution at 1 *fwhm*. The accuracy of the determined oscillator strengths scales was estimated to be better than ±5%. In the photon energy range 80-200 eV, we have used their measured total valence photoabsorption oscillator strength data and above 200 eV the same was extrapolated by Thomas-Reiche-Kuhn sum rule within ±10% uncertainty (see for instance Refs. 10). Thus the derived total photoabsorption cross sections have been distributed into ionic fragments using the constant ionization efficiency above the dipole breakdown limit of 30eV. The collision parameter $C_i$ and energy parameter $\varepsilon_0$ were evaluated as for other polyatomic molecules [1-2]. The vertical onsets or the ionization potentials corresponding to the various cations are given elsewhere [10]. In the present evaluations of cross sections, the estimated uncertainty is more or less the same as for the measurement of the photoionization cross sections.

The double differential cross sections as a function of energy and angle were evaluated by the differentiation of equation (1) with respect to the angle θ. The double differential calculations require an energy dependent asymmetric parameter $\beta$ (see Ref. 1 for detailed discussion). Its evaluation is difficult due to the lack of wave functions for the polyatomic molecules. However an attempt was made to evaluate $\beta$ for the inner-shell photoionization of $CCl_4$ in the high energy range using a quantum mechanical treatment [11]. In the valance-shell ionization of $CCl_4$, we have computed $\beta$ as the ratio of the Bethe spectral transitions $S_i(W)$ to the dipole matrix squared $M_i^2(W)$ [12].

The partial ionization cross section is obtained by the integration of the energy dependent single differential cross sections (equation 1) over the entire energy loss W and the total ionization cross section is obtained by summing over all types of ions i.

## 3. Results and discussion

The partial single differential cross sections as a function of secondary electron energy in terms of energy loss W at fixed incident electron energy of 100 eV are presented in Fig.1. In the figure, we have also shown the behaviour of the double differential cross sections with angle varying from 10° to 180° at constant secondary electron energy of 10 eV and the fixed primary electron energy of 100eV. The calculations as a function of energy loss W, at constant angle 30° and at the same incident electron energy have also been presented. A 3D profile of the total double differential cross section varying with the secondary electron energy and angle is also shown in the figure. To the best of our knowledge, no experimental data are available to compare the present results for the differential cross sections. However, the qualitative behaviour of the cross sections are the same as for other molecules investigated in this laboratory. The energy dependent cross sections are symmetric at W/2 where the energies of primary and the secondary electrons are almost equal. The present calculations account for the exchange effects through the non-dipole cross section term (second part) of the formulation at about W/2. The differential cross section profiles clearly show the weight contribution of the molecular and atomic cations. The atomic photoionization cross sections include the contribution of the structures and many body states produced near onsets which are speculated in the present calculations for the energy dependent differential cross sections [10].

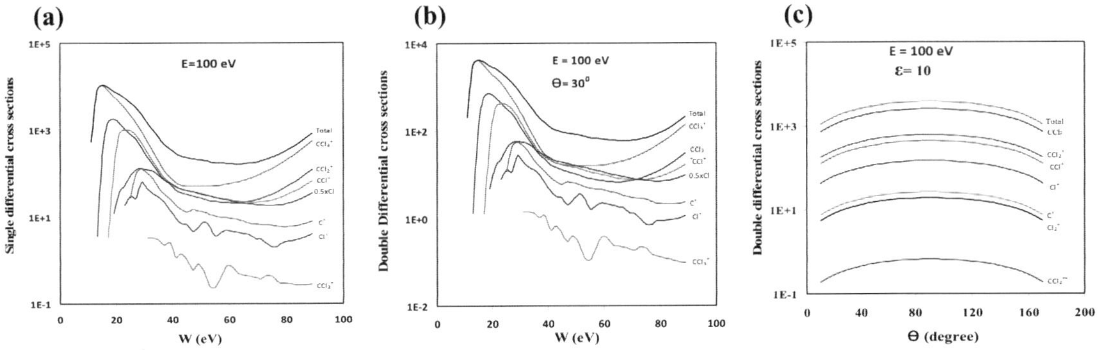


**(d)**

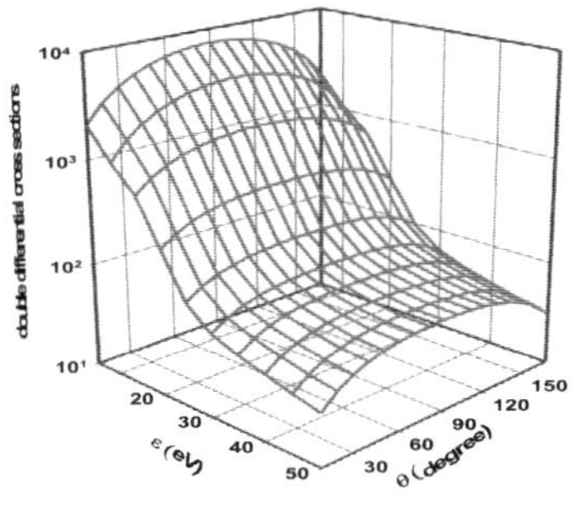


**Fig.1:** (a) The single differential cross sections (in the units of $10^{-20}$ $cm^2/eV$) as a function of energy loss W for the production of cations from electron impact ionization of $CCl_4$ at constant electron impact energy of 100 eV. (b) Same as (a) but for the double differential cross sections (in the units of $10^{-20}$ $cm^2/eV\text{-}sr$) at ($\theta$ =30° and E=100eV). (c) same as (b) but as a function of angle at constant secondary electron energy $\varepsilon$=10 eV. (d) Same as (b) but as for the total double differential cross sections as a function of angle $\theta$ and energy $\varepsilon$ (in the range of 0 to W/2).

Fig.2 shows the comparison of our evaluated partial and the total ionization cross section profiles with the recent experimental data [5-6].The experimental measurement of Leiter et al. [4] suffered a systematic error in the determination method, possible in the measured $CCl_4$ absolute pressure and hence not shown in the comparison. In the threshold and asymptotic regions the calculated results show a satisfactory agreement with the both experimental data [5-6]. In the case of quantitatively dominating molecular ions $CCl_3^+$ and $CCl_2^+$, our results are in satisfactory agreement, within experimental error bars, with the recent measurements [5-6]. However, our results for $CCl_3^+$ underestimate to the experimental data and for $CCl_2^+$ overestimate to the same experimental data sets. In the case of other smaller ions, our results underestimate than the experimental measurements [5-6]. Generally, a large disagreement has been noticed in the case of atomic ions and dications. The inter-comparison of the experimental data for atomic ions and dications also reveal a large disagreement among themselves. However, the modern experiments claim 100% ion collection efficiency yet may suffer with experimental error and mass discrimination effects. In addition to this, the calibration of the experimental results with standard results of inert gases may also be responsible for disagreement. Nevertheless, considering that this is a comparison with the absolute data where error bars of these sets of data are easily in the 5% to 25% regime (in particular taking into account the facts that the calculations depend on the accuracy of the experimental input parameters) the agreement is acceptable. The partial cross sections, when added up to a total cross section reveal a good agreement with the experimental data sets [5-7] alongwith the DM calculations, the MAR calculations and the modified BEB calculations carried out by Martinez et al. [8]and the complex potential model calculations of Joshipura et al. [9], in the complete energy range covered in the present calculations. The present calculations for the partial and the total ionization cross sections satisfy the necessary consistency checks to access their consistency and reliability [2].

For the first time, we have evaluated the secondary electron energy and angle dependent differential cross sections at a fixed impinging electron energy corresponding to the formation of various cations in electron impact dissociative ionization for the $CCl_4$ molecule by employing a semiempirical formalism based on the Jain-Khare approach. The corresponding derived partial ionization cross sections revealed a disagreement with the experimental data as expected. The total ionization cross sections were in excellent agreement with the established experimental and theoretical data. For the sake of accuracy and the consistency of the calculated data, further experimental measurements are undoubtedly needed.

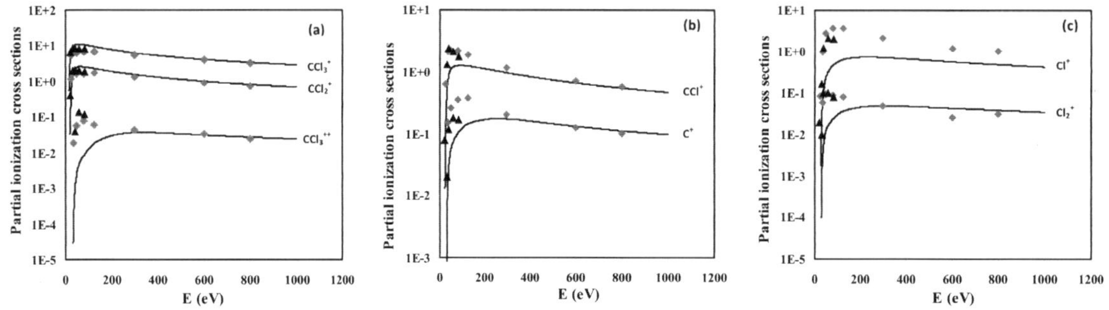


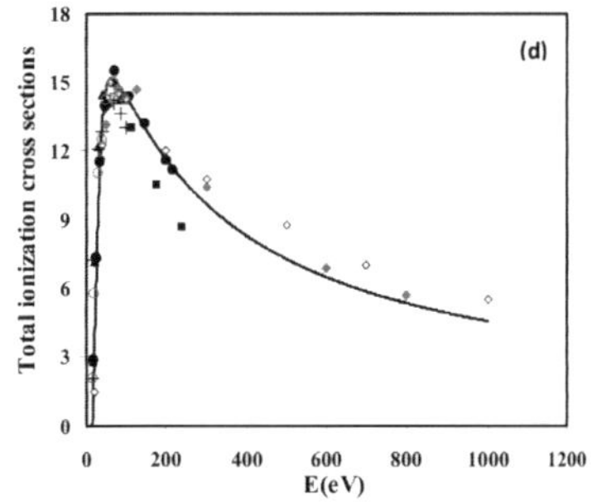


**Fig.2:** Ionization cross sections (in the units of $10^{-16}$ $cm^2$) and rate coefficients for electron impact ionization of $CCl_4$ (designated by solid line) in comparison with the various experimental data sets designated by:♦– Lindsay et al. [5], ▲- Sierra et al. [6], and ● – Bart et al. [7] and the theoretical data sets designated by: (o - modified BEB calculations , ■ – DM calculations and + – MAR) [8] and ◊ – Complex potential model calculations [9].

**Acknowledgements**
NK is thankful to the UGC for awarding project associateship with Grant No. F.30-10(2004)(SR). Anshu thanks the DAE for JRF award, through Grant No. 2007/37/13/BRNS.

**References**

[1] Pal S, Kumar J and Märk T D 2004 *J. Chem. Phys.* **120** 4658
[2] Pal S 2008 *Phys. Scr.* **77** 055304 and references therein
[3] *"Film Deposition by Plasma Techniques"* Konuma M 1992 (Springer-Verlag: Berlin Heidelberg)
[4] Leiter K, Stephan K, Märk E and Märk T D 1984 *Plasma Chem. Plasma Process.* **4** 235
[5] Lindsay B G, McDonald K F, Yu W S, Stebbings R F and Yousif F B 2004 *J. Chem. Phys.* **121** 1350
[6] Sierra B, Martínez R, Redondo C and Castańo F 2005 *Int. J. Mass. Spectrom.* **246** 105
[7] Bart M, Harland P W, Hudson J E and Vallance C 2001 *Phys. Chem. Chem. Phys.* **3** 800
[8] Martínez R, Sierra B, Redondo C, Sánchez Rayo M N and Castańo F 2004 *J. Chem. Phys.* **121** 11653
[9] Joshipura K N, Vinodkumar M, Limbachiya C G and Antony B K 2004 *Phys. Rev.* A**69** 022705
[10] Burton G R, Chan W F, Cooper G and Brion C E 1994 *Chem. Phys.* **181** 147
[11] Natalense A P P, Brescansin L M and Lucchese R R 2003 *Phys. Rev.* A**68** 032701
[12] Inokuti M 1971 *Rev. Mod. Phys.* **43** 297

# The first test experiment performed at the electron cooler of storage rings in Lanzhou

**L J Meng[1,2], X Ma[1], H P Liu[1], X D Yang[1], J W Xia[1], H S Xu[1], Z G Hu[1], X L Zhu[1], M Wang[1], R S Mao[1], D C Zhang[1], L J Mao[1], J Li[1], G H Li[1], Y Liu[1], J C Yang[1], Y J Yuan[1], J H Zheng[1], X T Yang[1], G Q Xiao[1], W L Zhan[1]**
[1] Institute of Modern Physics, Chinese Academy of Sciences, Lanzhou, 730000, China
[2] The Graduate University of the Chinese Academy of Sciences, Beijing, 100049, China

E-mail: x.ma@impcas.ac.cn

**Abstract.** The cooler storage ring (CSR) project was launched in 2000 at the Institute of Modern Physics, Chinese Academy of Sciences, Lanzhou. In 2007, the installation was completed and the commissioning of CSRs gained great success, a new highly precise generation of collision experiments will become accessible even for the heaviest ion species. A commissioning RR experiment was performed at the electron cooler with $Ar^{18+}$ ions, the results are reported. And the further development of the experiments at cooler will be discussed.

## 1. Introduction

Recombination between electrons and ions is one of the most fundamental atomic collision processes for all kinds of plasmas in the universe. The cross sections and the rate coefficients are required for understanding of astrophysical processes and fusion plasma. For the investigation of these processes, the experimental environment at the electron cooler storage ring is in particular well suited. Here the interaction of cooled heavy ions with electrons can be studied under well controlled conditions via the spectroscopy of projectile x-rays and by the analysis of projectile charge-exchange. In the electron cooler environment, two processes compete: dielectronic recombination (DR) and radiative recombination (RR). The rate coefficient, most important parameter in plasma modelling, can be obtained for various electronic configurations and in addition, detailed information on the atomic structure can be deduced. Theoretical investigations of the DR process show that DR might be a powerful tool for the investigation of the influences of nuclear effects on the atomic structure and may even be used to obtain model independent information on the nuclear structure, e.g. the nuclear charge radius. This has just been confirmed experimentally at the experimental storage ring (ESR) in Darmstadt [1].

## 2. The cooler storage rings and experimental devices

The cooler storage ring (CSR) project was launched in 2000 at the Institute of Modern Physics, Chinese Academy of Sciences, Lanzhou. The project consists of two rings, the main cooler storage ring (CSRm) and the experimental cooler storage ring (CSRe), as shown in figure 1. The CSRm works as a synchrotron which can accelerate the ions energy up to GeV/u (for light ions) and up to 500MeV/u (for heavy ions) [2]. Both CSR rings are equipped with electron cooling devices, the third generation of electron cooler, which can produce different electron density distributions [3]. In 2007, the installation was completed and the commissioning of CSRs gained great success, a new highly precise generation of collision experiments will become accessible even for the heaviest ion species. The CSR provides unique and unprecedented conditions for experiments based on the use of highly charged ions and in particular for research in the realm of atomic and nuclear physics. The interaction of the brilliant beams of cooled high-Z ions with low-dense gaseous matter as well as with electrons can be now studied under almost completely background free experimental conditions and with highest luminosity [4].

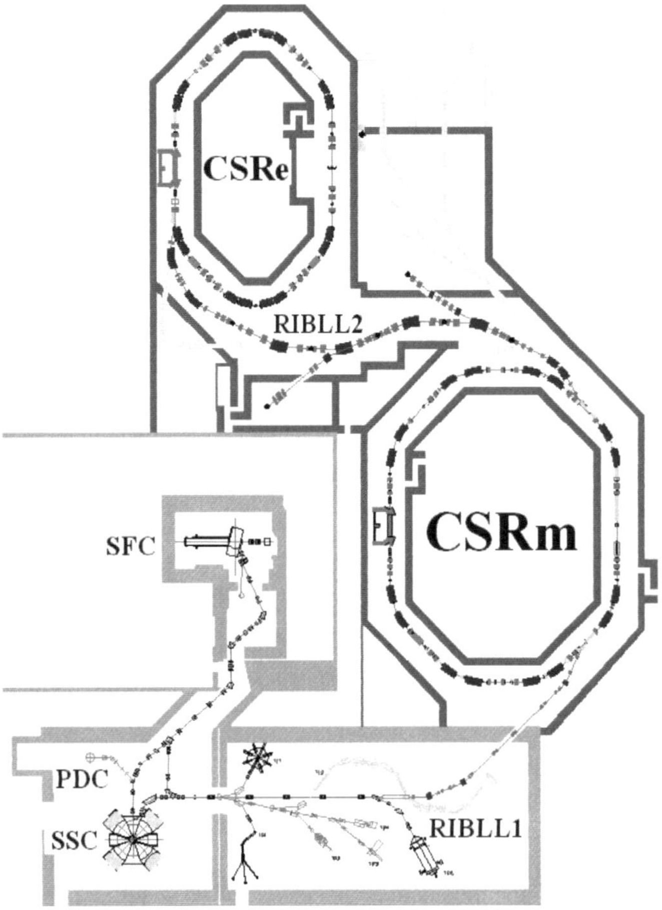


**Figure 1.** Overview of the Cooler Storage Rings and the heavy ion research facility at Lanzhou.

Between CSRm and CSRe a radioactive beam line was constructed, defined as RIBLL2 in figure 1. The ions in CSRm are accelerated to energy range of 100–500 MeV/u and then extracted fast to produce radioactive ion beams by nuclear fragmentation or highly charged heavy ions by thin foil stripping. The secondary beams (RIB or highly charged heavy ions) will be accepted and stored by the experimental ring, and RR and DR experiments can be performed at the CSRe cooler. DR experiments at CSRe will open a novel way for studying the ground-state properties of nuclei far from stability.

The recombination experiments were carried out at the electron cooler, the charge changed ions were detected by a particle detector located down stream after the dipole magnet, see figure 2 for details. In the present experiment, a plastic scintillator is employed as a particle detector. The particle detector is installed in a detector pocket which isolates the detector from ultra-high vacuum environment by a

30μm stainless steel window. The position of the particle detector together with the pocket can be controled by a step motor through computer network.

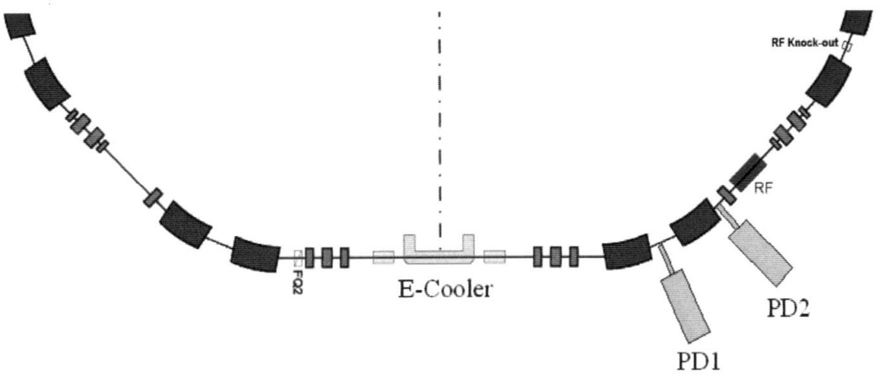


**Figure 2.** Layout of the electron cooler section at CSRm and configuration of the particle detectors.

## 3. Experimental results

The first test experiment was carried out at CSRm cooler with $Ar^{18+}$ ions at an energy of 21.44 Mev/u. The electron beam energy was set at the cooling point (11.765 keV) and the electron beam was always switched on during injection. The cooling of the beam is very effective, figure 3 shows the coasting beam lifetime in the CSRm, where electron beam intensities were set to 200mA, 150mA, 100mA, 50mA, and 0mA in the experiments. As can be seen from the fig. 3, for higher electron beam intensities cooling is better. The stored beam is lost quickly when the cooling electron beam was switched off right after the injections. This indicates that the interactions between stored ions and residual gas, the instability in magnetic fields may be the dominant contributions.

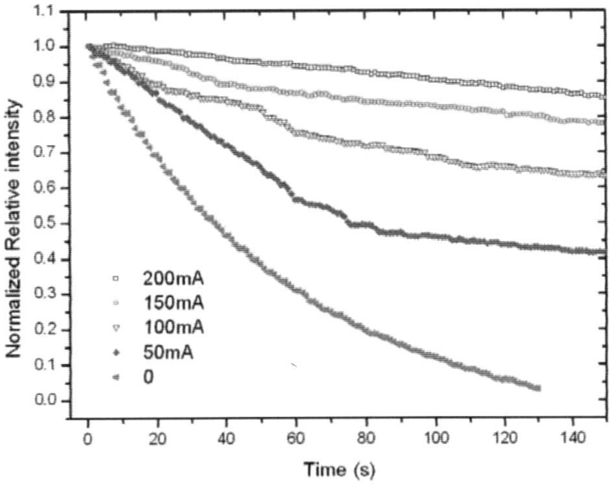


**Figure 3.** Stored ion beam lifetime measured at different cooling electron beam intensities.

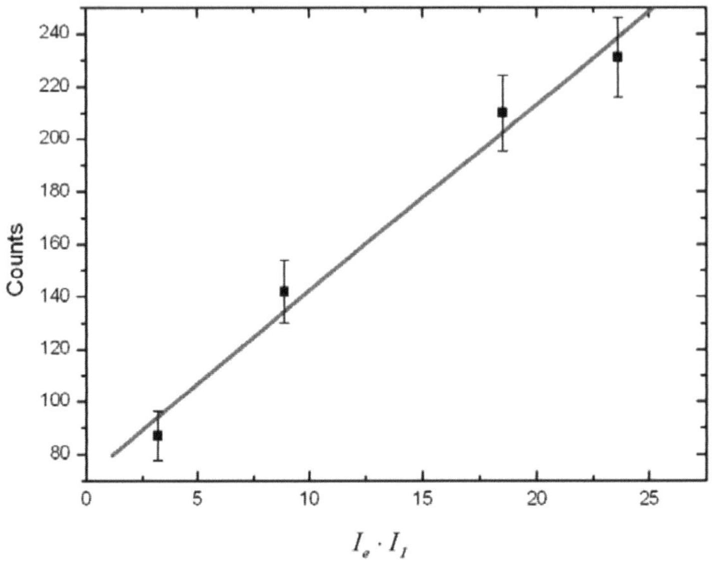


**Figure 4.** Registered ion counts as a function of the product of the electron beam intensity $I_e$ and the coasting beam intensity $I_I$ in units [$mA^2$]. The data points are measured at different electron beam intensities. The solid line represents a linear fit to the data points.

In the electron cooler the ions recombine with electrons resulting in the reduction of one charge unit of the projectile. The ions deviate the coasting beam orbit in the magnetic field and are recorded by the particle detector. Figure 4 shows the registered ion counts versus the electron beam intensities and the coasting beam intensities. It is clear that the recombination rates are proportional to the products of the electron beam intensities and the coasting beam intensities.

## 4. Outlook

In the commissioning of the CSR $C^{6+}$, $Ar^{18+}$ and $Xe^{44+}$ ions were injected, accelerated, and stored, and the first radiative recombination experiments was performed at the electron cooler. It is a good opportunity to study the interactions of highly charged ions with matter at the CSR in Lanzhou. Further RR and DR experiments are being under preparations. A program has been launched for fast fine detuning of the electron beam energy relative to the stored ion beam in order to perform high precision DR experiments.

This work would not be possible without the help of CSR team. The authors would like to express thanks to the CSR team.

**References**

[1] C. Brandau, C. Kozhuharov, Z. Harman, et. al., Phys. Rev. Lett. 100 (2008) 073201

[2] J. W. Xia, W.L. Zhan, B.W. Wei, et. al., Nucl. Instr. Meth. A 488 (2002) 11

[3] X. D. Yang, V.V. Parkhomchuk, W.L. Zhan, J.W. Xia, et al., Proceedings of COOL2007, Sept. 9-14, 2007, Bad Kreuznach, Germany, pp59 (http://accelconf.web.cern.ch/AccelConf/cl07/papers/tum1i02.pdf)

[4] J. W. Xia, W. L. Zhan, Y. J. Yuan, et. al., Proc. of 2007 Asian Particle Accelerator Conference (APAC), Jan 29-Feb 2, Indore, India (http://accelconf.web.cern.ch/AccelConf/a07/PAPERS/THXMA03.PDF)

# Angular momentum transfer and polarization degree of ions with one-valence electron by electron impact

**Kenichi Akita, Takeshi Kai*, Shinobu Nakazaki, Akinori Igarashi**

Department of Applied Physics, Faculty of Engineering, University of Miyazaki, Miyazaki 889-2192, Japan

(*) Quantum Beam Science Directorate, Japan Atomic Energy Agency 8-1 Umemidai Kizugawa-city Kyoto 619-0215, Japan

E-mail: akita@phys.miyazaki-u.ac.jp

**Abstract.** We carry out the $R$-matrix calculations for electron-impact excitations of ions with one valence electron. The integral cross sections and polarization degree are obtained for the excitation process from the ground state to the first $^2P^o$ state of $Li^{2+}$, $B^{2+}$ and $Al^{2+}$ as functions of electron incident energy. The differential cross sections and angular momentum transfer are also shown at non-resonant low-energy points. As for the angular momentum transfer ($L_\perp$) at small scattering angles, they are negative for $B^{2+}$ and $Al^{2+}$, while it is positive for $Li^{2+}$. Thus $L_\perp$ of doubly charged ions with one-valence electron is not simple.

## 1. Introduction

Excitation cross sections for positive ions by electron impact are required in order to interpret spectra of astrophysical plasmas and high-temperature laboratory plasmas. The cross sections from the ground S state to the lowest P state are of fundamental importance.

When velocity distribution of electrons in plasma is anisotropy, ions in excited states have unequal distribution for magnetic sublevels owing to collisions with such electrons. Radiation emitted from ions with unbalanced sublevel population is polarized. The detection of polarization degree is available for a plasma diagnostic [1].

In magnetic sublevels of final P state of atoms and ions formed in an electron-impact excitation from an initial S state, the expectation value of component of perpendicular to the scattering plane in angular momentum is called *angular momentum transfer* $L_\perp$. We pay attention to $L_\perp$ at small scattering angles. In electron-atom scattering, the value of $L_\perp$ near forward direction is positive for S $\rightarrow$ P transition [2]. It hold without exception in reported works. Srivastava *et al.*[3] have shown using the distorted wave calculation that $L_\perp$ is also positive for electron collisions with H-like ions (Z=2, 6, $\infty$) at X=2 and 4, where X is the incident electron energy in threshold unit. Kai *et al.*[4] have found a case where $L_\perp$ was negative in low-energy electron scattering with Mg-like Si ion using the $R$-matrix method.

In order to obtain acquaintances about $L_\perp$ in electron-impact excitation of doubly charged ions, we have applied the $R$-matrix calculation in electron scattering for ions with one valence electron such as H-like Li ($Li^{2+}$), Li-like B ($B^{2+}$) and Na-like Al ($Al^{2+}$). For the excitation processes of $Li^{2+}$(1s $\rightarrow$ 2p), $B^{2+}$(2s $\rightarrow$ 2p) and $Al^{2+}$(3s $\rightarrow$ 3p), the integral cross section (ICS) and polarization degree ($P$) are calculated, and the differential cross section (DCS) and $L_\perp$ are presented at selected energies in non-resonant region.

**Table 1.** Target states in the present R-matrix calculations.

| Symmetry | $Li^{2+}$(RM13) | $B^{2+}$(RM11) | $Al^{2+}$(RM11) |
|---|---|---|---|
| $^2S^e$ | 1s, 2s, 3s, 4s, 5s | 2s, 3s, 4s, $5\bar{s}$ | 3s, 4s, 5s |
| $^2P^o$ | 2p, 3p, 4p, 5p | 2p, 3p, 4p, $5\bar{p}$ | 3p, 4p, 5p |
| $^2D^e$ | 3d, 4d, 5d | 3d, 4d | 3d, 4d, 5d |
| $^2F^o$ | 4f | 4f | 4f, 5f |

**Table 2.** Excitation energies (eV) and oscillator strengths for $Li^{2+}$, $B^{2+}$ and $Al^{2+}$.

| | | Excitation energy | | Oscillator strength | | | |
|---|---|---|---|---|---|---|---|
| Target | State | Present | Other | Present | | Other | |
| | | | | Length | Velocity | Length | Velocity |
| $Li^{2+}$ | 2p $^2P^o$ | 91.839 | 91.841[a] | 0.4162 | 0.4162 | | |
| $B^{2+}$ | 2p $^2P^o$ | 6.0508 | 5.9984[a]<br>5.9913[b] | 0.3745 | 0.4089 | 0.36327[b] | 0.36325[b] |
| $Al^{2+}$ | 3p $^2P^o$ | 6.6089 | 6.6751[a]<br>6.6124[c] | 0.889 | 0.844 | 0.890[c] | 0.847[c] |

[a] NIST [8]. [b] Godefroid *et al.* [9]. [c] Dufton and Kingston [10].

## 2. Theory

### 2.1. Target wavefunctions and collision calculations

Target states used in the present R-matrix calculations are listed in Table 1. The target wave functions of $Li^{2+}$ are exact. The target states of $B^{2+}$ and $Al^{2+}$ are constructed in terms of configuration interaction expansions. The number after the letter "RM" in Table 1 indicates the number of coupled targets states in the calculations. The 1s and 2s orbitals of the $B^{2+}$ are the Hartree-Fock wave functions given by Clementi and Roetti [5] for the ground state $1s^2 2s$ $^2S^e$.The 2p, 3s, 3p, 3d, 4s, 4p, 4d and 4f orbitals of $B^{2+}$ are optimized on the energies of the 2p $^2P^o$, 3s $^2S^e$, 3p $^2P^o$, 3d $^2D^e$, 4s $^2S^e$, 4p $^2P^o$, 4d $^2D^e$ and 4f $^2F^o$ states, respectively, using the CIV3 atomic structure program of Hibbert [6]. The $5\bar{s}$ and $5\bar{p}$ pseudo-orbitals of the $B^{2+}$ are also determined by optimizing on the energy of the 2s $^2S^e$ and 2p $^2P^o$, respectively.For the orbitals of $Al^{2+}$, we adopt the orbitals used by Kimura *et al.* [7]. Table 2 shows excitation energies and oscillator strengths between the ground state and the first excited state in $^2P^o$ symmetry of $Li^{2+}$, $B^{2+}$ and $Al^{2+}$. Present excitation energies are in good agreement with those in [8, 9, 10]. There is a difference between length and velocity forms of our results. Our length forms compare reasonably well with the values reported by Godefroid *et al.* [9] and Dufton and Kingston [10]. Our wavefunctions of $B^{2+}$ are less accurate compared with those of Godefroid *et al.*, since much more configurations are used in their multiconfiguration Hartree-Fock method.

We perform scattering calculations for target $Li^{2+}$, $B^{2+}$ and $Al^{2+}$ using the $R$-matrix computer program of Berrington *et al.* [11].

### 2.2. The angular momentum transfer and polarization degree

For electron-impact excitation $^2S^e \rightarrow {}^2P^o$ in the present case, the angular momentum transfer $L_\perp$ at a given scattering angle is given by $L_\perp = -Im[f_1^0 f_0^{0*} + 3f_1^1 f_0^{1*}]/(\sqrt{2}\sigma)$ [12], where $f_M^S$ is the scattering amplitude for a magnetic sublevel with $M$ of the $^2P^o$ state and for the total spin $S$, and $\sigma$ is the DCS for the $^2P^o$ state. The ICS for the $^2P^o$ state is given by $Q = Q_0 + 2Q_1$. The $Q_M$ is the ICS for the excitation into a magnetic sublevel with $M$ of the $^2P^o$ state. The

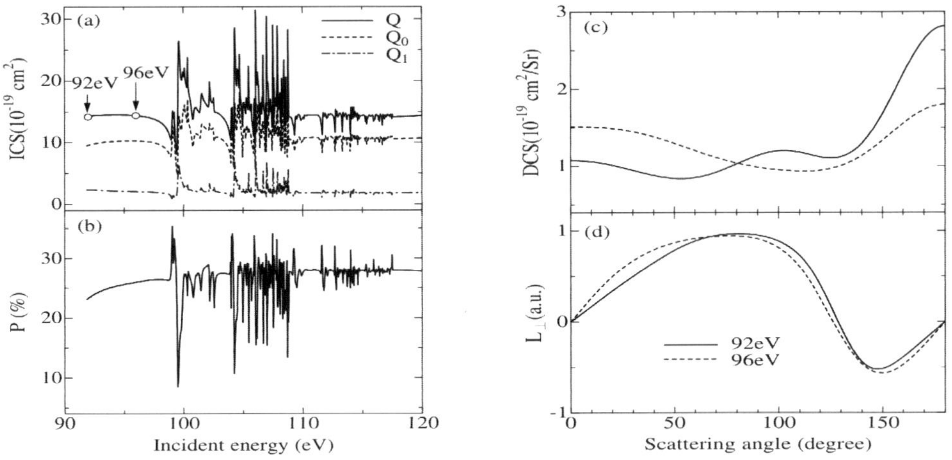


**Figure 1.** (a) ICS, (b) $P$, (c) DCS and (d) $L_{\perp}$ for the 1s $^2$S$^e \rightarrow$ 2p $^2$P$^o$ transition of $Li^{2+}$ above the threshold.

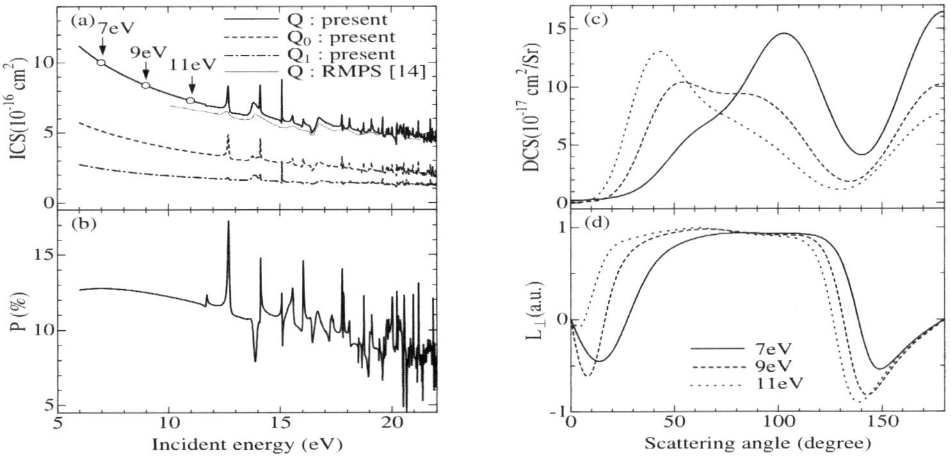


**Figure 2.** The same as Fig. 1 but for the 2s $^2$S$^e \rightarrow$ 2p $^2$P$^o$ transition of $B^{2+}$.

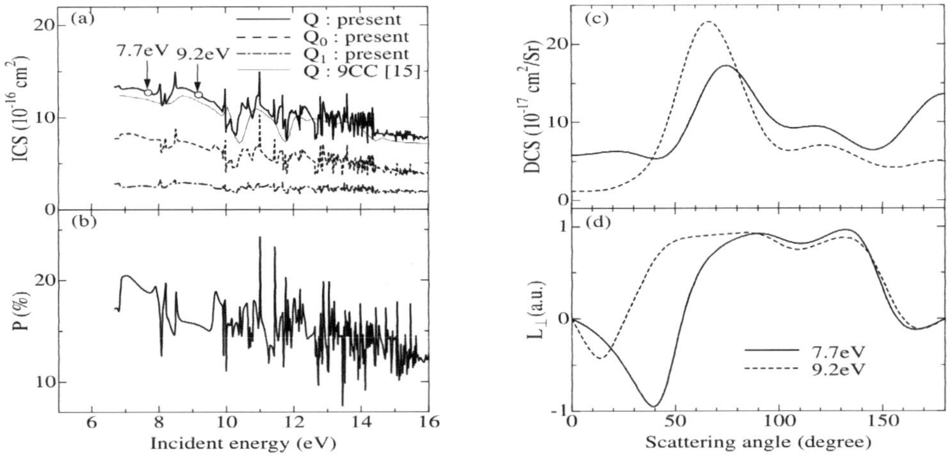


**Figure 3.** The same as Fig. 1 but for the 3s $^2$S$^e \rightarrow$ 3p $^2$P$^o$ transition of $Al^{2+}$.

polarization degree for the $^2S^e \to {}^2P^o$ transition is given by $P = 3(Q_0 - Q_1)/(7Q_0 + 11Q_1) \times 100$ [13]. The quantization axis is along the incident direction.In the case of neutral atom target, the emission line from the $^2P^o \to {}^2S^e$ transition is linearly polarized and $P$ is about 43 % at threshold [13], which is derived by setting $Q_1 = 0$ in the above expression.When the target is an ion, $Q_1$ does not vanish at the threshold and $P$ deviates from 43 %.

## 3. Results and discussion

Figure 1(a) and (b) show present results for ICS and $P$ of $Li^{2+}$. The $P$ is about 23 % at the threshold. The DCS and $L_\perp$ at incident energies 92 and 96 eV are shown in Fig. 1(c) and (d). As the incident energy increases, the DCS's increases at small angles and decreases at large angles. $L_\perp$'s are positive at small angles, which is consistent with the result of Srivastava *et al.* for hydrogenic ions [3].

The results of $B^{2+}$ are shown in Figure 2. Fig. 2(a) includes the ICS calculated by Marchalant *et al.* [14] using the $R$-matrix method with pseudo-states (RMPS), which used nine physical states and sixteen pseudo states. Though the calculation of Marchalant *et al* is more elaborate than the present one, the agreement of the two ICS's is good. The $P$ is about 13 % at the threshold (Fig. 2(b)) and monotonic up to 11.5 eV. We show the DCS's at energies around 7 – 11 eV in Fig. 2(c). The DCS near backward direction decreases with energies, while it remains small near forward angle. $L_\perp$ in Fig. 2(d) is negative at small angles.

The results for $Al^{2+}$ are shown in Figure 3. Mitroy and Norcross [15] have reported the ICS in a close-coupling calculation with nine states (9CC) of five atomic state and four pseudo states. In Fig. 3(a), the present ICS is larger than that of Mitroy and Norcross. This is due to difference of states adopted in the calculations. The $P$ is about 17 % at the threshold and shows a rapid increase due to a resonance just above the threshold in Fig. 3(b). We show the DCS (Fig. 3(c)) and $L_\perp$ (Fig. 3(d)) at incident energies 7.7 and 9.2 eV. The DCS values of 9.2 eV at forward and backward directions is smaller than those of 7.7 eV. $L_\perp$ is negative at small angles like the $L_\perp$ of $B^{2+}$ case.

To summarize, the ICS, $P$, DCS and $L_\perp$ for $Li^{2+}$, $B^{2+}$ and $Al^{2+}$ are calculated using $R$-matrix method. The DCS and $L_\perp$ are calculated at selected energies in non-resonant region. The $L_\perp$ at small angles are positive for $Li^{2+}$ and negative for $B^{2+}$ and $Al^{2+}$, and $L_\perp$ of doubly charged ions with one-valence electron is not simple. Further studies are required to understand the physical mechanism.

## References

[1] T. Kawamura, T. Kai, F. Koike, S. Nakazaki, Y. Inubushi, H. Nishimura, Phys. Rev. Lett. **99** 115003 (2007).
[2] N. C. Steph, D. E. Golden, Phys. Rev. A **21**, 1848 (1980)
[3] R. Srivastava, Y. Itikawa and K. Sakimoto, Phys. Rev. A **43**, 4736 (1991).
[4] T. Kai, R. Srivastava and S. Nakazaki, J. Phys. B: At. Mol. Phys. **37**, 2045 (2004).
[5] E. Clementi and C. Roetti, At. Data Nucl. Data Tables **14**, 177 (1974).
[6] A. Hibbert, *Comput. Phys. Commun.* **9**, 141 (1975).
[7] E. Kimura, A. Ohsaki, S. Nakazaki and Y. Itikawa, Astron. Astrophys. Suppl. **132**, 99 (1998).
[8] See http://physics.nist.gov/cgi-bin/AtData/levels_form
[9] M. Godefroid, C. Froese Fischer and P. Jönsson, J. Phys. B: At. Mol. Phys. **34** 1079 (2001).
[10] P. L. Dufton and A. E. Kingston, J. Phys. B: At. Mol. Phys. **20** 3899 (1987).
[11] K. A. Berrington, W. B. Eissner and P. H. Norrington, *Comput. Phys. Commun.* **92**, 290 (1995).
[12] N. Andersen and K. Bartschat, *Polarization, alignment, and orientation in atomic collisions* (Springer-Verlag New York, Inc. 2001).
[13] I. C. Percival and M. J. Seaton, Phil. Trans. Roy. Soc. A **251**, 113 (1958).
[14] P. J. Marchalant, K. Bartschat and I. Bray, J. Phys. B: At. Mol. Phys. **30** L435 (1997).
[15] J. Mitroy and D. W. Norcross, Phys. Rev. A **39** 537 (1989).

# Effective collision strengths for Mg-like iron peak ions

**P H Norrington and C E Hudson**

Department of Applied Mathematics and Theoretical Physics, The Queen's University of Belfast, Belfast BT7 1NN, UK

E-mail: p.norrington@qub.ac.uk

**Abstract.** Berrington *et al* [1] have calculated effective collision strengths for Fe XV using both the Breit-Pauli and Dirac $R$-matrix methods. The calculation was undertaken in the spirit of a case study to lay to rest assertions that the Dirac method is superior for such ions. They found good agreement between the two methods. Current development of the DARC program [2] is aimed at improving the quality of the target states. As a first step we have included the Breit interaction directly in the bound-bound continuum hamiltonians. DARC is based on the Dirac-Coulomb hamiltonian and in order to treat heavier systems correctly it is necessary to include the Breit interaction. We have also used Fe XV as a case study for this code development and in addition examined Cr XIII and Ni XVII. The effects on the effective collision strengths for the iron peak ions are as expected small. The work illustrates the extent to which the DARC program can be automated for isoelectronic series.

## 1. Introduction

The $R$-matrix method [3] is the tool of choice for the production of high quality collisional atomic data. Beginning with the publication [4] of the RMATRX code in 1974 a large number of FORTRAN programs have implemented this approach. These are used widely in producing data and also are continually being modified to incorporate new physics. An example of the latter is the use of target states built from relativistic gaussian-type-orbitals in a many-body perturbation theory framework [5]. This gives a fairly compact target representation of much higher accuracy. The effect of this modification on the calculational time is not apparent from the paper. A comparable level of accuracy can be achieved with the existent programs by using a large configuration interaction expansion. This requires considerable computational resources but the viability of such calculations is exemplified by a recent calculation for electron scattering by Fe V [6]. The large size of this calculation was not due to the configuration interaction expansion but rather to the number of low-lying and overlapping states in this ion. It should be possible to produce data of a high accuracy. Robust and reliable is how these programs are sometimes described, rarely are they described as easy to use.

'Users' of the atomic data, in for example astrophysical modelling, would ideally like the data to be of the highest quality. In conversation some admit that 'anything would do' and it seems in practice that medium quality data would suffice. Placing error bounds on the theoretical data is often difficult. $R$-matrix papers sometimes claim an accuracy of 10% for the data without much justification. As the target model increases in complexity there should be convergence in the atomic data and this will give some feel for the accuracy. The place of the high quality calculation is secure as a benchmark.

In principle with the free availability of the $R$-matrix programs [7] it should be possible for 'users' to generate their own atomic data. This study was undertaken in part to assess how feasible that would be. There is a benchmark calculation for Mg-like Fe XV [1]. This paper mainly demonstrates the expected near equivalence of the Breit-Pauli and Dirac hamiltonians for a system with this atomic number. In addition it has useful comments on the pitfalls of such a calculation. We have carried out simpler calculations for the ions Cr XIII, Fe XV and Ni XVII. The only essential change in the input data for the three ions was the atomic number. Our main purpose is obtain a overview of the complete calculation from target generation through to effective collision strength for a case of moderate size.

As part of code development to improve the target representation we have examined the inclusion of the Breit interaction in the bound-bound part of the continuum hamiltonian. This is only a partial introduction of the Breit interaction as it is not included in any radial integrals involving the continuum function. This is expected to be important for heavier systems.

## 2. Calculations

The configurations $3s^2$, 3s3p, $3p^2$, 3s3d, 3p3d and $3d^2$ were included in the target. This gives rise to 35 fine-structure levels. The target energies and orbitals were calculated using grasp0 [8]. The EAL (extended average level) MCDF (multi-configurational Dirac Fock) method was employed. We have not here included higher order radiative corrections because these corrections cannot be treated consistently in the present scattering calculation. The accuracy is at the 1 to 2% when compared to NIST data. So it is not at the highest accuracy but we are not seeking that here.

The $R$-matrix calculations used DARC (Dirac Atomic $R$-matrix Code) [2] interfaced with the UK APAP asymptotic and utility codes [7].

Limitations in these calculations are the target representation (missing $n = 4$ orbitals and resultant configuration interaction) and a modest energy range for resolution of the resonance structure.

Results are shown in Figure 1 for transitions between the ground-state and the first 4 excited levels. For Fe XV a comparison is made with results taken from [1].

Some observations:

- Figure 1(c) shows a failure to resolve the resonance structure for Fe XV. The low energy contribution is not showing.
- Figure 1(d) shows a failure at the high energy for Fe XV. This is an allowed transition ($3s^2$ $^1S_0$ — 3s3p $^1P_1$) and there may be an inconsistency in the contribution from high partial waves (top-up).

Overall there is reasonable agreement given the limitations. The behaviour of the other ions looks consistent. The discrepancies point us towards how we can improve the data within the context of maintaining moderate accuracy.

Figure 2 shows the effect for Fe XV of including the Breit interaction in the bound-bound part of the continuum hamiltonian. The effect here is to modify the thresholds and move the resonance structure. As the change in the energies for this ion is small we only see corresponding small effects in the low energy effective collision strengths. For the allowed transition for which the resonance structure is not dominant the two calculations are nearly identical.

## 3. Conclusions

The calculation of the effective collision strengths involved the following sequence of FORTRAN programs

grasp0 → darc0 → darc1a → darc1b → darc2 → dto3 → stg3r → stgf → omadd → adasexj

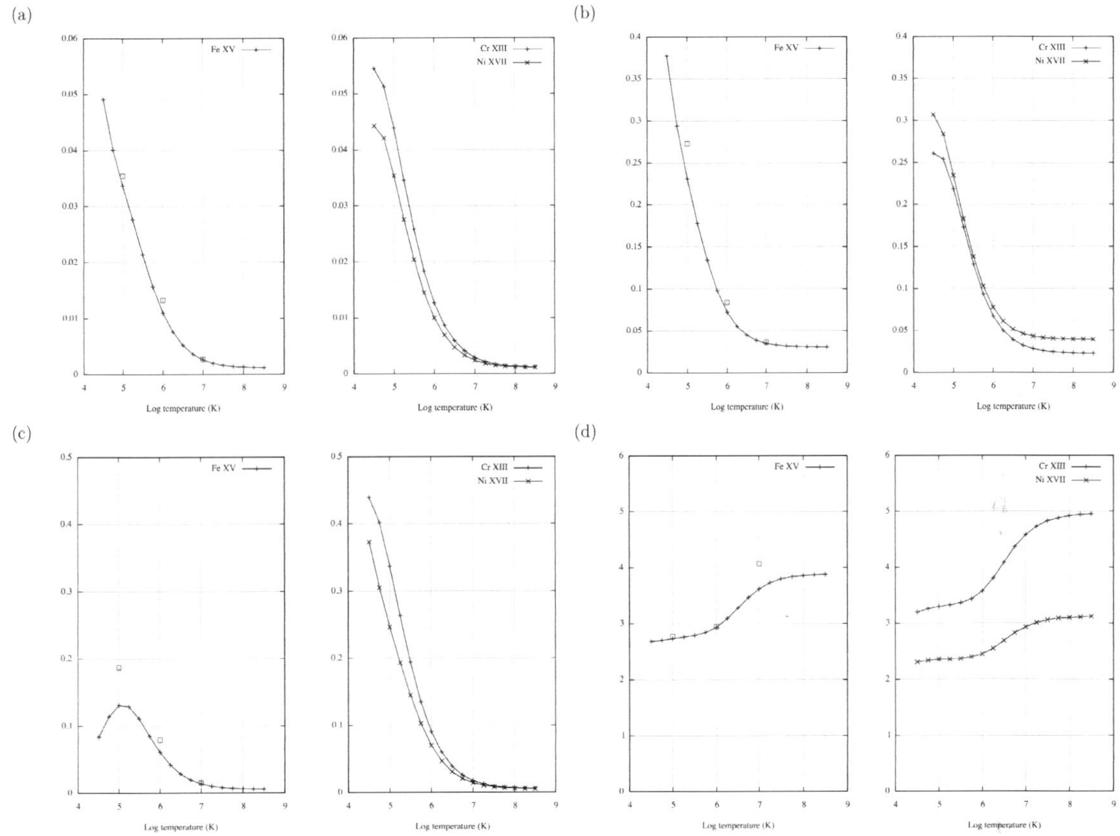


**Figure 1.** Effective collision strengths from 3s$^2$ $^1$S$_0$ to (a) 3s3p $^3$P$_0$; (b) 3s3p $^3$P$_1$; (c) 3s3p $^3$P$_2$; (d) 3s3p $^1$P$_1$. □ denotes the Fe XV data of [1].

The proliferation of programs is historic and in part due to computational limitations as the size of the problems increased. There are two many programs, each with its own input data and pitfalls, some of which we fell into. While it is true that each program can be run in a near default mode, there are hidden options and parameters. These parameters are sometimes solutions to numerical problems that have been discussed in the literature. Alternatively they may be adding new physics. Either way they reflect the long-time development of these programs. Supplied documentation is sparse.

The programs can be grouped into

| | |
|---|---|
| target | grasp0 → darc0 |
| *R*-matrix | darc1a → darc1b → darc2 → dto3 → stg3r |
| collision strength | stgf |
| rate coefficient | omadd → adasexj |

and this would make it more manageable. If we wish these programs to be a tool for 'users' to generate their own atomic data then we need such a merger together with a more consistent interface and documentation.

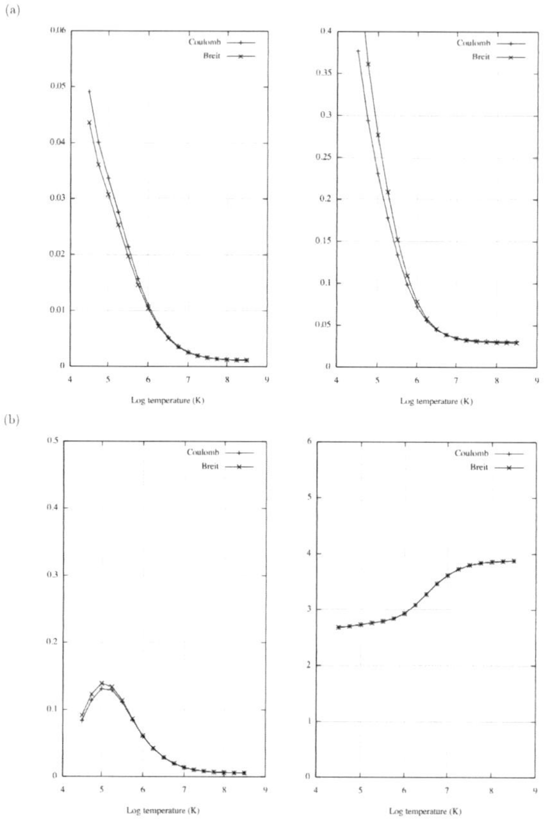


**Figure 2.** Effective collision strengths from $3s^2\ ^1S_0$ to (a) 3s3p $^3P_0$ and 3s3p $^3P_1$, (b) 3s3p $^3P_2$ and 3s3p $^1P_1$, in Fe XV.

A more detailed account of these calculations will be made available at the website www.am.qub.ac.uk/apa.

## Acknowledgments
One of us (PN) visited NIFS, Tajimi, last year and would like to express grateful thanks for the hospitality of Professor Takako Kato and her colleagues.

## References
[1] Berrington K A, Ballance C P, Griffin D C and Badnell N R 2005 *J. Phys. B: At. Mol. Opt. Phys.* **38** 1667-78
[2] Ait-Tahar S, Grant I P and Norrington P H 1997 *Phys. Rev. Let.* **79** 2955-58
[3] Burke P G and Berrington K A 1993 *Atomic and Molecular Processes, an R-matrix Approach* (Bristol: Institute of Physics Publishing)
[4] Berrington K A, Burke P G, Chang J J, Chivers A T, Robb W D and Taylor K T 1974 *Comput. Phys. Commun.* **8** 149
[5] Vilkas M J and Ishikawa Y 2007 *Physical Review* A **75** 062508
[6] Ballance C P, Griffin D C and McLaughlin B M 2007 *J. Phys. B: At. Mol. Opt. Phys.* **40** F327-35
[7] UK APAP Network website: `http://amdpp.phys.strath.ac.uk/UK_RmaX/`
[8] Grant I P, McKenzie B J, Norrington P H, Mayers D F and Pyper N C 1980 *Comput. Phys. Commun.* **21** 207

# Dielectronic and Radiative Recombination of Si- to N-like Tungsten Ions

**C Biedermann[1], R Radtke[1], R Seidel[1], and E Behar[2]**

[1]Institut für Physik der Humboldt-Universität zu Berlin, Lehrstuhl Plasmaphysik, Newtonstraße 15, 12489 Berlin, and Max-Planck-Institut für Plasmaphysik, EURATOM Association, 17491 Greifswald, Germany

[2]Technion Israel Institute of Technology, Physics Department, Haifa 32000, Israel and NASA / Goddard Space Flight Center, Greenbelt MD 20771, USA

E-mail: Christoph.Biedermann@ipp.mpg.de

**Abstract.** We have performed x-ray spectroscopic measurements of the dielectronic recombination resonance strength for the LMn (n=3, ..., 10) series of Si-like $W^{60+}$ to N-like $W^{67+}$ tungsten ions. Highly charged tungsten ions were produced, stored and excited with the Berlin electron beam ion trap and the emitted radiation was analyzed with a solid state detector. Information on the charge state abundance in the trap was extracted from a fit of the theoretical radiative recombination intensity to measured values. The fit procedure was only feasible when the fine structure, angular momentum of the recombination channels is taken into account. Our measurement of x-rays from n=2-3, 2-4 and higher DR resonance transitions was compared to relativistic calculations of the DR cross sections and rate coefficients calculated with the Hebrew University Lawrence Livermore Atomic Code (HULLAC). The previous theoretical predictions for Ne-like tungsten ($W^{64+}$) [3] were extended with calculations for ions in adjacent charge states and compare well with the observed DR resonance structure.

## 1. Introduction

Dielectronic recombination (DR) is an important recombination process for high-temperature laboratory and astrophysical plasmas. DR is the resonant electron capture of a free electron, where simultaneously a bound electron of the ion is excited. The resulting doubly excited intermediate state stabilizes by photon emission. Spectroscopy gathers detailed information on these signatures and allows conclusions on the specific conditions of excitation in the plasma.

Present and future fusion devices are and will be armored with the heavy element tungsten as wall material made possible by the excellent thermal properties and long erosion lifetime to withstand high power fluxes of reactor plasmas [1]. In spite of these advantages tungsten may be sputtered and ionized by the high temperature plasma. The resulting radiative transitions, as for example from DR processes, contribute to the radiation energy loss in fusion plasma, change the charge state balance and satellite lines can alter line shapes, widths and intensities of parent transitions. However, understanding the radiation pattern of highly-charged heavy element ions can provide profound diagnostics of high-temperature plasma conditions and test atomic structure calculations [2].

Calculations of DR cross sections and rate coefficients for Ne-like $W^{64+}$ tungsten have been published by Behar et al. [3] and observation is mentioned briefly from the Tokyo EBIT [4]. At the Berlin EBIT we have investigated the DR of $W^{60+}$ to $W^{67+}$ in detail and compared the spectroscopic information to relativistic predictions applying the HULLAC suite of codes.

## 2. Experiment

Highly charged tungsten ions are produced, stored and excited using the Berlin EBIT [2]. Electrons extracted from a heated cathode are accelerated to 20 kV, compressed by the 3-Tesla magnetic field generating a space-charge well for radial ion confinement and forming an electron beam with 32 eV energy width. The sectioned drift-tube assembly to which the accelerating potential is applied, provides axial trapping. Tungsten is introduced as $W(CO)_6$ vapor and ionized by successive electron impact. After breeding for 500 ms to ionize and trap highly charged $W^{60+ - 67+}$ ions the beam energy is rapidly ramped within 15 ms between 1.6 keV and 8 keV to excite and probe the ion ensemble. X radiation emitted from the ions is registered with a Ge-solid state detector. The ionization/DR-probe cycle is repeated many times recording the x-ray energy and the electron beam energy.

## 3. Radiative recombination

One specific feature of EBIT-x-ray spectra is the appearance of the Radiative Recombination (RR) process manifested as distinct lines. Fig. 1 shows the detail of the n=3 RR x-ray spectrum from highly charged tungsten ions at a fixed beam energy of 20 keV during the last 100 ms of ionization period before starting the fast DR ramp. The x-ray energy of the registered RR-feature is given by the well-defined electron beam energy and the binding energy of the capture state of the recombining ion. For each tungsten ion the fine structure is taken into account by calculating the binding energy for each *n=3* subshell (angular momentum state), the RR cross section [5] to determine the *3l* relative contribution for each ion and the electron beam properties (geometry and emission characteristic) to derive a relative theoretical RR intensity distribution. The colored lines in Fig. 1 present the contributions for the individual tungsten ions. A fit of the theoretical distribution for all relevant ion states to the experimental RR-spectrum yields the charge state abundance. The charge state distribution is displayed in Fig. 3.

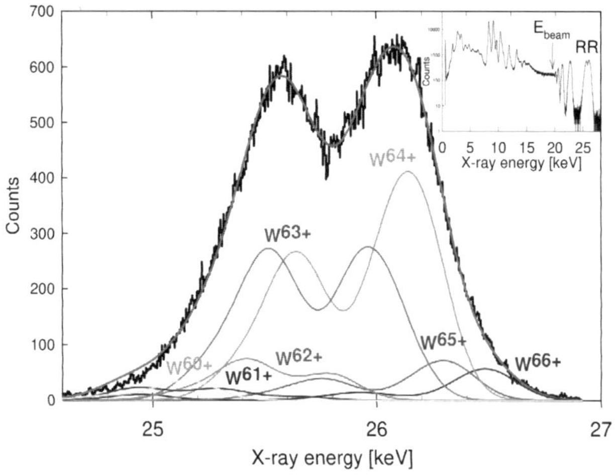


**Figure 1:** X-ray spectrum showing the detail of n=3 radiative recombination of highly charged tungsten ions. The black quivering line marks the observed intensity and is fitted by the ***nl*** contribution of $W^{60+}$ to $W^{67+}$ ions (colored lines) which sum up to the red line and gives the charge state distribution. The inset presents the wide range spectrum produced in EBIT at 20 keV electron beam energy with a logarithmic intensity scale. The spectrum is dominated by n=2-3 direct excitation lines between 8 and 13 keV rising above the Bremsstrahlung background.

## 4. Dielectronic recombination

Another characteristic target of EBIT investigation is the process of dielectronic recombination. The measurement of DR ramps records the x-ray emission energy as function of the electron beam energy. This information can be displayed as scatter plot (see e.g. [4]) or, emphasizing a particular DR detail, as excitation function of x-ray emission plotted against the electron interaction energy shown in Fig. 2

for the LMM DR resonance. Fig. 2 presents the excitation function for the case when an L shell electron is excited to the M shell leaving a $2p^{-1}{}_{3/2}$ hole in the *(2s2p)* configuration complex simultaneous to the resonant recombination of a free electron to the M shell followed by the emission of a 9.11 keV x-ray during the relaxation of the intermediate doubly-excited state by a $2p_{3/2} - 3d_{5/2}$ transition. Another possible LMM DR channel is the excitation leaving a $2p^{-1}{}_{1/2}$ hole in the L-shell and relaxation of the $(2s2p)^7\ 3l\ 3d_{3/2}$ doubly-excited state (here noted for Na-like $W^{63+}$) by a 11.8 keV photon originating from a $2p_{1/2}$-$3d_{3/2}$ transition. In this manner the excitation function for many other LMn (n=3 up to 10) DR resonances have been recorded in the experiment. Apart from the relaxation of the doubly excited intermediate state, x-rays from the deexcitation of the singly excited spectator electron are detected. The observed resonance structure is compared to theoretical predictions from multiconfiguration relativistic calculations using the Hebrew University Lawrence Livermore Atomic Code [3], which is based on the parametric potential and the isolated-resonance, distorted wave approximations [6]. Extensive level-by-level computation found that the population of the $(2s2p)^7$ *3l3l'* doubly excited configuration complex is the dominating DR channel. The original calculations of DR resonance strengths and rate coefficients for Ne-like $W^{64+}$ [3] have now been extended to the adjacent charge states ranging from Si-like $W^{60+}$ to N-like $W^{67+}$ LMn DR. All doubly-excited $2p^5\ 3l^x$ *3l' nl"* configurations with n=3, 4 were included in the computation for the DR of Na-like through Si-like W (x=1-4) and all of the $2p^x$ 3l nl' configurations with n=3, 4 were included for the DR of N-like through Ne-like W (x=2-5). Configuration mixing, which alters the atomic structure and can appreciably affect the DR rates was taken into account. The lower plot of Fig. 2 reveals how the contributions of the different charge states of tungsten ions to the LMM DR cross section shift with decreasing charge state to larger excitation energy. Cross sections are obtained by folding the calculated resonance strength with a 32-eV-wide Gaussian profile of the experimental electron-beam energy. Weighing the cross section of the relevant tungsten ions with the charge state abundance determined by the n=3 RR measurement results in the solid line overlaying the experimental data points in the upper part of Fig. 2. The radiation is observed at $90^o$ to the electron beam and is in general anisotropic [7]. For electric-dipole radiation predominantly emitted in the DR process the angular correction amounts here up to about 17%. However, this factor has not been included in the graph.

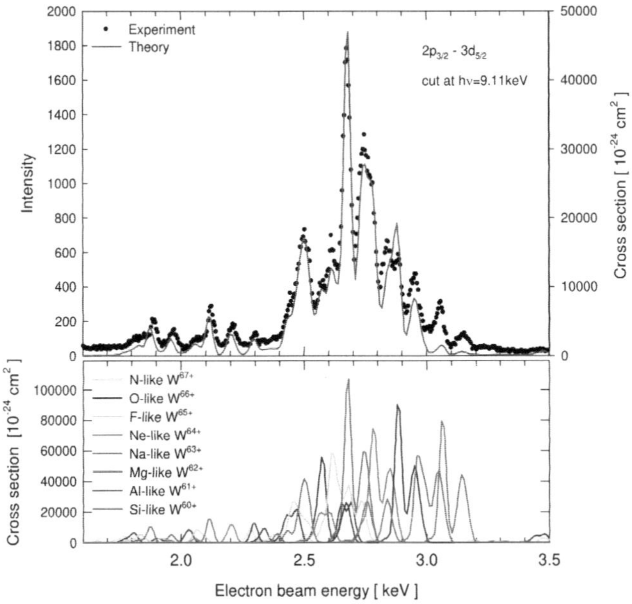


**Figure 2.** Comparison of measured and calculated LMM dielectronic recombination resonance cross section of highly charged tungsten ions (q=60+ to 67+). Shown here is the part of the LMM DR process during which a $2p_{3/2}$-electron is excited to *3l*, while a free electron recombines to *3l'*. The doubly excited state relaxes by emission of an x-ray following a $2p_{3/2}$ - $3d_{5/2}$ transition.

The detailed structure of the excitation function can be explained by contributions of individual resonances from certain highly charged tungsten ions.

## 5. Charge state abundance

To interpret the x-ray emission of EBIT it is necessary to gain information on the charge state abundance of the trapped ions. In section 3 we have demonstrated one possible estimation of the charge state distribution by fitting theoretical RR spectrum including the fine structure of the recombination states to the measured RR x-ray intensity. Similarly to this method we can exploit the good resolution of the measured DR emission pattern and fit the theoretical DR cross section to it. The resulting charge state distribution is plotted in Fig. 3. The shift of the distribution to lower charge states reflects the fact that the process of dielectronic recombination changes the ion balance of a plasma.

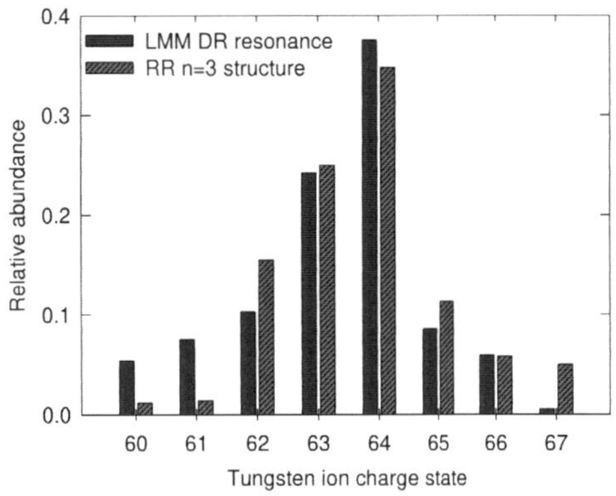


**Figure 3:** Charge state abundance of tungsten ions determined from the measured radiative recombination intensity at 20 keV, 100 ms before the DR ramp (red, hatched) and from a fit of the theoretical resonance strength to the observed LMM DR intensity.

## References

[1] Neu R, 2006 *Phys. Scr.* **T123** 33
[2] Radtke R, Biedermann C, Schwob J L, Mandelbaum P and Doron R 2001 Phys. Rev. **A64** 012720
[3] Behar E, Mandelbaum P and Schwob J L 1999 *Phys. Rev.* **A59** 2787
[4] Watanabe H, Nakamura N, Kato D, Nakano T and Ohtani S 2007 *Plasma and Fusion Res.* **2** 027
[5] Kim Y S and Pratt R H, 1983 *Phys. Rev.* **A27** 2913
[6] Bar-Shalom A, Klapisch M and Oreg J 2001 *J. Quant. Spectr Radiat. Transfer* **71** 169
[7] Fuchs T, Biedermann C, Radtke R, Behar E, Doron R 1998 Phys. Rev. **A58** 4518

# Effects of electron-impact ionization on the damage to biomolecules irradiated by XFEL

**Takeshi Kai and Kengo Moribayashi**

Japan Atomic Energy Agency, 8-1, Umemidai, Kizugawa, Kyoto, 619-0215, Japan

E-mail: kai.takeshi@jaea.go.jp

**Abstract.** The damage to biomolecules irradiated by an X-ray free electron laser is investigated with time-dependent rate equations, considering photoionization, Compton scattering, Auger decay and electron-impact ionization of the C atom and its ions. Averaged energies of secondary electrons, which are produced by electron-impact ionization, are calculated with the binary encounter dipole model (Kim Y-K. and Rudd M. E., 1994 Phys. Rev. A, **50** 3954) to study electron-energy distribution in the biomolecule. We found that those energies for the C atom, $C^{1+}$ and $C^{2+}$ ions are about 20, 60 and 150 eV, respectively, at an incident electron energy of 12keV, while those energies are in the order of tens of electronvolts when the incident electron energy is about 250 eV. We also found that the importance of electron-impact ionization increases as the X-ray flux decreases.

## 1. Introduction

In X-ray free electron laser (XFEL) applications, the analysis of three-dimensional structures of single biomolecules is a critical issue [1, 2]. When the biomolecule is irradiated by the XFEL, the electron density distribution of the biomolecule (three-dimensional structure) is obtained from the diffraction patterns of elastic-scattered photons. It is important to estimate intensities of the diffraction patterns, since a large number of X-ray photons will be needed for this measurement. The target samples, however, are damaged by photoionization, Auger decay, Compton scattering and electron-impact ionization processes. Therefore, we need to investigate the damage caused by those processes that influence the diffraction patterns.

Here, we concentrate on a description of the atomic processes leading to damage in the previous reports [1, 2]. The electron-impact ionization was not considered in Neutze's paper [1]. On the other hand, the Compton scattering was not considered in Hau-Riege's paper [2]. In the present study, photoionization, Auger decay, Compton scattering and electron-impact ionization are all taken into account. In addition, we also address here the electron-impact ionization, and study energy distribution of free electrons in the biomolecules. When a biomolecule is irradiated by the XFEL with a photon energy of 12 keV (or wavelength of $\sim$ 0.1 nm), the free electrons in the biomolecules have different energy components ($\sim$ 12 keV for photoelectrons, $\sim$ 250 eV for Auger electrons [1] and $\sim$ 270 eV for Compton recoil electrons). Moreover, to describe the damage, the energies of the secondary (or ejected) electrons produced by the processes listed above should also be estimated to describe the damage.

Time-dependent rate equations are solved with the atomic processes of the C atom and its ions. The population dynamics of the C atom and its ions, and the contributions of the previously mentioned atomic processes to ionization, are investigated at the X-ray flux from 1 $\times$ $10^{19}$ to 1 $\times$ $10^{22}$ photons/pulse/mm$^2$.

## 2. Atomic data and model simulation

For our model simulation, we assumed that the target biomolecules can be treated as non-bonded carbon clusters with radius $r$ [2, 4, 5]. Photoionization cross sections were calculated using the Flexible Atomic Code (FAC) [6]. Auger rates were taken from previous results [5]. Compton cross sections and Compton recoil electrons energies are calculated by the Klein-Nishina formula.

For the electron-impact ionization process, the energy distributions and averaged energies of secondary electrons were obtained to calculate the single differential cross section (SDCS). The SDCS of the C atom and its ions are calculated by BED model [3] given by

$$\frac{d\sigma}{dW} = \frac{S}{B(t+u+1)}\left\{\frac{N_i/N-2}{t+1}\left(\frac{1}{w+1}+\frac{1}{t-w}\right)\right.$$

$$\left.+\left[2-\frac{N_i}{N}\right]\left[\frac{1}{(w+1)^2}+\frac{1}{(t-w)^2}\right]\frac{\ln t}{N(w+1)}\frac{df(w)}{dw}\right\}, \tag{1}$$

with

$$t = T/B\ ,\ w = W/B\ ,\ u = U/B\ ,\ S = 4\pi a_0^2 N(R/B)^2\ ,\ N_i = \int_0^\infty \frac{df(w)}{dw}dw\ , \tag{2}$$

where $B$ is the binding energy of the ejected electron, $T$ is the nonrelativistic kinetic energy of the incident electron, $W$ is the kinetic energy of the ejected electron, $U = \langle p^2/2m\rangle$ is the orbital-electron-kinetic energy with the target electron momentum $\vec{p}$ and the electron mass $m$, $a_0$ is the Bohr radius and $R$ is the Rydberg energy. In equation (1), $df(w)/dw$ is the differential oscillator strength as a function of $w$. Hartree-Fock wavefunctions of the C atom and its ions [7] were used to calculate the orbital electron kinetic energy $U$. The differential oscillator strengths of the C atom and its ions were calculated with the FAC [6]. The integral cross section (ICS) $\sigma$ and averaged energy of secondary electron $\langle W\rangle$ are given by

$$\sigma = \int_0^{(T-B)/2}\frac{d\sigma}{dW}dW\ ,\ \langle W\rangle = \frac{\int_0^{(T-B)/2} W\frac{d\sigma}{dW}dW}{\int_0^{(T-B)/2}\frac{d\sigma}{dW}dW}\ . \tag{3}$$

The treatment of free electrons in this model simulation can be described as follows:. The Compton recoil electrons and the Auger electrons are trapped in a target sample because those kinetic energies are sufficiently low. The photoelectrons, however, escape from the target sample after $\sim$ 0.15 fs in the case of $r$ = 10 nm since these have high kinetic energies. Therefore, the target sample is ionized by the photoelectrons when these are within a target sample of radius $r$. The time-dependent rate equation is solved by applying this atomic model. A pulse X-ray flux with the energy of 12 keV was employed. The pulse had Gaussian distribution and a width of 10 fs.

## 3. Results and Discussion

To verify the results of the energies of the secondary electrons, we present the integral cross section (ICS) of a C atom for the BED calculation. Figure 1 shows the comparison of our calculated results with the recommended data [8] for a C atom from the ionization energy to 20 keV. Our result is in good agreement with the recommended data [8]. Note that our results for C ions for the BED calculation are also in good agreement with the recommended data [8].

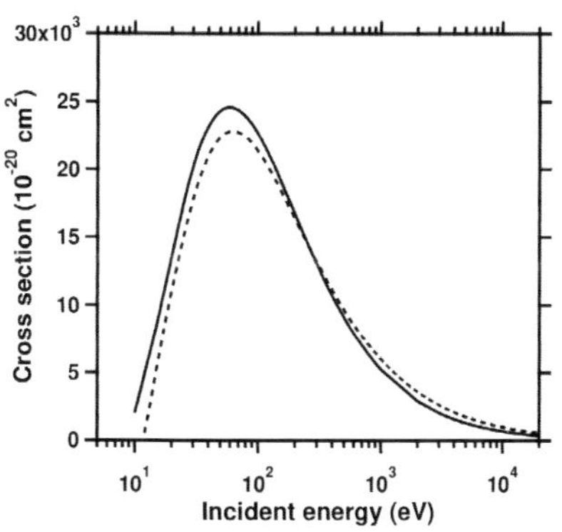


**Figure 1.** Ionization cross section of C atom from the ground state to all the $1s^22s^22p$ and $1s^22s2p^2$ states. ——, Present; - - - -, Recommended data [8].

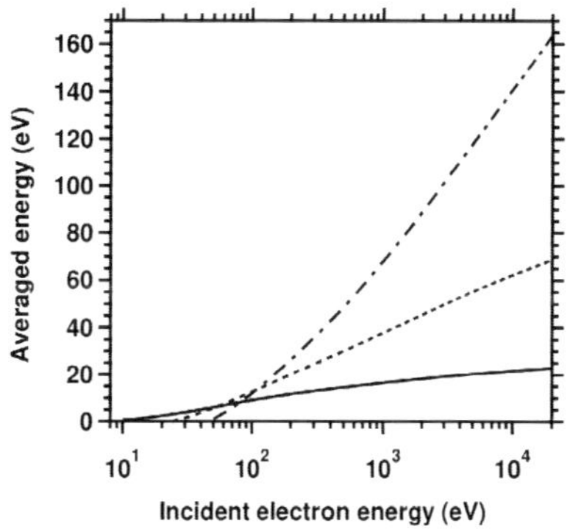


**Figure 2.** Averaged energies of secondary electron of C atom, $C^{1+}$ and $C^{2+}$ ions. ——, C atom (2p-orbital ionized); - - - -, $C^{1+}$ ion (2p-orbital ionized); — · —, $C^{2+}$ ion (2s-orbital ionized).

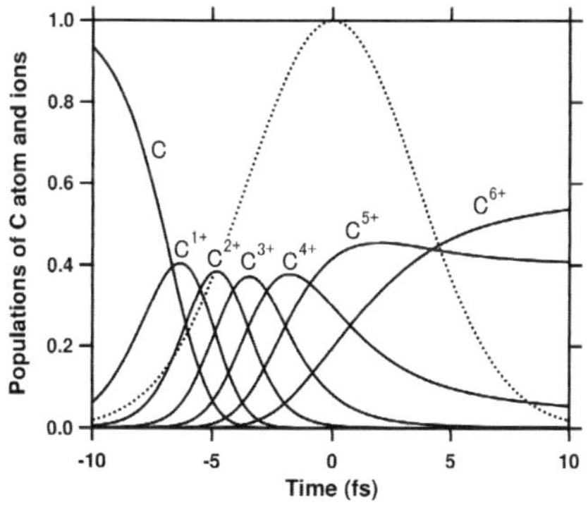


**Figure 3.** Normalized populations of C atom and ions at $1 \times 10^{22}$ photons/pulse/mm$^2$, pulse width 10 fs, target radius 10 nm. - - - -, Normalized X-ray flux.

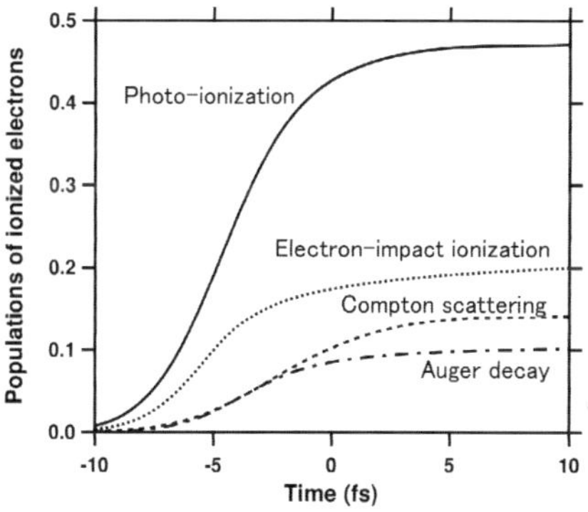


**Figure 4.** The contributions from the different atomic processes to ionization at $1 \times 10^{22}$ photons/pulse/mm$^2$, pulse width 10 fs, target radius 10 nm.

Figure 2 shows the results of the averaged energies of secondary electrons for the C atom, $C^{1+}$ and $C^{2+}$ ions. Those energies for the C atom, $C^{1+}$ and $C^{2+}$ ions are about 20, 60 and 150 eV, respectively, at the incident electron energy of 12 keV (photoelectron impact). Those energies are in the order of tens of electronvolts from figure 2 when the incident electron energy is about 250 eV (Compton recoil and Auger electron impact). Those results suggest that the ionization effects of the secondary electrons, which are produced by Compton recoil electrons and Auger electrons impact, are minor because those cross sections are small in this energy region.

Time-dependent rate equations were solved to investigate the population dynamics of the C atom and its ions. Figure 3 shows the resulting populations as a function of time (femtosecond). In this simulation, we chose a pulse width of 10 fs, and an X-ray flux of $1 \times 10^{22}$ photons/pulse/mm$^2$ and a target radius of 10 nm. The results show that the populations of $C^{5+}$ become dominant at the peak intensity of the X-ray flux. It is clear that the damage to the biomolecule is severe under these conditions. Figure 4 shows the contributions from the different atomic processes involved in ionization under the same conditions, as in figure 3. We found that the photoionization becomes dominant at an X-ray flux of $1 \times 10^{22}$ photons/pulse/mm$^2$, and

**Table 1.** The contributions of the each atomic processes to ionization at the peak intensity of X-ray flux (%).

| photons/pulse/mm$^2$ | (A) | (B) | (C) | (D) | (E) |
|---|---|---|---|---|---|
| $1 \times 10^{22}$ | 42.8 | 17.4 | 10.2 | 8.50 | 79.0 |
| $1 \times 10^{21}$ | 13.4 | 12.1 | 1.67 | 2.80 | 29.9 |
| $1 \times 10^{20}$ | 1.61 | 1.78 | 0.16 | 0.28 | 3.86 |
| $1 \times 10^{19}$ | 0.16 | 0.18 | 0.02 | 0.03 | 0.39 |

(A) Photo-ionization. (B) Electron-impact ionization. (C) Compton scattering. (D) Auger decay. (E) Total.

electron-impact ionization becomes a more important process in the investigation of the damage when the X-ray flux decreases, see table 1.

## 4. Conclusions

The atomic data of the photoionization, electron-impact ionization and Compton scattering of the C atom and its ions were calculated using the FAC, BED model and Klein-Nishina formula, respectively. We found that the averaged energies of secondary electrons for C atom, $C^{1+}$ and $C^{2+}$ ions were about 20, 60 and 150 eV, respectively, at an incident electron energy of 12keV, while those energies were in the order of tens of electronvolts when the incident electron energy was about 250 eV. The time-dependent rate equations were solved to investigate the damage to the target sample. We also found that the electron-impact ionization became a more important process as the X-ray flux decreased. To understand the damage evaluation, it is necessary to evaluate the damage as well as the intensities of the diffraction patterns at the same time. In the future, we will investigate the intensities of diffraction patterns in connection with the damage to the biomolecules for the various XFEL parameters.

## Acknowledgments

We wish to thank Professor N. Go, Drs H. Kono, A. Tokuhisa, J. Taka, T. Otobe, Y. Fukuda, T. Nakamura and M. Yamagiwa for their useful discussion. This study has been supported by the 'X-ray Free Electron Laser Utilization Research project' of the Ministry of Education, Culture, Sports, Science and Technology, Japan (MEXT)

## 5. References

[1] Neutze R, Wouts R, Spoel D, Weckert E, and Hajdu J, 2000 Nature, **406**, 752.
[2] Hau-Riege S P, London R, and Szoke A, 2004 Phys. Rev. E, **69**, 051906.
[3] Kim Y-K and Rudd M E, 1994 Phys. Rev. A, **50** 3954.
[4] Moribayashi K and Kai T, 2008 J.Phys: Conference series submitted.
[5] Moribayashi K, 2008 J. Phys. B: At. Mol. Opt. Phys., **41**, 085602.
[6] Gu M F, 2003 Astrophys. J, **582**, 1241.
[7] Clementi E and Roetti C, 1974 Atomic Data and Nuclear Data Tables, **14** 177.
[8] Suno H and Kato T, 2006 Atomic Data and Nuclear Data Tables, **92**, 407.

# HIGH-ENERGY ELECTRON-IMPACT EXCITATION CROSS SECTIONS OF HYDROGENLIKE IRON AND NICKEL IONS

**D B Thorn[1], P Beiersdorfer[1], G V Brown[1], R L Kelley[2], C A Kilbourne[2] and F S Porter[2]**

[1]Lawrence Livermore National Laboratory, Livermore CA 94550, USA.
[2]NASA/Goddard Space Flight Center, Greenbelt MD 20770, USA.

E-mail: dbthorn@ucdavis.edu

**Abstract.** We present a measurement of the cross section for electron-impact excitation of Lyman-$\alpha_1$ in hydrogen-like iron (Z=26) and nickel (Z=28) over a broad range of energies using the Lawrence Livermore National Laboratory SuperEBIT electron beam ion trap facility. The measurement was performed with electron beam energies between 35 keV and 85 keV. The hydrogen-like spectrum of iron and nickel was observed, and fully resolved, with the XRS/EBIT x-ray microcalorimeter spectrometer, which allowed for an absolute cross section to be measured by normalizing to the RR x-ray emission. These results are compared to theory.

## 1. Introduction

Electron-impact excitation (EIE) cross sections of K-shell ions are used in interpreting the spectra from astrophysical and laboratory-produced high-temperature plasmas. Of special importance are K-shell cross sections of iron and nickel as they have relatively high elemental abundances in celestial sources as well as being present in many laboratory plasmas. In K-shell systems the cross section for EIE rises till roughly 3 times the threshold energy. To construct a good model for high-Z K-shell emission from a high-temperture plasma, cross sections for electron energies up to 10 times threshold need to be included. Furthermore, for electron energies in excess of 100 keV relativistic effects are expected to play a role in the excitation process and modify the cross section [1]. Yet, for energies well above excitation threshold, no experimental data exist to benchmark and guide theory [2].

The emitted intensity of a EIE produced spectral line can be written as $I = n_e n_i < v_e \sigma_{EIE} >$, where $n_e$ is the density of electrons, $n_i$ is the density of the ions, $v_e$ is the velocity of the electrons, and $\sigma_{EIE}$ is the EIE cross section. The difficulty in making EIE measurements comes from the uncertainty in the knowledge of the density of the ions and electrons. However with a source that is mono-energetic there exists a way by which to remove the influence of the uncertain terms. That is by normalizing the direct excitation cross section to the cross section for radiative recombination (RR) since RR can be calculated to a much greater accuracy.

Radiative recombination is produced by the capture of an electron into a vacant atomic level. The photon that is produced has the energy of the binding energy of the atomic level plus the energy of the free electron. The mathematical expression for the intensity of a RR spectral feature has the same form as for EIE. Dividing the intensity of a given line from the EIE spectrum by that of a feature produced by RR forms a ratio of the two cross sections, and the terms with large uncertainties cancel out, leaving the EIE cross-section related to the RR cross section (which in turn has been shown to be calculated to an accuracy of 3%-5% [3]).

In this paper we present a measurement of the EIE cross sections for the Lyman-$\alpha_1$ line in hydrogenlike nickel and iron. The result of the measurement is compared to theory.

## 2. Experiment

Hydrogenlike iron ions were produced by a 35, 65, and 75 keV energy electron beam and the nickel ions were produced by a 65 and 85 keV energy electron beam in the SuperEBIT electron beam ion trap. The iron and nickel atoms were injected as metallo-organic gases (nickelocene and iron pentacarbonyl) by way of a ballistic gas injector. The hydrogenlike spectrum from the two elements was observed at 90° with a 6 mm thick coaxial EG&G IGLET-X high purity germanium detector and by the XRS/EBIT x-ray microcalorimeter spectrometer [4].

For iron and nickel, the K-shell RR spectrum sits roughly 10 keV higher in energy then the electron beam energy. The XRS/EBIT, which has a 6 eV full width half maximum resolution at 6 keV, has a low quantum efficiency (QE) at these high energies, and thus, cannot observe the RR spectrum (which is already 100 times less intense than the EIE emission). For that reason, the IGLET was used, as the QE of the Ge detector is above 85 % for photons under 100 keV[1]. On the other hand, the Ge detector cannot resolve the DE hydrogenlike emission because of its poor resolution (roughly 200 eV at 6 keV). Thus a hybrid measurement system was used, whereby the K-shell emission was recorded with both the Ge detector and the XRS/EBIT, but only the Ge detector recorded the RR spectrum, as shown in Figs. 1 and 2.

In our setup the XRS/EBIT is used to resolve the K-shell spectrum, and to relate the number of counts in the Lyman-$\alpha_1$ and Lyman-$\alpha_2$ lines to the intensity of the single spectral feature observed with the Ge detector, see Fig. 2. The Ge detector spectrum (as shown in Fig. 1), then, with the proper weightings of the lines, is used as the basis of the RR normalization. This allows for elimination of systematic error due to uncertainty in the geometry of the experimental setup. The spectral peaks recorded with both detectors are fit with gaussian fitting functions. This procedure is similar to that employed in [5] except that we employ the XRS/EBIT instead of a crystal spectrometer. It should be noted, that a new x-ray microcalorimeter spectrometer has been installed at EBIT after the completion of the measurements that negates the need for a Ge detector as it has good QE for photons under 100 keV [6].

In EBIT there are several processes that modify the intensity of the observed Lyman-$\alpha$ lines. These processes are polarization, cascades from higher levels, and charge exchange[2]. Polarization, affects the geometric emission of photons for Lyman-$\alpha_1$, which is important for relating the differential cross section (at 90°) to the total cross section for comparison to theory. Cascades can produce extra Lyman-$\alpha_1$ photons by the decay of higher lying atomic levels (like the n=3 level) down to the n=2 level, which, if not taken into account would cause the experiment to measure a larger cross section. Charge exchange occurs when a neutral atom is stripped of one of its valence electrons. The electron is captured into a high lying level, which may cascade down, and possibly make a Lyman-$\alpha_1$ photon. To take into account polarization and cascades, calculations were made with the Flexible Atomic Code [8] (FAC) to adjust the data. Since the contribution to the Lyman-$\alpha_1$ line for cascades and polarization is less then 10 % (using an estimated uncertainty of 20 %) the effect of including calculations only minimally affects the

[1] The absorption percentage was calculated using the mass attenuation coefficient taken from the NIST webpage http://physics.nist.gov/PhysRefData/XrayMassCoef/cover.html.

[2] RR capture into the n=2 shell can also effect the emission but the cross section is weaker as compared to the EIE cross sections by over a factor of 100, and thus, is not considered.

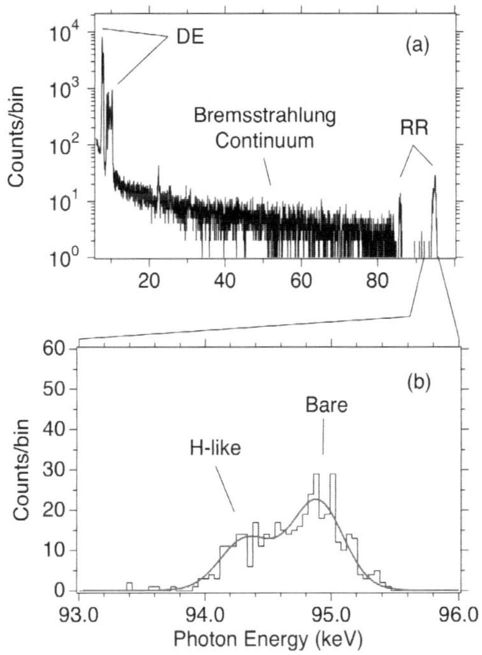


**Figure 1.** Graph (a) is the spectrum recorded with the IGLET for nickel at an electron beam energy of 85 keV. The DE spectrum as well as the L-shell and K-shell RR features are seen. Graph (b) is the K-shell RR peaks for H-like and bare nickel at 85 keV.

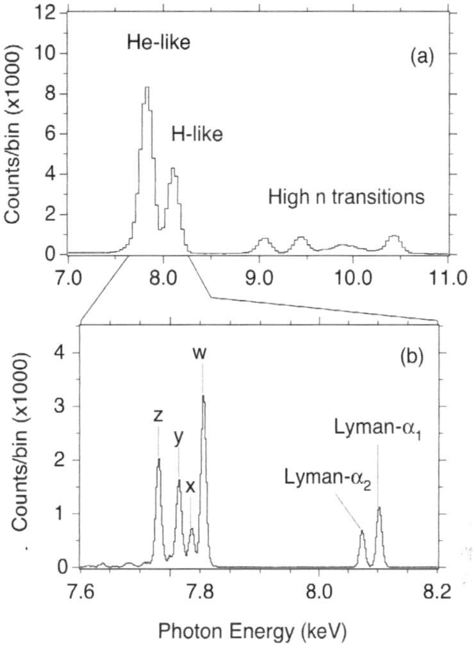


**Figure 2.** K-shell spectrum of nickel at an electron beam energy of 85 keV. Graph (a) is the spectrum recorded with the IGLET. Graph (b) is the spectrum recorded with the XRS/EBIT. The He-like charge state is labeled according to [7].

end result. We estimate that charge exchange effects the intensity of Lyman-$\alpha_1$ on the order of a few %[3]. Even with a possible order of magnitude uncertainty in the neutral density and the collision energy, the effect is still small and as such is not included in the analysis.

When taking into account the absorption percentage for RR photons, cascade, and polarization modifications to the DE, the formula for obtaining the EIE cross section takes the form,

$$\sigma_{EIE} = \frac{I_{DE}}{I_{RR}} \sigma_{RR} A_{RR} G(E), \tag{1}$$

where $\sigma_{EIE}$ is the cross-section for EIE, $\sigma_{RR}$ is the calculated cross section for RR, $I_{DE}$ is the number of counts in the DE line, $I_{RR}$ is the number of counts observed in the RR peak, $A_{RR}$ is the QE for RR photons in the Ge detector, and $G(E)$ is a factor that relates to the polarization and cascade contributions.

The RR cross sections used here are provided by Jim Scofield [11] and are not from FAC. Figure 3 shows the cross sections for the various energies for both nickel and iron ions and compared to FAC calculations. Each point is the average of several measurements (except for

[3] The cross-section for charge exchange is estimated from Müller and Salzborn and Salzborn and Müller [9, 10] to be the ion charge times $10^{-15}$ cm$^2$, the ion collision energy is estimated as 10 eV/amu, and the neutral density is around $3{\cdot}10^6$ cm$^{-3}$. Compared with the EIE cross section for Lyman-$\alpha_1$ of $10^{-22}$ cm$^2$, the electron collision energy of 35 keV to 85 keV, and the electron density around $5{\cdot}10^{11}$ cm$^{-3}$.

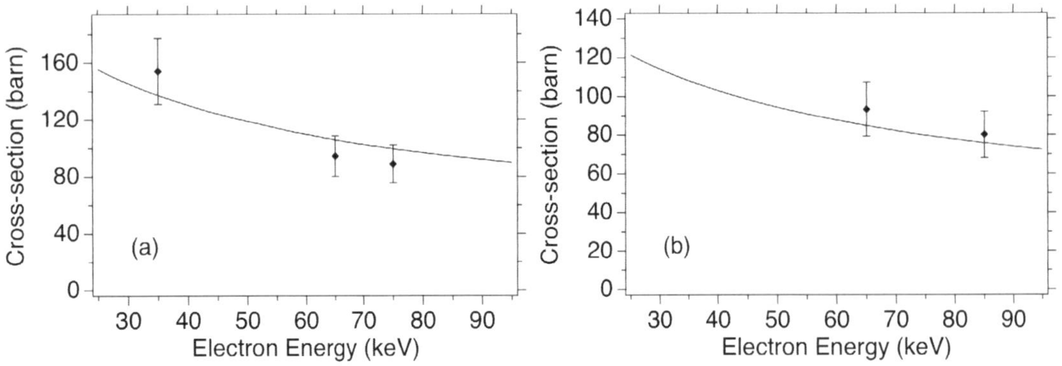


**Figure 3.** Experimental EIE cross sections for the Lyman-$\alpha_1$ line in hydrogenlike iron and nickel. Graph (a) represents the EIE cross sections for iron at energies of 35 keV, 65 keV and 75 keV. Graph (b) represents EIE cross sections for nickel at energies of 65 keV and 85 keV. The solid line is calculated by FAC.

the 35 keV and 75 keV measurements done with iron). The 15 % error bars are dominated by the approximately 10% statistical error associated with the determination of the number of counts in the H-like RR peak and, take into account the uncertainty in the calculation of the cascades, polarization and QE.

## 3. Conclusion

The cross sections measured here are the highest-energy cross sections that have been measured for a high-Z hydrogenlike ion. The results show that for the hydrogenlike ion, theory at high electron energies agrees well with experiment to roughly the 15 % level. Future studies may include a focus on heliumlike ions and go to higher-Z elements where QED effects become significant in the calculations of EIE cross sections [12].

## Acknowledgments

We would like to thank Ed Magee and Phil Dantonio for their technical support as well as John Gygax, Kevin Boyce, and Jonathan King for their work on design and fabrication of the XRS/EBIT. This work was performed under the auspices of the U. S. Department of Energy by Lawrence Livermore National Laboratory under Contract DE-C52-07NA27344 and was supported by NASA grants to LLNL and GSFC.

## References

[1] Bartiromo R, Bombarda F and Giannella R 1985 *Phys. Rev. A* **32** 531–537
[2] Chen H and Beiersdorfer P 2008 *Can. J. Phys.* **86** 55–71
[3] Saloman E B, Hubbell J H and Scofield J H 1988 *Atomic Data and Nuclear Data Tables* **38** 1–196
[4] Porter F S, Brown G V, Boyce K R, Kelley R L, Kilbourne C A, Beiersdorfer P, Chen H, Terracol S, Kahn S M and Szymkowiak A E 2004 *Rev. Sci. Instrum.* **75** 3772
[5] Wong K L, Beiersdorfer P, Reed K J and Vogel D A 1995 *Phys. Rev. A* **51** 1214–1220
[6] Porter F S, Gygax J, Kelley R L, Kilbourne C A, King J M, Beiersdorfer P, Brown G V, Thorn D B and Kahn S M 2008 *Rev. Sci. Instrum.* **79** 10E307
[7] Gabriel A H 1972 *Mon. Not. R. Astr. Soc.* **160** 99
[8] Gu M F 2008 *Can. J. Phys.* **86** 675
[9] Salzborn E and Müller A 1980 ed Oda N and Takayangi K (Amsterdam: North-Holland) p 407
[10] Müller A and Salzborn E 1976 *Phys. Lett.* **59A** 19
[11] Scofield J *private communication*
[12] Fontes C J, Sampson D H and Zhang H L 1993 *Phys. Rev. A* **47** 1009–1022

# Angular momentum transfer and polarization degree of ions with two-valence electrons by electron impact

**Kenichi Akita, Takeshi Kai*, Shinobu Nakazaki, Akinori Igarashi**

Department of Applied Physics, Faculty of Engineering, University of Miyazaki, Miyazaki 889-2192, Japan
(*) Quantum Beam Science Directorate, Japan Atomic Energy Agency 8-1 Umemidai Kizugawa-city Kyoto 619-0215, Japan

E-mail: akita@phys.miyazaki-u.ac.jp

**Abstract.** We study for electron-impact excitation of ions with two valence electrons ($Be^{2+}$, $C^{2+}$ and $Si^{2+}$) from the ground state to the first $^1P^o$ state using the $R$-matrix method. The integral cross sections and polarization degree for this transition of each ion are obtained. The differential cross sections and angular momentum transfer are also shown at a few energies in the non-resonant region. The present angular momentum transfer for $Be^{2+}$ and $C^{2+}$ at small scattering angles has positive values, while it is negative for $Si^{2+}$.

## 1. Introduction

Excitation cross section for electron impact is important to explain spectra of astrophysical plasmas and high-temperature laboratory plasmas. The radiation emitted from ions with unbalanced magnetic sublevel population by electron impact is polarized and its detection is available as a plasma diagnostic tool [1].

For excitation process from the ground S state to the excited P state, the expectation value of angular momentum component which is perpendicular to the scattering plane is called *angular momentum transfer* $L_\perp$. According to the propensity rule at high energies using the second-Born approximation method, the value of $L_\perp$ for this transition in electron-atom collision is positive at small scattering angles [2]. Srivastava *et al.* [3] reported using the distorted wave calculation that the values of $L_\perp$ for electron collisions with He-like ions (Z= 3, 8, $\infty$) are positive at small scattering angles at the incident electron energies of two and four times the excitation threshold energy. Kai *et al.* [4] have found a case where $L_\perp$ is negative in low-energy electron scattering with Mg-like Si ion ($Si^{2+}$) using the $R$-matrix method. The same method is applied to electron scatterings by doubly charged ions with one valence electron such as H-like Li ($Li^{2+}$) , Li-like B ($B^{2+}$) and Na-like Al ($Al^{2+}$) [5].

Here we treat the electron-impact excitations of doubly charged ions with two-valence electrons such as He-like Be ($Be^{2+}$), Be-like C ($C^{2+}$) and Mg-like Si ($Si^{2+}$). The integral cross section (ICS), polarization degree ($P$), the differential cross section (DCS) and $L_\perp$ are presented.

**Table 1.** Excitation energies (eV) and oscillator strengths for $Be^{2+}$ and $C^{2+}$.

| Target | State | Excitation energy | | Oscillator strength | | | |
|---|---|---|---|---|---|---|---|
| | | Present | Other | Present | | Other | |
| | | | | Length | Velocity | Length | Velocity |
| $Be^{2+}$ | 1s2p $^1P^o$ | 124.210 | 123.669[a] | 0.5595 | 0.5265 | 0.5515[a] | |
| $C^{2+}$ | 2s2p $^1P^o$ | 14.12 | 12.69[a] 13.03[b] | 0.832 | 0.831 | 0.788[b] | 0.873[b] |

[a] NIST [8]. [b] Glass [9].

## 2. Theory

### *2.1. Target wavefunctions and collision calculations*

The configuration interaction method is adopted to represent atomic states. To describe scattering wavefunctions, 19, 26 and 28 target states are used for $Be^{2+}$, $C^{2+}$ and $Si^{2+}$, respectively. In this work, the 1s orbital of $Be^{2+}$ and the 1s and 2s orbitals of $C^{2+}$ are the Hartree-Fock wave functions given by Clementi and Roettii [6] for the ground state $1s^2\ ^1S^e$ and $1s^22s^2\ ^1S^e$, respectively. The optimization of other high-lying orbitals are performed using the CIV3 atomic structure program of Hibbert [7]. The 2p, 3s, 3p, 3d, 4s, 4p, 4d and 4f orbitals of $Be^{2+}$ are optimized on the energies of the 1s2p $^1P^o$, 1s3s $^1S^e$, 1s3p $^1P^o$, 1s3d $^1D^e$, 1s4s $^1S^e$, 1s4p $^1P^o$, 1s4d $^1D^e$ and 1s4f $^1F^o$ states, respectively. The 2p, 3s, 3p and 3d orbitals of $C^{2+}$ are optimized on the energies of the 1s2p $^1P^o$, 1s3s $^1S^e$, 1s3p $^1P^o$ and 1s3d $^1D^e$ states, respectively. For the orbitals of $Si^{2+}$, we adopted the orbitals optimized by Kai *et al.* [4]. Table 1 shows the excitation energies and oscillator strengths between the ground state and the first excited $^1P^o$ state. The present excitation energies agree with the data of [8] within $\sim 0.5$ % for $Be^{2+}$ and with those of [8, 9] within $\sim 12$ % for $C^{2+}$. Our values of length and velocity oscillator strengths accord within $\sim 6$ % for $Be^{2+}$ and within less than 0.1 % for $C^{2+}$. The oscillator strengths are in fairly good agreement with the values of [9].

The $R$-matrix computer program of Berrington *et al.* [10] is used for the scattering calculation, where the experimental excitation energies [8] are adopted rather than the calculated values in Table 1.

### *2.2. The angular momentum transfer and polarization degree*

The angular momentum transfer for $^1S^e \to {}^1P^o$ transitions is given by $L_\perp = -2\sqrt{2}Im[f_1 f_0^*]/\sigma$ [11], where $\sigma$ is the DCS and $f_M$ is the scattering amplitude for a magnetic sublevel with $M$ of the $^1P^o$ state. The ICS for the $^1P^o$ state is given by $Q = Q_0 + 2Q_1$ ($Q_1 = Q_{-1}$). The $Q_M$ is the ICS for the excitation into a magnetic sublevel with $M$ of $^1P^o$ state. The polarization degree for $^1S^e \to {}^1P^o$ transitions is given by $P = (Q_0 - Q_1)/(Q_0 + Q_1) \times 100$ [12, 13]. The quantization axis is along the incident direction. In a case of neutral atom target, $p$ wave is dominant at the threshold, and $Q_1$ vanishes. Hence the emission line from the $^1P^o \to {}^1S^e$ transition is 100 % linearly polarized at threshold [12]. On the other hand, when the target is an ion, the $Q_1$ does not vanish at the threshold. Other partial waves as well as $p$ wave contribute to scattering owing to the attractive Coulomb force.

## 3. Results and discussion

Figure 1 shows the present results of the ICS, $P$, DCS and $L_\perp$ for $Be^{2+}(1s^2 \to 1s2p)$. $P$ is about 58 % at the threshold. DCS and $L_\perp$ at 124eV and 128eV are shown in Fig. 1 (c) and (d). With increase of energies, DCS increases at small scattering angles and decreases at large angles. The value of $L_\perp$ is positive at small angles and the trend is similar to the result for He-like ions of

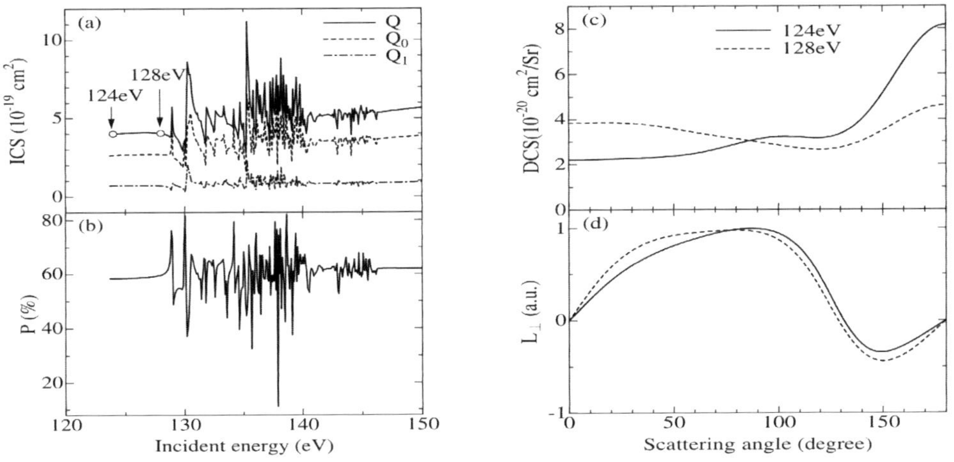


**Figure 1.** (a) ICS, (b) $P$, (c) DCS and (d) $L_\perp$ for the $1s^2\ ^1S^e \rightarrow 1s2p\ ^1P^o$ transition of $Be^{2+}$ above the threshold.

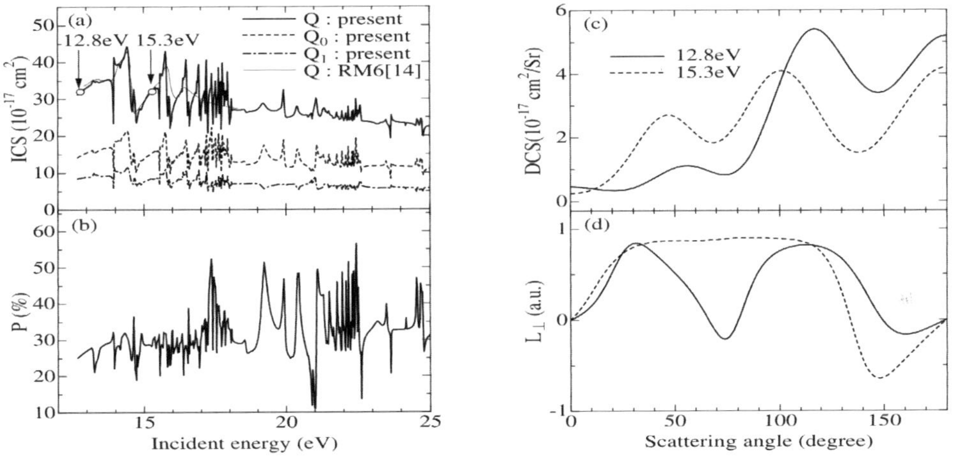


**Figure 2.** The same as Fig. 1 but for the $2s^2\ ^1S^e \rightarrow 2s2p\ ^1P^o$ transition of $C^{2+}$.

Srivastava *et al.* [3]. Figure 2 shows the present results for $C^{2+}(2s^2 \rightarrow 2s2p)$. Berrington *et al.* [14] have reported the collision strength for $C^{2+}$ using the $R$-matrix calculation of 6 target states constructed with many configurations. The collision strength of Berrington *et al.* is transformed to ICS and compared with the present one in Fig. 2(a). The agreement of the two ICS's are reasonable, but the resonance structure in the present ICS is more pronounced than that of the 6 states calculation. $P$ at the threshold is about 25 % in Fig. 2(b). Figure 2 (c) and (d) shows DCS and $L_\perp$ at 12.8 and 15.3 eV. The DCS near backward direction decreases with energies, while it remains small at small angles. Near 12.8 eV, $L_\perp$ is positive at small angles and it takes a negative value around 70°. Figure 3 shows the results for $Si^{2+}(3s^2 \rightarrow 3s3p)$. We have reproduced the $Si^{2+}$ results of Kai *et al.* [4], where the data except for $P$ is already reported. $P$ is about 48 % at the threshold in Fig. 3(b). The values of $L_\perp$ at small angles are negative differently from those of $Be^{2+}$ and $C^{2+}$ in Fig. 3(d).

The excitations of doubly charged ions with two valence electrons of $Be^{2+}(1s^2 \rightarrow 1s2p)$, $C^{2+}(2s^2 \rightarrow 2s2p)$ and $Si^{2+}(3s^2 \rightarrow 3s3p)$ are analogous to those with one valence electron of $Li^{2+}(1s \rightarrow 2p)$, $B^{2+}(2s \rightarrow 2p)$ and $Al^{2+}(3s \rightarrow 3p)$, respectively. The excitations of one valence

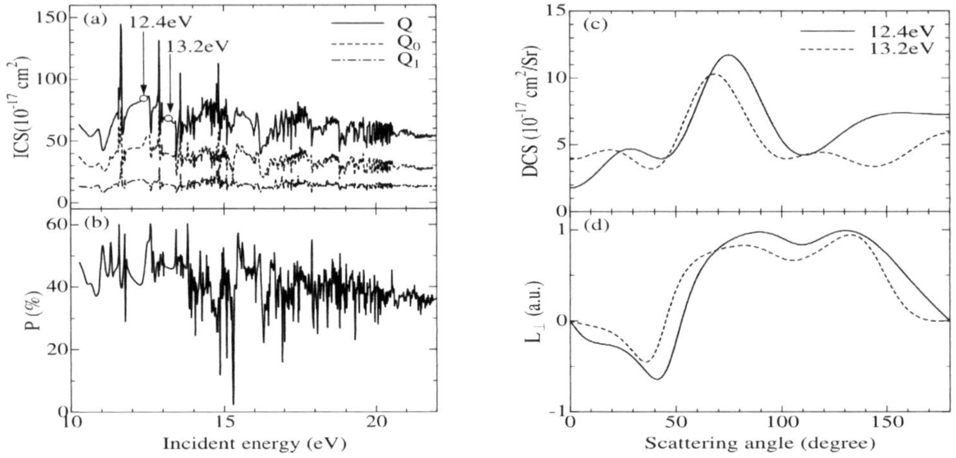


**Figure 3.** The same as Fig. 1 but for the $3s^2\ ^1S^e \rightarrow 3s3p\ ^1P^o$ transition of $Si^{2+}$.

electron such as $Li^{2+}$, $B^{2+}$ and $Al^{2+}$ are reported in [5]. It is interesting to compare the counterparts of one- and two-valence electron(s). The angular distributions of DCS are similar to each other. Similarities are also found for the sign of $L_\perp$ at small angles, namely, they are positive for $Li^{2+}$ and $Be^{2+}$ and negative for $Al^{2+}$ and $Si^{2+}$. However, disagreement occurs for a pair of $B^{2+}$ (negative) and $C^{2+}$ (positive). Thus, comparing with analogue ions of one valence electron, the sign of $L_\perp$ does not necessarily agree.

To summarize, the ICS, $P$, DCS and $L_\perp$ are calculated for electron-impact excitations of $Be^{2+}$, $C^{2+}$ and $Si^{2+}$ with two valence electrons for the transition from the ground state ($^1S^e$) to the first $^1P^o$ excited state. $L_\perp$ at small scattering angle is positive for $Be^{2+}$ and $C^{2+}$ but negative for $Si^{2+}$. Comparing with analogous ions of one valence electron, the sign of $L_\perp$ agrees for pairs $Be^{2+}$ – $Li^{2+}$ and $Si^{2+}$ – $Al^{2+}$, but disagrees for a pair $C^{2+}$ – $B^{2+}$. Hence the sign of $L_\perp$ is a collision parameter sensitive to target ions. To give the physical reason is a problem to be cleared in future.

## References

[1] T. Kawamura, T. Kai, F. Koike, S. Nakazaki, Y. Inubushi, H. Nishimura, Phys. Rev. Lett. **99** 115003 (2007).
[2] K. Bartschat, N. Andersen and D. Loveall, Phys. Rev. Lett. **83**, 5254 (1999).
[3] R. Srivastava, Y. Itikawa and K. Sakimoto, Phys. Rev. A **43**, 4736 (1991).
[4] T. Kai, R. Srivastava and S. Nakazaki, J. Phys. B: At. Mol. Phys. **37**, 2045 (2004).
[5] K. Akita et al. 'Angular momentum transfer and polarization degree of ions with one-valence electron by electron-impact' submitted to "Proceedings of HCI2008".
[6] E. Clementi and C. Roetti, At. Data Nucl. Data Tables **14**, 177 (1974).
[7] A. Hibbert, *Comput. Phys. Commun.* **9**, 141 (1975).
[8] See http://physics.nist.gov/cgi-bin/AtData/levels_form
[9] R. Glass, J. Phys. B: At. Mol. Phys. **12** 1633 (1979).
[10] K. A. Berrington, W. B. Eissner and P. H. Norrington, *Comput. Phys. Commun.* **92**, 290 (1995).
[11] N. Andersen and K. Bartschat, *Polarization, alignment, and orientation in atomic collisions* (Springer-Verlag New York, Inc. 2001).
[12] I. C. Percival and M. J. Seaton, Phil. Trans. Roy. Soc. A **251**, 113 (1958).
[13] T. Kai, S. Nakazaki, K. A. Berrington, Nucl. Instrum. Methods Phys. Res. B, **235**, 249 (2005)
[14] K. A. Berrington, P. G. Burke, P. L. Dufton and A. E. Kingston, J. Phys. B: At. Mol. Phys. **10** 1465 (1977).

# Excitation rates for transitions in Kr XXXII

**K M Aggarwal[1], F P Keenan[1] and K D Lawson[2]**

[1]Astrophysics Research Centre, School of Mathematics and Physics, Queen's University Belfast, Belfast BT7 1NN, Northern Ireland, UK
[2]EURATOM/UKAEA Fusion Association, Culham Science Centre, Abingdon, Oxforshire OX 14 3DB, UK

E-mail: K.Aggarwal@qub.ac.uk

**Abstract.** In this paper we report our results for collision strengths and effective collision strengths for transitions among the lowest 125 levels of the $2s^22p$, $2s2p^2$, $2p^3$, $2s^23\ell$, $2s2p3\ell$, and $2p^23\ell$ configurations of Kr XXXII. For our calculations both the FAC and DARC codes have been employed.

## 1. Introduction

Krypton is used in a variety of experiments in tokamak fusion plasmas where high temperatures ($> 10^6$ K) give rise to many of its ionisation stages. Therefore, atomic data (namely energy levels, oscillator strengths or radiative decay rates, excitation rates, etc.) are required for many ions in order to estimate the power loss from the impurities. In a recent paper [1] we reported energy levels, lifetimes, and radiative rates for four types of transitions, namely electric dipole (E1), electric quadrupole (E2), magnetic dipole (M1), and magnetic quadrupole (M2) for five Kr ions (Kr XXXII - Kr XXVIII). Here we report our results for excitation rates for transitions in boron-like Kr XXXII. These results along with those already reported [1] will be helpful in the modelling of plasmas.

Earlier calculations for Kr XXXII ([2], [3]) included only 20 levels among the $2s^22p$, $2s2p^2$, $2p^3$, and $2s^23\ell$ configurations, reported values of collision strengths ($\Omega$) at a few energies above thresholds, and did not include resonances which can significantly contribute to the determination of excitation rates, even at the high temperatures found in fusion plasmas, at least for some of the transitions. Therefore, in the present work we include 125 levels belonging to the $2s^22p$, $2s2p^2$, $2p^3$, $2s^23\ell$, $2s2p3\ell$, and $2p^23\ell$ configurations of Kr XXXII.

For our calculations of wavefunctions the GRASP (general-purpose relativistic atomic structure package) code has been adopted, and for calculating collision strengths ($\Omega$) and subsequently the effective collision strengths ($\Upsilon$), the *Dirac atomic R-matrix code* (DARC) is employed. Additionally, in order to make an accuracy assessment of our results and to estimate the contribution of resonances, we have also performed calculations from the *Flexible Atomic Code* (FAC) of Gu [4], available from the website `http://kipac-tree.stanford.edu/fac`. This is a fully relativistic code and provides a variety of atomic parameters. Furthermore, the code yields results comparable to those obtained from other atomic structure codes such as CIV3 and GRASP - see, for example, Aggarwal et al. [5] and references therein, and scattering codes such as *R*-matrix - see, for example, Aggarwal et al. [6] and references therein. Furthermore, FAC is based on the well known and widely used *distorted-wave* (DW) method and yields results for collision strengths ($\Omega$) comparable with those from the $R$-matrix method, particularly for the allowed transitions, as demonstrated in several of our earlier papers on a variety of ions, including many of iron, i.e. Fe IX - Fe XXVI. However, resonances have not been resolved in the calculations from FAC.

## 2. Collision strengths and effective collision strengths

In our calculations from DARC, the $R$-matrix radius is adopted to be 1.48 au, and 15 continuum orbitals are included for each channel angular momentum for the expansion of the wavefunction. This allows us to compute values of $\Omega$ up to an energy of 500 Ryd, sufficient to determine the values of $\Upsilon$ up to a temperature of $10^{7.3}$ K. Furthermore, in order to obtain convergence of $\Omega$ for all transitions and at all energies, we have included all partial waves with angular momentum $J \leq 40$. To account for the inclusion of higher neglected partial waves, we have also included a top-up, based on Coulomb-Bethe approximation for allowed transitions and geometric series for others. Resonances in the thresholds region have been resolved in a narrow energy mesh, and results for effective collision strengths ($\Upsilon$) have been obtained after averaging the values of $\Omega$ over a Maxwellian distribution of electron velocities. Detailed results for all 7750 transitions over a wide temperature range below $10^{7.3}$ K are reported in a separate paper [7].

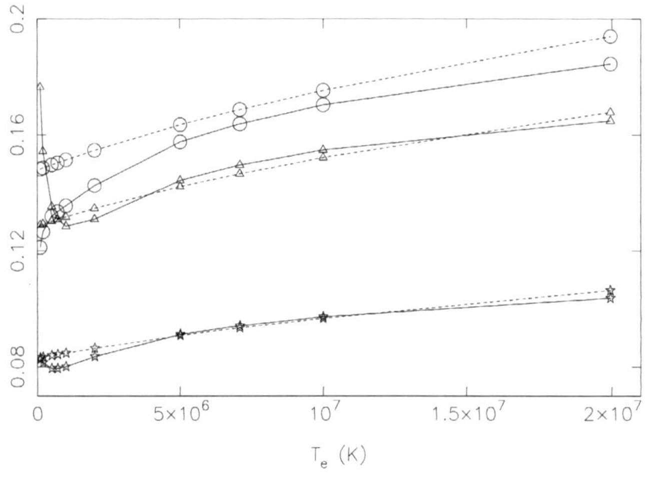


**Figure 1.** Comparison of effective collision strengths from DARC (continuous curves) and FAC (broken curves) for the 1-6 (circles: $2s^22p\ ^2P^o_{1/2}$ - $2s2p^2\ ^2D_{3/2}$), 2-8 (triangles: $2s^22p\ ^2P^o_{3/2}$ - $2s2p^2\ ^2D_{5/2}$), and 3-11 (star: $2s2p^2\ ^4P_{1/2}$ - $2p^3\ ^4S^o_{3/2}$) allowed transitions of Kr XXXII.

Our calculations from the FAC code cover a wider energy range up to $\sim$ 1600 Ryd, and hence allow us to calculate values of $\Upsilon$ up to $T_e = 10^9$ K. Furthermore, the energy levels and A-values obtained from FAC are comparable for a majority of levels (transitions), as discussed and demonstrated by Aggarwal et al. [1]. Similarly, values of $\Omega$ obtained from FAC are generally in good agreement with those from DARC for a majority of transitions, particularly at higher energies. In Fig. 1 we compare our results of $\Upsilon$ from DARC and FAC for three *allowed* transitions, namely 1-6 ($2s^22p\ ^2P^o_{1/2}$ - $2s2p^2\ ^2D_{3/2}$), 2-8 ($2s^22p\ ^2P^o_{3/2}$ - $2s2p^2\ ^2D_{5/2}$), and 3-11 ($2s2p^2\ ^4P_{1/2}$ - $2p^3\ ^4S^o_{3/2}$). Since the oscillator strength (f- values) and collision strengths for these three (and many other) transitions are comparable between the GRASP/DARC and FAC calculations, so are the values of $\Upsilon$, particularly at higher temperatures, as shown in Fig. 1. At temperatures towards the lower end, some allowed transitions do have the higher values of $\Upsilon$ from DARC, because of the resonances, as may be noted for the 2-8 transition in Fig. 1. However, the effect of resonances is not significant for a majority of the *allowed* transitions, and particularly at the high temperatures achieved in fusion plasmas. Similarly, $\Upsilon$ values from FAC are slightly higher at lower temperatures for some transitions, such as 1-6. This is due to the fact that the values of $\Omega$ are higher from FAC than from DARC at lower energies. Such differences are partly due to the differences in the $R$-matrix and DW approaches adopted in the two calculations, and mainly due to the interpolation and extrapolation techniques employed in the FAC code, which is designed to generate a large amount of atomic data in a comparatively very short period of time, and without too much loss of accuracy. Therefore, sometimes a problem of a few anomalies may arise from the calculations from FAC, but overall we observe no major discrepancy with our calculations performed from the DARC code, as also observed for many other ions, such as those of iron - see, for example, Aggarwal et al. [6] and the references therein.

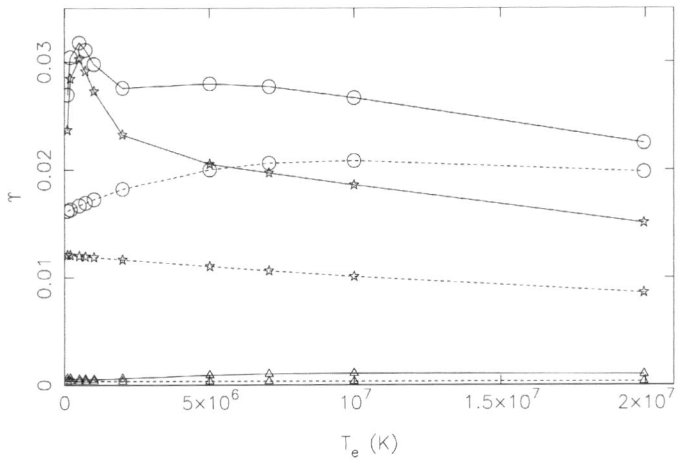


**Figure 2.** Comparison of effective collision strengths from DARC (continuous curves) and FAC (broken curves) for the 1-2 (circles: $2s^22p\ ^2P^o_{1/2}$ - $2s^22p\ ^2P^o_{3/2}$), 2-11 (triangles: $2s^22p\ ^2P^o_{3/2}$ - $2p^3\ ^4S^o_{3/2}$), and 3-4 (star: $2s2p^2\ ^4P_{1/2}$ - $2s2p^2\ ^4P_{3/2}$) forbidden transitions of Kr XXXII.

Comparisons of $\Upsilon$ values from DARC and FAC for three *forbidden* transitions, namely 1-2 ($2s^22p\ ^2P^o_{1/2}$ - $2s^22p\ ^2P^o_{3/2}$), 2-11 ($2s^22p\ ^2P^o_{3/2}$ - $2p^3\ ^4S^o_{3/2}$), and 3-4 ($2s2p^2\ ^4P_{1/2}$ - $2s2p^2\ ^4P_{3/2}$), are shown in Fig. 2. For all these three (and many other) forbidden transitions, our $\Upsilon$ values from DARC are invariably higher than from FAC. This is partly due to corresponding differences in the $\Omega$ values, but for only some transitions, mainly the weaker ones. Most of the differences are due to the inclusion of resonances in the calculations from DARC. For the same reason the differences in the $\Upsilon$ values from the two independent calculations decrease with increasing temperature. Nevertheless, the importance of resonances is clear from Fig. 2, as the enhancement in the $\Upsilon$ values is up to a factor of three, even at the highest common temperature of $2\times10^7$ K - see, for example, transition 2-11. In fact there are some resonance transitions, such as 1-15 ($2s^22p\ ^2P^o_{1/2}$ - $2p^3\ ^2P^o_{3/2}$) and 1-48 ($2s^22p\ ^2P^o_{1/2}$ - $2p^23s\ ^4P_{1/2}$), for which our values of $\Upsilon$ from DARC are higher than from FAC by over an order of magnitude over the entire range of temperature. Similar enhancements in the values of $\Upsilon$ have also been observed for transitions among the excited levels. Thus the values of $\Upsilon$ adopted with the inclusion of resonances may affect the modelling of plasmas.

A major limitation of our calculations from DARC is the limited range of energy (below 500 Ryd), and hence the temperature, i.e. below $2{\times}10^7$ K, whereas higher temperatures may prevail in fusion plasmas. Extending the temperature range is currently not feasible with the DARC, as the calculations become too lengthy with the resources available with us. Therefore, we have calculated values of $\Upsilon$ up to $T_e = 10^9$ K, but only from FAC which do not include the contribution of resonances. A complete set of data for all transitions from both calculations and over a wide range of temperatures are reported in a separate paper [7]. We hope those results along with the already reported A- values will be highly useful in the modelling of fusion plasmas for the under progress ITER project. Finally, we would like to state that similar calculations, both from DARC and FAC, are in progress for F-like Kr XXVIII, and the excitation rates among 113 levels belonging to the $2s^22p^5$, $2s2p^6$, $2s^22p^43\ell$, $2s2p^53\ell$, and $2p^63\ell$ configurations will be reported in a future publication.

## Acknowledgments

This work has been partly funded by the EPSRC, UK, and partly by the European Communities under the contract of Associations between EURATOM and UKAEA. FPK is grateful to AWE Aldermaston for the award of a William Penney Fellowship.

## References

[1] Aggarwal K M, Keenan F P and Lawson K D 2008 *At. Data Nucl. Data Tables* **94** 323
[2] Bhatia A K, Feldman U and Seely J F 1986 *At. Data Nucl. Data Tables* **35** 319
[3] Zhang H L and Sampson D H 1994 *At. Data Nucl. Data Tables* **56** 41
[4] Gu M F 2003 *Astrophys. J.* **582** 1241
[5] Aggarwal K M, Tayal V, Gupta G P and Keenan F P 2007 *At. Data Nucl. Data Tables* **93** 615
[6] Aggarwal K M, Hamada K, Igarashi A, Jonauskas V, Keenan F P and Nakazaki S 2008 *Astron. Astrophys.* **484** 879
[7] Aggarwal K M, Keenan F P and Lawson K D 2009 *At. Data Nucl. Data Tables* in press

# Structure deformation dynamics of acetylene molecules following electron loss and capture collisions of 6 MeV $O^{4+}$ ions

**T. Mizuno[1], T. Yamada[1], H. Tsuchida[1], Y. Nakai[2] and A. Itoh[1]**

[1]Department of Nuclear Engineering, Kyoto University, Kyoto 606-8501, Japan
[2]RIKEN Nishina center, Wako Saitama 351-0198, Japan

E-mail: mizuno@nucleng.kyoto-u.ac.jp

**Abstract.** We investigated structure deformation dynamics of acetylene molecules in charge-changing collisions of 6 MeV $O^{4+}$ ions by means of a momentum 3D imaging technique. We obtained an angular correlation between fragment ions of $H^+$ and $C^+$ produced in ($H^+$+$C^+$+$C^{2+}$+H), ($H^+$+$C^+$+$C^+$+H) and ($H^+$+$C^+$+$CH^+$) measured for single electron loss and capture collisions. It was found that in fragmentation channel of ($H^+$+$C^+$+$CH^+$) bending motion of $C_2H_2$ occurs strongly compared with ($H^+$+$C^+$+$C^{2+}$+H) and ($H^+$+$C^+$+$C^+$+H). We concluded that the structure deformation is caused by electronic and vibrational excitation.

## 1. Introduction

In the last decade, structure deformation and bond rearrangement dynamics of molecules have been extensively studied by using a 3D momentum imaging technique [1-10]. This method enables us to obtain information about molecular structure deformation prior to its dissociation since the momentum vectors of product ions are sensitive to the structure. Hishikawa *et. al.* demonstrated visualization of ultrafast migration of hydrogen in deuterated double charged acetylene ion using intense ultrashort laser pulses [7]. They used a pump-probe method in combination with a 3D momentum imaging technique. They found the deuterium atom to move at first from one carbon site to the other in a short time scale (90 fs) and then to migrate back to the original carbon site by 280 fs, in competition with the molecular fragmentation. Osipov *et. al.* studied hydrogen migration of $C_2H_2$ by inner-core ionization of carbon K shell [8]. They also found the hydrogen atom to move to the other carbon site within 60 fs. By contrast, little is known about structure and bond arrangements of multiply charged molecules for which the deformation process may be different greatly because the liftime of the molecule prior to dissociation is expected to be shorter than neutral or lowly charged molecules.

In order to understand the deformation process of multiply charged molecules more clearly, we performed collision experiments of 6 MeV $O^{4+}$ ions in a $C_2H_2$ molecular target. This is because the swift heavy ion impact can easily produce highly charged molecules [4, 10] and serves as a useful tool to probe the structure deformation of multiply charged molecules. We used a momentum 3D imaging method to obtain information about the structure before its fragmentation for individual fragmentation channel. As the charge-changing collision of

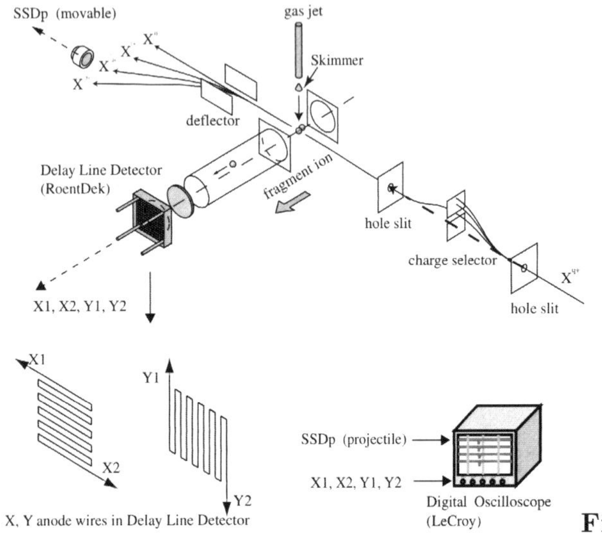


**Figure 1.** Experimental setup

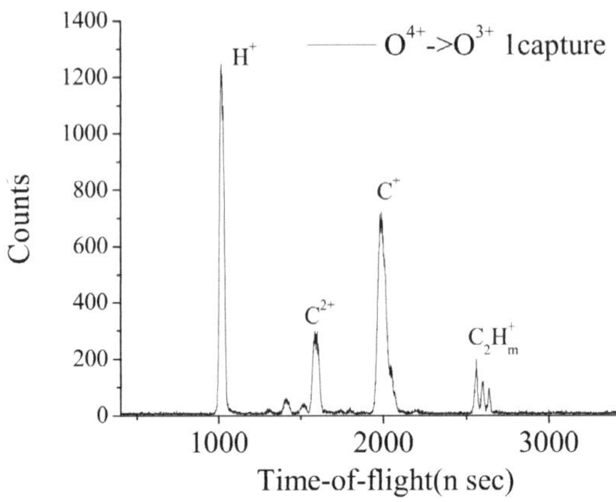


**Figure 2.** TOF spectrum of fragment ions obtained in 1e-capture collisions of 6 MeV $O^{4+}$ ions.

projectiles plays an essential role in ionization and fragmentation of target molecules [10], we also measured the charge state of a outgoing particle.

## 2. Experiment

Experiments were performed at the QSEC 1.7-MV tandem accelerator facility of Kyoto University. A schematic diagram of the experimental setup is shown in Figure 1. A beam of 6 MeV $O^{4+}$ ions was carefully collimated (<1 mm in diameter) and was crossed with an effusive molecular beam target of $C_2H_2$. A base pressure of the collision chamber was kept below $1 \times 10^{-5}$Pa. After collisions, final charge states of outgoing projectile particles were analyzed by an electrostatic beam deflector and detected by a movable solid state detector (SSD). Fragment ions as well as intact ions produced in collisions were measured with a position-sensitive delay line detector (DLD) [11]. Mass distributions of fragment ions were obtained by a time-of-flight (TOF) technique under a Wiley-Mclaren spatial-focusing condition. Signals of fragment ions were recorded event-by-event in a digital oscilloscope (Wavepro7000, LeCroy) in coincidence with outgoing projectile ions. In this way, we studied the following single electron (1e) capture and loss collisions.

$$O^{4+} + C_2H_2 \rightarrow O^{3+} + C_2H_2^{r+*} + (r-1)e^- \quad \text{(1e-capture)} \qquad (1)$$

$$\rightarrow O^{5+} + C_2H_2^{r+*} + (r+1)e^- \quad \text{(1e-loss)} \qquad (2)$$

The momentum vectors of fragment ions were deduced from positions $(x, y)$ and flight times $t$ of these products.

## 3. Results and discussion

Figure 2 shows a time-of-flight spectrum obtained in 1e-capture collisions of 6 MeV $O^{4+}$ ions. Peaks are mostly fragment ions like $H^+$, $C^+$, $C^{2+}$ and $C_2H_m^+$ (m=0,1,2) ions, and no other ions such as $CH_2^+$ and $H_2^+$ produced from bond rearrangement reactions are observed.

The angular correlations between $H^+$ and $C^+$ measured for 1e-capture collisions are shown in Figure 3 (a)-(c), and those for 1e-loss collisions are in (d)-(f). The abscissa $\theta$ is the angle between the momenta of $C^+$ and $H^+$, defined as, $P(C^+) \cdot P(H^+) = |P(C^+)||P(H^+)| \cos\theta$. The top figures

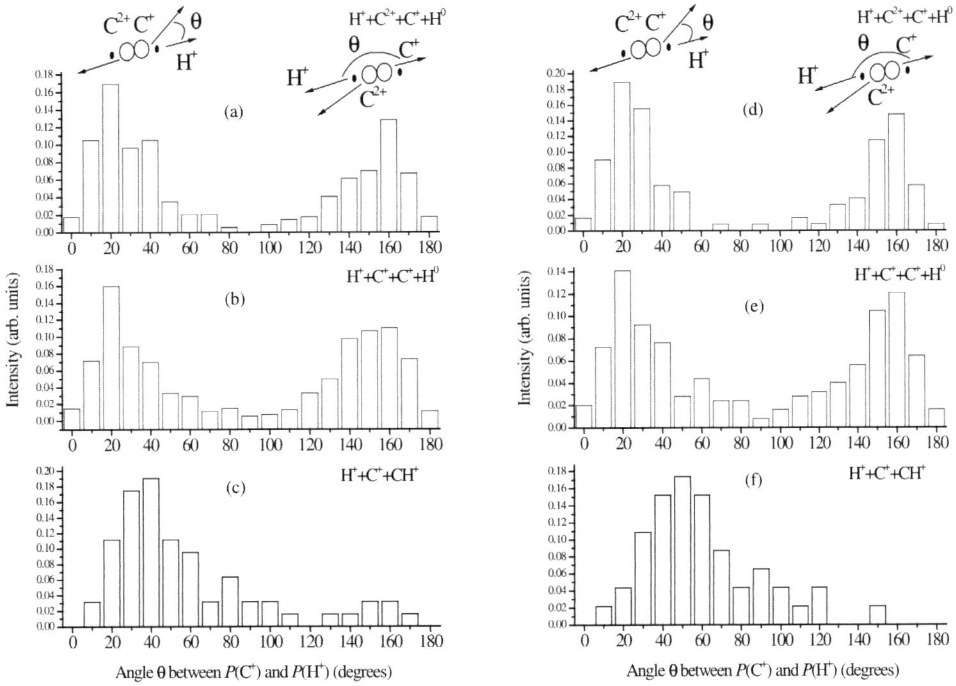


**Figure 3.** Angular correlations between two momenta of $P(C^+)$ and $P(H^+)$ obtained in single electron capture ((a)-(c)) and loss ((d)-(f)) collisions of 6 MeV $O^{4+}$ ions.

((a) and (d)) are produced from intact parent ions $(C_2H_2)^{4+*}$ and the middle ones ((b) and (e)) are from $(C_2H_2)^{3+*}$. The bottom figures ((c) and (f)) show the case of three-body fragmentation ($H^+$+$C^+$+$CH^+$). The angular correlations shown in the top and middle figures exhibit bimodal distributions peaking at about 20 and 160 degrees. These are interpreted as follows. In case of fragmentation of $(C_2H_2)^{4+*}$, resulting in the same final products (H,$H^+$,$C^+$,$C^{2+}$), there are two dissociation schemes of ($HC^{2+}-C^+H^+$) and ($H^+C^{2+}-C^+H$) prior to dissociation. In the former case, both $C^+$ and $H^+$ move in the same direction after dissociation and separate from each other afterwards as depicted in the figure, giving rise to the correlation angle of 20 degrees. In the latter case, $C^+$ and $H^+$ move in the opposite direction, giving rise to the peak of 160 degrees. As for the case of fragmentation of $(C_2H_2)^{3+*}$, there is only one dissociation scheme of ($H^+C^+$-$C^+H$) resulting in the final form of ($H^+$+$C^+$+$C^+$+H) as drawn in the figures. Hence, the bimodal pattern appears also in this case due to the same reason described above. On the other hand, the angular correlation exhibits single and broader peak structure in case of three-body fragmentation of ($H^+$+$C^+$+$CH^+$), as shown in the bottom figures. This is due to the reason that both $C^+$ and $H^+$ move in the same direction after dissociation.

The broader $\theta$-distribution in three-body fragmentation indicates evidently that the bending motion of $C_2H_2$ is enhanced more strongly in comparison with four-body fragmentation. This can be explained as in the following way. The coulomb repulsion force as well as excitation energy in four-body fragmentation is expected to be larger than three-body fragmentation. Thus, the lifetime of intact parent ions leading to four-body fragmentation may be shorter than three-body fragmentation, because highly charged and/or excited states are more unstable. As a result, the four-body fragmentation channels dissociate immediately, leading to near ground state structures. By contrast, the three-body fragmentation with relatively longer lifetimes has enough time for relaxation to bending structure states.

Molecular deformation may be caused by the following three mechanisms. First is the Renner-Teller effect following C 1s→ $\pi^*$ transition. Saito *et. al.* showed that this transition causes a bending motion more easily compared with C 1s ionization [9]. In the present collision system, however, the energy deposition to C 1s-electrons is smaller than those of valence electrons, as calculated from the Casp code [12]. Therefore, the Renner-Teller effect may be neglected in the present work. Second is the effect of valence electron excitation. For example, Zyubina *et al.* [13] reported that, in the first excited triplet state ($1^3\Pi$) of $(C_2H_2)^{2+}$, the linear structure $(HCCH)^{2+}$ is unstable but the vinylidene configuration $(CCH_2)^{2+}$ is stable, and they predicted the structure deformation to the vinylidene configuration from $(C_2H_2)^{2+}$. As fast ion collision is known to produce a large number of electronic excited states, the structure deformation due to this effect is possibly expected to play a role in our collision system. Third is the effect of vibrational excitation as reported by Rajgara *et al.* who investigated the molecular deformation of $CS_2$ by 120 keV $Ar^{8+}$ impacts [6]. Although the vibrational excitation would hardly occur in our fast collisions, relaxation from electronic to vibrational excitation may certainly take place because the three-body fragmentation has a long lifetime.

We found that the $\theta$-distribution in the four-body fragmentation is essentially the same for both electron loss and capture collisions. On the other hand, there is somewhat large difference in case of the three-body fragmentation, showing broader distribution in loss collisions compared to capture collisions. These results seem also support our remarks described above. Namely, the four-body fragmentation channels dissociate more rapidly without accompanying molecular deformation and do not depend greatly on the electron loss and capture collisions. On the other hand, the three-body fragmentation with relatively longer lifetimes can lead to bending structure more easily. Consequently, the degree of its deformation depends on the excited state population which is probably different for electron loss and capture collisions.

## 4. Summary

Fragmentation dynamics of $C_2H_2$ molecules by 6 MeV $O^{4+}$ impacts is studied by means of a 3D momentum imaging technique. The decay channel associated with $(C_2H_2)^{4+*}$ was identified for the first time to ($HC^{2+}$-$C^+H^+$) and ($H^+C^{2+}$-$C^+H$) prior to its dissociation. We also found that in the three-body fragmentation the bending motion is enhanced strongly in comparison with four-body fragmentation. It is concluded that the structure deformation is certainly induced by valence electronic excitation and vibrational excitation but not by inner shell excitation.

## Acknowledgments

This work was supported by a Grant-in-Aid for Scientific Research (B), No. 16360474, from the Japan Society for the Promotion of Science.

## References

[1] Ullrich J *et. al.* 2003 *Rep. Prog. Phys.* **66** 1463
[2] De S, Rajput J, Roy A, Ghosh P N, and Safvan C P 2008 *Phys. Rev. A.* **77** 022708
[3] De S, Rajput J, Roy A, Ghosh P N, and Safvan C P 2007 *J. Chem. Phys.* **127** 051101
[4] Adoui L *et.al.* 2006 *Nucl. Instr. and Meth. B* **245** 94
[5] Werner U, Siegmann B, Lebius H, Huber B, Lutz H O, 2003 *Nucl. Instr. and Meth. B* **205** 639
[6] Rajgara F A, Krishnamurthy M, Mathur D, Nishide T, Kitamura T, Shiromaru H, Achiba Y and Kobayashi N 2001 *Phys. Rev. A.* **64** 032712
[7] Hishikawa A, Matsuda A, Fushitani M, and Takahashi E J 2007 *Phys. Rev. Lett.* **99** 258302
[8] Osipov T *et.al.* 2003 *Phys. Rev. Lett.* **90** 233002
[9] Saito N, Nagoshi M, Machida M, Koyano I, De Fanis A, Ueda K 2004 *Chem. Phys. Lett.* **393** 295
[10] Mizuno T, Majima T, Tsuchida H, Nakai Y, and Itoh A 2007 *J. Phys.: Conf. Ser.* **58** 173
[11] Sobottka S and Williams M, 1988 *IEEE Trans. Nucl. Sci.* **35** 348
[12] Grande P L and Schiwietz G, 1998 *Phys. Rev. A.* **58** 3796
[13] Zyubina T S *et.al.* 2005 *J. Chem. Phys.* **123** 134320

Journal of Physics: Conference Series **163** (2009) 012040 doi:10.1088/1742-6596/163/1/012040

# Systematic multiconfiguration Dirac-Fock method study of the K-X-ray spectra of silicon

**Katarzyna Słabkowska and Marek Polasik**
Faculty of Chemistry, Nicolaus Copernicus University, 87-100 Toruń, Poland
E-mail: Katarzyna.Slabkowska@uni.torun.pl

**Abstract.** Extensive multiconfiguration Dirac-Fock calculations with the inclusion of the transverse Breit interaction and QED corrections have been performed for silicon to explain the influence of removing electrons from $L$ and $M$ shells on the shapes and parameters (the average $K\alpha$ and $K\beta$ transition energies and the values of $K\beta/K\alpha$ intensity ratio) of its $K$-X-ray spectra. For $K\alpha$ diagram lines and for certain $L$- and $M$-satellite lines the theoretical stick spectra (line positions with their relative intensities) have been presented. Moreover for each type of lines two theoretical spectra have been synthesized: one being a sum of the Lorentzian natural line shapes and the other one being a convolution of the sum of the Lorentzian natural line shapes with the Gaussian instrumental response.

## 1. Introduction

In recent years, several theoretical models for reliable description of very complex X-ray spectra accompanying the ionization (of both targets and projectiles) in collision processes based on the multiconfiguration Dirac-Fock (MCDF) method have been developed [1–10]. These models have been successfully applied to the interpretation of very complex X-ray spectra generated in near-central collisions of various light and heavy projectiles with many target atoms [11–25].

Recently the $K\alpha$ X-ray emission spectra of the low-density $SiO_2$ aerogel target bombarded by $^{48}Ca^{6+}$ ions with initial energy of 11.4 MeV/u have been measured with a high spectral and spatial resolution along the ion beam stopping path [25]. The structure of the observed spectrum [25] indicates that the reliable interpretation of $K$-X-ray spectra of silicon induced by heavy projectiles requires detailed and extensive theoretical studies of the influence of removing electrons from $L$ and $M$ shells on the shapes and parameters of its $K$-X-ray spectra. It is worth to be noticed that owing to the open-like character of the electronic ground state configuration of neutral silicon atom even for the diagram $K\alpha$ and $K\beta$ lines the number of transitions achieves the large value. Therefore the detailed theoretical studies on the shapes and parameters of the $K$-X-ray spectra for multiply ionized silicon atom presented in this paper require performing advanced and exhausting MCDF calculations.

## 2. MCDF calculations

The methodology of MCDF calculations performed in the presented studies is similar to the published earlier, in many papers (e.g. [5, 26, 27]). Within the MCDF scheme the effective Hamiltonian for an N-electron system is expressed by

$$H = \sum_{i=1}^{N} h_D(i) + \sum_{j>i=1}^{N} C_{ij}, \tag{1}$$

where $h_D(i)$ is the Dirac operator for $i$-th electron and the terms $C_{ij}$ account for electron-electron interactions. The latter are a sum of the Coulomb interaction operator and the transverse Breit operator. In the MCDF method an atomic state function (ASF) with the total angular momentum $J$ and parity $p$ is assumed in the form

$$\Psi_s(J^p) = \sum_m c_m(s)\Phi(\gamma_m J^p), \tag{2}$$

where $\Phi(\gamma_m J^p)$ are configuration state functions (CSF), $c_m(s)$ are the configuration mixing coefficients for state $s$, $\gamma_m$ represents all information required to uniquely define a certain CSF. In this work the modified special average-level version of MCDF calculations, MCDF-MSAL [5], has been used. Apart from the transverse Breit interaction two types of quantum electrodynamics (QED) corrections (self-energy and vacuum polarization) have been included.

## 3. Results and discussion

In order to explain the influence of removing electrons from $L$ and $M$ shells on the shapes and parameters (the average $K\alpha$ and $K\beta$ transition energies and the values of $K\beta/K\alpha$ intensity ratio) of the silicon $K$-X-ray spectra the extensive MCDF calculations including the transverse Breit interaction and QED corrections have been performed for various electronic configurations of silicon ions. The results of these theoretical studies have been given in Table 1.

**Table 1.** Effect of removing electrons from various subshells on the number of $K\alpha$ and $K\beta$ transitions, the relative average energy positions (with respect to diagram lines) of $K\alpha$ and $K\beta$ lines (in eV) and the values of $K\beta/K\alpha$ intensity ratio for silicon. The electronic ground state configuration of neutral silicon atom is $1s^2 2s^2 2p^6 3s^2 3p^2$.

| Initial hole configuration | The number of $K\alpha$ transitions | The $K\alpha$ energy shift | The number of $K\beta$ transitions | The $K\beta$ energy shift | The $K\beta/K\alpha$ intensity ratio |
|---|---|---|---|---|---|
| $1s^{-1}$ | 129 | 0.0 | 14 | 0.0 | 0.0232 |
| $1s^{-1}3p^{-1}$ | 30 | 0.5 | 2 | 3.9 | 0.0171 |
| $1s^{-1}3s^{-1}$ | 454 | 0.4 | 47 | 4.7 | 0.0277 |
| $1s^{-1}3s^{-2}$ | 129 | 1.0 | 14 | 10.3 | 0.0327 |
| $1s^{-1}3s^{-1}3p^{-1}$ | 102 | 1.1 | 6 | 9.5 | 0.0195 |
| $1s^{-1}3s^{-2}3p^{-1}$ | 30 | 2.0 | 2 | 16.0 | 0.0224 |
| $1s^{-1}2p^{-1}$ | 1345 | 12.0 | 284 | 33.0 | 0.0351 |
| $1s^{-1}2p^{-2}$ | 3818 | 25.6 | 1345 | 68.4 | 0.0541 |
| $1s^{-1}2p^{-3}$ | 3837 | 41.9 | 2190 | 106.1 | 0.0800 |
| $1s^{-1}2p^{-4}$ | 1345 | 60.0 | 1340 | 145.3 | 0.1321 |
| $1s^{-1}2p^{-5}$ | 129 | 75.9 | 285 | 190.1 | 0.4260 |
| $1s^{-1}2s^{-1}$ | 454 | 10.5 | 47 | 27.9 | 0.0331 |
| $1s^{-1}2s^{-2}$ | 129 | 22.3 | 14 | 57.8 | 0.0418 |
| $1s^{-1}2s^{-1}3p^{-1}$ | 102 | 11.4 | 6 | 32.9 | 0.0223 |
| $1s^{-1}2s^{-2}3p^{-1}$ | 30 | 23.6 | 2 | 63.8 | 0.0268 |

Moreover, in Figures 1 and 2 for $K\alpha$ diagram line and for certain $L$- and $M$-satellite lines both stick (calculated line positions with their relative intensities) and synthesized spectra of silicon have been presented. The synthesized spectra for each case have been constructed as follows: one (dotted lines) being the sum of the Lorentzian natural line shapes with the appropriate natural linewidth of 0.48 eV [28] and, to better simulate the experimental spectra, the other one (solid lines) being a convolution of the sum of the Lorentzian natural line shapes with the Gaussian instrumental response having a width of 1.0 eV.

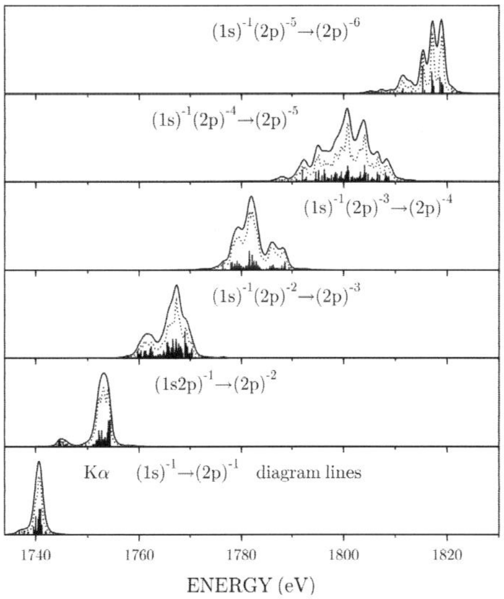


Figure 1: Theoretical $K\alpha_{1,2}$ X-ray spectra for diagram and $L$-satellite transitions in silicon.

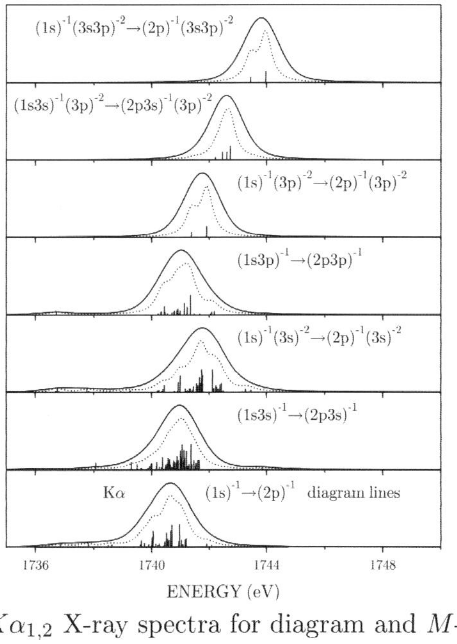


Figure 2: Theoretical $K\alpha_{1,2}$ X-ray spectra for diagram and $M$-satellite transitions in silicon.

Due to the open-like type of the electronic ground state configuration of neutral silicon atom even the diagram $K\alpha$ and $K\beta$ lines origin from many different transitions. As can be seen from Table 1 the largest number of transitions appears for the $2p^{-1}$, $2p^{-2}$, $2p^{-3}$, $2p^{-4}$ configurations for both $K\alpha$ and $K\beta$ lines. It is worth to be noticed that for all initial hole configurations the energy shifts (with respect to diagram lines) in the case of $K\beta$ lines are much larger than for $K\alpha$ lines. Moreover, the very large shifts of $K\alpha$ and $K\beta$ transition energies are typical in the cases of the ionization of the $2p$ and $2s$ subshells while the energy shifts in the cases of the ionization of the $3p$ and $3s$ subshells are many times smaller. The calculated $K\beta/K\alpha$ intensity ratio is influenced, first of all, by the ratio of the number of electrons in $3p$ and $2p$ subshells, and, secondly, by ionization degree of $2s$ and $3s$ subshells. It is very important to note that the effects of removal of electrons from the particular subshells on the $K\alpha$ and $K\beta$ transition energies are strong nonadditive, i.e. the $K\alpha$ and $K\beta$ transition energies increase much faster than linearly with the number of holes in a given subshell. Moreover, the effects of removal of electrons from different ($L$ and $M$) shells on the $K$-X-ray spectra parameters are also strongly nonadditive. As can be seen (Figs. 1 and 2) in most cases the stick spectra have very rich structure and they are much more scattered (and shifted with respect to diagram lines) in the case of $L$-shell than $M$-shell $K\alpha$ satellite lines. Moreover, in most cases the subsequent theoretical spectra being the sum of the Lorentzian natural line shapes are rich and complex.

It is worth to be noticed that for all types of $L$- and $M$-shell $K\alpha$ satellite lines the convolution of the sum of the Lorentzian natural line shapes with the Gaussian instrumental response smoothes the resultant spectrum. One can observe (see Fig. 1) that for $L$-shell $K\alpha$ satellite lines the spectra being the convolution of the sum of the Lorentzian natural line shapes with the Gaussian instrumental response retain a very rich structure. Therefore in the case of $L$-shell $K\alpha$ satellite lines none of them can be represented as a single Voigt function, even in a very crude approximation. However as can be seen from Fig. 2 in the case of $M$-shell $K\alpha$ satellite lines the spectra being the sum of the Lorentzian natural line shapes are rather smooth. Owing to this the spectra being the convolution of a sum of the Lorentzian line shapes with the Gaussian instrumental response are non-structured and highly symmetrical. Therefore they can be quite well represented as the Voigt function in all cases.

The results of presented theoretical study will be applied to perform the reliable interpretation of the $K\alpha$ X-ray emission spectra of the low-density $SiO_2$ aerogel target bombarded by $^{48}Ca^{6+}$ ions (with initial energy of 11.4 MeV/u) measured with a high spectral and spatial resolution along the ion beam stopping path [25]. They will also be helpful in interpreting various silicon target $K$-X-ray spectra induced by photons and different light and heavy projectiles.

## Acknowledgments

This work was supported by the Polish State Committee for Scientific Research under Grant No. N N202 1465 33.

[1] Polasik M 1989 *Phys. Rev.* A **39** 616-627
[2] Polasik M 1989 *Phys. Rev.* A **39** 5092-5097
[3] Polasik M 1989 *Phys. Rev.* A **40** 4361-4368
[4] Polasik M 1990 *Phys. Rev.* A **41** 3689-3697
[5] Polasik M 1995 *Phys. Rev.* A **52** 227-235
[6] Polasik M 1998 *Phys. Rev.* A **58** 1840-1845
[7] Polasik M *et al.* 1999 *J. Phys. B: At. Mol. Opt. Phys.* **32** 3711-3725
[8] Słabkowska K and Polasik M 2006 *Radiat. Phys. Chem.* **75** 1471-1476
[9] Polasik M and Słabkowska K 2006 *Radiat. Phys. Chem.* **75** 1497-1502
[10] Słabkowska K and Polasik M 2007 *J. Phys.: Conf. Ser.* **58** 263-266
[11] Carlen M W *et al* 1992 *Phys. Rev.* A **46** 3893-3903
[12] Carlen M W *et al* 1994 *Phys. Rev.* A **49** 2524-2534
[13] Ludziejewski T *et al* 1995 *Phys. Rev.* A **52** 2791-2803
[14] Ludziejewski T *et al* 1996 *Phys. Rev.* A **54** 232-240
[15] Herren Ch *et al* 1998 *Phys. Rev.* A **57** 235-245
[16] Mauron O *et al* 2000 *Phys. Rev.* A **62** 062508-1-11
[17] Kravchuk V L *et al* 2001 *Phys. Rev.* A **64** 062710/1-6
[18] Raboud P-A *et al* 2002 *Phys. Rev.* A **65** 062503-1-8
[19] Majewska U 2002 *J. Phys.* B **35** 1941-1957
[20] Rzadkiewicz J *et al* 2003 *Phys. Rev.* A **68** 032713/1-14
[21] Czarnota M *et al* 2003 *Nucl. Instr. and Meth. in Phys. Res.* B **205** 133-138
[22] Czarnota M *et al* 2003 *Radiat. Phys. Chem.* **68** 121-125
[23] Braziewicz J *et al* 2004 *Phys. Rev.* A **69** 0627105-1-10
[24] Czarnota M *et al* 2006 *Braz. J. Phys.* **36** 546-549
[25] Rzadkiewicz J *et al* 2007 *High Energy Density Physics* **3** 233-236
[26] Grant I P 1984 *Int. J. Quantum Chem.* **25** 23-46
[27] Dyall K G, Grant I P, Johnson C T, Parpia F A and Plummer E P 1989 *Comput. Phys. Commun.* **55** 425-456
[28] Campbell J L and Papp T 2001 *At. Data Nucl. Data Tables* **77** 1-56

# Electron angular distributions in He single ionization impact by $H_2^+$ ions at 1 MeV

**Shaofeng Zhang[1,2,5], J Suske[2], D Fischer[2,4], K U Kuehnel[2], S Hagmann[3], A Voitkiv[2], B Najjaril[2], A Krauss[2], R Moshammer[2], J Ullrich[2], Xinwen Ma[1,5]**

[1]Institute of Modern Physics, CAS, Lanzhou 730000, China
[2]Max-Planck-Institut für KernPhysik, Saupfercheckweg 1, 69117 Heidelberg
[3]Gesellschaft für Schwerionenforschung, Planckstr. 1, 64291 Darmstadt, Germany
[4]Stockholm University, AlbaNova University Centre, 10691 Stockholm, Sweden
[5]Graduate University of Chinese Academy of Sciences, Beijing 100049 China

zhangshf@impcas.ac.cn

**Abstract**. For the first time we investigated in a kinematically complete experiment the ionization of helium in collisions with $H_2^+$-molecular ions at 1 MeV. Using two separate detectors, the orientation of the projectile $H_2^+$-molecular ions was determined at the instance of the collision. The electron angular distribution was measured by a "Reaction Microscope". The observed structures are found in agreement with theoretical calculations, indicating that the ionized electron of He shows a slight preferential emission direction parallel to the molecular axis.

## 1. Introduction

Studies of particle-induced dissociation of $H_2$, the simplest molecule composed of two atoms, have been reported extensively last decade. Particular attention is devoted to spatial effects in target ionization since the two centers are indistinguishable, where the contributions from each center add coherently. Early studies of collisionally induced fragmentation of $H_2$ mainly focused on the processes of electron capture[1, 2] and photoionization[3-5]. The following theoretical work was developed quite rapidly[6], whereas more experimental work mainly focused on the heavier molecules with synchrotron radiation[7, 8]. Up to now, little has been done for the collision experiments with $H_2^+$ as projectiles, which is the simplest molecular system in the nature, in despite of the advances in the collision experimental[9, 10] and theoretical work[11-13] with $H_2$ as a target. In a kinematically complete experiment the spatial effects in the ionization of helium induced by the molecular ion $H_2^+$ are investigated. In this work the orientation and internuclear distance of the molecular ion at the instance of the collision could be determined by momentum imaging technique. The phenomena resulting from the two-center potential of the molecular projectile in the dissociation channel were found.

## 2. Experimental setup

In our experiment, we employed a Recoil-Ion Momentum Spectroscopy (RIMS)[14] , which is suitable for the investigation of any kind of atomic reaction dynamics. The features of the experimental setup concerning molecular ion impact are presented. Briefly, the $H_2^+$ ion beam (1 MeV) was delivered by a linear accelerator at Max-Planck institute for nuclear physics. After collimation the ion beam entered the interaction chamber through a 0.5 mm diameter hole and crossed a cold localized gas-jet target. The outgoing projectile was charge selected by a dipole magnet, the neutral and the proton produced in the dissociation channel were delivered to different directions and detected by two separate position sensitive detectors respectively, while the primary beam was directed into a Faraday cup. The ejected electrons and the recoils produced in target ionization were extracted and accelerated by a 3 V/cm electric field from the interaction region, and then they were projected onto the detectors, respectively. In order to achieve a high acceptance together with a good resolution for both electrons and recoil-ions in coincidence, a weak homogenous magnetic field (generated by a pair of large Helmholtz coils) was superimposed along the spectrometer axis effectively confining the electron motion in space (figure 1).

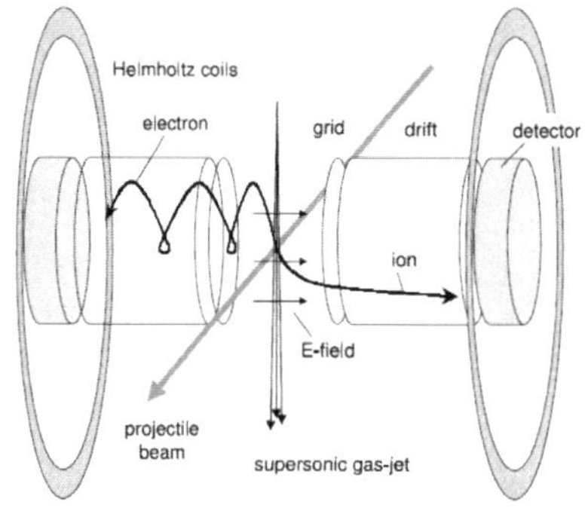


**Figure 1. Schematics of an reaction microscope**

The reaction channel of interest in the present experiment is the target ionization and simultaneous dissociation of the projectile molecular ion,

$$H_2^+ + He \rightarrow H + p + He^+ + e \qquad (1)$$

Here, four-fold coincidence measurements between the ionized electron, the recoil ion $He^+$, the projectile fragments (the neutral hydrogen atom H and the proton p) were performed, the data were recorded in event-by-event mode.

## 3. Results and discussion

The momenta of the four particles in eq. (1) were reconstructed according to their time and position on the detectors. The kinetic energy release (KER) of the projectile molecular $H_2^+$, were illustrated in the left panel in figure 2. By selecting specific KER the nuclear distance of the molecular ion can be determined. The right panel in figure 2 shows the projection of the spherical shell of the molecular ion with the KER condition around 10 eV.

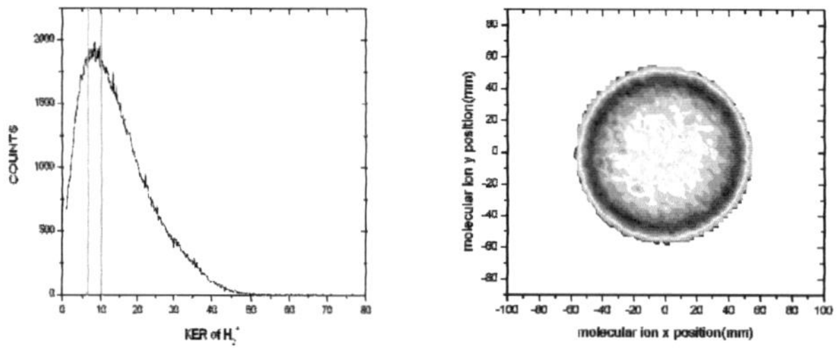


**Figure 2.** Left: The kinetic Energy Release (KER) spectrum of the $H_2^+$ ion in dissociation channel; Right: The projection of spherical shell of the fragments when the events between the two straight lines of the left plot are selected.

The theoretical electron angular distribution with respect to the molecular axis was obtained under the axial recoil approximation[15], which holds that when a diatomic molecular ion is formed in a dissociative state, the atoms produced in the dissociation process will move outward along the straight line defined by the internuclear axis of the molecule. From the estimation below it is clear that this approximation works well in our case. If we treat the diatomic molecule as a rotating dumbbell composed of two point masses with a distance R, its rotational energy is given by

$$\frac{1}{2}I_0\omega_0^{\,2} = J(J+1)\frac{\hbar^2}{2I_0} \tag{2}$$

where $\omega_0$ is the initial angular velocity of the molecular ion, and $I_0$ is its moment of inertia. Taking the rotational temperature of the gas to be 300 K, the final rotation angle can be written as below, by neglecting other aspects which can contribute to the rotational angle[16].

$$\Delta\varphi = \frac{8R_0\sqrt{E_\omega E_c}}{q^2} \tag{3}$$

where $E_c$ is the coulomb potential energy, $E_\omega$ the rotational energy, $R_0$ the initial nuclear distance, and $q$ the charge of the fragment. In our case, $R_0$=1, $q$=1, $E_c$=0.2, and $E_\omega$=0.001 (in atomic units), we finally get $\Delta\varphi$=6.3°, which is sufficiently accurate to account for the data in our work.

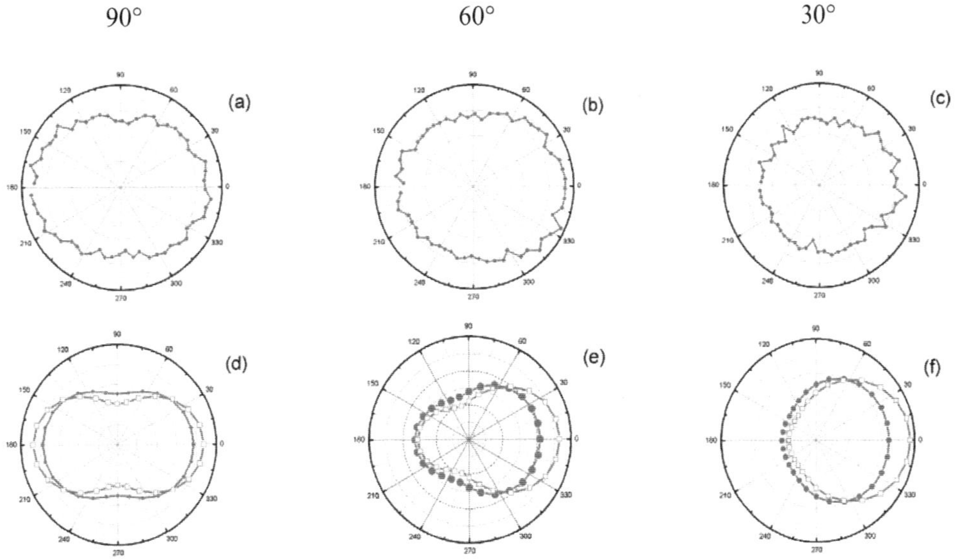


**Figure 3.** Electron angular distributions corresponding to three different molecular axis angles with respect to the projectile direction (from left to right the angles are 90°, 60°, and 30°, respectively). Upper row: experimental (a, b, c), lower row: theoretical (d, e, f).

Since the orientation of the $H_2^+$ axis can be reconstructed for each event, the electron angular distributions with respect to molecular axis can be obtained. Both experimental (upper row) and theoretical (lower row) results of the electron angular distributions are shown in figure 3. The projection plane in figure 3 is perpendicular to the beam propagation direction (the propagation direction is looking down into the figure). Plot (a) shows the electron angular distribution when the molecular axis is perpendicular to the beam propagation and plot (d) shows the theoretical results under the same situation. Plot (b) and (e) as well as plot (c) and (f) present the experimental and theoretical electron angular distributions for the molecular axis angle 60° and 30° with respect to projectile direction, respectively.

The experimental data show that the emission of the ionized electron from the target has a slight preference along the molecular axis. Theoretical calculations from the quantum plane-wave Born approximation show more pronounced structure variation while the direction of the molecular axis changes. This phenomenon indicates the structured projectile does have some impacts on the collision reaction.

## 4. Summary

The first successful experiment for projectile-orientation-reconstruction in high energies (1 MeV) collisions was performed. The experiment was precisely organized to study the spatial effects in ionization of helium by the hydrogen molecular ion impact in dissociation channel: $H_2^+ + He \rightarrow H + p + He^+ + e$. The experimental results show less pronounced structure effects than the quantum plane-wave Born approximation predictions.

**Acknowledgement**

This work has been supported by the Doctoral Training Program between Chinese Academy of Sciences and the Max-Planck Society. Zhang and Ma were also supported by NSFC (No. 10434100).

**References**

[1] Tuan T F and Gerjuoy E 1960 *Physical Review* **117** 756
[2] Joulakian B, Hanssen J, Rivarola R and Motassim A 1996 *Physical Review A* **54** 1473
[3] Cohen H D and Fano U 1966 *Physical Review* **150** 30
[4] Dörner R, Bräuning H, Jagutzki O, Mergel V, Achler M, Moshammer R, Feagin J M, Osipov T, Bräuning-Demian A, Spielberger L, McGuire J H, Prior M H, Berrah N, Bozek J D, Cocke C L and Schmidt-Böcking H 1998 *Physical Review Letters* **81** 5776
[5] Walter M and Briggs J 1999 *Journal of Physics B: Atomic, Molecular and Optical Physics* **32** 2487-2501
[6] Corchs S E, Busnengo H F and Rivarola R D 1999 *Nuclear Instruments and Methods in Physics Research Section B: Beam Interactions with Materials and Atoms* **149** 247-256
[7] Heiser F, Geßner O, Viefhaus J, Wieliczek K, Hentges R and Becker U 1997 *Physical Review Letters* **79** 2435
[8] Mills J D, Sheehy J A, Ferrett T A, Southworth S H, Mayer R, Lindle D W and Langhoff P W 1997 *Physical Review Letters* **79** 383
[9] Stolterfoht N, Sulik B, Gulyás L, Skogvall B, Chesnel J Y, Frémont F, Hennecart D, Cassimi A, Adoui L, Hossain S and Tanis J A 2003 *Physical Review A* **67** 030702
[10] Stolterfoht N, Sulik B, Hoffmann V, Skogvall B, Chesnel J Y, Rangama J, Frémont F, Hennecart D, Cassimi A, Husson X, Landers A L, Tanis J A, Galassi M E and Rivarola R D 2001 *Physical Review Letters* **87** 023201
[11] Galassi M E, Rivarola R D, Fainstein P D and Stolterfoht N 2002 *Physical Review A* **66** 052705
[12] Nagy L, Kocbach L, Pora K and Hansen J P 2002 *Journal of Physics B: Atomic, Molecular and Optical Physics* **35** L453-L459
[13] Sarkadi L 2003 *Journal of Physics B: Atomic, Molecular and Optical Physics* **36** 2153-2163
[14] Ullrich J, Moshammer R, Dorn A, Dörner R, Schmidt L P H and Schmidt-Böcking H 2003 *Reports on Progress in Physics* **66** 1463-1545
[15] Zare R N 1967 *The Journal of Chemical Physics* **47** 204-215
[16] Wood R M, Zheng Q, Edwards A K and Mangan M A 1997 *Review of Scientific Instruments* **68** 1382-1386

# State-selective differential cross section measurements for the one-electron capture processes in the $F^{4+}$ - He, Ne, Ar systems at $E_{lab}$ = 45 eV

**Y Itoh[1] and K Ohtsuki[2]**

[1]Physics Laboratory, Josai University, Sakado, Saitama 350-0295, Japan
[2]Department of Applied Physics and Chemistry, The University of Electro-Communications, Chofu, Tokyo 182-8585, Japan

yitoh@josai.ac.jp

**Abstract**. Using a crossed-beam apparatus, we measured systematically the relative state-selective differential cross sections for the charge-transfer processes at very low collision energy. The cross section obtained in $F^{4+}$ + He system shows a peak around 0° angle, while those in $F^{4+}$ + Ne and Ar collisions at the same collision energy show clear angular thresholds. The strong peak observed in $F^{4+}$ + He system is interpreted as due to the glory scattering, and the cross section maximum in $F^{4+}$ + Ne and Ar collisions at non-zero angles is due to the rainbow scattering.

## 1. Introduction

There exists various state-selective measurements for the charge-transfer reactions involving multiply-charged ions performed at a few hundred eV or higher energy region [1]. The purpose of the study is to obtain information on the selection rules and the transition probabilities as well as to obtain the absolute value of the integral cross sections for each reaction channel.

The deflection angle of the scattered ions is small in such energy regions, hence it is almost impossible to determine the angular dependence of the cross sections unless an apparatus with an extremely high angular-resolution is used [2]. When the collision energy is lowered, the angular scattering becomes more prominent. The angular differential cross sections are very sensitive to the shape of the potential curves concerning the process. Therefore, experimental results can be analyzed in detail in comparison with the theoretical ones based on *ab initio* potentials. However, only few experimental results have so far been reported at very low energy region, because it is still not an easy task to produce highly-charged ions and to decelerate them [3 - 6]. We introduced a compact electron-beam ion-source to a conventional crossed-beam apparatus to measure state-selective differential cross sections below 100 eV region. In this paper, present status of the measurements and the analysis based on the theoretical potentials is reported.

## 2. Experimental

Details of the crossed-beam apparatus used were described previously [7]. It consists of a small electron-beam ion-source, a Wien-filter as a mass selector, an energy selector and a one-dimensional position-sensitive ion-detector. The energy- and mass-selected primary ion beam was crossed at a right

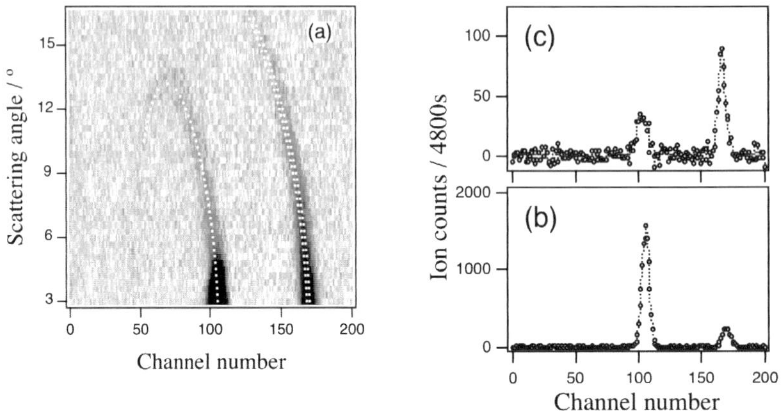


**Figure 1.** Energy spectra obtained in $F^{4+}$ - He system. (a) A two dimensional graph. Calculated energy-positions are shown by dots. (b) A spectrum obtained at $\theta_{lab} = 3.0^o$, and (c) at $\theta_{lab} = 6.0^o$.

angle with a supersonic nozzle beam. An in-plane geometry was used to detect the product ions. The base pressure of the scattering chamber was about $4 \times 10^{-6}$ Pa, and the background pressure was about $1 \times 10^{-4}$ Pa with the target beam running.

The intensity of the ion beam after the mass selection was on the order of $10^{-11}$ A, and the energy spread of the ion beam was about $1.5 \times q$ eV (FWHM), where $q$ indicates the charge number of the ions. After the mass selection, the ions were decelerated to the transmission energy of the double hemispherical energy selector. The transmission energy was 42 eV, and the energy width of the primary beam was reduced to about $0.25 \times q$ eV. The intensity of the primary beam at the collision center was less than 1 pA for $F^{4+}$ beam at $E_{lab} = 44$ eV. The overall angular resolution of the detector system was determined to be better than ±1° (FWHM) by measuring the angular distribution of the primary beam.

State-selective differential cross sections were obtained by measuring the kinetic energy of the product ions. The energy spectra of the scattered ions at different scattering angles were recorded by rotating the energy analyzer.

Accurate determination of the collision energy is important for the assignment of reaction channels from the energy spectrum of the product ions. For this purpose, we measured the elastically scattered ions simultaneously, because the energy and angular dependence of the elastic scattering are uniquely determined by the scattering geometry. The target beam energy in the isentropic limit was estimated to be $E = (5/2)kT$ ( $T$ = nozzle temperature: 300K), and this value was used in the calculation of the kinematics.

In figure 1(b) and 1(c), the energy spectra of the scattered ions obtained in $F^{4+}$ + He system are shown. In figure 1(a), all the measured spectra are shown in a 2-dimention plot. In figure 1(b) and 1(c), two peaks are clearly observed. The left side peak corresponds to the elastically scattered signals, and the right side one is due to the charge transfer reactions. Changing the value of the impact energy, we calculated the angular and energy dependence of the elastic scattering so as to reproduce the measurements, and determined the initial energy to be 44 ± 1 eV. The calculated positions for the elastic scattering and the following charge transfer reactions are shown by dots in figure 1(a):

$$F^{4+}(2s^2 2p\ ^2P) + He \rightarrow F^{3+}(2s^2 2p3s\ ^1P) + He^+ + 10.0\ eV, \quad (1)$$

$$\rightarrow F^{3+}(2s^2 2p3s\ ^3P) + He^+ + 10.9\ eV. \quad (2)$$

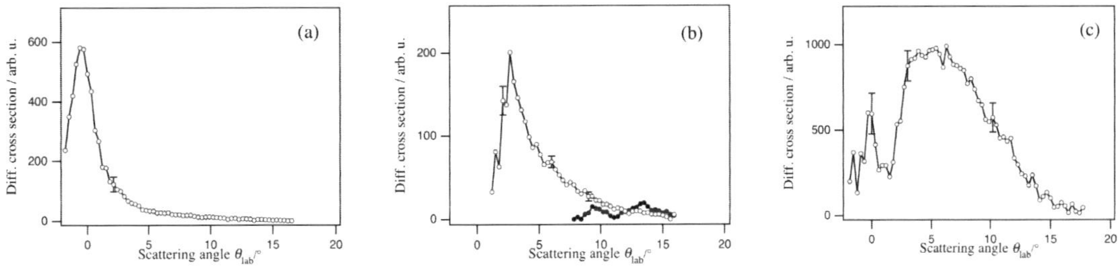


**Figure 2.** Measured differential cross sections. (a) $F^{4+}$ - He, ○ : the sum of channel (1) and (2). (b) $F^{4+}$ - Ne, ○ : the sum of channel (3) and (4), •: channel (5). (c) $F^{4+}$ - Ar , ○ : channel (6).

We observed the $F^{3+}$ ions around 0° angle even when the target beam was stopped. Therefore, special care had to be taken to ensure a correct background subtraction. We have carefully determined the amount of background counts using the method reported previously [7].

## 3. Results

In figure 2(a), relative differential cross section in the laboratory frame for the reaction channels (1) and (2) are shown. In the present measurements, energy resolution was not sufficient to separate these channels, so the obtained cross sections are the sum of the channel (1) and (2). Note that the definition of the differential cross section is $d\sigma/d\Omega$, not $d\sigma/d\theta = 2\pi \sin\theta \; d\sigma/d\Omega$. The cross section shows a peak around 0° and decreases monotonically with the increase of the scattering angle.

Results obtained in $F^{4+}$ + Ne system are shown in figure 2(b). The open circle corresponds to the following reactions:

$$F^{4+}(2s^2 2p\ ^2P) + Ne \rightarrow F^{3+}(2s^2 2p3p\ ^3D) + Ne^+ + 9.6\ \text{eV}, \qquad (3)$$

$$\rightarrow F^{3+}(2s^2 2p3p\ ^1P) + Ne^+ + 10.0\ \text{eV}. \qquad (4)$$

The cross section increases steeply with the increase of the scattering angle, shows peak at around 3°, and decreases monotonically. We noticed that another peak starts to appear in the energy spectra at around 8°. This process could be uniquely assigned as,

$$F^{4+}(2s^2 2p\ ^2P) + Ne \rightarrow F^{3+}(2s^2 2p3s\ ^1P) + Ne^+ + 13.1\ \text{eV}. \qquad (5)$$

Due to poor statistics for this channel, a fitting program was used to derive the cross section value. The cross section for channel (5), displayed in filled circle in figure 2(b), shows two maxima at around 9.5° and 13.5°.

The $F^{3+}$ signal observed in $F^{4+}$ + Ar collision is assigned to the charge transfer to the excited state: $F^{3+}$ ($2s^2 2p3d$ ) with the exothemicity of about 9 eV:

$$F^{4+}(2s^2 2p\ ^2P) + Ar \rightarrow F^{3+}(2s^2 2p3d) + Ar^+ + \sim 9\ \text{eV}. \qquad (6)$$

As this electronic configuration results in the closely lying terms of $^1F$, $^3F$, $^1D$ and $^3D$, the cross section shown in figure 2(c) is considered to be the sum to all these states. The cross section shows a peak at 0° angle and has maximum at around 5°.

## 4. Discussion

In a separate measurement for $N^{2+}$ + He collision, which is the isoelectronic system of the $F^{4+}$ + He, a similar angular dependence was observed in the reaction $N^{2+}(2s^2 2p\ ^2P) + He \rightarrow N^+(2s^2 2p^2\ ^3P) + He^+ + 5.0$ eV. The interaction potentials for $NHe^{2+}$ system have been reported by Lafyatis *et al* [8]; we also calculated them again and fitted the results to an appropriate analytical function to obtain classical

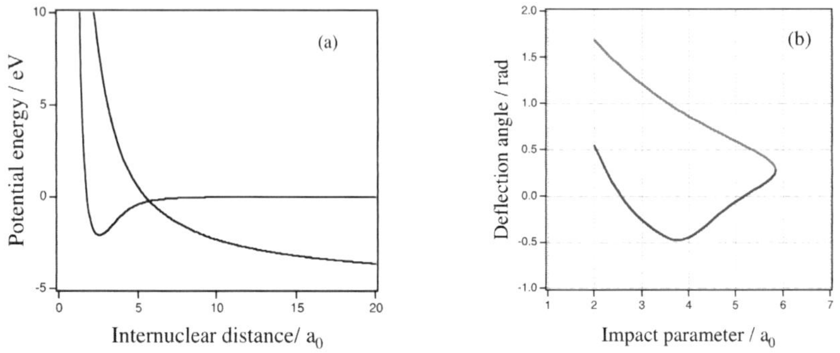


**Figure 3.** (a) Interaction potentials of the $^2\Pi$ states of $NHe^{2+}$. (b) Deflection functions for $N^{2+}$ + He → $N^+$ + $He^+$ + 5.0 eV collisions at $E_{lab}$ = 20.5 eV. The upper-half of the curve corresponds for the reaction in the incoming channel, and the lower-half of it for the reaction in the outgoing channel.

deflection function within a two-state model. Results are shown in figure 3(a), and 3(b).

When the charge transfer reaction occurs in the incoming part of the trajectory, the trajectory will be determined mainly by the repulsive Coulomb force.

While if the reaction occurs in the outgoing part of the trajectory, the ion trajectory before the reaction will be determined by the attractive or repulsive force depending on the collision systems, then the trajectory will be controlled by the repulsive Coulomb force. The $NHe^{2+}$ system has a deep well of about 2 eV, therefore the ion trajectory is determined by the strong attractive force before the reaction. When the deflection angle comes to 0 at a certain impact parameter, the classical differential cross section, $d\sigma/d\Omega = b\ /\ [\sin\theta\,(d\theta/db)]$, diverges due to the factor $1/\sin\theta$. This is known as the glory scattering.

We temporarily assigned the strong forward peak observed in the charge transfer reaction to the glory scattering both for $N^{2+}$ + He and $F^{4+}$ + He systems.

The minimum in the lower-half of the deflection function, $d\theta/db = 0$, gives another singularity to the classical differential cross section, known as the rainbow scattering. When the interaction potentials have not strong attractive part or have only repulsive character, the glory scattering will not occur, and the rainbow scattering will come to be seen clearly. The observed structure in $O^{2+}(2s^22p^2\ ^3P)$ + He → $O^+(2s^22p^3\ ^2P)$ + $He^+$ + 5.5 eV collisions [7], which appears at the same reduced collision angle, was tentatively assigned to the rainbow scattering.

We again consider that the maximum observed in $F^{4+}$ + Ne and Ar collisions at a certain angle may be due to the rainbow effect. Detailed theoretical analysis is in progress concerning these assignments.

## 5. Acknowledgements

This work was supported in part by the Grant-in-Aid for Scientific Research (C) （No. 17540373） .

## References

[1] Seredyuk B *et al* 2005 *Phys. Rev.* A **71** 022705
[2] Lindsay B G, Yu W S and Stebbings R F 2005 *Phys. Rev.* A**71** 032705
[3] Friedrich B *et al* 1986 *J. Chem. Phys.* **84** 807
[4] Biedermann C *et al* 1990 *Phys. Rev.* A **41** 5889
[5] Hoshino M *et al* 2007 *Phys. Rev.* A **75** 012716
[6] Kamber E Y, Abu-Haija O and Wardwell J A 2008 *Phys. Rev.* A **77** 012701
[7] Itoh Y 2002 *J. Phys.* B **35** 3217
[8] Lafyatis G P, Kirby K and Dalgano A 1993 *Phys. Rev.* A **48** 321

# Recoil Momentum of Molecular Ions in Collisions of $Ar^{6+}+N_2$ at Energies below 300 eV/u

**Y Inoue, K Ishii[1] and H Ogawa[1]**

Graduate school of Humanities and Sciences, Nara Women's University, Nara, 630-8506, Japan
[1]Dept. of Physics, Nara Women's University, Nara, 630-8506, Japan

E-mail: bay.inoue@cc.nara-wu.ac.jp

**Abstract.** To elucidate the collision energy dependence of the recoil momentum of molecular ions after electron transfer, we measure the recoil momentum of non-dissociated molecular ions for single electron capture collisions of $Ar^{6+}$ with $N_2$ at collision energies from 7.5 to 240 eV/u. The measured collision energy dependence of the recoil momentum is compared with those obtained by a theoretical calculation using a deflection function with polarization potential. We obtain fairly good agreement between the measured and calculated results.

## 1. Introduction

Charge transfer processes involving ionization, electron capture, and molecular fragmentation in collisions of multiply charged ions (MCIs) with atoms and molecules play important roles not only in atomic physics but also in astrophysics and plasma physics [1, 2]. In particular, in the last few decades, the field of fragment processes of diatomic molecules in electron capture collisions of MCIs at energies above 1 keV/u has grown rapidly. In molecular fragmentation, it is understood that a diatomic molecular target undergoes Coulomb explosion after electron capture, and then, the two fragments fly in opposite directions, their kinetic energies given by the dissociation energy [3]. Kaneyasu et al.[4] and Ohyama-yamaguchi et al.[5] conducted studies on molecular fragmentation processes induced by MCI collisions at low energies below 1 keV/u. Kaneyasu et al. measured two fragment ions in a direction perpendicular to a projectile beam axis in coincidence with a projectile after double and triple electron capture. As a result, an interesting observation was reported that recoil momenta of non-dissociated molecular ions depend on collision energy.

In this study, in order to investigate the collision energy dependence of recoil momentum, we measure recoil momenta of the target of a $Ar^{6+}+N_2$ collision system at energies from 7.5 to 240 eV/u. The energy dependence of the measured recoil momentum is compared with those obtained by theoretical calculations performed using the Coulomb potential and the polarization potential.

## 2. Experiment

The experiment was performed using a mini-EBIS (electron beam ion source) atomic collision facility of Nara Women's University [6, 7]. MCI beams extracted from the mini-EBIS were made to collide with effusive $N_2$ gas in the collision region. Outgoing projectile ions from the collision region were analyzed using a parallel-plate electrostatic analyzer with a two-dimensional

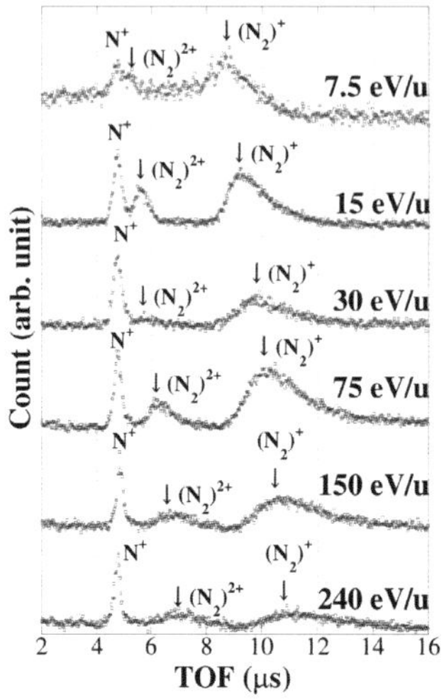


**Figure 1.** TOF spectra of molecular ions and fragment ions in collisions of $Ar^{6+}+N_2$ at energies from 7.5 to 240 eV/u.

position sensitive detector. Two fragment ions or non-dissociated molecular ions were detected separately by two time-of-flight (TOF) analyzers installed at 90° and -90°, respectively, with respect to the projectile beam axis. It should be noted that no-extraction fields in the collision region made it possible to identify the recoil momentum of target ions after collisions at low energy. In this experiment, we measured the time difference between the projectile ions and the target ions. All signals were directly recorded as a single event by a PC, using digitizers.

## 3. Results and Discussion

Single electron capture (SC) processes of $Ar^{6+}$ + $N_2$ are given as follows:

$$\begin{aligned} Ar^{6+} + N_2 &\rightarrow Ar^{5+*}(nl) + {N_2}^+ + Q_1, & (1)\\ &\rightarrow Ar^{5+*}(n'l') + N^+ + N + Q_2, & (2)\\ &\rightarrow Ar^{5+*}(n''l'') + {N_2}^{2+} + e^- + Q_3, & (3)\\ &\rightarrow Ar^{5+*}(n'''l''') + N^+ + N^+ + e^- + Q_4. & (4) \end{aligned}$$

Reaction(1) indicates that $Ar^{6+}$ ions capture an electron into the energy state of $Ar^{5+}(nl)$ from a $N_2$ molecule with reaction energy $Q_1$. Reaction(2) represents a molecular dissociative process occurring via reaction(1). Reactions(3) and (4) represent transfer ionization processes occurring via double electron capture with non-dissociative and dissociative processes of $N_2^{2+}$ ions, respectively.

Measured TOF spectra of SC processes in collisions of $Ar^{6+}$ with $N_2$ at energies from 7.5 to 240 eV/u are shown in Figure 1. At first glance, the peak positions of $N^+$ fragment ions appear to be independent of the collision energy but those of non-dissociated ${N_2}^+$ and ${N_2}^{2+}$ molecular ions appear to depend on the collision energy. The correlation diagram (Newton Diagram) of the velocity vectors of projectile ions and targets in the laboratory frame and the center of mass (C-M) frame is shown in Figures 2(a) and 2(b). In the Newton Diagram, vectors $\vec{v_i}$ and $\vec{v_t}$ are the velocities of projectile and target molecules, respectively, after inelastic collisions in the laboratory frame. Vectors $\vec{v}_{CM}$ and $\vec{v_r}'$ are the velocity for the center of mass and the relative velocity between projectile and target ions in the C-M frame, respectively. Experimentally, the

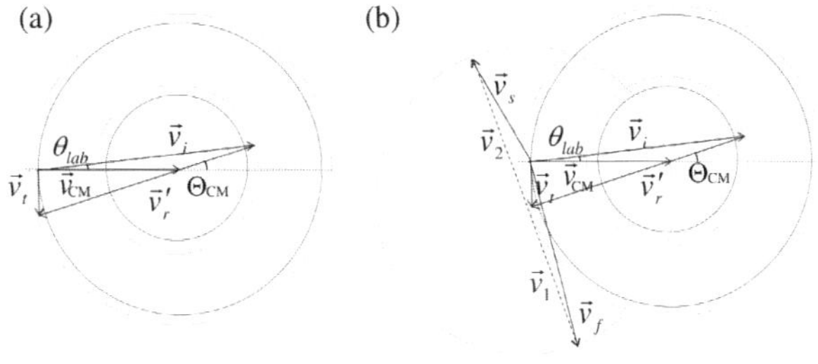


**Figure 2.** Correlation diagram (Newton Diagram) of velocity vectors in the laboratory frame and the C-M frame for non-dissociative process (a) and dissociative process (b). The $\theta_{lab}$ and $\Theta_{CM}$ are scattering angles in the laboratory frame and the C-M frame, respectively.

vectors for $N_2^+$ and $N_2^{2+}$ molecular ions correspond to $\vec{v_t}$ and the vector for the $N^+$ fragment ion corresponds to $\vec{v_f}$ or $\vec{v_s}$. Here, $\vec{v_f}$ and $\vec{v_s}$ are given by $\vec{v_f} = \vec{v_t} + \vec{v_1}$ and $\vec{v_s} = \vec{v_t} + \vec{v_2}$, respectively, where $\vec{v_1}$ and $\vec{v_2}$ are the dissociation velocity vectors due to Coulomb explosion. In the case of a homonuclear diatomic molecule, $\vec{v_1}$ is equal to -$\vec{v_2}$. In general, the velocity vectors of molecular ions, $\vec{v_t}$, depend on the collision energy, but those of fragment ions, $\vec{v_1}$ and $\vec{v_2}$, are independent of the collision energy. Furthermore, the dissociation velocity vectors $\vec{v_1}$ and $\vec{v_2}$ are considerably larger than the velocity vector $\vec{v_t}$ of the non-dissociated molecular ion, i.e., $\vec{v_f} \cong \vec{v_1}$ and $\vec{v_s} \cong \vec{v_2}$. Therefore, the energy dependence of the experimental TOF spectra can be qualitatively explained by using the velocity vectors in the Newton Diagram.

In order to elucidate the collision energy dependence of recoil momenta, we compare experimental recoil momenta with theoretical ones. The experimental recoil momentum($P_\perp$) is estimated as $P_\perp = \mu v_r' \sin\Theta_{CM}$ (see Figures 2(a) and 2(b)), where $\mu$, $v_r'$, and $\Theta_{CM}$ are the reduced mass, relative velocity, and scattering angle, respectively, in the C-M frame. Assuming $\sin\Theta_{CM} \approx \Theta_{CM}$ and noting the energy conservation before and after the collision, $E'_{CM}= E_{CM} + Q$, and the recoil momentum is expressed as $P_\perp \propto \sqrt{E'_{CM}}\ \Theta_{CM}$. We calculate the energy dependence of recoil momentum by using the deflection function $d\varphi/dr$ in the classical trajectory equation with interaction potential $V(r)$, which is given as

$$\frac{d\varphi}{dr} = \pm\frac{b}{r^2}\sqrt{1 - \frac{b^2}{r^2} - \frac{V(r)}{E'_{CM}}}\ , \tag{5}$$

where $\varphi$, $r$, and $b$ are the deflection angle, internuclear distance, and impact parameter, respectively [8]. In the case of $V(r)$ in powers of $r$, $V(r) = C/r^s$, we obtain $\Theta_{CM} \propto (b^s\ E'_{CM})^{-1}$. It is well known that electron transfer occurs at some particular internuclear distance $R$, and in this study, we have used $R$ of the closest approach. Practically, $R$ is calculated by the classical over-barrier model [9, 10] using the Atomic Spectra Database of the NIST [11]. According to this model, $R$ is given by $R = (2\sqrt{qt} + q)/I_t$, where $q$ is the charge state of the incident ion and $I_t$ is the $t$-th ionization potential of the target. The relation between $b$ and $R$ is given by $b^2 = R^2(1 - V(R)/E'_{CM})$.

In the present calculation, we have theoretically estimated the energy dependence of recoil momentum using two types of potentials– the Coulomb potential and the polarization potential. In the case of the Coulomb potential ($s = 1$), assuming $b \approx R$, $P_\perp$ is expressed as follows:

$$P_\perp \propto (\sqrt{E'_{CM}}\ R)^{-1}. \tag{6}$$

On the other hand, for the polarization potential ($s = 4$), $V(r) = -(\alpha q^2)/(2r^4)$, and $P_\perp$ is

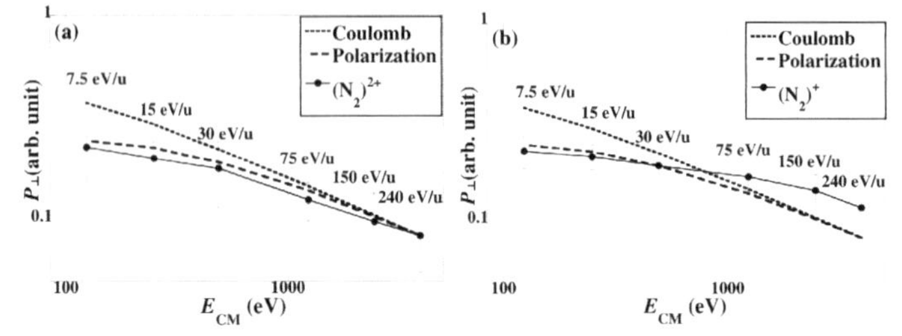


**Figure 3.** Collision energy dependence of recoil momenta of $N_2^{2+}$ (a) and $N_2^{+}$ (b). The circles show experimental recoil momenta of non-dissociated molecular ions. The dotted and dashed lines show the theoretical collision energy dependence of recoil momentum.

expressed as follows:

$$P_\perp \propto \left(\sqrt{E'_{CM}}\, R^4\left[1+\frac{\alpha q^2}{2R^4E'_{CM}}\right]^2\right)^{-1}, \tag{7}$$

where $\alpha$ is the polarizability of $N_2$. Figures 3(a) and 3(b) show the experimental and theoretical energy dependences of recoil momenta of molecular ions. The experimental dependence for $N_2^{2+}$ is good agreement with the theoretical dependence obtained using the polarization potential rather than that using the Coulomb potential. The relative dependence in the case of $N_2^{+}$ agrees with the theoretical dependence obtained using the polarization potential only at low energies below 100 eV/u. This suggests that the polarization potential plays a crucial role in the low-energy region. The disagreement between experimental and theoretical results of $N_2^{2+}$ at high energies above 100 eV/u may be due to the fact that internuclear distances vary with an increase in the collision energy, although we have used energy-independent $R$ in this model.

## 4. Conclusion

We have experimentally measured the recoil momentum of molecular ions for SC collisions and compared it with the theoretical one in order to elucidate the collision energy dependence of the recoil momentum. The experimental energy dependence of recoil momenta of $N_2^{2+}$ and $N_2^{+}$ are in agreement with the theoretical dependence obtained using polarization potential at energies below 100 eV/u. It is concluded that polarization potential plays a crucial role in the low-energy region. To gain a better understanding of the energy dependence of recoil momentum, coincidence measurements of TOF spectra with energy gain spectra of the projectile after the collision are necessary.

## References

[1] Tawara H, Kato T and Nakai Y 1985 *Atomic Data and Nucl. Data Tables* **32** 235
[2] Janev R K, Phaneuf R A and Hunter H 1988 *Atomic Data and Nucl. Data Tables* **40** 249
[3] Barat M and Roncin P 1992 *J. Phys. B: At. Mol. Phys.* **25** 2205
[4] Kaneyasu T, Azuma T and Okuno K 2005 *J. Phys. B: At. Mol. Phys.* **38** 1342
[5] Ohyama-yamaguchi T and Ichimura A 2006 *J. Phys. B: Conf. Ser.* **34** 245
[6] Okuno K 1986 *J. Phys. Soc. Jpn.* **55** 1504
[7] Ishii K, Inoue Y, Ogawa H, Itoh A and Sakamoto N 2006 *J. Phys. B: Conf. Ser.* **34** 245
[8] Bransden B H and McDowell M R C 1992 P.21 Charge exchange and the theory of ion-atom collisions, Oxford, New York
[9] Ryufuku H, Sasaki K and Watanabe T 1980 *Phys. Rev. A* **21** 745
[10] Niehaus A 1986 *J. Phys. B: At. Mol. Phys.* **19** 2925
[11] Ralchenko Yu, Kramida A E and NIST ASD Team 2008 *NIST Atomic Spectra Database* (version 3.1.5) (*http://physics.nist.gov/asd3*)

# Interference effects in electron emission spectra for 3 MeV/u $H^+$ + $O_2$ collisions

**M Winkworth[1], P D Fainstein[2], M E Galassi[3], J Baran[1], S Das[1], B S Dassanayake[1], A Kayani[1] and J A Tanis[1,*]**

[1]Department of Physics, Western Michigan University, Kalamazoo, MI 49008, USA
[2]CNEA, Centro Atómico Bariloche, Avenida E. Bustillo 9500, 8400 Bariloche, Argentina
[3]Instituto de Física Rosario (CONICET-UNR) and Facultad de Ciencias Exactas, Universidad Nacional de Rosario, Avenida Pellegrini 250, 2000 Rosario, Argentina

[*]Corresponding author's e-mail address: john.tanis@wmich.edu

**Abstract.** The present work extends the study of oscillatory structures associated with electron emission from diatomic molecular targets to 3 MeV/u $H^+$ + $O_2$. Electrons ejected from $O_2$ were detected in the energy range 5 to 400 eV and for observation angles 30°, 60°, 90° and 150°. Experimental molecular $O_2$ cross sections normalized to theoretical molecular $O_2$ cross sections reveal oscillatory structures suggestive of secondary interferences as evidenced by their independence on the observation angle. Contrary to results for 1, 3 and 5 MeV/u $H^+$ + $H_2$, no obvious evidence for primary Young-type interferences was seen, a result similar to that found for $H^+$ + $N_2$ collisions.

## 1. Introduction

Quantum mechanical interference effects are a well-known consequence of the wavelike behavior of particles, and have generated considerable interest among experimentalists and theoreticians for decades. Early predictions and observations of two-center interference effects in the photoionization of $N_2$ and $O_2$ [1] and of $H_2$ [2] were followed about thirty years later by detailed theoretical investigations of similar phenomena for photon collisions with $H_2^+$ [3]. In the latter work the molecular orbitals are represented by a combination of two atomic orbitals, the continuum electron is described by a plane wave, and the nuclei do not move (fixed-in-space molecules). Moreover, experiments involving collisions of ions and electrons with $H_2$ [4,5,6,7,8,9,10,11] uncovered structures analogous to the interferences observed in Young's famous two-slit experiment. Corresponding theoretical calculations $H_2$ [12,13,14,15] were generally in good agreement with the experimental measurements. In addition to Young-type interferences, more detailed analysis of the measured electron spectra in [6,9] revealed second-order oscillations superimposed on the primary interference structures with about 2-3 times higher frequency.

Recently, the work of Hossain *et al.* [9] for 1-5 MeV/u $H^+$ + $H_2$ was extended to proton collisions with more complex diatomic molecular targets. In contrast to the Young-type and secondary oscillatory structures identified in the ejected electron spectra of $H_2$, results for 1-5 MeV/u $H^+$ + $N_2$

collisions revealed apparently only secondary oscillations [16]. Similar findings were recently reported for 1.9 MeV/u $O^{5,8+} + O_2$ [17].

In order to gain new insight into the interference patterns reported for the above mentioned collisions, experimental investigations have been performed for 3 MeV/u $H^+ + O_2$. As a result of the larger internuclear separation of 2.28 a.u. for $O_2$, compared with 2.1 a.u. for $N_2$ and 1.4 a.u. for $H_2$, the primary Young-type interference structures would be expected to have higher frequencies. Additionally, due to the similar projectile energies but different molecular targets, the results for $H^+ + O_2$ collisions allow quantitative comparison with previous results for $H^+ + N_2$ [16].

## 2. Theory

Theoretical cross section calculations using different models [12,13,14,15,18] follow quite closely the general behavior of the experimental findings, and contribute to the interpretation of interferences resulting from coherent electron emission from the identical molecular centers of $H_2$. A description of the observed oscillatory structures is offered in the work of Ref. [4] by the relation:

$$(\sigma_{\mathrm{H}_2})_{norm} = \frac{\mathrm{d}^2\sigma_{\mathrm{H}_2}}{\mathrm{d}\Omega \mathrm{d}\varepsilon} \bigg/ \frac{\mathrm{d}^2\sigma_{2\mathrm{H}}}{\mathrm{d}\Omega \mathrm{d}\varepsilon} = \int \frac{\mathrm{d}^3\sigma_{2\mathrm{H}}}{\mathrm{d}q \mathrm{d}\Omega \mathrm{d}\varepsilon}\left[1 + \frac{\sin|k - q|d}{|k - q|d}\right] \times \mathrm{d}q \bigg/ \int \frac{\mathrm{d}^3\sigma_{2\mathrm{H}}}{\mathrm{d}q \mathrm{d}\Omega \mathrm{d}\varepsilon} \mathrm{d}q \qquad (1)$$

where $\sigma_{\mathrm{H}_2}$ is the cross section for two-center emission, $\sigma_{2\mathrm{H}}$ the cross section for emission from H, $\Omega$ is the solid angle, and $\varepsilon$ the energy of the outgoing electron. The sinusoidal term, where $|k - q|$ is the difference between the outgoing electron momentum $k$ and the momentum transfer $q$, and $d$ is the internuclear distance, describes the interference caused by coherent emission from the two nuclei.

For more complex systems, despite the difficulties involved in providing accurate theoretical calculations, efforts proved to be successful in identifying interference effects for 1, 3 and 5 MeV/u $H^+ + N_2$, (see Ref.[16]). In this latter work, the theoretical atomic N cross sections were calculated using a first-order impact parameter method. Hartree-Fock wavefunctions were used to describe the initial state of the N atom, while that corresponding to the ejected electron has been calculated in the field of the residual ion. In the present work, motivated by the results for $H^+ + N_2$ [16], theoretical calculations for molecular $O_2$ cross sections were performed using one-center wavefunctions, following the methods described in Ref. [19].

## 3. Experimental procedure

The measurements were performed at Western Michigan University using the 6-MV tandem Van de Graaff accelerator. A collimated beam of 3 MeV/u $H^+$ was directed into the scattering chamber onto an $O_2$ gas target supplied by a gas nozzle with the flow rate set to maintain a pressure of $\sim 4 \times 10^{-5}$ Torr inside the chamber, for which single collision conditions were found to prevail. The incident beam intensity was measured downstream in a Faraday cup. Electrons emitted from the target were energy analyzed using an electrostatic parallel-plate spectrometer and counted with a channel-electron multiplier for ejection energies 5-400 eV and observation angles of 30°, 60°, 90° and 150° with respect to the incident beam direction. Electron spectra were measured without the target gas at a background pressure of $\sim 2 \times 10^{-6}$ Torr to correct for residual gas events. Shielding with a μ-metal liner was used to minimize stray magnetic fields inside the scattering chamber.

## 4. Results and discussion

Following corrections for contributions from the background gas the ejected electron yields, which are proportional to the cross sections, were normalized to the calculated theoretical cross sections for 3 MeV/u $H^+ + O_2$ by setting the experimental value equal to the theoretical cross section near the middle of the spectrum (on a logarithmic scale) as shown in Figures 1a, b. As the theoretical cross sections decrease faster than the experimental yields for all electron observation angles, the experimental-to-theoretical $O_2$ cross section ratios exhibit rising trends as shown in Figures 1c,d, where the ratios are plotted as a function of the ejected electron velocity *k* (results are shown only for two of the four measured angles). A similar behavior was observed in the work of Refs. [4,16].

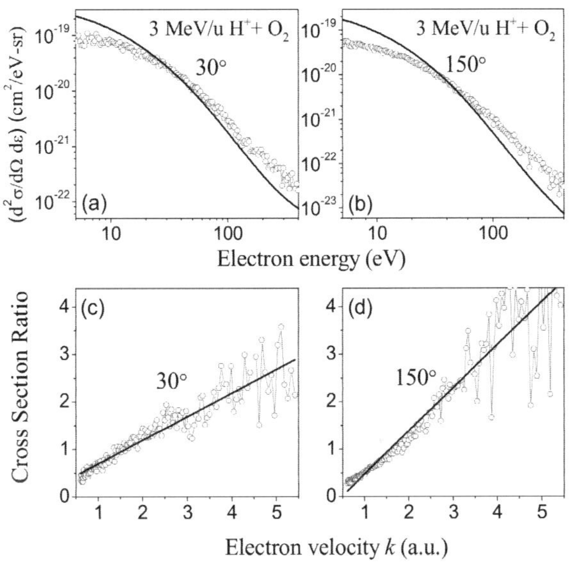


Figure 1: Experimental (open circles) and theoretical (solid lines) $O_2$ cross sections for electron emission in 3 MeV/u $H^+$ + $O_2$ collisions as a function of electron energy at the observation angles of 30° (a) and 150° (b) with respect to the incident beam direction. Experimental-to-theoretical cross section ratios for 3 MeV/u $H^+$ + $O_2$ plotted as a function of electron velocity *k* at the observation angles of 30° (c) and 150° (d), respectively.

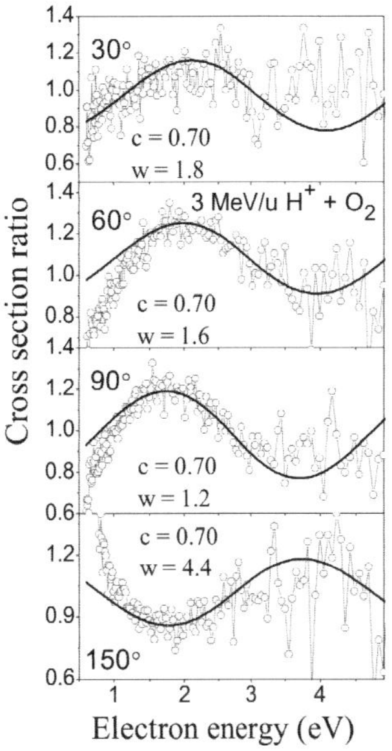


Figure 2: Experimental-to-theoretical cross section ratios (open circles) obtained by dividing the ratios from Figures 1c,d by the corresponding linear fits. The smooth curves are the results of fitting the phase-shifted sinusoidal function given by Equation 2 to the ratios.

The overall increase of the above mentioned ratios was removed by fitting a straight line to the ratios, clearly revealing the oscillatory structures as shown in Figure 2. The modified ratios have statistical fluctuations associated with experimental uncertainties that tend to become larger, at backward angles and higher electron velocities. The frequencies of these structures show no obvious dependence on the observation angle contrary to the behaviour expected for Young-type interferences [4,5,9]. To quantify the oscillations, a phase shifted sinusoidal function was used to fit the ratios:

$$f(k) = A[\sin(kcd - w)] + B; \tag{2}$$

a method that was first used in Ref. [6]. In Equation 2 *d* is the internuclear distance (2.28 a.u.), *c* is an adjustable frequency parameter; *w* is an adjustable phase shift, and A and B are constants. The resulting frequency *c* and phase shift *w* parameters are shown in Figure 3. The values obtained for 1.9 MeV/u $O^{5,8+}$ + $O_2$ [17] are added for comparison. The frequency parameter *c* remains nearly constant for all electron observation angles, a characteristic that is suggestive of secondary interferences [6,9], and was also seen for 1-5 MeV/u $H^+$ + $N_2$ Ref.[16] and 1.9 MeV/u $O^{5,8+}$ + $O_2$ collisions [17].

With an apparent minimum at 90°, the phase parameter *w* displays a similar pattern to that reported for $H^+$ + $N_2$ [16] and $O^{5,8+}$ + $O_2$ [17]. Notably, an oscillation interval of $\Delta k \sim 4$ a.u. is found as in Ref. [17], a value two times larger than for 3 MeV/u $H^+$ + $N_2$ ($\Delta k \sim 2$ a.u.). The reason for this difference is presently not known but may be due to the differences in the $N_2$ and $O_2$ wavefunctions.

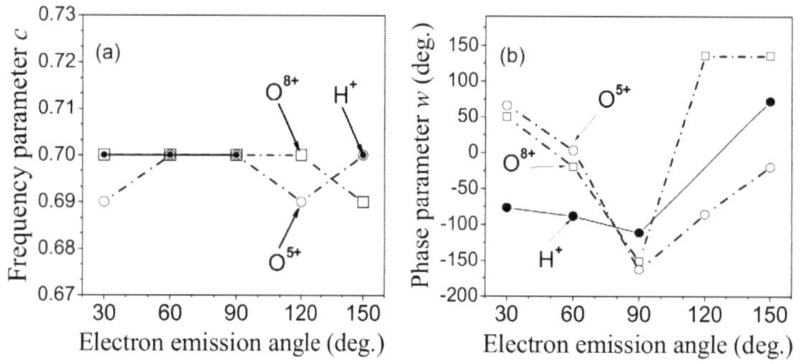


Figure 3. Frequency $c$ and phase shift $w$ parameters for 3 MeV/u $H^+$ + $O_2$ (solid circles) determined from fitting Equation 2 to the normalized cross section ratios shown in Figure 2. Additionally, $c$ and $w$ from previous measurements for 1.9 MeV/u $O^{q+}$ + $O_2$ are shown (open circles: $O^{5+}$; open squares: $O^{8+}$) [17].

## 5. Conclusions

Evidence for interference effects in the electron emission spectra of $O_2$ by 3 MeV/u $H^+$ impact is revealed. In qualitative agreement with the results obtained for $H^+$ + $N_2$ [16], the observed oscillatory structures are independent of the observation angle, and consequently are identified as second-order interferences from intramolecular scattering. In contrast with the oscillatory structures reported for $H^+$ + $H_2$ [9], no primary Young-type interferences are seen suggesting that interference effect for heavy targets differ fundamentally from those for $H_2$. To better understand these differences, theoretical calculations are needed to provide a quantitative understanding of the frequency and magnitude of the secondary oscillations.

## Acknowledgment

We acknowledge D. Cassidy and T. E. Elkafrawy for assistance in the data collection.

## References

[1] Cohen H D and Fano U 1966 *Phys. Rev.* **150** 30
[2] Kaplan I G and Markin A P 1969 *Sov. Phys. Dokl.* **14** 36
[3] Walter M and Briggs J 1999 *J. Phys. B* **32** 2487
[4] Stolterfoht N *et al* 2001 *Phys. Rev. Lett.* **87** 023201
[5] Stolterfoht N *et al* 2003 *Phys. Rev.* A **67** 030702
[6] Stolterfoht N *et al* 2004 *Phys. Rev.* A **69** 012701
[7] Misra D, *et al 2004 Phys. Rev. Lett.* **92** 153201
[8] Kamalou O, *et al* 2005 *Phys. Rev.* A **71** 010702(R)
[9] Hossain S et *al* 2005 *Phys. Rev.* A **72** 010701(R)
[10] Tanis J A *et al* 2006 *Phys. Rev.* A **74** 022707
[11] Misra D, *et al* 2006 *Phys. Rev.* A **74** 060701
[12] Galassi M E *et al* 2002 *Phys. Rev.* A **66** 052705
[13] Nagy L *et al* 2002 *J. Phys.* B: *At. Mol. Opt. Phys.* **35** L453 - L459
[14] Sarkadi L 2003 *J. Phys.* B **36** 2153
[15] Galassi M E *et al* 2004 *Phys. Rev.* A **70** 032721
[16] Baran J L *et al* 2007 *J. Phys.: Conf. Ser.* **58** 215 – 218 ; Baran J L *et al* 2008 *Phys. Rev.* A **78** 012710
[17] Winkworth M *et al* 2008 *Nucl. Instrum. and Meth., Phys. Res.* B, in press.
[18] Póra K and Nagy L 2005 *Nucl. Instrum. and Meth., Phys. Res.* B **233** 293–297
[19] Galassi M E *et al* 2000 *Phys. Rev.* A **62** 022701

# Electron capture from excited hydrogen atoms by highly charged beryllium and carbon ions

**N Shimakura[1], N Kobayashi[1], M Honma[1], T Nakano[2] and H Kubo[2]**

[1] Faculty of Science, Niigata University, Niigata 950-2181, Japan
[2] Naka Fusion Institute, Japan Atomic Energy Agency, 801-1 Mukoyama, Naka, Ibaraki, Japan

E-mail: shima@shimakura.gs.niigata-u.ac.jp

**Abstract**. The cross sections for electron capture in collisions of bare and helium-like ions of Be and C with excited H atoms ($n$ = 2) have been calculated by a molecular-state close-coupling method in a collision energy range of 60 eV/amu–6 keV/amu. Calculated results show that these cross sections are remarkably larger than those for collisions with ground-state H atoms. For example, for collisions of $Be^{2+}$, $Be^{4+}$, $C^{4+}$, and $C^{6+}$ ions with H(2s) atoms at ~1 keV/amu, the cross sections are 190 × $10^{-16}$ cm$^2$, 610 × $10^{-16}$ cm$^2$, 920 × $10^{-16}$ cm$^2$, and 720 × $10^{-16}$ cm$^2$, respectively. In contrast, the cross sections for collisions with ground-state H atoms are 25 × $10^{-16}$ cm$^2$, 35 × $10^{-16}$ cm$^2$, 30 × $10^{-16}$ cm$^2$, and 40 × $10^{-16}$ cm$^2$, respectively.

## 1. Introduction

Thus far, a number of experimental and theoretical studies have been conducted on electron capture in collisions of highly charged ions with ground-state hydrogen atoms. However, despite the known importance of such processes in understanding the impurity behaviour in tokamak divertors, very few studies have been conducted on electron capture cross sections of electronically excited hydrogen atoms. To our knowledge, only two theoretical estimations of such processes have been carried out in the case of $C^{4+}$ + H($n$ = 2) collisions. One is the classical trajectory Monte Carlo calculation carried out by Zaniol *et al.* [1] at a collision energy of 1 eV/amu; the other is the molecular-state close-coupling (MSCC) calculation carried out by Shimakura *et al.* [2].

Macek and Ovchinnikov [3] have developed a simple model for electron capture from high-$n$ states of H atoms by multiply charged ions and have predicted that the cross sections depend on $n$ as $n^5$. They have concluded that this high-$n$ dependence explains why a small fraction (less than 0.1%) of H atoms in high-$n$ states contribute 30% of the capture yield at 50 eV/amu during the $O^{5+}$ + H collisions. Combining this model with the cross sections for ground-state H atoms, we can estimate the cross sections for excited H atoms.

In this study, the cross sections for electron capture in collisions of $Be^{2+}$, $Be^{4+}$, $C^{4+}$, and $C^{6+}$ ions with electronically excited H($n$ = 2) atoms are calculated in the range of 60 eV/amu–6 keV/amu using an MSCC method [4].

## 2. Calculation method

Since details of the theoretical treatment in this paper have already been described elsewhere [4], only specific information used for the present calculation is given below.

*2.1. Electronic states and couplings*

The electronic structures were calculated by a modified valence-bond configuration-interaction methods. In order to represent the $Be^{2+}(1s^2)$ and $C^{4+}(1s^2)$ cores, an *l*-dependent Gaussian-type pseudopotential was employed; hence, only one active electron was treated explicitly in all collision systems considered in this paper. The pseudopotential parameters for $Be^{2+}(1s^2)$ and $C^{4+}(1s^2)$ ions were taken from papers by Wetmore *et al.* [5] and Kimura and Olson [6]. The orbital exponents of the Slater-type orbitals were obtained by optimizing the energies of the respective electronic states. The accuracy of the electronic state calculations performed in this study with respect to spectroscopic values [7] is better than 0.05 eV, except for the states whose contribution to the electron capture processes was insignificant.

We calculated radial and rotational couplings with atomic-type electron translation factors (ETFs) to solve the MSCC equations. The ETFs were included to the first order in velocity.

*2.2. Collision dynamics*

A semi-classical MSCC method was employed. The coupled equations were solved numerically under the assumption of straight-line trajectories of the heavy particles. By squaring the resulting amplitude, the transition probability was obtained as a function of collision energy and impact parameter. The cross sections were obtained by integrating the impact-parameter-weighted transition probability over the impact parameter.

## 3. Results and discussion

*3.1. $Be^{2+}$ + H collision system*

The total cross sections for the $Be^{2+}$ + H(1s) collisions depend weakly on the collision energy; their maximum value is ~20 × $10^{-16}$ $cm^2$ at 1.68 keV/amu. The electron in H(1s) is dominantly captured into the $Be^+$(2s) state.

The calculated cross sections for electron capture in the $Be^{2+}$ + H(2s) collisions are shown in figure 1. The total cross sections are ten times larger than those for collisions with H(1s) at ~1 keV/amu. At the lowest and the highest ends of the collision energy region covered in this work, the electron is mainly captured into the $Be^+(4l)$ states (~90%). At intermediate energies, the contribution of the $Be^+(3l)$ states increases.

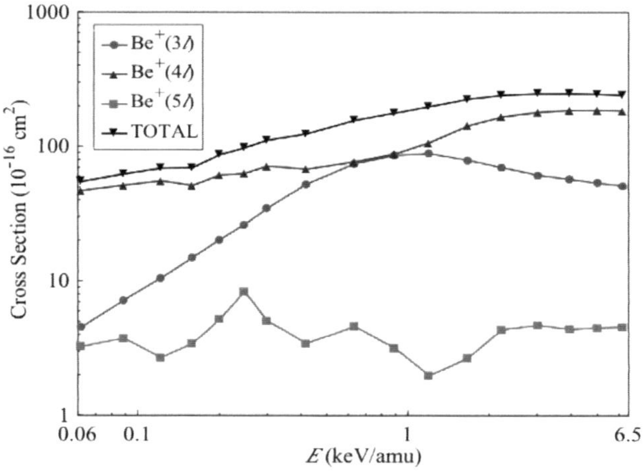


**Figure 1.** Total and partial cross sections for electron capture into $Be^+(nl)$ states in the $Be^{2+}$ + H(2s) collisions.

The total cross sections for the $Be^{2+}$ + H(2p) collisions are approximately five times larger than those for collisions with H(1s) targets at ~1 keV/amu. At the lowest and the highest collision energies, the electron is mainly captured into the $Be^+(3l)$ (~80%) and $Be^+(4l)$ states (~70%), respectively.

*3.2. $Be^{4+}$ + H collision system*

The cross sections for the $Be^{4+}$ + H(2s) collisions are shown in figure 2. The calculations were carried out for values of Λ from 0 to 3, namely, for Σ, Π, Δ, and Φ symmetries, where Λ is the electronic angular momentum of the total system projected onto the internuclear axis. The solid curves in figure 2 represent the cross sections including all the symmetries up to Λ = 3, whereas the broken curves include only up to $\Lambda = 2$. This figure proves that the contributions from $\Lambda = 3$ and above too, perhaps, are negligibly small. The total cross sections are $500 \times 10^{-16}$ cm$^2$ ~ $700 \times 10^{-16}$ cm$^2$. The $n = 6$ states are the most dominant over the collision energy range covered in this study; at the lowest and highest collision energies, the contributions of the $n = 6$ states are ~85% and ~65%, respectively. The second dominant are the $n = 7$ states. The cross sections for electron capture into these states increase with the collision energy.

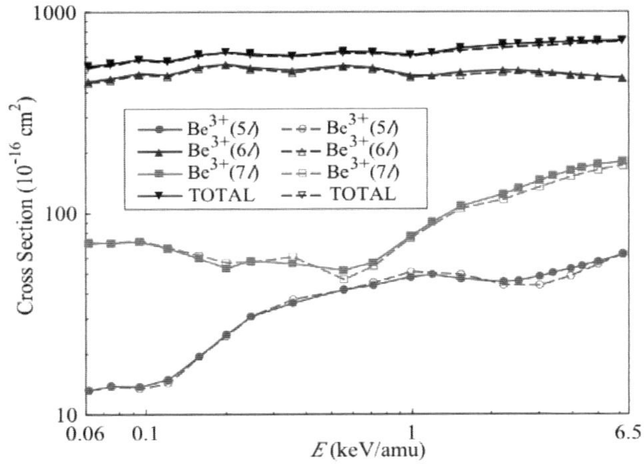


**Figure 2.** Total and partial cross sections for electron capture into $Be^{3+}(nl)$ states in the $Be^{4+}$ + H(2s) collisions.

The total cross sections for the $Be^{4+}$ + H(2p) collisions are smaller than those for the $Be^{4+}$ + H(2s) collisions. At the lowest and the highest collision energies, the cross sections are $200 \times 10^{-16}$ cm$^2$ and $500 \times 10^{-16}$ cm$^2$, respectively. The electron is mainly captured into the $Be^{3+}(6l)$ states ($200 \times 10^{-16}$ cm$^2$ ~ $330 \times 10^{-16}$ cm$^2$). The second dominant at the lowest and the highest collision energies are the $Be^{3+}(5l)$ and $Be^{3+}(7l)$ states, respectively. The cross section for capture into $Be^{3+}(7l)$ takes the largest value of $100 \times 10^{-16}$ cm$^2$ at the highest energy.

### *3.3. $C^{4+}$ + H collision system*

Experimental and theoretical cross sections for the $C^{4+}$ + H(1s) collisions are found in the literature. All have a maximum value of ~$30 \times 10^{-16}$ cm$^2$ at ~250 eV/amu. The total cross section for the $C^{4+}$ + H(2s) collisions obtained by the CTMC method [1] is ~$380 \times 10^{-16}$ cm$^2$ at 1 eV/amu. Our calculated cross sections for the latter collisions are shown in figure 3. The solid curves are for the calculations including the $\Lambda = 0 - 3$ symmetries and the broken curves only up to $\Lambda = 2$. The total cross sections are $800 \times 10^{-16}$ cm$^2$ ~ $930 \times 10^{-16}$ cm$^2$. The electron is mainly captured into the $C^{3+}(6l)$ states.

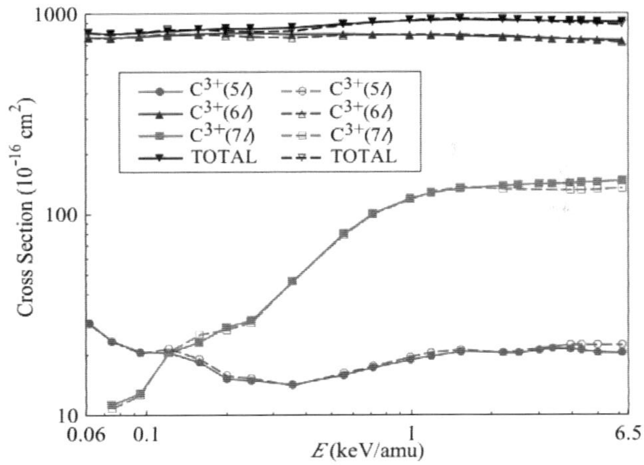


**Figure 3.** Total and partial cross sections for electron capture into $C^{3+}(nl)$ states in the $C^{4+}$ + H(2s) collisions.

The total cross sections for the $C^{4+}$ + H(2p) collisions ($580 \times 10^{-16}$ cm$^2$ ~ $650 \times 10^{-16}$ cm$^2$) are smaller than those for the $C^{4+}$ + H(2s) collisions. Capture into the $n = 6$ states ($550 \times 10^{-16}$ cm$^2$ ~ $600 \times 10^{-16}$ cm$^2$) is the most dominant over the collision energy range covered in this study; its contribution is ~80%. The second dominant are the $n = 7$ states, except at low collision energies; their maximum contribution is ~10%.

### *3.4. $C^{6+}$ + H collision system*

Thus far, two theoretical and one experimental reports are found in the literature for the $C^{6+}$ + H(1s) collisions. We have also calculated the electron capture cross sections using the MSCC method. Our results are similar to the previously reported ones; the cross sections increase with the collision energy.

Our calculated cross sections for the $C^{6+}$ + H(2s) collisions are shown in figure 4. The total cross sections are $550 \times 10^{-16}$ cm$^2$ ~ $750 \times 10^{-16}$ cm$^2$. The electron is mainly captured into the $C^{5+}(8l)$ states; the cross sections are in a range of $450 \times 10^{-16}$ cm$^2$ ~ $490 \times 10^{-16}$ cm$^2$. The contribution is 82% at the lowest collision energies. The $C^{5+}(9l)$ and $C^{5+}(7l)$ states are the second dominant states at the lowest and highest energies, respectively; the cross sections for capture into the $C^{5+}(9l)$ states are $77 \times 10^{-16}$ cm$^2$ ~ $107 \times 10^{-16}$ cm$^2$.

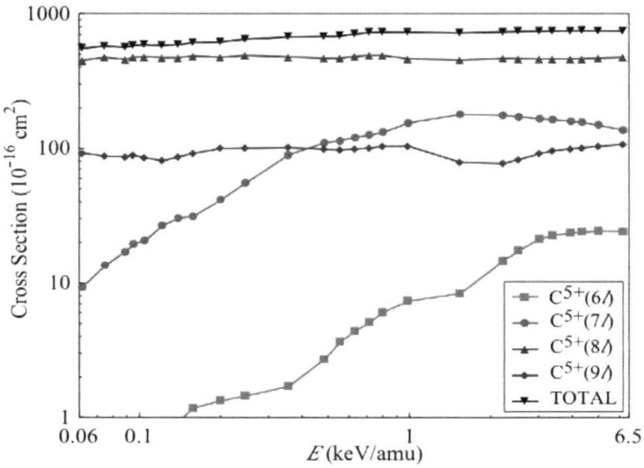


**Figure 4.** Total and partial cross sections for electron capture into $C^{5+}(nl)$ states in the $C^{6+}$ + H(2s) collisions.

The total cross sections for the $C^{6+}$ + H(2p) collisions are $430 \times 10^{-16}$ cm$^2$ ~ $560 \times 10^{-16}$ cm$^2$. The $n = 8$ states are the most dominant, the cross section being $310 \times 10^{-16}$ cm$^2$ ~ $360 \times 10^{-16}$ cm$^2$; their contribution is ~80% at the lowest collision energies. The second dominant are the $n = 7$ states; their contribution is 10% ~ 18%.

*3.5. Test of the $n^5$ rule on the MSCC cross sections*

Table 1 shows the calculated cross sections for the $Be^{2+}$, $Be^{4+}$, $C^{4+}$, and $C^{6+}$ + H(1s, 2s) collisions at ~1keV/amu along with the H(2s) cross sections estimated from the $n^5$ rule as $2^5$ times the H(1s) cross sections. A good agreement is found for the $C^{4+}$+ H system, but a large discrepancy is seen for the $Be^{2+}$ + H collisions.

**Table 1.** Test of the Macek-Ovchinnikov $n^5$ rule on our cross sections (in $10^{-16}$ cm$^2$) calculated at ~1keV/amu.

| | $Be^{2+}$ | $Be^{4+}$ | $C^{4+}$ | $C^{6+}$ |
|---|---|---|---|---|
| H(1s) | 25 | 35 | 30 | 40 |
| H(2s) | 190 | 610 | 920 | 720 |
| $2^5$×H(1s) | 800 | 1100 | 960 | 1300 |

## 4. Summary and conclusion

Our calculations show that the electron capture cross sections for collisions with electronically excited hydrogen atoms are remarkably larger than those for collisions with ground-state hydrogen atoms. This can be explained in terms of the effective distances at which electron capture occurs. The H(2*l*) states are more diffusely distributed than the H(1s) state, and hence, the $Be^{2+}$, $Be^{4+}$, $C^{4+}$, and $C^{6+}$ ions can capture the electron in H(2*l*) at longer distances than that in H(1s).

**Acknowledgment**

This study was supported in part by the Japan Atomic Energy Agency.

**References**

[1] Zaniol B, Isler R C, Brooks N H, West W P and Olson R E 2001 *Phys. Plasmas* **8** 4386
[2] Shimakura N, Honma M and Kubo H 2006 *J. Plasma Fusion Res. Ser.* **7** 199
[3] Macek J and Ovchinnikov S Y 1992 *Phys. Rev. Lett.* **69** 2357
[4] Kimura M and Lane N F 1989 *Advances in Atomic, Molecular, and Optical Physics* vol 26, ed by Bates D R and Bederson B (New York :Academic) pp 79
[5] Wetmore A E, Cole H R and Olson R E 1986 *J. Phys. B: At. Mol. Phys.* **19** 1515
[6] Kimura M and Olson R E 1984 *J. Phys. B: At. Mol. Phys.* **17** L713
[7] Bashkin S and Stoner J R Jr 1975 *Atomic Energy levels and Grotrian Diagrams* vol I (Amsterdam: North-Holland)

# Isotope effect in the dissociation of methanol dications produced by collision with $Ar^{8+}$

**K Hayakawa, J Matsumoto, H Shiromaru and Y Achiba**

Department of Chemistry, Tokyo Metropolitan University, Hachioji 192-0397, Japan

shiromaru-haruo@tmu.ac.jp

**Abstract.** The H/D isotope effect in the ion-pair dissociation processes of doubly charged methanol formed by the impact of low-energy $Ar^{8+}$ was investigated. By multi-hit position-sensitive time-of-flight measurements, the branching ratios of dissociation channels yielding H, $H_2$ or $H_3$ fragment ions, or H/D substituted ions, and the kinetic energy release distributions of these dissociation channels were determined for methanol isotopomers.

## 1. Introduction

Experiments examining the collision of highly charged ions with molecules have addressed a variety of targets, from diatomics to polyatomics, including organic molecules and clusters. Isotope substitution is useful for investigating the complex dissociation dynamics of multiply ionized polyatomic molecules. Strong isotope effects have been reported for ion-pair dissociation of the water dication; the reaction $HDO^{2+} \rightarrow H^+ + OD^+$ takes preference over the reaction $HDO^{2+} \rightarrow D^+ + OH^+$. These channels also differ considerably in their kinetic energy release (KER) distributions [1, 2]. In the case of the HDO molecule, the D-substitution breaks the $C_{2v}$ symmetry of water, which may be a key factor of the strong isotope effects. On the other hand, isotopomers of methanol may present isotope effects while keeping the $C_s$ symmetry.

Eland and coworkers investigated the ion-pair dissociation of methanol dications by the photoelectron-photoion-photoion coincidence (PEPIPICO) method, also using isotopomers to distinguish between OH and CH bond breakups [3]. They reported that the peaks in the KER distributions were the same for all isotopomers within experimental uncertainty (0.4~0.6 eV). A coincidence study on the dissociation of collisionally prepared methanol dications was reported by De et al. [4], while the details of the isotope effects remain an open question. In the present study, the branching ratios and KER distributions of various ion-pair dissociation channels of methanol dications were measured, focusing on the effects of H/D substitution.

## 2. Experimental setup

A projectile beam, 120 keV $Ar^{8+}$, extracted from a 14-GHz electron cyclotron resonance-ion source at Tokyo Metropolitan University (TMU-ECRIS), hit the target molecule with a 90° crossed-beam configuration. Methanol ($CH_3OH$) or methanol-d ($CD_3OH$: 99.8% isotope enrichment, or $CD_3OD$: 99.8% isotope enrichment) targets prepared using a conventional degassing procedure were introduced into the beam-source chamber through a pin-pole 0.6 mm in diameter, without seed gas. The molecular beam, collimated by a skimmer, crossed the projectile beam and was caught in a liquid nitrogen trap placed in the collision chamber. The typical operating pressure was $1\times10^{-3}$ Pa in the beam-source chamber and $1\times10^{-6}$ Pa in the collision chamber.

The position-sensitive time-of-flight (PSTOF) measurement procedure reported previously [5] was employed, except that a fast ADC card (Acqiris DC270) was used for recording PSTOF output signals. Briefly, the fragment ions were accelerated by an electric field (4.90 V/mm) in the drift tube (449 mm) and hit the aluminum plate. Secondary electrons guided by an inhomogeneous magnetic field were accelerated in the opposite direction by this electric field, while maintaining positional information, and were detected by a time- and position-sensitive detector [6]. The TOF measurements were triggered by detection of the projectile passing through the collision area.

## 3. Results and discussion

### 3.1. TOF coincidence map

Hereafter, the ion-pair dissociation channel yielding fragment $A^+$ and $B^+$ is denoted by ($A^+$, $B^+$), and the number of evaporated atoms associated with the ion-pair dissociation is denoted by $\Delta$. Typical coincidence maps for $CH_3OH$ and $CD_3OH$ targets are shown in figures 1(a)-(g), where islands indicate events in which $H^+$ (or $D^+$) and heavier counterpart ions were detected in coincidence. Figures 1(a)-(c) show examples of the two-body ion-pair dissociation channels of the dications ($\Delta$=0), figures (d) and (e) are those accompanied with evaporation of neutral species ($\Delta$=1), and figures (f) and (g) are those for the dissociation of multiply ionized methanol, $(CD_3OH)^{n+}$; $n \geq 3$.

The islands in figures (a)-(c) are narrow lines, as the fragment TOFs are well correlated. The intensity ratio of the islands in figures (b)-(c) is 1:20. It is clear that the C-D bond breakup takes preference over the O-H bond breakup, beyond the statistical weight (1:3), while the ratio is less disproportionate than in the case of 30.4-nm photoionization (1:52) [3]. The islands in figures (d) and (e) are broadened due to the momentum shared by the neutral fragment. The propensities of evaporation were derived from figures 1(b)-(e); $H^+$ ejection tends to induce evaporation of D, whereas this is not the case for $D^+$ ejection. As will be discussed later, the ratio of $\Delta$=1 to $\Delta$=0 is considered a measure of the average excess energy of the ion-pair dissociation. The propensity given by

$$\frac{\{I(D^+,CHDO^+)+I(D^+,CD_2O^+)\}/I(D^+,CHD_2O^+)}{I(H^+,CD_2O^+)/I(H^+,CD_3O^+)}$$

is 0.08, which is also less disproportionate than in the case of photoionization (0.033).

The shapes of the ($H^+$, $CO^+$) and ($D^+$, $CO^+$) islands are quite different from those of the other islands. The minimum in the middle indicates that $H^+$ or $D^+$ emitted perpendicular to the TOF axis is missing due to relatively high initial kinetic energy given by more violent Coulomb repulsion in

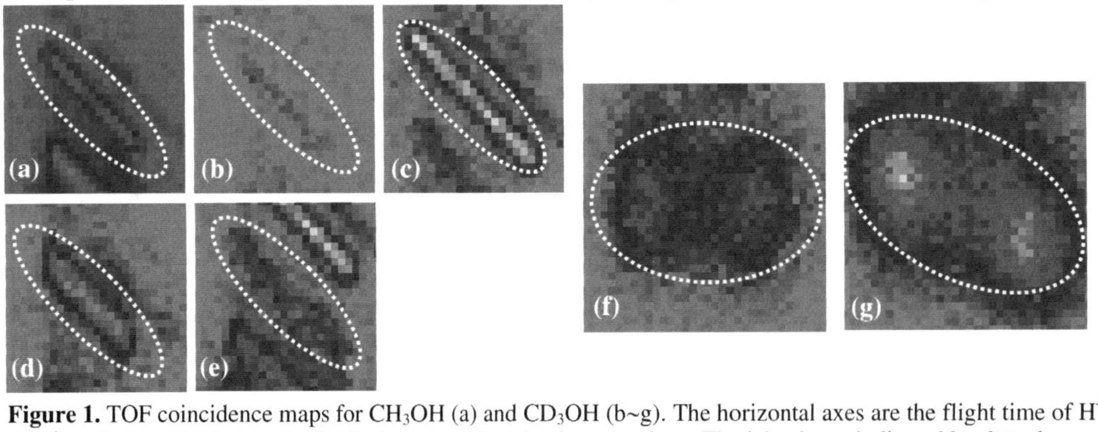


**Figure 1.** TOF coincidence maps for $CH_3OH$ (a) and $CD_3OH$ (b~g). The horizontal axes are the flight time of $H^+$ or $D^+$, and the vertical axes are the flight time of heavier fragment ions. The islands are indicated by dotted ellipsoid, (a)($H^+$,$CH_3O^+$), (b) ($H^+$,$CD_3O^+$), (c)($D^+$,$CHD_2O^+$), (d) ($H^+$,$CD_2O^+$), (e) ($D^+$,$CD_2O^+$), (f) ($H^+$,$CO^+$), (g) ($D^+$,$CO^+$). The full scale is 250 ns for (a)~(e), and 330 ns for (f) and (g).

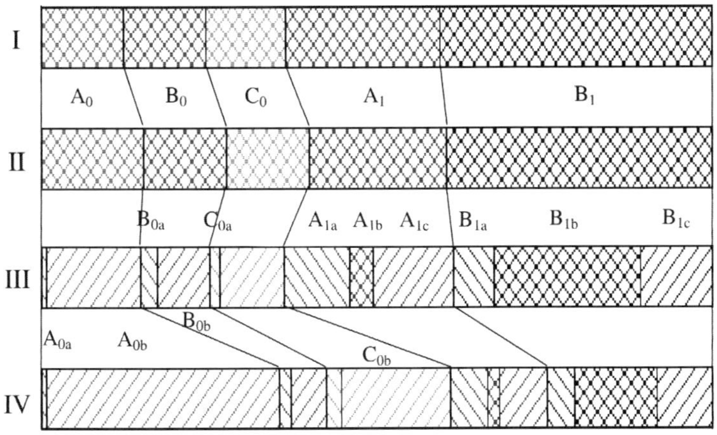


| CH$_3$OH, CD$_3$OD | | CD$_3$OH | |
|---|---|---|---|
| $A_0$ | ($H^+$, $CH_3O^+$) | $A_{0a}$ | ($H^+$, $CD_3O^+$) |
| | ($D^+$, $CD_3O^+$) | $A_{0b}$ | ($D^+$, $CHD_2O^+$) |
| $B_0$ | ($H_2^+$, $CH_2O^+$) | $B_{0a}$ | ($HD^+$, $CD_2O^+$) |
| | ($D_2^+$, $CD_2O^+$) | $B_{0b}$ | ($D_2^+$, $CHDO^+$) |
| $C_0$ | ($H_3^+$, $CHO^+$) | $C_{0a}$ | ($HD_2^+$, $CDO^+$) |
| | ($D_3^+$, $CDO^+$) | $C_{0b}$ | ($D_3^+$, $CHO^+$) |
| $A_1$ | ($H^+$, $CH_2O^+$) | $A_{1a}$ | ($H^+$, $CD_2O^+$) |
| | ($D^+$, $CD_2O^+$) | $A_{1b}$ | ($D^+$, $CD_2O^+$) |
| | | $A_{1c}$ | ($D^+$, $CHDO^+$) |
| $B_1$ | ($H_2^+$, $CHO^+$) | $B_{1a}$ | ($HD^+$, $CDO^+$) |
| | ($D_2^+$, $CDO^+$) | $B_{1b}$ | ($D_2^+$, $CDO^+$) |
| | | $B_{1c}$ | ($D_2^+$, $CHO^+$) |
| $A_2$ | ($H^+$, $CHO^+$) | $A_{2a}$ | ($H^+$, $CDO^+$) |
| | ($D^+$, $CDO^+$) | $A_{2b}$ | ($D^+$, $CDO^+$) |
| | | $A_{2c}$ | ($D^+$, $CHO^+$) |

**Figure 2.** Branching ratios for the ion-pair dissociation with $\Delta = 0$, indicated by the subscript "0", and $\Delta = 1$, indicated by the subscript "1". Target molecules are I: $CH_3OH$, II: $CD_3OD$, III: $CD_3OH$. The results of the photoionization study for $CD_3OH$ are shown in IV. Assignment of the channels is listed in the right. The channels for $\Delta = 2$ are certainly identified in the coincidence map ($A_2$ and $A_{2a\text{-}c}$ in the list), however, the branching ratios are not determined due to serious overlap with the neighbouring islands.

highly ionized methanol. The round shape of the islands is consistent with the assignment, as the coulombic force of the undetected fragment ions would considerably affect the TOFs. For the ($H^+$, $CO^+$) channel, the undetected ion is $D^+$, whereas for the ($D^+$, $CO^+$) channel, the rest is $H^+$ or $D^+$. The Coulomb repulsion of $D^+$ and $H^+$ acting on $CO^+$ would be partly canceled making the island horizontal. No cancellation is expected for the reaction $CO^+ + D^+ + D^+$.

The relative intensities of the islands identified for the ion-pair dissociation channels with $\Delta = 0$, 1 of the three isotopomers are shown in figure 2, together with the results of the photoionization study [3]. The channels for ion-pair dissociation of dications with $\Delta$=0-2 are listed on the right; those with $\Delta$=3 were observed only for the ions in the higher charge states. Previously, we have shown that $\Delta$ values are a good measure of the excess energy of benzene dications [5]. If this is also true for methanol, the observation of various channels yielding $H^+ \sim H_3^+$ isotopomers (groups $A_0$, $B_0$, $C_0$) for $\Delta$=0 indicates that all the reactions may proceed with low excess energies. With the larger excess energies associated with $\Delta$=1, the channels yielding $H_3^+$ isotopomers are closed. A reasonable conjecture is that the production of trihydrogen ions requires somewhat "mild" conditions. If the excess energies are very large, the channels yielding $H_2^+$ isotopomers are also closed, *i.e.*, only the channels yielding $H^+$ or $D^+$ survive for $\Delta$=2.

The ($HD^+$, $CD_2O^+$) and ($HD^+$, $CDO^+$) channels were clearly observed, and a weak ($HD_2^+$, $CDO^+$) channel was also seen for the $CD_3OH$ target, indicating that intramolecular proton (or hydrogen) migration occurs, as demonstrated previously in photoionization and ion-impact studies [3, 4]. In the dissociation of $CD_3OD$ dication, the evaporation of neutrals was slightly suppressed relative to $CH_3OH$ dication, which may be rationalized by considering the increase in the bond dissociation energy for heavier isotopes.

In comparison with the photoionization study, it is notable that collisional processes induce more evaporation of neutrals ($A_1$ and $B_1$). This trend can be readily understood, as collisional ionization generally allows access to a wide variety of ionic states and enhances vibrational excitations, in contrast to photoionization, which is governed by selection rules and Franck-Condon factors. The aforementioned enhancement of the minor process of O-H bond breakup is consistent with this trend.

3.2. KER of two-body ion-pair dissociation

The KER distributions of the two-body ion-pair dissociation channels for three methanol isotopomers yielding $H^+ \sim D_3^+$ are shown in figures 3(a)-(c). A general trend observed in these plots is that the KER distributions for the $CH_3OH$ and the $CD_3OD$ dissociation channels are similar, especially for $H_2^+$, $D_2^+$ formation and $H_3^+$, $D_3^+$ formation channels, whereas those for the $CD_3OH$ dissociation channels are shifted to lower energy. The peak values in figure 3(a) coincide with those reported in the photoionization study within experimental uncertainty, whereas those in figure 3(c) are considerably lower.

As mentioned in section 3.1, the intensity of the ($H^+$, $CD_3O^+$) channel was very low, and thus, the two-body charge separation with O-H bond breakup is most likely a minor channel for all the isotopomers. The peaks of the KER distribution differed slightly among the isotopomers, $E_{(H+,\ CD3O+)} < E_{(D+,\ CD3O+)} < E_{(H+,\ CH3O+)}$. For the minor channel ($H^+$, $CD_3O^+$), which would be energetically unfavorable, showed low-energy shift in KER, similar to the case of photoionization of methanol and the dissociation of $HDO^{2+}$ [1].

For the channels yielding $H_2^+$ isotopomers, the KER distributions varied with H/D substitutions, *i.e.*, $E_{(HD+,\ CD2O+)} < E_{(D2+,\ CHDO+)} < E_{(D2+,\ CD2O+)} \approx E_{(H2+,\ CH2O+)}$. This trend was also observed for the channels yielding $H_3^+$ isotopomers. The shift induced by H/D substitution was about 0.2-0.5 eV.

It should be noted that the isotope effect was not prominent when all the H atoms were substituted by D atoms, but was most enhanced by partial substitution. This observation for different target isotopomers is the outcome of a different type of isotope effect from that reported previously for water isotopomers, in which a larger shift (~1 eV) was observed between two ion-pair dissociation channels of $HDO^{2+}$. Further extension of the target molecules is needed to clarify whether these are specific phenomena for methanol isotopomers or are generally true for organic molecules.

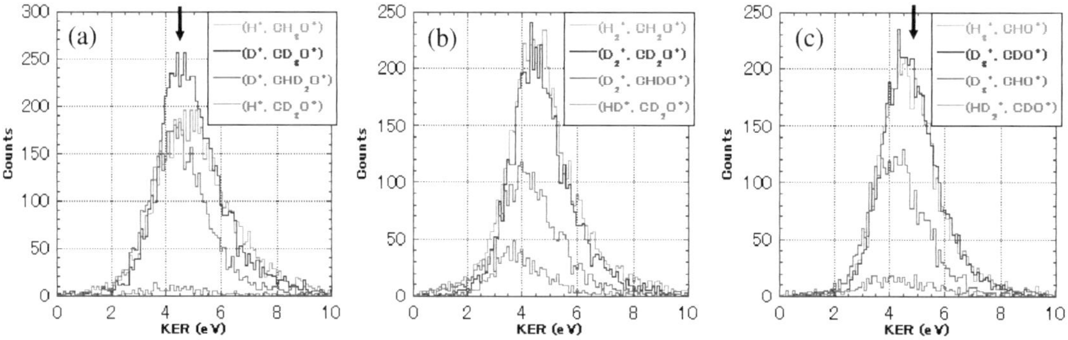


**Figure 3.** KER distribution of two-body ion-pair dissociation channels for three isotopomers. Since the plots are not normalized by $Ar^{8+}$ beam intensity nor accumulation time, the intensity ratio is meaningful only between the two channels for $CD_3OH$. The vertical arrows indicate peaks reported by photoionization study. (a) $A_0$ group, (b) $B_0$ Group, (c) $C_0$ group.

## 4. Acknowledgement
The authors thank Mr. R. Hatsuda for his contribution in the early stage of the methanol dication experiments.

## References
[1] Legendre S, Giglio E, Tarisien M, Cassimi A, Gervais B and Adoui L 2005 *J. Phys. B: At. Mol. Opt. Phys.* **38** L233
[2] Sayler A M, Leonard M, Carnes K D, Cabrera-Trujillo R, Esry B D and Ben-Itzhak I 2006 *J. Phys. B: At. Mol. Opt. Phys.* **39** 1701
[3] Eland J H D and Treves-Brown B J 1992 *Int. J. Mass. Spectrom. Ion Processes.* **113** 167
[4] De S, Rajput J, Roy A, Ghosh P N, and Safvan C P 2006 *Phys. Rev. Lett.* **97** 213201
[5] Veshapidze G, Shiromaru H, Achiba Y and Kobayashi N 2004 *Int. J. Mass Spectrom.* **239** 27
[6] Veshapidze G, Nishide T, Shiromaru H, Kobayasi N and Mizogawa T 2002 *Jpn. J. Appl. Phys.* **41**, 871

# Analysis of charge-asymmetric Coulomb explosion of $N_2$ molecules with slow $Kr^{8+}$ ions

**Tomoko Ohyama-Yamaguchi[1] and Atsushi Ichimura[2]**

[1] Tokyo Metropolitan College of Industrial Technology, Shinagawa, Tokyo 140-0011, Japan
[2] Institute of Space and Atronautical Science, JAXA, Sagamihara, 229-8510, Japan

E-mail: [1]`yamaguti@tokyo-tmct.ac.jp`, [2]`ichimura@isas.jaxa.jp`

**Abstract.** We investigate a nontrivial asymmetric population observed in a triple-coincidence measurement for charge-asymmetric Coulomb explosion of $N_2$ molecules in collisions with a slow (10 — 100 eV/amu) $Kr^{8+}$ ions. The three-center over-the-barrier model is developed to describe collision dynamics including many active electrons as the formation of a triatomic quasi-molecule and its decay into three moving atomic ions. The present model predicts an asymmetry effect consistent with the measurement, where molecular bond elongation during a collision is crucial. Electron localization processes in the quasi-molecular decay are examined in detail in relation to two saddle points in the three-center Coulomb potential.

## 1. Introduction

Much attention has been called to the Coulomb explosion of molecules in a variety of collision processes on a basis of the momentum imaging technique. Of particular interest is a triple coincidence measurement done by Kaneyasu and coworkers [1] for collisions of slow (10 — 200 eV/amu) $Kr^{8+}$ ions with $N_2$ molecules; they detected, in a coplanar geometry, the dissociating ion pair $(N^{Q+}, N^{Q'+})$ in transverse to the beam axis and the scattered ion in forward left and right directions. A remarkable result thereby was unequal strengths of charge-asymmetric fragmentation between the *near* and *far* sites relative to the projectile trajectory. For a given nonequivalent charge pair $Q \neq Q'$, the asymmetry parameter is obtained as $A(Q,Q') = (P_> - P_<)/(P_> + P_<)$ through $P_>$ and $P_<$, where $P_>$ ($P_<$) denotes a coincident population for $Q_{\rm far} > Q_{\rm near}$ ($Q_{\rm far} < Q_{\rm near}$), with $Q_{\rm far}$ and $Q_{\rm near}$ being the far and near fragment charges. The experiment generally indicated a striking positive asymmetry $A$; the far site is populated more by a higher charge than by a lower charge. This asymmetry was interpreted [1] to come from target electron polarization by the projectile charge during a collision.

We have made some analyses in previous works [2, 3] to get insights into the observations [1] by developing the three-center Coulombic over-the-barrier model. In the first paper [2], we took a conventional but questionable two-step picture, *i.e.*, 'electron removal followed by Coulomb explosion'. In this picture, residual electrons in the molecule are polarized by a receding projectile and get localized into two atomic sites on the way of fragmentation. Although leading to a correct sign of $A$, the polarization effect was found too small to account for the measurement. Hence, in the second paper [3], we extended the model so as to describe an *inseparable* process of electron transfer and Coulomb explosion; we treated the formation of a three-center quasi-molecule and its decay into three (one scattering and two dissociating) moving atomic ions. The large asymmetry observed was suggested to reflect bond elongation during a collision.

In the present paper, we further develop the model to describe the decay of a three-center quasi-molecule including many active electrons. The electron localization probabilities into the near and far sites along with their dependence on the impact-parameter vector are analyzed.

## 2. Three-center Coulombic over-the-barrier model

We address collisions of a slow ($v \ll 1$ au) highly charged ion $A^{q+}$ with a homonuclear diatomic molecule BC. The molecular bond-length is not fixed to be the equilibrium one but allowed to be elongated during a collision.

The many-electron dynamics is treated as formation and decay of a triatomic quasi-molecule,

$$A^{q+} + BC \quad \to \quad ABC^{q+} \quad \to \quad A^{(q-r)+} + B^{Q_B+} + C^{Q_C+}, \tag{1}$$

where $r = Q_B + Q_C$ is the number of removal electrons from the molecule. According to the original over-the-barrier model by Niehaus [4], the dynamics is described by a sequence of quasi-molecule formations of respective electrons in numerical order of 'rank' $t$ $(= 1, 2, \ldots)$ in the way in of a collision, and by a sequence of their decays in the reverse order in the way out. One electron dynamics in respective steps is described with a three-center Coulomb potential $U(\boldsymbol{r})$ determined by ion core charges $q_A$, $q_B$, and $q_C$. This potential has two saddle points ($U_{\rm high}$ and $U_{\rm low}$) in contrast with the two-center potential containing one saddle point (see Fig. 1).

The quasi-molecule is formed at a critical distance $R_t$ determined by a classical *over-the-barrier* criterion, $E = U_{\rm high}$, with the electron diabatic energy $E$ and the barrier height $U_{\rm high}$. Here the $t$-th electron is assumed initially bound in the target with its $t$-th ionization potential.

The quasi-molecule decays via two-step electron localization into three atomic sites associated with the two saddle points. This decay may occur in three processes (labeled by I, II, and III in Fig. 1) according to the nuclear configuration formed by three ion cores A, B, and C; the configuration space is divided into three domains labeled by $\mathcal{D}_{A-BC}$, $\mathcal{D}_{C-AB}$, and $\mathcal{D}_{B-CA}$ in view of classical electron motion when $U_{\rm high} > E > U_{\rm low}$. For the three processes, one-electron localization probabilities into three atomic sites are respectively given by

$$\text{process I} \; : \; p_A = p^{A}_{\rm high}, \qquad p_B = p^{BC}_{\rm high} \times p^{B}_{\rm low}, \qquad p_C = p^{BC}_{\rm high} \times p^{C}_{\rm low}, \tag{2}$$

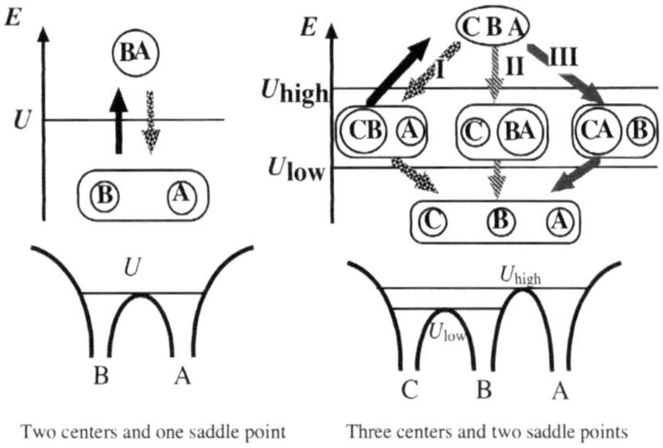


**Figure 1.** Formation (upward arrows) and decay (downward arrows) of a quasi-molecule. In the decay of a three-center quasi-molecule, an electron is localized into three atomic sites via two steps associated with higher and lower saddle points.

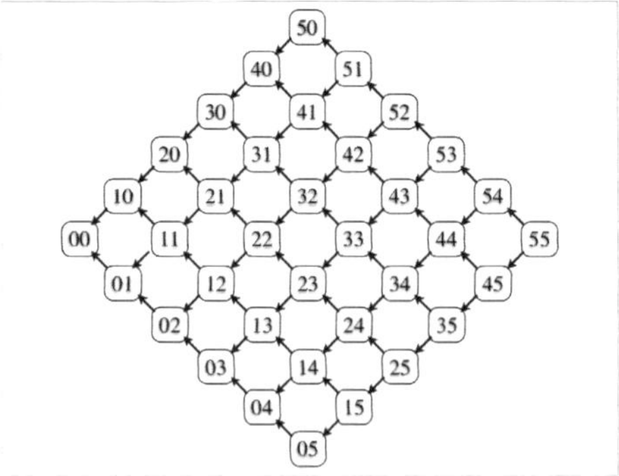


**Figure 2.** Network of the charge state $(q_B, q_C)$ evolution in the decay of a three-center quasi-molecule with 10 active electrons.

$$\text{process II} \;:\; p_{\rm A} = p^{\rm AB}_{\rm high} \times p^{\rm A}_{\rm low}, \quad p_{\rm B} = p^{\rm AB}_{\rm high} \times p^{\rm B}_{\rm low}, \quad p_{\rm C} = p^{\rm C}_{\rm high}, \tag{3}$$

$$\text{process III} \;:\; p_{\rm A} = p^{\rm CA}_{\rm high} \times p^{\rm A}_{\rm low}, \quad p_{\rm B} = p^{\rm B}_{\rm high}, \quad p_{\rm C} = p^{\rm CA}_{\rm high} \times p^{\rm C}_{\rm low}, \tag{4}$$

through the branching probabilities calculated with phase-space volumes at $U_{\rm high}$ and $U_{\rm low}$,

$$p^{\Gamma}_{\rm s} = \frac{\rho^{\Gamma}_{\rm s}}{\rho^{\Gamma}_{\rm s} + \rho^{\Delta}_{\rm s}}, \qquad \rho^{\Gamma,\Delta}_{\rm s} = \int_{\Gamma,\Delta} {\rm d}\boldsymbol{p}{\rm d}\boldsymbol{r}\,\delta(U_{\rm s} - \boldsymbol{p}^2/2 - U(\boldsymbol{r}; q_{\rm A}, q_{\rm B}, q_{\rm C})). \tag{5}$$

Here a suffix 's' labels higher or lower saddle point, which divides the available phase space into two separable regions denoted by $\Gamma$ and $\Delta$.

Considering all the electrons, we obtain the probability distribution for dissociating ion-pair,

$$P(Q_{\rm B}, Q_{\rm C}) = \sum_{\rm string} \prod_{t=1}^{t_{\max}} p_{j(t)}(q_{\rm A}(t), q_{\rm B}(t), q_{\rm C}(t);\ \text{string}), \tag{6}$$

where $t_{\max}$ denotes the number of active electrons determined by the impact parameter $b$ and the critical distances $R_t$. In Eq. (6) $p_j$ indicates one-electron localization probability given by Eqs. (2—5). The suffix $j$ (= 0, 1, 2) labels A, B or C according to a 'string' [4], $\{j(t);\ t = 1, 2, \cdots, t_{\max}\}$, representing evolution of ion-core charges $(q_{\rm A}, q_{\rm B}, q_{\rm C})$ due to successive electron localization in descending order of $t$. The evolution starts at $t = t_{\max}$, where we take $q_{\rm A} = q$ (incident ion charge) and $(q_{\rm B}, q_{\rm C}) = (t_{\max}/2,\ t_{\max}/2)$ for even $t_{\max}$. For odd $t_{\max}$, we consider two kinds of evolution starting from $((t_{\max}+1)/2,\ (t_{\max}-1)/2)$ and from $((t_{\max}-1)/2,\ (t_{\max}+1)/2)$, and take their average in Eq. (6). The summation in Eq. (6) is taken over all possible 'strings' leading to ion pair $(Q_{\rm B}, Q_{\rm C})$. Figure 2 shows a network of the charge-state evolution projected onto a $(q_{\rm B}, q_{\rm C})$ plane starting from (5,5).

## 3. Results and Discussion

We apply the present model to collisions of $\mathrm{Kr}^{8+} + \mathrm{N}_2$ at a velocity of $v = 0.04$ au (*i.e.*, 40 eV/amu) and consider 10 covalent electrons in the $\mathrm{N}_2$ molecule as active electrons. We assume linear trajectories of incidence and of dissociation retaining the beam axis ($\hat{\boldsymbol{v}}$) and the initial molecular orientation ($\hat{\boldsymbol{d}}$), satisfying $\hat{\boldsymbol{d}} \perp \hat{\boldsymbol{v}}$.

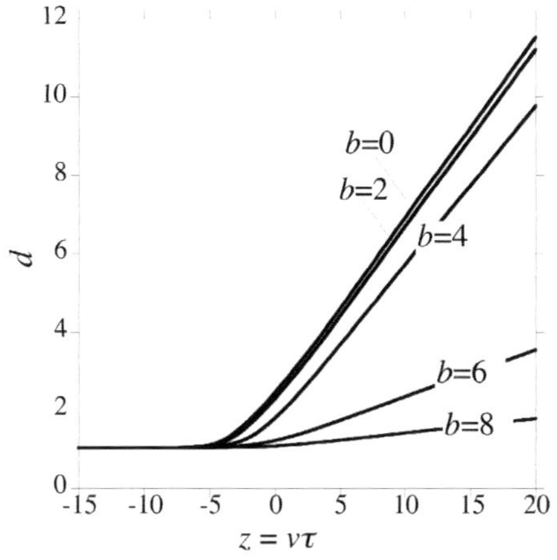


**Figure 3.** Time $\tau$ evolution of half the bond length $d$ at $v = 0.04$ au for impact parameters $b = 0$, 2, 4, 6 and 8 au.

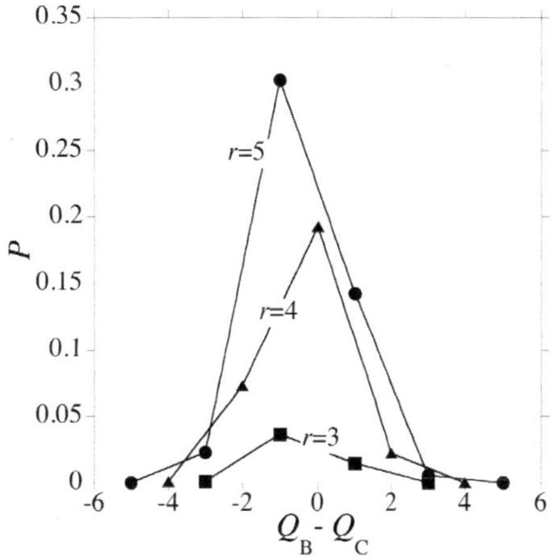


**Figure 4.** Probability distributions over charge-asymmetry $Q_{\rm B} - Q_{\rm C}$ when $r$ electrons are removed at $v = 0.04$ au, $b = 3$ au and $\phi = 0\,$deg.

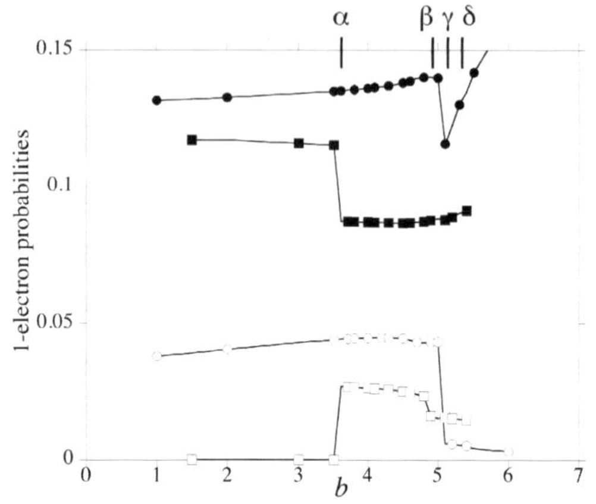


**Figure 5.** Impact parameter $b$ dependence (its direction being $\boldsymbol{b} \parallel \boldsymbol{d}$) of one-electron localization probabilities when $q_{\mathrm{A}} = 8$. Plotted are $p_{\mathrm{B}} + p_{\mathrm{C}}$ for ion-core charges of $(q_{\mathrm{B}}, q_{\mathrm{C}}) = (2,2)$ (closed circles) and the average for (2,1) and (1,2) (closed squares). Also plotted are $p_{\mathrm{B}} - p_{\mathrm{C}}$ for $(q_{\mathrm{B}}, q_{\mathrm{C}}) = (2,2)$ (open circles) and the average for (2,1) and (1,2) (open squares). Vertical bars on the upper horizontal axis indicate the positions of discontinuities, one labelled by $\alpha$ and by $\beta$ coming from (1,2), $\gamma$ from (2,2) and $\delta$ from (2,1).

Time $\tau$ evolution of half the bond length, $d$, is calculated and shown in Fig. 3, where the horizontal scale is taken $z = v\tau$ representing the projectile position along the incident trajectory. In this calculation, we have solved Newton equation by taking a sum of empirical N—N interatomic potential and Coulomb repulsion between dissociating ions with charges of $q_{\mathrm{B}} = q_{\mathrm{C}} = t/2$ after passing the $t$-th critical distance. It is seen from the figure that the molecule dissociates more rapidly as the impact parameter $b$ decreases, because the number, $t_{\mathrm{max}}$, of active electrons increases. In particular, the dissociation velocity changes strikingly in a region of $6 > b > 4$, where $t_{\mathrm{max}}$ increases from 4 to 10.

With these trajectories, we calculate the probability disribution $P(Q_{\mathrm{B}}, Q_{\mathrm{C}})$ through Eqs. (2—6) by assuming that the quasi-molecule decays at $\tau = 500$ au ($z = 20$ au). Figure 4 shows the result when $r = 3$ — 5 electrons are removed from the target BC. As the figure indicates, the distributions are asymmetric with respect to the sign of $Q_{\mathrm{B}} - Q_{\mathrm{C}}$, leading to an asymmetry parameter of $A > 0$ consistent with the measurement [1]. Note here that B (C) is defined as the near (far) site relative to the incident trajectory of A.

To clarify a mechanism causing this asymmetry effect, we examine the sum and the difference, $p_{\mathrm{B}} \pm p_{\mathrm{C}}$, of one-electron localization probabilities into sites B and C. They are plotted in Fig. 5 for ion-cores of $(q_{\mathrm{B}}, q_{\mathrm{C}}) = (2,2)$, (2,1) and (1,2). It is seen from the figure that an inequality, $p_{\mathrm{B}} \geq p_{\mathrm{C}}$, holds universally if one simultaneously considers a contribution from 'ion-core exchange', such as one from $(q_{\mathrm{B}}, q_{\mathrm{C}}) = (1,2)$ in addition to one from $(2,1)$. It is also seen that the localization probabilities are discontinuous with respect to the impact parameter $b$, *i.e.*, with respect to the nuclear configuration at the time of quasi-molecular decay. This behavior is due to alternation of the decay process (see Fig. 1). A discontinuity labelled by $\alpha$ in Fig. 5 arises from alternation as process III $\rightarrow$ II, while $\beta$, $\gamma$ and $\delta$ from I $\rightarrow$ II. As the figure indicates, process I (*i.e.*, nuclear domain $\mathcal{D}_{\mathrm{A-BC}}$) is only responsible for large impact parameters ($b > 5$); it gives vanishingly small values of $p_{\mathrm{B}} - p_{\mathrm{C}}$, which comes from electron polarization in the target molecule by the projectile ion. Hence, process II ($\mathcal{D}_{\mathrm{C-AB}}$) gives a dominant contribution to the asymmetry effect. Note that the process III ($\mathcal{D}_{\mathrm{B-CA}}$) is only responsible for $(q_{\mathrm{B}}, q_{\mathrm{C}}) = (1,2)$ in the figure. It gives negative values of $p_{\mathrm{B}} - p_{\mathrm{C}}$ as a result of rapid bond-elongation at small impact parameters such as leading to $d \gg b$ (see Fig. 3). This negative contribution from process III for $(q_{\mathrm{B}}, q_{\mathrm{C}}) = (1,2)$ cancels a positive contribution from process II for $(2,1)$ at $b < 3.5$.

We have found through the analysis that the present three-center model is explainable for the tendencies of measured charge-asymmetry [1]. The asymmetry effect reflects all the three localization processes illustrated in Fig. 1, among which process II is dominant.

## References

[1] Ehrich M, Werner U, Lutz H O, Kaneyasu T, Ishii K, Okuno K and Saalmann U 2002 *Phys. Rev.* A **65** 030702(R); Kaneyasu T, Azuma T and Okuno K 2005 *J. Phys. B: At. Mol. Opt. Phys.* **38** 1341
[2] Ohyama-Yamaguchi T and Ichimura A 2005 *Nucl. Instr. and Meth.* B **235** 382
[3] Ohyama-Yamaguchi T and Ichimura A 2007 *J. Phys.: Conf. Series* **58** 247
[4] Niehaus A 1986 *J. Phys. B: At. Mol. Opt. Phys.* **19** 2925

# Coster-Kronig electrons from $N^{q+}$ ($q$=1-3) and $O^{q+}$ ($q$=1,2) Rydberg states produced in high-energy collisions with He

**K Kawatsura[1,2], K Takahiro[1], M Sataka[3], M Imai[4], H Sugai[3], K Ozaki[1], K Kawaguchi[1], H Shibata[4] and K Komaki[5]**

[1] Department of Chemistry and Materials Technology, Kyoto Institute of Technology, Sakyo, Kyoto 606-8585, Japan
[2] Theoretical Radiation Research Laboratory, 7-11-401, Shoden-cho, Matsugasaki, Sakyo, Kyoto 606-0966, Japan
[3] Advanced Science Research Center and Nuclear Science Research Institute, Japan Atomic Energy Agency, Tokai, Ibaraki 319-1195, Japan
[4] Department of Nuclear Engineering, Kyoto University, Sakyo, Kyoto 606-8501, Japan
[5] National Center for University Entrance Examination, Meguro, Tokyo 153-8501, Japan

E-mail: nh9h5z@bma.biglobe.ne.jp (K Kawatsura)

**Abstract.** Coster-Kronig (C-K) electrons ejected through autoionization decay of high-Rydberg states in high-energy collisions of $N^{q+}$ ($q$ = 1-3) and $O^{q+}$ ($q$ = 1,2) with He have been measured with high resolution by using zero-degree electron spectroscopy. Autoionizing lines are observed corresponding to decays from $1s^22p(^2P)nl$ Rydberg states for 21 MeV $N^{3+}$ + He, from $1s^22s2p(^3P)nl$ Rydberg states for 21 MeV $N^{2+}$ + He and from $1s^22s2p^2(^4P)nl$ Rydberg states for 14 MeV $N^+$ + He, respectively. Angular momentum distributions of these three series of C-K decaying Rydberg states for $N^{q+}$ ($q$ = 1-3) projectiles are also discussed, where the highly excited states are formed by single-electron excitation. Similarly, series of the autoionizing Rydberg states are observed for two charge states of incident O ions. The C-K spectra observed can be assigned to decays from $1s^22s2p^2(^4P)nl$ states for 30 MeV $O^{2+}$ + He and from $1s^22s2p^3(^5S)nl$ states for 15 MeV $O^+$ + He.

## 1. Introduction

Beryllium is the simplest atom with two closed shells. However, compared to helium it has only been studied marginally. Be-like ions are also interesting to the theorists and experimentalists continuously. Due to their relative simplicity, this four-electron system is an excellent test case for atomic theory as it can be treated in a rigorous manner [1-3]. It is still a labor to study a large amount of various states of Be-like ions [1,4,5]. The Coster-Kronig (C-K) transitions for Be-like N, O, Si, S and Ar ions in high-energy collisions with He-gas and C-foil targets have been widely investigated experimentally and theoretically by our group [6-12]. Furthermore, it is well known that multiply charged ions of low-Z elements such as carbon, nitrogen, and oxygen are important species existing in the terrestrial atmosphere, in the interstellar medium, as well as in the envelopes of many stars. In the present study, we have systematically studied high-resolution C-K electron spectra from high-Rydberg states produced in Be- to C-like $N^{q+}$ projectile ions ($q$ = 1-3) in high-energy collisions with He, where the highly excited states were formed by single-electron excitation. We also present the first measurements for C-K electron spectra from high-Rydberg states produced in high-energy collisions of $O^{q+}$ ($q$ = 1,2) with He for comparison with the collision system of $N^{q+}$ ($q$ = 1-3) + He and for extension of previous measurements for 32 MeV $O^{q+}$ ($q$ = 3,4) + He [11,13].

## 2. Experiment

The experiments were performed at the tandem accelerator facility at the Japan Atomic Energy Agency (JAEA) of Tokai, using the 20 MV tandem accelerator. The experimental setup has been presented before [6] so that only a brief outline is given here. The primary $N^{q+}$ ($q$ = 1-3) and $O^{q+}$ ($q$ = 1,2) ion beams were produced by using ECRIS installed at the high-energy terminal of the tandem accelerator [14], and then accelerated up to 14 MeV for $N^+$ and 21 MeV for $N^{q+}$ ($q$ = 2,3) ions, and 15 MeV for $O^+$ and 30 MeV for $O^{2+}$ ions. The projectile beams passed through a He gas target under single collision conditions. The C-K electrons produced in the target cell were measured under an observation angle of 0° relative to the incident direction with a tandem-type 45° parallel plate electron spectrometer. The beam currents were 0.3~30 nA and were collected in the Faraday cup placed right after the spectrometer. All data were normalized to the same target pressure and ion fluence. The electron spectra presented here are transformed from the laboratory into the projectile rest frame.

## 3. Results and discussion

### *3.1. Angular momentum (l) distribution for Coster-Kronig spectra of Be-, B- and C-like $N^{q+}$ (q = 1-3)*

We have measured for the first time Coster-Kronig electrons ejected form high-Rydberg states produced in high-energy collisions of $N^{q+}$ ($q$ = 1-3) + He [12]. In such high-energy collisions using a He target, the ejected electron spectrum is expected to be pure and simple as the "needle ionization" is realized. There, we have calculated the C-K electron energies by using following equation,

$$E_n = \Delta E - Q^2 R_y / n^2, \tag{1}$$

where $E_n$ is the C-K electron energy, $n$ is the principal quantum number, $\Delta E$ is the energy difference between the initial and final states of the ion core configuration, taken from the compiled transition energy tables [15], $Q$ is the effective charge of the ion core (assumed to be $q$+1 for this case), and $R_y$ is the Rydberg energy (13.606 eV). It has been found that the C-K spectra for $N^{q+}$ projectiles ($q$ = 1-3) show some structures reflecting the $l$ disributions in the low-energy side of the C-K electron energies calculated by using equation (1).

In the present study, the kinetic energies of C-K electrons from initial levels, $1s^22p(^2P)nl$, $1s^22s2p(^3P)nl$, and $1s^22s2p^2(^4P)nl$ are estimated with $l$-dependent version of the equation (1) including the quantum defect correction as following ,

$$E_{nl} = \Delta E - Q^2 R_y / (n - \mu_l)^2, \tag{2}$$

where $E_{nl}$ is the C-K electron energy, $l$ is the angular momentum, and $\mu_l$ is the quantum defect [16]. Figure 1 shows C-K electron spectrum obtained for 21 MeV $N^{3+}$ + He. It is noted that the C-K transitions for $n < 5$ are energetically forbidden. In figure 1, the representative peaks are assigned to a decay series from $1s^22p(^2P)nl$ ($n$ = 5,6) states. For high-energy collisions with gas targets such as $H_2$ and He, single-electron excitation/ionization process is dominant [17]. So, $1s^22p(^2P)nl$ configurations are produced by single-electron excitation of the $1s^22s2p(^3P^o)$ metastable state for Be-like $N^{3+}$ ions, but not by double-electron excitation of the $1s^22s^2(^1S_0)$ ground state. The kinetic energies of $1s^22p(^2P)5l$ and $6l$ C-K electrons obtained using the quantum defect theory, *i.e.*, equation (2) are indicated by the vertical bars in figure 1. The averaged energies of electrons emitted via C-K transitions $1s^22p5l$ – $1s^22s\varepsilon l'$ were calculated to be 0.236, 0.794, 1.233 and 1.298 eV for $l$ = 0, 1, 2 and 3, respectively, by averaging the electron energies of $1s^22p(^2P_{1/2}, {}^2P_{3/2})$ – $1s^22s(^2S_{1/2})$ for $N^{4+}$ over $J$ [15]. These values are qualitatively in agreement with the calculated ones by Glans *et al.* [18]. We obtained the averaged C-K electron energies for $1s^22p6l$ – $1s^22s\varepsilon l'$ to be 3.35, 3.67, 3.92 and 3.96 eV for $l$ = 0, 1, 2 and 3,

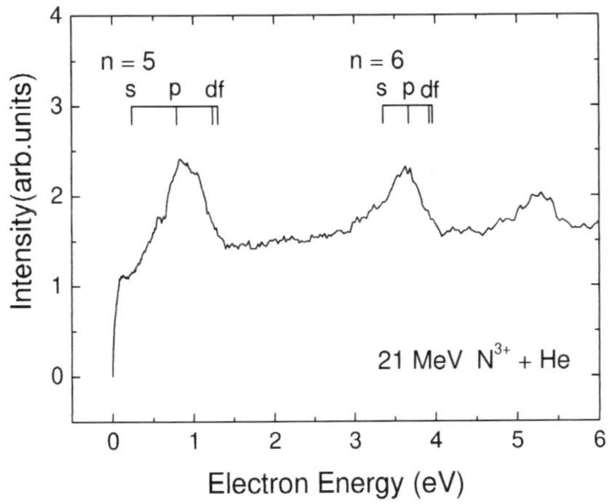


**Figure 1.** Coster-Kronig electron spectrum from $1s^22p(^2P)nl$ states. The vertical bars indicate energy positions calculated by equation (2).

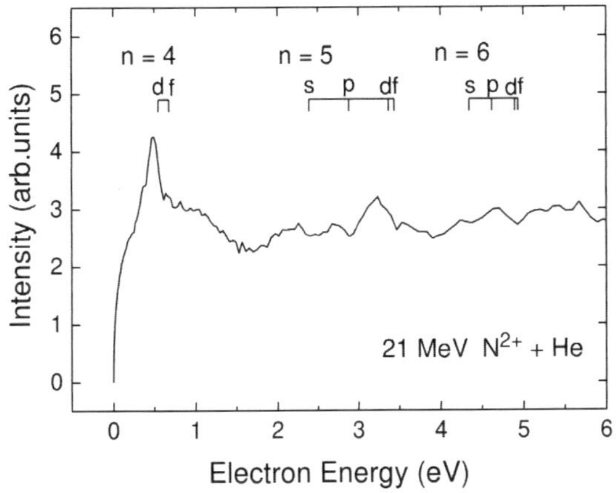


**Figure 2.** Coster-Kronig electron spectrum from $1s^22s2p(^3P)nl$ states. The vertical bars indicate energy positions calculated by equation (2).

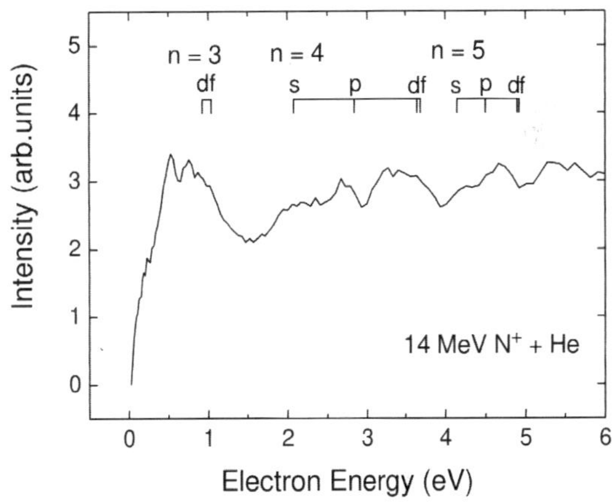


**Figure 3.** Coster-Kronig electron spectrum from $1s^22s2p^2(^4P)nl$ states. The vertical bars indicate energy positions calculated by equation (2).

respectively. By comparing the experimental and calulated results, we can identify the most intense peak to a low angular momentum state of the initial configuration $1s^22pnl$ ($n$ = 5,6) with $l$ = 1, which obeys dipole selection rules. In the previous experiment, we have determined the C-K electron energies of the configurations $1s^22p6l$ for 32 MeV $O^{4+}$ + He collisions and shown that the line due to the low angular momentum $l$ = 1 is the most intense [11]. On the other hand, Meyer *et al.* [19] have investigated the angular momentum ($l$) distribution of the lines associated with the configurations $1s^22pnl$ ($n$ = 6,7) for 30-105 keV $O^{6+}$ + He collisions and found that the lines due to the low angular momenta $l$ = 0 and 1 are rather weak and the maximum intensity is attributed to the high angular momentum states $l \geq 2$. This was pointed out to be a Stark deformation of the Rydberg orbit by the Coulomb field of the receding ion in double electron capture (DEC) by low-energy highly charged ions [20].

Figure 2 shows Coster-Kronig electron spectrum obtained for 21 MeV $N^{2+}$ + He. The transition energies of $1s^22s2p(^3P)nl$ – $1s^22s^2(^1S)\varepsilon l'$ ($n$ = 4-6) C-K electrons are obtained by using equation (2). In the figure, the representative peaks are assigned to decays of $1s^22s2p(^3P)nl$ ($n$ = 4-6) states. It is noted that the C-K transitions for $l < 2$ for $n$ = 4 are energetically forbidden. The averaged energies of electrons emitted via C-K transitions $1s^22s2p4l$ – $1s^22s^2\varepsilon l'$ were determined to be 0.55 and 0.67 eV for $l$ = 2 and 3. Those for $1s^22s2p5l$ – $1s^22s^2\varepsilon l'$ were determined to be 2.39, 2.89, 3.37 and 3.44 eV for $l$ =

0, 1, 2 and 3, respectively. These energies agree qualitatively with the peak positions of 2.35, 2.75 and 3.3 eV, respectively, as seen in figure 2. Those for $1s^2 2s2p6l$ were determined to be 4.34, 4.62, 4.90 and 4.94 eV for $l$ = 0, 1, 2 and 3, respectively. In the case of $N^{2+}$ ions, $1s^2 2s2pnl$ states, which are mainly populated in $n$p and $n$d states, are formed by 2s – $nl$ and 2p – $nl$ transitions from the ground states of $1s^2 2s^2 2p(^2P)$ and metastable states of $1s^2 2s2p^2(^4P)$, respectively. Spectrum in the low-energy region, where the $1s^2 2s2pnl$ ($n$ = 5,6) states contribute, shows that the peaks due to the angular momenta $l \geq 2$ are the most intense. This result for the $l$ distributions is very similar to that for the high-energy collisions of $O^{3+}$ + He [13], but different from those of dielectronic recombination (DR) processes [21].

Figure 3 shows C-K electron spectrum for 14 MeV $N^+$ + He. The C-K electron energies for $1s^2 2s2p^2(^4P)nl$ – $1s^2 2s^2 2p(^2P)\varepsilon l'$ ($n$ = 3-5) are obtained by using equation (2). In the figure, the representative peaks are assigned to decays of $1s^2 2s2p^2(^4P)nl$ ($n$ = 3-5) states. It is noted that the C-K transitions for $n$ < 3 and for $l$ = 0 and 1 of $n$ = 3 are energetically forbidden. Therefore, the line structure of the C-K transitions for $n$ = 3 is rather simple and those for $n$ = 4 and 5 show several structures between 1.5 and 5.0 eV. The averaged energies of electrons emitted via C-K transitions $1s^2 2s2p^2 3l$ – $1s^2 2s^2 2p\varepsilon l'$ were determined to be 0.93 and 1.05 eV for $l$ = 2 and 3 respectively. Those for $1s^2 2s2p^2 4l$ and $5l$ – $1s^2 2s^2 2p\varepsilon l'$ were determined to be 2.09, 2.85, 3.65 and 3.70 eV and 4.15, 4.50, 4.90 and 4.92 eV for $l$ = 0, 1, 2 and 3, respectively, which are in agreement with the line structures of the Rydberg series seen in figure 3. For $N^+$ ions, the spectrum for the $1s^2 2s2p^2 nl$ ($n$ =4,5) states shows that the peaks due to the angular momenta $l \geq 2$ are the most intense, similar to the case for $N^{2+}$ ions.

### *3.2. Coster-Kronig spectra of C- and N-like $O^{q+}$ (q = 1,2)*

Recently we have reported on Coster-Kronig electron spectra in high-energy collisions of 32 MeV $O^{q+}$ ($q$ = 3,4) ions with a He gas target [11,13]. For Be-like $O^{4+}$ ions, a series of $1s^2 2p(^2P)nl$ - $1s^2 2s(^2S)\varepsilon l'$ ($n$ = 6-11) C-K transitions was assigned. The high-resolution spectrum for the $1s^2 2p6l$ state has shown that the peak due to the low angular momentum $l$ = 1 is the most intense, which obeys dipole selection rules. This result for the $l$ distribution has been found to be different from those of the DEC processes [19,22]. For B-like $O^{3+}$ ions, a series of $1s^2 2s2p(^3P)nl$ - $1s^2 2s^2(^1S)\varepsilon l'$ ($n$ = 5-11) C-K transitions was assigned. The sharp lines on both sides of the cusp peak have been confirmed to be due to the very low energy transition from the $1s^2 2s2p(^3P)5s$ state as suggested by Zouros *et al.* [23]. It is also found that states with the angular momenta of $l$ = 0-3 are populated and the maximum intensity is attributed to the states with higher angular momenta $l \geq 2$. The C-K spectra were compared with those for the DR processes [21].

In the present study, we have systematically measured C-K electrons from high-Rydberg states produced in high-energy collisions of 30 MeV $O^{2+}$ + He and 15 MeV $O^+$ + He for comparison with the results for the collision system of $N^{q+}$ ($q$ = 1-3) + He and for extending the investigation of C-K electron spectra past highly charged $O^{q+}$ ($q$ = 3,4) ions. Figure 4 shows high-resolution spectrum of electrons ejected in the collision of 30 MeV $O^{2+}$ ions with a He target. For this collision system, projectile $O^{2+}$ ions have 6 electrons which are more than in the previous experiments for $O^{3+}$ and $O^{4+}$ ions. To our knowledge there is no report on C-K electron spectra for such complicated collision systems. Indeed, figure 4 shows the complicated C-K electron spectrum from $O^{2+}$ $1s^2 2s2p^2 nl$ states and other excited states which were produced by electron excitation/ionization processes during 30 MeV $O^{2+}$ + He collision. Here, we calculate ejected electron energies arising from the possible C-K transitions using equation (1), where $Q$ is the effective charge of the highly exited $O^{2+}$ ion core (assumed to be +3 for this case) and the values of $\Delta E$ are taken from the compiled transition energy tables [15]. In the figure, the representative peaks are assigned to a series of $1s^2 2s2p^2(^4P)nl$ – $1s^2 2s^2 2p(^2P)\varepsilon l'$ ($n \geq 4$) C-K transitions and vertical bars in figure 4 indicate the line positions obtained by equation (1). In this spectrum, the transitions to the $^2P_{1/2}$ and $^2P_{3/2}$ final states are not observed but some structures are recognized at each value of $n$, to which the angular momentum ($l$) distribution should be responsible.

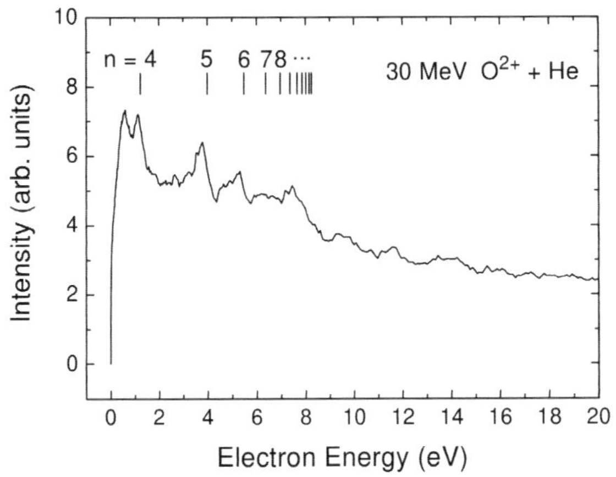


**Figure 4.** Coster-Kronig electron spectrum ejected at 0° in collisions of 30 MeV $O^{2+}$ on He. The vertical bars indicate the positions of the $1s^22s2p^2(^4P)nl$ Rydberg series calculated by equation (1).

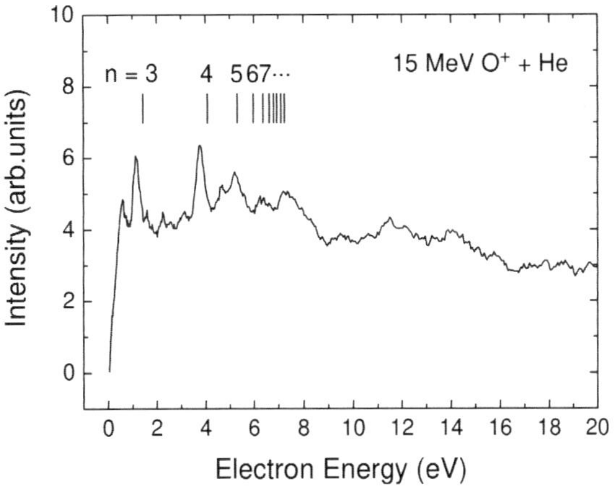


**Figure 5.** Coster-Kronig electron spectrum ejected at 0° in collisions of 15 MeV $O^+$ on He. The vertical bars indicate the positions of the $1s^22s2p^3(^5S)nl$ Rydberg series calculated by equation (1).

Figure 5 shows the measured electron spectrum resulting from the collisions of 15 MeV $O^+$ ions on He. For this collision system, projectile $O^+$ ions have 7 electrons which are even more than the above-mentioned collision system of 30 MeV $O^{2+}$ with a He target. This figure shows the complicated C-K electron spectrum ejected from $O^+$ $1s^22s2p^3nl$ states and other excited states which were produced by electron excitation/ionization processes during 15 MeV $O^+$ + He collision. Here, we calculate ejection energies for the possible C-K transitions by using equation (1), where $Q$ is the effective charge of the highly exited $O^+$ ion core (assumed to be +2 for this case), adopting the values of $\Delta E$ given by Ref. [15]. In the figure, the representative peaks are assigned to a series of $1s^22s2p^3(^5S)nl - 1s^22s^22p^2(^3P)\varepsilon l'$ ($n \geq 3$) C-K transitions and vertical bars indicate line positions obtained by equation (1). In this spectrum, difference among the transitions to final $^3P_0$, $^3P_1$ and $^3P_2$ states is not observed but structures in the peaks for each principal quantum number ($n \geq 3$), up to the series limit of 7.48 eV ($n \rightarrow \infty$), seem to reflect the angular momentum distribution. Moreover, some structures present in the higher energy region between 8.0 and 16 eV could be assigned to a series of $1s^22s2p^3(^3D)nl - 1s^22s^22p^2(^3P)\varepsilon l'$ and/or $1s^22s2p^3(^3P)nl - 1s^22s^22p^2(^3P)\varepsilon l'$ C-K transitions, whose series limits are 14.88 eV and 17.65 eV, respectively.

## 4. Summary

We have performed high-resolution measurements for Coster-Kronig electron spectra from high-Rydberg states produced in collisions of 14-21 MeV $N^{q+}$ ($q$ = 1-3) + He. It is found that the low angular momentum states of the initial configuration $1s^22p(^2P)nl$ ($n$ = 5,6) with $l$ = 1 are highly populated for $N^{3+}$ projectile ions. It shows that the high-Rydberg states are produced by 2s – $n$p electron excitation and these transitions obey dipole selection rules, which agrees with the results for $O^{4+}$ projectile ions [11]. On the other hand, it is found that the states with higher angular momenta are populated in the cases of $1s^22s2p(^3P)nl$ ($n$ = 5,6) for $N^{2+}$, and $1s^22s2p^2(^4P)nl$ ($n$ = 4,5) for $N^+$ ions. High-Rydberg states are possibly produced by 2s – $n$p and 2p – $n$d one-electron dipole excitation for $N^{2+}$ and $N^+$ projectile ions in ground state and metastable state configurations with more than one 2p electron, respectively. The 2p – $n$d transitions could contribute to increase the population of high angular momentum states. It is also suggested that both one-electron excitation processes from the metastable and ground states and decay processes from the high-Rydberg states play an important role in the *l* distribution for C-K electron spectra in high-energy collisions of $N^{q+}$ ($q$ = 1-3) + He.

We also report on the recent efforts to extend the measurements of C-K electron spectra in high-energy collisions of $O^{q+}$ ($q$ = 3,4) + He to 15-30 MeV $O^{q+}$ ($q$ = 1,2) + He. The C-K spectra observed can be assigned to decays from the $1s^22s2p^2(^4P)nl$ states for $O^{2+}$ ion and the $1s^22s2p^3(^5S)nl$ states for $O^+$ ions, respectively.

However, a comprehensive analysis for high-resolution measurements of C-K electron spectra as well as precise theoretical calculations are needed to obtain more detailed information about the production of high-Rydberg states and their decay processes in the high-energy collision regime.

**Acknowledgments**
The authors would like to express their thanks to the staff of the Accelerator division of JAEA for their kind operation and to A Terai and S Oizumi for their assistance during these experiments. This work is supported by the Common Use Program of JAEA.

**References**

[1] Krause M O and Caldwell C D 1987 *Phys. Rev. Lett.* **59** 2736
[2] Kim D. S, Tayal S S, Zhou H-L and Manson S T 2000 *Phys. Rev.* A **61** 062701
[3] Chen M-K 2005 *Nucl. Instrum. Methods* B **235** 165
[4] Hasegawa S *et al.* 2006 *Phys. Rev. Lett.* **97** 023001
[5] Olalde-Velasco P, Mendez-Martinez E, Jimenez-Mier J, Wehlitz R and Whitfield S B 2007 *Phys. Rev.* A **76** 032701
[6] Kawatsura K *et al.* 1990 *Nucl. Instrum. Methods* B **48** 103
[7] Kawatsura K *et al.* 1997 *Nucl. Instrum. Methods* B **124** 381
[8] Sataka M *et al.* 2002 *J. Phys. B: Atom. Mol. Opt. Phys.* **35** 267
[9] Sataka M *et al.* 2002 *Phys. Rev.* A **65** 052704
[10] Imai M *et al.* 2006 *Brazil. J. Phys.* **36** 541
[11] Kawatsura K, Takahiro K, Imai M, Sataka M, Komaki K and Shibata H 2003 *Nucl. Instrum. Methods* B **205** 528
[12] Kawatsura K *et al.* 2008 *Nucl. Instrum. Methods* B (submitted)
[13] Kawatsura K, Takahiro K, Sataka M, Imai M, Komaki K, Sugai H and Shibata H 2006 *Nucl Instrum. Methods* B **245** 44
[14] Matsuda M, Kobayashi C and Takeuchi S 1999 *Heavy Ion Accelerator Technology* ed K W Shepard (New York: AIP) p. 65 and this conference
[15] NIST Atomic Spectra Database: http://www.physics.nist.gov/PhysRefData/ASD/index.html
[16] Theodosiou C D, Inokuchi M and Manson S T 1986 *At. Data Nucl. Data Tables* **35** 473
[17] Sulik B *et al.* 1995 *Phys. Rev.* A **52** 387
[18] Glans P, Lindroth E, Badnell N R, Eklow N, Zong W, Justiniano E and Schuch R, 2001 *Phys. Rev.* A **64**, 043609
[19] Meyer F W, Griffin D C, Havener C C, Huq M S, Phaneuf R A, Swenson J K and Stolterfoht N 1988 *Phys. Rev. Lett.* **60** 1821
[20] Bordenave-Montesquieu A, Moretto-Capelle P, Gonzalez A, Benhenni M, Bachau H and Sanchez I 1994 *J. Phys. B: Atom. Mol. Opt. Phys.* **27** 4243
[21] Fogle M, Badnell N R, Glans P, Loch S D, Madzunkov S, Abdel-Naby Sh A, Pindzola M S and Schuch R 2005 *Astrnom. Astrophys.* **442** 757
[22] Stolterfoht N, Sommer K, Griffin D C, Havener C C, Huq M S, Phaneuf R A, Swenson J K and Meyer F W 1989 *Nucl. Instrum. Methods* B **40/41** 28
[23] Zouros T J M, Toth G, Liao C, Montenegro E C, Hagmann H, Richard P, Grabbe S, Bhalla C P and Wong K L 1995 *Nucl. Instrum. Methods* B **99** 27

# Theoretical predictions of the structure of M-X-ray lines of heavy atoms

**P Matuszak, K Kozioł, M Polasik, K Słabkowska**
Faculty of Chemistry, Nicolaus Copernicus University, 87-100 Toruń, Poland
E-mail: mpolasik@uni.torun.pl

**Abstract.** Extensive multiconfiguration Dirac-Fock calculations with the inclusion of the transverse (Breit) interaction and QED corrections have been carried out on gold and thorium to elucidate the structure of various satellite (additional vacancies in $N$ and/or $O$ shells) and hypersatellite (additional vacancies in $M$ or $M$ and $N$ shells) $M\alpha_{1,2}$ ($M_5N_{6,7}$) and $M\beta_1$ ($M_4N_6$) lines in its X-ray spectra. For every calculated type of $M\alpha_{1,2}$ and $M\beta_1$ lines the theoretical stick spectra (line positions with their relative intensities) have been presented. Moreover for each type of lines two theoretical spectra have been predicted: one being a sum of the Lorentzian natural line shapes and the other one being a convolution of the sum of the Lorentzian natural line shapes with the Gaussian instrumental response. The obtained results will be very helpful in quantitative interpretation of a very complex origin of $M\alpha_{1,2}$ and $M\beta_1$ line structures in various high-resolution X-ray spectra of heavy atoms induced by different light and heavy projectiles.

## 1. Introduction

The multiple ionization of various shells of target atoms in collisions with heavy ions makes the origin of observed $K$-, $L$- and $M$-X-ray lines very complex. Therefore the results of theoretical study [1–13] on the effect of the ionization of various shells on the positions and shapes of particular X-ray lines are absolutely essential for reliable and detailed interpretation of the measured X-ray spectra. The profound importance of such theoretical research has been revealed in numerous cases of the thorough analysis of observed $K$-, $L$- and $M$-X-ray lines [2, 7, 14–28]. In the present work in order to explain the structure of various satellite (additional vacancies in $N$ and/or $O$ shells) and hypersatellite (additional vacancies in $M$ or $M$ and $N$ shells) $M\alpha_{1,2}$ and $M\beta_1$ lines in the X-ray spectra of heavy atoms the extensive multiconfiguration Dirac-Fock (MCDF) calculations have been carried out on gold and thorium.

## 2. MCDF calculations

The methodology of MCDF calculations performed in the presented studies is similar to the published earlier in many papers (e.g. [5, 29, 30]). The effective Hamiltonian for an N-electron system is expressed by

$$H = \sum_{i=1}^{N} h_D(i) + \sum_{j>i=1}^{N} C_{ij}, \tag{1}$$

where $h_D(i)$ is the Dirac operator for $i$-th electron and the terms $C_{ij}$ account for electron-electron interactions. The latter are a sum of the Coulomb interaction operator and the transverse Breit operator. An atomic state function (ASF) with the total angular momentum $J$ and parity $p$ is assumed in the form

$$\Psi_s(J^p) = \sum_m c_m(s)\Phi(\gamma_m J^p), \tag{2}$$

where $\Phi(\gamma_m J^p)$ are configuration state functions (CSF), $c_m(s)$ are the configuration mixing coefficients for state $s$, $\gamma_m$ represents all information required to uniquely define a certain CSF. In this work the modified special average-level version of MCDF calculations, MCDF-MSAL [5], has been used. Apart from the transverse Breit interaction two types of QED corrections (self-energy and vacuum polarization) have been included.

## 3. Results and discussion

In Table 1 the number of transitions for each considered transition type and the theoretical relative "average" positions of each group of satellite and hypersatellite $M\alpha_{1,2}$ and $M\beta_1$ lines (with respect to $M\alpha_{1,2}$ and $M\beta_1$ diagram lines, denoted $3d^{-1} \to 4f^{-1}$) have been given for gold (Z=79) and thorium (Z=90). One can see that the hypersatellite lines have a substantial average energy shift (in the range of 35-43 eV for gold and of 42-53 eV for thorium), which is the biggest when the hole is in the $3d$ subshell, the second biggest when it is in the $3s$ subshell, whereas the smallest shifts are observed for a $3p$ hole. For the $N$-shell satellite lines the effect is much smaller (4.6-7.7 eV for gold and 7.5-10.4 eV for thorium) and for the $O$-shell satellite lines it is very small (0.1-0.4 eV for gold and 0.7-1.2 eV for thorium).

**Table 1.** The number of transitions for each transition type and the theoretical relative "average" positions of each groups of satellite and hypersatellite $M\alpha_{1,2}$ and $M\beta_1$ lines (with respect to $M\alpha_{1,2}$ and $M\beta_1$ diagram lines) for gold and thorium.

| Transition | The number | Energy shifts (eV) | | | |
|---|---|---|---|---|---|
| | | $M\alpha_{1,2}$ | | $M\beta_1$ | |
| type | of transitions | Au | Th | Au | Th |
| $3d^{-1} \to 4f^{-1}$ | 3 | 0.0 | 0.0 | 0.0 | 0.0 |
| $(3s3d)^{-1} \to (3s4f)^{-1}$ | 11 | 35.7 | 43.5 | 37.6 | 45.5 |
| $(3p3d)^{-1} \to (3p4f)^{-1}$ | 83 | 35.0 | 41.9 | 36.4 | 43.1 |
| $3d^{-2} \to (3d4f)^{-1}$ | 81 | 42.9 | 51.1 | 42.0 | 53.5 |
| M-shell hypersatellites | 175 | 39.2 | 47.0 | 39.5 | 48.8 |
| $(3d4s)^{-1} \to (4s4f)^{-1}$ | 11 | 7.7 | 10.4 | 7.6 | 10.3 |
| $(3d4p)^{-1} \to (4p4f)^{-1}$ | 83 | 6.9 | 9.2 | 6.9 | 9.3 |
| $(3d4d)^{-1} \to (4d4f)^{-1}$ | 171 | 6.8 | 9.2 | 6.1 | 8.7 |
| $(3d4f)^{-1} \to 4f^{-2}$ | 108 | 5.1 | 8.3 | 4.6 | 7.5 |
| N-shell satellites | 373 | 6.2 | 8.9 | 5.7 | 8.4 |
| $(3d5s)^{-1} \to (4f5s)^{-1}$ | 11 | 0.4 | 1.2 | 0.4 | 1.2 |
| $(3d5p)^{-1} \to (4f5p)^{-1}$ | 83 | 0.3 | 1.0 | 0.3 | 1.0 |
| $(3d5d)^{-1} \to (4f5d)^{-1}$ | 171 | 0.1 | 0.7 | 0.1 | 0.7 |
| O-shell satellites | 265 | 0.2 | 0.8 | 0.2 | 0.9 |

In Figs. 1-3, both stick (calculated line positions with their relative intensities) and synthesized spectra for all considered types of the transitions of gold are presented together with the summary spectra (analogical presentation of theoretical spectra for thorium has been already published in [12]). Two synthesized spectra for each case have been constructed: one (dotted lines) being the sum of the Lorentzian natural line shapes with the appropriate natural linewidth [31] and, to better simulate the experimental spectra, the other one (solid lines) being a convolution of the sum of the Lorentzian natural line shapes with the Gaussian instrumental response having a width of 1.0 eV [25]. For the better observation of the differences between shapes of two types of synthesized spectra the bigger peak height for convoluted than non-convoluted spectra has been assumed.

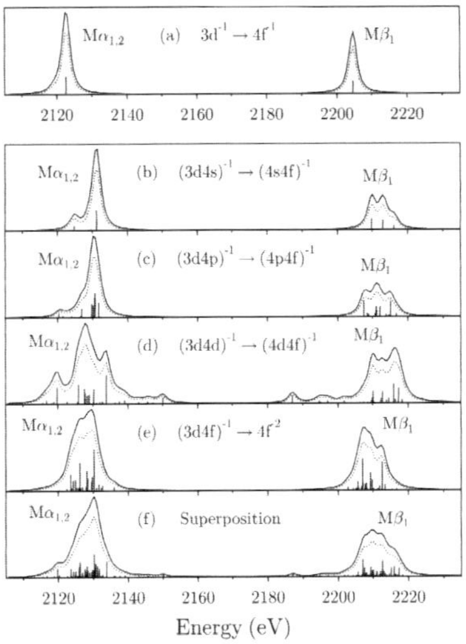


**Figure 1.** Calculated stick and predicted spectra for the $M\alpha_{1,2}$ and $M\beta_1$ diagram [spectrum (a)] and $N$-shell satellite transitions [(b)-(f)] in gold. Spectrum (f) is the summary spectrum [(b)+ (c)+(d)+(e)].

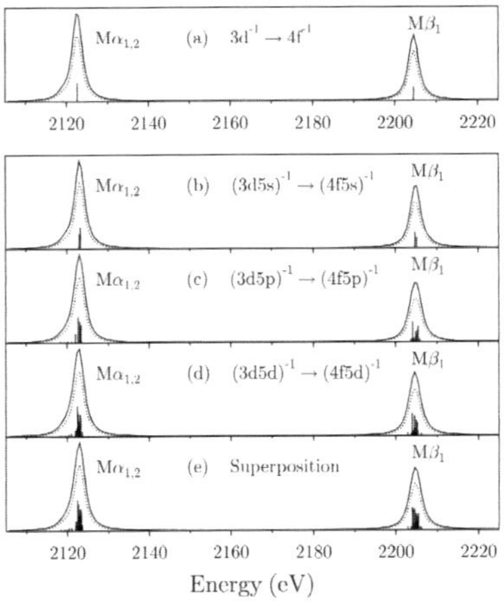


**Figure 2.** As in Figure 1, but for the $M\alpha_{1,2}$ and $M\beta_1$ diagram and $O$-shell satellite transitions [(b)-(e)] for gold. Spectrum (e) is the summary spectrum [(b)+ (c)+(d)].

It can be observed from Figs. 1-3 that for each discussed case the structures of stick spectra for the $M\alpha_{1,2}$ and for the $M\beta_1$ satellite and hypersatellite lines of gold are different. In the case of the $O$-shell satellite lines (Fig. 2) the sticks are close to each other and the shapes of all the synthesized spectra (both the sum of the Lorentzian natural line shapes and the convolution of a sum of the Lorentzian line shapes with the Gaussian instrumental response) are similar to the shapes of the corresponding diagram lines (though the satellite bands are of course wider than the diagram ones).

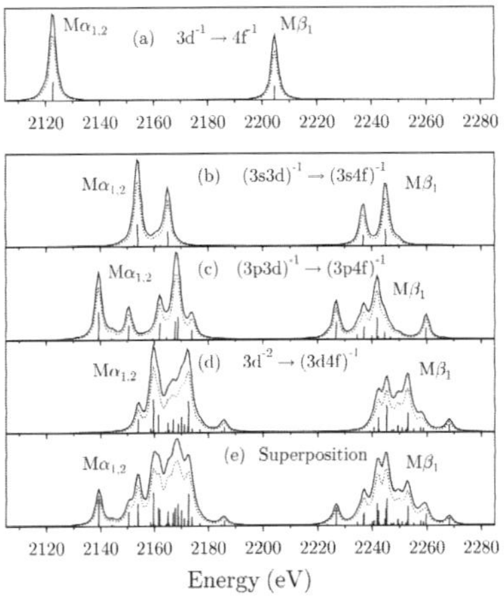


**Figure 3.** As in Figure 1 and Figure 2, but for the $M\alpha_{1,2}$ and $M\beta_1$ diagram and $M$-shell hypersatellite transitions [(b)-(e)] for gold. Spectrum (e) is the summary spectrum [(b)+ (c)+(d)].

For the $N$-shell satellite lines (Fig. 1) the sticks are noticeably scattered in the rage of 40 eV and the corresponding stick spectra and both kinds of the synthesized spectra differ considerably from each other. In the case of the $M$-shell hypersatellite lines (Fig. 3) stick spectra are widely distributed in the range of 60 eV and both kinds of the synthesized spectra are mostly rich in structure. Generally it can be noticed that the shapes of the predicted $M\alpha_{1,2}$ and $M\beta_1$ lines of gold can be quite well represented as single Voigt functions only in the case of the $O$-shell satellite transitions.

The obtained theoretical results will be very helpful in reliable and quantitative interpretation of the experimental satellite and hypersatellite $M\alpha_{1,2}$ and $M\beta_1$ lines of heavy atoms such as gold and thorium bombarded by different light and heavy projectiles. It is worth noting that the preliminary results of this and previous [12] studies enabled for the first time in case of heavy atoms the identification and quantitative interpretation of the $N$-shell satellite as well as the $M$-shell hypersatellite in the observed $M\alpha_{1,2}$ X-ray lines of thorium excited by an impact of energetic oxygen and neon ions measured by means of a high-resolution von Hamos crystal spectrometer having the energy resolution of about 1 eV [25].

## Acknowledgments
This work was supported by the Polish State Committee for Scientific Research under Grant No. N N202 1465 33.

[1] Polasik M 1989 *Phys. Rev.* A **39** 616-627
[2] Polasik M 1989 *Phys. Rev.* A **39** 5092-5097
[3] Polasik M 1989 *Phys. Rev.* A **40** 4361-4368
[4] Polasik M 1990 *Phys. Rev.* A **41** 3689-3697
[5] Polasik M 1995 *Phys. Rev.* A **52** 227-235
[6] Polasik M 1998 *Phys. Rev.* A **58** 1840-1845
[7] Polasik M *et al.* 1999 *J. Phys. B: At. Mol. Opt. Phys.* **32** 3711-3725
[8] Polasik M and Lewandowska-Robak M 2004 *Phys. Rev.* A **70** 052502-1-15
[9] Polasik M and Lewandowska-Robak M 2005 *J. Phys. B: At. Mol. Opt. Phys.* **38** 2407-2425
[10] Polasik M and Lewandowska-Robak M 2006 *J. Phys. B: At. Mol. Opt. Phys.* **39** 1169-1186
[11] Słabkowska K and Polasik M 2006 *Radiat. Phys. Chem.* **75** 1471-1476
[12] Polasik M and Słabkowska K 2006 *Radiat. Phys. Chem.* **75** 1497-1502
[13] Słabkowska K and Polasik M 2007 *J. Phys.: Conf. Ser.* **58** 263-266
[14] Carlen M W *et al* 1992 *Phys. Rev.* A **46** 3893-3903
[15] Carlen M W *et al* 1994 *Phys. Rev.* A **49** 2524-2534
[16] Ludziejewski T *et al* 1995 *Phys. Rev.* A **52** 2791-2803
[17] Ludziejewski T *et al* 1996 *Phys. Rev.* A **54** 232-240
[18] Herren Ch *et al* 1998 *Phys. Rev.* A **57** 235-245
[19] Mauron O *et al* 2000 *Phys. Rev.* A **62** 062508-1-11
[20] Kravchuk V L *et al* 2001 *Phys. Rev.* A **64** 062710/1-6
[21] Raboud P-A *et al* 2002 *Phys. Rev.* A **65** 062503-1-8
[22] Majewska U 2002 *J. Phys.* B **35** 1941-1957
[23] Rzadkiewicz J *et al* 2003 *Phys. Rev.* A **68** 032713/1-14
[24] Czarnota M *et al* 2003 *Nucl. Instr. and Meth. in Phys. Res.* B **205** 133-138
[25] Czarnota M *et al* 2003 *Radiat. Phys. Chem.* **68** 121-125
[26] Braziewicz J *et al* 2004 *Phys. Rev.* A **69** 0627105-1-10
[27] Czarnota M *et al* 2006 *Braz. J. Phys.* **36** 546-549
[28] Rzadkiewicz J *et al* 2007 *High Energy Density Physics* **3** 233-236
[29] Grant I P 1984 *Int. J. Quantum Chem.* **25** 23-46
[30] Dyall K G, Grant I P, Johnson C T, Parpia F A and Plummer E P 1989 *Comput. Phys. Commun.* **55** 425-456
[31] Campbell J L and Papp T 2001 *At. Data Nucl. Data Tables* **77** 1-56

# Influence of changes in the valence electronic configuration on the structure of L-X-ray spectra of molybdenum

**M Polasik[1], K Kozioł[1], K Słabkowska[1], M Czarnota[2], M Pajek[2]**

[1] Faculty of Chemistry, Nicolaus Copernicus University, 87-100 Toruń, Poland
[2] Institute of Physics, Jan Kochanowski University, 25-406 Kielce, Poland

E-mail: mpolasik@uni.torun.pl

**Abstract.** Extensive multiconfiguration Dirac-Fock (MCDF) calculations with the inclusion of the transverse (Breit) interaction and QED corrections have been carried out on molybdenum to explain the dependence of the structure of $L\alpha_{1,2}$ and $L\beta_1$ lines on the changes in configurations of the valence electrons belonging to two different configuration types: three open-shell $4d^{6-r}5s^r$ (r=2,1,0) configurations and one closed-shell $4d^4_{3/2}\,5s^2$ configuration. It has been found that the MCDF predictions for open-shell valence configurations ($4d^4\,5s^2$, $4d^5\,5s^1$, $4d^6\,5s^0$) much better reproduce observed structure of $L\alpha_{1,2}$ lines in X-ray spectra of molybdenum than closed-shell $4d^4_{3/2}\,5s^2$ valence configuration. The influence of changes in the valence electronic configuration on the structure of $L$-X-ray spectra of molybdenum is noticeable. Moreover, the observation of the shapes of $L$-X-ray spectra seems to be very good method to investigate the changes of the valence electronic configuration caused by the chemical environment.

## 1. Introduction

Numerous studies based on the interpretation of the $K$-, $L$- and $M$-X-ray spectra [1–15] have been performed in order to investigate the ionization in collision processes. The multiple ionization of various shells of target atoms makes the origin of the measured X-ray spectra very complex. In order to obtain from these X-ray spectra the information about the $L$- or $M$-shell ionization probability it is vital to make use of the results of theoretical calculations [16–28] of the positions and intensities of various X-ray lines.

Recently the $L\alpha_{1,2}$ ($L_3M_{4,5}$) and $L\beta_1$ ($L_2M_4$) lines in the X-ray spectra of molybdenum induced by X-ray photons, 279-MeV oxygen ion and 178-MeV neon ion beams [12] have been measured by means of a high-resolution von Hamos crystal spectrometer having the energy resolution of about 1 eV. The structure of the observed spectrum [12] induced by X-ray photons (especially specific shape of $L\alpha_{1,2}$ diagram line) indicates that the reliable interpretation of $L$-X-ray spectra of molybdenum requires detailed theoretical studies of the influence of changes in the valence electronic configuration on its spectra structure. Therefore in the present work the structures of the $L\alpha_{1,2}$ and $L\beta_1$ X-ray spectra for various valence electronic configurations of molybdenum, predicted using multiconfiguration Dirac-Fock (MCDF) method [16–20], have been presented and analyzed with regards to their parameters like shapes and energy shifts.

## 2. Theoretical background

The methodology of MCDF calculations performed in the presented studies is similar to the published earlier, in many papers (e.g. [20, 29, 30]). The effective Hamiltonian for an N-electron system is expressed by

$$H = \sum_{i=1}^{N} h_D(i) + \sum_{j>i=1}^{N} C_{ij}, \tag{1}$$

where $h_D(i)$ is the Dirac operator for $i$-th electron and the terms $C_{ij}$ account for electron-electron interactions. The latter are a sum of the Coulomb interaction operator and the transverse Breit operator. An atomic state function (ASF) with the total angular momentum $J$ and parity $p$ is assumed in the form

$$\Psi_s(J^p) = \sum_m c_m(s)\Phi(\gamma_m J^p), \tag{2}$$

where $\Phi(\gamma_m J^p)$ are configuration state functions (CSF), $c_m(s)$ are the configuration mixing coefficients for state $s$, $\gamma_m$ represents all information required to uniquely define a certain CSF. In this work the modified special average-level version of MCDF calculations, MCDF-MSAL [20], has been used. Apart from the transverse Breit interaction two types of quantum electrodynamics (QED) corrections (self-energy and vacuum polarization) have been included.

## 3. Results and discussion

In the present work in order to explain the dependence of the structure of $L\alpha_{1,2}$ and $L\beta_1$ lines on the changes in the valence electronic configurations of the extensive MCDF calculations with the inclusion of the transverse (Breit) interaction and QED corrections have been carried out on molybdenum for two different configuration types: three open-shell $4d^{6-r}5s^r$ (r=2,1,0) configurations and one closed-shell $4d^4_{3/2}\ 5s^2$ configuration. The results of calculations of $L\alpha_{1,2}$ and $L\beta_1$ line positions and $L\alpha_{1,2}$ and $L\beta_1$ energy shifts for these valence configurations of molybdenum have been given in Table 1.

**Table 1.** Comparision of $L\alpha_{1,2}$ and $L\beta_1$ line positions (in eV) for four valence configurations. The $4d^5\ 5s^1$ configuration have been taken as a reference.

| Valence electronic configuration | The number of transitions | $L\alpha_{1,2}$ transition energy | $L\beta_1$ transition energy | $L\alpha_{1,2}$ energy shift | $L\beta_1$ energy shift |
|---|---|---|---|---|---|
| $4d^4\ 5s^2$ | 22426 | 2292.26 | 2394.67 | 0.09 | 0.15 |
| $4d^5\ 5s^1$ | 120395 | 2292.17 | 2394.52 | 0 | 0 |
| $4d^6\ 5s^0$ | 22257 | 2292.13 | 2394.47 | -0.04 | -0.05 |
| $4d^4_{3/2}\ 5s^2$ | 3 | 2292.36 | 2394.00 | 0.19 | -0.52 |

Generally, in the presented cases the number of transitions depends very strongly on the configuration type (see Table 1). For closed-shell $4d^4_{3/2}\ 5s^2$ valence configuration there are only 2 initial states of the type $2p^{-1}$, 2 final states of the type $3d^{-1}$ and only 3 transitions. For all open-shell valence configurations the number of transitions is much larger than for the closed-shell configuration case. The most complicated case is for the $4d^5 5s^1$ configuration, in which occupation of $4d$ shell and $5s$ shell is half-and-half. For this case there are 417 initial states, 647 final states and 120395 transitions. From Table 1 it can be seen that the position of $L\alpha_{1,2}$ and $L\beta_1$ X-ray lines are different for different valence configurations. However, these differences are small and it seems to be difficult to detect this effect experimentally.

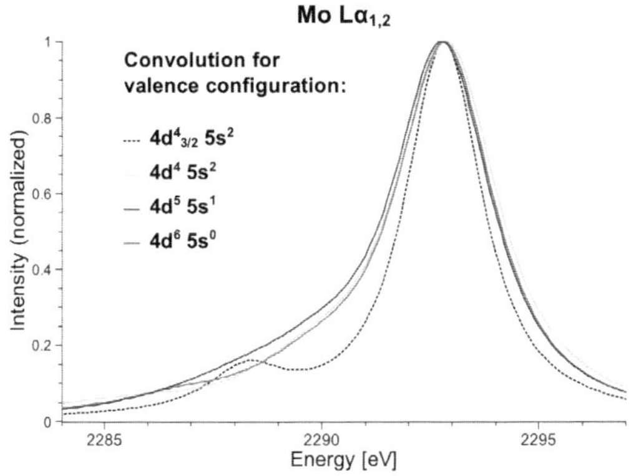


**Figure 1.** Influence of changes in the valence configuration on the structure of $L\alpha_{1,2}$ X-ray spectra of molybdenum.

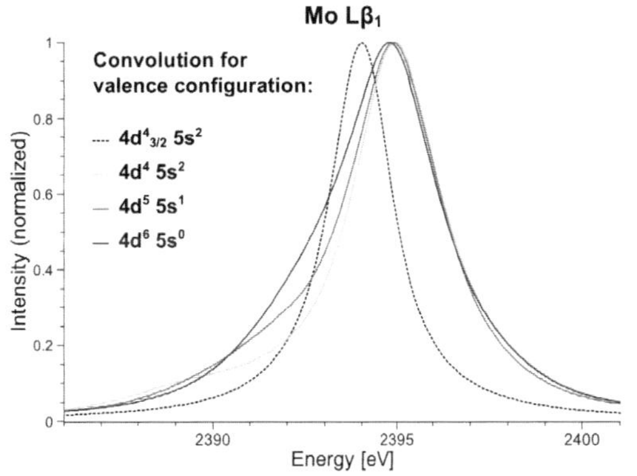


**Figure 2.** Influence of changes in the valence configuration on the structure of $L\beta_1$ X-ray spectra of molybdenum.

The theoretical simulations for four valence configurations (open-shell: $4d^4\ 5s^2$, $4d^5\ 5s^1$, $4d^6\ 5s^0$ and closed-shell – $4d^4_{3/2}\ 5s^2$) of molybdenum have been shown in Figure 1 (for $L\alpha_{1,2}$ X-ray lines) and Figure 2 (for $L\beta_1$ X-ray lines). The lines represent convolution of sum of the Lorentzian line shapes [31] with the Gaussian instrumental response. As it can be observed in Figures 1 and 2, the change of shapes of the $L\alpha_{1,2}$ and $L\beta_1$ X-ray lines for different valence configurations is distinct. In the case of $L\alpha_{1,2}$ lines larger changes are observed for energy range of 2284-2290 eV (due to $L\alpha_2$ X-ray line). Moreover, it has been found that the MCDF predictions for open-shell valence configurations ($4d^4\ 5s^2$, $4d^5\ 5s^1$, $4d^6\ 5s^0$) much better reproduce observed structure of $L\alpha_{1,2}$ lines in X-ray spectra of molybdenum [12] than closed-shell $4d^4_{3/2}\ 5s^2$ valence configuration. From Figure 2, it can be seen that the larger changes of shapes $L\beta_1$ line are observed for energy range of 2387-2394 eV.

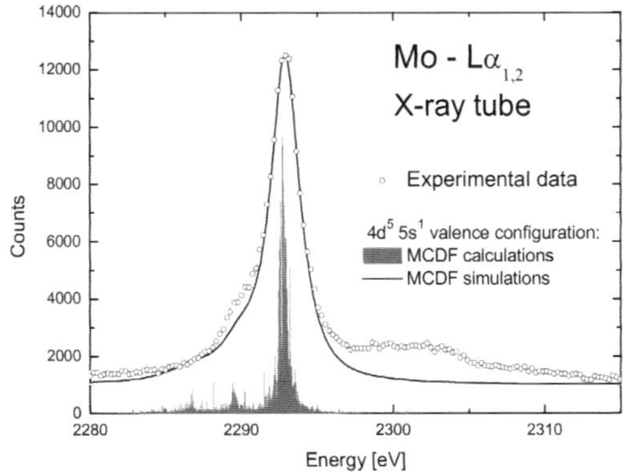


**Figure 3.** Experimental $L$-X-ray spectrum of molybdenum and the MCDF predictions for $4d^5 5s^1$ valence configuration.

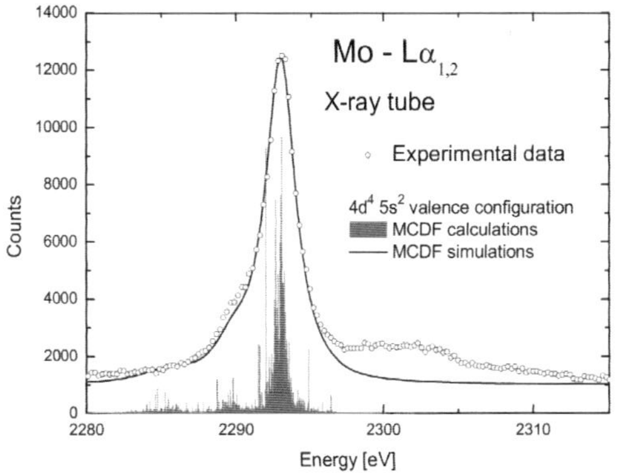


**Figure 4.** Experimental $L$-X-ray spectrum of molybdenum and the MCDF predictions for $4d^4 5s^2$ valence configuration.

Comparisons of the MCDF predictions with the observed structure of $L\alpha_{1,2}$ lines in the X-ray spectrum of molybdenum induced by X-ray photons [12] have been shown in Figure 3 (theoretical predictions for $4d^5 5s^1$ configuration) and in Figure 4 (theoretical predictions for $4d^4 5s^2$ configuration). As it can be seen the MCDF predictions for both open-shell valence configurations ($4d^4\ 5s^2$ and $4d^5\ 5s^1$) reproduce very well the observed structure of $L\alpha_{1,2}$ diagram lines in X-ray spectra of molybdenum. It is important to note that the bump for energy range of 2297-2310 eV originates from $M$-shell satellite $L\alpha_{1,2}$ transitions not considered in this work.

## 4. Summary and conclusions

In order to explain the dependence of the structure of $L\alpha_{1,2}$ and $L\beta_1$ lines on the changes in configurations of the valence electrons the extensive MCDF calculations with the inclusion of the transverse (Breit) interaction and QED corrections have been carried out on molybdenum for two different configuration types [three open-shell $4d^{6-r}5s^r$ (r=2,1,0) configurations and one closed-shell $4d^4_{3/2}$ $5s^2$ configuration]. It has been found that the changes of shapes of the $L\alpha_{1,2}$ and $L\beta_1$ X-ray lines for different valence configurations are discernible. Moreover, the MCDF predictions for open-shell valence configurations ($4d^4$ $5s^2$, $4d^5$ $5s^1$, $4d^6$ $5s^0$) much better reproduce observed shape of $L\alpha_{1,2}$ diagram lines of molybdenum [12] than closed-shell $4d^4_{3/2}$ $5s^2$ valence configuration. The obtained results are very helpful in reliable and quantitative interpretation of the structure of $L\alpha_{1,2}$ and $L\beta_1$ lines having a very complex origin, in various high-resolution X-ray spectra of $4d$ transition-metals induced by X-ray photons and different light and heavy projectiles. Moreover, the observation of the shapes of $L$-X-ray spectra seems to be very good method to investigate electronic valence structure, especially for the study of the influence of changes in the valence electronic configuration caused by the chemical environment [32].

### Acknowledgments

This work was supported by the Polish State Committee for Scientific Research under Grant No. N N202 1465 33.

[1] Carlen M W *et al* 1992 *Phys. Rev.* A **46** 3893-3903
[2] Carlen M W *et al* 1994 *Phys. Rev.* A **49** 2524-2534
[3] Ludziejewski T *et al* 1995 *Phys. Rev.* A **52** 2791-2803
[4] Ludziejewski T *et al* 1996 *Phys. Rev.* A **54** 232-240
[5] Herren Ch *et al* 1998 *Phys. Rev.* A **57** 235-245
[6] Mauron O *et al* 2000 *Phys. Rev.* A **62** 062508-1-11
[7] Kravchuk V L *et al* 2001 *Phys. Rev.* A **64** 062710/1-6
[8] Raboud P-A *et al* 2002 *Phys. Rev.* A **65** 062503-1-8
[9] Braziewicz J *et al* 2004 *Phys. Rev.* A **69** 0627105-1-10
[10] Majewska U 2002 *J. Phys.* B **35** 1941-1957
[11] Rzadkiewicz J *et al* 2003 *Phys. Rev.* A **68** 032713/1-14
[12] Czarnota M *et al* 2003 *Nucl. Instr. and Meth. in Phys. Res.* B **205** 133-138
[13] Czarnota M *et al* 2003 *Radiat. Phys. Chem.* **68** 121-125
[14] Czarnota M *et al* 2006 *Braz. J. Phys.* **36** 546-549
[15] Rzadkiewicz J *et al* 2007 *High Energy Density Physics* **3** 233-236
[16] Polasik M 1989 *Phys. Rev.* A **39** 616-627
[17] Polasik M 1989 *Phys. Rev.* A **39** 5092-5097
[18] Polasik M 1989 *Phys. Rev.* A **40** 4361-4368
[19] Polasik M 1990 *Phys. Rev.* A **41** 3689-3697
[20] Polasik M 1995 *Phys. Rev.* A **52** 227-235
[21] Polasik M 1998 *Phys. Rev.* A **58** 1840-1845
[22] Polasik M *et al.* 1999 *J. Phys. B: At. Mol. Opt. Phys.* **32** 3711-3725
[23] Polasik M and Lewandowska-Robak M 2004 *Phys. Rev.* A **70** 052502-1-15
[24] Polasik M and Lewandowska-Robak M 2005 *J. Phys. B: At. Mol. Opt. Phys.* **38** 2407-2425
[25] Polasik M and Lewandowska-Robak M 2006 *J. Phys. B: At. Mol. Opt. Phys.* **39** 1169-1186
[26] Słabkowska K and Polasik M 2006 *Radiat. Phys. Chem.* **75** 1471-1476
[27] Polasik M and Słabkowska K 2006 *Radiat. Phys. Chem.* **75** 1497-1502
[28] Słabkowska K and Polasik M 2007 *J. Phys.: Conf. Ser.* **58** 263-266
[29] Grant I P 1984 *Int. J. Quantum Chem.* **25** 23-46
[30] Dyall K G, Grant I P, Johnson C T, Parpia F A and Plummer E P 1989 *Comput. Phys. Commun.* **55** 425-456
[31] Campbell J L and Papp T 2001 *At. Data Nucl. Data Tables* **77** 1-56
[32] Liu Z, Yuge K, Kawai J 2004 *Spectrochimica Acta Part B* **59** 93-99

# X-ray signatures of charge exchange in L-shell iron and sulfur

**M Frankel[1,2], P Beiersdorfer[1,3], G V Brown [1], J Clementson[1,2], M F Gu[1,3], L Schweikhard[4]**

[1]Lawrence Livermore National Laboratory, Livermore, CA 94551, USA
[2]Physics Department, Lund University, SE-221 00 Lund, Sweden
[3]Space Science Laboratory, University of California, Berkeley, 94720, USA
[4]Institut für Physik, Ernst-Moritz-Arndt Universität, D-17487 Greifswald, Germany

E-mail: frankel4@llnl.gov

**Abstract.** The X-ray signatures of L-shell charge exchange in sulfur and iron were studied in the laboratory. Charge states from $S^{11+}$ to $S^{14+}$ and $Fe^{18+}$ to $Fe^{24+}$ were created in electron beam ion traps (EBITs) and were left to interact via charge exchange with neutral gases. The measurements were monitored with a high-resolution microcalorimeter spectrometer for sulfur and a moderate-resolution SiLi solid-state detector for iron. Comparison of the charge exchange (CX) spectra with those obtained under electron-impact excitation showed marked differences. In particular, we show that CX shifts the centroid energy of the dominant $n = 3 \rightarrow n = 2$ emission. We explain this by an enhancement of the intensity of the 3s $\rightarrow$ 2p transitions and a shift towards lower energy of the 3d $\rightarrow$ 2p peak.

## 1. Introduction

Laboratory L-shell charge exchange (CX) emission data are important for understanding CX as an X-ray production mechanism. CX emission has proven to be of great importance in the solar wind interaction with both cometary comae and planetary atmospheres [1]. However, X-ray emission from CX is poorly understood. Assumptions made in the early X-ray emission models of the K-shell CX emission were shown to be inadequate when compared to laboratory data [2]. Even more updated, sophisticated models [3, 4, 5, 6] can still not fully reproduce many of the important features in K-shell CX emission [7].

L-shell CX emission is even harder to model than K-shell CX emission, and less laboratory data exist. Only EUV emission from low-Z ions (C, N, O ) [8, 9, 10] and some X-ray spectra from high-Z ions (Kr, U) [11, 12] have been measured. Laboratory data of L-shell CX are essential in order to guide the development of theory [13].

We here present two recent measurements of L-shell CX emission: moderate-resolution results from the L-shell iron ions, $Fe^{17+}$ to $Fe^{23+}$; and high-resolution results from the L-shell sulfur ions, $S^{10+}$ to $S^{13+}$.

## 2. Experimental setup

CX emission from L-shell iron was created on EBIT-II [14] at the Livermore EBIT facility. The sulfur CX measurements were performed on SuperEBIT [14], also at Livermore. The highly charged ions were produced, excited and trapped using the electron-trapping mode [15]. When the charge-state equilibria had been reached, the beam was turned off and the ions could CX with neutral gas in the magnetic-trapping mode [15]. A typical timing pattern for CX, which was used in the iron measurement, is shown in Fig. 1. In the iron measurement the neutral targets were background gases in EBIT, such as $O_2$, $N_2$, $H_2O$, and $CO_2$ [16]. In the sulfur measurement, most of the CX reactions were with neutral $SF_6$ gas, which was continuously leaked into the trap. However, there is always background gases present in the trap and there could also be some molecular fragments of $SF_6$. These gases could act as neutral targets in the CX reactions and there is no way for us to distinguish whether the CX is with $SF_6$ or with background gases. The iron and sulfur measurements, respectively, were monitored with a moderate-resolution SiLi solid-state detector (resolving power $E/\Delta E \approx 10$, at around 2 keV) [16] and a high-resolution microcalorimeter spectrometer (resolving power $E/\Delta E \approx 80$, at around 400 eV) [17]. The interacting iron ions $Fe^{18+}$ to $Fe^{24+}$ were produced at electron beam energies between 1.65 keV and 2.45 keV and the sulfur ions $S^{11+}$ to $S^{14+}$ were produced at 3.09 keV.

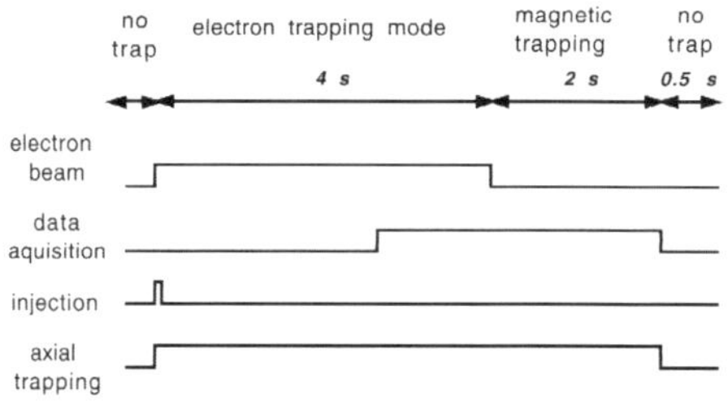


**Figure 1.** Experimental timing sequence for the L-shell iron measurement.

## 3. Measurement of iron L-shell charge exchange

In the electron-impact excitation spectra, the strongest feature is the peak due to the $n = 3 \rightarrow n = 2$ transitions. At electron beam energies of 1.65 keV and 1.85 keV, this peak can be observed at 990 eV and 1025 eV, respectively. The corresponding energies for the $n = 4 \rightarrow n = 2$ peak are 1290 eV and 1380 eV, respectively. Also the $n = 5 \rightarrow n = 2$ peak is visible but is significantly weaker than the $n = 4 \rightarrow n = 2$ peak.

Whereas for direct excitation the peak intensity decreases as a function of $n$, CX shows a different pattern. The intensity observed with the SiLi detector at the location of the $n = 4 \rightarrow n = 2$ transitions is strongly enhanced compared to the $n = 3 \rightarrow n = 2$ peak. The $n = 5 \rightarrow n = 2$ peak is of similar intensity as the $n = 4 \rightarrow n = 2$ peak, and hence shows an even greater enhancement. The unresolved CX emission from the levels of electron captures, $n_c$ = 8-10 $\rightarrow n = 2$, can also be observed. This peak is weaker than the $n = 5 \rightarrow n = 2$ peak, showing a similar or lower enhancement. Fig. 2 shows the electron-impact excitation spectrum overlaid with the CX spectrum. The peak around 0.5 keV is due to oxygen, which is one of the background gases in EBIT [16]. The factors of enhancement in CX is about 2 for the $n = 4 \rightarrow n = 2$ peak, ranges between 3 and 4 for the $n = 5 \rightarrow n = 2$ peak and ranges between 2 and 4 for the $n_c$ = 8-10 $\rightarrow n = 2$ peak [16].

Fig. 3 shows an overlaid direct excitation spectrum with a CX spectrum, where the $n = 3 \rightarrow n = 2$ emissions of both spectra are normalized to each other and the oxygen peak, that was visible in figure 2, has been subtracted. The enhancement in the flux from $n \geq 4$ is clearly visible. However, it is also observable that the $n = 3 \rightarrow n = 2$ peak is slightly shifted towards lower energy in CX. Unfortunately, the charge states are only roughly the same in the two overlaid spectra [16]. However, if this shift is significant, it would suggest that the transition 3s

→ 2p has been enhanced in the CX spectrum, as it has the lowest energy of the $n = 3 \rightarrow n = 2$ transitions [16].

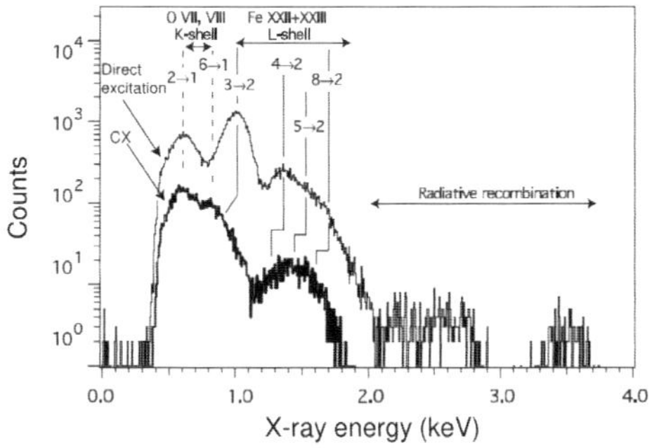


**Figure 2.** Overlaid Fe X-ray spectra from direct excitation with charge exchange at electron beam energy 1.85 keV.

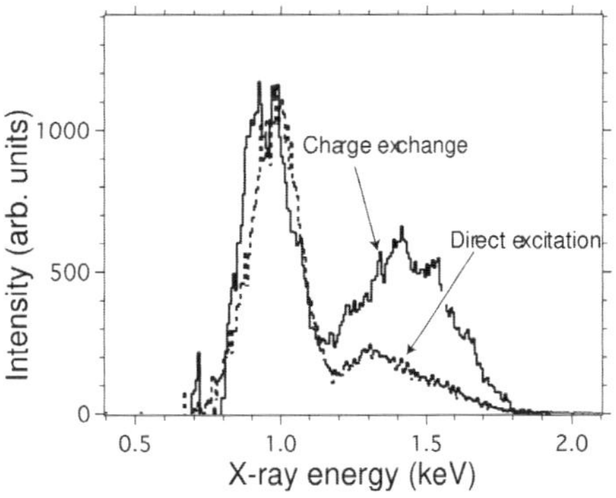


**Figure 3.** Comparison of Fe X-ray spectra produced by direct excitation at at electron beam energies 1.75 keV in direct excitation and 1.85 keV in CX. The oxygen peak has been subtracted.

## 4. Measurement of sulfur L-shell charge exchange

In the L-shell sulfur CX spectra the individual transitions in the $n = 3 \rightarrow n = 2$ peak can be observed due to the significantly improved resolution of the micro-calorimeter [17]. The results are, in fact, the first high-resolution data of L-shell CX, with a resolution just below 5eV in the measurement. As a result, we can investigate the potential change in the $n = 3 \rightarrow n = 2$ peak, as intimated by the iron data. We will here focus mainly on the lithium-like lines, even though lines from lower charge states are present in the spectra.

In the direct excitation spectra, we see a similar pattern as in the iron measurement. The spectrum is dominated by the n = 3 → $n = 2$ transitions; for S XIV that is 3d → 2p, 3p → 2s and 3s → 2p in order of decreasing intensity. These lines were found between 369 eV and 407 eV, as shown in Fig. 4. This is an overlaid spectrum of the $n = 3 \rightarrow n = 2$ transitions of lithium-like sulfur in direct excitation and CX, where the respective 3d→2p peaks have been normalized to each other. A significant change in these transitions is the enhancement in the 3s → 2p peak in CX, as well as enhancement in the 3p → 2s peak . In addition, the 3d → 2p peak, which is however blended with lower charge states, also shifts to lower energy. Since the 3s → 2p peak has the lowest energy of the $n = 3 \rightarrow n = 2$ transitions, and the 3d → 2p shifts towards lower energy, this means that the centroid of the $n = 3 \rightarrow n = 2$ emission shifts to lower energy when produced by CX. This observation agrees with the centroid shift intimated by our iron measurements.

## 5. Discussion and summary

The principal quantum number of the electron capture can be approximated by the Janev and Winter approximation [18]. For iron ions ($q \leq 24$) this gives $n_c \approx$ 8-10. This agrees quite well with the experimental data. For helium-like sulfur ($q = 14$) reacting with $SF_6$, the approximation gives $n_c \approx 8$. This agrees with what we observe in the spectra from the sulfur measurement (full analysis in progress).

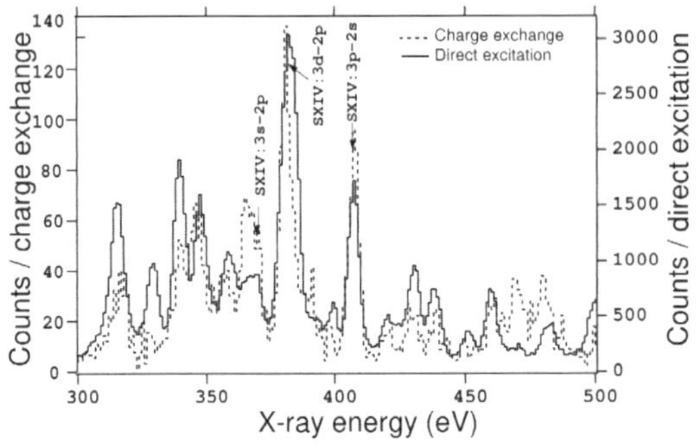


**Figure 4.** Comparison of the SXIV $n = 3 \rightarrow n = 2$ transitions X-ray spectra produced by direct excitation and CX at electron beam energy 3.09 keV.

We have detected a shift in the $n = 3 \rightarrow n = 2$ centroid energy during the CX excitation of iron. It seems plausible, if this shift is significant, that the 3s→2p transition has been enhanced in CX. In the high- resolution results from lithium-like sulfur, Fig. 4, this specific enhancement is actually resolved. In addition, the blended 3d-2p peak shifts to lower energy. This suggests that the $n = 3 \rightarrow n = 2$ centroid energy shift, in CX, might be an L-shell CX feature.

For L-shell ions, an electron captured into a level $n_c$ can only decay to the 2s or 2p states if it is captured in an angular momentum state s, p, or d due to the $\Delta$l=1 selection rule which governs the electric dipole transitions. For charge transfer to higher l-states the electron typically cascades down by a series of transitions, changing its l-value with 1 for every step [7]. The enhanced emission from $n = 4$ and 5 could either be populated by a single electron capture into $n_c$, followed by cascades to $n = 4$ and 5, or by multi-electron capture where one electron drops from the capture level while the second electron autoionizes. Which of these paths the electron takes depends among other things on the collision energy of the electron. Low collision energy favors low l-state capture [19]. Since we observe a strong enhancement of the high-n flux, we may assume that low l-levels were populated in the charge transfer, which suggests that low-l double electron capture is more probable than high-l capture into $n_c$ followed by cascades. Hopefully, detailed radiative cascade models will someday be able to discern which of these processes is more probable [16] .

## Acknowledgments

Part of this work was performed under the auspices of the U. S. Department of Energy by Lawrence Livermore National Laboratory under Contract DE-AC52-07NA27344 and was supported in part by NASA grants to LLNL, SAO, UC Berkeley and GSFC as well as by LLNL Laboratory Directed Research and Development project OG-ERD-010.

## References

[1] Bhardwaj A, Elsner R F, Randall Gladstone G, Cravens T E, Lisse C M, Dennerl K, Branduardi-Raymont G, Wargelin B J, Hunter Waite J, Robertson I, stgaard N, Beiersdorfer P, Snowden Steven L and Kharchenko V 2007 *Planetary and Space Science.* **55** 1135–1189

[2] Beiersdorfer P, Behar E, Boyce K R, Brown G V, Chen H, Gendreau K C, Graf A, Gu M F, Harris C L, Kahn S M, Kelley R L, Lepson J K, May M J, Neill P A, Pinnington E H, Porter F S, Smith A J, Stahle C K, Szymkowiak A E, Tillotson A, Thorn D B, Trabert E and Wargelin B J 2003 *Nuclear Instruments and Methods in Physics Research Section B* **205** L99–L102

[3] Kharchenko and Dalgarno 2001 *The Astrophysical Journal* **554** 173–177

[4] Perez J A and Olson R E 2005 *Nuclear Instruments and Methods in Physics Research Section B* **241** 134–138

[5] Kharchenko V, Rigazio M, Dalgarno A and Krasnopolsky V A 2003 *The Astrophysical Journal* **585** L73–L75

[6] Otranto S, Olson R E and Beiersdorfer P 2006 *Physical Review A* **73**(2)
[7] Wargelin B J, Beiersdorfer P, Neill P A, Olson R E and Scofield J H 2005 *The Astrophysical Journal* **634** 687–697
[8] Bliek F W, Woestenenk G R, Hoekstra R and Morgenstern R 1998 *Physical Review A (Atomic, Molecular, and Optical Physics)* **57** 221–226
[9] Lubinski G, Juhsz Z, Morgenstern R and Hoekstra R 2000 *Journal of Physics B: Atomic, Molecular, and Optical Physics* **33** 5275–5296
[10] Lubinski G, Juhsz Z, Morgenstern R and Hoekstra R 2001 *Physical Review Letters* **86** 616–619
[11] Beiersdorfer P, Olsen R, Schweikhard L, Liebisch P, Brown G, Crespo Lopez-Urrutia J, Harris C, Neill P, Utter S and Widmann K 2000 *CP500, The physics of Elelctronic and Atomic Collisions., published by American Institute of Physics, Melville, NY, 2000, edited by Itikawa, Y et al.*
[12] Tawara H, Richard P, Safronova U I, Vasilyev A A, Hansen S and Shlyaptseva A S 2002 *Physical Review A* **65**(4)
[13] Wargelin B J, Beiersdorfer P and Brown G V 2008 *Canadian Journal of Physics* **86** 151–169
[14] Beiersdorfer P 2008 *Canadian Journal of Physics* **86** 1–10
[15] Beiersdorfer P, Schweikhard L, Crespo Lòpez-Urrutia J and Widmann K 1996 *Review of Scientific Instruments* **67** 3818–3826
[16] Beiersdorfer P, Schweikhard L, Liebisch P and Brown G V 2008 *The Astrophysical Journal.* **672** 726–732
[17] Porter F S, Beck B R, Beiersdorfer P, Boyce K R, Brown G V, Chen H, Gygax J, Kahn S M, Kelley R L, Kilbourne C A, Magee E and Thorn D B 2008 *Canadian Journal of Physics* **86** 231–240
[18] Janev R K and Winter H 1985 *Physics Reports* **117** 265–387
[19] Ryufuku H and Watanabe T 1979 *Physical Review A* **20** 1828–1837

# STUDIES OF X-RAY PRODUCTION FOLLOWING CHARGE EXCHANGE RECOMBINATION BETWEEN HIGHLY CHARGED IONS AND NEUTRAL ATOMS AND MOLECULES

**G. V. Brown†, P. Beiersdorfer†, H. Chen†, J. Clementson†, M. Frankel†, M. F. Gu†, R. L. Kelley♯, C. A. Kilbourne♯, F. S. Porter♯, D. B. Thorn†, B. J. Wargelin‡**

†Lawrence Livermore National Laboratory, Livermore, CA 94550
♯NASA/Goddard Space Flight Center, Greenbelt, MD 20770
‡Harvard-Smithsonian Center for Astrophysics, Cambridge, MA 02138

E-mail: `gregbrown@llnl.gov`

**Abstract.** We have used microcalorimeters built by the NASA/Goddard Space Flight Center and the Lawrence Livermore National Laboratory SuperEBIT electron beam ion trap to measure X-ray emission produced by charge exchange reactions between highly charged ions colliding with neutral helium, hydrogen, and nitrogen gas. Our measurements show the spectral dependence on neutral species and also show the distinct differences between spectra produced by charge exchange reactions and those produced by direct electron-impact excitation. These results are part of an ongoing experimental investigation at the LLNL SuperEBIT facility of charge exchange spectral signatures and can be used to interpret X-ray spectra produced by a variety of laboratory and celestial sources including cometary and planetary atmospheres, the Earth's magnetosheath, the heliosphere, and tokamaks.

The radiationless transfer of one or more electrons to a projectile ion from a target neutral atom or molecule is known as charge exchange recombination (CX). Photon emission is produced by CX when the transferred electron is captured into an excited state and that state radiatively decays. CX plays an important role in a variety of plasmas and is believed to be a significant X-ray line formation process in many non-equilibrium astrophysical objects. For example, X-ray emission from CX has been identified or hypothesized as a source of X-ray production in planetary atmospheres, including the aurora of Jupiter [1, 2] and the Earth's magnetosheath [3], supernova remnants [4], and the galactic center and ridge [4, 5]. CX is also believed to contribute significantly to the cosmic soft X-ray background. In addition, CX produces a significant amount of radiation in some laboratory plasmas, such as during neutral beam heating in tokamak plasmas. Identification and interpretation of the X-ray spectral signatures from charge exchange in complex sources, however, has been challenging because little targeted laboratory data are available, and while in some cases agreement between theory and measurement is good [6], in many significant discrepancies still exist [7, 8, 9]. To help better understand charge exchange recombination and to provide benchmarks for theory, we have used the SuperEBIT electron beam ion trap (EBIT) facility at Lawrence Livermore National Laboratory (LLNL) and a NASA/Goddard Space Flight Center microcalorimeter instrument to measure the X-ray spectral signatures of charge exchange. Here we present a brief description of the measurement technique and selected results.

The LLNL EBIT facility, home of the original EBIT [10, 11], is well tested and has been used in numerous atomic physics and laboratory astrophysics experiments. Most basically, it consists of an electron beam, compressed by a 3 Tesla magnetic field, that creates and excites ions, and a trap region defined by three drift tubes that confine the ions along the axis of the electron beam. The LLNL EBIT can operate in two modes: the electron trapping mode, and the magnetic trapping mode [12]. In the magnetic trapping mode, the electron beam is turned off and the ions are trapped radially by the 3 Tesla magnetic field. This differs from the electron trapping mode, most often used in EBIT experiments, where the beam is on and the ions are trapped radially by the potential created by the electron beam's space charge. For CX experiments, EBIT is operated in a phase that includes both trapping modes. Ions are created in the electron trapping mode, then the X-ray emission from CX is measured in the magnetic trapping mode. Neutral targets are introduced continuously throughout the EBIT phase using a ballistic gas injector. Figure 1 shows a typical EBIT phase diagram denoting the electron and magnetic trapping modes and the spectra acquired during each mode.

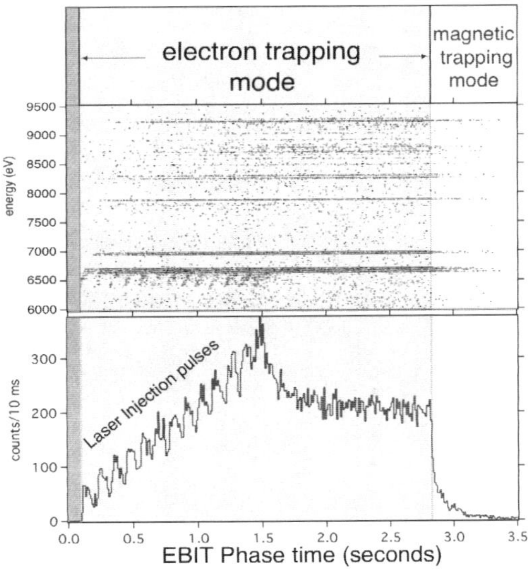


**Figure 1.** Energy (upper frame) and total counts (lower frame) versus EBIT phase time. The electron and magnetic trapping mode portions of the phase are both shown. The X-ray emission from charge exchange between highly charged Fe ions and $N_2$ takes place during the magnetic trapping mode. For this experiment, Fe was injected into EBIT using a laser ablation injection system [13]. Each of the laser injection pulses can be seen in both frames. Between 0 and 0.1 seconds the ions are dumped from the trap and no X-ray emission occurs.

For these experiments, photons were detected using one of the microcalorimeter instruments developed and built at the NASA/Goddard Space Flight Center [14, 15] and operated at LLNL. In particular, the second generation XRS/EBIT was used[1]. The XRS/EBIT has an energy

[1] For a more complete description of the microcalorimeter instruments operated at the LLNL EBIT facility, see Porter et al. these proceedings.

resolution of 5 – 6 eV at 6 keV and is able to measure time-resolved spectra. Time resolution is necessary in order to "cut out" the electron trapping mode portion of the EBIT phase so that the X-ray spectra produced solely by charge exchange can be analyzed separately. High energy resolution is necessary to resolve high-$n$ Rydberg states (see below).

Figures 2 and 3 show spectra produced by charge exchange between hydrogenic $Fe^{25+}$ and bare $Fe^{26+}$ projectiles and neutral $H_2$, $N_2$, and He targets producing X-ray emission from helium-like $Fe^{24+}$ and hydrogenic $Fe^{25+}$, respectively. Figure 2 shows the He-like $Fe^{24+}$ spectrum produced by CX compared to the spectrum from $Fe^{24+}$ produced by direct electron impact excitation. In the CX case, the 1s2s $^3S_1 \rightarrow 1s^2\ ^1S_0$ forbidden line "z" is much stronger than the 1s2p $^1P_1 \rightarrow 1s^2\ ^1S_0$ resonance line "w", in stark contrast to the case of direct electron impact excitation. This well known spectral signature [4, 6, 8] can be used to identify the presence of CX as an X-ray production mechanism, even using a relatively low-resolution spectrometer, such as a CCD or germanium detector [16, 17].

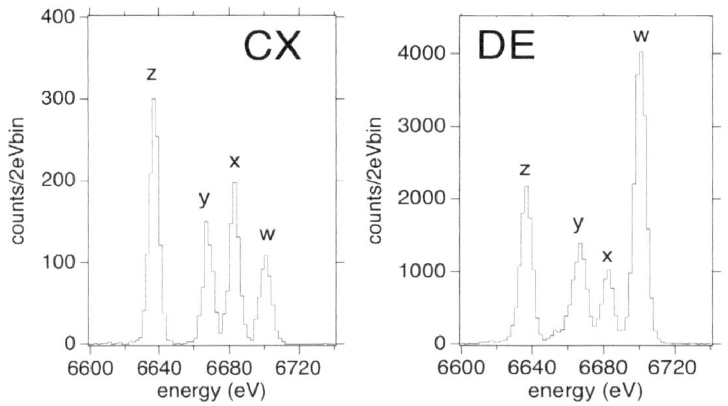


**Figure 2.** Comparsion of X-ray emission from helium-like $Fe^{24+}$ produced by charge exchange (CX) between hydrogenic $Fe^{25+}$ and $N_2$ and produced by direct electron impact excitation (DE).

Figure 3 compares the high-$n$ Rydberg X-ray lines produced by CX between $Fe^{26+}$ and neutral $H_2$, $N_2$, and He. Note that the strongest X-ray line in the case of He is $n_{max} = 12 \rightarrow 1$, while in the case of $H_2$ and $N_2$ it is $n_{max}$=14 $\rightarrow$ 1. The lower $n_{max}$ for the He targets is caused by the fact that the principal quantum number, $n$, of the state of the ion with the maximum capture cross section depends on the ionization potential of the neutral material [4]. Although these results do follow the trend predicted by theory, i.e., $n_{max}$ decreases as the ionization potential of the neutral target increases, the measured $n_{max}$ is about 2 levels higher than predicted by the usual scaling formula. Our results show that, given high enough energy resolution, CX spectra can be used to distinguish among neutral target material.

Significant advances have been made in understanding the X-ray production from CX; however, much work remains to fully realize its diagnostic potential, particularly in the case of charge exchange with L-shell ions [18, 19]. Studies of CX will be especially useful in the analysis of data acquired by future satellite missions, such as Astro-H, which will carry a microcalorimeter similar to the one used in these experiments, and that will for the first time provide high resolution, high signal-to-noise spectra of extended sources.

**Acknowledgments**

Part of this work was performed under the auspices of the U. S. Department of Energy by Lawrence Livermore National Laboratory under Contract DE-AC52-07NA27344 and is also supported by NASA grants to LLNL, SAO, Stanford University, and the NASA/GSFC.

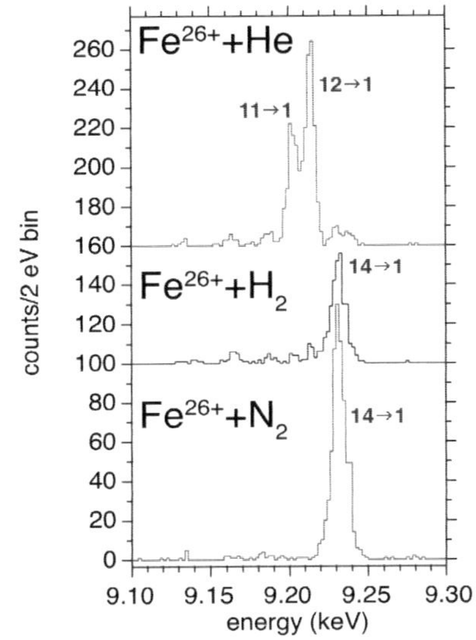


**Figure 3.** Comparison of X-ray emission from high-$n \rightarrow 1$ Rydberg transitions following CX between bare $Fe^{26+}$ and He, $H_2$, and $N_2$. The ionization potential for He, $H_2$, and $N_2$ are 24.6 eV, 15.4 eV, and 15.6 eV, respectively. The predicted value for $n$ for the state of maximum capture for the He targets is 9, and for $H_2$, and $N_2$ is 12 or 13. The collision energy for these reactions is $\approx$ 10 eV/amu.

## References

[1] Gladstone G R, Waite J H, Grodent D, Lewis W S, Crary F J, Elsner R F, Weisskopf M C, Majeed T, Jahn J M, Bhardwaj A, Clarke J T, Young D T, Dougherty M K, Espinosa S A and Cravens T E 2002 **415** 1000
[2] Elsner R F, Lugaz N, Waite J H, Cravens T E, Ford P, Grodent D, Bhardwaj A, Macdowell R J, Desch M D and Majeed T
[3] Fujimoto R, Mitsuda K, Mccammon D, Takei Y, Bauer M, Ishisaki Y, Porter S F, Yamaguchi H, Hayashida K and Yamasaki N Y 2007 *pasj* **59** 133–140
[4] Wargelin B J, Beiersdorfer P and Brown G V 2008 *Can. J. Phys.* **86** 151–169
[5] Beiersdorfer P, Bitter M, Marion M and Olson R E 2005 *pra* **72** 032725–+
[6] Beiersdorfer P, Olson R E, Brown G V, Chen H, Harris C L, Neill P A, Schweikhard L, Utter S B and Widmann K 2000 *Phys. Rev. Lett.* **85** 5090
[7] Beiersdorfer P, Lisse C M, Olson R E, Brown G V and Chen H 2001 *Astrophys. J. Lett.* **549** L147 – L150
[8] Otranto S, Olson R and beiersdorfer P 2008 *Can. J. Phys.* **86** 171–174
[9] Beiersdorfer P 2008 *Can. J. Phys.* **86** 1–10
[10] Marrs R 2008 *Can. J. Phys.* **86** 11–18
[11] Beiersdorfer P, Schweikhard L, López-Urrutia J C and Widmann K 1996 *Rev. Sci. Instrum.* **67** 3818–3826
[12] Niles A M, Magee E W, Thorn D B, Brown G V, Chen H and Beiersdorfer P 2006 *rsi* **77**
[13] Porter F S, Beiersdorfer P, Boyce K R, Brown G V, Chen H, Gygax J, Kahn S M, Kelley R L, Kilbourne C A, Magee E and Thorn D B 2008 *Can. J. Phys.* **86** 231–240
[14] Porter F S, Beiersdorfer P, Boyce K R, Brown G V, Chen H, Kelley R L and Kilbourne C A 2004 *Rev. Sci. Instrum.* **75** 3772
[15] Koyama K, Hyodo Y, Inui T, Nakajima H, Matsumoto H, Tsuru T G, Takahashi T, Maeda Y, Yamazaki N Y, Murakami H, Yamauchi S, Tsuboi Y, Senda A, Kataoka J, Takahashi H, Holt S S and Brown G V 2007 *pasj* **59** 245–255
[16] Wargelin B J, Beiersdorfer P, Neill P A, Olson R E and Scofield J H 2005 *apj* **634** 687–697

# Vacuum ultraviolet spectra in charge transfer collisions of multiply charged Sn ions

**S Suda[1], H Ohashi[1], H Tanuma[1], S Fujioka[2], H Nishimura[2] and K Nishihara[2]**

[1]Department of Physics, Tokyo Metropolitan University, Hachioji, Tokyo 192-0397, Japan
[2]Institute of Laser Engineering, Osaka University, Suita, Osaka 565-0871, Japan

E-mail: suda-shintaro@ed.tmu.ac.jp

**Abstract.** We measured the vacuum ultraviolet (VUV) emission spectra following charge transfer collisions of $Sn^{q+}$ ($q$ = 3–6) with neutral rare gas targets at collision energies of 50–100 keV. We identified some of emission lines from reported energy levels. Several lines were observed in collisions of different charge states of ions. This finding means that the multiple-electron capture processes can take place effectively in our experiment. Significant target dependences were observed in the spectra with the projectile of $Sn^{4+}$ and $Sn^{6+}$.

## 1. Introduction

Nowadays the wavelength of the light for the semiconductor lithography is 193 nm, which is emitted from an ArF excimer laser. Since shorter wavelength is necessary for higher integration of semiconductor devices, the extreme ultraviolet (EUV) light, ~10 nm, will be applied to the next-generation semiconductor lithography. The EUV light is easily absorbed by the air or optical lenses, so the EUV exposure system must be in vacuum and transmissive lenses are not available. The Mo/Si multilayer mirrors having peak reflectivity around 13.5 nm in the EUV region will be used for light focusing instead of optical lenses. Therefore light sources which emit the 13.5 nm light had been searched for actively [1-3]. Consequently, the laser-produced Sn plasma was selected as a light source [4-6]. However, spectroscopic data about multiply charged Sn ions were limited; therefore, it was hard to identify emission lines from Sn plasmas. To provide the EUV spectroscopic data of multiply charged Sn ions, we have measured the charge-selected EUV emission spectra by means of charge exchange spectroscopy [7].

The emission of 13.5 nm with 2% band width is called "in-band" emission. However, not only the "in-band" light around 13.5 nm, but also the "out-of-band" light with wavelengths longer than 130 nm can be reflected by the Mo/Si mirror efficiently [8]. Hence, VUV spectroscopic data of Sn ions were also required. In this work, we measured VUV emission spectra by charge exchange spectroscopy as we have done for the EUV spectra.

## 2. Experiment

Multiply charged Sn ions were produced as before [7]. We used a 14.25 GHz electron cyclotron resonance ion source (ECRIS) at Tokyo Metropolitan University. We put sintered tin oxide ($SnO_2$) pellets inside the plasma chamber at the ECRIS. Multiply charged Sn ions produced in the ECRIS were extracted with an electric potential of 20 kV for $q$ = 4–6 or 17.5 kV for $q$ = 3,

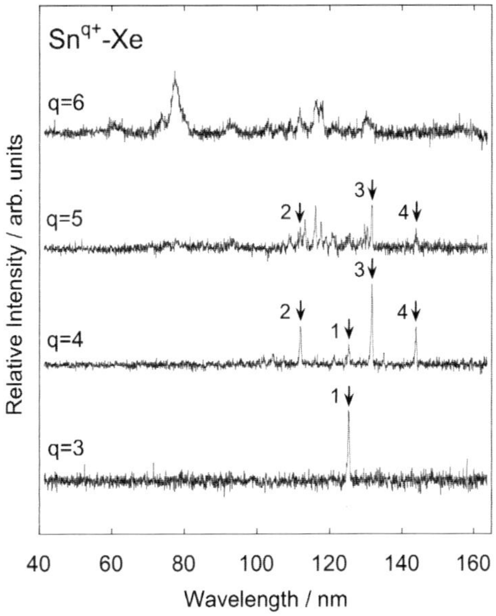


**Figure 1.** Emission spectra in collisions of $Sn^{q+}$ ($q = 3$–$6$) with Xe. The emission lines labeled as 1–4 are listed in table 1.

and were charge-separated with a magnetic field. We mention that collision energy dependences of emission cross sections are not significant in such energy region.

Charge-selected incident ions were made to collide with neutral gas targets in a collision chamber. The projectile beam current was measured with a Faraday cup at the end of beamline: it was typically 1 $\mu$A for $q = 4$–$6$ or 0.5 $\mu$A for $q = 3$. The target gas jet was ejected from a multi-capillary plate to cross the ion beam. During the experiment, the gas pressure was low enough to avoid multiple collisions.

The emitted light following charge transfer collisions was observed by a Seya-Namioka spectrometer equipped with a Peltier-cooling CCD camera. The spectrometer was set up vertically to the ion beam. The measured wavelength range was from 40 nm to 225 nm. However, in the range of 145 nm to 225 nm, no emission line was observed. The resolution was determined by a slit which was located at the entrance of the spectrometer: FWHM was 0.4 and 1.0 nm for the slit width of 100 and 300 $\mu$m, respectively.

## 3. Result

### *3.1. Charge State Dependence*

Figure 1 shows emission spectra in collisions of $Sn^{q+}$ with Xe. The resolution is 0.4 nm at $q = 3$–$5$ and 1.0 nm at $q = 6$. The emission intensity is normalized with the incident ion current and the target gas pressure. We estimated wavelengths of emission lines from reported energy levels of Sn ions so that we could identify the transition lines [9]. We indicate some of identified transition lines in table 1. The emission lines labeled as 1–4 in figure 1 were listed in table 1.

In the $Sn^{3+}$ - Xe collision, we observed only the $5s^2$ ($^1S_0$)–5s5p ($^1P^o_1$) emission line. At $q = 4$ we observed several lines, and identified almost all of them. We can see many emission lines at $q = 5$; most of them are the $4d^95s$–$4d^95p$ transitions, in which states have fine structures. In case of $q = 6$, we could not identify emission lines because of the absence of information about energy levels and wavelengths.

Generally single electron capture (SEC) is dominant in low energy collisions of highly charged ions with neutral targets [11]:

$$\mathrm{Sn}^{q+} + \mathrm{Xe} \longrightarrow \mathrm{Sn}^{(q-1)+*} + \mathrm{Xe}^{+} \longrightarrow \mathrm{Sn}^{(q-1)+} + \mathrm{h}\nu + \mathrm{Xe}^{+}. \tag{1}$$

In addition, transfer ionization (TI) or autoionizing double capture (ADC) play an important role [12]:

$$\mathrm{Sn}^{q+} + \mathrm{Xe} \longrightarrow \mathrm{Sn}^{(q-2)+**} + \mathrm{Xe}^{2+} \longrightarrow \mathrm{Sn}^{(q-1)+*} + \mathrm{e}^{-} + \mathrm{Xe}^{+}. \tag{2}$$

However, in cases of low charged ions, the cross section of the multiple electron capture might become comparable to that of the SEC because of the energy matching [13]. In figure 1, there exist several lines which were observed simultaneously in different spectra. The emission lines labeled as 1 and 2–4 were observed in the spectra of $q = 3$ and 4, and in those of $q = 4$ and 5, respectively. These lines observed in larger $q$ should be caused by the true double electron capture (TDC), i.e.

$$\mathrm{Sn}^{q+} + \mathrm{Xe} \longrightarrow \mathrm{Sn}^{(q-2)+**} + \mathrm{Xe}^{2+} \longrightarrow \mathrm{Sn}^{(q-2)+} + \mathrm{h}\nu + \mathrm{Xe}^{+}. \tag{3}$$

### *3.2. Target Dependence*

Figure 2 shows the target dependence of the emission spectra in collisions of $\mathrm{Sn}^{4+}$ with rare gas targets. The resolution is 1.0 nm for Ne target and 0.4 nm for others. The emission intensity is normalized with the incident ion current and the target gas pressure. The emission lines labeled as 1–6 are listed in table 1.

The electron configuration of $\mathrm{Sn}^{4+}$ is $4\mathrm{d}^{10}$ for the ground state. After the SEC, electron configuration would be $4\mathrm{d}^{10}nl$. On the other hand, it would be $4\mathrm{d}^{10}nln'l'$ in the cases of TDC. The emission lines 1 and 5, which are identified as the $5\mathrm{s}^2$–5s5p and the 5s5p–5s5d of Sn III, are due to the TDC, whereas the others are due to the SEC.

We can see the strong target dependence in the intensity of the $5\mathrm{s}^2$–5s5p of Sn III and in the intensity-ratio of the 5s–5p($^2\mathrm{P}^{\mathrm{o}}_{3/2}$) to the 5s–5p($^2\mathrm{P}^{\mathrm{o}}_{1/2}$). The emission line of the $5\mathrm{s}^2$–5s5p appears for Ar, Kr and Xe targets. The emission intensity of this line is the strongest for Kr target. Although the intensity-ratio of the 5s–5p($^2\mathrm{P}^{\mathrm{o}}_{3/2}$) to the 5s–5p($^2\mathrm{P}^{\mathrm{o}}_{1/2}$) should be 2 : 1 if the population of the 5p states simply obeys the statistical weight, it shifts away especially in the case of Ar target.

Figure 3 shows a series of emission spectra in collisions of $\mathrm{Sn}^{6+}$ with rare gas targets. The resolution was 1.0 nm in these spectra. A significant target dependence can be seen. However, as we mentioned before, we could not identify emission lines. A precious calculation for transition energies considering electron correlations in detail is desired to identify emission lines.

**Table 1.** A list of some identified lines. The line numbers correspond to labels in figure 1 and 2. $^{a)}$The wavelengths obtained from the NIST database [10]. $^{b)}$The wavelengths deduced from the reported energy levels [9].

| Line no. | Wavelength / nm | | | Charge state | Lower state | Upper state |
|---|---|---|---|---|---|---|
| | this work | NIST$^{a)}$ | Moore$^{b)}$ | | | |
| 1 | 125.2 | 125.992 | 125.14 | III | $5\mathrm{s}^2$ ($^1\mathrm{S}_0$) | 5s5p ($^1\mathrm{P}^{\mathrm{o}}_1$) |
| 2 | 112.0 | 112.068 | 111.93 | IV | $4\mathrm{d}^{10}$ 5p ($^2\mathrm{P}^{\mathrm{o}}_{3/2}$) | $4\mathrm{d}^{10}$ 5d ($^2\mathrm{D}_{5/2}$) |
| 3 | 131.5 | 131.455 | 131.45 | IV | $4\mathrm{d}^{10}$ 5s ($^2\mathrm{S}_{1/2}$) | $4\mathrm{d}^{10}$ 5p ($^2\mathrm{P}^{\mathrm{o}}_{3/2}$) |
| 4 | 143.8 | 143.752 | 143.75 | IV | $4\mathrm{d}^{10}$ 5s ($^2\mathrm{S}_{1/2}$) | $4\mathrm{d}^{10}$ 5p ($^2\mathrm{P}^{\mathrm{o}}_{1/2}$) |
| 5 | 134.8 | —— | 134.76 | III | 5s5p ($^1\mathrm{P}^{\mathrm{o}}_1$) | 5s5d ($^1\mathrm{D}_2$) |
| 6 | 107.4 | 107.341 | —— | IV | —— | —— |

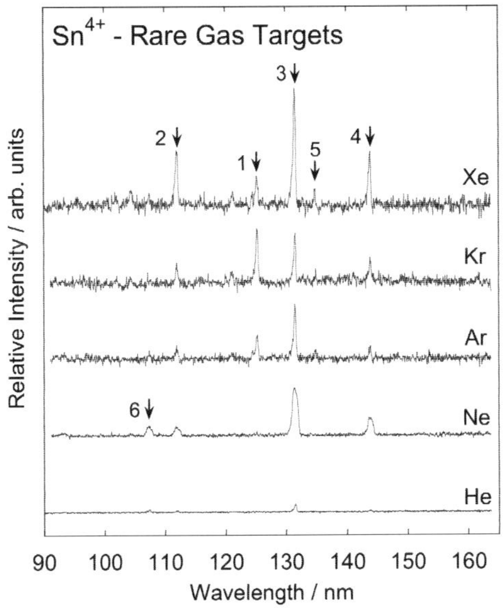


**Figure 2.** Emission spectra in collisions of $Sn^{4+}$ with rare gases. The emission lines labeled as 1–6 are listed in table 1.

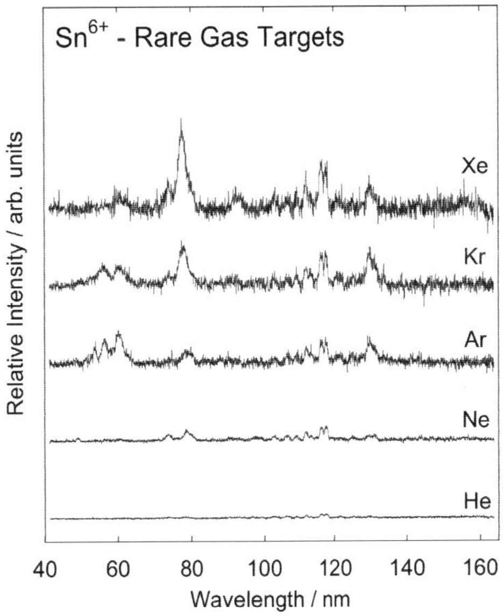


**Figure 3.** Emission spectra in collisions of $Sn^{6+}$ with rare gases.

## 4. Conclusion

The vacuum ultraviolet (VUV) emission spectra after charge transfer collisions of $Sn^{q+}$ ($q = 3$–6) with rare gas targets were measured. Emission lines following both single and double capture were observed. Significant target dependences were observed in the spectra with the projectile of $Sn^{4+}$ and $Sn^{6+}$. A precious calculation for transition energies is desired to identify emission lines especially for the projectile of $Sn^{6+}$.

## 5. Acknowledgments

This work was financially supported in part by MEXT (Ministry of Education, Culture, Sports, Science, and Technology, Japan) under construct subject" Leading Project for EUV lithography source development" . And a part of this work was performed under the collaboration program of the Institute of Laser Engineering, Osaka University.

## 6. References

[1] Jin F and Richardson M 1995 *Appl. Opt.* **34** 5750
[2] Spitzer R C, Orzechowski T J, Phillion D W, Kauffman R L and Cerjan C 1996 *J. Appl. Phys.* **79** 2251
[3] Shimoura A, Mochizuki T, Miyamoto S, Amano S and Uyama T 1999 *Appl. Phys. Lett.* **75** 2026
[4] Aota T and Tomie T 2005 *Phys. Rev. Lett.* **94** 015004
[5] Fujioka S *et al* 2005 *Phys. Rev. Lett.* **95** 235004
[6] Okuno T *et al* 2006 *Appl. Phys. Lett.* **88** 161501
[7] Ohashi H, Tanuma H, Fujioka S, Nishimura H, Sasaki A and Nishihara K 2007 *J. Phys. Conf.* **58** 235
[8] Bakshi V 2006 *EUV source for lithography* (Washington: SPIE Press Book) p 38
[9] Moore C E 1958 *Atomic Energy Levels Vol. III* (Washington: NBS-Cir) 467
[10] National Institute for Science and Technology (NIST), http://www.nist.gov
[11] Justiniano E, Cocke C L, Gray T J, Dubois R, Can C and Waggoner W 1984 *Phys. Rev.* A **29** 1088
[12] Cederquist H *et al* 1992 *Phys. Rev.* A **46** 2592
[13] Ishii K, Okuno K and Itoh A 2004 *Phys. Rev. A* **70** 042716

# Dissociative and non-dissociative charge-changing processes in 1.0-2.0 MeV/u $O^{5+} + O_2$ collisions

**D P Cassidy, E Y Kamber, A Kayani and J A Tanis**[1]

Department of Physics, Western Michigan University, Kalamazoo, MI 49008 USA

E-mail: john.tanis@wmich.edu

**Abstract.** In the present work molecular ionization and dissociation correlated with projectile single electron capture and loss is investigated as a function of energy for $O^{5+} + O_2$ collisions. Coincidence recoil-ion spectra as well as total projectile charge-changing yields were measured for projectile energies of 1.0, 1.5, and 2.0 MeV/u. Charge states for dissociative and non-dissociative recoiling products of the target molecular oxygen were identified. Ratios of charged molecular fragment yields to single non-dissociative ionization yields as a function of projectile energy were determined as well as total cross sections for projectile single electron capture and loss.

## 1. Introduction

Fundamental interactions of molecular oxygen during collisions with charged particles are relevant to numerous fields of study including but not limited to astrophysics, atmospheric physics, biophysics, and plasma physics [1]. These interactions are especially relevant to the life sciences as both atomic and molecular oxygen are abundant in most living matter, as well as being one of the essential components of many organic molecules. The generation of ion-radiation induced reactive oxygen and how it affects surrounding cellular tissue is a primary example [2]. Within the field of atomic physics results for the ionization of molecular oxygen by fast ions are important for analyzing the fragmentation of $H_2O$ and other molecules. Similar work has been the subject of recent attention for protons, neutral hydrogen, and carbon projectiles [3] [4] colliding with $H_2O$. In an effort to gain additional understanding of these and other such interactions, dissociative and non-dissociative ionization of $O_2$ associated with single electron capture, single electron loss, and direct ionization have been investigated for 1.0, 1.5 and 2.0 MeV/u $O^{5+} + O_2$ collisions.

## 2. Experimental setup

The measurements were conducted at Western Michigan University using the 6 MV tandem Van de Graaff accelerator. An $O^{5+}$ beam was collimated to a diameter of about 2 mm before entering the target region. In the target region the $O^{5+}$ projectile beam impacted an $O_2$, Ne, or He gas target contained within a differentially pumped target cell 4.0 cm in length. The pressure in the target cell, monitored by a capacitance monometer system, was varied between 0 - 2 mTorr.

Time-of-flight techniques were used to detect coincidences between recoiling charged molecular and atomic oxygen fragments and individual outgoing projectile charge states. The recoiling

[1] To whom any correspondence should be addressed.

target ions were extracted transverse to the beam direction by an electric field created by a pair of electrodes biased at ±1000 volts respectively. Following extraction the fragments drifted in a field-free region at a velocity determined by their charge-to-mass ratio and then were detected using a set of coupled microchannel plates. Meanwhile the outgoing projectile beam was magnetically analyzed and the individual charge-changed components were detected using silicon surface-barrier detectors to determine the respective coincidence and total charge-changing yields. Simultaneously the unchanged primary beam was collected in a Faraday cup for normalization.

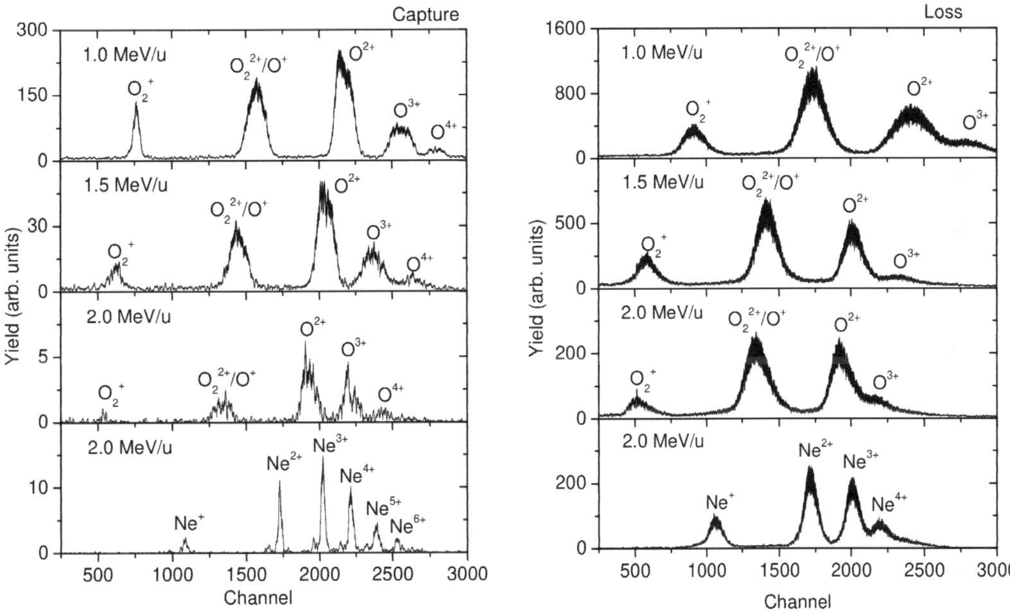


**Figure 1.** Relative yields for ionization and fragmentation of molecular oxygen associated with projectile single electron capture (left panel) and loss (right panel) for 1.0, 1.5, and 2.0 MeV/u $O^{5+}$ + $O_2$ collisions normalized to the incident beam current. The bottom panel in each case shows the corresponding normalized spectrum for 2.0 MeV/u $O^{5+}$ + Ne.

## 3. Results and data analysis

Typical coincidence spectra are shown in figure 1. The spectra show recoiling products of the target molecular oxygen with charge states as high as 4+ for capture and 3+ for loss. All of these spectra were collected at a pressure of 2.0 mTorr and are normalized to the total beam current and as such show the changing fragmentation yields of measured coincidences as the beam energy increases.

To ensure the validity of the $O^{5+}$ + $O_2$ data and to determine the recoil-ion products, additional recoil-ion spectra for He and Ne, as well as total projectile charge-changing yields, were measured for projectile energies of 1.0, 1.5, and 2.0 MeV/u. For each of the three target gases at all energies a minimum of three pressures was used, typically with values of 0.5, 1.0, 1.5, and 2.0 mTorr. The pressure dependence was used to determine the fractional yields and to check that the experiment was within single collision conditions. In addition spectra for no target gas were measured for purposes of background correction.

For $O^{5+}$ + He the fractional yields for total projectile capture and loss were determined by normalizing the measured numbers of counts to the total incident beam current and plotting the results as a function of the target gas pressure. Then the He total capture and loss cross sections were determined from

$$\sigma = \frac{\Delta F / \Delta P}{KL\varepsilon} \qquad (1)$$

where F is the fractional yield, P is the target gas pressure, (ΔF/ΔP is the slope of the linear regression of F vs. P), K = 3.3 × $10^{13}$ atoms/mTorr-$cm^3$, L is the target length in cm, and ε is the detection efficiency (assumed to be unity for the surface-barrier detectors). The calculated He cross sections for 1.0 and 1.5 MeV/u $O^{5+}$ were compared to previous results [5] [6] and subsequently used to check for systematic error, which was found to be about 20%. The He electron capture cross sections also provided an overall normalization factor, which was used to determine the total capture and loss cross sections for $O^{5+}$ + $O_2$ as well as the cross sections for recoil-ion production.

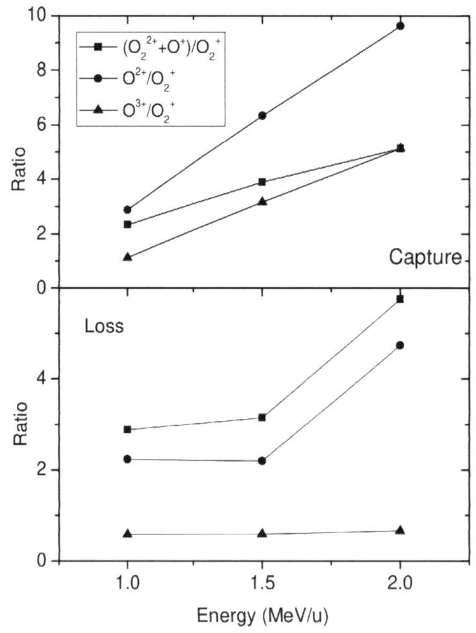


**Figure 2.** Ratios of charged molecular fragment yields to single non-dissociative ionization yields of molecular oxygen associated with projectile single electron capture and loss as a function of projectile energy.

The recoiling molecular and atomic oxygen peaks in the coincidence spectra were identified by comparison with the spectra for $O^{5+}$ + Ne (see figure 1) for which the peaks are readily identified [7]. The observed peaks for oxygen were integrated, the background subtracted, and the ratios of the charged molecular fragment yields to the single non-dissociative ionization yields as a function of energy were calculated as shown in figure 2. The ratios associated with electron capture increase nearly uniformly over the energy range measured, while those associated with electron loss increase non-uniformly. The fragment ratios associated with electron capture are generally larger than those associated with electron loss likely because capture has a higher probability to be accompanied by excitation or ionization. While the underlying mechanisms for the observed dependences are unknown, more data are clearly needed over a broader range of energies to better define the behaviours with collision energy.

The fractional yields for total electron capture and loss as a function of pressure were also determined for $O_2$, and the corresponding cross sections were obtained using equation 1 and by normalizing to the known He cross sections [5] [6]. These cross sections with their systematic plus statistical uncertainties are listed in table 1.

**Table 1.** Total single capture and loss cross sections for 1.0, 1.5, and 2.0 MeV/u $O^{5+}$ incident on molecular oxygen with absolute uncertainties. Scaled estimates from other work are given for comparison.

| Energy (MeV/u) | $\sigma_{q,q-1}$ ($\times 10^{-18}$ cm$^2$) | Ref. [8][a] ($\times 10^{-18}$ cm$^2$) | $\sigma_{q,q+1}$ ($\times 10^{-17}$ cm$^2$) | Refs. [5] [9][b] ($\times 10^{-17}$ cm$^2$) |
|---|---|---|---|---|
| 1.0 | 10±2 | 25 | 3.5±0.7 | 4.2 |
| 1.5 | 2.9±0.6 | 9.3 | 4.4±0.9 | 8.0 |
| 2.0 | 0.29±0.06 | 4.2 | 23±5 | 15 |

[a] Calculated from Ref. [8] assuming that the cross section for $O_2$ is twice that for O.
[b] Estimated by scaling results for Ne targets in Refs. [5] [8]

## 4. Conclusion

Ratios of charged molecular fragment yields associated with projectile charge-changing have been investigated for 1.0-2.0 MeV/u $O^{5+}$ + $O_2$ collisions. Measured yields for the molecular and atomic recoil-ion products were compared to the yields for singly-charged non-dissociative $O_2^+$. Additionally, total projectile electron capture and loss cross sections were determined for $O^{5+}$ + $O_2$. This study of molecular oxygen ionization and dissociation correlated with projectile single electron capture and loss for $O^{5+}$ + $O_2$ collisions will be extended to a wider range of incident energies. This will help to better determine the energy dependence of these ratios. Other projectile ion charge states will also be investigated, particularly fully-stripped and hydrogen-like ions. This information combined with data already available can be used to further investigate the more complex system of $O^{q+}$ + $H_2O$.


[1] Tachni C A, Galassi M E and Rivarola R D 2007 *J. Phys.: Conf. Ser.* **58** 271

[2] Leach J K, Van Tuyle G, Lin P-S, Schmidt-Ullrich R, Mikkelsen R B 2001 *Cancer Res.* **61** 3894

[3] Luna H et al 2007 *Phys. Rev. A* **75** 042711

[4] Montenegro E C, Shah M B, Luna H, Scully S W J, de Barros A L F, Wyer J A, Lecointre J 2007 *Phys. Rev. Lett.* **99** 213201

[5] Boman S A, Bernstein E M and Tanis J A 1989 *Phys. Rev. A* **39** 4423

[6] Price R N 1988 *M. A. Thesis,* Western Michigan University, unpublished

[7] Maurer R J, Can C and Watson R L 1987 *Nucl. Instrum. Meth. Phys. Res. A* **262** 99

[8] Schlachter A S, Stearns J W, Graham W G, Berkner K H, Pyle R V and Tanis J A 1983 *Phys. Rev. A* **27** 3372

[9] Houck J H, Závodszky P A and Tanis J A 1997 *Phys. Rev. A* **56** 1954

# Charge transfer processes in collisions of slow highly charged ions with polar molecules CO and $C_3H_8$

**M V Khoma[1], Makoto Imai[2], O M Karbovanets[1], Y Kikuchi[2], M Saito[3], Y Haruyama[3], M I Karbovanets[1], I Yu Kretinin[4], A Itoh[2,5] and R J Buenker[6]**

[1] Department of Theoretical Physics, Uzhgorod National University, Uzhgorod 88000, Ukraine
[2] Department of Nuclear Engineering, Kyoto University, Kyoto 606-8501, Japan
[3] Kyoto Prefectural University, Kyoto 606-8522, Japan
[4] Voronezh State University, University Sq. 1, 394006 Voronezh, Russia
[5] Quantum Science and Engineering Center, Kyoto University, Kyoto 611-0011, Japan
[6] Theoretische Chemie, Bergische Universität Wuppertal, D-42097 Wuppertal, Germany

E-mail: m.khoma@gmail.com

**Abstract.** Charge transfer processes resulting from low energy collisions of polar molecules CO and $C_3H_8$ with highly charged $Be^{2+}$ and $B^{2+}$ ions have been investigated experimentally and theoretically. The potential for the one–electron exchange interaction between a polar molecule and highly charged atomic ion were obtained in closed analytical form in the framework of a semiclassical approach. Obtained results were used for close–coupling calculations of total cross sections for single electron transfer in the studied reactions.

## 1. Introduction

Electron transfer processes in collisions of neutral molecules with slow highly charged ions are of great importance both for theoretical atomic physics and also for applied fields such as fusion plasmas and astrophysics. Much attention has been very recently called to the charge exchange processes involving polar molecules [1, 2]. In the present work we have studied single and double electron capture reactions in low energy collisions of atomic ions $Be^{2+}$ and $B^{2+}$ with polar molecules CO and $C_3H_8$ experimentally as well as charge transfer reactions,

$$C_3H_8 + Be^{2+} \rightarrow C_3H_8^+ + Be^+, \tag{1}$$

$$CO + A^{Z_A+} \rightarrow CO^+ + A^{(Z_A-1)+}, \tag{2}$$

where: (a) $A^{Z_A+} = B^{2+}(1s^2 2s)$, and (b) $A^{Z_A+} = Be^{2+}(1s^2)$, theoretically. Our theoretical analysis is carried out as an extension of the asymptotic theory of electron transfer in slow ion–atomic collisions [3] to processes involving a non–central target particle with a permanent dipole moment.

## 2. Experiment

A portion of the experimental cross sections reported here has been published in our previous paper [4], and more recently cross section data have been measured with almost the same

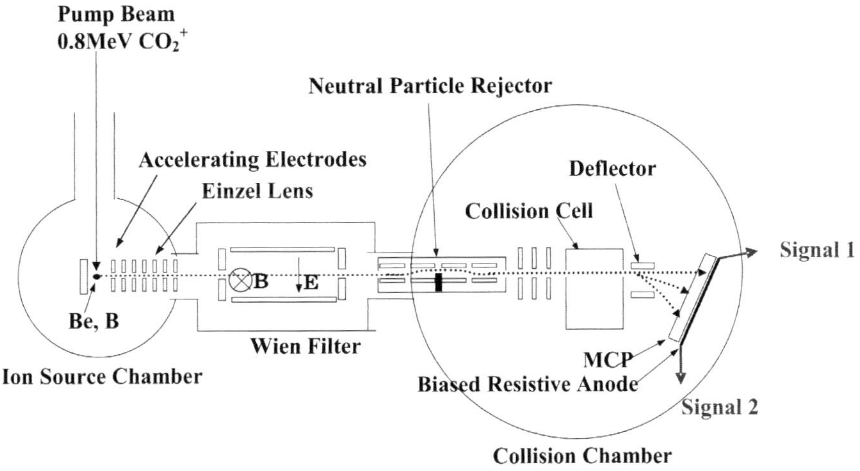


**Figure 1.** The experimental setup.

procedure [5]. Only major parameters are therefore reported here. The experiment was performed using the Van de Graaff Accelerator Facility of the Quantum Science and Engineering Center, Kyoto University. A schematic view of the experimental apparatus is given in figure 1. Incident ions were obtained by an Ion–Impact–Ion–Source [6], in which a pump beam of 0.8 MeV $CO_2^+$ impacts a target of glassy beryllium or boron containing alloy in wire shape, and fractions of sputtered ions were extracted and accelerated to desired energies of 15–32 keV. Extracted ions were velocity-analyzed by a combination of an E × B filter (Wien filter) and a 'neutral particle rejector', and collided with a target gas (CO and $C_3H_8$) in a gas cell, whose pressure was monitored with a high–sensitivity Pirani gauge calibrated with a capacitance manometer (MKS Baratron). Outgoing ions were electrostatically separated and detected with a Position–Sensitive Detector utilizing Micro–Channel Plate (MCP–PSD). Cross sections were derived by the growth method under so–called single–collision condition with target gas pressure of $10^{-2} \sim 10^{-1}$ Pa, while the base pressure of the collision chamber was kept below $1 \times 10^{-5}$ Pa. The measured cross sections $\sigma^{(21)}$ and $\sigma^{(20)}$ are presented in table 1. Total uncertainties are estimated to be 9 – 20 %, which mainly come from the effect of residual gases and the uncertainty of Pirani gauge readings.

**Table 1.** Single $\sigma^{(21)}$, and double $\sigma^{(20)}$ charge transfer cross sections for doubly–charged beryllium and boron ions colliding with CO and $C_3H_8$ targets.

| incident ion | target gas | collision energy (keV) | $\sigma^{(21)}$ ($cm^2$) | $\sigma^{(20)}$ ($cm^2$) | Ref. |
|---|---|---|---|---|---|
| $Be^{2+}$ | CO | 15 | $1.16 \times 10^{-15}$ | | [5] |
| | | | | $4.16 \times 10^{-17}$ | present work |
| $Be^{2+}$ | $C_3H_8$ | 16 | $2.1 \times 10^{-15}$ | $1.1 \times 10^{-16}$ | [4] |
| | | 24 | $1.9 \times 10^{-15}$ | $1.4 \times 10^{-16}$ | [4] |
| | | 32 | $1.8 \times 10^{-15}$ | $1.4 \times 10^{-16}$ | [4] |
| $B^{2+}$ | CO | 20 | $1.2 \times 10^{-15}$ | $4.3 \times 10^{-16}$ | [4] |

## 3. Theoretical method

Further we shortly outline the theoretical model used to describe the process of one–electron transfer. The theory for two–electron processes will be represented in our further publications.

We shall consider the charge transfer process within the 'one–electron approximation'. The active electron is considered as moving in the combined field of the molecular core $M^{Z_M+}$, which has permanent dipole moment $\mathbf{d}$, and the atomic ion $A^{Z_A+}$, respectively. Thus, the corresponding Schroedinger equation reads

$$\left(-\frac{\Delta}{2}+V_A(|\mathbf{r}-\mathbf{R}|)+V_M(\mathbf{r})-\frac{\mathbf{d}\cdot\mathbf{r}}{r^3}-E\right)\Psi(\mathbf{r})=0, \tag{3}$$

where $\Psi(\mathbf{r})$ and $E$ are the electronic wave function and energy, $\mathbf{r}$ and $\mathbf{R}$ are the position vectors of the electron and atomic ion $A^{Z_A+}$ with respect to the centre of mass of the molecule. We denote the energy and the wave function by $E_M$ and $\Psi_M$ when the electron is bound to core $M^{Z_M+}$ (quasimolecule $M^{(Z_M-1)+}+A^{Z_A+}$) and by $E_A$ and $\Psi_A$ when it is bound to atomic center $A^{Z_A+}$ (quasimolecule $M^{Z_M+}+A^{(Z_A-1)+}$), respectively. The effective potentials $V_A$ and $V_M$ of the electron interaction with corresponding core $M^{Z_M+}$ and $A^{Z_A+}$ have the following asymptotic behavior $V_A(|\mathbf{r}-\mathbf{R}|)=-Z_A/|\mathbf{r}-\mathbf{R}|$, and $V_M(\mathbf{r})=-Z_M/r$. In the limit $R\to\infty$, the eigenvalue spectrum of system $[AM]^{(Z_A+Z_M-1)+}$ is the sum of the eigenenergies $E_M^{(0)}$ or $E_A^{(0)}$ of the isolated $M^{(Z_M-1)+}$ or $A^{(Z_A-1)+}$ particles, respectively, the corresponding wave functions being $\Psi_M^{(0)}$ and $\Psi_A^{(0)}$. The exchange interaction $H_{if}$ responsible for electron transfer processes (1), (2) can be given in terms of a surface integral [3]

$$H_{if}=\frac{1}{2}\int\limits_S d\mathbf{S}\left(\Psi_A^*\nabla\Psi_M-\Psi_M^*\nabla\Psi_A\right), \tag{4}$$

where the $S$-plane divides the electronic location in the initial and final channels of the reactions (1), (2). For calculation of the wave functions $\Psi_A$ and $\Psi_M$ we employ a standard semiclassical approach [3]. In the tunneling region between interacting particles the approximate analytical representation of the wave function $\Psi_M$ can be expressed as [5]

$$\Psi_M(\mathbf{r})\underset{r\sim R/2}{=}r^{-1}Q(\mathbf{r})\sum_{\ell=|m_1|}^{\infty}\sum_{k=-\ell}^{+\ell}a_\ell^{m_1}(\mathrm{d})D_{km_1}^{\ell}(0,\beta,0)Y_\ell^k(\theta,\varphi), \tag{5}$$

where $m_1$ is the projection of the orbital momentum of the tunneling electron on the axis $\mathbf{d}$, expansion coefficients $a_\ell^{m_1}$ are given in [5], $D_{km_1}^{\ell}$ are the Wigner $D$–functions, $\beta$ being the angle between vectors $\mathbf{R}$ and $\mathbf{d}$, and $\{\theta,\varphi\}$ are the spherical angles of the vector $\mathbf{r}$ in the spherical coordinate system with axis $z$ oriented along the vector $\mathbf{R}$. Introduce the cylindrical coordinates $\{\rho,z,\varphi\}$ ($\rho=r\sin\theta$). The main contribution to the electron capture probability gives a small region of the surface $S$ near axis $\mathbf{R}$, where $\rho\ll z$.

We shall obtain the wave function $\Psi_M$ in this classically forbidden (under–barrier) region of the electronic motion along the coordinate $z$, ($z_1\le z\le z_2$, where $z_1$, $z_2$ are the turning points). In terms of the quasi momentum $p^2(\mathbf{r})\approx p^2(z)=2\left[|E|-Z_M/z-Z_A/(R-z)\right]$, the semiclassical expression for $Q(\mathbf{r})$ has the following form

$$Q(\mathbf{r})\underset{\rho\ll z}{=}\frac{n_1^{-1}}{\sqrt{(n_1Z_M)!p(z)}}\left(\frac{n_1Z_M}{e}\right)^{n_1Z_M}\exp\left(-\frac{\rho^2p(z)}{2z}-\int\limits_{z_1}^{z}\left|p\left(z'\right)\right|dz'\right),\quad e=2.718.... \tag{6}$$

The asymptotic expansion for barrier integral in equation (6) is given in [3]. The wave function $\Psi_A$ can be obtained in the same manner. Calculating the surface integral in (4) using the expressions for the one–electron wave functions $\Psi_M$, $\Psi_A$ one can obtain the final expression for exchange interaction $H_{if}$ [5].

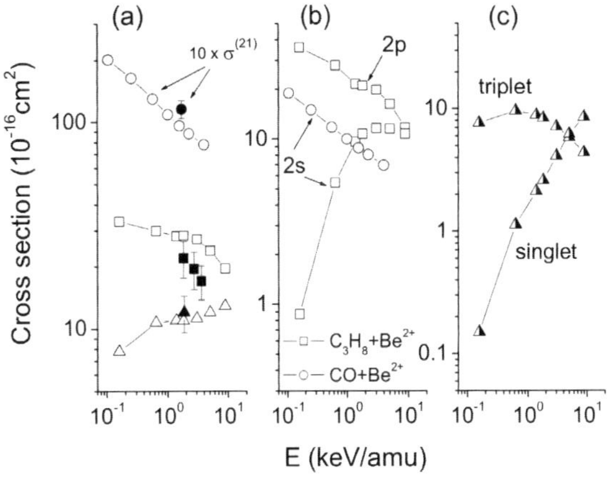


**Figure 2.** (a). Measured cross sections for single electron capture $\sigma^{(21)}$, ■ : $C_3H_8$ + $Be^{2+}$; ● : CO + $Be^{2+}$; ▲: CO + $B^{2+}$. Solid lines with open symbols: present calculations. (b). Present calculation for partial single capture cross sections for collisions CO + $Be^{2+}$ and $C_3H_8$ + $Be^{2+}$. (c). Present calculation for summed single capture cross sections $\sigma^{(21)}$ for collision CO + $B^{2+}$ into singlet and triplet states.

## 4. Results, discussion and concluding remarks

The dynamical treatment of the collision was carried out using the impact parameter method with straight–line trajectories. The resulting cross sections were then averaged over the different orientations of the dipole vector **d** with respect to the incident velocity vector. The measured and calculated results for total cross sections $\sigma^{(21)}$ are plotted in figure 2(a). Reactions (1) and (2b) take place without any changes in the $^1$S ionic core $Be^{2+}(1s^2)$, and we treated these colliding systems within the approximation of a single active electron interacting with frozen ionic core and corresponding polar ion. The most important products of the reaction (1) is $Be^+(2s)$ ions and for reaction (2b) are $Be^+(2s)$ and $Be^+(2p)$ ions. Corresponding partial cross sections are plotted in figure 2(b). The CO + $B^{2+}(1s^2 2s)$ system [reaction (2a)] was considered in the two–electron approximation (active 'transient' molecular electron and the 2s electron of the atomic core). Since no spin interaction was considered, the total spin was conserved and thus the calculations were carried out separately for the spin singlet and spin triplet symmetries. The total cross section were obtained by summation of the singlet and triplet cross sections with 1/4 and 3/4 weighting factors, respectively. Corresponding total cross sections are plotted in figure 2(c). For reaction (1) our calculations reproduced the measured energy behavior of the total cross section, but slightly ($\sim$ 15–20 %) overestimated the experimental data. Measured cross sections for reactions (2a) and (2b) are in good accord with the present calculations.

## References

[1] Otranto S and Olson R E 2008 *Phys. Rev.* A **77** 022709
[2] Bodewits D and Hoekstra R 2007 *Phys. Rev.* A **76** 032703
[3] Chibisov M I and Janev R K 1988 *Phys. Rep.* **166** 1; Chibisov M I 1981 *Zh. Eksp. Teor. Fiz.* **51** 470
[4] Imai M, Shirai T, Saito M, Haruyama Y, Itoh A, Imanishi N, Fukuzawa F and Kubo H 2006 *J. Plasma Fusion Res. Series* **7** 323
[5] Khoma M V, Imai M, Karbovanets O M, Kikuchi Y, Saito M, Haruyama Y, Karbovanets M I, Kretinin I Yu, Itoh A, and Buenker R J 2008 *Chem. Phys.* **352** 142
[6] Kusakabe T, Hanaki H, Nagai N, Horiuchi T and Sakisaka M 1983 *Physica Scripta* **T3** 191

# Cross Sections for Single electron capture in collisions of $N^{2+}$ with He at low energies

**K Soejima, S Tanaka and A Danjo**

Niigata University, Faculty of Science, Department of Environmental Science
8050 Ikarashi-ninocho, Nishi-ku, Niigata 950-2181, Japan

E-mail: soejima@env.sc.niigata-u.ac.jp

**Abstract**. The absolute cross sections of single electron capture during the collision of $N^{2+}$ with He have been measured over the energy range 0.1–300 eV/u. From the plot showing the dependence of cross sections on energy, the following three remarkable observations can be made: (1) the cross sections exhibit an oscillatory structure below 1eV/u, (2) a bend in the cross sections appears at approximately 1 eV/u and (3) there is a dip in the cross sections at approximately 70 eV/u. The formation of an oscillatory structure is attributed to the glory effect. The bend in the cross sections occurs owing to the attractive potential part of a $1^2\Pi$ state of quasimolecule $NHe^{2+}$. The dip in the cross sections appears as a result of opening a new reaction path above 70eV/u. The glory effect can be observed for the first time on a total cross section even for slow ion-neutral collisions. The objective of this study is to explore the possibilities of providing a detailed explanation of energy dependence measurements in order to study reaction dynamics for a charge changing process.

## 1. Introduction

Charge-changing processes involving multiply charged ions are not only fundamentally significant but are also widely used in various fields such as upper atmospheric physics, interstellar science, and fusion plasma science [1, 2]. It is essential to estimate the absolute values of the total charge-changing cross sections in order to successfully carry out charge-changing processes in the abovementioned fields. At present, however, there is a scarcity of experimental and theoretical data on the energy region below 1 keV/u.

Maier *et al.* [3] have measured single electron capture cross sections for $N^{2+}$-He ions over the low energy range 0.7–6 eV/u. Lennon *et al.* [4] have also measured the cross sections over the energy range 100–360 eV/u. Even more recently, Ishii *et al.* [5] have measured the cross sections over the energy range 0.2–260 eV/u. Theoretically, Lafyatis *et al.* [6] have calculated the potential energy curves for $NHe^{2+}$ using a multiconfiguration self-consistent field method and also estimated the cross sections using the Landau-Zener (LZ) model.

Recently, Ito [7] performed differential cross section measurements; the results of these measurements suggested that the single electron capture cross section could be enlarged at 0 °, which is a result of glory effect. The cross sections mentioned above are not affected by the glory effect. During slow neutral-neutral collisions, an oscillatory structure is observed in total cross sections due to the glory effect [8]. Does the oscillatory structure appear in the total cross sections even during ion-atom collisions? This study investigates the reason for the formation of the oscillatory structure. The absolute values of the total cross sections have been obtained for single electron capture during the collision of $N^{2+}$ with He over the energy range 0.1–300 eV/u. The present measurements have been performed over a wide energy range with fine energy step, which provide a detailed explanation of the dependence of cross sections on low energy. The dependence of cross sections on energy during ion-atom collisions was observed for the first time as an evidence of the glory effect.

## 2. Experimental Setup

The experimental apparatus and measuring procedure have been previously described in detail [9]. This section only summarizes the main features. Figure 1 shows the schematic diagram of the experimental setup.

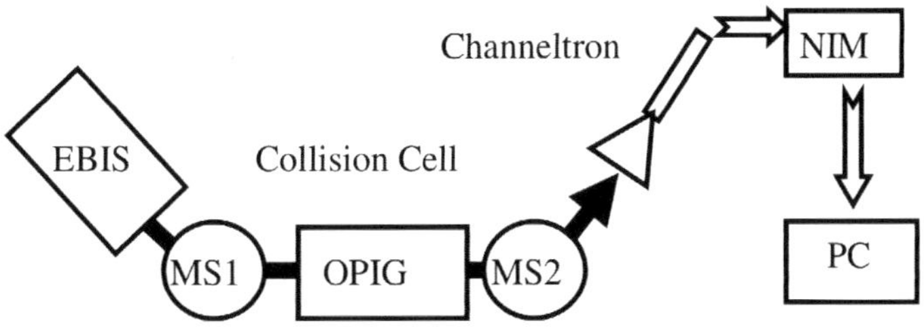


**Figure 1.** Schematic diagram of the experimental setup.

A 300 eV/u mass-selected beam of $N^{2+}$ ions was obtained from an electron beam ion source (EBIS). This beam, which has a measured full width and half maximum (FWHM) energy spread of 0.1 eV/u, was then focused and decelerated to the required energy before it entered a collision cell. The collision cell was equipped with an octupole ion beam guide (OPIG), which prevents the slow moving ions from diverging by means of an RF field [10]. The OPIG has a length of 160 mm, inner diameter of 7 mm, and an RF field strength of ~100 V at 14 MHz. Ion energy in the collision cell is defined as the difference between the voltages of the ion source and collision cell. The target gas pressure, which was low enough to ensure single collision conditions at all times, was determined using an MKS baratron pressure gauge, type 690A. Primary and product ions were reaccelerated after the beam passed through the collision cell, and their m/q was determined using the second electromagnet MS2. These ions were detected by a channeltron multiplier. The signal transmitted from the channeltron and fed into an electronic circuit consists of NIM modules, which were transferred to a personal computer.

For low energy collision, it is important to consider the possible effect of large angle scattering, because electron capture processes are exoergic and both products are positively charged. However, the OPIG used in this study can accommodate all the projectiles scattered in the forward direction. Furthermore, in order to check whether back scattering had a significant effect on the observed cross sections, attenuation cross sections were determined over the entire collision energy range. For all cases, the attenuation cross sections were found to be in good agreement with the single electron capture cross sections, within the estimated experimental error. It can be concluded that backward scattering did not have a significant effect on the observed cross sections at all energies. The overall uncertainty in the measured cross sections was estimated to be approximately $\pm 20\%$. The target gas pressure and collision length were the parameters that significantly contributed to this uncertainty. The uncertainty of the target gas pressure was estimated to be 10%. For the determination of gas pressure gradients, the collision path length was estimated to be $9.0 \pm 0.5$ cm.

It is also important to consider the possible effect of metastable $N^{2+}$ ions on the primary beam, since such contaminants are known to have significant and dramatic effects on the cross sections [11]. However, it is difficult to produce metastable ions from the EBIS-type ion source used in this experiment [12]. In addition, the pressure in the EBIS under operating collisions and the background pressure outside the source region were of the orders of $10^{-7}$ Pa and $10^{-8}$ Pa, respectively. At such low pressures, metastable ions, even if they are produced, will be quenched since they are produced as a result of multielectron impact collisions. Therefore we conclude that our measurements are valid only for ground state $N^{2+}$ ions.

### 3. Results and Discussion

Single electron capture cross sections during the collision of $N^{2+}$ with He over the energy range 0.1–300 eV/u, are shown in figure 2. Absolute errors of the cross sections are estimated to be ±20%.

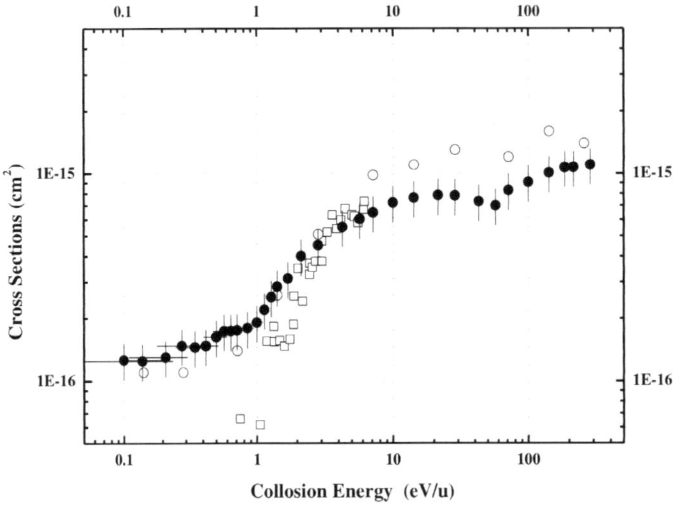


**Figure 2.** Single electron capture cross sections during $N^{2+}$-He collisions: ●, present results; ○, results obtained by Ishii *et al.*; □, results obtained by Maier II and Stewart.

Figure 2 shows that the results of our experiment are in good agreement with the previous measurements. The energy dependence of the cross sections has interesting structures.

A bend and a dip in the cross sections were clearly observed at collision energy of 1 and 70 eV/u, respectively. The cross sections estimated by Lafyatis *et al.* [6] using the LZ calculations results in the following reaction paths:

$$N^{2+}(2p,{}^2P)+He \rightarrow N^+(2p^2,{}^3P)+He^+ +5.0eV \quad (1)$$

$$\rightarrow N^+(2p^2,{}^1D)+He^+ +3.0eV \quad (2).$$

The ground state of quasimolecule $NHe^{2+}$ is $1^2\Pi$, which separates to the $N^+(2p^2,{}^3P)+He^+$ (1) limit. The potential energy curve of $1^2\Pi$ has an avoided crossing near 2 Å with the $2^2\Pi$ state and has a potential well of 1.9 eV at an equilibrium separation of 1.3 Å [6]. Above 0.1 eV/u, the reaction path via the avoided crossing represents the main path. In the case of $1^2\Sigma^+$ for $N^+(2p^2,{}^1D)+He^+$ (2) limit, the potential energy curve has an avoided crossing near 5 Å with the $2^2\Sigma^+$ and has a potential well of 0.4 eV at 1.8 Å. Below 0.1 eV/u, the reaction path changes to a path via the avoided crossing. Therefore, it is hard to consider that a cause of the bend in the cross sections at 1 eV/u is changing the main path from (2) to (1) with an increase in collision energies. The bend probably appears as a result of increasing the contribution of an attractive potential of the $1^2\Pi$ state to the reaction below 1 eV/u. It is well known that a dependence of total cross sections on energy for an attractive potential is different from that for a repulsive potential [13]. The dip in the cross sections at 70 eV/u is probably caused by opening a new reaction path via an avoided crossing which is located at a small internuclear distance.

An oscillatory structure was observed in the cross sections below 1eV/u, which can be attributed to the glory effect. If the differential cross section at 0 ° exhibits an oscillatory structure, the total cross section σ also exhibits an oscillatory structure according to the following relation:

$$\sigma \cong \frac{4\pi}{k}\sqrt{q(0)}$$

, where $q(0)$ is the differential cross section at 0 ° and $k$ is the wave number. At a collision energy of 1.5 eV/u, an enlargement of the differential cross section at 0 ° for single electron capture has been observed by Ito [7]. However, there is no clearly evidence of an oscillatory structure in our total cross sections around 1.5 eV/u. It is need to observe an oscillatory structure in total cross sections that the differential cross section exhibits a supernumerary rainbow. Peaks of a supernumerary rainbow move to 0 ° and disappear with an increase in collision energies; this is a cause of an oscillatory structure in the $q(0)$. The Peaks increase with a decrease in collision energies. Therefore, it becomes easier to observe an oscillatory structure in total cross sections at rather low energies. Since the collision energy of 1.5 eV/u is not enough low to observe a supernumerary rainbow, the oscillatory structure could not be observed. This can be confirmed to measure differential cross sections below 1 eV/u.

## 4. Conclusion

In this study, the absolute values of the total cross sections have been obtained for single electron capture during the collisions of $N^{2+}$ with He over the energy range 0.1–300 eV/u. The present measurements have been performed over a wide energy range with fine energy step, which provide a detailed explanation of the dependence of cross sections on low energies. Our results are in good agreement with those of the previous measurements. The energy dependence of the cross sections has interesting structures. A bend in the cross sections was observed at a collision energy of approximately 1eV/u; this probably appears as a result of increasing a contribution of an attractive part of $1^2\Pi$ potential curve below 1 eV/u. A dip in the cross sections was clearly observed at a collision energy of approximately 70 eV/u; this can be attributed to the changing of the main path. Below 1eV/u, an oscillatory structure was observed in the cross sections, which can be attributed to the glory effect.

The detailed measurements of energy distribution of total cross sections can serve as a basis for reaction dynamics.

## Acknowledgment

This study was supported by the Uchida Energy Foundation.

## References

[1] Fawcett B C 1991 *Z. Phys. D Suppl.* **21** S1
[2] Tawara H and Fritsch W 1989 *Phys. Scr.* T **28** 69
[3] Maier II W B and Stewart B 1978 *J. Chem. Phys.* **68** 4228
[4] Lennon M, McCullough R W and Gilbody H B 1983 *J. Phys.* B **16** 2191
[5] Ishii K, Itoh A and Okuno K 2004 *Phys. Rev.* A **70** 042716
[6] Lafyatis G P, Kirby K, and Dalgarno A 1993 *Phys. Rev.* A **48** 321
[7] Ito Y Private Communication
[8] Neumann W and Pauly H 1968 *Phys. Rev. Lett.* **20** 357
[9] Okuno K, Soejima K and Kaneko Y 1991 *Nucl. Instrum. Methods* B **53** 387
[10] Teloy E and Gerlich D 1974 *Chem. Phys.* **4** 417
[11] Irvane A D and Latimer C J 1991 *J. Phys.* B **24** L145
[12] Okuno K, Tawara H, Iwai T, Kaneko Y, Kimura M, Kobayashi N, Matsumoto A, Ohtani S, Takagi S and Tsurubuchi S 1983 *Phys. Rev.* A **28** 127
[13] Pauly H 1975 *Phys. Chem.* **6B** 553

Journal of Physics: Conference Series **163** (2009) 012057 doi:10.1088/1742-6596/163/1/012057

# Gross and partial ionization cross sections in 6-MeV/amu bare-ion collisions with methane

**Y. Ohno[1], T. Matsuo[2], T. Kohno[3], T. Nakazato[4], S. Watanabe[2] and T. Murakami[1]**

[1]National Institute of Radiological Sciences, 4-9-1 Anagawa, Inage-ku, Chiba, 263-8555, Japan
[2]Kanagawa Institute of Technology, 1030 Shimo-ogino, Atsugi, 243-0292, Japan
[3]Tokyo Institute of Technology, 4259 Nagatsuta-cho, Midori-ku, Yokohama, 226-8502, Japan
[4]Institute of Laser Engineering, Osaka University, 2-6 Yamadaoka, Suita, 565-0871, Japan.

E-mail: matuo@rm.kanagawa-it.ac.jp

**Abstract**. Gross and partial ionization cross sections of $CH_4$ molecules were measured in 6-MeV/amu $H^+$, $He^{2+}$, $C^{6+}$, $Ne^{10+}$ and $Ar^{18+}$ ion impact. The gross ionization cross sections in light $H^+$ and $He^{2+}$ ion impact were found to be proportional to square of projectile charge $q$ in accord with the first Born approximation. As the projectile charge was increased, the dependence became weaker than $q^2$. From mass-spectroscopic measurements, the cross sections for $CH_4^+$ and $CH_3^+$ ion production were also found to obey the $q^2$ dependence in low charge ion impact. In high charge ion impact, their dependence, however, became weak compared with $q^2$ scaling. The fragmented ions, such as $CH_2^+$, $CH^+$ and $C^+$, showed charge dependence steeper than $CH_4^+$ and $CH_3^+$ ions. Multiply charged carbon ions up to $C^{5+}$ were also observed.

## 1. Introduction

Ionization cross sections in energetic (>MeV/amu) heavy ion collisions are important for understanding the ionization mechanisms and fundamentals of astrophysics, cosmic ray physics, plasma physics and so forth. We have studied ionization of rare gases, hydrocarbons, and water in collisions of 6 MeV/amu bare-ions by measurements of absolute cross sections and relative intensities of secondary ions. Ionization processes of $CH_4$ have been studied under impact of various projectiles; 5-4000 keV protons [1], 1 MeV/amu -$F^{4,6-9+}$, -$Si^{5,6,8-12+}$, -$C^{3-6+}$ [2], 1-12 MeV protons [3], up to 3keV electrons [4], 50-6000keV protons and antiprotons [5], and 0.5-3.5 MeV proton and electron [6]. However, a limited number of experiments have been reported in collisions of energetic highly-charged ions. This paper discusses gross and partial ionization cross sections of $CH_4$ in 6-Mev/amu-$H^+$, -$He^{2+}$, -$C^{6+}$, -$Ne^{10+}$ and -$Ar^{18+}$ ion collisions.

## 2. Experiment

The experiments were carried out at the *Middle-Energy beam course* of Heavy Ion Medical Accelerator in Chiba (HIMAC) of the National Institute of Radiological Sciences (NIRS) in Japan. The gross (net) ionization cross sections were measured with a condenser-plate assembly, which was installed in a collision cell. When projectile ions enter the collision cell, recoil ions are produced. These ions are in turn collected with a parallel-plate electrode by the electric field. The cross sections were calculated from the ratios of total recoil ion current *Ir* to the projectile ion current *Ip*, (*Ir*/ *Ip*), measured as a function of the target-gas pressure, which were measured with a capacitance manometer. The systematic uncertainties in the gross ionization cross section were estimated to be 8.7%, and the statistical error was within 1.2% [7].

A mass-spectroscopic technique was applied to separate recoiled ions to get partial ionization cross sections. Secondary ions produced in collisions were extracted with an electrostatic lens, mass/charge analyzed by a double focusing sector magnet and finally detected by a Channeltron detector. Detailed description of the experimental setup and procedure have been reported in previous papers [7, 8].

## 3. Results and discussion

### 3.1. Gross ionization cross section

The measured gross ionization cross sections are summarized in table 1 and also shown in figure 1. In figure 1, they were plotted as a function of projectile charge $q$.

**Table 1.** Gross ionization cross sections for $CH_4$ in bare ion impact ( in $10^{-16}$ $cm^2$ )

| $H^+$ | $He^{2+}$ | $C^{6+}$ | $Ne^{10+}$ | $Ar^{18+}$ |
|---|---|---|---|---|
| 0.403 | 1.62 | 12.9 | 30.9 | 83.7 |

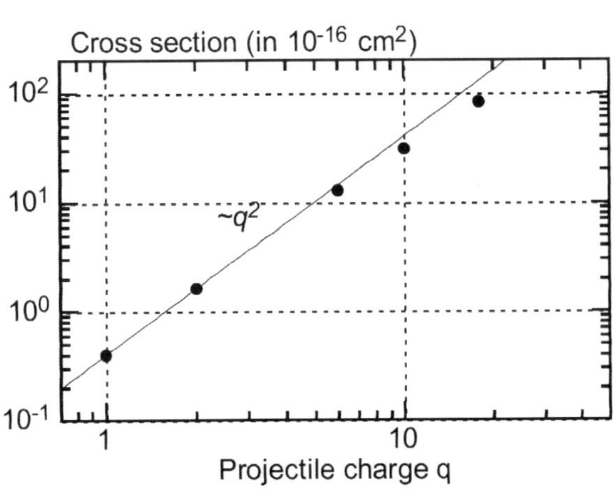


**Figure 1.** Projectile charge $q$ dependence of gross ionization cross sections for $CH_4$. Filled circles show the present results. Solid line represents the $q^2$ scaling.

For light projectiles, $H^+$ and $He^{2+}$, the gross ionization cross sections are well on a line showing $q^2$ dependence, as predicted by the first Born approximation. However the $q$ dependence became weaker as the projectile charge became increased, and the cross sections calculated from $q^2$ dependence results in 1.56 times larger than the experimental value in $Ar^{18+}$ impact, indicating a failure of the first Born approximation for high charge projectiles even in the present collision energy of 6 MeV/amu.

*3.2. Partial ionization cross sections*

In figure 2, mass/charge spectra for $CH_4$ in $He^{2+}$ and $Ar^{18+}$ impact are shown for comparison. They are characterized by several peaks corresponding to fragmented ions as well as the most prominent peak for parent $CH_4^+$ ions. As can be noted in figure 2, the intensity of $CH_4^+$ and $CH_3^+$ peaks are almost the same in $He^{2+}$ and $Ar^{18+}$ projectiles. Differences in peak intensity appear for fragmented ions, such as $CH_2^+$, $CH^+$ and $C^+$. The intensities of these fragmented ions relative to parent $CH_4^+$ ion increased as the projectile charge increased. In $Ar^{18+}$ impact, multiply charged $C^{n+}$ (n=2-4) are clearly seen. Production of $C^{5+}$ ions was also confirmed in a separate measurement. Partial ionization cross sections were calculated from the relative intensities of peaks in the mass spectra and the gross ionization cross sections in table 1.

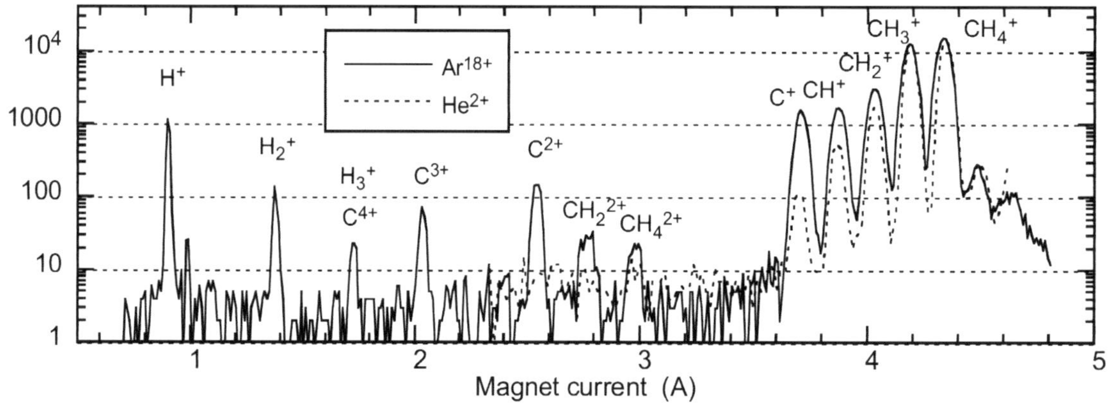


**Figure 2.** Mass/charge spectra for $CH_4$ in $He^{2+}$ and $Ar^{18+}$ impact.

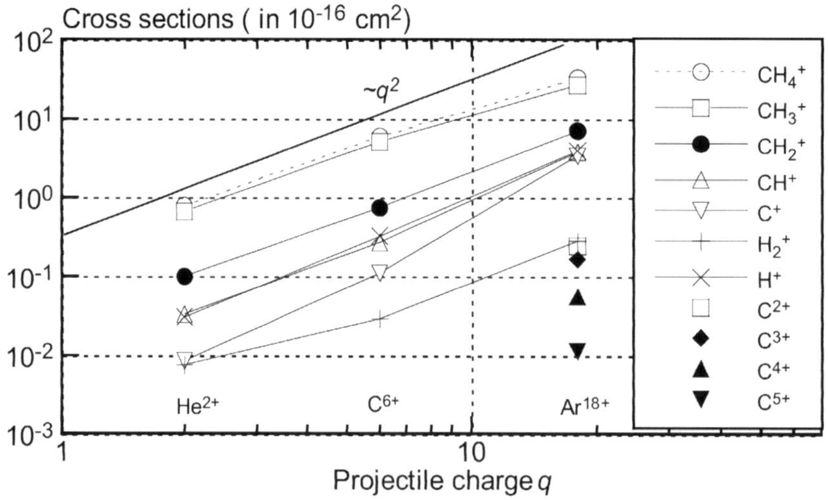


**Figure 3.** Partial ionization cross sections for $CH_4$ as a function of projectile charge $q$.

Figure 3 shows the obtained partial cross sections as a function of projectile charge $q$. The cross sections for $CH_4^+$ and $CH_3^+$ ions are prominent, and they increase with a slope of around 2 in a log-log plot. This indicates that they are mainly produced at distant collisions, where the first Born approximation would be applicable. In addition, the intensity ratios of fragmented $CH_3^+$ ions to parent $CH_4^+$ ion were almost independent of $q$, indicating that these ions are mainly produced through two-step processes as described by the quasiequilibrium theory (QET) by Rosenstock [9]: An excited state molecular $CH_4^{+*}$ ion is first produced in the impact, and this ion then de-excites or fragments statistically with a fragmentation pattern which depend only on the internal energy of the excited $CH_4^{+*}$ ion and does not depend on the ionization cross section.

On the other hand, fragmented ions, such as $CH_2^+$, $CH^+$ and $C^+$, showed charge dependence steeper than the $q^2$ scaling, indicating deviation from the first Born approximation and that other mechanisms, such as multiple ionization and inner-shell ionization, might contribute to their production. Moreover, the intensity ratios of these ions to parent $CH_4^+$ ions were found to change considerably with projectile charge $q$, indicating deviation from the quasiequilibrium theory.

## 4. Summary

Gross ionization cross sections for $CH_4$ molecule in light $H^+$ and $He^{2+}$ ion impact were found to show $q^2$ dependence, as predicted by the first Born approximation. As the projectile charge was increased, the charge dependence became weaker than $q^2$, indicating a failure of the first Born approximation for high charge projectiles such as $C^{6+}$, $Ne^{10+}$, and $Ar^{18+}$ even in the present collision energy of 6MeV/amu. The mass/charge spectra of recoil ions showed that the most prominent product was parent $CH_4^+$ ions in all projectiles studied. Highly fragmented $CH_m^+$ (m=0,1 and 2) ions and multiply-charged $C^{i+}$ (i=up to 5) ions were detected with considerable intensities in high charge $Ar^{18+}$ impact. Deviation from quasiequilibrium theory was observed in production of these fragmented ions.

## Acknowledgements

We are grateful to the members of NIRS-HIMAC for their skillful operation of the accelerator and their support during this work. This work was performed as part of the research project with Heavy Ions at NIRS-HIMAC.

## References

[1] Rudd M E, DuBois R D, Toburen L H, Ratcliffe C A, and Goffe T V, 1983, *Phys. Rev.* A 28 3244-3257
[2] Malhi N B, Ben-Itzhak I, Gray T J, Legg J C, Needham V, Carnes K, and McGuire J H, 1987, *J. Chem. Phys.* 87 (11), 6502-6506
[3] Ben-Itzhak I, CarnesK D, Johnson D T, Norris P J, and Weaver O L, 1994, *Phys. Rev.* A 49 881-888
[4] Nishimura H and Tawara H, 1994, *Phys. B: At. Mol. Opt. Phys.* 27 2063-2074
[5] Knudsen H, Mikkelsen U, Paludan K, Kirsebom K, Moller S P, Uggerhoj E, Slevin J, Charlton M, and Morenzoni E, 1995, *Phys. B: At. Mol. Opt. Phys.* 28 3569-3592
[6] Luna H, Cavalcanti E G, Nickles J, Sigaud G M and MontenegroE C, 2003 *Phys. B: At. Mol. Opt. Phys.* 36 4717-4729
[7] Matsuo T, Tonuma T, Kumagai H, and Tawara H, 1994 *Phys. Rev.* A 50 1178-1183
[8] Matsuo T, Kohno T, Makino S, Mizutani M, Tonuma T, Kitagawa A, Murakami T and Tawara H 1999 *Phys. Rev.* A 60 3000-07
[9] Rosenstock H M, Wallenstein M B, Wahrhaftig A L, and Eyring H 1952 *Proc. Natl, Acad. Sci. U.S.A.* 38 667

Journal of Physics: Conference Series **163** (2009) 012058 doi:10.1088/1742-6596/163/1/012058

# High resolution low-energy dielectronic recombination rate coefficients of beryllium-like germanium: QED test bench for two-valence-electron systems

**D. A. Orlov[1], C. Krantz[1], D. Bernhardt[2], C. Brandau[3], J. Hoffmann[1], A. Müller[2], T. Ricsóka[2], S. Schippers[2], A. Shornikov[1], and A. Wolf[1]**

[1] Max-Planck-Institut für Kernphysik, D-69117 Heidelberg, Germany

[2] Institut für Atom- und Molekülphysik, Universität Giessen, D-35392 Giessen, Germany

[3] Gesellschaft für Schwerionenforschung, D-64291 Darmstadt, Germany

Email: `orlov@mpi-hd.mpg.de`

**Abstract**. Low-energy dielectronic recombination resonances of beryllium-like $Ge^{28+}$ have been studied by electron collision spectroscopy in the heavy-ion storage ring TSR using a twin-electron beam technique combined with an ultracold photoelectron source. Rich resonance structure below 1.7 eV is observed which mainly arises from resonant capture of the free electron to $9l$ Rydberg states accompanied by excitation of inner valence electrons from the $2s^2(^1S_0)$ ground state to $2s2p(^1P_1)$. The positions of resonances below 0.3 eV are measured with an accuracy better than 0.5 meV. Accurate calculations of Rydberg binding energies for these resonances are still required in order to derive a precise value for the $^1S_0-^1P_1$ transition energy, which may give the opportunity for high-accuracy testing of QED corrections to atomic energy levels for a two valence electron system.

## 1. Introduction

In a dielectronic recombination (DR) process the capture of a free electron into a Rydberg state is accompanied by excitation of inner electrons. Radiative decay of this intermediate autoionizing state completes the recombination. DR is a resonant process which takes place when the core excitation energy equals the sum of electron-ion collision energy and Rydberg binding energy. In many cases of lithium-like and beryllium-like systems, binding energies of high Rydberg states can be calculated with high accuracy [1]. In these cases, measurements of the DR rate coefficient at high energy resolution are tools to precisely derive the transition energies of the core valence electrons. The highest precision is reached when studying resonances of low energies (ideally below ~ 0.1 eV). In this case the experimental resolution reaches an optimal value about equal to the transverse electron temperature $kT_{\perp}$, which is better than 1 meV for the TSR's ultra cold electron target.

Measurements of low-energy DR resonances of lithium-like $Sc^{18+}$, performed at the Heidelberg Test Storage Ring (TSR) with a cold photoelectron target, allowed us to determine the $2s_{1/2}$-$2p_{3/2}$ transition energy with an accuracy of 4.6 ppm, less than 1% of the few-body effects on radiative corrections [2]. Following these studies we have investigated beryllium-like heavy ions with low-energy DR resonances where quantum-electrodynamical (QED) corrections to atomic energy levels are large, making the systems promising candidates for accurate QED tests with two valence electrons.

In this paper we present high resolution measurements of the electron-ion recombination rate coefficient of beryllium-like germanium. The study revealed a series of sharp low-energy resonances. In the range from about 0.1 eV up to 1.8 eV dielectronic recombination is governed by $(2s2p\,^1P_1)9l$ resonances. Additional dielectronic $(2s2p\,^3P_0)14l$ and trielectronic $(2p^2\,^1D_2)7l$ resonances can also contribute to the low-energy part of the recombination rate spectrum. Accurate calculations of Rydberg binding energies for these resonances are still required to derive a precise value for the $^1S_0 - {}^1P_1$ transition energy.

## 2. Experimental overview

The experiment at the Heidelberg TSR was carried out with beryllium-like $^{74}Ge^{28+}$. The MPIK tandem accelerator was used to inject ions of kinetic energies of about 340 MeV into the TSR. The TSR's twin-beam electron facility was used for these measurements [3]. Phase space cooling of ions was provided by a collinear electron beam of high density ($1.6\times10^7$ cm$^{-3}$) from the electron cooler at a cathode voltage of about 2.56 kV, yielding a stored ion beam with a relative momentum spread below $10^{-4}$ and of cross-section diameter ~ 1 mm. The cooling time was below 1 s, the lifetime of the stored ion beam, limited by radiative recombination in the cooler, was about 45 s. For the measurements a continuous injection scheme with a repetition rate of about 1-3 s was used, leading to a stored ion current of 5-10 μA after a few minutes of accumulation.

The ultra cold electron target was used for collision spectroscopy, employing a low electron beam density of $5.2\times10^5$ cm$^{-3}$ from the cryogenic photoelectron source. The photocathode gun delivered a magnetically guided dc electron beam of about 350 μA, emitted in space charge limited mode by a p-GaAs photocathode of diameter 3 mm cooled by liquid nitrogen. Details of the cryogenic photocathode gun can be found elsewhere [4-5]. The transverse energy of the electron beam was reduced by adiabatic magnetic expansion (guiding field ratio: 20) and the electron temperatures of the target beam after acceleration to the final energy were about 1.5 meV and 0.04 meV in the transverse and longitudinal direction respectively. The electron beam overlapped the stored $Ge^{28+}$ beam in an interaction region of length 1.5 m. For recombination spectrum measurements the electron-ion collision energies in the c.m. frame are varied by detuning the acceleration voltage of the electron target from the cooling voltage [2]. The zero collision energy corresponds to matched electron and ion velocities at the cooling voltage.

The recombination rate of electrons and $Ge^{28+}$ ions was measured by counting the $Ge^{27+}$ product ions using a detector behind the first ion-bending dipole magnet downstream of the interaction region of the electron target. The measured rate spectrum allowed to derive the recombination rate coefficient as a function of electron-ion collision energy using the measured ion and electron beam currents, the electron target voltage, and the well-understood electron space charge corrections to the latter.

## 3. Results and discussion

The overview spectrum of the $Ge^{28+}$ recombination rate coefficient in the energy range of 0-8.5 eV is shown in figure 1a. Calculated energy positions of $l$-series limits for dielectronic $2s2p\,^1P_19l$ (1.70 eV), $^3P_014l$ (0.22 eV), $^3P_114l$ (7.76 eV), $^3P_211l$ (4.34 eV) and trielectronic $2p^2\,^1D_27l$ (7.70 eV) capture resonances are also shown. Resonances from other dielectronic and trielectronic terms lie at higher energies and not considered here. To estimate the positions of $l$-series limits, the excitation energies from the $2s^2$ ground state to $^1P_1$ (133.46±0.04 eV), $^3P_1$ (62.25±0.01 eV) and $^3P_2$ (92.53±0.015 eV) were taking from spectroscopic measurements in tokamak plasmas [6]. The transition energies for $^3P_0$ (54.64 eV) and $^1D_2$ (225.58 eV) are Z-interpolated data of theoretical values obtained by a

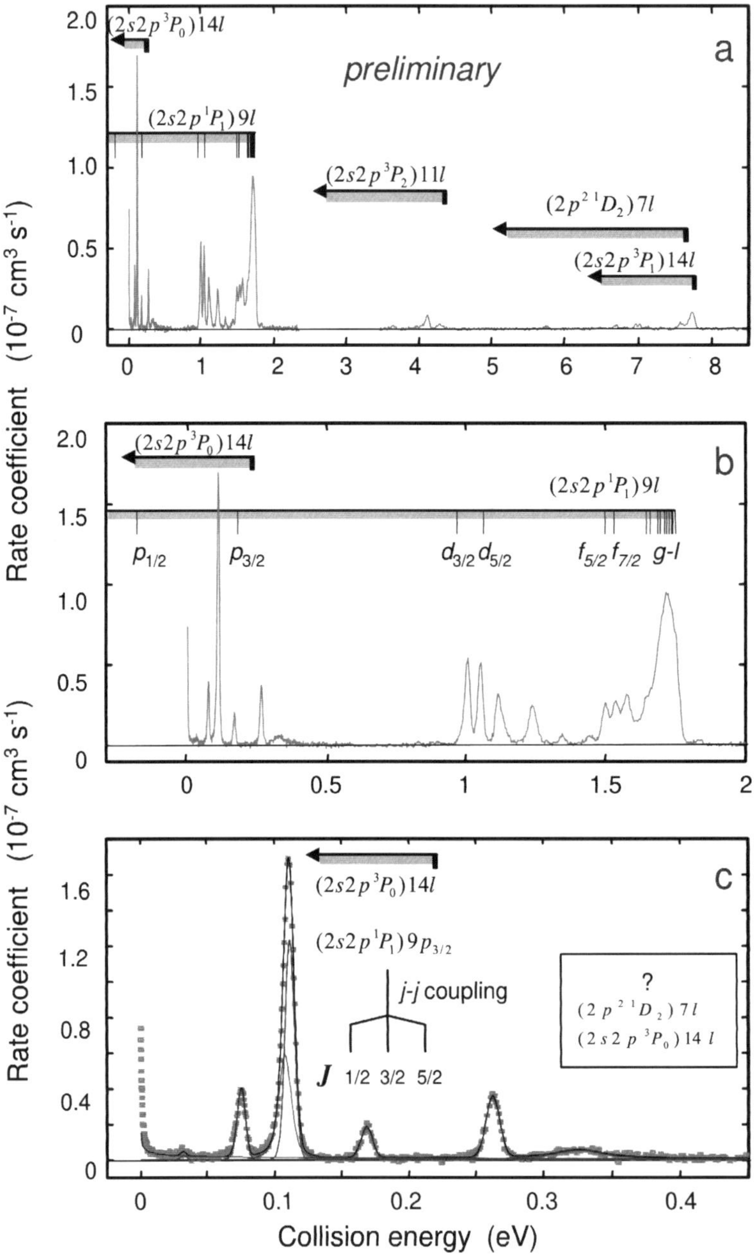


**Figure 1.** High resolution recombination rate coefficients of beryllium-like $^{74}Ge^{28+}$. Energy positions of $l$-series limits for dielectronic $2s2p$ $^1P_19l$, $^3P_014l$, $^3P_114l$, $^3P_211l$ and trielectronic $2p^2$ $^1D_27l$ resonances are also shown. The low-energy part of the spectrum below 0.5 eV ("c") is fitted with seven resonances convoluted with an electron energy distribution ($kT_\perp$=1.7 meV, $kT_\parallel$=0.47 meV).

multiconfiguration Dirac-Fock technique for excitation energies of beryllium-like ions with different nuclear charge $Z$ [7]. The same interpolation procedure was also applied for $^1P_1$, $^3P_1$ and $^3P_2$ to estimate the accuracy of the interpolation (about 0.5 eV) by comparison to spectroscopic data [6]. Rydberg binding energies were calculated using the H-like formula with $Z_{eff}=Z-4$, which should be good enough to estimate $l$-series limits when the effect of the smeared potential from the core electrons is small for high orbital angular momentum quantum numbers.

The rich structure below 2 eV (see figure 1b) is governed by resonant capture of a free electron to the $9l$ Rydberg state accompanied by transition of a valence electron from the $2s^2$ ground state to the $2s2p$ orbital forming a $^1P_1$ singlet state. The estimated positions of $^1P_1$ $9l$ resonances, calculated in a Dirac-Fock potential from $1s^22s^2$ [8], are also shown. While the $9s$ and $9p_{1/2}$ binding energies are below the $2s2p$ $^1P_1$ excitation energy, $9p_{3/2}$ is very closely above.

The most "attractive" resonances lie at low energies (see Figure 1c), where the highest resolution can be obtained. Below 0.4 eV six resonances (without counting the radiative recombination peak at zero energy) are clearly observed. The fitting of these resonances (see Figure 1c) was done by convolution with an electron energy distribution with electron temperatures of $kT_{\perp}$=1.7 meV and $kT_{\parallel}$=0.47 meV, the latter being a fitting parameter. The strongest resonance lies at 0.1132 eV. The fitting suggests that there is one more broad resonance close to the strongest one. The widths of this resonance as well as of a resonance at 0.328 eV were fitting parameters. The remaining resonances were assumed to be δ-peaks, and only their positions and strengths were fitted. However, to get precise (better than 0.5 meV) positions of these low-energy resonances it is important to assign all peaks and to calculate their natural width. In the $^1P_19l$ series the splitting of $9p_{3/2}$ resonance due to $j$-$j$ coupling, with total angular momentum $J$=1/2, 3/2 or 5/2, is expected to give three resonances split by typically [1] about 10-100 meV. The assignment of the fine structure levels and of the other resonances is still an open question (calculation in progress [8]). Candidates are $(2s2p\ ^3P_0)14l$ or $(2p^2\ ^1D_2)7l$ dielectronic resonances. Despite the fact that the $(2p^2\ ^1D_2)7l$ series limit lies far above threshold at 7.7 eV the $l$-resonances from this term can be shifted strongly to low energies (comparable to $^3P_114l$ and $^3P_211l$) due to their small principal quantum number $n$ and because of their orbital configuration being $2p^2$ rather than $2s2p$.

In conclusion, high resolution measurements of low-energy electron capture recombination rate coefficient have been performed for beryllium-like germanium. Rich resonance structure at low energies is found. Dielectronic $2s2p$ $(^1P_19p_{3/2})_{J=1/2,\ 3/2,\ 5/2}$ resonances are supposed to give the main contribution to the low-energy part of the spectrum. Contributions from other terms $(2s2p\ ^3P_0)14l$ and $(2p^2\ ^1D_2)7l$ are also expected. To derive the $^1S_0 - ^1P_1$ transition energy, accurate calculation of the Rydberg binding energies of the excited ions is still required. The obtained data may give the opportunity to test QED corrections with high accuracy for atomic energy levels for two valence electron systems.

**Acknowledgments**
The authors wish to thank Dr. E. Lindroth for fruitful discussions. The great support of this experiment by the TSR team, in particular from behalf of Dr. M. Grieser and Dr. R. Repnow, is gratefully acknowledged.

**References**
[1] Tokman M, *et al.* 2002 *Phys. Rev. A* **66** 012703
[2] Lestinsky M, *et al.* 2008 *Phys. Rev. Let.* **100** 033001
[3] Sprenger F, *et al.* 2004 *NIM A* **532** 298
[4] Orlov D A, *et al.* 2004 *NIM A* **532** 418
[5] Orlov D A, *et al.* 2005 *J.Phys.: Conf. Series* **4** 290
[6] Hinnov E, *et al.* 1986 *J. Opt. Soc. Am. B* **3** 1288; http://physics.nist.gov/PhysRefData/
[7] Cheng K T, Kim Y K, Deslaux J P 1979 *At. Data and Nucl. Data Tabl.* **24** 111
[8] Lindroth E (Stockholm) *private communication*

Journal of Physics: Conference Series **163** (2009) 012059 doi:10.1088/1742-6596/163/1/012059

# Low-lying dielectronic resonances of Fe XXII for accurate energy determination of autoionising Rydberg levels on boronlike cores

**C Krantz[1], D A Orlov[1], D Bernhardt[2], C Brandau[3], J Hoffmann[1], A Müller[2], T Ricsóka[2], S Ricz[2], S Schippers[2], and A Wolf[1]**

[1] Max-Planck-Institut für Kernphysik, 69117 Heidelberg, Germany
[2] Institut für Atom- und Molekülphysik, Justus-Liebig Universität, 35392 Gießen, Germany
[3] Gesellschaft für Schwerionenforschung, 64291 Darmstadt, Germany

E-mail: `claude.krantz@mpi-hd.mpg.de`

**Abstract.** We present high resolution measurements of dielectronic recombination (DR) of Fe XXII forming doubly excited Fe XXI. These measurements were performed at the Test Storage Ring (TSR) of the Max-Planck-Institute of Nuclear Physics (MPIK). Low-lying DR resonances allow precise energy determination of the doubly excited autoionising levels in Fe XXI and, with suitable future calculations of Rydberg binding energies ($n \geq 7$), of $2s^22p$ to $2s2p^2$ excitation energies of the boronlike Fe XXII core.

## 1. Introduction

Experiments at the TSR have shown that highly precise DR measurements can compete with optical spectroscopy at determining the core excitation energy of a highly charged lithiumlike ion if the autoionising state is sufficiently low-energetic [1, 2]. Such determination of the core excitation energy from DR resonances relies on accurate calculation of the highly-excited Rydberg electron eigenstates. As higher order effects like QED corrections are negligible for the Rydberg electron due to its large distance to the atomic nucleus, its binding energy can be computed within the framework of many-body perturbation theories (MBPT) [1].

While highly precise MBPTs exist for helium- and lithiumlike ionic cores, work on more complex systems with several valence electrons is still underway. We measured low-lying DR resonances of boronlike Fe XXII at an energy resolution much better compared to a previous measurement performed on this system. Our data can thus serve as a testbench for future atomic structure calculations aiming at accurately determining the total energy of the corresponding doubly excited autoionising states.

## 2. Overview of the experiment

The MPIK's tandem-booster accelerator facility was used to produce a beam of $^{56}Fe^{21+}$ at a total kinetic energy of 250 MeV. This beam was injected and stored in the TSR at a magnetic rigidity of about 1.1 Tm. The TSR's electron cooler was used to cool the ion beam down to a cross-section diameter of about 1 mm. Beam cooling with the electron cooler was practically instantaneous, thus a continuous injection scheme could be used during the measurements, in contrast to a more common mode of operation, where, following an ion injection, data taking has to be interrupted for several seconds in order to give the beam time to cool down. While

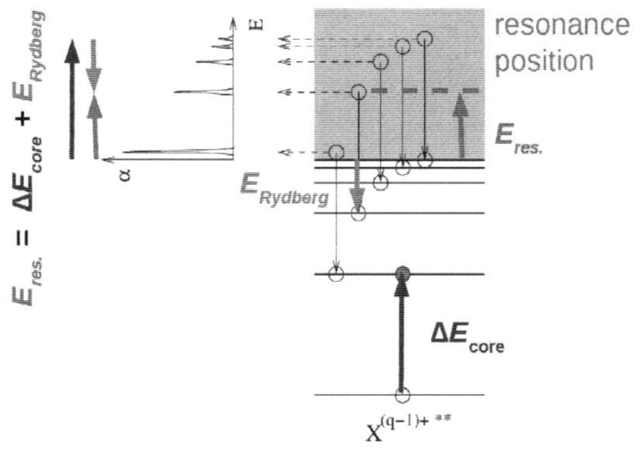


**Figure 1.** DR is the resonant capture of a free electron by an atomic ion through an intermediate, doubly-excited state. The incident electron is bound in a high Rydberg state while its energy is absorbed into an ionic core excitation. Subsequently the autoionising state stabilises radiatively, completing the recombination process.

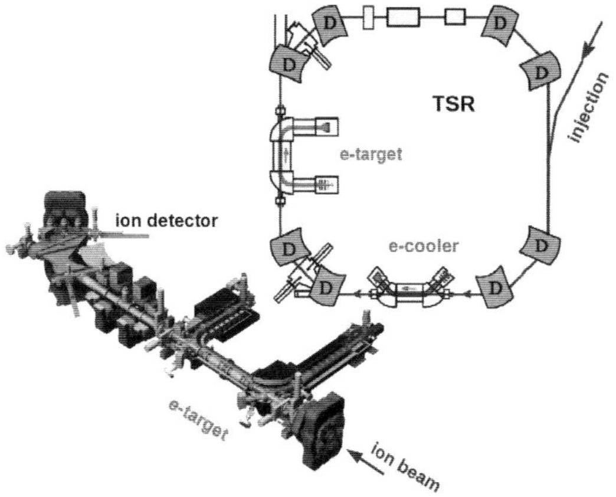


**Figure 2.** Schematic view of the heavy ion storage ring TSR and close-up view of the electron target. In a straight section of the TSR, the stored ions are collided with a cold electron beam. Recombination products are separated from the stored beam at the TSR's bending magnet following the electron target and are counted by a scintillation detector.

circulating ion currents as high as 65 $\mu$A could be achieved through multiturn-injection and electron cooler stacking, low currents of about 5...10 $\mu$A were used in the experiment in order to avoid heating of the ion beam by intra-beam scattering.

In another straight section of the storage ring, the TSR's second merged electron beam – referred to as the "electron target" – was used to collide the stored ions with a cold electron beam of well defined energy [3]. The collision energy arises from velocity detuning of the electron beam with respect to the ions. At the TSR bending magnet following downstream of the electron target, the charge-changed recombination products are separated from the stored beam, and the electron-ion recombination rate is measured using a scintillation counting detector.

## 3. Achieving high energy resolution

In order to achieve the highest possible energy resolution in a merged-beams DR measurement several aspects are of importance.

### *3.1. Low-lying DR resonances*

It is a fundamental property of the merged-beams technique that its resolution is highest close to the condition of vanishing electron-ion collision energy. The collision energy arises from the velocity difference $\Delta v$ between electrons and ions. In a non-relativistic approximation on can write

$$E_{\rm coll} \approx \frac{m_{\rm e}}{2}\Delta v^2 = \left(\sqrt{E_{\rm e}} - \sqrt{\frac{m_{\rm e}}{m_{\rm i}}E_{\rm i}}\right)^2 \approx \frac{m_{\rm i}}{m_{\rm e}}\frac{(\Delta E_{\rm e})^2}{4E_{\rm i}}\,, \tag{1}$$

where $E_{\rm e}$ and $E_{\rm i}$ are the kinetic energies, and $m_{\rm e}$ and $m_{\rm i}$ the masses of electrons and ions respectively. $\Delta E_{\rm e}$ is the lab-frame energy detuning of the electron target beam. One sees that the collision energy $E_{\rm coll}$ in the centre-of-mass frame of electron and ion is approximately proportional to the lab-frame energy detuning squared. Thus for small detunings, the resolution in the collision energy $E_{\rm coll}$ improves linearly with $\Delta E_{\rm e}$ as one approaches the point of velocity matching between the two beams.

Journal of Physics: Conference Series **163** (2009) 012059 doi:10.1088/1742-6596/163/1/012059

### *3.2. Independent cooler and target beams*

The TSR is the world's only heavy-ion storage ring featuring two independent electron beam merging sections. For DR measurements the electron beam has to be detuned from velocity matching with the ion-beam. In the case of a single electron beam, the ions velocity will change as they feel a friction force in the velocity-detuned electron bath. While this can to some extent be prevented by fast switching of the electron beam between cooling and collision modes, electron-induced changes of the ion velocity remain the dominant resolution limiting factor for collision energies below 1 eV for DR measurements performed with a single electron merging section [1].

However, usage of the TSR's independent electron target section completely eliminates this limitation, as the ion velocity continues to be defined by the dedicated electron cooler, even for very small detunings of the electron target beam corresponding to collision energies of only a few meV [3].

### *3.3. Ultra-cold electron target beam*

The target beam is produced by a cryogenic photoelectron gun based on GaAs-photocathodes [4, 5]. This electron source provides a free electron gas of an initial thermal energy spread $k_BT = 10\ldots15$ meV [4]. Magnetic expansion of the electron beam by guiding field ratios of typically 20 to 40 reduce the kinetic energy spread transverse to the beam propagation direction down to $k_BT_\perp = 1\ldots2$ meV. Subsequent acceleration to ion velocity kinematically reduces the longitudinal velocity spread down to values $k_BT_\| \approx$ few $10^{-5}$ eV in the centre-of-mass frame.

## 4. Experimental results

The experimentally determined recombination rate coefficient of Fe XXII is shown in figure 3. The improvement in energy resolution compared to an earlier measurement also performed at the TSR, but employing the electron cooler alone [6], is significant. Our measurement with the

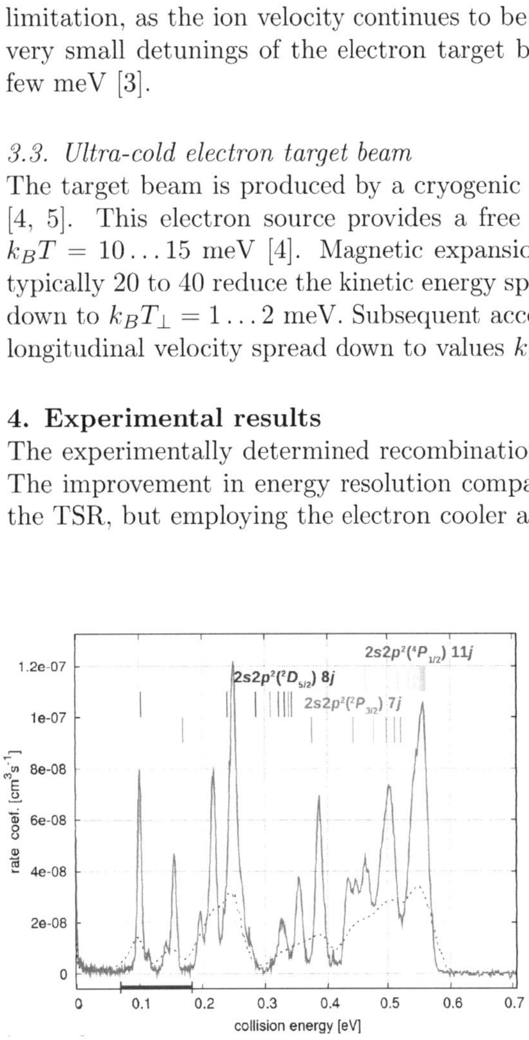


**Figure 3.** Our experimental recombination rate coefficient for Fe XXII forming Fe XXI (solid red) and indicative DR resonance positions, calculated assuming a hydrogenic potential for the Rydberg electron and using core excitation energies taken from [7]. The dotted black curve is an earlier TSR measurement performed using only the electron cooler [6].

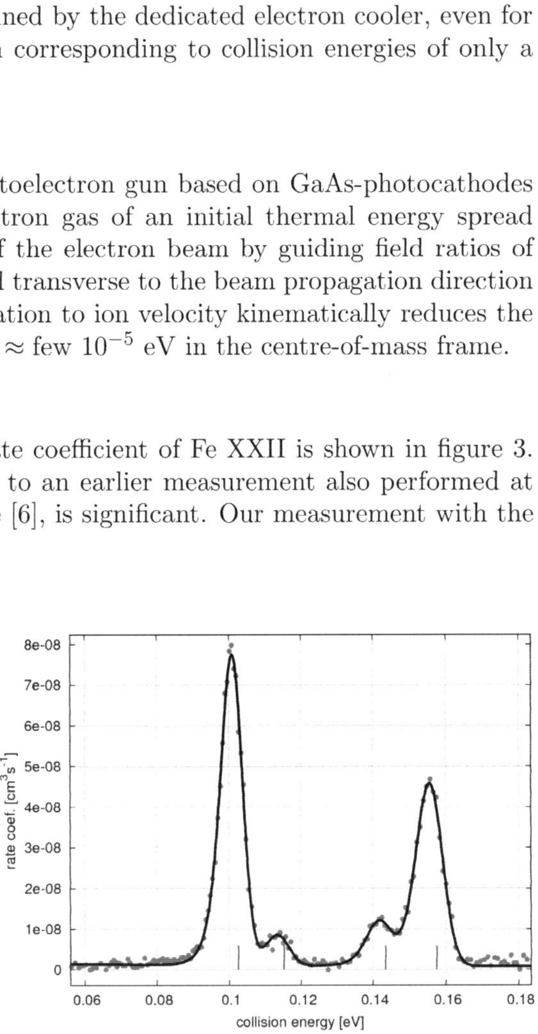


**Figure 4.** Blow-up of the highlighted part of the experimental DR spectrum shown in figure 3 (red dots). A fit to the data (solid black) suggests that the four lowermost lines in the observed spectrum are individually resolved. The fit assumes delta-shaped DR resonances convoluted with the electron velocity distribution in the target beam. The resonance positions are indicated in blue.

Journal of Physics: Conference Series **163** (2009) 012059 doi:10.1088/1742-6596/163/1/012059

electron target unveils complex resonance structures at electron-ion collision energies of 100 meV to 600 meV which were previously washed-out by the limited energy resolution of the electron cooler.

The origin of the magnitude difference between both measurements is unclear. It could arise from the fact that we used – in an effort to achieve the lowest possible beam temperature – low ion currents of 5...10 $\mu$A. At such low currents the TSR's build-in beam diagnostics is estimated to be accurate to around 40%. Thus we might underestimate the ion-current which would lead to a systematically too high recombination rate coefficient. It should be noted however, that the energy scale calibration of the measurement is ion-current-independent, i.e. an uncertainty in the ion-current measurement does not affect the spectroscopic reliability of the data.

A simple calculation of the Rydberg binding energies in the Dirac approximation indicates that 3 excitations of the Fe XXII core from its $2s^2 2p(^2P_{1/2})$ ground state [7] can lead to doubly excited Rydberg states of energies below 0.6 eV:

$$\mathrm{Fe}^{21+}\left[2s^2 2p(^2P_{1/2})\right] + \mathrm{e} \rightarrow \begin{cases} \mathrm{Fe}^{20+}\left[2s2p^2(^2P_{3/2})\ 7j\right] & , \Delta E_{\mathrm{core}} = 123.032\ \mathrm{eV} \\ \mathrm{Fe}^{20+}\left[2s2p^2(^2D_{5/2})\ 8j\right] & , \Delta E_{\mathrm{core}} = 94.130\ \mathrm{eV} \\ \mathrm{Fe}^{20+}\left[2s2p^2(^4P_{1/2})\ 11j\right] & , \Delta E_{\mathrm{core}} = 50.158\ \mathrm{eV} \end{cases} \quad (2)$$

The respective core excitation energies $\Delta E_{\mathrm{core}}$ are taken from [7] which reports an absolute uncertainty of $\pm 37$ meV for the three transitions. Each principal number $n$ of the Rydberg electron leads to a $j$ series of DR resonances (c.f. figure 3). Due to angular momentum coupling between Rydberg and core shell electrons, each $(n, j)$-state itself splits into fine structure components of total angular momentum $J$. In a more accurate calculation the states of low $j$ are expected to be shifted to lower energies as the quantum defect rises for orbitals of small angular momentums. Some states might even be shifted down to negative energies and thus not contribute to dielectronic recombination.

Figure 4 shows a blow-up of the lowest-lying lines observed in the recombination rate coefficient in the energy range 100 meV ...160 meV. A fit to the data assuming 4 delta-shaped DR resonances convoluted with the electron velocity distribution in the target beam is also shown. The quality of the fit suggests that these four lines are individually resolved. They appear thus suitable to accurately derive the energies of the corresponding autoionising Rydberg levels $(2s2p^2\ nj)_J$ for $n = 7, 8, 11$ once they can be identified on the basis of theoretical calculations.

## 5. Conclusions

Employing the Heidelberg TSR's twin merged beam setup and ultracold electron target, low-energy dielectronic resonances on boron-like iron ions have been measured at highest spectroscopic resolution. In the energy range below 0.6 eV, the number of observed resonances is at least doubled compared to earlier measurements [6]. A fit to the data suggests that the four lowermost autoionising Rydberg states with a boron-like excited core are individually resolved and thus promise sensitive probing of awaited theoretical predictions for these energy levels.

## References


[1] Kieslich S *et al.* 2004 *Phys. Rev. A* **70**, 042714
[2] Lestinsky M *et al.* 2008 *Phys. Rev. Lett.* **100**, 033001
[3] Sprenger F *et al.* 2004 *Nucl. Instr. and Meth. A* **532**, 298
[4] Orlov D A *et al.* 2004 *Nucl. Instr. and Meth. A* **532**, 418
[5] Orlov D A *et al.* 2005 *J. Phys. Conf. Series* **4**, 290
[6] Savin D W *et al.* 2004 *Astrophys. J. Suppl. Ser.* **147**, 421
[7] NIST 2008 `http://physics.nist.gov`

Journal of Physics: Conference Series **163** (2009) 012060 doi:10.1088/1742-6596/163/1/012060

# Effective collision strengths for transitions of Ni XVII

**C. E. Hudson, P. H. Norrington and C. A. Ramsbottom**

School of Mathematics and Physics, The Queens University of Belfast, Belfast BT7 1NN, Northern Ireland, U.K.

E-mail: c.hudson@qub.ac.uk

**Abstract.** Electron impact excitation collision strengths are required for the analysis and interpretation of stellar observations. This calculation aims to provide fine structure effective collision strengths for the Ni XVII ion using a method which includes contributions from resonances. A DARC calculation has been performed, involving 37 $J\pi$ states. The effective collision strengths are calculated by averaging the electron collision strengths over a Maxwellian distribution of electron velocities. The non-zero effective collision strengths for transitions between the fine structure levels are given for electron temperatures ($T_e$) in the range $\log_{10} T_e(\mathrm{K}) = 4.5 - 8.5$. Data for several transitions from the ground state are discussed in this paper.

## 1. Introduction

Ni XVII has been recently detected by the EIS instrument onboard Hinode [1]. The 249.18 Å line corresponding to the $3s^2\,{}^1S_0$ - $3s3p\,{}^1P_1^o$ transition is analogous to the 284.16 Å line of the iso-electronic Fe XV ion. Since this line is unblended, Young et al [1] indicate that it will be a useful probe of the hot cores of active regions.

Collision strength data for this ion is sparse. Christensen et al [2] carried out a calculation using the distorted wave approximation with 16 states and collision strengths being determined at 7 energies (though values are only given at 3 energies in their paper). The results were determined in $LS$-coupling and transformed to intermediate-coupling using the JAJOM program [3; 4]. These collision strengths were then used to produce Maxwellian-averaged effective collision strengths by Pradhan [5] for electron temperatures $\log_{10} T_e(\mathrm{K}) = 5.0 - 7.8$. However, this method does not include contributions from resonances, which can significantly affect the effective collision strength, hence an R-matrix calculation for this ion is needed.

## 2. Target Description

Wavefunctions have been calculated using the fully relativistic GRASP (General-purpose Relativistic Atomic Structure Package) code [6; 7]. Ten orbitals have been employed in the description i.e. up to 4f. The 37 $J\pi$ levels included in the calculation have configurations $3s^2$, $3p^2$, $3d^2$, 3s3p, 3s3d, 3p3d and 3s4s. The energy levels determined are given in Table 1 and are compared to the values given by NIST [8]. Christensen et al [2] included 16 levels which equate to levels 1-14, 36 & 37 of the current calculation, with their corresponding energies also being displayed in Table 1. We find that the current calculation agrees to within 1 per cent of the values of NIST [8].

**Table 1.** Energy levels (Ry) for the 37 $J\pi$ levels included in the calculation, relative to the Ni XVII ground state. Observed values from NIST [8] and the values calculated by Christensen et al [2] are also shown (differences from NIST are shown in parentheses).

| Index | Configuration | NIST | Current | Christensen |
|---|---|---|---|---|
| 1 | $3s^2\,{}^1S_0$ | 0.0000 | 0.0000 (0.0000) | 0.0000 (0.0000) |
| 2 | $3s3p\,{}^3P^o_0$ | 2.4097 | 2.3998 (0.0099) | 2.4569 (0.0472) |
| 3 | $3s3p\,{}^3P^o_1$ | 2.4844 | 2.4771 (0.0073) | 2.5255 (0.0411) |
| 4 | $3s3p\,{}^3P^o_2$ | 2.6763 | 2.6724 (0.0038) | 2.6559 (0.0204) |
| 5 | $3s3p\,{}^1P^o_1$ | 3.6569 | 3.7072 (0.0503) | 3.6800 (0.0231) |
| 6 | $3p^2\,{}^3P_0$ | 5.7220 | 5.7480 (0.0260) | 5.8183 (0.0963) |
| 7 | $3p^2\,{}^1D_2$ | 5.8214 | 5.8372 (0.0158) | 5.8236 (0.0022) |
| 8 | $3p^2\,{}^3P_1$ | 5.8668 | 5.8941 (0.0273) | 5.8685 (0.0017) |
| 9 | $3p^2\,{}^3P_2$ | 6.1013 | 6.1302 (0.0290) | 6.0995 (0.0018) |
| 10 | $3p^2\,{}^1S_0$ | 6.8756 | 6.9496 (0.0740) | 6.8924 (0.0168) |
| 11 | $3s3d\,{}^3D_1$ | 7.0283 | 7.0661 (0.0378) | 7.0754 (0.0471) |
| 12 | $3s3d\,{}^3D_2$ | 7.0437 | 7.0830 (0.0393) | 7.0792 (0.0355) |
| 13 | $3s3d\,{}^3D_3$ | 7.0675 | 7.1089 (0.0414) | 7.1067 (0.0392) |
| 14 | $3s3d\,{}^1D_2$ | 7.8776 | 7.9830 (0.1055) | 7.9500 (0.0276) |
| 15 | $3p3d\,{}^3F^o_2$ | 9.6042 | 9.6348 (0.0306) | —— |
| 16 | $3p3d\,{}^3F^o_3$ | 9.7381 | 9.7730 (0.0349) | —— |
| 17 | $3p3d\,{}^1D^o_2$ | 9.8519 | 9.8932 (0.0413) | —— |
| 18 | $3p3d\,{}^3F^o_4$ | 9.8973 | 9.9360 (0.0386) | —— |
| 19 | $3p3d\,{}^3D^o_1$ | 10.1654 | 10.2255 (0.0601) | —— |
| 20 | $3p3d\,{}^3D^o_2$ | 10.1915 | 10.2508 (0.0593) | —— |
| 21 | $3p3d\,{}^3D^o_3$ | 10.3317 | 10.3973 (0.0656) | —— |
| 22 | $3p3d\,{}^3P^o_0$ | 10.3496 | 10.4094 (0.0598) | —— |
| 23 | $3p3d\,{}^3P^o_1$ | 10.3531 | 10.4143 (0.0612) | —— |
| 24 | $3p3d\,{}^3P^o_2$ | 10.3579 | 10.4212 (0.0634) | —— |
| 25 | $3p3d\,{}^1F^o_3$ | 11.0174 | 11.1418 (0.1245) | —— |
| 26 | $3p3d\,{}^1P^o_1$ | 11.1423 | 11.2718 (0.1294) | —— |
| 27 | $3d^2\,{}^3F_2$ | 14.1966 | 14.2648 (0.0682) | —— |
| 28 | $3d^2\,{}^3F_3$ | 14.2202 | 14.2922 (0.0719) | —— |
| 29 | $3d^2\,{}^3F_4$ | 14.2506 | 14.3254 (0.0748) | —— |
| 30 | $3d^2\,{}^1D_2$ | 14.5331 | 14.6320 (0.0990) | —— |
| 31 | $3d^2\,{}^3P_0$ | 14.5566 | 14.6467 (0.0901) | —— |
| 32 | $3d^2\,{}^3P_1$ | 14.5702 | 14.6570 (0.0869) | —— |
| 33 | $3d^2\,{}^3P_2$ | 14.5960 | 14.6884 (0.0924) | —— |
| 34 | $3d^2\,{}^1G_4$ | 14.5931 | 14.7035 (0.1105) | —— |
| 35 | $3d^2\,{}^1S_0$ | —— | 15.5667 | —— |
| 36 | $3s4s\,{}^3S_1$ | 19.9349 | 19.9218 (0.0131) | 19.9655 (0.0306) |
| 37 | $3s4s\,{}^1S_1$ | —— | 20.1689 | 20.2367 |

## 3. Scattering Calculation

Collision strengths are generated using DARC (Dirac Atomic R-matrix Code) [9], interfaced with the UK APAP asymptotic and utility codes [10]. Twenty continuum orbitals per angular momentum are used, with the $R$-matrix boundary radius being 4.0 atomic units. Collision strengths have been generated in the energy range from threshold up to 120 Rydbergs in order to compare with the data of Christensen et al [2], but only values up to 60 Rydberg ($\sim 3\times$ energy of last target level) are used to calculate the Maxwellian-averaged effective collision strengths.

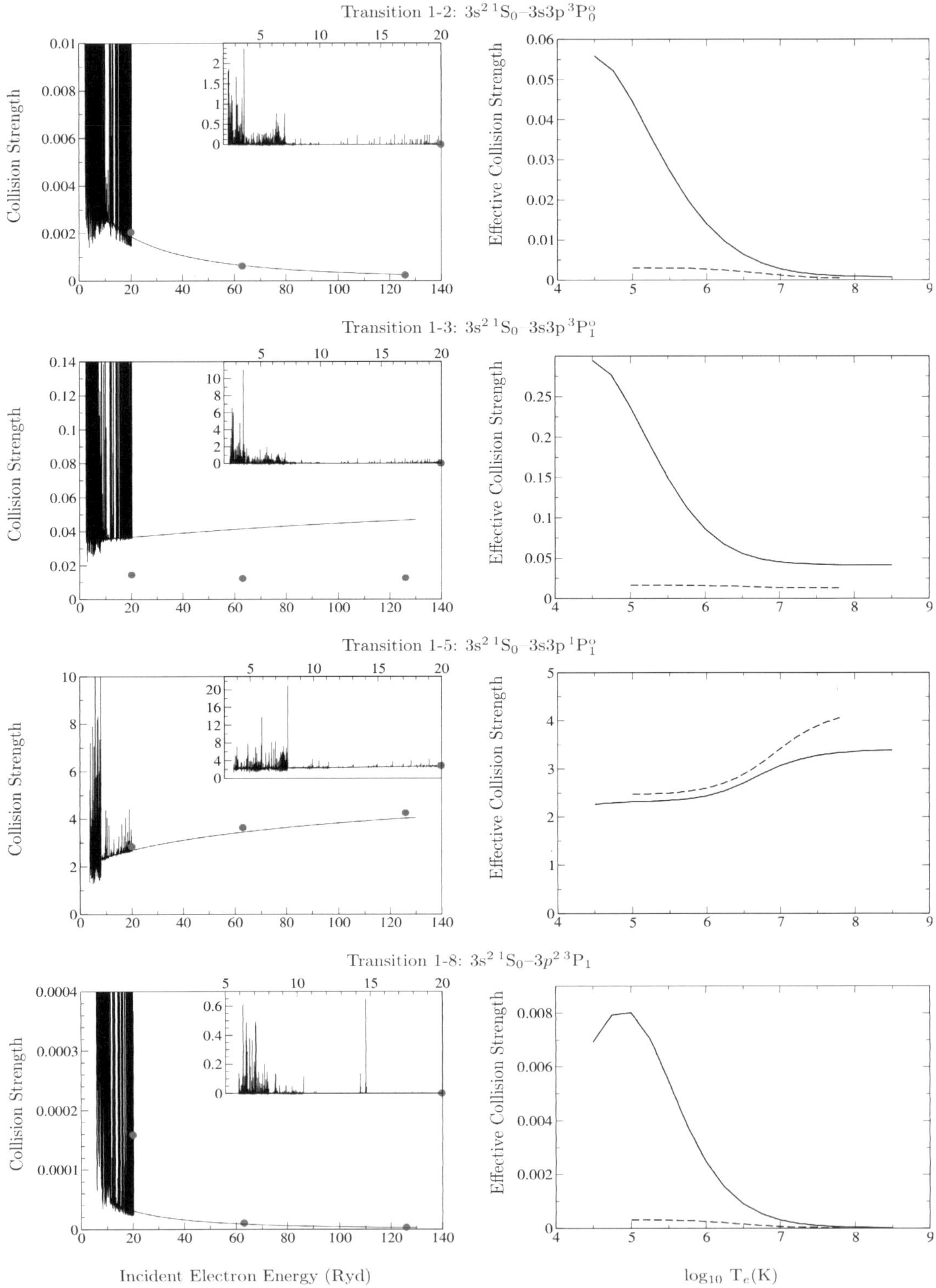


**Figure 1.** Collision strengths and Maxwellian-averaged effective collision strengths. —— Current ($R$-matrix) ● Christensen et al [2] - - - - Pradhan [5] (both Distorted Wave).

Approximately 70,000 points are used to properly delineate the resonance structure with an energy mesh of $\sim 2.5\times10^{-4}$ Rydberg. All partial waves up to $2J = 45$ are included. For allowed transitions, to take account of neglected higher partial waves, a 'top-up' procedure is applied. The resulting collision strengths ($\Omega$) between initial state 'i' and final state 'j' are then averaged over a Maxwellian distribution of electron velocities to give the effective collision strengths ($\Upsilon$)

$$\Upsilon_{ij}(T_e) = \int_0 \Omega_{ij}(E_f)\exp(-E_f/kT_e)d(E_f/kT_e). \quad (1)$$

## 4. Results and Discussion

In Figure 1 we show a few selected collision strengths and their corresponding effective collision strengths. The graphs show the kind of agreement/differences we see between a distorted wave calculation and an R-matrix calculation. In general there is good agreement between the background level of the collision strengths, but since the distorted wave approximation only determines values at a few energies the contribution from resonances is neglected.

The resonances, which can be quite large, feed in to the effective collision strengths and have the ability to enhance the effective collision strength at lower temperatures. At the high temperature extremity the effective collision strengths generally converge. Transitions 1-2, 1-3 and 1-8, all follow this pattern, though the agreement in background level of the collision strength at 20 Rydbergs in transition 1-8 is not as good as for the other transitions and for transition 1-3 the background level is a factor of 2-3 higher than the distorted wave values.

For the allowed transition 1-5, the distorted wave collision strength is slightly higher than our $R$-matrix results. The resonance structure does not dominate this collision strength (see insert on graph), meaning there is little low temperature enhancement. This along with a high-energy tail which increases, gives an effective collision strength which increases with temperature and one where the distorted wave values are higher for all values of $T_e$ in the present range.

## 5. Conclusions

Collision strengths and effective collision strengths for all 666 possible transitions have been generated. A few selected transitions are discussed here. Differences exist between the current values and those from distorted wave evaluations, in general we find resonance contributions in our collision strengths greatly enhance the effective collision strengths. More details of this work and additional data will be available at `www.am.qub.ac.uk/apa`.

## Acknowledgments

This work has been supported by PPARC, under the auspices of a Rolling Grant.

## References

[1] Young P R *et al.* 2007 *Publ. Astron. Soc. Japan* **59** S857
[2] Christensen R B, Norcross D W and Pradhan A K 1986 *Phys. Rev.* A **34** 4704
[3] Saraph H E 1978 *Comput. Phys. Commun.* **15** 247
[4] Saraph H E 1972 *Comput. Phys. Commun.* **3** 256
[5] Pradhan A K 1988 *Atomic Data and Nuclear Data Tables* **40** 335
[6] Grant I P, McKenzie B J, Norrington P H, Mayers D F and Pyper N C 1980 *Comput. Phys. Commun.* **21** 207
[7] Dyall K G, Grant I P, Johnson C T, Parpia F A and Plummer E P 1989 *Comput. Phys. Commun.* **55** 425
[8] NIST Atomic Spectroscopic Data URL `http://physics.nist.gov/PhysRefData/ASD/index.html`
[9] Ait-Tahar S, Grant I P and Norrington P H 1997 *Phys. Rev. Lett.* **79** 2955
[10] UK APAP Network URL `http://amdpp.phys.strath.ac.uk/UK_RmaX/`

# Dielectronic recombination into Mg-like ions

**I Murakami[1], U I Safronova[2] and T Kato[1]**

[1] National Institute for Fusion Science, Toki, Gifu, 509-5292, Japan
[2] Department of Physics, University of Nevada, Reno, NV 89557, USA

Email: murakami.izumi@nifs.ac.jp

**Abstract**. Dielectronic recombination (DR) rate coefficients for Na-like ions forming Mg-like ions of Fe, Zn, Kr, and Mo are calculated. Energy levels, radiative transition probabilities, and autoionization rates for $1s^22s^22p^63l'nl$, $1s^22s^22p^64l'nl$, and $1s^22s^22p^53l''3l'nl$ states are calculated by using the Hatree-Fock-relativistic method (Cowan code). Total and state-selective DR rate coefficients are derived as a function of electron temperature. Configuration mixing plays an important role for DR rates of $1s^22s^22p^63snl$ states. Energy level structure near the first 3s threshold is important for DR rate coefficients behaviour at low temperature.

## 1. Introduction

Dielectronic recombination (DR) is important process governing ionization state of ions in various plasmas. Many theoretical and experimental studies on DR have been done for K-shell and L-shell ions, but not so many for M-shell ions. Recently Netzer [1] and Kraemer et al. [2] suggested importance of M-shell Fe ions and the data needs for astrophysical plasmas such as photoionized plasma of active galactic nuclei. The solar observational satellite HINODE was launched in 2006 and has EUV Imaging Spectrometer (EIS) to observe the sun with EUV spectral lines of Fe M-shell ions [3].Plasma diagnostics by using spectral lines of Fe ions will provide information on the active region and the transition region of the sun to study the coronal heating problem. Reliable data of total and state-selective DR rate coefficients are needed for determining ionization state of ions and level populations to estimate spectral line intensities for plasma diagnostics.

Here we focus on DR process forming Mg-like ions. The DR rate coefficients are calculated including $1s^22s^22p^63l'nl$, $1s^22s^22p^64l'nl$, and $1s^22s^22p^53l''3l'nl$ states in Mg-like iron ($Fe^{14+}$), Mg-like Zn ($Zn^{18+}$), Mg-like Kr ($Kr^{24+}$) and Mg-like Mo ($Mo^{30+}$) ions. The energy levels, transition probabilities and autoionization rates required for calculating the DR rate coefficients are determined. The present paper continues our efforts on calculation of the low-temperature DR rate coefficients for CI [4], CII [5,6], CIII [7], OIII [8], OIV [9], OV [10], Ne VII [11], and Fe XV [12]. In paper [12] we did not include $1s^22s^22p^53l''3l'nl$ states which contribute to the DR rate coefficient at high temperature. Hereafter we omit the core $1s^22s^2$ from the configuration notation.

## 2. Atomic data

Atomic parameters for intermediate and final states of DR processes are needed to determine the DR rate coefficients. We calculated energy levels, radiative transition probabilities, and autoionization rates for $2p^63l'nl$ ($n$ =3-$n_l$, $l$<8), $2p^64l'nl$ ($n$ =4-7, $l$<$n$), and $2p^53l''3l'nl$ ($n$ =3-4, $l$<$n$) states in $Fe^{14+}$, $Zn^{18+}$, $Kr^{24+}$ and $Mo^{30+}$ ions by using the Hartree-Fock-Relativistic method (Cowan code [13,14]). The upper limit of $n$ for calculating $2p^63l'nl$ states, $n_l$, is set as 12 for $Fe^{14+}$ and $Zn^{18+}$, 13 for $Kr^{24+}$ and 14 for $Mo^{30+}$. Due to computational issues, calculation of the DR parameters involving $2p^64l'nl$ states was performed with account of $2p^63l'nl$ states with $n \leqslant 8$ only, and calculation of the DR parameters involving $2p^53l''3l'nl$ states, with account of $2p^63l'nl$ states with $n \leqslant 5$ only. We use one scaling factor (0.85) for all electrostatic integrals in the Cowan code. Radiative transition probabilities with $A_r < 10^5$ $s^{-1}$ were omitted from consideration for simplicity.

Autoionizing states are the $2p^63pn_2l$ states with $n_2 \geqslant 10$ ($Fe^{14+}$), 11 ($Zn^{18+}$), or 12 ($Kr^{24+}$ and $Mo^{30+}$) and the $2p^63dn_3l$ states with $n_3 \geqslant 7$ ($Fe^{14+}$ and $Zn^{18+}$), 8 ($Kr^{24+}$), or 9 ($Mo^{30+}$). Some of $2p^64l'4l$ states listed in Table 1 are located below the first 3s threshold and are not autoionizing. All $2p^64l'nl$ states with $n$ = 4-7 are below the third 3d threshold for these ions.

**Table 1.** $2p^64l'4l$ states located below the 3s threshold.

| Ion | $I_{3s}$ ($cm^{-1}$) | states |
|---|---|---|
| $Fe^{14+}$ | 3,686,000 | $2p^64s^2\ ^1S_0$ |
| $Zn^{18+}$ | 5,670,000 | $2p^64s^2\ ^1S_0$, $2p^64s4p\ ^3P_{0,1,2}$, $^1P_1$ |
| $Kr^{24+}$ | 9,315,400 | $2p^64s^2\ ^1S_0$, $2p^64s4p\ ^3P_{0,1,2}$, $^1P_1$, $2p^64p^2\ ^3P_{0,1,2}$, $2p^64s4d\ ^3D_{1,2,3}$, $^1D_2$ |
| $Mo^{30+}$ | 13,920,000 | $2p^64s^2\ ^1S_0$, $2p^64s4p\ ^3P_{0,1,2}$, $^1P_1$, $2p^64p^2\ ^3P_{0,1,2}$, $^1D_2$, $^1S_0$, $2p^64s4d\ ^3D_{1,2,3}$, $^1D_2$, $2p^64s4f\ ^3F_{2,3,4}$, $^1F_3$, $^1S_0$, $2p^64p4d\ ^3P_2$, $^3D_1$, $^3F_{2,3}$ |

## 3. Dielectronic recombination rate coefficients

The DR rate coefficients are obtained by the following equations:

$$\alpha_d(i_0, j) = 3.3 \times 10^{-24} \left( \frac{I_H}{kT_e} \right)^{3/2} \frac{1}{g_0} \sum_i Q_d(j,i) \exp\left( -\frac{E_s(i)}{kT_e} \right), \tag{1}$$

$$Q_d(j,i) = \frac{g(i) A_a(i,i_0) A_r(j,i)}{\sum_{i'_0} A_a(i,i'_0) + \sum_k A_r(k,i)}, \tag{2}$$

where $i_0$ indicates initial state, $j$ final state, $i$ intermediate autoionizing state. $I_H$ is hydrogen ionization potential energy, $T_e$ electron temperatnre, $E_s(i)$, energy from ionization threshold, $g_0$ and $g(i)$ statistical weights of initial state and $i$ state, $A_a$ autoionization rate, and $A_r$ radiative transition probability.

The contributions from the high $n$ autoionizing states to the DR rate coefficients are estimated by using empirical scaling laws [4] for atomic parameters. The one-electron dipole transitions, eg., 3s-$n$p, 3p-$n$s, 3p-$n$d, 3d-$n$p, and 3d-$n$f, are taken into account. Details are described in [12]. In addition 2p-$nl$ transitions are also taken into account in this paper. The $1/n^3$ scaling law for $A_a$ and $A_r$ is used. The high $n$ state contributions up to $n$ =1000 are calculated.

The state-selective and total DR rate coefficients are derived as a function of electron temperature. Figures 1(a)-(d) show the total DR rate coefficients of Mg-like Fe, Zn, Kr and Mo ions. Contributions of $2p^63lnl'$, $2p^64lnl'$, and $2p^53ln'l'n''l''$ states to the DR rate coefficients are also shown. Figure 2 shows examples of state-selective DR rate coefficients in Mg-like Fe. Contributions of $2p^53ln'l'n''l''$ states are seen at $T_e$ >100eV.

Configuration mixing for $2p^6$ [3$s$$n$s + 3$p$$n$p + 3d$n$d], $2p^6$ [3s$n$p + 3p$n$s + 3p$n$d + 3d$n$p] and $2p^6$ [3s$n$d + 3p$n$p + 3p$n$f + 3d$n$s + 3d$n$g] states plays an important role for the DR rate coefficients of 3s$nl$ levels with low $n$ at low temperature. For example, $2p^6$ [$3s^2$ + $3p^2$ + $3d^2$] (all), $2p^6$ [3s6p + 3p5d + 3p6s + 3d5p] (Mg-like Fe), $2p^6$ [3s7s + 3p6p + 3d5d] (Mg-like Zn), $2p^6$ [3s5d + 3p5p + 3p5f + 4d5s] (Mg-like Kr), and $2p^6$ [3s8d + 3p7p + 3d6s] (Mg-like Mo) show configuration mixing.

The total DR rate coefficients agree well with previously published data by Gu [15], Altun *et al.* [16], and Fournier *et al.* [17] (Mg-like Mo) except for low temperature region. The behavior of the DR rate coefficients at low temperature strongly depends on the energy level structure. DR processes through autoionizing levels just above the first 3s threshold contribute to the DR rate coefficients at low temperature. For Mg-like Fe, $2p^63p10l$ and $2p^64s4p$ states are located near the 3s threshold and DR processes through these states contribute to the DR rate at low temperature. The $2p^63d7l$ states for Mg-like Zn, the $2p^63d8l$ and $2p^63p12l$ states for Mg-like Kr and the $2p^64p4l$ states for Mg-like Mo locate close to the 3s threshold, respectively and contribute to the DR rate coefficients at low temperature. These different level structures cause different behavior of the DR rate coefficients.

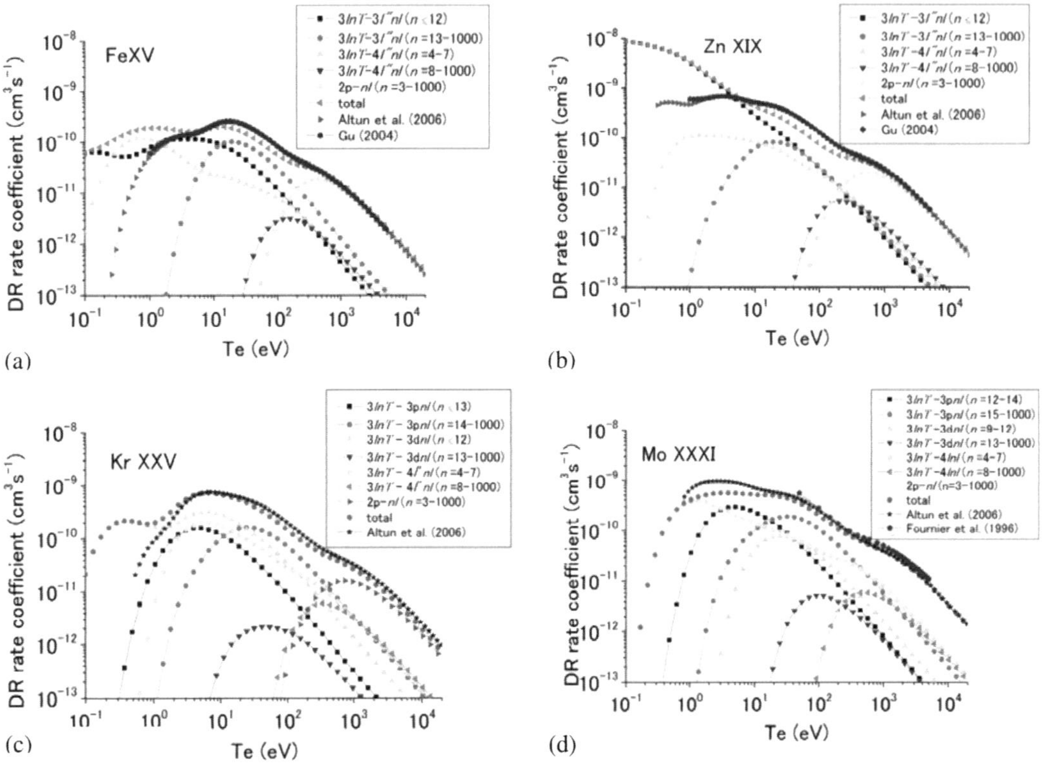


**Figure 1.** Dielectronic recombination rate coefficient in Mg-like Fe (a), Zn, (b), Kr (c), and Mo (d) as a function of electron temperature. Contributions of transitions through $2p^63lnl$, $2p^64lnl$ and $2p^53lnln'l'$ states are shown. Results by Altun et al. [16], Gu [15] and Fournier et al. [17] are also shown for comparison.

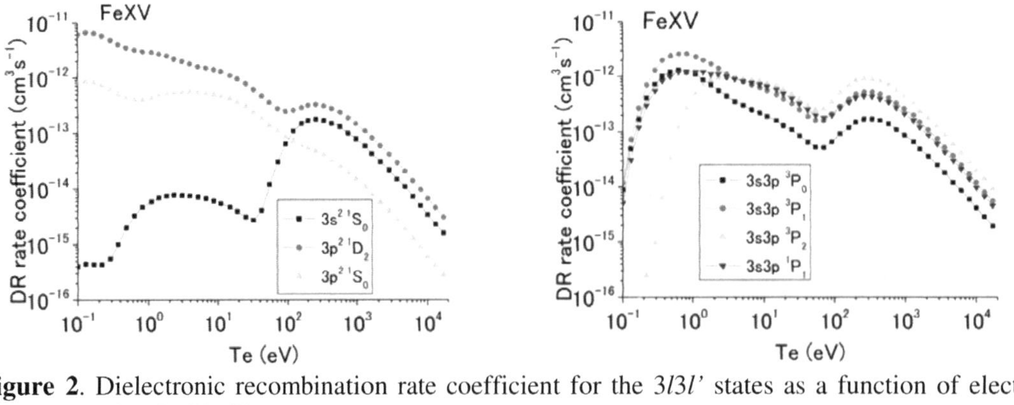


**Figure 2**. Dielectronic recombination rate coefficient for the *3l3l'* states as a function of electron temperature in Mg-like Fe.

Contributions of high *n* states become important at $T_e$ >10-30eV and contributions of 2p-*nl* transitions through $2p^5 3ln'l'n''l''$ become important at $T_e$ > 200-600eV.

## 4. Conclusion

In this paper we calculate a large set of atomic data related to DR processes of Na-like ions into Mg-like ions of Fe, Zn, Kr and Mo. Energy levels, radiative transition probabilities and autoionization rates are calculated for Mg-like Fe, Zn, Kr, and Mo ions by using the Hatree-Fock-relativistic method (Cowan code). The calculated atomic data are used to obtain the DR rate coefficients. We take into account doubly excited states $2p^6 3lnl'$ and $2p^6 4lnl'$ and inner-shell excited states $2p^5 3ln'l'n''l''$ as intermediate states with *n* up to 1000 to calculate the DR rate coefficients.

Configuration mixing plays an important role for the DR rate coefficients of 3s*nl* states with low *n*. The autoionizing states which located close to the first 3s threshold contribute to the DR rate coefficient at low temperature, which strongly depends on the energy level structure of each ion. The DR rate coefficients are in good agreement with previously published data at medium and high temperature.

The state selective rate coefficients can be used in collisional-radiative modelling for population kinetics and plasma diagnostics in recombining plasmas, which is important to examine behaviour and properties of non-equilibrium plasma such as the transition region of the sun.

## Acknowledgement

This work was supported in part by Matsuo Foundation. We would like to thank to Dr. A Kramida for supervising the use of his version of the Cowan code.

## References

[1] Netzer H 2004 *Astrophysical J.* **604** 551
[2] Kraemer S B, Ferland G L and Gabel J R 2004 *Astrophys. J.* **604** 556
[3] Culhane J L 2007 *Solar Phys.* **243** 19
[4] Safronova U I and Kato T 1998 *J. Phys. B* **31** 2501
[5] Safronova U I and Kato T 1996 *Phys. Scr.* **52** 461
[6] Kato T, Safronova U I and Ohira M 1997 *Phys. Scr.* **55** 185
[7] Safronova U I, Kato T and Ohira M 1997 *J. Quant. Spectr. Radiat. Transfer* **58** 193
[8] Safronova U I, Ralchenko Yu, Murakami I, Kato T and Kato D 2006 *Phys. Scr.* **73** 143
[9] Murakami I, Safronova U I, Vasilyev A A and Kato T 2005, *At. Data Nucl. Data Tables* **90** 1
[10] Murakami I, Safronova U I and Kato T 2002 *Can. J. Phys.* **80** 1525
[11] Murakami I, Safronova U I and Kato T 1999 *J. Phys. B* **32** 5351
[12] Murakami I, Kato T, Kato D, Safronova U I, Cowan T E and Ralchenko Yu 2006 *J. Phys. B* **39** 2917
[13] Cowan R D 1981 *The Theory of Atomic Structure and Spectra* (Berkeley: Univ. of California Press)
[14] Kramida A 1998 URL= http://das101.isan.troitsk.ru/cowan.htm
[15] Gu M F 2004 *Astrophys. J. Suppl.* **153** 389
[16] Altun Z, Yumak A, Badnell N R, Loch S D and Pindzola M S 2006 *Astron.Astrophys* **447** 1165
[17] Fournier K B, Cohen M, Goldstein W H, Osterheld A L, Finkenthal M, May M J, Terry J L, Graf M A and Rice J 1996 *Phys. Rev. A* **54** 3870

# High-precision measurement of the spectral width of the nickel-like molybdenum x-ray laser

**N. Hasegawa[1], T. Kawachi[1], A. Sasaki[1], H. Yamatani[1], M. Kishimoto[1], Y. Ochi[1], M. Tanaka[1], M. Nishikino[1], Y. Kunieda[1], K. Nagashima[1], and H. Yoneda[2]**

[1]Quantum Beam Science Directorate, Japan Atomic Energy Agency, 8-1 Umemidai, Kizugawa, Kyoto, 619-0215, Japan

[2]Institute for Laser Science, University of Electro-Communications, 1-5-1, Chofugaoka, Chofushi, Tokyo 182, Japan

E-mail:hasegawa.noboru@jaea.go.jp

**Abstract**. In this study, we performed the first measurement of the spectral width of a nickel-like molybdenum x-ray laser ($\lambda$ = 18.895 nm) by use of a high-resolution spectrometer. The intrinsic and amplified spectral width were measured to be 36 and 18 mÅ under the substantial lasing conditions, respectively. It corresponds to a temporal coherence of the amplified x-ray laser of 0.34 ps.

## 1. Introduction

In recent years, the advent of high-performance ultra short pulse lasers has made it possible to generate the high-brightness compact x-ray lasers (XRL) [1-3] based on laser-produced plasmas. The gain media of the plasma x-ray lasers are highly charged ions in high-density plasmas. Only a few measurements of the spectral width of the XRL have been reported [4,5], because the spectral width of the XRL is so narrow that the required spectral resolution is quite high, *e.g.* $\Delta\lambda/\lambda \sim 10^{-4}$. The precise knowledge about the wavelength and the spectral width of the lasing line is important for the atomic physics and applications of the XRL. First, the spectral width is a good benchmark for the atomic code because it depends on the electron collisional impact excitation and de-excitation rate coefficient. Second, the spectral width shows the temporal coherence of the XRL. Temporal coherence is important for the applications such as the holography or diffraction imaging [6,7]. Third, the spectral width is an important parameter to determine the strength of the magnetic field required for the generation of the circularly polarized XRL. The circularly polarized XRL will be useful for the light source of the circular dichroism measurement [8]. The details of the circularly polarized XRL are described in section 5.

In this study, we choose the Ni-like Mo XRL ($3d^9 4d\ {}^1S_0 - 3d^9 4p\ {}^1P_1$) as an example and measured the spectral width by using a high-resolution spectrometer. Reasons for using this laser are: First, this laser line is in the wavelength region in which Mo/Si multi-layer mirrors have substantial reflectivity; therefore, this laser line is important for applications. Second, the strong amplification of this laser line has been obtained in previous work [2], and the high precision measurement can be obtained. Third, since the Ni-like Mo XRL media is the high electron density plasma, it is expected that the influence of stark broadening becomes dominant. In addition, we produced the XRL by using the grazing incident pumping scheme (GRIP) [3]. GRIP scheme can control the electron density of the gain region of the x-ray laser. It implies that the experimental result can be compared with the theoretical one easily.

## 2. Experimental set-up

The experimental set-up is shown in figure 1. The nickel-like molybdenum x-ray laser was measured by the high-resolution spectrometer, HIREFS made by Hettrick scientific [9].

### 2.1. Pumping laser system

We used the GRIP scheme [3] for the focusing system of the Nd:glass laser at a wavelength of 1053 nm. GRIP scheme can control the electron density ($n_e$) of the laser absorption region by the grazing angle ($\theta$) to the target surface. The electron density of the laser absorption region is described to be $n_e = n_{cr} \sin^2 \theta$. $n_{cr}$ is the critical electron density and it is $10^{21}$ cm$^{-3}$ for the laser with the wavelength of 1053 nm. The grazing incidence angle was 14 degree, and the electron density of the laser absorption region was $6 \times 10^{19}$ cm$^{-3}$. The focal width and length were measured to be 70 μm and 5 mm, respectively. Nd:glass laser light consists of two pulses, the pre-pulse and the main pulse. The total energy on the target was 10 J, and the energy ratio of these pulses was 1 : 4. The pulse separation was 1.1 ns, and each duration was 7ps.

### 2.2. High-resolution spectrometer (HIREFS).

HIREFS has a spherical and elliptical mirror to transfer the image of the slit on to the detector. The magnification of the horizontal plane (= the direction of the plasma expansion) and the vertical plane (= the direction of the wavelength dispersion) were 2.4 and 3.2, respectively. The back-illuminated CCD (Princeton, PI-SX:1K, pixel size 13μm) was used for the detector.

The carbon spectrum was measured to decide the inverse linear dispersion of HIREFS. Carbon Balmer α line (182.20 Å) and Ni-like Mo XRL line (188.95 Å) [10] were used for calibration. The inverse linear dispersion was measured to be 785 mÅ /mm on the CCD surface. The resolution of the HIREFS was estimated to be 12.7 mÅ ($d\lambda / \lambda = 7 \times 10^{-5}$) from the inverse linear dispersion, CCD pixel size, slit size (= 3 μm) and magnification of HIREFS (= 3.2).

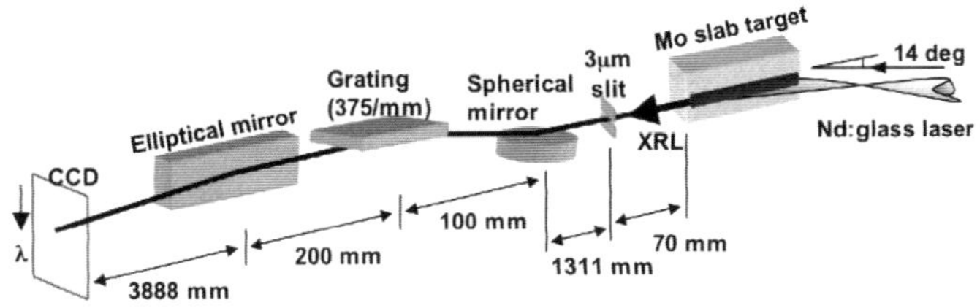


**Figure 1.** Experimental set-up.

## 3. Experimental result

Figure 2 shows the spectral profiles of the nickel-like molybdenum x-ray laser line at gain length product ($gl$) were 3.1 (a) and 6.9 (b). Each dot and the solid curve show the data point and the fitting curve with the Voigt profile [4], respectively. The detail of the Voigt fitting curve is described in section 4. The spectral width of the XRL was measured to be 36 (a) and 18 mÅ (b) at the center wavelength of 18.895 nm.

Figure 3 shows the gain narrowing behavior of the spectral width of the Ni-like Mo XRL. Each XRL spectra were fitted by the Voigt profile. The spectral width was clearly increased at the gain length product of less than 4.

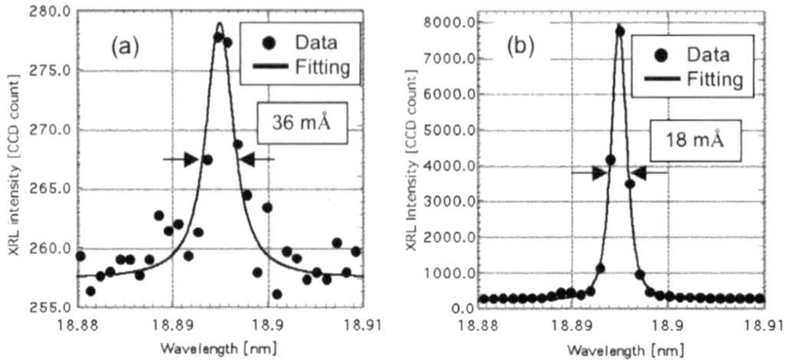


**Figure 2.** XRL spectra at $gl = 3.1$ (a) and 6.9 (b).

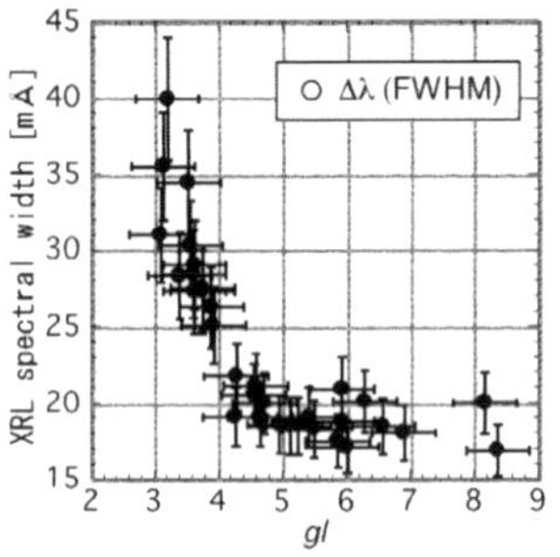


**Figure 3.** Gain narrowing behavior.

## 4. Discussion

The intrinsic spectral profile of the XRL is determined by convolution of the homogeneous and inhomogeneous broadening, *e.g.* Voigt profile. Homogeneous broadening (= $\Delta\omega_L$, (full width at half maximum: FWHM)) is determined by the depopulation rate of the lasing levels. Inhomogeneous component is due to Doppler broadening (= $\Delta\omega_D$ (FWHM)). The spectral profile, $f(\omega)$, was described by the convolution of these broadening.

$$f(\omega) = \frac{2\ln 2\Delta\omega_L}{\pi^{3/2}\Delta\omega_D}\int_{-\infty}^{+\infty}\frac{\exp(-4\ln 2(\omega_0-\omega_\xi)^2/\Delta\omega_D^2)}{(\omega_\xi-\omega)^2+(\Delta\omega_L/2)^2}d\omega_\xi \qquad (1)$$

$\omega$ and $\omega_0$ were angular frequency and angular frequency of the wavelength center, respectively. The Doppler broadening is written by the ion temperature ($T_i$), ion mass ($m_i$) and light velocity ($c$); $\Delta\omega_D/\omega_0 = (8\ \ln 2\ kT_i\ /\ (m_i\ c^2))^{0.5}$. The typical plasma parameters in the present experiment were calculated by the 1-D hydrodynamic code HYADES [11]. The ion temperature ($T_i$), electron temperature ($T_e$) and electron density ($n_e$) were 60 eV, 400 eV and $6 \times 10^{19}$ cm$^{-3}$, respectively. It implied the Doppler broadening ($\Delta\lambda_D$) was 12 mÅ. We supposed the spectral profile at $gl = 3.1$ to be the intrinsic spectrum of the lasing line without gain narrowing effect. The intrinsic spectrum (figure 2 (a)) was fitted by the Voigt profile with homogeneous width ($\Delta\lambda_L$) of 30 mÅ. The amplified XRL spectrum (figure 2 (b)) was fitted by the Voigt profile with $\Delta\lambda_D$ = 6 mÅ and $\Delta\lambda_L$ = 15 mÅ. From this result, the temporal coherence of the intrinsic and the amplified XRL dominated by the homogeneous broadening were about 0.17 ps and 0.34 ps, respectively. In the previous experiment, the duration of the Ni-like Mo XRL was measured to be 2.5 ps [12], it implies the coherent photon ratio of the amplified XRL (= temporal coherence / duration) was 0.34 / 2.5 = 0.14.

Now we considered the theoretical spectral width of the homogeneous broadening ($\Delta\lambda_{L\_calculation}$) from the result of the calculation for the Ni-like Ag x-ray laser line by HULLAC code ($\Delta\lambda_{L\_Ag}$ = 18 mÅ at $\lambda$ = 13.9 nm) [13,14]. Homogeneous broadening mainly depended on the depopulation rate of the lasing levels. They were estimated from the value for the Ni-like Ag case using the scaling low [15]; $C_z = Z^{-3}\ C^H$, where $C_z$ is the collisional rate coefficient for ion $Z$ and $C^H$ is the reduced collisional rate coefficient for $Z = 1$. In this calculation, $C_{Ag} = 6.53 \times 10^{12}$ s$^{-1}$ at $n_{e\_Ag} = 4.0 \times 10^{20}$ cm$^{-3}$. It red to be the theoretical width of Voigt profile was 20 mÅ with homogeneous width ($\Delta\lambda_{L_calculation}$) of 13 mÅ and inhomogeneous width of 12 mÅ. The theoretical broadening was small compared with the experimental result. The cause of the difference might be that the XRL was generated in high electron density region.

## 5. Future plan, generation of circularly polarized x-ray laser

The nickel-like ion x-ray laser (*4p* ($J$ = 1) - *4d* ($J$ = 0)) has three degenerated states (magnetic sublevel: $m$ = -1, 0, 1) in the lower level. If the quantization axis is parallel to the longitudinal axis of the XRL media due to a strong magnetic field, the polarization of the XRL can be selected automatically. Left handed circularly ($\sigma'$, $\Delta m$ = -1) and right handed circularly polarized ($\sigma$, $\Delta m$ = 1) x-ray laser will be generated. Each circular polarization can be selected by a spectrometer because of the different wavelength of each polarization by the Zeeman effect under influence of the strong magnetic field. The value of the Zeeman split ($\Delta\varepsilon$) is described to be $\Delta\varepsilon = 2.8\ \mu_B B$, where $\mu_B$ is the Bohr magnetron (= $5.8 \times 10^{-5}$ eV/T), and $B$ is the strength of the magnetic field. From the measurement of the spectral width of Ni-like Mo XRL, the strength of the magnetic field required for the splitting of the circular polarization component was estimated to be 40 T. The pulse-power magnet system [16] for the next experiment is now under construction.

## 6. Summary

We have measured the precise spectral width of the nickel-like molybdenum x-ray laser. The intrinsic and amplified spectral width were 36 mÅ and 18 mÅ at the wavelength of 18.895 nm, respectively. It implies the temporal coherence of the intrinsic and amplified XRL were 0.17 ps and 0.34 ps, respectively. From this result, the strength of the magnetic field for the generation of circularly polarized x-ray laser was obtained (40 T). In the next work, we will try the generation of the circularly polarized x-ray laser with the pulse power magnet.

## Acknowledgments

This work was supported by a Grand-in-Aid for Young Scientists (B) (No. 19740350) from the Japan Society for the Promotion of Science.

We would like to thank that we got the use of HIREFS from Professor Nishimura in Osaka University.

## References

[1] Kalachnikov M P, Nickles P V, Schnürer M, Sandner W, Shlyaptsev V N, Danson C, Neely D, Wolfrum E, Zhang J, Behjat A, Demir A, Tallents G J, Warwick P J and Lewis C L S, 1998 *Phys. Rev. A* **57**, 4778

[2] Luther B M, Wang Y, Larotonda M A, Alessi D, Berrill M, Marconi M C, Rocca J J and Shlyaptsev V N, 2005 *Opt. Lett.* **30**, 165

[3] Keenan R, Dunn J, Shlyaptsev V N, Smith R F, Patel P K and Price D F, 2003 *Proc. SPIE* Vol. 5197, 213

[4] Koch J A, MacGowan B J, Silva L B D, Matthews D L, Underwood J H, Batson PJ, Lee R W, London R A and Mrowka S, 1994 *Phy. Rev. A* **50**, 1877

[5] Hasegawa N, Kawachi T, Utsumi T, Sasaki A, Tanaka M, Kado M, Sukegawa K, Lu P, Kishimoto M, Tai R, Nagashima K, Koike M, Daido H and Kato Y 2004 *Jpn. J. Appl. Phys.* **43**, 2519

[6] Wachulak P W, Bartels R A, Marconi M C, Menoni C S, Rocca J J, Lu Y and Parkinson B, 2006 *Opt. Exp.* Vol. 14, No 21, 9636

[7] Miao J, Charalambous P, Kirz J and Sayre D, 1999 *Nature*, 400, 342

[8] Tanaka M, Nakagawa K, Agui A, Fujii K and Yokoya A, 2005 *Phys. Scr.* T115, 873-876

[9] Hettrick M C, Underwood J H, Batson P J, and Eckart M J, 1988 *Appl. Opt.* Vol. 27, 200

[10] Li Y, Nilsen J, Dunn J and Osterheld A L, 1998 *Phys. Rev. A*, **58**, R2668

[11] Pert G J, 1983 *J. Fluid Mech.*, **131**, 401

[12] Ochi Y, Kawachi T, Hasegawa N, Sasaki A, Nagashima K, Sukegawa K, Kishimoto M, Tanaka M, Nishikino M and Kado M, 2004 *Appl. Phys. B* **78,** 961

[13] Kawachi T, Kado M, Tanaka M, Sasaki A, Hasegawa N, Kilpio A V, namba S, Nagashima K, Lu P, Takahashi K, Tang H, Tai R, Kishimoto M, Koike M, daido H and Kato Y, 2002 *Phys. Rev. A* **66** 33815

[14] Klapisch M and Bar-Shalom A, 1997 *J. Quant. Spectrosc. Radiat.Transf.* **58**, 687

[15] T. Fujimoto, 1985 *J. Phys. Soc. Jpn.* Vol. 54, No. 8, 2905

[16] Kindo K, 2001 *Physica B* 294-295, 585

# Isomer effect on ionization processes in collisions of 6-MeV/amu bare ions with $C_3H_6$ molecules

T Nakazato[1], T Matsuo[2], T Kohno[3], Y Ohno[4], S Watanabe[2] and T Murakami[4]

[1] Institute of Laser Engineering, Osaka Univ., Suita-shi, 565-0871 Osaka, Japan
[2] Kanagawa Institute of Technology, Atsugi-shi, 243-0292 Kanagawa, Japan
[3] Tokyo Institute of Technology, Yokohama-shi, 226-8502 Kanagawa, Japan
[4] National Institute of Radiological Sciences, Inage-ku, 263-8555 Chiba, Japan

e-mail: matuo@rm.kanagawa-it.ac.jp

**Abstract.** Gross and partial ionization cross sections for $C_3H_6$ isomers, propene and cyclopropane, have been measured under impact of 6 MeV/amu fully stripped ions with charge state $q$=1, 2, 6, 10, and 18. The gross ionization cross sections have been found to depend weakly on the projectile charge $q$ compared with the $q^2$ dependence, especially for highly-charged projectiles. The cross sections for cyclopropane have been found to be slightly larger than those for propene in all projectiles studied. The most prominent product was parent $C_3H_6^+$ ions in cyclopropane, whereas $C_3H_5^+$ ion dominated $C_3H_6^+$ ion in propene, indicating an isomer effect on fragmentation processes. Mechanisms relevant to the production of fragmented ions are also discussed.

## 1. Introduction

Collision processes involving hydrocarbon molecules are known to be important in a number of applications such as material, plasma, astrochemical, and medical sciences. We have been studying ionization of small hydrocarbons in 6-MeV/amu fully stripped-ion impact through measurements of absolute cross sections and relative intensities of individual secondary ions. For hydrocarbons, as the number of carbon atoms increases, the number of isomers tends to increase. It has been recognized that isomers show apparent differences in various physical and chemical properties. In the present study, the target was chosen to be $C_3H_6$ isomers, cyclopropane (denoted by $(CH_2)_3$) and propene $C_3H_6$, to study the effect of molecular structure on the ionization and fragmentation processes under energetic heavy ion impact.

## 2. Experimental method

The experiment was carried out using the 6-MeV/amu-$H^+$, -$He^{2+}$, -$C^{6+}$, -$Ne^{10+}$, and -$Ar^{18+}$ projectiles provided by the Heavy Ion Medical Accelerator in Chiba (HIMAC) of the National Institute of Radiological Sciences (NIRS). The absolute values of gross ionization cross sections were measured using the parallel-plate-condenser method, detail of which was described in a previous paper [1]. The systematic uncertainties in the gross ionization cross section were estimated to be 8.7%, and the statistical error was within 1.2%. The mass/charge spectra of secondary ions were measured using a double-focusing sector magnet [2].

## 3. Results and discussion

### *3.1 Ionization cross section*

The measured gross ionization cross sections are listed in table 1. To illustrate how the cross sections depend on the projectile charge $q$, they were plotted as a function of $q$ in figure 1.

**Table 1.** Measured gross ionization cross sections (in $10^{-16}$ $cm^2$)

| | $H^+$ | $He^{2+}$ | $C^{6+}$ | $Ne^{10+}$ | $Ar^{18+}$ |
|---|---|---|---|---|---|
| $C_3H_6$ | 0.956 | 3.77 | 30.2 | 71.3 | 179 |
| $(CH_2)_3$ | 0.993 | 3.88 | 31.1 | 72.7 | 183 |

As can be seen, the cross sections for $H^+$ and $He^{2+}$ projectiles are well on a line showing $q^2$ dependence, indicating validity of the perturbation treatment such as first Born approximation for light projectiles. On the other hand, in higher charge projectiles such as $Ne^{10+}$and $Ar^{18+}$, the cross sections are considerably smaller than those predicted by the $q^2$ dependence. This causes that the cross sections for these molecules are about 40% smaller than the prediction from $q^2$ dependence under $Ar^{18+}$ impact, indicating failure of the first Born approximation for high charge projectiles even in the present collision energy of 6 MeV/amu.

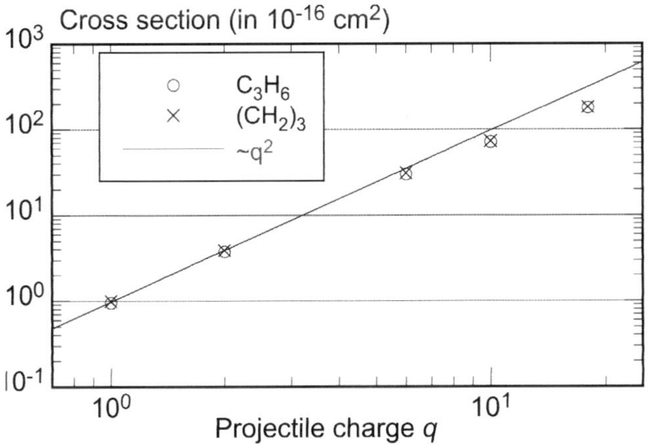


**Figure 1.** Projectile charge $q$ dependence of gross ionization cross sections for propene $C_3H_6$ and cyclopropane $(CH_2)_3$.

In table 1 we can also note that the cross sections of cyclopropane are 2-3% larger than those of propene for all projectiles studied here. On the contrary, the electron impact experiments for $C_3H_6$ isomers showed that the cross sections of propene were larger than those of cyclopropane by 2-4% in the energy range between 10 and 3000 eV [3]. The cause of such differences might be explained by the deviation from the first Born approximation. For impact of fast projectiles, perturbation theory is valid, and the cross section for a specified process can be written as a Born series,

$$\sigma = a_1(V_p)q^2 + a_2(V_p)q^3 + a_3(V_p)q^4 + \dots\dots , \tag{1}$$

where $q$ is the projectile charge and $Vp$ the projectile velocity. At very high velocity, the cross section is given to high accuracy by the first term. It is proportional to $q^2$ and independent of projectile mass and sign (±) of charge as well. The second term might be responsible for the difference between ion and electron impact.

*3.2 Mass/charge spectra of recoil ions*

Mass/charge spectra of recoiled ions were measured under $He^{2+}$, $C^{6+}$, and $Ar^{18+}$ collisions. Figure 2 shows the recoil ion spectra for propene and cyclopropane in $Ar^{18+}$ ion impact. It should be noted that the most prominent peak is the parent molecular $(CH_2)_3^+$ ion in cyclopropane, whereas $C_3H_5^+$ ion predominates $C_3H_6^+$ in propene, indicating an isomer effect in fragmentation. Doubly charged molecular ions such as $C_3H_6^{2+}$, $C_3H_5^{2+}$, $C_3H_4^{2+}$and $C_3H_3^{2+}$ are found to be appreciably produced in these molecules. The apparent peak shift appearing in the left panel is mainly due to a hysteresis of the analyzing maget used in the present experiment.

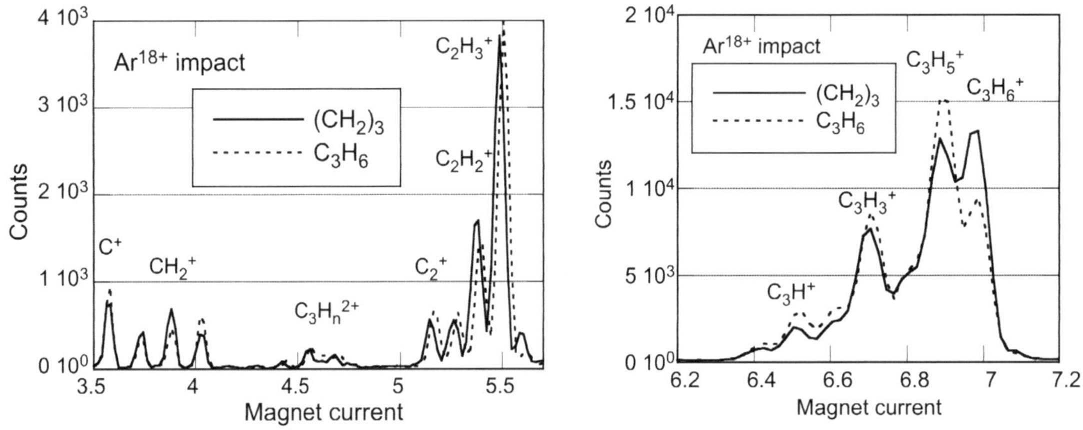


**Figure 2.** Mass/charge spectra for $(CH_2)_3$ and $C_3H_6$ molecules in $Ar^{18+}$ impact

*3.3 Partial ionization cross sections*

From the measured ionization cross sections and relative intensities of recoil ions, we can calculate partial ionization cross sections for individual ion species. The obtained partial cross sections for $(CH_2)_3$ target are plotted as a function of charge $q$ in figure 3.

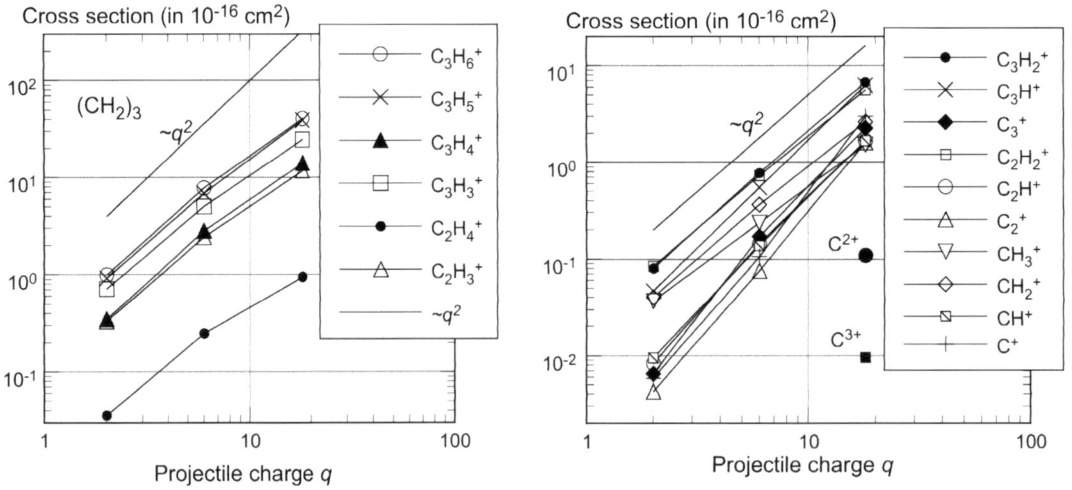


**Figure 3.** Partial ionization cross sections for cyclopropane as a function of projectile charge $q$

The charge dependence of cross sections could be divided tentatively into two groups. As shown in the left panel of figure 3, one consists of parent $(CH_2)_3^+$ ion and fragmented $C_3H_5^+$, $C_3H_3^+$, $C_2H_4^+$, and $C_2H_3^+$ ions, charge dependence of which is quite similar to the gross ionization cross sections and slightly weak compared with the $q^2$ dependence, indicating that they are mainly produced through distant collisions. Moreover, the intensity ratios of fragmented ions to parent $(CH_2)_3^+$ ion were independent of $q$ within experimental error. This result suggests that these fragments are mainly produced through two-step processes as described by the quasiequilibrium theory (QET) by Rosenstock [4]: An excited state molecular $(CH_2)_3^{+*}$ ion is first produced in the ion impact, and this ion then fragments statistically with a fragmentation pattern which depend only on the internal energy of the excited $(CH_2)_3^{+*}$ ion.

As the right panel of figure 3 shows, in highly fragmented ions such as $C_3H_2^+$, $C_3H^+$, $C_3^+$, $CH_2^+$ and so on, cross sections show charge dependence steeper than $q^2$, indicating that intimate collisions might take part in their production. In addition, the intensity ratios of these ions to parent $(CH_2)_3^+$ ion were found to change considerably with projectile charge $q$. In production of such highly fragmented ions, multiple ionization, which would yield immediate dissociation of molecular ion, might contribute to considerable extent. Multiply charged $C^{2+}$ and $C^{3+}$ ions were also detected with remarkable cross sections of $10^{-17} \sim 10^{-18}$ cm$^2$ in the $Ar^{18+}$ impact. Quite similar results have also been obtained in propene molecule.

## 4. Summary

Gross ionization cross sections for $C_3H_6$ and $(CH_2)_3$ molecule in collision of 6-MeV/amu-$H^+$, -$He^{2+}$, -$C^{6+}$, -$Ne^{10+}$, and -$Ar^{18+}$ ions were found to depend weakly on the projectile charge $q$ compared with the $q^2$ dependence predicted by the first Born approximation. The cross sections for cyclopropane in ring shape was found to be slightly larger than those for propene, in contrast to the electron impact ionization, in which cross sections were reported to be larger in propene. These results might indicate the failure of the first Born approximation for high charge projectiles even in the present collision energy. The mass/charge spectra of recoil ions showed that the most prominent product was fragmented $C_3H_5^+$ ion in propene target, while $C_3H_6^+$ ion dominated $C_3H_5^+$ in cyclopropane, indicating an isomer effect in fragmentation. Deviation from quasiequilibrium theory was observed in production of highly fragmented ions.

## Acknowledgement

This work was performed as part of the Research Project with Heavy Ions at NIRS-HIMAC. We would like to thank the NIRS-HIMAC crew for providing excellent beam and their support during this work.

## References

[1] Matsuo T, Kohno T, Makino S, Mizutani M, Tonuma T, Kitagawa A, Murakami T and Tawara H 1999 *Phys. Rev.* A 60 3000-07

[2] Tonuma T, Shibata H, Be S H, Kumagai H, Kase M, Kambara T, kohno I, Osaki A and Tawara H, 1986 *Phys. Rev.* A 33 3047-53

[3] Nishimura H and Tawara H, 1994 *J. Phys. B: At. Mol. Opt. Phys.* 27 2063-74

[4] Rosenstock H M, Wallenstein M B, Wahrhaftig A L, and Eyring H 1952 *Proc. Natl, Acad. Sci. U.S.A.* 38 667

Journal of Physics: Conference Series **163** (2009) 012064 doi:10.1088/1742-6596/163/1/012064

# Differential Experimental Investigation of Electron Capture for $He^{2+}$ on Helium at Low Energy

**X. L. Zhu[1], X. Ma[1,2], B. Li [1,2], S. F. Zhang[1,2], W. T. Feng[1,2], H. P. Liu[1],L. F. Chen[1,2], S. P. Cao[1,2], D. B. Qian[1,2], D. C. Zhang[1,2]**

1 Institute of Modern Physics, Chinese Academy of Sciences, Lanzhou 730000 China
2 Graduate University of Chinese Academy of Sciences, Beijing 100049 China

E-mail: zhuxiaolong@impcas.ac.cn

**Abstract.** State-selective single- and double-electron capture in $He^{2+}$-He collisions at energies ranging from 20 to 40 keV have been studied experimentally by means of cold target recoil ion momentum spectroscopy. The differential cross sections have been obtained by measuring the longitudinal and transverse momenta of recoil ions. The results show that single electron capture into L shell is the dominant reaction channel, and double-electron capture into the ground state of projectile is by far the dominant process in the incident energy range. The measured state-selective cross sections agree well with the semi-classical close-coupling calculations.

## 1. Introduction

The collision processes of multiply charged ions with atom/molecules are of fundamental interest because they involve many-body dynamics. Charge transfer processes during keV collisions of multiply charged ions with various atomic and molecular targets play an important role in various fields, such as astrophysical plasmas, accelerator physics [1-4]. The symmetrical $He^{2+}$-He collision system is the simplest two-electron system in which coulomb ion–electron and nuclear-electron interactions, and the correlation between electrons can be studied. This collision system has been studied heavily in experimental [5-14] and theoretical [14-17] works. However, only a few measurements or calculations of differential angular scattering cross sections have been carried out for the $He^{2+}$-He collision system in the energy range from 15 keV to 150 keV.

In this paper, we present measured differential cross-sections for single electron capture (SC),

$$He_p^{2+} + He_t \rightarrow He_p^{+}(n) + He_t^{+}(n') \qquad (1)$$

and double-electron capture (DC),

$$He_p^{2+} + He_t \rightarrow He_p^{0}(nn') + He_t^{2+} \qquad (2)$$

at 20, 30 and 40keV by using cold target recoil ion momentum spectroscopy (COLTRIMS). Results were compared with theoretical calculations.

## 2. Experimental setup

The experiments have been performed in the atomic physics platform at the Institute of Modern Physics (IMP) in Lanzhou. Details of the experimental set-up are described in references [18,19].

Briefly, a continuous beam of $He^{q+}$ is extracted from the ECR source, whose extraction voltage of the source was varied from 5 to 23 kV, and the charge state is analysed by a magnet. The ion beam enters the collision chamber through the two-pair collimators with four adjustable slits. The beam-spot size of below 0.4×0.4 $mm^2$ at collision point was obtained. The $He^{2+}$ ion beam is crossed with a supersonic He gas target. 20 cm downstream of the interaction point an electrostatic deflector analyses the final projectile charge states. Then, a two-dimensional position-sensitive detector (PSD) with delay-line anode detects the projectile products. The recoil ions generated in the collisions were extracted by an electric field and accelerated toward the recoil PSD. After passing the field-free time-of-flight region, the ions were post-accelerated to a few keV to increase the detection efficiency. Coincidence technique was used between recoil ions and projectiles. Coincident event was recorded by PC-based data acquisition system in even-by-event mode. Figure 1 shows the two-dimension coincidence map of recoil ions with projectiles ions at 30keV $He^{2+}$-He collisions. From the map, the different reaction channels can be easily identified.

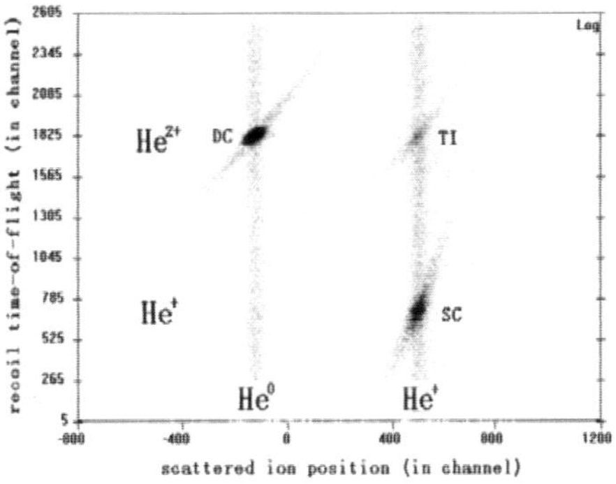


**Figure 1.** The two-dimensional coincidence map of recoil ions with the projectiles ions at 40 keV $He^{2+}$+He collisions.

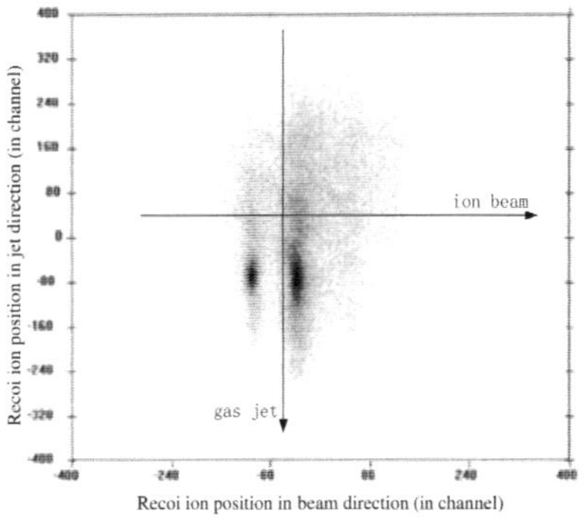


**Figure 2.** Recoil ion position image on the detector for SC process at 40 keV $He^{2+}$-He collisions.

## 3. Results and discussion

The COLTRIMS collects two classes of information: energy-gain spectra and angular differential cross sections. The former one gives the state-selective information on charge transfer, and the later one gives the information on the dynamics of the populated states. For pure electron captures, according to momentum and energy conservation laws, the longitudinal momentum and transverse momentum of the recoil ion was deduced as follows:

$$P_{//R} = -Q / v_p - n_c \cdot v_p / 2 \tag{3}$$

$$P_{\perp R} = m_p \cdot v_p \cdot \theta \tag{4}$$

where, $v_p$ is the projectile velocity, $m_p$ is the projectile mass, $n_c$ is the number of captured electrons, Q is the change in binding energy of the active electrons, and θ is the scattering angle of the projectile products.

### *3.1.1. Single electron capture*

Figure 2 shows the distribution of $He^+$ recoil ions on the recoil ion detector for SC processes at 40 keV impact energy. From figure 2, electron capture into the different final bound states was clearly distinguished. The single electron capture into L shell is the dominant reaction channel. The experimental state-selective cross sections are deduced by normalizing the total single electron capture cross sections to the data of references [8,9].

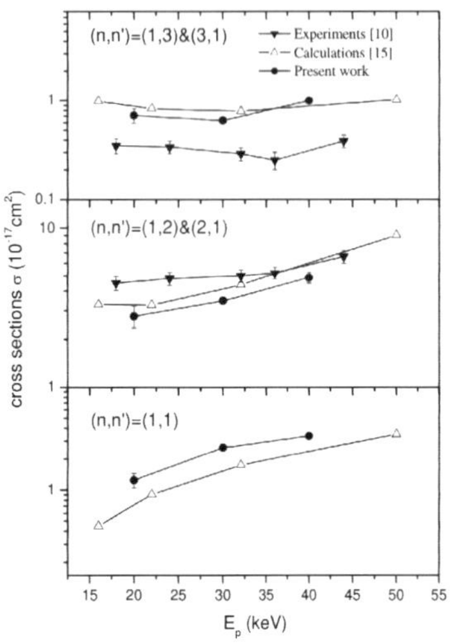


**Figure 3.** State-selective capture cross sections for SC process of $He^{2+}$ on He collisions.

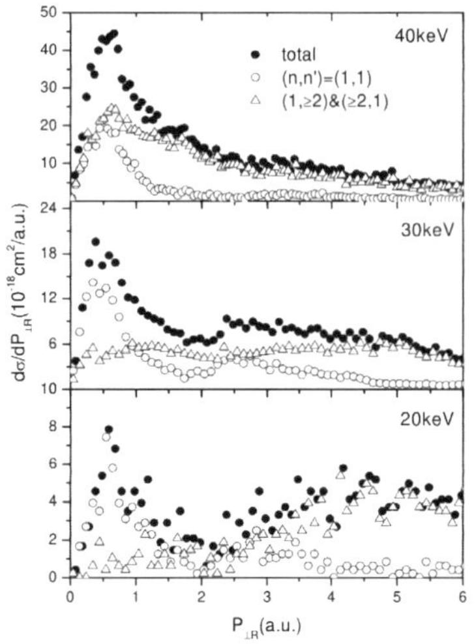


**Figure 4.** Differential state-selective capture cross sections for SC process of $He^{2+}$ on He collisions.

Figure 3 shows the energy dependence of state-selective capture cross sections, which are compared with calculations given by Fritsch [15] and other experimental results [10]. From figure 3, we can observe that with increasing projectile energy the state-selective cross section for single electron capture into K and L shells increases clearly more than for electron capture into M shell. The experimental data are reasonably described by semi-classical close-coupling calculations [15].

From the evaluation of the transverse momentum of recoil ion, the angular differential cross sections are obtained and presented in Figure 4. Total differential state-selective cross sections exhibit that the single electron capture into ground state mainly contributes at small scattering angles, and the single electron capture into excited states contribute at large scattering angles. The single electron capture into K shell exhibits an oscillating structure, while the capture to excited states shows a smooth decrease at larger transverse momenta. The oscillating structures can be understood by the interference of molecular orbits during the collisions processes [20].

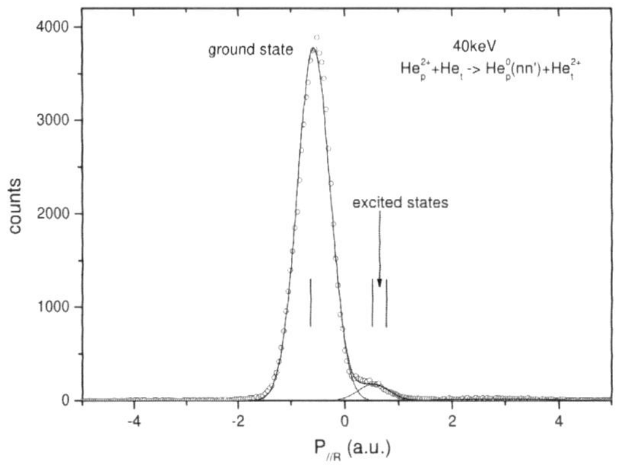


**Figure 5.** The longitudinal momentum spectra of recoil $He^{2+}$ ion for DC processes.

*3.1.2. Double electron capture*

Figure 5 shows the longitudinal momentum spectrum obtained for double-electron capture by $He^{2+}$ ions on helium at 40 keV. According to Eq. (3), the peak in figure 5 can easily be identified. The left peak in the spectrum corresponds to the double-electron capture into the ground state of the projectile; the right peak corresponds to the case of one of the two electrons captured into the ground state and the other one captured into excited states. It is clearly found that double electron captured into the ground state is by far the dominant reaction channel.

## 4. Summary

We have experimentally studied the state-selective single- and double-electron capture processes for slow $He^{2+}$ on helium collisions between 20 keV and 40 keV. COLTRIMS technique was utilized. Differential cross sections have been obtained by the measurement of the recoil ion momentum vector. At low energies between 20 and 40 keV, the experimental results show that single electron capture into L shell is the main reaction channel, and double-electron capture into the ground state of projectile is by far the dominant process. The measured data is reasonably described by semi-classical close-coupling calculations.

## Acknowledgments

This work is supported by the Natural Science Foundation of China (Grant No. 10434100).

## References

[1] M. Barat, P. Roncin, 1992 *J. Phys.* B **25** 2205
[2] D. Mathur, 1988 *Phys. Rep.* **225** 193
[3] M. Ehrich, B. Siegmann, U. Werner and H. Lebius, 2003 *Radiat. Phys. Chem.* **68** 127
[4] R. K. Janev, T. Kato and J. Wang, 2000 *Phys. Plasmas* **7** 4364
[5] V. V. Afrosimov, A. A. Basalaev, G. A. Leiko andM. N. Panov, 1978 *Sov. Phys. JETP* **47** 837
[6] M. E. Rudd, T. V. Goffe and A. Itoh, 1985 *Phys. Rev. A* **32** 2128
[7] R. D. Dubois, 1986 *Phys. ReV. A* **33** 1595
[8] R. D. Dubois, 1987 *Phys. Rev. A* **36** 2585
[9] M. B. Shah, P. McCallion and H. B. Gillbody, 1989 *J. Phys. B: At. Mol. Opt. Phys.* **22** 3037
[10] H. O. Folkerts, F. W. Bliek,L. Meng, R. E. Olson, R. Morgenstern, M. V. Von Hellermann, H. P. Summers and R. Hoekstra, 1994 *J. Phys. B: At. Mol. Opt. Phys.* **27** 3475
[11] V. Mergel, R. Dörner, J. Ullrich, O. Jagutzki, S. Lencinas, S. Nüttgens, L. Spielberger, M. Unverzagt, C. L. Cocke, R. E. Olson, M. Schulz, U. Buck, E. Zanger, W. Theisinger, M. Isser, S. Geis and H. Schmidt-Böcking, 1995 *Phys. Rev. Lett.* **74** 2200
[12] R. Dörner, V. Mergel, L. Spielberger, O. Jagutzki, J. Ullrich and H. Schmidt-Böcking, 1998 *Phys. Rev. A* **57** 312
[13] R. Dörner, V. Mergel, O. Jagutzki, L. Spielberger, J. Ullrich, R. Moshammer and Schmidt-Böcking, 2000 *Phys. Rep.* **330** 95
[14] R. S. Gao, C. M. Dutta, N. F. Lane, K. A. Smith, R. F. Stebbings and M. Kimura, 1992 *Phys. Rev. A* **45** 6388
[15] W. Fritsch, 1994 *J. Phys. B: At. Mol. Opt. Phys.* **27** 3461
[16] Ashok Jain, C. D. Lin and W. Fritsch, 1989 *Phys. Rev. A* **39** 1741
[17] Wu Yong, Yan Bin, Liu Ling and Wang Jian-guo, 2007 *Chin. Phys. Lett.* **24** 1999
[18] Zhu Xiao-long, Ma Xin-wen, Li Bin, Liu Hui-ping, Chen Lan-fang, Zhang Shao-feng, Qian Dong-bin, Feng Wen-tian, Cao Shi-pin, Sha Shan and Zhang Da-cheng, 2006 *Chin. Phys. Lett.* **23** 587
[19] Ma Xin-wen, Zhu Xiao-long, Liu Hui-ping, Li Bin, Zhang Shao-feng, Cao Shi-pin, Feng Wen-tian, Xu Shen-yue, 2008 *Science in China Series G: Physics, Mechanics & Astronomy* **51** 755
[20] W. C. Keever and E. Everhart, 1966 *Phys. Rev.* **150** 43

# ENERGY GAIN MEASUREMENTS OF $Xe^{9+}$ IONS IN COLLISIONS WITH He AND Ar

**K. Ishii, K. Yoshimura[1], Y. Inoue[1] and H. Ogawa**

Department of Physics, Nara Women's University, Nara 630-8506, Japan
[1]Graduate school of Humanities and Sciences, Nara Women's University, Nara 630-8506, Japan

E-mail: ishii@cc.nara-wu.ac.jp

**Abstract.** To clarify a collision dynamics of multiply charged Xe ions with rare gas target, we have measured energy gain spectra of $Xe^{9+}$ in collisions with He and Ar at 200 $q$ eV and have compared with the theoretical energy gain peak calculated by the classical over barrier model. Using energy levels of excited $Xe^{8+}(nl)$ ions referred to the NIST atomic spectra database, we have determined capture states of $Xe^{9+}$ ions after single electron transfer.

## 1. Introduction

The study of collision dynamics of multiply charged ions with atoms and molecules at low energies is important not only in atomic physics but also in various applied fields. Recently, high temperature plasmas of heavy elements such as Xe and Sn have attracted great attention in developing extreme ultra-violet (EUV) light sources ($\lambda$ =13.5 nm) for the technology of the lithography in next generation[1]. In order to clarify emitting processes of EUV lights from the Xe or Sn plasmas, we should understand basic charge transfer processes in these plasmas since the EUV light is emitted as one of the de-exciting processes of excited states after the charge transfer.

An energy gain spectroscopy technique firstly developed by Ohtani et al is very useful to determine the final states of multiply charged ions in collisions with atoms or molecules[2]. For the collision systems involving Xe, Hvelplund et al have measured energy gain spectra of multiply charged Xe ions with rare gas targets. They identified the states into which an electron was captured for $Xe^{q+}$ ions based on classical over barrier model[3, 4]. Recently, Tanuma et al have measured EUV emission spectra from charge selected Xe and Sn ions extracted from ECR ion source with several targets[5]. They found that most of 13.5 nm EUV is mostly emitted from excited Xe ions for $q$=10 and Sn ions for $q \geq 10$.

To determine the states into which an electron is captured by multiply charged Xe ions with the energy gain spectra, precise energy levels for ground and excited states of multiply charged ions are necessary. Saloman has compiled available energy levels and observed spectral lines at all stages of ionization of Xe ions [6]. As a database of energy levels of excited multiply charged ions, it is well known that the NIST atomic spectra database is quite useful [7]. In the case for $Xe^{9+}$ ion, the 61 levels for the ground and excited $Xe^{8+}$ ions are presented in the NIST database which include 59 levels summarized by Saloman.

In this work we have measured energy gain spectra of $Xe^{9+}$ in collisions with He and Ar at 200 eV/$q$. Present results are compared with the predicted energy gain by the classical over barrier calculation. We have also calculated potential curve crossings based on Coulomb repulsion using energy levels from NIST atomic spectra database and have determined capture levels after single electron transfer collisions.

## 2. Experimental

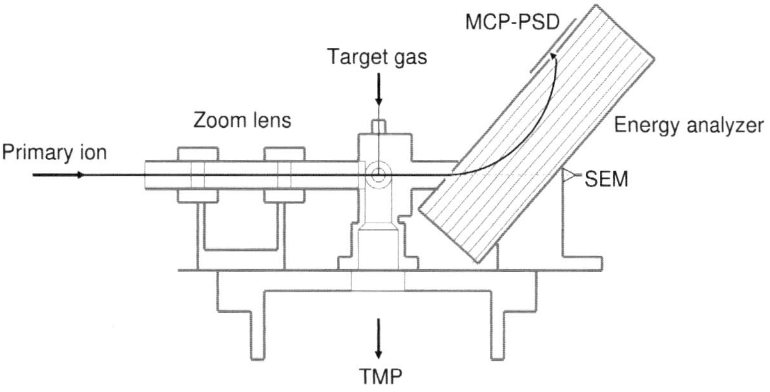


**Figure 1.** Schematic diagram of experimental setup.

The experiment was performed using the mini-EBIS atomic collision facility at Nara Women's University. Fig. 1 shows a schematic diagram of our experimental setup[8]. Projectile ions extracted from the mini-EBIS ion source were mass-separated by an electromagnet analyzer and decelerated before entering collision region. The target gas flow is effused from a nozzle placed above 1 mm from the collision center. After charge transfer at the collision center, projectile ions are energy-analyzed by an electrostatic analyzer of a parallel plate type and are detected by a position sensitive detector system.

In the energy gain spectroscopy, the reaction energy, $Q$-value, can be determined by the energy gain $\Delta E$ of the projectile ion after the charge transfer. The energy gain can be given by following,

$$\begin{aligned}\Delta E &= (\frac{m_1}{m_1+m_2})^2 E_0 \\ &\times \left\{\cos\theta + \left[\left(\frac{m_2}{m_1}\right)^2\left(1+\frac{m_1+m_2}{m_2}\frac{Q}{E_0}\right) - \sin^2\theta\right]^{1/2}\right\}^2 - E_0,\end{aligned}$$

where $m_1$ and $m_2$ are the masses of the projectile and the target, respectively and $E_0$ and $\theta$ are the collision energy and the scattering angle of the projectile ions in the laboratory frame, respectively. Assuming the conditions $\theta \cong 0$ and $E_0 \gg Q$, $\Delta E \sim Q$.

We have routinely checked long-lying excited ions in the primary ion beam by means of a beam attenuation method [8, 9]. The mini-EBIS was operated with a DC mode at a pressure around $10^{-8}$ Pa under liquid nitrogen temperature, so that practically no long-lying excited ions were produced via electron capture collisions with residual gases inside the ion source [10]. It is noted that most long-lying excited ions, if any, may be quenched easily inside the ion source because the confinement time of about 1 ms is sufficiently long to quench such ions.

## 3. Results and discussion

Measured energy gain spectra of $Xe^{9+}$ in collisions with He and Ar at 200 eV/$q$ are shown in Figure 2. The energy resolution of the energy analyzer is around 0.01. The energy width of primary ion beam was measured to be around 10 eV at full width at half maximum (FWHM). In the measured energy gain spectra, both spectra have similar structures and there are single peaks at about 27 eV and 20 eV for He and Ar, respectively. Each energy width of energy gain spectra for both targets are about 20 eV. This indicates that many final states of $Xe^{8+}(nl)$ after the single electron transfer contribute in the present collision systems. Note that small structures around 40 eV and 55 eV for both targets are regarded as contributions from transfer ionization processes via autoionizing double electron capture states.

In the classical over the barrier frame, the peak energy of energy gain spectra for single electron transfer can be predicted as follows,

$$Q = \frac{(q-1)}{2\sqrt{q}+1} I_1, \tag{1}$$

where $q$ is charge state of projectile and $I_1$ is the ionization potential of the target. In the case for single electron capture of $Xe^{9+}$ ions, the predicted values are 28.1 eV and 18.0 eV for He and Ar, respectively. These results are fairly consistent with measured ones. That is, the classical over barrier model reasonably works out in these collision systems since it is considered that many excited states of multiply charged Xe ions exist almost continuously.

To determine final states of projectile after the charge transfer, we have used energy levels for ground and excited states of $Xe^{8+}$ ions referred to NIST Atomic Spectra Database[6]. Figure 3 shows calculated potential curves using these energy levels. Noting that the results of energy gain spectra correspond to the energy differences between initial and final states in the potential curves, we can find capturing energy levels after single electron transfer. The major channels for these collision systems can be estimated as follows,

$$\begin{aligned} Xe^{9+} + He &\rightarrow Xe^{8+}(4d^9 5f^1, 4d^9 6p^1) + He^+, \\ Xe^{9+} + Ar &\rightarrow Xe^{8+}(4d^9 6f^1, 4d^9 7p^1) + Ar^+. \end{aligned}$$

It should be noted that many energy levels between $Xe^{8+}(5f)$ and $Xe^{8+}(7p)$ in the figure 3 must exist almost continuously although such energy levels are not found in the NIST atomic spectra database. The contributions of major channels can be determined when the database are fulfilled. It is required for precise analysis to fulfill energy levels of the highly charged Xe ions.

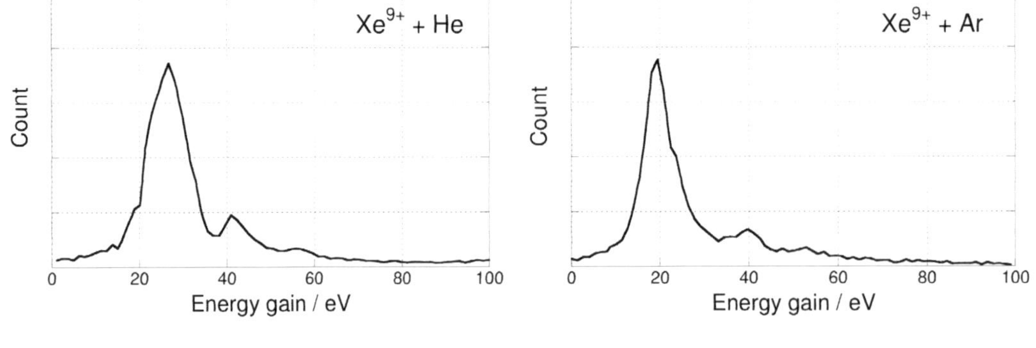


**Figure 2.** Measured energy gain spectra of $Xe^{9+}$ ions in collisions with He and Ar at 200 eV/$q$.

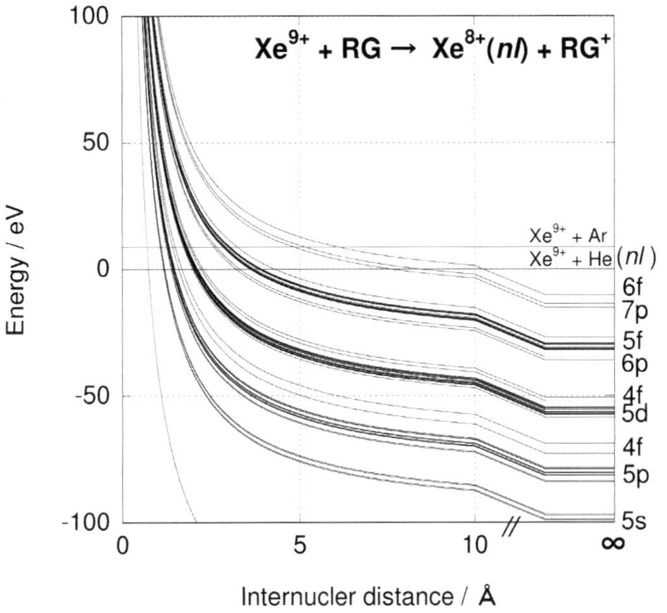


**Figure 3.** Potential curve crossing of $Xe^{9+}$ in collisions with He and Ar. The energy levels of excited $Xe^{8+}(nl)$ ions are referred from the NIST atomic spectra database.

## 4. Conclusion

We have measured energy gain spectra of $Xe^{9+}$ in collisions with He and Ar at 200 eV/$q$. Predictions by the classical over barrier model are fairly consistent with experimental results. Using energy levels from the NIST Atomic Spectra Database, we have determined capturing state of $Xe^{9+}$ ions after single electron transfer. Fulfilling database for highly charged ions are required.

## Acknowledgments

The authors would like to thank Junya Karimata of Nara Women's University. This work was partly supported by a Grant-in-Aid for Scientific Research from the Ministry of Education, Culture, Sports, Science and Technology of Japan and by Nara Women's University Intramural Grant for Project Research.

## References

[1] Bakashi V 2006 *EUV source for Lithography* (Washington: SPIE press)
[2] Ohtani S, Kaneko Y, Kimura M, Kobayashi N, Iwai T, Matsumoto A, Okuno K, Takagi S, Tawara H and Tsurubuchi S 1982 *J. Phys. B* **15**, L533
[3] Hvelplund P 1987 *J. Phys. B: At. Mol. Opt. Phys.* **20** 2515
[4] Niehaus A. 1986 *J. Phys. B: At. Mol. Opt. Phys.* **19** 2925
[5] Tanuma H, Ohashi H, Shibuya E, Kobayshi N, Okuno T, Fujioka S, Nishimura H and Nishihara K 2005 *Nucl. Instr. Meth. B* **235** 331
[6] Saloman E B 2004 *J. Phys. Chem. Ref. Data* **33** 765
[7] Ralchenko, Yu., Kramida, A.E., Reader, J., and NIST ASD Team (2008). NIST Atomic Spectra Database (version 3.1.5): http://physics.nist.gov/asd3
[8] Ishii K, Inoue Y, Ogawa H, Itoh A and Sakamoto N 2007 *J. Phys. Conf. Ser.* **58** 275
[9] Okuno K 1986 *J. Phys. Soc. Japan* **55** 1504
[10] Ishii K, Itoh A and Okuno K 2004 *Phys. Rev. A* **70** 042716

Journal of Physics: Conference Series **163** (2009) 012066 doi:10.1088/1742-6596/163/1/012066

# Differential cross sections for single-electron capture by low-energy $O_2^{2+}$ ions from Ne and $O_2$

**E Y Kamber**

Physics Department, Western Michigan University, Kalamazoo, MI 49008-5252, USA

E-mail: emanuel.kamber@wmich.edu

**Abstract**. State-selective differential cross sections for single-electron capture processes in low-energy collisions of $O_2^{2+}$ recoil ions with Ne and $O_2$ at impact energy of 100 eV and scattering angles between 0° and 6° have been studied using a differential energy-gain spectrometer. In all of the collision systems studied here, contributions from processes commencing with long-lived metastable states in the incident $O_2^{2+}$ ion beam are detected. The energy-gain spectra are interpreted qualitatively in terms of the reaction windows, which are calculated using the Landau-Zener model and the extended version of the classical over-the-barrier model.

## 1. Introduction

A large variety of molecular ions have been identified in cold astrophysical environments and in planetary atmospheres. Molecular oxygen ions are of particular interest due to their potential influence on the properties of the ionosphere and their involvement in many atmospheric phenomena. Furthermore, electron capture involving multiply charged molecular ions has been proposed as a mechanism for generating ions in the earth's upper atmosphere [1,2].

Most doubly-charged molecules are not stable because the Coulomb repulsion energy between the positive holes exceeds the molecular binding energy. Only a few highly bound covalent molecules have been found, for example, $N_2^{2+}$, $O_2^{2+}$, $NO^{2+}$, $CO^{2+}$, $CO_2^{2+}$ and $NH_3^{2+}$, whose lifetimes may depend on the ionization method [1]. Recently we have extended our study of ion-atom collisions to include single-electron capture by $N_2^{2+}$ and $CF_2^{2+}$ ions from atomic and molecular targets by means of translational energy spectroscopy [3,4]

In the present work, a differential energy-gain spectrometer [5], capable of measuring simultaneously the scattering angle and the energy-gain of projectile products in ion-atom collisions, has been used for the study of state-selective non-dissociative single-electron capture in collisions of $O_2^{2+}$ recoil ions with Ne and $O_2$ at an impact energy 100 eV and scattering angles between 0° and 6°.

Unfortunately, the peak of doubly charged molecular ion $O_2^{2+}$ coincides with the atomic ion $O^+$ peak in mass spectrum at m/q =16. No attempt was made to estimate the fraction of $O_2^{2+}$ in the $O^+$ beam. However, we were able to easily separate the $O_2^+$ products resulting from electron capture by using an electrostatic analyzer (ESA) voltage which was double that required to transmit the elastically scattered ion beam. To identify the reaction channels involved, the energy-gain spectrum for $N_2^{2+}$ - Ne collision system that were previously observed in the same apparatus [3], were used as a standard to calibrate the energy scale (i.e., Q scale). The energy levels for $O_2^{2+}$ and $O_2^+$ ions used in calculating the energy defect of the reaction were taken from photoelectron spectroscopy data [6] and other sources [7-9]. The energies were calculated assuming that the projectile, molecular target, and their product ions are at the lowest vibrational level ($\nu = 0$) and calculated on the basis of the Franck-Condon principle, i.e. the ionization energies are those for vertical processes.

Following the nomenclature described by Kamber et al. [10], the observed reaction channels are denoted in the following fashion: the designations I, II, and III represent, respectively, the ground, first, and second electronically excited states of $O_2^{2+}$; α, β, γ, represent the ground and subsequent electronically excited states of $O_2^+$ ion; X, represents the ground state of the target product.

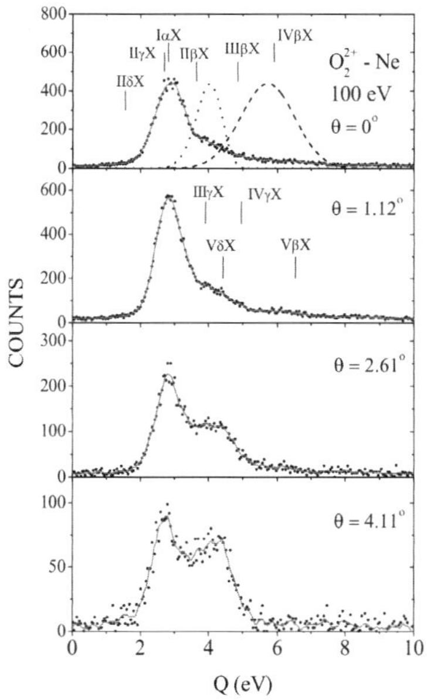


Figure 1. Translational energy-gain spectra for single-electron capture by 100 eV $O_2^{2+}$ ions from Ne at different projectile laboratory scattering angles. Also shown are reaction windows calculated on the basis of a single-crossing LZ model (dotted curve) and the ECOB model (dashed curve). Smooth lines are drawn to guide the eye.

## 2. Results and discussion

2.1. *$O_2^{2+}$ + Ne collisions.* Figure 1 shows the translational energy-gain spectra obtained for single-electron capture by 100 eV $O_2^{2+}$ ions from Ne at different scattering angles. At 0° scattering angle, the energy-gain spectrum indicates that the dominant reaction channel is due to non-dissociative single-electron capture into the ground state (X $^2\Pi_g$) of the product $O_2^+$ from the ground state incident $O_2^{2+}$ (X $^1\Sigma_g^+$) ions with contributions due to capture from the low-lying metastable state (A $^3\Sigma_u$) of the $O_2^{2+}$ into the (A $^2\Pi_u$) state of $O_2^+$. There are smaller contributions due to capture into the a $^4\Pi_u$, A $^2\Pi_u$, and b $^4\Sigma_g^-$ states from the metastable states W $^3\Delta_u$, B $^3\Pi_g$, and B' $^3\Sigma_g^-$ of the $O_2^{2+}$ ions via reaction channels IIβX, IIIγX, IIIβX, VδX, IVγX, IVβX, and VβX. A comparison with the measurements of Pederson [11] at 100 eV, and Hamdan and Brenton [12] at 6 keV shows good agreement with the present measurements.

As the scattering angle is increased, capture into the X $^2\Pi_g$ state (the IαX channel) remains dominant but contributions from reaction channels IIβX, IIIγX, and VδX, respectively, due to capture from the metastable states W $^3\Delta_u$ and B' $^3\Sigma_g^-$ of the $O_2^{2+}$ ions into the a $^4\Pi_u$, A $^2\Pi_u$, and b $^4\Sigma_g$ states of $O_2^+$, increase with increasing the scattering angles. This indicates that the angular distribution for the IαX channel is more strongly peaked in the forward direction than for the reaction channels IIβX, IIIγX, and VδX. Figure 1 also shows our calculated reaction windows for 100 eV $O_2^{2+}$ - Ne collisions, using both a single-crossing Landau-Zener (LZ) model [13] and the extended version of the classical over-the-barrier (ECOB) model [14]. Calculated peak values have been normalized to our observed peak values in the energy spectrum. The reaction based on a single-crossing LZ model does not describe the position of the dominant reaction channel, while the reaction window based on the ECOB model maximizes at about 6 eV and underestimates the contribution of the IαX channel.

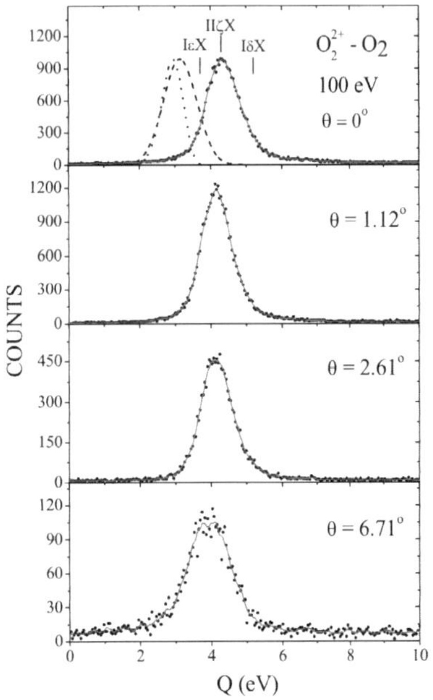


Figure 2. Translational energy-gain spectra for single-electron capture by 100 eV $O_2^{2+}$ ions from $O_2$ at different projectile laboratory scattering angles. Also shown are reaction windows calculated on the basis of a single-crossing LZ model (dotted curve) and the ECOB model (dashed curve). Smooth lines are drawn to guide the eye.

### 2.2. *$O_2^{2+}$ + $O_2$ collisions.*

Figure 2 shows the translational energy-gain spectra for single-electron capture by 100 eV $O_2^{2+}$ ions from $O_2$ at different projectile laboratory scattering angles. The observed spectra clearly indicate that non-dissociative single-electron capture into the $O_2^+$ (c $^4\Sigma_u$) state from the low-lying metastable state (A $^3\Sigma_u$) of the $O_2^{2+}$ is the predominant channel, although capture by the ground state (X $^1\Sigma_1$) of $O_2^{2+}$ ions into b $^4\Sigma_g$, and B $^2\Sigma_g$ states are also significant. As the scattering angle is increased, a relative shift of about 0.45 eV in the energy gain of the IIζX channel is observed going from scattering angle of 0° to 6.71°. This is attributed to the energy given to the target and is in accordance with energy and momentum conservation rules. In addition, the shift in the energy-gain of the dominant peak could be also due to the increasing contributions from the reaction channel IεX as the scattering angle is increased. The reaction windows favor Q values smaller than those observed and are positioned near the IεX channel.

### 2.3. *Differential cross sections*

The experimental total differential cross sections ($d\sigma/d\Omega$) for single-electron capture by 100 eV $O_2^{2+}$ ions from Ne and $O_2$ are shown in Figure 3. The differential cross sections were determined using the translational energy-gain technique, by calculating the area under the peaks in the energy-gain spectra at different projectile laboratory scattering angles using a curve fitting program. The general features of the distributions are qualitatively explained in terms of semi-classical model based on Coulomb potential curves which have been described in detail by Andersson et al. [15]. The traditional two-state model has been used to estimate the critical angle $\theta_c$, which corresponds to capture at an impact parameter equal to the crossing radius, by assuming that capture occurs at a localized curve crossing between the potential energy curves for entrance and exit channels. For small angles, $\theta_c = Q/2E$, where Q is the exoergicity of the collision and E is the collision energy. This angle separates the events scattered at smaller angles due to capture on the way-out of the collision and events scattered at larger angles due to capture on the way-into the collision.

For $O_2^{2+}$ - $O_2$ collisions, the distribution is peaked in the forward direction inside the critical angle $\theta_c$ = 1.32°, which corresponds to the IIζX capture channel. This distribution represents contributions from electron capture that takes place on the way-out of the collisions. For $O_2^{2+}$ - Ne collisions, the measurements show that the projectile products which correlate with capture into the (X $^2\Pi_g$) state are distributed forward inside the critical angle $\theta_c$ = 0.8°, indicating that capture takes place on the way-out of the collision.

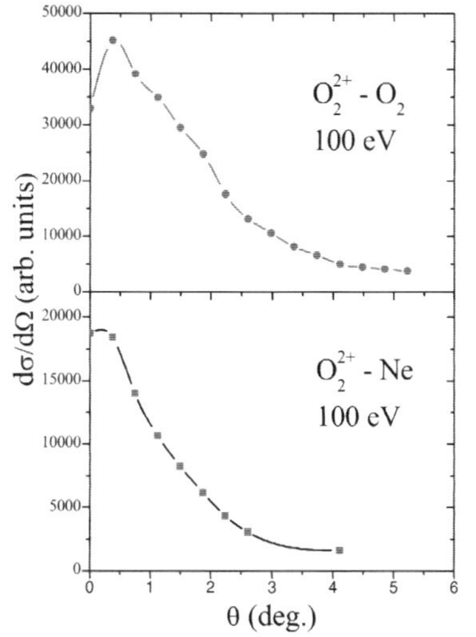


Figure 3. Experimental differential cross sections for single-electron capture by $O_2^{2+}$ ions from Ne and $O_2$. Spline lines are drawn to guide the eye.

In summary, doubly differential cross sections, in energy and angle, for single-electron capture by 100 eV $O_2^{2+}$ ions from Ne and $O_2$ have been studied by means of translational energy gain spectroscopy. Translational energy gain spectra for single-electron capture by $O_2^{2+}$ ions from Ne indicated that dominant reaction channel was correlated with capture into the (X $^2\Pi_g$) state, while for the $O_2$ the dominant channel was due to capture into the $O_2^+$ (c $^4\Sigma_u$) state from the low-lying metastable state (A $^3\Sigma_u$) of the $O_2^{2+}$. The energy-gain spectra were interpreted qualitatively in terms of the reaction windows, which were calculated using the single-crossing LZ model and the ECOB model. The reaction windows did not describe the position of dominant processes. We have also studied differential cross sections for single-electron capture processes, the angular distribution spectra contain a mean peak lying just inside a critical angle $\theta_c$, corresponding to capture at an impact parameter equal to the crossing radius of the dominant reaction channels. The peaks are qualitatively explained by a two-state model and are attributed to a capture process on the way-out from the collision. No data on single-electron capture by $O_2^{2+}$ ions from $O_2$ are, however, available for comparison.

**References:**

[1] Mathur D 1993 *Phys Rep* **225** 193
[2] Prasad S S and Furman D R 1975 *J GeoPhys* **80** 1360
[3] Kamber E Y, Akgungor K, Safvan C and Mathur D 1996 *Chem Phys Lett* **258** 336
[4] Abu-Haija O, Kamber E Y and Mathur D 2005 *Chem Phys Lett* **408** 5
[5] Yaltkaya S, Kamber E Y and Ferguson S M 1993 *Phys Rev* A **48** 382
[6] Turner D W, Baker A D, Baker C and Brundle C R 1970 *Molecular Photoelectron Spectroscopy* (Wiley, London)
[7] Richard-Viard M, Dutuit O, Lavollee M, Govers T, Guyon P M and Durup J 1985 *J Chem Phys* **82** 4054
[8] Ellis K, Hall R J, Avaldi L, Dawber G, McConkey A G, Andric L and King G C 1994 *J Phys* B **27** 3415
[9] Dawber G, McConkey A G, Avaldi L, MacDonald M A, King G C and Hall R J 1994 *J Phys* B **27** 2191
[10] Kamber E Y, Mathur D and Hasted J B 1982 *J Phys* B **16** 3025
[11] Pedersen J O P 1986 M.S. Thesis University of Aarhus, Aarhus, Denmark
[12] Hamdan M and Brenton A G 1989 *Chem Phys Lett* **164** 413
[13] Olson R E and Salop A 1976 *Phys Rev* A **14** 579
[14] Niehaus A 1986 *J Phys* B **19** 2925
[15] Andersson L R, Danared H and Barany A 1987 *Nucl Instrum and Meth* B **23** 54

# Non-perturbative investigation of the interference effects in the ionization of the $H_2$ by charged particle impact

**S. Borbély, K. Póra and L. Nagy**

Faculty of Physics, Babeş-Bolyai University, str. Kogălniceanu nr. 1, 400084 Cluj-Napoca, Romania

E-mail: lnagy@phys.ubbcluj.ro

**Abstract.** The interference effects in the double differential ionization cross section of $H_2$ are studied theoretically. Theoretical, non-perturbative models (Volkov model, numerical solution of the TDSE) developed for the ionization of atomic systems by intense ultrashort laser pulses are applied with success to study the ionization by fast charged particle impact. These results are compared with those obtained by a perturbative impact parameter method.

## 1. Introduction

Interference effects in the ionization of the hydrogen molecule due to the two-center character of the target have been observed experimentally and studied theoretically both for charged particle [1, 2, 3, 4] and photon [5, 6, 7, 8, 9] impact. Just like in the Young experiments in optics, the atoms in the molecule can be considered as separate coherent sources of electrons, leading to interference patterns in the ionization cross sections. The presence of such interference effects in the ionization spectra of diatomic molecules by highly charged particle impact was observed experimentally for the first time by Stolterfoht et al [1].

Experiments usually do not detect the orientation of the molecular axis. However, recently it has become possible to observe experimentally the alignment dependence of the ionization in the case of diatomic homonuclear molecules [10] and important experimental and theoretical work is done to achieve better alignments of the molecular targets [11]. These efforts suggest, that in the near future there will be available angular resolved ionization cross sections for given molecular alignments. Unlike in the case of photoionization, there are only a few theoretical studies [4, 12, 13] in the case of the ionization by charged particle impact or by ultrashort laser pulses, which produce reliable angular resolved theoretical data for fixed molecular orientation.

In the present paper we use a method developed for the ionization by short laser pulses [14] for the ionization by charged particle impact and compare with the previous results obtained with a perturbative impact parameter method [2].

## 2. Theory

The Coulomb interaction between the projectile and electron is expanded into multipole series. It is known that at high velocities of the projectile the ionization is dominated by distant collisions, and in the multipole series of the interaction the dipole term is the most important [15]. The electric field is decomposed into a component parallel and a component perpendicular relative to the projectile trajectory. After basic calculations can be proved, that for high projectile velocities (more than 50 a.u. in the present calculations) the overall momentum transfer toward the electron is along the perpendicular direction, because the parallel components for the approaching and departing parts of the trajectory cancel out. The electron system does not change significantly in the short time interval of the interaction. In these conditions the electron transitions are mainly driven by the perpendicular component of the electric field, and it is safe to neglect the parallel component during our calculations. This approximation for high projectile velocities is valid also for atoms, and in case of molecules for any orientation of the molecular axis. The remaining perpendicular component can be approximated by a sine-square half-cycle electromagnetic pulse [16].

The time evolution of the active electron in the $H_2$ in the presence of an external electromagnetic field is described using the time dependent Schrödinger equation. The screening effect of the passive electron is included into the Coulomb potential $V(\vec{r})$, which acts between the active electron and the nuclei of the molecule. In the present approach the time dependent wave function is expressed as a superposition of Volkov wave functions [14]

$$\Psi(\vec{r},t) = \int d\vec{p}c(\vec{p},t)\Psi_V(\vec{p},t). \tag{1}$$

Using the time dependent Schrödinger equation for the expansion coefficients, the following integro-differential equation is obtained (for details see [14])

$$\frac{\partial}{\partial t}c(\vec{p},t) = -\frac{i}{(2\pi)^3}\int d\vec{s}c(\vec{s}+\vec{p},t)e^{-\frac{i}{2}\int_0^t dt'\vec{s}[\vec{s}+2\vec{p}+2\vec{A}(t')]}\int d\vec{r}V(\vec{r})e^{i\vec{s}\vec{r}}, \tag{2}$$

where $\vec{A}(t) = -\int_0^t \vec{E}(t')dt'$ is the vector potential of the external electromagnetic field. The initial condition for the expansion coefficients is obtained from the initial state wave function $\psi_i(\vec{r})$. The molecular orbital describing the the initial state of the active electron is written as a linear combination of two atomic orbitals with effective charge $\alpha$. Denoting by $S$ the overlap integral between the two atomic orbitals, the initial condition is

$$c^0(\vec{p}) = \sqrt{\frac{\alpha^5}{2\pi^5(1+S)}\frac{2\cos(\vec{p}\vec{R}_0)}{(\alpha^2+p^2)^2}}, \tag{3}$$

where $\vec{R}_0$ is the half of the nuclear axis. The simplest approximate solution of Eq. 2 is obtained by neglecting the Coulomb potential ($V(\vec{r}) = 0$). In this case for the expansion coefficient we have $c_V(\vec{p},t) = c^0(\vec{p})$. This approximation, also called Volkov model, provides good results only for high field intensities. In improving this approximation a possibility is the first order iterative solution of Eq. 2, which in the case of atomic systems was used successfully [14], however the most accurate results can be obtained by the full numerical solution. In the present work Eq. 2 is solved on a FEDVR (finite element discrete variable representation) numerical grid using fourth order Runge-Kutta method. The obtained numerical solution further on will be called TDSE. From the time dependent wave functions calculated in the framework of Volkov or TDSE model the transition probability amplitude to a plane wave final state can be calculated as $a_{if} = (2\pi)^{3/2}c(\vec{k}-\vec{A}(\tau),\tau)$. We compare these results with those obtained by a perturbative impact parameter method, developed previously [2]. In this model we consider the interaction of the projectile with the active electron as a perturbation. In the calculation of the the transition amplitude the peaking approximation was applied. For the transition amplitude we get [2]

$$a(b,\hat{R}_0) = \frac{2\sqrt{2}iZ_pN\alpha}{vk^2\gamma\sqrt{\pi}}e^{-i\vec{b}\vec{k}_\perp}\left[b_+K_1(b_+\gamma)e^{-i(q-k_\parallel)R_{0\parallel}} + b_-K_1(b_-\gamma)e^{i(q-k_\parallel)R_{0\parallel}}\right], \tag{4}$$

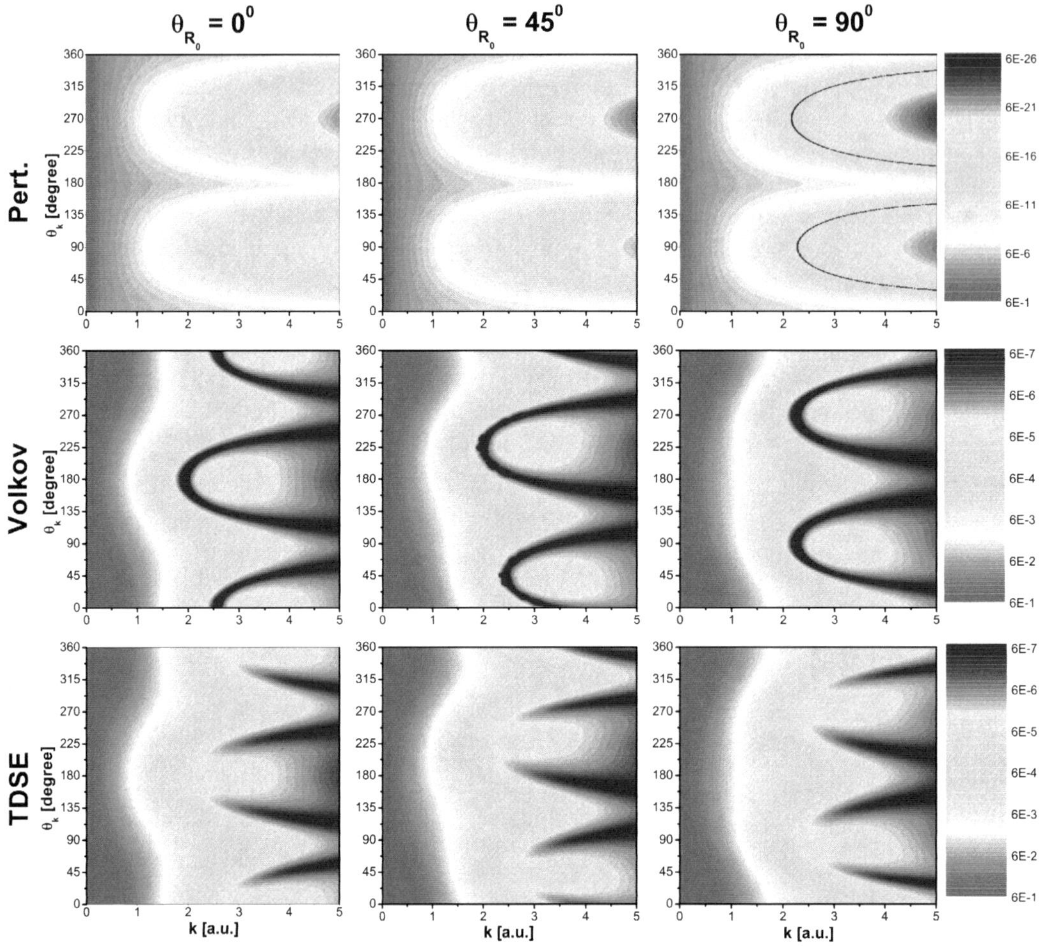


**Figure 1.** Ionization probability densities in the collisional plane as a function of electron momentum ($k$) and electron ejection angle ($\theta_k$ is measured relative to the direction of the perpendicular momentum transfer) for different molecular axis orientations ($\theta_{R_0}$) represented on a logarithmic scale.

$$\gamma = \sqrt{\alpha^2 + (q - k_{\|})^2}, \tag{5}$$

Here $b_+$ and $b_-$ are the impact parameters relative to the two nuclei, while $R_{0\|} = R_0 \cos \beta_{R_0}$, $\beta_{R_0}$ being the angle between the molecular axis and the projectile beam direction. The parallel component of the ejected electron momentum $k$ is $k_{\|} = k \cos \beta_k$, with $\beta_k$ the electron ejection angle. $K_1(x)$ denotes the modified Bessel function, $N$ a normalization factor. $q = \Delta E/v$ is the minimum value of the momentum transfer, where $\Delta E = E_i + k^2/2$ is the energy transfer to the electron.

## 3. Results and discussions

Calculations were performed for 68 MeV/u ($v = 52.18$ a.u.) $Kr^{33+}$ projectile at fixed impact parameter ($b = 4a_0$) value. The electric field corresponding to these parameters has a strength of $E_0 = -Z_p/b^2 = 2.0625$ a.u. and duration of $\tau = 4b/v = 0.306$ a.u. The probability densities were obtained for different molecular axis orientations ($\theta_{R_0}$ is measured relative to the direction of the perpendicular momentum transfer) in the framework of the Volkov, TDSE and perturbative models. Their values in the collisional plane are presented on Fig. 1. A good agreement between Volkov and TDSE results is observed. For both models well defined deep minima in the probability densities may be observed, which appear due to the interference. The minimum is defined by the zero values of expression $\cos[(\vec{k} - \Delta\vec{p})\vec{R}_0]$, where $\Delta\vec{p} = \vec{A}(\tau)$ is the momentum transferred from the projectile to the electron. The same formula was also obtained empirically by Stolterfoht et. al. [1]. The good agreement between the TDSE and Volkov models is due to the short duration and high intensity of the electric pulse, i.e. the sudden approximation applied in the Volkov model is valid in accordance with the findings of Duchateau et al [17].

For the parallel orientation of the molecular axis relative to the projectile trajectory (i.e. $\theta_{R_0} = 90^0$) the same minima are obtained also with the perturbative model. For this orientation $b_+ = b_-$ and the amplitude given by Eq. 4 is proportional to $\cos[(k_\parallel - q)R_0]$, leading to a similar dependence as for the other two methods (3). For the other orientations of the molecular axis the impact parameters relative to the two nuclei ($b_+$ and $b_-$) are different, causing one term of the amplitude (4) being much larger than the other. Consequently, in this model, the interference is unobservable for molecular orientations $\theta_{R_0} = 0^0$ and $\theta_{R_0} = 45^0$.

Another difference between the models is, that in the case of the non-perturbative models the electrons are predominantly ejected in the direction of the momentum transfer, while in the perturbative model also the ejection in the opposite direction has the same probability, leading to a dipole character.

We assume, that in the calculated cross sections these important differences between the models will diminish, because of the integration over all possible impact parameters. The calculation of the differential cross section with the Volkov and TDSE models will be the topic of our future work.

## Acknowledgments

This work was supported by the Romanian Academy of Sciences (grant No. 31/2008) and by the Romanian National Plan for Research (PNII) under contract No. ID 539.

## References

[1] Stolterfoht N et al 2001 *Phys. Rev. Lett.* **87** 023201.
[2] Nagy L, Kocbach L, Póra K and Hansen J P 2002 *J. Phys. B: At. Mol. Opt. Phys.* **35** L453.
[3] Sarkadi L 2003 J. Phys. B **36** 2153.
[4] Laurent G et al 2002 *J. Phys. B: At. Mol. Opt. Phys.* **35** L495.
[5] Cohen H D and Fano U 1966 *Phys. Rev.* **150** 30.
[6] Walter M and Briggs J 1999 *J. Phys. B: At. Mol. Opt. Phys.* **32** 2487.
[7] Nagy L, Borbély S and Póra K 2004 Phys. Lett. A **327** 281.
[8] Fojón O A et al 2004 *J. Phys. B: At. Mol. Opt. Phys.* **37** 3035.
[9] Borbély S and Nagy L 2007 *Rad. Phys. Chem.* **76** 516.
[10] Baier S, Becker A and Plaja L 2008 *Phys. Rev. A* **78** 013409.
[11] Sakai H, Minemoto S, Nanjo H, Tanji H and Suzuji T 2003 *Phys. Rev. Lett.* **85** 083001.
[12] Nicolas Sisourat, Jérémie Caillat, Alain Dubois and Pablo D. Fainstein 2007 *Phys. Rev. A* **76** 012718.
[13] Póra K and Nagy L 2008 *Nucl. Instr. Meth. B*, doi:10.1016/j.nimb.2008.10.059, *in press*
[14] Borbély S, Tőkési K and Nagy L 2008 *Phys. Rev. A* **77** 033412.
[15] Rodriguez V D and Falcon C A 1990 *J. Phys. B: At. Mol. Opt. Phys.* **23** L547.
[16] Rodriguez V D, Macri P, Gayet R 2005 *J. Phys. B: At. Mol. Opt. Phys.* **38** 2775.
[17] Duchateau G, Illescas C, Pons Z, Cormier E and Gayet R 2000, *J. Phys. B: At. Mol. Opt. Phys.* **33** L571.

# A four-body fully distorted wave - eikonal initial state model for ionization of He targets by ion impact

**J. M. Monti†, O. A. Fojón†, J. Hanssen‡and R. D. Rivarola†**

† Instituto de Física Rosario (CONICET-UNR) and Facultad de Ciencias Exactas, Ingeniería y Agrimensura, Universidad Nacional de Rosario, Pellegrini 250, 2000 Rosario, Argentina
‡ Institut de Physique, Laboratoire de Physique Moléculaire et des Collisions, Université Paul Verlaine - Metz, 1 Bv. Arago, 57078 Metz Cedex 3, France.

E-mail: jmonti@ifir.edu.ar

**Abstract.** A four body-distorted wave model is introduced to study collisions between swift bare ions and dielectronic atomic targets. Both electrons are considered as active, being one of them ionized while the other one remains bound to the residual target. The relevance of electron correlation on the resulting emission electron spectra is investigated for the case of protons impacting on He atoms.

## 1. Introduction

The reaction of single electron ionization of atomic and simple molecular targets by impact of swift heavy ions has been a matter of active research in the last forty years. An increasing interest on this subject was produced following the development of a new experimental technique known as cold target recoil ion momentum spectroscopy (COLTRIMS) [1]. Employing this technique a complete mapping of the momentum of all particles resulting from the reaction can be obtained. From the theoretical point of view, *one*-active electron models were commonly used to describe this multiple particle process. The reduction of this single ionization problem to a one electron treatment was formally given by Fainstein *et al.* [2] assuming that the time evolution of the active electron is produced independently of the other ones (passive electrons), which are supposed to remain as frozen during the collision. This implies the concept of independent electrons: the active one and the core of passive electrons of the residual target. So, the multiple body reaction (projectile, active and passive electrons, and target nucleus) is described using a three-body model (projectile, active electron and residual target). A certain success was obtained in representing single ionization experimental data for numerous collision systems [3-6]. Moreover, in the case of a *multiple*-electron process where all electrons are considered as active ones, when the independent electron model (IEM) is used, each one of the electrons is considered to evolve in time independently of the dynamical evolution of the other ones. So, in such representation electrons are assumed to be *statically* and *dynamically* decorrelated. By static electron correlation we understand the interaction among the active electrons in the asymptotic initial state and by dynamical correlation the fact that all electrons evolve simultaneously at the 'same' time.

Focussing our interest in the case of a bare swift ion impacting on a two-electron atomic target, a four-body theoretical representation appears as necessary to include static and dynamic correlations. We propose thus the treatment of the four-body reaction as a real four-body problem, avoiding the limitations of a three-body representation. It implies to deal with harder computational difficulties, than the ones corresponding to the simpler IEM. Studies using four-body models for ionization of two-electron atomic targets are scarce in the existing literature (for a review on these type of models applied to different electronic reactions, see Ref. [7]). The continuum distorted wave - eikonal initial state approximation was used in a four-body formalism (4B-CDW-EIS) to analyse single electron detachment of negative atomic hydrogen $H^-$ by proton impact [8], showing the important role played by the static electron correlation on ionization total cross sections. Correct boundary conditions were preserved in both the entry and exit channels and the electron remaining in the residual H atom was supposed to occupy its fundamental state. However, the distortion of the two-electron initial and final wavefunctions provoked by the projectile potential was considered to affect only the ionized electron, in an unequal treatment of both target electrons.

In the present work, the influence of the projectile potential on the initial wavefunction is considered on equal footing for both electrons, so that two associated distortion continuum eikonal phases are included describing the interaction of the impinging bare ion with each one of these electrons. In the exit channel the non-ionized electron is distorted by a projectile continuum eikonal phase according to the symmetric eikonal (SE) model employed with success to study electron excitation [9]. In the final wavefunction, a product of the continuum target wavefunction and a projectile continuum factor is proposed to represent the ionized electron moving in the combined fields of the residual target and projectile, according to the CDW-EIS approximation for ionization [11]. We call the present model, the four-body fully distorted wave - eikonal initial state approximation (4B-FDW-EIS).

Atomic units will be used in the text, except otherwise stated.

## 2. Theory

Let us consider the following reaction produced by the impact of a bare ion of nuclear charge $Z_P$ on a He atom of nuclear charge $Z_T$:

$$Z_P + (Z_T, 2e^-) \longrightarrow Z_P + (Z_T, e^-)^* + e^- \qquad (1)$$

where one of the electrons is ionized and the other one remains bound to the target nucleus.

An initial distorted wave function is chosen as

$$\chi_\alpha^+ = \varphi_\alpha(\mathbf{x}_1, \mathbf{x}_2)\exp(-i\varepsilon_\alpha t)L_\alpha^+(\mathbf{s}_1, \mathbf{s}_2)\exp[i\frac{Z_P Z_T}{v}\ln(vR + \mathbf{v}.\mathbf{R})] \qquad (2)$$

In (2), $\varphi_\alpha$ is the initial two-electron bound state, $\mathbf{v}$ is the collision velocity, $\mathbf{R}$ is the position of the projectile with respect to the target nucleus and $\varepsilon_\alpha$ the initial atomic energy. Also, $\mathbf{x}_j$ (with $j = 1, 2$) and $\mathbf{s}_j$ (with $j = 1, 2$) give the positions of the $jth$-electron seen from a reference frame fixed on the target nucleus and on the projectile, respectively. The term $L_\alpha^+$ is the distortion function given by

$$L_\alpha^+(\mathbf{s}_1, \mathbf{s}_2) = \exp[-i\frac{Z_P}{v}\ln(vs_1 + \mathbf{v}.\mathbf{s}_1)]\exp[-i\frac{Z_P}{v}\ln(vs_2 + \mathbf{v}.\mathbf{s}_2)] \qquad (3)$$

Eq. (3) indicates that the eikonal forms of the continua of both electrons in the projectile field are included into the initial channel. The last multiplicative factor in (2) represents the elastic scattering between the nuclei.

The final distorted wavefunction is chosen as

$$\chi_\beta^- = \varphi_\beta(\mathbf{x}_1, \mathbf{x}_2)\exp(\varepsilon_\beta)N^*(\frac{Z_P}{p})\,{}_1F_1(-i\frac{Z_P}{v}, 1, -i(ps_1 + \mathbf{p}.\mathbf{s}_1))$$
$$\exp[i\frac{Z_P}{v}\ln(vs_2 - \mathbf{v}.\mathbf{s}_2)]\exp[-i\frac{Z_P Z_T}{v}\ln(v - \mathbf{v}.\mathbf{R})] \qquad (4)$$

where $\varphi_\beta$ represent a final target state with one of the electrons bound and the other one traveling in the continuum of this three-body subsystem, and $\varepsilon_\beta = -i\frac{k^2}{2}t - i\varepsilon_2 t$ the corresponding energy, with $\mathbf{k}$ the momentum of the ejected electron with respect to the target nucleus and $\varepsilon_2$ the orbital energy corresponding to the bound electron. Also, ${}_1F_1$ is a Kummer's confluent hypergeometric function representing the continuum factor of an electron moving with momentum $\mathbf{p}$ in the projectile field, as seen from a reference frame fixed on this nucleus. $N(\frac{Z_P}{p})$ is the normalization constant of the Kummer's function. Thus, according to expression (4), the ejected electron moves in the combined field of the residual target and the projectile, while the other electron is considered to be bound to the target nucleus but at same time moving in a projectile continuum (given by its eikonal approximation). The last multiplicative factor in (4) describes the relative internuclear movement.

## 3. Results and conclusions

In this first work, the case of proton impact is analyzed considering that a large amount of experimental double differential cross sections (DDCS) exist. For multiply charged projectiles the influence of dynamical correlation could be expected to be even more relevant than for the proton case, because the collision time at equal projectile velocities will be larger, allowing thus dynamical correlation to play a role during the collision. Different initial $\varphi_\alpha(\mathbf{x}_1, \mathbf{x}_2)$ and final $\varphi_\beta(\mathbf{x}_1, \mathbf{x}_2)$ wavefunctions are chosen. For the initial channel we take a product of 1s1s configuration of single $Z$-functions (with $Z = 1.6875$) or a two-electron Hylleraas-Eckart wavefunction [11, 12] with two variational $Z$-charges (1s1s' configuration) with $Z_1 = 2.15$ and $Z_2 = 1.19$. These wavefunctions take into account approximately 80% and 90% of the radial static correlation, respectively. In the exit channel we consider $\varphi_\beta = \phi_{1s}(\mathbf{x}_2)\phi_c(\mathbf{x}_1)$, where $\phi_{1s}(\mathbf{x}_2)$ is the wavefunction corresponding to the dominant residual $He^+(1s)$ state and $\phi_c(\mathbf{x}_1)$ is taken as a Coulomb continuum wavefunction in the residual target field. An effective charge is considered in this continuum state: $\zeta = \sqrt{-2\varepsilon_1}$, with $\varepsilon_1 = -0.91795$ the initial orbital energy of the ionized electron as given by a Roothaan-Hartree-Fock calculation [13]. We have verified that the use of an asymptotic $\zeta = 1$ does not produce noticeable changes in DDCS for all cases treated in this work. So, in the following only DDCS computed employing the effective charge are presented.

In Fig. 1, DDCS are shown as a function of the emitted electron energy at fixed ejection angles of 0°, 30°, 50° and 90° for impact of 1 MeV protons. 4B-FDW-EIS results obtained employing both initial configurations indicated above are compared with 3B-CDW-EIS calculations [2] and with existing experiments for 30°, 50° and 90°. Present calculations using the simplest 1s1s configuration are in close agreement with 3B-CDW-EIS predictions, except in the binary encounter region, where the last ones coincides with 4B-FDW-EIS results corresponding to the 1s1s' configuration. In general, DDCS obtained for the initial 1s1s' configuration overestimate, at high enough emitted electron energies, the other calculations shown in figure, which seem to fit better the experimental data. The largest difference between 4B-FDW-EIS corresponding to 1s1s' and 1s1s configurations appears for the 0° case, where unhappily no experiments are available. In order to study this behavior, we present in Fig. 2 DDCS at 0°-emission angle but for energies of 50 keV, 100 keV and 1.5 MeV, for which experimental data exist. At emission energies lower than the ones corresponding to the capture to continuum peak, experiments place under the 4B-FDW-EIS predictions for a 1s1s' configuration and above the other theoretical results. In the binary encounter region, for the lower impact energies considered, experiments are in better agreement with 3B-CDW-EIS and 4B-FDW-EIS (1s1s configuration) calculations. On the contrary, for 1.5 MeV impact energy, 3B-CDW-EIS and 4B-FDW-EIS (1s1s' configuration) are in better accordance.

In conclusion, a four body treatment of the single ionization reaction for impact of bare ions on dielectronic targets is presented. It has been applied to the case of the $H^+ + He(1s^2)$

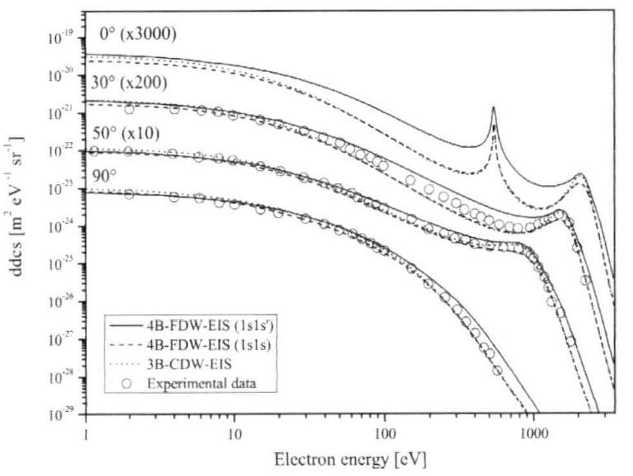


**Figure 1.** DDCS for electron emission in collisions of 1 MeV $H^+$ with He as a function of electron energy at fixed ejection angles. Theory: ——, 4B-FDW-EIS 1s1s' configuration; - - - -, 4B-FDW-EIS 1s1s configuration; ······, 3B-CDW-EIS [2]. Experimental data: ○ , [14]

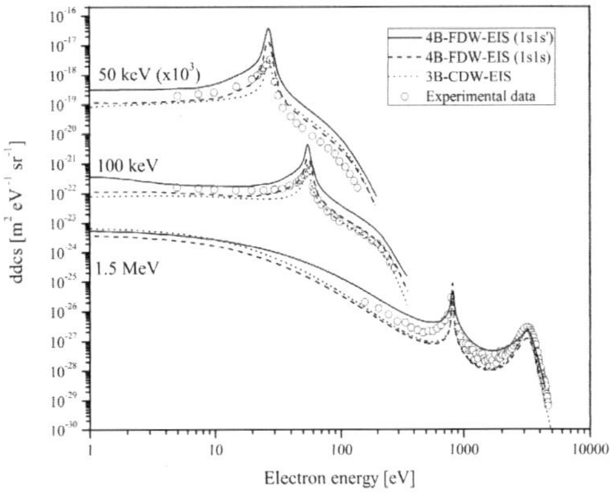


**Figure 2.** DDCS at 0° electron emission angle in collisions of 50 keV, 100 keV and 1.5 MeV $H^+$ with He. ——, 4B-FDW-EIS 1s1s' configuration; - - - -, 4B-FDW-EIS 1s1s configuration. 50 keV and 100 keV: ······, 3B-CDW-EIS calculations [15]; ○ , experimental data [16]. 1.5 MeV: ······, 3B-CDW-EIS [17]; ○ , experimental data [18]

collision. Comparisons are given with existing DDCS experimental data and with three body calculations. The role played by electron correlation is analyzed. The use of more elaborated initial wavefunctions including radial and angular correlations is being the matter of our present research. Preliminary 4B-FDW-EIS results obtained in this way show a promissory close agreement with experiments at all ejection energies. We are now focusing our interest on the application of this four-body theory to multiply charged projectile impact on He atoms.

## 4. References

[1] Moshammer R *et al.* 1994 *Phys. Rev. Lett.* **73** 3371.
[2] Fainstein P D, Ponce V H and Rivarola R D 1988 *J. Phys. B: At. Mol. Phys.* **21** 2989.
[3] Stolterfoht N, DuBois R D and Rivarola R D 1997 *Electron Emission in Heavy Ion – Atom Collisions* (Berlin: Springer).
[4] Rivarola R D and Fainstein P D 2003 *Nucl. Instr. Meth. B* **205** 448.
[5] Foster M *et al.* 2004 *J. Phys. B: At. Mol. Opt. Phys.* **37** 3797 *and references therein.*
[6] Ciappina M F and Cravero W R 2006 *J. Phys. B: At. Mol. Opt. Phys.* **39** 2183 *and references therein.*
[7] Belkic Dz, Mancev I and Hanssen J 2008 *Rev. Mod. Phys.* **80** 249.
[8] Belkic Dz 1997 *Nucl. Instr. Meth. B* **124** 365.
[9] Deco G R, Fainstein P D 1986 *J. Phys. B: At. Mol. Phys.* **19** 213.
[10] Crothers D S F and McCann J F 1983 *J. Phys. B: At. Mol. Phys.* **16** 3229.
[11] Hylleraas E A 1929 *Z. Physik* **54** 347.
[12] Eckart C 1930 *Phys. Rev.* **36** 878.
[13] Clementi E and Roetti C 1974 *Atomic Data and Nuclear Data Tables* **14** 177.
[14] Rudd M E, Toburen L H and Stolterfoht N 1976 *At. Data Nucl. Data Tables* **18** 413
[15] M F Ciapina, W R Cravero and C R Garibotti 2004 *Phys. Rev. A* **70** 062713
[16] G C Bernardi *et al.* 1989 *Phys. Rev. A* **40** 6863
[17] P D Fainstein, V H Ponce and R D Rivarola 1991 *J. Phys. B: At. Mol. Phys.* **24** 3091
[18] D H Lee *et al.* 1990 *Phys. Rev. A* **41** 4816

Journal of Physics: Conference Series **163** (2009) 012069 doi:10.1088/1742-6596/163/1/012069

# New classical CTMC approaches to $A^{q+}$ + He processes

**F. Guzmán[1], L.F. Errea[1], L. Méndez[1], B. Pons[2] and A. Riera[1]**

[1]Departamento de Qumica, Universidad Autnoma de Madrid, 28049 Madrid (Spain)
[2]CELIA,UMR 5107 du CNRS, Universit de Bordeaux-1,351 Cours de la Libration, 33405 Talence (France)

E-mail: francisco.guzman@uam.es

**Abstract.** Any small perturbations in the classical stable Helium atom may produce unphysical autoionization that prevents its classical dynamical study. Previous attempts have required barrier potentials to limit the electron energy. As a first approach we consider two sets of $N$ classical electron trajectories, whose evolution obey the Hamilton equations in the full Hamiltonian. Each classical electron of one set feels the average repulsion due to the classical electrons of the other set. Smoothing of the evaluation of the classical electronic repulsion is proposed assuming that each electron can be described by Gaussian functions centred on the position of the corresponding electrons

## 1. Introduction

Previous attemps to describe clasically (CTMC) two electron systems [1, 2, 3] have required new potential energy terms that depend on both space and momentum coordinates.

To minimize the non central components of repulsive forces that appear just before the onset of unphysical autoinization, we included two sets of $N$ classical electrons that simulate the corresponding quantal electronic densities, averaging repulsive forces that act over each electronic trajectory. We have also implemented time dependent model potentials in target and projectile centers in a different approach. Both methods are hightlighted here.

## 2. Averaged Collective Repulsion approach (ACR)

We use two coupled sets of $N$ classical electron trajectories which evolve collectivelly because of the repulsion term according to the Hamilton equations [4], under the Hamiltonian:

$$H_i = \frac{p_i^2}{2m} - \frac{Z_T}{r_i^T} - \frac{Z_P}{r_i^P} + \frac{1}{N}\sum_j^N \frac{1}{|\vec{r}_i - \vec{r}_j|} \qquad (1)$$

for each trajectory $i$ of one set with $j$ running over all trajectories of the other set and where $T$ and $P$ refers to target and projectile.

### 2.1. Gaussian kernel approximation

Evaluation of the repulsion potential is performed between electrons described by Gaussian functions centered on the positions of the corresponding classical trajectories at each time in a

Journal of Physics: Conference Series **163** (2009) 012069 doi:10.1088/1742-6596/163/1/012069

kernel representation [5]. For example, the repulsive potential felt by the electron $i$ of the set 1 due to the whole set of electrons 2 is:

$$V_{i_1}^{\text{rep}}(\boldsymbol{r}_{i_1}) = \frac{1}{(2\pi)^{\frac{3}{2}}\sigma^3 N}\left[\sum_{j_2}\int \mathrm{d}\vec{r}e^{-\frac{(\boldsymbol{r}-\boldsymbol{r}_{j_2}(t))^2}{2\sigma^2}}\frac{1}{|\boldsymbol{r}_{i_1}-\boldsymbol{r}|}\right] \tag{2}$$

where $\boldsymbol{r}_{i_1}$ and $\boldsymbol{r}_{j_2}$ are the positions of the electronic trajectories and $\sigma$ is the smoothing parameter.

*2.2. Dynamics*

We have tested the stability of the He atom against small electrostatic perturbations obtaining stability with respect to autoinization.

To carry out a collision, we first generate a microcanonical distribution with an effective charge. Then an elestrostatic model potential generated by the previous microcanonical distribution is obtained and finally, spacial and momentum coordinates of the classical trajectories that define the actual initial distribution are generated using the tranformation given by C.O. Reinhold and C.A. Falcon (1986) [6] under the generated model potential.

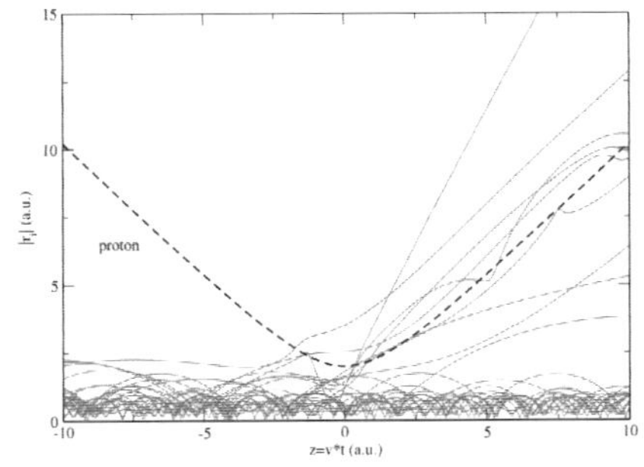


**Figure 1.** Representation of the target distance (module) of a collision at impact parameter $b$ = 2a.u. and velocity $v$ = 1a.u. with two sets of N=250 and a smoothing parameter $\sigma = 0.28$.

At the end of the collision, a number of trajectories will be captured by the projectile, a different number will remain in the target and the rest will be ionized as can be seen in figure 1. As each electron of the first set is correlated to each electron of the second set and vice versa, an $N^2$ statistics is performed to obtain the probabilities for each process.

## 3. Dynamic Screening

In an alternative way we have modified and extended to the projectile the dynamic screening model developed by V. J. Montemayor and G. Schiwietz (1989) [3] which consisted of independent pairs of trajectories evolving on time dependent central model potentials with magnitudes that depend on the bond energy of the other electron in each pair according to:

$$H = H_0 + V_{ee,1}^{A}(\vec{r}_1, E_{0,2}) + V_{ee,2}^{A}(\vec{r}_2, E_{0,1}) + V_{ee,1}^{B}(\vec{r}_1, E_{0,2}) + V_{ee,2}^{B}(\vec{r}_2, E_{0,1}) \tag{3}$$

where $H_0$ is the sum of the monoelectronic Hamiltonians and $V_{ee,i}^{(A,B)}$ are time-dependent model potentials of the form, for $i \neq j$:

$$V_{ee,i} = N_c(E_{0,j})\frac{1-(1+\alpha r_i)e^{-2\alpha r_i}}{r_i} \tag{4a}$$

$$N_c(E_{0,j}) = \begin{cases} 0 & \text{for } E_{0,j} \geq 0 \\ g(\mu)^2 & \text{for } I_{1s^2} < E_{0,j} < 0 \\ 1 & \text{for } E_{0,j} < I_{1s^2} \end{cases} \tag{4b}$$

$$g(\mu) = \frac{1.25^{1.25/2}\mu}{(\mu^2 + 1.25 - 1)^{1.25/2}} \tag{4c}$$

$$\mu(E) = |\frac{-E}{|E_{lim}|} + \frac{1}{2}| - |\frac{-E}{|E_{lim}|} - \frac{1}{2}| \tag{4d}$$

$$E_{0,j} = -\frac{Z}{r_j} + \frac{p_j^2}{2}(+V_{ee,j}) \tag{4e}$$

$I_{1s^2}$ is the ionization potential. $E_{lim}$ is a constant that refers to the point where the model potential drops smoothly. Here we use $N_c$ as the time-dependent screening parameter varying between 1 and 0, unlike [3] where this variation was included in the paramater $\alpha$. Collisions were carried out using a microcanonical distribution in a model potential initial conditions [6].

## 4. Results and Conclusions

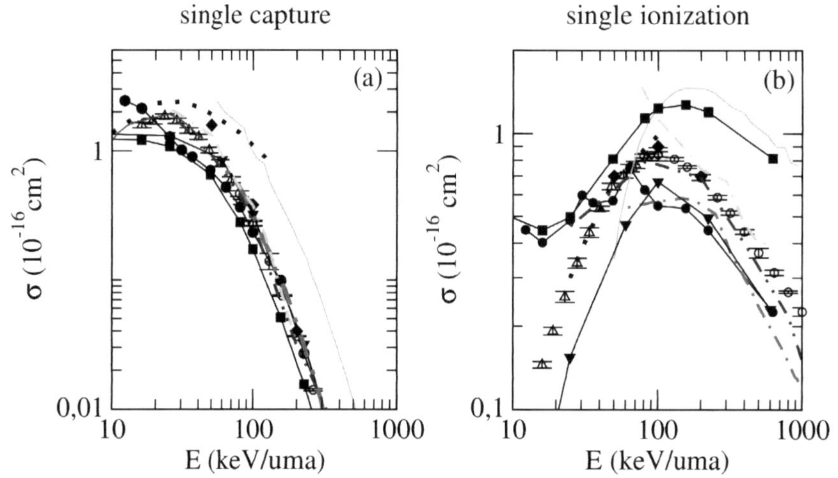


**Figure 2.** Single capture (a) and single ionization (b) processes for $He(1s^2)+H^+$ collision. Own results: ACR (——●), Dynamic Screening (——■), Independent Gaussian Kernel pairs (——▼). Previous results, theory: *ab initio* [7] (——), Semiclassical *Close Coupling* IPM [8, 9] (······), AO [10] (♦),CTMC $V_{mod}$ IPM [6](— · —),CTMC-EB [2] (— · · —),CTMC-HC [11] (——) [12] (——); experiment: [13](△), [14] (○).

Our ACR results for the collisions $He(1s^2)+H^+$ are illustrated in figures 2 for single capture and ionization, and in figures 3 for double processes compared with previous theooretical and experimental results. Our calculations fairly agree with experimental values of single capture with a good convergence at high energies, and only slightly understimate those of single ionization in figure 2b, although they present the correct shape as also happens with the results from independent pairs of electronic trajectories using a gaussian kernel approximation for the repulsion. Double process results (figure 3) overstimate the experimental data but improve the previous Heisemberg Core CTMC calculations [11]. Dynamic screening results for single ionization and double processes are above the experimental ones except for single capture. To understand this behaviour a further study is required.

We have developed a new method using an average of repulsion terms between the two set of electrons that represent the quantum electronic cloud of Helium atom. A Gaussian kernel

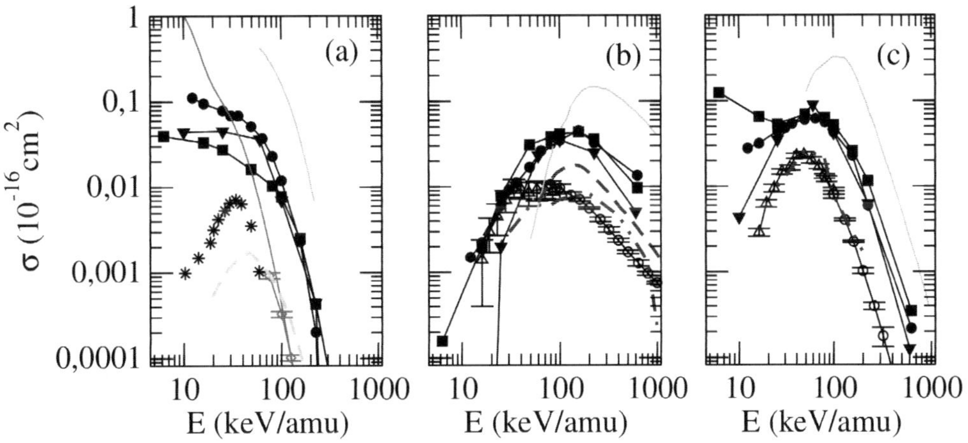


**Figure 3.** Double capture (a), double ionization (b) and capture-ionization (c) processes for He($1s^2$)+$H^+$ collision. Own results: ACR (——●), Dynamic Screening (——■), Independent Gaussian Kernel pairs(——▼). Previous results, theory: CDW [15] (——), C.D. Lin [16] model ( ),CTMC-EB [2] (— · · —),CTMC-HC [11] (——), CTMC $V_{mod}(t)$ [3] – – –; experiment: [17](∗), [18] (◇), [13](△), [14] (○).

approximation has been employed to eliminate unphysical ionization. We have also extended the dynamic screening method to the projectile modifying the time dependent parameter. We have obtained good results for single processes while double processes overestimate experimental data whith both methods. Further work on the optimization of the time-dependent model parameters and new calculations employing larger statistics must be performed.

## Acknowledgements

This work has been partially supported by project ENE2004-62934 from Dirección General de Investigación (Spain). F.G. acknowledges also to Fundacion Avelino López de Castro and Ciemat by their support.

## References

[1] Kirschbaum C L and Wilets L 1980 *Phys. Rev. A* **21** 834
[2] Cohen J S 1996 *Phys. Rev. A* **54** 573
[3] Montemayor V J and Schiwietz G 1989 *Phys. Rev. A* **40** 6223
[4] Goldstein H 1954 *Classical Mechanics* (Addison-Wesley Academic Press, Orlando)
[5] Donoso A and Martens C C 2002 *J. Chem Phys.* **116** 10598
[6] Reinhold C O and Falcón C A 1986 *Phys. Rev. A* **33** 3859
[7] Errea L F, Méndez L and Riera A 1989 *Z. Phys. D* **14** 229
[8] Slim H, Heck E, Bransden B and Flower D 1991 *J. Phys. B: At. Mol. Opt. Phys.* **24** 1683
[9] Slim H, Heck E, Bransden B and Flower D 1991 *J. Phys. B: At. Mol. Opt. Phys.* **24** L421
[10] Winter T 1991 *Phys. Rev. A* **44** 4353
[11] Wetmore A E and Olson R E 1988 *Phys. Rev. A* **38** 5563
[12] Zajfman D and Maor D 1986 *Phys. Rev. Lett.* **56** 320
[13] Shah M B, McCallion P and Gilbody H B 1989 *J. Phys. B: At. Mol. Opt. Phys.* **22** 3037
[14] Shah M B and Gilbody H B 1985 *J. Phys. B: At. Mol. Phys.* **18** 899
[15] z Belkić D and Mančev I 1993 *Phys. Scr.* **47** 18
[16] Lin C D 1979 *Phys. Rev. A* **19** 1510
[17] Williams J 1966 *Phys. Rev.* **150** 7
[18] Toburen L and Nakai Y 1969 *Phys. Rev.* **177** 191

Journal of Physics: Conference Series **163** (2009) 012070 doi:10.1088/1742-6596/163/1/012070

# A new MOTRIMS apparatus for high resolution measurements in ion-atom collisions and trapped atoms studies

**J Blieck[1], X Fléchard[1,3], A Cassimi[2], H Gilles[2], S Girard[2], and D Hennecart[2]**

[1] LPC Caen, ENSICAEN, Université de Caen, CNRS/IN2P3, Caen, France

[2] CIMAP, ENSICAEN,CNRS, CEA, Université de Caen, Caen, France

E-mail: flechard@lpccaen.in2p3.fr

**Abstract**. A new MOTRIMS (magneto-optical trap-target recoil ion momentum spectroscopy) apparatus has been built at the LPC-CAEN, and recently tested with a beam of 2 keV $Na^+$ projectiles colliding on a trapped Rb target. In this work, we demonstrate the capability of the setup to provide, with a very high signal over background ratio, fully differential cross sections in scattering angle, initial state, and final state of the system. The fine control of the trapping lasers, the extraction of the recoil ions transverse to the ion beam axis, and the fast switching for the MOT magnetic field, are three characteristics promising for new studies including collisions with oriented targets or requiring electron detection.

## 1. Introduction

Two experimental techniques developed in the last decades, MOTs (magneto-optical traps), and COLTRIMS (cold target recoil ion momentum spectroscopy), have recently been merged by using trapped atoms as a cold target for the COLTRIMS methodology [1-3]. This new development, called MOTRIMS, added several advantages to the standard COLTRIMS technique. First, it provided an additional set of atomic species such as alkali or alkaline-earth atoms. Using such targets, measurements with different initial states, including aligned or oriented states, can now be performed. A second advantage of MOTRIMS is the very low temperature of the atomic target (of the order of 100 μK) that no longer limits the resolution in recoil momentum.

Using this novel and promising technique, various aspects of the electron capture process, MOT populations dynamic, and photoassociation in cold atoms have already been studied [4]. It is in this context that the LPC-CAEN and the CIMAP laboratories built jointly a new MOTRIMS set-up, devoted in a first step to ion-atom collisions study. Since its technical details have already been described [5], the main section of this paper focuses only on specific instrumental aspects. The performance of the apparatus, deduced from test experiments involving $Na^+$ projectiles at 2 keV colliding with a target of trapped $^{87}Rb$ atoms, are then presented. Possible modifications of the set-up and future MOTRIMS experiments are discussed in the last section.

## 2. Experimental set-up

A simplified schematic of the experimental set-up is shown in Fig. 1. It comprises three main parts: i) the magneto-optical trap with three pairs of counter-propagating laser beams and anti-Helmholtz magnetic coils, ii) a projectile beam-line with a charge state analyzer and, iii) the recoil ion spectrometer.

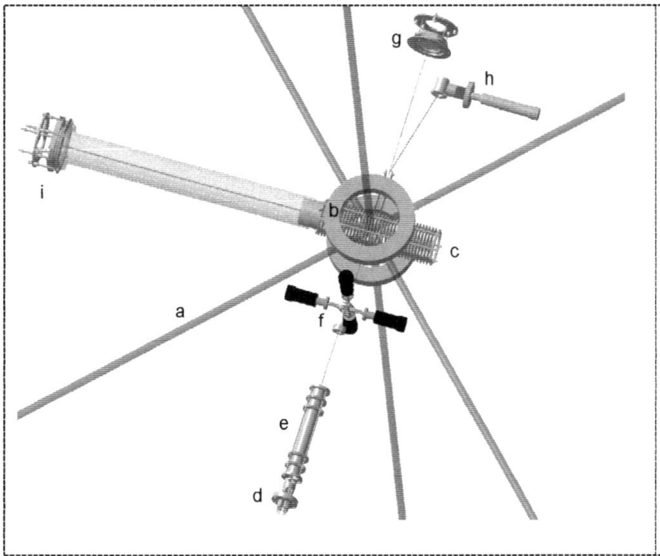


**Figure 1.** Schematic view of the experimental set-up. With the combination of six laser beams (a) and the magnetic field resulting from an anti-Helmholtz pair of coils (b), the atomic target is trapped in the middle of the spectrometer (c). The beam of projectile ions is provided by an ion gun (d). After focussing (e) and collimation (f), the beam enters the collision chamber, passes through the cloud of cold trapped atoms, and is charge-state analyzed at the exit of the spectrometer. The portion of the beam having captured an electron is guided towards the projectile position sensitive detector (g), while the primary beam is collected by a Faraday-cup (h). The recoil ions resulting from charge transfer are extracted from the collision region with 100% efficiency towards the recoil ion position sensitive detector (i).

### *2.1. The magneto-optical trap*

The $^{87}$Rb MOT requires two laser frequencies: one tuned a few MHz below the cooling/trapping transition $5^2S_{1/2}(F=2) - 5^2P_{3/2}\ (F'=3)$ and a second tuned on the repumping transition $5^2S_{1/2}(F=1)$ - $5^2P_{3/2}\ (F'=2)$. We use for these transitions two external cavity diode lasers (DL), each providing an output power of 100 mW at 780 nm. A small amount of the laser power is first sent toward a frequency stabilization set-up, while the main laser beams are merged through a polarizing beam splitting cube (PBSC). The merged beams are then sent to a 110 MHz acousto-optic modulator (switch-AOM) used as a fast beam shutter. The negative first-order diffracted beam is enlarged with a two lens telescope and split into three beams of equal power with a set of half-wave plates and PBSCs. The three beams enter the trapping chamber with orthogonal directions, and are retro-reflected with gold coated mirrors on the opposite side of the chamber. Six quarter-wave plates insure adequate circular polarization for the three pairs of counter-propagating beams. The 6 laser beams have a 1.4 cm diameter and deliver to the trapping chamber a total laser power of 65 mW at the trapping frequency. When the RF power of the first AOM is switched-off, the negative first-order diffracted beam of a second 110 MHz AOM can be used to provide a polarized beam for orientation or alignment of the trapped atoms. The frequency stabilization scheme chosen for the laser diodes makes also use of an AOM [6]. The modulation for the lock-in detection is applied to the AOM radio frequency and does not affect the laser light used for the experiment. This laser control scheme provides a wide range of frequency tuning, and a very stable frequency lock with drifts below 100 kHz. This allows, for instance, to switch easily from $^{87}$Rb to $^{85}$Rb, and to vary easily and accurately the detuning of the trapping laser without affecting the laser power sent to the MOT. The magneto optical trap requires a strong inhomogeneous magnetic field (typical gradients of 10 Gauss/cm) not compatible with the recoil ion momentum spectroscopy. It is then necessary to operate the MOT in a pulsed mode where the B-field is switched off periodically for data acquisition. This B-field switch has to be fast enough to limit the cloud expansion and the trapped atoms losses. For this purpose, a fast current switch for high current and high inductance has been developed. It allows to ramp up and down the 2 A current of the coils in less than 300 μs within a simple timing sequence where the B-field is *off* during 3 ms and *on* during 9 ms. The size, the density, and the temperature of the trapped atomic cloud are measured using a fast triggered CCD-camera. The cloud size and position, and the total rate of emitted photons $\gamma_{tot}$ are monitored permanently while operating the MOT. The number of trapped atoms $N$ can then be deduced by dividing $\gamma_{tot}$ by the scattering rate per atom $\gamma_p$, given by the spontaneous decay rate $\gamma$= 38.11 MHz times the excited fraction $f$ of the cloud population. The temperature measurement is performed using the ballistic expansion technique. A peak density of 1.5 x $10^{11}$ atoms per $cm^3$ has been obtained with a cloud temperature of ~200 μK and a typical cloud diameter of 0.5 mm (FWHM). This cloud temperature of 200 μK corresponds to a standard deviation $\sigma_p$ = 0.01 a.u. for the three components of the target atoms momentum.

### *2.2. The recoil ion spectrometer*

The recoil ion spectrometer is made of three main elements: a stack of electrodes providing an extraction field transverse to the projectile beam axis, a field free flight tube, and a 2-dimensional position sensitive detector (2D-PSD). We use for the recoil ion detection a commercial 83 mm MCP 2D-PSD from Roentdek. A post-acceleration region is defined by an 85 % transmission grid placed 3 mm upstream from the MCPs. The grid is at the same potential as the field free region and the front area of the detector is typically at - 4 kV. This provides the additional acceleration eventually needed to ensure maximum detection efficiency.To maximize the recoil momentum resolution, the spectrometer extraction field is shaped to achieve the *three dimensional focussing* (3D-focussing) of the collision region in the detection plane. The inhomogeneous extraction field ensuring this focussing, generated by the simple succession of a weak field region with a strong field region, was designed with SIMION. The polarization scheme of the spectrometer was chosen so that the potential of the collision region is set to the ground. This requires a polarized field free region, defined by a stainless steel tube connecting the last ring electrode to the 2D-PSD post acceleration grid. That way, the access holes needed for the trapping lasers have negligible effect on the 3D-focusing properties. In addition, this polarization scheme is well adapted to collisions involving very low energy projectile ions. The focussing properties of the spectrometer reduce by a factor of 5 the effect of the collision region size.

## 3. Sample results

To test the apparatus, we chose the 2 keV $Na^+$ + $^{87}$Rb(*5s, 5p*) collision system, which has already been studied using the MOTRIMS technique [7]. For each detected event, the full momentum of the recoil ions could be reconstructed, and the event flagged with both the status of the switch-AOM and the status of the B-field. Doubly differential cross sections in $P_{//}$ and $P_{\perp}$ for the *laser on* and *laser off* cases are displayed in Fig. 2. The relative cross sections of the different capture channels can thus be extracted along with their associated scattering angle distributions. It is to be noticed that due to the transverse extraction geometry and to simple filters applied to the data (on the projectile position and recoil ion time of flight), the signal over backgroud ratio obtained here is excellent and allows the observation of very weak electron capture channels.

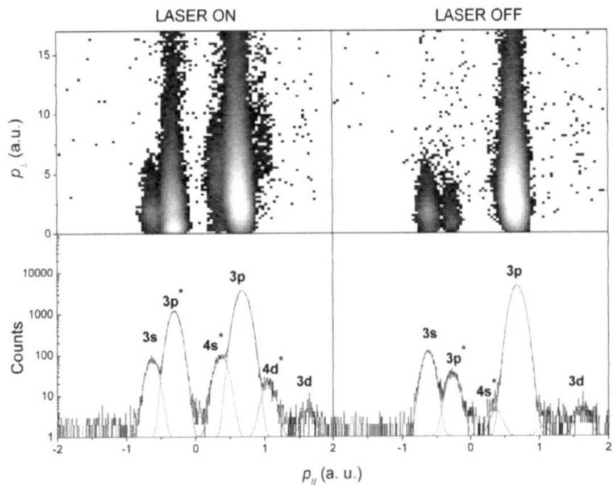


**Figure 2.** Doubly differential cross sections in $p_{//}$ and $p_{\perp}$ (upper panels) and differential cross sections in $p_{//}$ (lower panels) for the laser *on* and laser *off* cases. The different electron capture channels are labelled by the final state of the Na projectile. An asterisk is added for the capture from the Rb(*5p*) excited state of the target.

With the ability to switch the MOT B-field combined with 3-D focusing, the resolution on the recoil ion momentum measurement is currently $\sigma_p = 0.07$ a.u., and was found to be only limited by the use of a post-acceleration grid in front of the recoil detector. Using the results from both cases *laser on / off* and the methodology described in [4], the excited fraction of the atomic target $f = 0.30(3)$ could also be deduced from the data. This gives direct access to the cross section ratios from the two possible initial states.

## 4. Future applications

In this work, a direct measurement of the excited fraction of the atomic cloud has been performed using charge exchange as a probe. The laser lock system of our apparatus is particularly well adapted for a systematic study of excited fractions as a function of the MOT parameters, like laser detuning and intensity. This study would provide a test of the simple models commonly used to determine the number of trapped atoms in a MOT.

The laser system was also designed to easily "prepare" the target in an anisotropic excited state. With the high resolution of MOTRIMS, the study of charge transfer with such oriented targets looks very promising. Similar experiments have been performed using energy gain spectroscopy, but not using the RIMS technique, which gives access to much smaller scattering angles.

To study ionization processes, future developments should include the detection of electrons with an additional detector. This is here made possible by the use of transverse extraction geometry combined with the ability to switch off the MOT B-field. This improvement would not require a large modification of the experimental set-up.

Last, the capability to control accurately the position of the atomic cloud, with a set of correction coils for the B-field and a CCD-camera, can give access to absolute cross sections measurements. Such measurements are usualy limited by an unsufficient knowledge of the target and ion beam densities in the interaction region. With our apparatus, the target density profile can be permanently monitored with the CCD-camera, while the correction coils are used to scan the interaction region by moving the atomic cloud through the projectile ion beam.

## 5. Conclusion

A new MOTRIMS apparatus using transverse extraction geometry has been designed and tested for the 2 keV $Na^+ + {}^{87}Rb$ (*5s, 5p*) collision system. Its capability to provide fully differential cross sections in scattering angle, initial state, and final state of the system has been demonstrated. Compared to previous experiments using longitudinal extraction, a very good signal over background ratio has been obtained. The resolution of the recoil ion momentum is currently limited by the use of a post-acceleration grid in front of the recoil detector. Regarding this point, the apparatus will be improved by combining several grids. Specific features such as the transverse extraction geometry, the ability to switch off the MOT B-field, and the fine control of the target using our laser system, will allow a large variety of studies involving ion-atom collisions.

## References

[1] M. van der Poel *et al*, Phys. Rev. Lett. **87**, 123201 (2001).
[2] J.W. Turkstra *et al*, Phys. Rev. Lett. **87**, 123202 (2001).
[3] X. Flechard *et al*, Phys. Rev. Lett. **87**, 123203 (2001).
[4] B.D. DePaola *et al*, Adv. In At. Mol. And Opt. Phys. **55**, 139 (2008)
[5] J. Blieck *et al*, Rev. Sci. Instrum. **79**, 103102 (2008).
[6] H.J. Lewandowski *et al*, Journal of Low Temperature Physics **132**, 309 (2003).
[7] T. G. Lee *et al*, Phys. Rev. A **66**, 042701 (2002).

Journal of Physics: Conference Series **163** (2009) 012071 doi:10.1088/1742-6596/163/1/012071

# Complementary spectroscopy of tin ions using ion and electron beams

**H Ohashi[1], S Suda[1], H Tanuma[1], S Fujioka[2], H Nishimura[2], K Nishihara[2], T Kai[3], A Sasaki[3], H A Sakaue[4], N Nakamura[5] and S Ohtani[5]**

[1] Department of Physics, Tokyo Metropolitan University, Hachioji, Tokyo 192-0397, Japan
[2] Institute of Laser Engineering, Osaka University, Suita, Osaka 565-0871, Japan
[3] Japan Atomic Energy Agency, Kizu, Kyoto 619-0215, Japan
[4] National Institute for Fusion Science, Toki, Gifu 509-5292, Japan
[5] Institute for Laser Science, The University of Electro-Communications, Chofu, Tokyo 182-8585, Japan

E-mail: `ohashi-hayato@ed.tmu.ac.jp`

**Abstract.** Extreme ultra-violet (EUV) emission spectra of multiply charged tin ions were measured in the wavelength range of 10–22 nm following charge exchange collisions of $Sn^{q+}$($q$ = 15–21) ions with rare gas targets at 20 keV/$q$ or the electron impact excitation of tin ions with electron energy of 312–613 eV. In charge exchange collisions, we observed both the resonance lines and the emission lines corresponding to the transitions between the excited states. On the other hand, we observed mainly the resonance lines in the electron impact experiments. We can distinguish the resonance lines from other emission lines in the charge exchange spectrum by comparison with the emission lines in the electron impact spectrum. The comparison of the complementary experimental results can be regarded as a new method of spectroscopy of multiply charged ions.

## 1. Introduction

Tin plasmas have been studied not only by experimental investigations but also theoretical calculations as a candidate for a powerful light source for next generation semiconductor lithography [1, 2]. This technology requires light with a wavelength around 13.5 nm in the EUV region because a Mo/Si multi-layer coated mirror has a maximum reflectivity at this wavelength. Various kinds of plasma, e.g. capillary discharge and vacuum spark, were studied for the experimental spectroscopy of multiply charged tin ions because tin plasmas have a favorable emission around 13.5 nm [3, 4]. These plasmas, however, usually contain various charge states of tin ions [5]. Although there are some reported emission spectra of several ions for individual charge states around 13.5 nm [6, 7], more spectroscopic data of tin ions are necessary to diagnose and to understand tin plasmas.

An EBIT (electron beam ion trap) is a powerful device to produce highly-ionized atoms by electron impact [8]. We can estimate the highest charge state in the EBIT from the electron beam energy and the ionization potentials of the ions for each charge state in the EBIT. Most emission from the EBIT can be considered as the resonance lines. In charge exchange collisions, the excited states of the ion are determined by the charge state of the ion, the ionization energy of the target atom and the collision energy [9]. In general, emission from excited states produced via charge exchange collisions is considered as due to both resonance lines and emission lines corresponding to transitions between excited states in charge exchange collisions.

In this study, we measured EUV emission from excited tin ions via both charge exchange collisions and electron impact to provide fundamental spectroscopic data and some information about the charge transfer processes of tin ions.

## 2. Experiment

In the charge exchange spectroscopy (CXS), multiply charged tin ions were produced in a 14.25 GHz ECR (electron cyclotron resonance) ion source at Tokyo Metropolitan University [10]. We inserted a rod of sintered tin oxide ($SnO_2$) in a plasma chamber and introduced $O_2$ gas to a pressure of about $1\times10^{-5}$ Pa as a support gas for the plasma. The $Sn^{q+}$($q$ = 15–21) ions were extracted with an electric potential of 20 kV and selected by a 110° double-focusing dipole magnet according to their mass-to-charge ratios. The ion-beam was directed into a collision chamber, where it intersected an effusive beam of target gas ejected from a multi-capillary plate. The background pressure in the collision chamber was $6\times10^{-6}$ Pa and the target gas pressure in the chamber was held at about $1\times10^{-3}$ Pa during the measurements. In this setup, we could not directly measure the target gas density in the collision volume. The target gas pressure, however, was low enough to maintain single-collision conditions. The primary ion-beam, which had about 6 mm diameter, had a typical electrical current of 0.03–0.15 $\mu$A as measured by a Faraday cup located behind the collision region. Optical radiation in the EUV region from the collision volume was observed at 90° to the ion-beam direction with a compact flat-field grazing-incident spectrometer (SSK-260, Shin Seiki Co.) equipped with a toroidal-type converging mirror and a variable line space (ca. 1200 lines/mm) grating blazed at 100 nm. A liquid nitrogen cooled CCD (charge coupled device) camera (C4880, Hamamatsu) was installed in the EUV spectrometer. Emissions in a 18 nm range of wavelengths were observed and accumulated simultaneously. A slit width of 200 $\mu$m between the converging mirror and the grating gave the wavelength resolution of about 0.03 nm. The calibration of the wavelength was performed by observing more than 20 lines of O VI and O VII in collisions of $O^{6+}$ and $O^{7+}$ with Xe gas. The uncertainty of the observed wavelength was estimated at 0.02 nm.

In the electron impact experiment, highly charged tin ions were produced by the Tokyo EBIT at the University of Electro-Communications [11]. A Knudsen cell with metallic tin was placed horizontally at the center level of the drift tube of the EBIT [12, 13]. Tin vapour was produced in the cell and was introduced to the drift tube. Electron energies between 312 and 693 eV were applied, an extremely low energy for this device as it was designed to operate at energies of up to 300 kV, and the electron beam currents were between 2.2 and 6.4 mA. The optical radiation from the ion trap region was observed at 90° to the electron beam direction with the same detection setup which was used in the CXS.

## 3. Results and Discussion

In low energy collisions of $Sn^{q+}$ with He, the following processes take place:

$$
\begin{aligned}
\mathrm{Sn}^{q+} + \mathrm{He} &\rightarrow \mathrm{Sn}^{(q-1)+}(n\ell) + \mathrm{He}^{+} && \text{: single electron capture (SC),}\\
&\rightarrow \mathrm{Sn}^{(q-2)+}(n\ell n'\ell') + \mathrm{He}^{2+} && \text{: true double electron capture (TDC),}\\
&\rightarrow \mathrm{Sn}^{(q-1)+}(n\ell) + \mathrm{He}^{2+} + e^{-} && \text{: transfer ionization (TI).}
\end{aligned}
$$

Emission from $Sn^{(q-1)+}$ and $Sn^{(q-2)+}$, therefore, must be considered. In slow collisions of multiply charged ions with He, generally speaking, SC is the dominant process and TI has a significant contribution, while TDC has very small cross section, as observed in $Xe^{q+}$($q$ = 10–43), $Ar^{q+}$($q$ = 8–16) and other multiply charged ions [14, 15, 16, 17]. Therefore, we can assume that the final charge states in the product Sn ions are $(q-1)+$ in collisions of $Sn^{q+}$ ions with He. On the other hand, in the EBIT under our experimental conditions, the following processes take place dominantly:

$$\begin{aligned} Sn^{q+} + e^- \;&\rightarrow\; Sn^{(q+s)+} + (s+1)e^- && : \text{electron impact ionization,} \\ &\rightarrow\; Sn^{q+*} + e^- && : \text{electron impact excitation,} \\ &\quad\hookrightarrow\; Sn^{q+} + h\nu && : \text{radiative de-excitation.} \end{aligned}$$

The distributions of both the charge states and the excited states depend on the electron beam energy. The highest charge state can be estimated by the ionization energies of tin ions, which had been obtained with SCF (self-consistent field) calculations [18]. Most of the radiation from the ions can be assumed to be due to resonance lines, not to transitions between excited states.

Figure 1 shows the EUV emission spectra in which we consider that most emissions come from the same charge state. Figure 1(a) shows the observed emission spectra in collisions of $Sn^{15+}$ with He and in the EBIT with an electron beam energy of 312 eV. Both of the spectra have a strong emission line at 13.3 nm, and we consider it to be a resonance line of $Sn^{14+}$. We also consider that the emissions between 13.5 to 14.3 nm are satellite lines corresponding to the 4d-4f transitions by comparison with the theoretical calculations [19]. Since the $4d^{k-1}n\ell$ configuration has quite a number of fine-structure levels, the 4d-$n\ell$ transitions make a band-like emission, which is called an UTA (unresolved transition array). In figure 1(b), some narrow

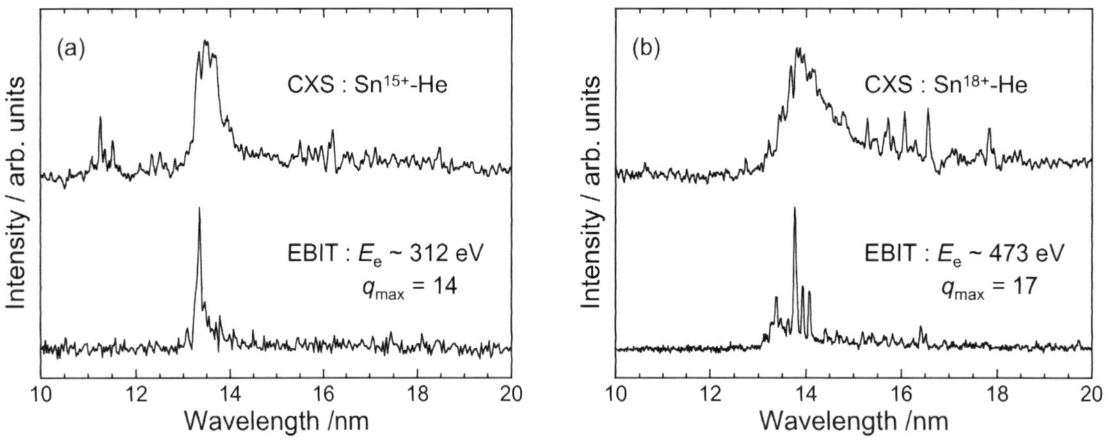


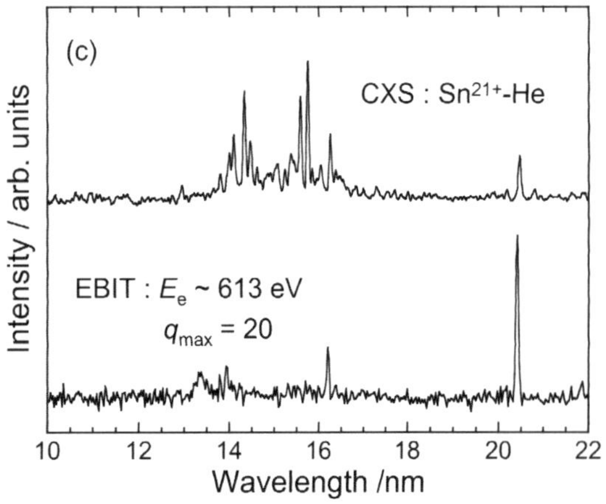


**Figure 1.** EUV emission spectra in collisions of $Sn^{q+}$ with He and in the electron impact collisions in the EBIT; (a) $Sn^{15+}$ and $E_e \sim 312$ eV, (b) $Sn^{18+}$ and $E_e \sim 473$ eV, (c) $Sn^{21+}$ and $E_e \sim 613$ eV. The $q_{max}$ is the highest charge state in the EBIT, which is estimated from the electron beam energy.

emission lines were observed in the EBIT spectrum, and are regarded as the resonance lines of $Sn^{17+}$. These lines were also observed as a part of the UTA in the CXS spectrum. In figure 1(c), the EBIT spectrum has strong emission lines at 16.2 and 20.4 nm. Although the CXS spectrum also has those two lines, we considered only the line at 20.4 nm to be a resonance line of $Sn^{20+}$. Most ions in the EBIT are estimated as $Sn^{20+}$, but the EBIT also traps some ions with other charge states. The emission line at 16.2 nm is considered as the emission from $Sn^{19+}$ by comparison with a calculation using the FAC (flexible atomic code). We can assume that double electron capture occurs in collisions of $Sn^{21+}$ with He. More stronger emission lines between 13.8 and 15.8 nm were observed in the CXS spectrum. We consider that emission lines are correspond to cascade transitions which are considered as the 4d-4f transitions.

Although we cannot identify many emission lines in the CXS spectra, we can distinguish the resonance lines from other emission lines by comparison with the EBIT spectra. This complementary research using the different experimental methods is an effective way to study the spectroscopic properties of multiply charged ions.

**Acknowledgments**
This work was financially supported in part by MEXT (Ministry of Education, Culture, Sports, Science, and Technology, Japan) under construct subject "Leading Project for EUV lithography source development". Part of this work was performed under the collaboration program of the Institute of Laser Engineering, Osaka University.

## References

[1] Bakshi V 2006 *EUV Source for Lithography* (Washington: SPIE press)
[2] Kieft E R, van der Mullen J J A M, Kroesen G M W, Banine V and Koshelev K N 2005 *Phys. Rev.* E **71** 026409
[3] Kieft E R, Graloff K, van der Mullen J J A M and Banine V 2005 *Phys. Rev.* E **71** 036402
[4] Azarov V I and Joshi Y N 1993 *J. Phys.* B **26** 3495
[5] Svendsen W and O'Sullivan G 1994 *Phys. Rev.* A **50** 3710
[6] Tanuma H, Ohashi H, Fujioka S, Nishimura H, Sasaki A and Nishihara K 2007 *J. Phys.: Conf. Series* **58** 231
[7] Ohashi H, Tanuma H, Fujioka S, Nishimura H, Sasaki A and Nishihara K 2007 *J. Phys.: Conf. Series* **58** 235
[8] Marrs R E, Levine M A, Knapp D A and Henderson J R 1988 *Phys. Rev. Lett.* **60** 1715
[9] Niehaus A 1986 *J. Phys.* B **19** 2925
[10] Tanuma H, Matsumoto J, Nishide T, Shiromaru H and Kobayashi N 2001 *J. Chin. Chem. Soc.* **48** 389
[11] Watanabe H *at al.* 1997 *J. Phys. Soc. Japan* **66** 3795
[12] Knudsen M 1908 *Ann. Phys.* (Leipzig) **28** 75
[13] Yamada C *et al.* 2007 *J. Phys.: Conf. Series* **58** 403
[14] Andersson H, Astner G and Cederquist H 1988 *J. Phys.* B **21** L187
[15] Selberg N, Biedermann C and Cederquist H 1996 *Phys. Rev.* A **54** 4127
[16] Vancura J, Marchetti J V, Perotti J J and Kostroun V O 1993 *Phys. Rev.* A **47** 3758
[17] Justiniano E, Cocke C L, Gray T J, Dubois R, Can C, Waggoner W, Schuch R, Schmidt-Böcking H and Ingwersen H 1984 *Phys. Rev.* A **29** 1088
[18] Carlson T A, Nestor C W, Jr., Wasserman N and McDowell J D 1970 *At. Data* **2** 63
[19] Sasaki A, Sunahara A, Nishihara K, Nishikawa T, Fujima K, Kagawa T, Koike F and Tanuma H 2007 *High Energy Density Phys.* **3** 250

Journal of Physics: Conference Series **163** (2009) 012072 doi:10.1088/1742-6596/163/1/012072

# Polarization studies of radiative electron capture into highly-charged uranium ions

**S. Hess[1,2], H. Bräuning[1], U. Spillmann[1,2], C. Brandau[1], S. Geyer[1,2], S. Hagmann[1,2], M. Hegewald[1,2], C. Kozhuharov[1] T. Krings[4], A. Kumar[1], R. Märtin[1,2], D. Protic[4], B. O'Rourke[1], R. Reuschl[1,2], M. Trassinelli[1], S. Trotsenko[1,2], G. Weber[1,3], D.F.A. Winters[1,3], and Th. Stöhlker[1,3]**

[1]Gesellschaft für Schwerionenforschung, 64291 Darmstadt, Germany
[2]Institut für Kernphysik, University of Frankfurt, 60486 Frankfurt, Germany
[3]Physikalisches Institut, University of Heidelberg, 69120 Heidelberg, Germany
[4]Forschungszentrum Jülich, 52428 Jülich, Germany

E-mail: s.hess@gsi.de, t.stoehlker@gsi.de

**Abstract.**
Recent advances in the development of 2D microstrip detectors open up new possibilities for hard x-ray spectroscopy, in particular for polarization studies. These detectors make ideal Compton polarimeters, which enable us to study precisely the polarization of hard x-rays. Here, we present recent results from measurements of Radiative Electron Capture into the K-shell of highly-charged uranium ions. The experiments were performed with a novel 2D Si(Li) Compton polarimeter at the Experimental Storage Ring at GSI. Stored and cooled beams of $U^{91+}$ and $U^{92+}$ ions, with kinetic energies of 43 MeV/u and 96 MeV/u respectively, were crossed with a hydrogen gasjet. The preliminary data analysis shows x-rays from the K-REC process, emitted perpendicularly to the ion beam, to be strongly linearly polarized.

## 1. Introduction

Modern position sensitive 2D detectors have good spatial, temporal, and energy resolution, and are thus perfectly suited as Compton polarimeters. These detectors uniquely enable studies of the polarization of x-rays emitted during Radiative Electron Capture (REC), and of characteristic transitions in heavy highly-charged ions [1,2]. The REC process is found to be quite sensitive to certain features of the collision system, such as the spin direction [3], which can be linked to the degree of polarization. Being the time-reversed process of photoionisation, REC (and radiative recombination - RR) can be used to improve our understanding of photoionisation in heavy highly-charged ions (see Fig. 1). In this paper, we report on new measurements of the polarization of x-rays emitted from REC into the K-shell of bare uranium ions, and compare these results with the unpolarized $K\alpha_2$ transition after electron capture. These polarization measurements were carried out with our novel 2D Si(Li) Compton polarimeter.

## 2. The Si(Li) Compton polarimeter

The Si(Li) Compton polarimeter has an active area of 64 x 64 mm$^2$ and a crystal thickness of 6.65 mm. The active area is surrounded by a guard ring of roughly 7 mm width. The

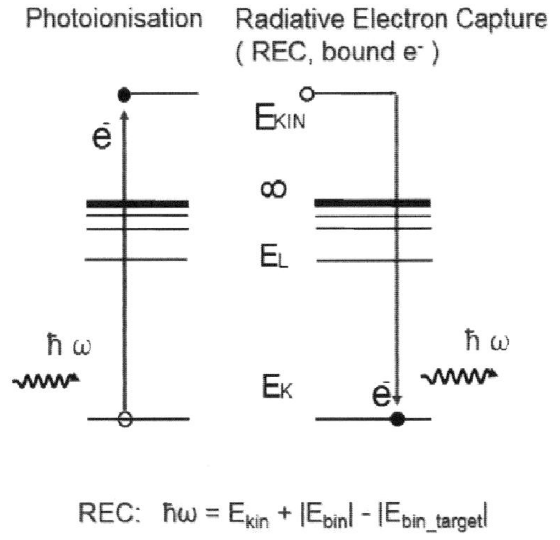


**Figure 1.** Radiative Electron Capture (left) and its time-reversed process: photoionisation (right).

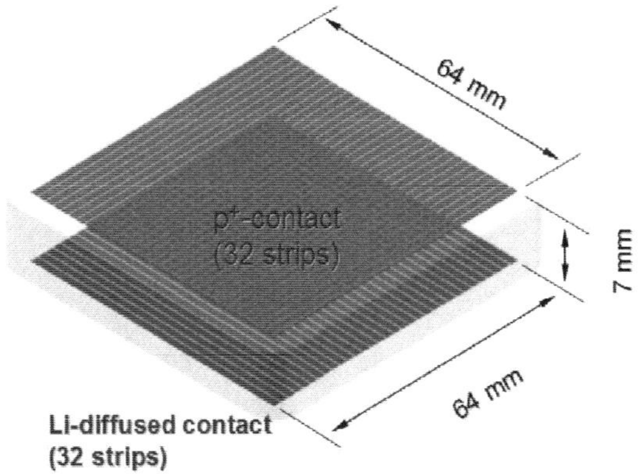


**Figure 2.** Schematic of the 2D strip detector (Compton polarimeter). The Si(Li) crystal is equipped with 32 front- and backside strips, each with a pitch of 2 mm.

commercially available silicon crystal was converted into a 2D detector by lithium-drifting and plasma-etching [7]. Both, positive and negative contact electrodes are divided into 32 strips, each with a pitch of 2 mm. The typical bias voltage applied to the detector crystal is about 1000 V positive. The beryllium entrance window covers the entire active area and has a thickness of 0.5 mm. Each strip is read out separately, and the data is written to harddisk in listmode file format.

Information on the degree of polarization can be obtained from the anisotropy of the Compton-scattering process (see Ref. [6]). The differential cross-section for Compton scattering (Klein-Nishina formula) for 100 % linearly polarized light is given by:

$$\frac{d\sigma}{d\Omega} = \frac{r_e}{2}\frac{\hbar\omega'^2}{\hbar\omega^2}\left(\frac{\hbar\omega'}{\hbar\omega}\frac{\hbar\omega}{\hbar\omega'} - 2\sin^2\theta\cos^2\varphi\right) \quad (1)$$

For a fractional amount of polarization, the Klein-Nishina function can be adapted by replacing $cos^2\varphi$ in eq. 1 with

$$P(d) = 1/2(1-d) + d\cos^2\varphi \quad (2)$$

where $d$ corresponds to the degree of linear polarization.

## 3. Experiment

The experiments were performed at the internal gasjet target of the Experimental Storage Ring (ESR) at GSI, Darmstadt (see Fig. 3). $U^{91+}$ and $U^{92+}$ ions, with energies of 43 MeV/u and 96 MeV/u, respectively, were crossed with a molecular hydrogen gasjet. Typical energy spectra of one of the frontside strips of the Si(Li) polarimeter are shown in Fig. 4. The x-rays

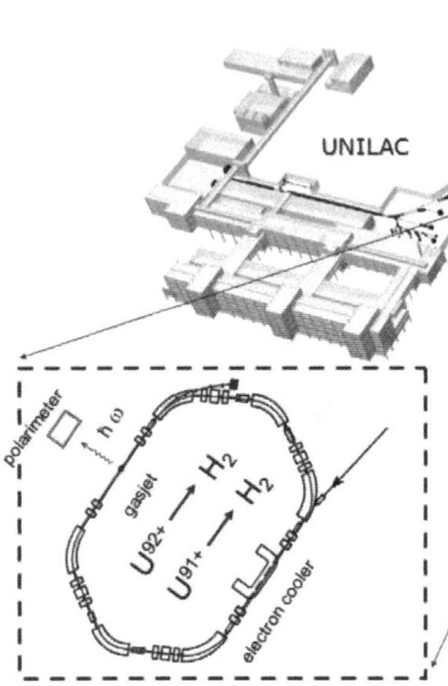


**Figure 3.** The experiments were performed at the internal gasjet in the experimental storage ring (ESR) at GSI, Darmstadt.

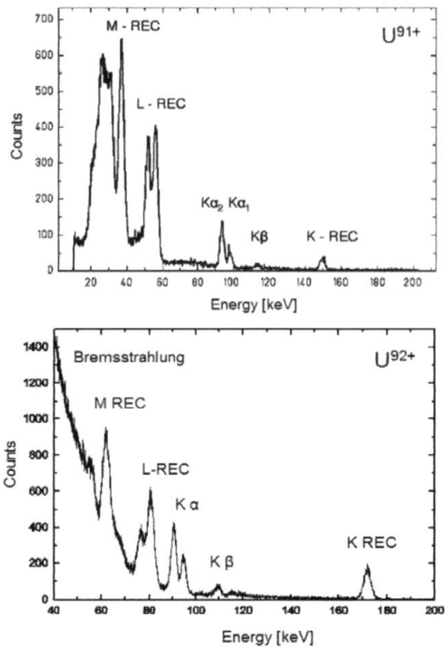


**Figure 4.** Typical energy spectra of one of the frontside strips of the Si(Li) Compton polarimeter.

from REC into the K-shell ($U^{92+}$) have an energy of 172 keV in the laboratory frame. For the polarization measurements, Compton scattering events in the detector were identified using appropriate energy- and multiplicity conditions. Fig. 3 shows a correlation plot, where exactly two events in different strips of the 2D detector are required (multiplicity 2). The energy of one strip is plotted against that of a second strip. The energy splitting between the Compton scattered electrons and Compton photons is indicated by the diagonal lines. At the observed x-ray energies, the recoil electrons have lower energies than the scattered photons, while the sum of both must be equal to the energy of the incident photon.

Finally, the spatial distribution of the scattered K-REC photons is shown in Fig. 6. Here, the scattered photons are plotted relative to the location of the identified Compton recoil electrons. The Compton scattering angle $\vartheta$, which is perpendicular to the detector plane, has been fixed to $90° \pm 5°$ by energy conditions corresponding to the kinematical relation between recoil electron energy and scattering angle $\vartheta$. For comparison, a Compton image of the unpolarized $K\alpha_2$ transition after capture into hydrogen-like uranium is also shown. In Fig. 4 (top) the energy of this transition is at 92 keV. Based on our preliminary data analysis, it can be stated that the observed anisotropy of the K-REC Compton image is in good agreement with the theoretical predictions for polarization of the K-REC transition into bare uranium [8]. In contrast, the Compton image of the $K\alpha_2$ x-rays is highly isotropic, as expected for unpolarized radiation. It can be concluded that, by these measurements, the novel 2D Si(Li) strip detector has proven it's capability as a highly-effective Compton polarimeter.

## Acknowledgments

Martino Trassinelli has been supported by the Alexander von Humboldt foundation and Ajay Kumar acknowledges support by DAAD.

i

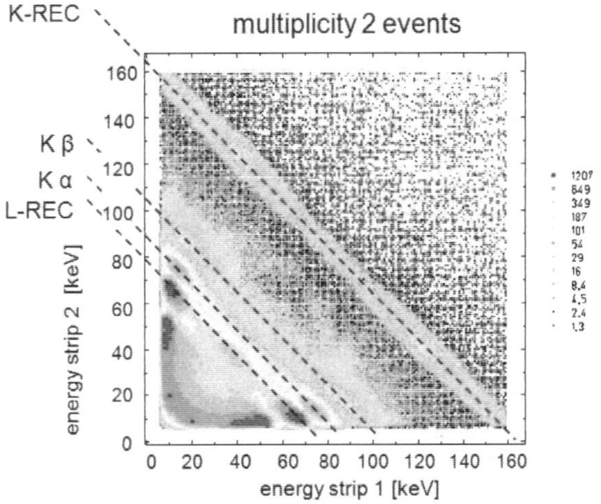


**Figure 5.** Multiplicity 2 correlation plot. The true Compton events are marked by the diagonal lines.

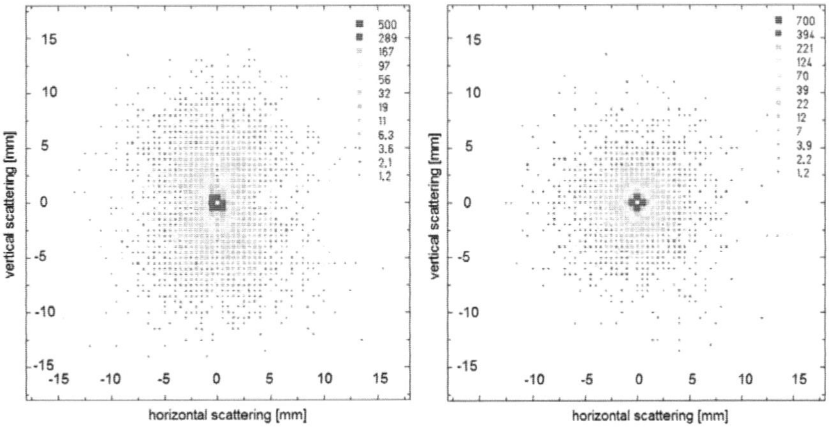


**Figure 6.** Compton image of the x-ray radiation originating from the K-REC transition into bare uranium (left panel). The anisotropy of the angular distribution corresponds to the Klein-Nishina differential cross section for linearly polarized light. In comparison, the panel on the right shows a Compton image of the unpolarized K$\alpha_2$ transition into hydrogen-like uranium.

## References

[1] D. Protic, T. Krings, *IEEE Transactions on Nuclear Science* **50**, (2003) 1008-1012
[2] S. Tashenov, Th. Stöhlker, et al., *Physical Review Letters* **97**, (2006) 223202
[3] A. Surzhykov, S. Fritzsche, T.Stöhlker et al., *Physical Review A* **68**, (2003) 022710
[4] J. Eichler und Th. Stöhlker, *Physical Review* **439**, (2007) 1
[5] F. Lei, J. Dean and G.L. Hills, *Space Science Review* **82**, (1997) 309-388
[6] U. Spillmann, H. Bräuning, S. Hess, Th. Stöhlker et al., *Review of Scientific Instruments* **79**, (2008) 083101
[7] D. Protic, T. Krings and R. Schleichert, *IEEE Trans. Nucl. Sci.* **49**, (2002) 1992-1998
[8] A. Surzhykov, S. Fritzsche, Th. Stöhlker, *Physcal Letters A* **289**, (2001) 213-218

Journal of Physics: Conference Series **163** (2009) 012073 doi:10.1088/1742-6596/163/1/012073

# Multiple electron scattering sequences in $C^{q+}$ + Ar collisions at energies and charge states relevant for carbon ion therapy

K. Tőkési and B. Sulik
Institute of Nuclear Research of the Hungarian
Academy of Sciences (ATOMKI), H-4001, Debrecen, Hungary, EU

E-mail: tokesi@atomki.hu

**Abstract.** Secondary electrons, created in biological tissues by high energy ion impact, may significantly contribute to the fragmentation of small molecules up to single- and double-strand brakes in DNA. Differential spectra of electrons emitted in the collisions of decelerating ions are very important for estimating ion impact radiation damages. In this work, we consider a specific ionization mechanism, the so called Fermi-shuttle acceleration, which can produce fast electrons even by slow ion impact. Focusing our interest to the distal region of the Bragg peak, we calculate electron spectra for singly and doubly charged carbon ions using the classical trajectory Monte Carlo method at 60 keV impact energy, where large fragmentation yields for water molecules have recently been observed experimentally.

## 1. Introduction

The interaction of an energetic heavy ion with dense matter follows a complex scenario with different types of collision events [1-4]. At a few tens of MeV/u or higher impact energies, the dominant process is ionization for both the projectile ion and the target atoms. Accordingly, the equilibrium charge state of the projectile ion is close to that of the bare projectile nucleus. After decelerated to a few hundred keV/u, excitation and charge transfer processes come into play and the mean charge state of the ions increases. Free electron production by direct ionization or autoionization, however, remains an important contribution to the energy transfer from ions to matter, in the entire energy range during the deceleration of the ions, down to their stopping. For practice, the most interesting impact energy region is where the energy transfer per unit length shows a sharp maximum, usually referred to as the Bragg peak [5]. For carbon ions, e.g., the maximum of the Bragg peak is located at ~300 keV/u, where the mean charge state is $q \approx 3.5$ [6]. Towards lower energies, the charge state further decreases; e.g., $q \approx 1$ at 10 keV/u projectile energy. In a recent work [6], it has been shown experimentally that at the low energy side (i.e., in the distal region) of the Bragg peak, the total water fragment ion production had a much flatter profile with projectile energy than would be expected if the water radical formation was assumed to follow the energy-loss profile. This seems to be an important region for understanding the details of the characteristic mechanisms of ion therapy.

A molecule in a biological surrounding can be fragmented in direct ion-molecule collisions or due to the impact of secondary particles (electrons, photons or molecular fragments) emitted in the direct collisions. We note here, that not all the particle-emitting direct collisions lead to a molecular fragmentation, and therefore the yield of secondary processes might be eventually higher than that of the direct fragmentation events. Most of the direct fragmentation processes start with electronic transitions, i.e., the transfer of electrons from the target to the projectile, or with target ionization.

When electrons are removed from the molecule, the repulsive forces between its positive and/or polarized cores may result in a Coulomb explosion [7].

Among the secondary particles emitted in direct collisions, secondary electrons are responsible for many of the further fragmentation events. This has been understood recently, following the focus of interest to radiation damages in biological tissues [8-10]. The majority of the secondary electrons are ballistic electrons ($E \leq 20$ eV). They easily create neutral radicals, ions and electrons. At low energies they may be captured at particular molecular sites or trapped as solvated electrons, contributing to single and double strand breaks in DNA [11]. Even in the sub-excitation energy range (0 –4 eV), resonant processes may enhance molecular dissociation enormously [12].

From the point of view of radiation induced damages, the higher energy part of the secondary electron spectrum is also very important [13,14]. The yield of double-strand breaks induced by electrons in the DNA shows a local maximum at energies of a few hundred eV. These fast electrons have been found to be very effective to segment the DNA molecule [13]. It has also been pointed out, that the fragmentation cross section of water molecules by electron impact maximizes between 50 and 300 eV electron energies [14]. Since free radicals are largely responsible for DNA strand brakes in real tissues, the role of few hundred eV electrons seems to be essential. Accurate knowledge of the amount and distribution of the emitted electrons is especially important in the region of the Bragg peak.

It has been shown recently, that a large amount of unexpectedly fast electrons can be emitted by the impact of slow ions. The responsible mechanism is the so-called Fermi-shuttle acceleration [15-19], where the electron is scattered forward and backward by the incoming heavy projectile ion and the target core before being ejected. Due to the repeated collisions, the electron can be accelerated to relatively high energies.

Up to now, most of the experiments were performed with dressed heavy projectiles and heavy target atoms, where the cross section is large enough for observing such multiple scattering processes. These systems are particularly challenging theoretically to describe the double differential ionization cross sections. So far, no quantum calculations were carried out at such complexity, only classical trajectory Monte-Carlo (CTMC) calculations have been performed for modeling these collisions. Rather surprisingly, for many collision systems quantitative agreement has been found compared with the experimental data [20]. On the other hand, in such a classical calculation, the analysis of the individual particle trajectories allows us to identify the multiple scattering events in a purely classical sense.

In this situation, we decided to perform a combined, systematic experimental and theoretical study of fast electron emission by Fermi-shuttle acceleration in the distal region of the Bragg peak. In the present work, we show the results of the first step in the systematic theoretical studies. We calculated the double differential cross section for electron emission from the collisions of singly and doubly charged carbon ions with argon atoms at 60 keV impact energy (5 keV/u). This energy falls in the distal region of the Bragg peak, where the fragmentation yield for the water molecule has been found to be large [6]. Argon ($Z$=18) is a target which provides a high cross section for the multiple electron scattering process, therefore it is easy to handle experimentally. Moreover, it can be considered as a rather good model for chlorine ($Z$=17) and calcium ($Z$=22) atoms. The latter two atoms are likely to play an important role in producing fast electrons in biological tissues.

## 2. Calculation

The CTMC simulations have been performed in the three-body approximation. In our CTMC model the three particles are the projectile ion core (P), one atomic active target electron (e), and the remaining target ion core (T). Both projectile and target cores include nucleus and electrons. The three "particles" are characterized by their masses and the interaction potentials between them. For the description of the interaction among the particles, central model potentials (based on Hartree-Fock calculations) have been used. For more details of the CTMC calculations see Ref. [20].

In the calculations, electron emission contributions originating from the target and projectile ionization were calculated separately. The sum of these two components, in the rest frame of the target, is the quantity to be compared with the planned experiments on an absolute scale.

Finally, a random sampling analysis of the individual CTMC trajectories has been performed. In the present stage of the work, only 0.01 percent of the trajectories have been analyzed individually above the energy of 15 eV, so this analysis is still preliminary. At 60 keV total energy (5 keV/u), the maximum energy of the single electron scattering ionization (the so called binary encounter peak) is 10.4 eV, and so 15 eV can be considered as an approximate upper limit of observing projectile-target (P-T) scattering events. The results clearly show that most of the high energy ($E$>15eV) electrons, emitted in these collision systems suffered multiple collisions with the target nucleus and the projectile core.

## 3. Results

Fig. 1 shows the calculated double differential electron emission spectra. Projectile and target ionization contributions are shown separately, for both singly and doubly charged carbon ions. It is clearly seen that a significant part of the calculated spectra falls above 15 eV. The projectile velocity ($V$=0.44 atomic units, with $\hbar=e=m_e=1$) corresponds to 2.6 eV electrons. The binary encounter peak at zero degree is located at $2V$ (10.4 eV). The maximum energy of the triple, projectile-target-projectile

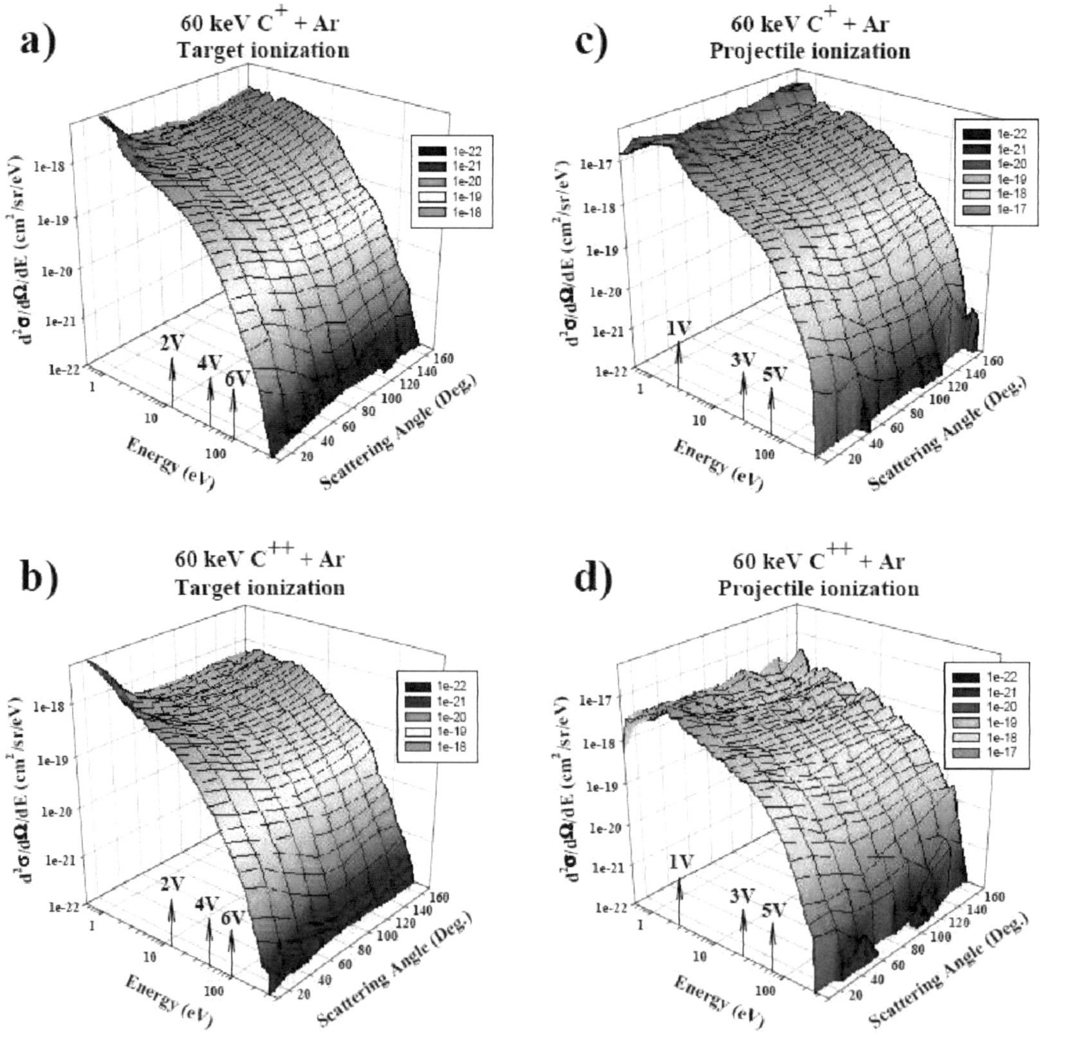


**Figure 1.** Double differential cross section for electron emission as a function of electron energy and observation angle (relative to the beam direction). *a*) and *b*): target ionization, *c*) and *d*): projectile ionization, *a*) and *c*): singly charged carbon ion projectile, *b*) and *d*): doubly charged carbon ion projectile.

(P-T-P) and quadruple (P-T-P-T) scattering at $4V$ is 41.6 eV, while that for 5-fold and 6-fold sequences it is 93.6 eV ($6V$). Looking at the figures 1*a* and 1*b* (target ionization), it is clear that the calculated cross sections indicate even 10-fold sequences ($10V$, 260eV), which seems to be unexpected, but such trajectories could really be found in the trajectory analysis. Only future experiments can decide whether this prediction is realistic or not. Similarly, 9-fold scattering sequences are predicted for the projectile ionization (figures 1*c* and 1*d*). These strange predictions are all supported by the preliminary random sampling analysis of the CTMC trajectories. We would like to draw attention here to some other features of the calculated spectra. One is that the $C^{++}$ cross sections are systematically smaller than those for the $C^{+}$ projectile ions. Moreover, for $C^{+}$, the low energy part of the projectile ionization cross section is higher than that for the target ionization. A more detailed analysis of these data will be provided, together with those from the presently running other calculations will be provided in a forthcoming publication.

In conclusion, our preliminary results indicate unexpectedly long sequences of multiple scattering in the studied projectile energy range. Since this is already the quasi-molecular collision regime, these classical predictions should be tested by both experiments and quantum calculations. The latter, however is not very easy. Quantum and classical calculations could be compared first for much simpler collision systems. According to our preliminary statistics, multiple (3-fold or higher order) scattering sequences dominantly contribute (with ~70%) to the electron emission above 15 eV energy. As a next step, the predicted double differential cross sections should be tested by experiment.

**Acknowledgement**

This work was supported by the Hungarian National Science Foundation, OTKA (Grant Nos: K073703, K72172), the grant ``Bolyai" from the Hungarian Academy of Sciences, the TeT Grant No. JP-21/2006, and the Hungarian National Office for Research and Technology.

## References

[1] M. E. Rudd and J. H. Macek, *Case Studies in Atomic Physics* **3** (1976) pp 47-136.
[2] R. A. Baragiola (ed.), *Ionization of Solids by Heavy Particles*, Nato ASI Series B: Physics, vol. 306 (Plenum Press, New York, 1993)
[3] N. Stolterfoht, R.D. DuBois and R.D. Rivarola, *Electron Emission in Heavy Ion-Atom Collisions* Springer, Berlin, 1997).
[4] H. Rothard, *Nucl. Instr. Meth.* B **146**, 1 (1998).
[5] Z. Deng, I. Bald, E. Illenberger, M. A. Huels, *Phys. Rev. Lett.* **95**, 153201 (2005).
[6] E. C. Montenegro, M. B. Shah, H. Luna, S. W. J. Scully, A. L. F. de Barros, J. A. Wyer, and J. Lecointre, Phys. Rev. Lett. **99**, 213201 (2007)
[7] Z.D.Pesic, J.-Y.Chesnel, R.Hellhammer, B.Sulik, N.Stolterfoht, *J. Phys* B **37**, 1405 (2004).
[8] G. Kraft and M. Krämer, Adv. *Rad. Biol.* **17** (1993) 1.
[9] M. Scholz and G. Kraft, *Rad. Prot. Dos.* **52** (1994) 29.
[10] Collins G P 2003 *Sci. Am.* **289**(3) 26
[11] Boudaïffa B, Cloutier P, Hunting D, Huels M A and Sanche L 2000 *Science* **287** 1658
[12] Abdoul-Carime H, Gohlke S and Illenberger E, 2004 *Phys. Rev. Lett.* **92** 168103
[13] Friedland W, Jacob P, Paretzke H G and Stork T, 1998 *Radiat. Res.* **150** 170
[14] S. W. J. Scully, V. Senthil, J. A. Wyer, E. C. Montenegro, and M. B. Shah, ICPEAC 2005, 24th International Conference on Photonic Electronic and Atomic Collisions Rosario, Argentina, July 20-26, 2005, *Abstarcts of Contributed Papers,* Edited by: F. D. Colavecchia *et al.*, p 324.
[15] S. Suarez, R. O. Barrachina, W. Meckbach, Phys. *Rev. Lett.* **77** (1996) 474.
[16] U. Bechthold, S. Hagmann, J. Ullrich, B. Bathelt, A. Bohris, R. Moshammer, U. Ramm, C. Bhalla, G. Kraft and H. Schmidt-Böcking, *Phys. Rev. Lett.* **79** (1997) 2034.
[17] G. Lanzanò, E. De Filippo, D. Mahboub, H. Rothard, S. Aiello, A. Anzalone, S. Cavallaro, A. Elanique, E. Geraci, M. Geraci, F. Giustolisi, A. Pagano, G. Politi, *Phys. Rev. Lett.* **83** (1999) 4518.
[18] B. Sulik , Cs. Koncz, K. Tőkési, A. Orbán and D. Berényi, *Phys. Rev. Lett.* **88** (2002) 073201.
[19] B Sulik, N Stolterfoht, R Hellhammer, Z Pesic, Cs Koncz, K Tőkési and D Berényi, *Nucl. Inst. and Meth.* B **212,** 32 (2003).
[20] B. Sulik and K. Tőkési, Adv. Quant. Chem. **52** (2007) 253.

# Young type interference in electron emission from $H_2$ and forward-backward asymmetry: a new approach

**S Chatterjee[1], D Misra[1], A H Kelkar[1], PD Fainstein[2] and L C Tribedi[1]**

[1]Tata Institute of Fundamental Research, Homi Bhabha Road, Colaba, Mumbai 400 005, India
[2]Centro Atomico Bariloche, 8400 Bariloche, Argentina

E-mail: `lokesh@tifr.res.in`

**Abstract.** The energy and angular distribution of electron double differential cross sections (DDCS) in collision of 95 MeV $F^{9+}$ ions with molecular hydrogen have been measured. We explain the observed distributions in terms of the two-center effect and the Young-type interference effect. The asymmetry parameter, derived from the cross sections for complementary forward ($30^0$) and backward ($150^0$) angles, shows an oscillatory behavior as a function of electron velocity. The measured energy and angular distributions as well as the asymmetry parameter are compared with a molecular CDW-EIS (continuum distorted wave-eikonal initial state) theory.

## 1. Introduction

The low energy electron emission spectrum from molecular hydrogen induced by heavy ion collision contains rich information about various ionization mechanisms. Since the two atoms in molecular hydrogen are indistinguishable, their contributions to the ionization probability add coherently and an interference effect might be expected in the single ionization of $H_2$. Such electron emission from $H_2$ may be closely related to the well-known Youngs double-slit experiment. It was predicted nearly forty years ago by Cohen and Fano in photoionization [1] and was first observed only recently in the electron spectra resulting from the ionization of molecular hydrogen by GeV energy heavy-ions in a pioneering investigation by Stolterfoht *et al.* [2] and by MeV energy heavy-ions [3, 4, 5]. The oscillations observed in the ratio of DDCS of $H_2$ and 2H, as a function of electron velocity i.e. de Broglie wavelength of electrons was termed as the signature of Young type interference. Such ratios were fully measured by Misra *et al.* [3] and Tribedi *et al.* [6, 7] by measuring the DDCS of $H_2$ as well as atomic H. Recently, the interference effect has been shown in the DDCS spectrum and in the ratio ($H_2$-to-H) spectrum as function of emission angle [8],which is an analogous situation of original Young's double-slit experiments with light or with single electrons, where the interference fringes are observed as a function of observation angle.

The two-center effect as well as non-Coulomb potential for two or multielectron targets are known to cause a large forward-backward angular asymmetry, which can be seen in the large difference in the DDCS values for small forward and large backward angles [9, 10, 11, 12, 13, 14, 15]. Apart from the methods of deriving interference pattern from the ratio of molecular DDCS to that of atomic one, another way to look into it comes from the analysis of the angular asymmetry of molecular DDCS, which is derived "only" from molecular DDCS and is free from any normalization. This procedure, which was introduced by Misra *et al.* [17] can be used self-consistently to study the interference.

We provide new measurements on the angular and energy distributions of the DDCS and show examples of interference effect by deriving the DDCS ratios ($H_2$/2H) using theoretical values of atomic H for 95 MeV $F^{9+}$ ions colliding with $H_2$. Finally we derive the forward-backward asymmetry parameter from the measured DDCS as a function of emission velocity. The measured DDCS, their angular distributions and the asymmetry parameter are compared with the state-of-the art model calculation based on the molecular CDW-EIS approximation.

## 2. The experiment

The measurement was carried out for a 95 MeV ($v_p \sim 14$ a.u.) $F^{9+}$ beam (obtained from BARC-TIFR Pelletron accelerator facility) colliding with $H_2$. The emitted electrons were detected using an electron spectrometer equipped with a hemispherical electrostatic energy analyzer and a channel electron multiplier. The energy resolution of the spectrometer was about 6% of the electron energy. Experiments were done by flooding the chamber with target gas keeping pressure of 0.15 mTorr for ejected electron energies 1-40 eV. For higher energy electrons the gas pressure was 0.2 mTorr. The front and exit apertures of the spectrometer were biased to a small voltages of +6V in order to help the lowest energy electrons to be detected. Background pressure was kept at $1\text{x}10^{-7}$ Torr. The DDCSs were studied for different angles ranging from $20^0$ to $160^0$ (in steps of $15^0$) and for electron energies between 1 and 300 eV. The further details of the experimental setup are described in [16].

## 3. Energy and angular distributions

Figures 1. (a) and (b) show the measured energy distribution of DDCS at two complementary angles $30^0$ and $150^0$ respectively for electron emission in 5 MeV/u $F^{9+}$+$H_2$ collision along with the CDW-EIS predictions. The DDCS varies over five orders of magnitude in an energy interval of 300 eV and reaches a maximum value for very low energy electrons. The angular distributions, one for low energy (15 eV) and one for high energy electrons (120 eV) are shown in figure 1(c,d). The measured data show a large forward-backward asymmetry caused by the two centre effect and is quite well reproduced by the CDW-EIS.

## 4. Interference effect in DDCS ratio spectrum

Since $H_2$ is inversion symmetric homonuclear diatomic molecule, the contributions from each atomic center to the ionization probability add coherently and this may lead to oscillatory structure in the electron emission spectrum representing the interference effect. In order to amplify the visibility of the oscillatory structure, the experimental DDCS for $H_2$ is divided by twice the DDCS corresponding to the two-effective H atom calculations. In figure 2(a) we show the ratio of the experimentally measured DDCS of $H_2$ to the twice the theoretically calculated DDCS of H for $30^0$ emission angle. The ratio goes through an oscillation around the fitted straight line (D1). This oscillation can be attributed to the Young type interference effect [3, 4]. To compare the oscillations with theory i.e., to produce the oscillations around a horizontal line the ratio is divided by D1. The normalized ratio is plotted in Fig. 2(b), which is in good agreement with the molecular CDW-EIS prediction. The normalized ratio is fitted with Cohen-Fano model function, which can be in found in ref. [2, 18].

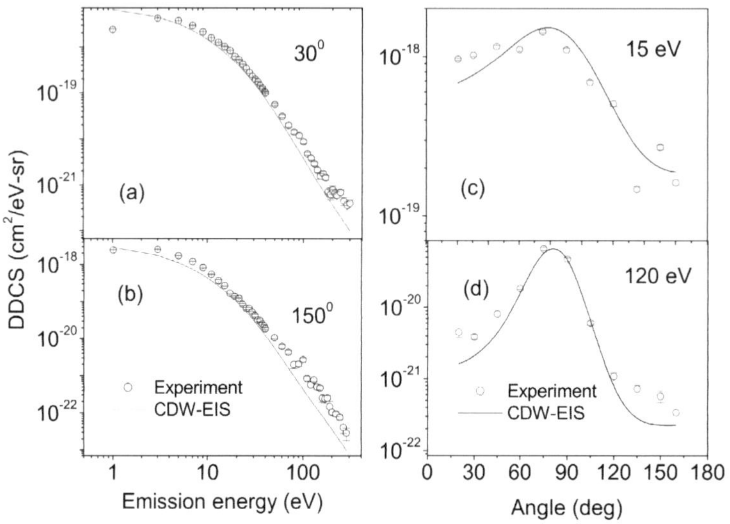


**Figure 1.** Electron energy distribution of DDCS for emission angle $30^0$ (a) and $150^0$ (b); angular distributions of DDCS for electrons with energy 15 eV (c) and 120 eV (d). Solid lines are CDW-EIS predictions for 95 MeV $F^{9+}$ + $H_2$ collision.

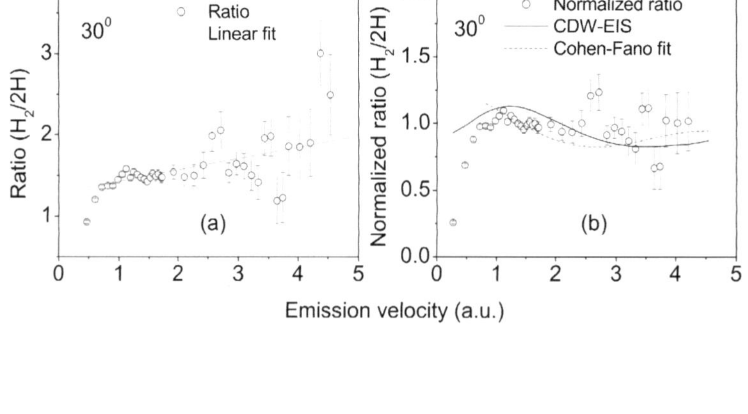


**Figure 2.** (a) DDCS ratio ($H_2$/2H) for 95 MeV $F^{9+}$ + $H_2$ collision. Solid line: linear fit to the data. (b) Symbols: the normalized ratio, Solid line: CDW-EIS and dashed line for Cohen-Fano model fit.

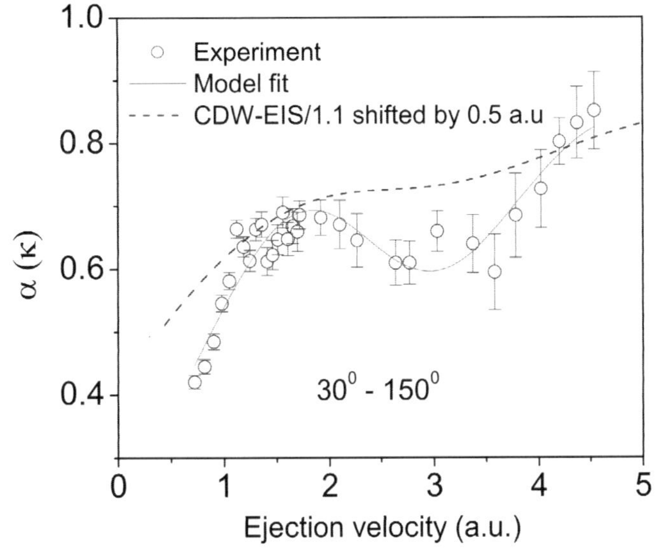


**Figure 3.** The asymmetry parameter for 95 MeV $F^{9+}$ collisions with $H_2$ including a model fit in solid line and dashed line for CDW-EIS theory [to be published elsewhere in details]. Two complementary forward-backward angles $30^0$ and $150^0$ are considered.

## 5. Forward-backward asymmetry and interference effect

Besides first-order effects derived from the molecular to atomic cross-section ratios, the interference mechanism has been investigated in asymmetry parameter of the electrons emitted from $H_2$ targets in collision with 95 MeV $F^{9+}$. The asymmetry parameter can be defined as [15, 17] $\alpha(k) = (\sigma(\theta, k) - \sigma(\pi - \theta, k))/(\sigma(\theta, k) + \sigma(\pi - \theta, k))$, where $\sigma$ refers to the DDCS, $\theta$ is any small forward angle and $k$ is the ejection velocity. The asymmetry parameter $\alpha(k)$ derived from the molecular DDCSs for emission angles $30^0$ and $150^0$ is plotted as a function of ejection velocity in Fig. 3. It can be seen that the molecular CDW-EIS (dashed line) predicts a similar oscillation in $\alpha$ with a slightly different phase. Such oscillation may stem from difference of frequencies of first-order oscillations of the two complementary angles as explained earlier in ref. [17] i.e., the frequencies of oscillation for backward angles are more than complementary forward angles, causing the oscillatory structure in $\alpha$. Fig. 3 also shows an oscillation starting at a value $\alpha$ of about 0.4 and the oscillation is seen to be around a straight line with a steep slope. The steep slope for heavy-ion collision on $H_2$ could be attributed to the post-collisional two-center effect [14, 15, 17] where influence of the field of the highly-charged positive ions on the final state of the ejected electron contributes towards a continuous increment of the yield in the forward direction compared to the backward angles. The effect increases with increasing ejection velocity in the velocity range considered. Therefore, the asymmetry parameter for $H_2$ reveals the combined effect of two-center post-collision and interference effect. Using Cohen-Fano type model for interference and considering the observed frequency difference in forward-backward angles a good fitting is obtained (solid line in Fig. 3). The details about the model fitting has been discussed in ref. [17].

## 6. Conclusion

We have studied energy and angular distribution of electron emission and derived the first-order interference as well as the forward-backward asymmetry parameter from the DDCS spectrum at $30^0$ and $150^0$. The CDW-EIS prediction is in good agreement with the experimental observations. It is shown that the asymmetry parameter which is based on the DDCS of $H_2$ only, can be used as self normalizing way of the obtaining the Young type interference. The simple model fitting based on Bessel function type term for DDCS and the frequency difference between forward and backward angles reproduces the oscillation quite well.

## References

[1] Cohen H D and Fano U 1966 Phys. Rev **150** 30
[2] Stolterfoht N *et al.* 2001 Phys. Rev. Lett. **87** 023201
[3] Misra D *et al.* 2004 Phys. Rev. Lett. **92** 153201
[4] Hossain S *et al.* 2003 Nucl. Instr. Meth. Phys. Res. B **205** 484
[5] Hossain S, Landers A, Stolterfoht N and Tanis J A 2005 Phys. Rev. A **72** 010701(R)
[6] Tribedi L C *et al.* 1998 J. Phys. B **31** L369
[7] Tribedi L C *et al.* 2001 Phys. Rev. A **63** 062723
[8] Chatterjee S *et al.* 2008 Phys. Rev. A (in press)
[9] Stolterfoht N *et al.* 1987 Euro. Phys. Lett. **4** 899
[10] Pedersen J O P, Hvelplund P, Petersen A and Fainstein P 1991 J. Phys. B **24** 4001
[11] Tribedi L C *et al.*1998 Phys. Rev. A **58** 3619
[12] Moshammer R *et al.* 1994 Phys. Rev. Lett. **73** 3371
[13] Tribedi L C *et al.* 1996 Phys. Rev. Lett. **77** 3767
[14] Fainstein P D, Ponce V H, and Rivarola R D 1996 J. Phys. B **24** 3091
[15] Fainstein P D, Gulyás L, Martín L, Salin L 1996 Phys. Rev. A **53** 3243
[16] Misra D *et al.* 2007 Phys. Rev. A **75** 052712
[17] Misra D *et al.* 2006 Phys. Rev. A **74** 060701(R)
[18] Nagy L *et al.* 2002 J. Phys. B **35** L453

# Loss mechanisms and lifetimes of heavy ion beams in HIRFL-CSRe

**Y. L. Xue, X. Cai, D. Yu, J. Shao, F. F. Ruan, D. J. Qi, M. W. Zhang, W. Wang**

Institute of Modern Physics, Chinese Academy of Sciences, Lanzhou, 730000 China

E-mail: caixh@impcas.ac.cn

**Abstract.** The loss mechanisms and lifetimes of ion beams in a cooler storage ring in collisions with residual gas, internal gaseous target, and electrons in an electron-cooler were studied. Partial beam lifetimes, which were due to various loss mechanisms, and total beam lifetimes of 50-500 MeV/u $^{12}C^{6+}$, $^{36}Ar^{18+}$, $^{132}Xe^{54+}$ and $^{238}U^{92+}$ ion beams stored in HIRFL-CSRe were also calculated. The calculations indicate that the beam lifetimes are restricted considerably by charge exchange between the beam ions and the internal gaseous target, as well as radiative recombination with the electrons. For heavy ion beams such as $^{132}Xe^{54+}$ and $^{238}U^{92+}$, the radiative recombination with electrons in the electron-cooler is the main loss mechanism.

## 1. Introduction

The loss mechanisms and lifetimes of ion beams circulating in a cooler storage ring are routinely considered in various cooler storage rings experimentally and/or theoretically [1-7]. The lifetime of an ion beam mainly depends on the interaction between the ions and the residual gas, internal gaseous target, and electrons in the electron-cooler. The main processes include single Coulomb scattering, multiple Coulomb scattering and charge exchange between the beam ions and the residual gas or internal gaseous target, as well as radiative recombination, which occurs during the interaction of the beam ions with the cooling electrons.

In this work the loss mechanisms and lifetimes of heavy ion beams circulating in HIRFL-CSRe—the experimental ring of the newly finished Cooler-Storage-Ring system HIRFL-CSR in Lanzhou [8] were studied. Partial beam lifetimes, which were due to various loss mechanisms, and total beam lifetimes of 50-500 MeV/u $^{12}C^{6+}$, $^{36}Ar^{18+}$, $^{132}Xe^{54+}$ and $^{238}U^{92+}$ ion beams stored in HIRFL-CSRe in collisions with the residual gas ($3.0\times10^{-11}$ mar, 85% $H_2$, 15% $N_2$), internal gaseous target ($N_2$: $5.0\times10^{12}$ atoms/cm$^2$) and electrons in the electron-cooler ($5\times10^7$ electrons/cm$^3$) were calculated.

## 2. Beam lifetime due to interaction with residual gas

For each beam loss mechanism due to the interaction with residual gas, the corresponding lifetime can be expressed as

$$\tau_{rj} = (\sigma_j \rho_r \beta_i c)^{-1}, \tag{1}$$

where $\sigma_j$ is the corresponding interaction cross section such as single Coulomb scattering [9], multiple Coulomb scattering [9-11], electron capture [12-25] or electron loss [1, 13, 20, 22, 26-32]. For simple molecules such as $H_2$, CO, $CO_2$ and $N_2$, $\sigma_j$ could be obtained by adding together the individual cross section of each atom in the molecule if the velocity of the ion beam is high. $\rho_r$ is the number density of the residual gas [33], and $\beta_i$ is the ion beam velocity in units of $c$ ($3\times10^8$ m/s).

If the residual gas consists of two components, $a$ and $b$, the corresponding beam lifetime is [11]

$$(\tau_{rj}) = \left((\sigma_{ja}\rho_{ra}\beta_i c) + (\sigma_{jb}\rho_{rb}\beta_i c)\right)^{-1}, \tag{2}$$

where $\rho_{ra}$ and $\rho_{rb}$ are the number densities of $a$ and $b$, respectively.

The total beam lifetime resulting from the interaction with the residual gas can be described as

$$(\tau_r) = (\sum_j (\tau_{rj})^{-1})^{-1}. \tag{3}$$

## 3. Beam lifetime due to interaction with internal gaseous target

The interaction processes between the beam ions and the internal gaseous target are similar to those with the residual gas, namely single Coulomb scattering, multiple Coulomb scattering, electron capture and electron loss. However, the beam lifetime related to each loss mechanism should be described as

$$\tau_{tj} = (\sigma_j \rho_t f_0)^{-1}, \tag{4}$$

where $\rho_t$ is the area number density of the target and $f_0$ is the revolution frequency of the ion beam. Additionally, the beam lifetime due to energy loss can be found in Boine-Frankenheim's work [6].

## 4. Beam lifetime due to interaction with electrons in the electron-cooler

To lower (or eliminate) the transverse and longitudinal emittance increasing in the cooler storage ring caused by Coulomb scattering and intra-beam scattering, the electron cooling is used. However, radiative recombination during the interaction of the beam ions with the cooling electrons can cause beam loss. The corresponding lifetime can be expressed as [11]

$$\tau_{rec} = \gamma_i^2 \alpha_r^{-1} n_e^{-1} \eta_{ec}^{-1}. \tag{5}$$

Here, $\gamma_i = (1-\beta_i^2)^{-1}$, $\alpha_r$ is the recombination coefficient in cm$^3$/s [34], $n_e$ is the number density of the electrons in cm$^{-3}$, and $\eta_{ec}$ is the ratio of the cooling length to the ring circumference.

## 5. Total beam lifetime

Under the influence of the residual gas, internal gaseous target and electrons in the electron-cooler, the total beam lifetime can be expressed by

$$\tau_{total} = \left(\tau_r^{-1} + \tau_t^{-1} + \tau_{rr}^{-1}\right)^{-1}, \tag{6}$$

where, $\tau_r$, $\tau_t$ and $\tau_{rr}$ are the partial beam lifetimes due to interaction with the residual gas, internal gaseous target and electrons in the electron-cooler, respectively.

We used the method in the reference [9] to predict the cross sections of single Coulomb scattering. Partial beam lifetimes due to multiple Coulomb scattering and energy loss were investigated according to the way published by [7,10] and [6], respectively. For fast ion beams such as $Xe^{54+}$ and $U^{92+}$ with the light target, the cross sections of the dominate electron capture channel, radiative electron capture, were obtained by interpolating in published data table II of Mclaughlin's work [24], and then timing the quasifree electrons in the light target. For $C^{6+}$ and $Ar^{18+}$, we used the equation (5) in Franzke's work [7] to calculate the electron capture cross sections. Then, the beam lifetime of $^{238}U^{92+}$ stored in ESR [4] were calculated and compared with their experimental data in table 1.

Furthermore, according to the main parameters of the HIRFL-CSRe, the partial lifetimes as well as the total lifetimes of 50-500 MeV/u $^{12}C^{6+}$, $^{36}Ar^{18+}$, $^{132}Xe^{54+}$ and $^{238}U^{92+}$ in collisions with the residual gas ($3.0\times10^{-11}$ mbar, 85% $H_2$, 15% $N_2$), internal gaseous target ($N_2$: $5.0\times10^{12}$ atoms/cm$^2$) and electrons in the electron-cooler ($5\times10^7$ electrons/cm$^3$) were calculated and listed in table 2.

Table 1. Calculated and experimental beam lifetime of $U^{92+}$ in ESR.

| Ion | Energy (MeV/u) | Target | Beam lifetime (s) Present work (not cooled) | Experimental data [4] (not cooled) |
|---|---|---|---|---|
| $U^{92+}$ | 49 | $N_2(5.0\times10^{11}$ mol./cm$^2)$ | 146 | 80 |
| | 68 | $N_2(5.0\times10^{11}$ mol./cm$^2)$ | 197 | 130 |
| | 220 | $N_2(5.0\times10^{11}$ mol./cm$^2)$ | 668 | 750 |
| | 358 | $N_2(5.0\times10^{11}$ mol./cm$^2)$ | 1230 | 1300 |

Table 2. Partial and total lifetimes of 50-500 MeV/u $^{12}C^{6+}$, $^{36}Ar^{18+}$, $^{132}Xe^{54+}$ and $^{238}U^{92+}$ stored in HIRFL-CSRe. $\tau_r(s)$, $\tau_t(s)$, $\tau_{rr}(s)$ are the partial lifetimes due to interaction with residual gas, internal gaseous target and electrons in the electron-cooler respectively, and $\tau_{total}(s)$ is the total lifetimes.

| Ion | Energy (MeV/u) | $\tau_r(s)$ | $\tau_t(s)$ | $\tau_{rr}(s)$ | $\tau_{total}(s)$ |
|---|---|---|---|---|---|
| $C^{6+}$ | 50 | $1.91\times10^5$ | $1.79\times10^2$ | $3.77\times10^3$ | $1.71\times10^2$ |
| | 100 | $8.73\times10^5$ | $7.68\times10^2$ | $4.17\times10^3$ | $6.48\times10^2$ |
| | 200 | $3.66\times10^6$ | $3.03\times10^3$ | $5.02\times10^3$ | $1.89\times10^3$ |
| | 300 | $7.55\times10^6$ | $6.28\times10^3$ | $5.94\times10^3$ | $3.05\times10^3$ |
| | 400 | $1.16\times10^7$ | $1.00\times10^4$ | $6.95\times10^3$ | $4.10\times10^3$ |
| | 500 | $1.52\times10^7$ | $1.39\times10^4$ | $8.03\times10^3$ | $5.09\times10^3$ |
| $Ar^{18+}$ | 50 | $5.62\times10^4$ | $5.52\times10^2$ | $3.54\times10^2$ | $2.15\times10^2$ |
| | 100 | $1.39\times10^5$ | $1.41\times10^2$ | $3.90\times10^2$ | $1.04\times10^2$ |
| | 200 | $8.00\times10^5$ | $7.98\times10^2$ | $4.70\times10^2$ | $2.96\times10^2$ |
| | 300 | $2.24\times10^6$ | $2.22\times10^3$ | $5.57\times10^2$ | $4.45\times10^2$ |
| | 400 | $4.42\times10^6$ | $4.48\times10^3$ | $6.51\times10^2$ | $5.68\times10^2$ |
| | 500 | $7.02\times10^6$ | $7.42\times10^3$ | $7.52\times10^2$ | $6.83\times10^2$ |
| $Xe^{54+}$ | 50 | $1.39\times10^5$ | $1.40\times10^2$ | 33.8 | 27.2 |
| | 100 | $2.99\times10^5$ | $3.05\times10^2$ | 37.3 | 33.2 |
| | 200 | $7.12\times10^5$ | $7.43\times10^2$ | 44.9 | 42.3 |
| | 300 | $1.24\times10^6$ | $1.33\times10^3$ | 53.2 | 51.1 |
| | 400 | $1.87\times10^6$ | $2.07\times10^3$ | 62.2 | 60.4 |
| | 500 | $2.57\times10^6$ | $2.95\times10^3$ | 71.9 | 70.2 |
| $U^{92+}$ | 50 | $3.38\times10^4$ | 35.3 | 10.9 | 8.3 |
| | 100 | $6.45\times10^4$ | 67.7 | 12.0 | 10.2 |
| | 200 | $1.34\times10^5$ | $1.42\times10^2$ | 14.5 | 13.1 |
| | 300 | $2.18\times10^5$ | $2.32\times10^2$ | 17.1 | 16.0 |
| | 400 | $3.17\times10^5$ | $3.40\times10^2$ | 20.0 | 18.9 |
| | 500 | $4.30\times10^5$ | $4.65\times10^2$ | 23.2 | 22.1 |

Comparing the calculations with the experimental data in table 1, we overestimate the lifetimes of $^{238}U^{92+}$ at low energy, but underestimate them at high energy. This is because we do not take the non-radiative electron capture into account, which becomes important in the collision between the highly charged heavy ions and the light target at low energy. The ratio of the calculated lifetime to the experimental data ranges from 0.9 to 1.9.

For 50-500 MeV/u $^{12}C^{6+}$, $^{36}Ar^{18+}$, $^{132}Xe^{54+}$ and $^{238}U^{92+}$ circulating in HIRFL-CSRe, the calculations indicate that:

1) The loss mechanisms resulting from the residual gas, with an average pressure of $3.0\times10^{-11}$ mbar, can be neglected.

2) With the increase of energy, the beam lifetimes become longer, and the light ion beams can be stored longer than the heavy ones with the same energy.

3) Lifetimes of $^{12}C^{6+}$and $^{36}Ar^{18+}$ are several tens of minutes, mainly due to charge exchange between the beam ions and the internal gaseous target, as well as radiative recombination with electrons in the electron-cooler.

4) $^{132}Xe^{54+}$ and $^{238}U^{92+}$ ion beams can be stored only for several tens of seconds in HIRFL-CSRe. The main loss mechanism is radiative recombination with the cooling electrons.

The calculations and conclusions presented here will be very helpful for the experiments at HIRFL-CSRe.

## 6. Acknowledgments

This work is supported by the National Scientific Project (HIRFL-CSR) and National Natural Science Foundation of China (10375080, 10775160).

## References

[1] Habs D, Baumann W, Berger J *et al* 1989 *Nucl. Instr. and Meth. B* **43** 390
[2] Franzke B, Beckert K, Bosch F *et al* 1992 *Procedding of the Third European Particle Accelerator Conference(Berlin)* (Berlin/Germany) p 444
[3] Ekström C, Fridé C J, Jansson A *et al* 1996 *Nucl. Instr. and Meth. A* **371** 572
[4] Stöhlker Th, Reich H, Beckert K *et al* 1997 *Hyperfine Interaction* **108** 29
[5] DuBois R D, de Lucio O, Thomason M *et al* 2007 *Nucl. Instr. and Meth. B* **261** 230
[6] Boine-Frankenheim O, Hasse R, Hinterberger F *et al* 2006 *Nucl. Instr. and Meth. A* **560** 245
[7] Franzke B 1992 *CERN* **92-01** 100
[8] Xia J W, Zhan W L, Wei B W *et al* 2002 *Nucl. Instr. and Meth. A* **488** 11
[9] Jackson J D 1975 *Classical Electrodynamics 2nd ed.* (New York:John-wiley and Sons Inc)
[10] Hinterberger F and Prasuhn D 1989 *Nucl. Instr. and Meth. A* **279** 413
[11] Poth H 1990 *Phys. Rep.* **196** 135
[12] Nikolaev V S 1965 *Sov. Phys. Usp.* **8** 269
[13] Franzke B 1981 *IEEE Trans. on Nucl. Sci.* **NS-28** 2116
[14] Schlachter A S, Stearns J W, Gramam W G *et al* 1983 *Phys. Rev. A* **27** 3372
[15] Dmitriev I S, Zaikov V P, Kral'kina E A *et al* 1986 *Nucl. Instr. and Meth. B* **14** 515
[16] Cornelius K R, 2006 *Phys. Rev. A* **73** 032710
[17] Anholt R and Eischer J 1985 *Phys. Rev. A* **31** 3505
[18] Eichler J 1985 *Phys. Rev. A* **32** 112
[19] Meyerhof W E and Anholt R 1985 *Phys. Rev. A* **32** 3291
[20] Anholt R, Xu X -Y, Stoller Ch *et al* 1988*Phys. Rev. A* **37** 1105
[21] Stöhlker Th, Mokler P H, Brckert K *et al* 1994 *Nucl. Instr. and Meth. B* **87** 64
[22] Scheidenberger C, Geissel H, Stöhlker Th *et al* 1994 *Nucl. Instr. and Meth. B* **90** 90
[23] Stobbe M 1930 *Ann. Phys.* **7** 661
[24] McLaughlin J D and Hahn Y 1991 *Phys. Rev. A* **43** 1313
[25] Stöhlker Th, Kozhuharov C, Mokler P H *et al* 1995 *Phys. Rev. A* **51** 2098
[26] Bates D R and Grifffing G W 1955 *Phys. Soc. A* **68** 90
[27] Anholt R 1979 *Phys. Rev. A* **19** 1004
[28] Anholt R 1985 *Phys. Rev. A* **31** 3579
[29] Anholt R 1986 *Phys. Lett. A* **114** 126
[30] Anholt R Meyerholf W E, Gould H *et al* 1985 *Phys. Rev. A* **32** 3302
[31] Shevelko V P, Böhne D and Stöhlker Th 1998 *Nucl. Instr. and Meth. A* **415** 609
[32] Shevelko V P, Tolstikhina I Yu and Stöhlker Th 2001 *Nucl. Instr. And Meth. B* **184** 295
[33] Mahner E 2007 *Vacuum* **81** 727
[34] Bell M and Bell J S 1982 *Part. Accel.* **12** 49

# Electron-impact excitation of $2p^5 3l \rightarrow 2p^5 3l'$ line emission of Fe XVII

**D Kato, H A Sakaue, I Murakami, T Kato, N Nakamura[1], S Ohtani[1], N Yamamoto[2], and T Watanabe[3]**

National Institute for Fusion Science, Toki, Gifu 509-5292, Japan
[1]ILS, The Univ. of Electro-Commun., Chofu, Tokyo 182-8585, Japan
[2]ILE, Osaka Univ., Suita, Osaka 565-0871, Japan
[3]National Astronomical Observatory of Japan, Mitaka, Tokyo 181-8588, Japan

kato.daiji@nifs.ac.jp

**Abstract.**
We present theoretical investigations on the EUV line emission of Fe XVII in an electron-beam-ion-trap. Calculations of atomic data and the collisional-radiative model, used to study population kinetics of $2p^5 3l$ levels of Fe XVII, are described. Intensities of the strongest EUV lines at 204.6 Å and 254.8 Å are found to be 2 orders of magnitudes smaller than the x-ray line intensities at 15 Å and 17 Å. Significant enhancement of the EUV lines around 250-400 Å due to resonance excitation is demonstrated.

## 1. Introduction

Ne-like Fe (Fe XVII) is abundant in many astrophysical and laboratory plasmas because of its L-shell closed electronic configuration of the ground state. Fe XVII lines of extreme ultraviolet wavelengths (EUV) have been observed in solar flares [1] and beam-foil spectra [2]. The lines were identified as transitions among $2p^5 3l$ excited levels [2-4].

$2p^5 3p$ ($^1S_0$) $\rightarrow$ $2p^5 3s$ $(1/2, 1/2)_1$ transition is known as an electron-impact pumping scheme to produce soft x-ray or EUV lasers. Laboratory experiments with laser produce plasmas [5, 6] have, however, shown very small gain of this transition for Se XXV; other transitions in the $2p^5 3l$ levels exhibited the largest gain in the experiments. This result has been puzzling, since collisional excitation models predicted that the $2p^5 3p$ ($^1S_0$) $\rightarrow$ $2p^5 3s$ $(1/2, 1/2)_1$ line should have one of the highest intensities. Later, Doschek et al. [7] reported that observed line intensities of Fe XVII in a solar flare were in good agreement with intensities calculated, taking account collisional excitations from the ground state of Fe XVII. In the both (observed and calculated) spectra, 204.6 Å and 254.8 Å lines, identified as $2p^5 3p$ ($^1S_0$) $\rightarrow$ $2p^5 3s$ $(3/2, 1/2)_1$ and $\rightarrow$ $2p^5 3s$ $(1/2, 1/2)_1$, respectively, had the strongest and almost the same intensities. They concluded therefore that inaccurate collision excitation rates in the model should not be the cause of the puzzle.

However, new observation of a solar flare by means of the EUV Imaging Spectrometer on *Hinode* satellite [8] has shown difficulty in reconciling observed Fe XVII line intensities with available atomic data. It was reported [9] that the observed intensity of the 254.8 Å line was factor of 2 smaller than that calculated by using CHIANTI code [10]. Although the discrepancy may be ascribed partially to calibration issues, re-investigation of the relevant atomic data seems to be necessary for a better understanding.

In this paper, we describe excitation cross sections related to the EUV line emission of Fe XVII with emphases on resonance excitation effects and particularity of the $2p^53p$ ($^1S_0$) state. To the best of our knowledge, the EUV line emission of Fe XVII has not been studied thoroughly with an electron-beam-ion-trap (EBIT), while there are studies on the x-ray lines at 15-17 Å [11-13]. We therefore present line emissivity and intensity ratios, calculated for experimental conditions of an EBIT.

## 2. Atomic data and collisional-radiative model of EBIT

In this section, we briefly describe calculations of atomic data and the collisional-radiative model used to study population kinetics of $2p^53l$ levels of Fe XVII in the EBIT.

### 2.1. Atomic data

HULLAC code [14] was used to obtain level energies, collision strengths of electron-impact excitation, and radiative-decay and auto-ionization rates. To account for strong valence-core correlation known for the $2p^53l$ excited states, configuration interaction with $2s2p^6nl'$ states was included. Nevertheless, a large error remained in the calculated transition energy (wavelength) of $2p^53p$ ($^1S_0$) → $2p^53s$ $(3/2, 1/2)_1$ and → $2p^53s$ $(1/2, 1/2)_1$. By taking account core polarization effects, the transition energy was improved to some extent, however the effects made the other transition energies less accurate. In earlier studies [15], multi-configuration Dirac-Fock calculations in the extended average mode gave inaccurate level energies of the $^1S_0$ state. To improve this, a separate optimization of radial wavefunctions for this state was required, due to a particular radial dependence of this state. However, this pragmatic remedy can not be adopted with HULLAC code, since, in the HULLAC, every state of a given electronic configuration is represented by a common set of the radial wavefunctions obtained with a single parametric potential. For the sake of convenience, we just take correct transition energies from the NIST database [16].

Another feature of $2p^53p$ ($^1S_0$) is the large direct excitation cross section and then the small resonance excitation effect compared with the other states involved in the EUV line emission. This is clearly demonstrated in Fig. 1. The resonance excitation cross sections were obtained with a Gaussian energy distribution function using FWHM of 10 eV.

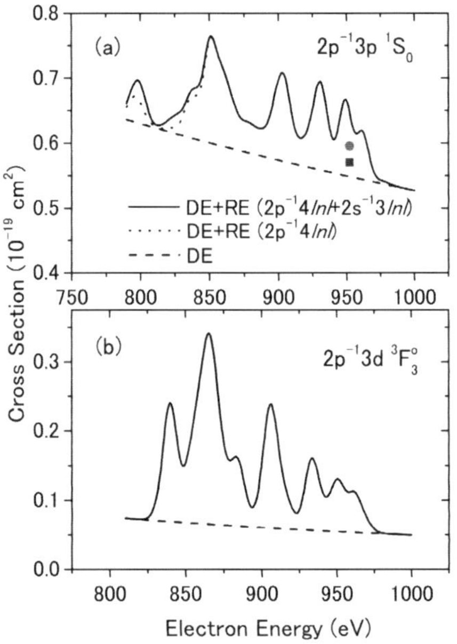


Fig.1 Excitation cross sections of (a) $2p^53p$ $^1S_0$ and (b) $2p^53d$ $^3F_3$ from the ground state $2p^6$ of Fe XVII. Solid circle is relativistic R-matrix calculations of Gupta et al. [17], and solid square relativistic distorted-wave calculations of Cornille et al. [18].

### 2.2. Collisional-radiative model of EBIT

In the EBIT, ions are excited with a quasi mono-energy electron beam. Recombination from higher ionization states can be avoided by setting the electron energy below the ionization threshold of Fe XVII (about 1260 eV). A collisional excitation model can therefore be used for the population kinetics of the excited levels. Because of low electron number densities ($N_e = 10^{10-11}$ cm$^{-3}$) in the EBIT, the corona model is also applied. In the quasi stationary condition, the fractional populations of the excited levels $n_i$ are written,

$$n_i = N_e(\alpha_{DE} + \alpha_{RE})n_g \Big/ \sum_{j<i} A_{ji} ,$$

where $\alpha_{DE}$ and $\alpha_{RE}$ are the direct excitation and resonance excitation rate coefficients, respectively, $A_{ji}$ is the Einstein's coefficients of spontaneous decay, and the population of the ground level $n_g = 1$.

## 3. Line emissivity and intensity ratio

In Fig.2, the line emissivity divided by the electron number density, $\Delta E_{ji} A_{ji} n_i / N_e$, is plotted for the electron beam energy of 850 eV. The lines in x-ray region are also shown for comparison. The strongest EUV lines at 204.6 Å and 254.8 Å have intensities 2 orders of magnitudes smaller than the x-ray line intensities at 15 Å and 17 Å. It suggests that the EUV lines are generally more difficult to observe in laboratory plasmas of low electron densities like the EBIT. It is seen in the figure that the two lines at 204.6 Å and 254.8 Å have almost the same intensities, because of strong intermediate coupling in the lower levels, for the two lines is the electric-dipole branching ratio close to unity. It is noted that many other EUV lines also appear around 250-400 Å as strong as the two lines. The resonance excitation contribution dominates these line emissions. The resonance excitation effect manifests itself more clearly in the energy dependence of line intensity ratios.

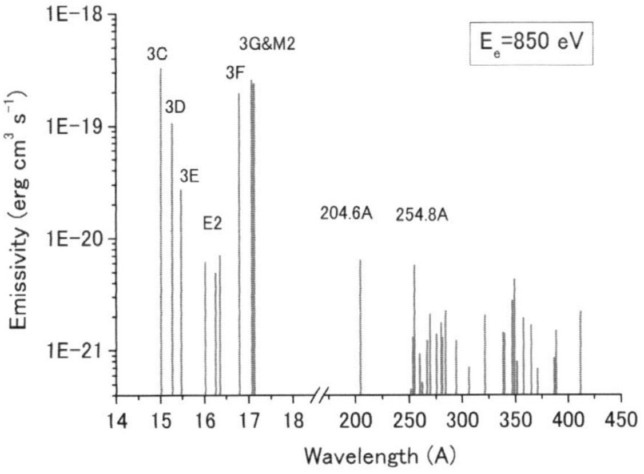


Fig. 2 Line emissivity of Fe XVII in EBIT at electron energy of 850 eV.

The line intensity ratios of 269.4 Å /254.8 Å are plotted at electron energies for 800-950 eV in Fig.3. In the figure, the intensity ratios, calculated by accounting for the direct excitation only, are also shown for comparison. It is quite obvious that the resonance excitation enhances the intensity ratio significantly. This is easily understood by recalling that the resonance excitation effect of the upper level of the 269.4 Å line, $2p^5 3d$ ($^3F_3$), is much larger than that of the upper level of the 254.8 Å line, $2p^5 3p$ ($^1S_0$) (see Fig. 1). Due to the narrow energy width of the electron beam, the intensity ratio shows rapid variation along the electron energy. At the highest peak, the intensity ratio is enhanced by about a factor of 3. It is noted that a similar enhancement of the intensity ratio of x-ray lines at 17 Å to the 15 Å line (3F+3G+M2 and 3C, respectively, as designated in Fig. 2) has been reported in earlier measurements with an EBIT [11].

## 4. Summary

Although the EBIT is know as the one of the best apparatus to examine detailed electron collision processes of highly charged ions, no observation of the EUV lines of Fe XVII has been reported with

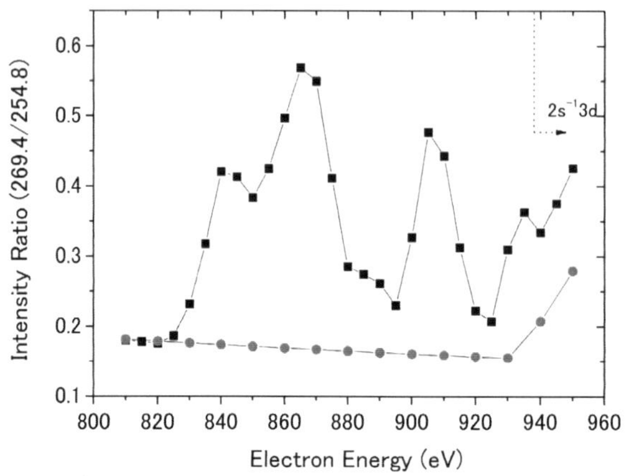


Fig. 3 Line intensity ratio of 269.4 Å /285.4 Å. Solid circles are results of direct excitation only. Excitation threshold of $2s2p^63d$ state is indicated by dotted line.

the EBIT to date. We investigated theoretically the EUV line emission in the EBIT by using the collisional excitation model with atomic data from HULLAC code. The intensities of the strongest EUV lines at 204.6 Å and 254.8 Å were found to be 2 orders of magnitudes smaller than the x-ray line intensities at 15 Å and 17 Å, suggesting that the EUV lines would be more difficult to observe with the EBIT. The present results demonstrate significant enhancement of other EUV lines in 250-400 Å due to resonance excitation.

**Acknowledgment**
This work is supported by the NIFS/NINS project of Formation of International Network for Scientific Collaborations.

**References**

[1] K.P Dere, Astrophys. J. **221**, 1062 1978.
[2] J.P. Buchet et al., Phys. Scr. **31**, 364 1985.
[3] C. Jupen, Mon. Not. R. Astr. Soc. **208**, short communication 1 1984.
[4] U. Feldman, G.A. Doschek and J.F. Seely, Mon. Not. R. Astr. Soc. **212**, 41 1985.
[5] R.A. London et al., J. Phys. B **22**, 3363 1989.
[6] C.J. Keane et al., J. Phys. B **22**, 3343 1989.
[7] G.A. Dochek, U. Feldman and A.K. Bhatia, Phys. Rev. A **43**, 2565 1991.
[8] http://solar-b.nao.ac.jp/index_e.shtml.
[9] H.P. Warren, U. Feldman and C.M. Brown (private communications).
[10] For example, E. Landi et al., Astrophys. J. Suppl. Ser. **162**, 261 2006.
[11] P. Beiersdofer et al., Astrophys. J. **576**, L169 2002.
[12] J.M. Laming et al., Astrophys. J. **545**, L161 2000.
[13] G.V. Brown et al., Phys. Rev. Lett. **96**, 253201 2006.
[14] A. Bar-Shalom et al., *The HULLAC Package Computer Set of Codes for Atomic Structure and Processes in Plasmas*, unpublished.
[15] J.A. Cogordan et al., Phys. Scr. **31**, 545 1985.
[16] http://physics.nist.gov/PhysRefData/ASD/index.html.
[17] G.P. Gupta, N.C. Den and A.Z. Msezane, At Data Nucl. Data Tables **74**, 257 2000.
[18] M. Cornille et al., Astron. Astrophys. Suppl. Ser. **105**, 77 1994.

# Fragmentation of meso-Tetraphenyl iron(III) porphyrin chloride (FeTPPCl) in collisions with $Kr^{8+}$ ions

**B Li[1,2], G Gutierrez[1], L Chen[1], R Brédy[1], J Bernard[1], G Montagne[1], X Ma[2], V Bernigaud[3], B Manil[3], L Maunoury[3], B Huber[3] and S Martin[1]**

[1] Université de Lyon, Université Lyon 1, CNRS, LASIM UMR5579, Bat Orion 43 Bd du 11 Novembre 1918, F69622, Villeurbanne, France

[2] Institute of Modern Physic, Chinese Academy of Sciences, Lanzhou 730000, China

[3] CIMAP (CEA-CNRS-ENSICaen-UCBN), Bd Henri Becquerel (BP 5133), F-14070 Caen Cedex, France

E-mail: chen@univ-lyon1.fr

**Abstract**. Fragmentation of FeTPPCl ($C_{44}H_{28}N_4FeCl$) induced by $Kr^{8+}$ impact at 80 keV has been investigated. Fragmentation schemes have been studied for different charge states of parent ions (r=1-3). The dominant low-energy-cost fragmentation channel is the emission of neutral chlorine independently to the charge of the molecule. For the monocharged FeTPPCl parent ions, a minor channel corresponding to the emission of a chlorine anion is observed with a measured rate of about 3% ± 1%. For the dication molecule, the evaporation of a neutral phenyl molecule is the dominant channel, whereas for the trication molecule the dominant channel is the asymmetrical fission by emission of a charged phenyl. These processes are discussed in terms of charge distribution of the molecule.

## 1. Introduction

Porphyrins are the most important and widespread class of natural pigment [1]. A large variety of tetrapyrrolic pigments exist with the similar ligand core, but differ in the metal center and the side groups attached to the porphyrin rings. The metal center can be an iron (heme), a magnesium (chlorophyll) or cobalt (vitamin B12). Porphyrins are important in numerous processes in biology, chemical catalysis and as geological markers. They are also used in photodynamic tumor treatment [2]. Single electron transfer reactions such as photosynthesis have been investigated using synthetic porphyrin. With their symmetry and remarkable structure, metal porphyrins have also large theoretical interests for physicists and chemists [3]. From an experimental point of view tetra phenyl porphyrin can be considered as a model system. This molecule has a robust macro cycle nucleus and can be easily prepared with different group surrounding the cycle (see Figure 1). After ionization, the positive charge is localized in the extended $\prod$ system of the macro cycle and the fragmentation occurs mainly by emission of molecules fixed on the ring system as well observed in many experiments via electron [4], photon [5], or ion [6] impact.

In the present work, we study the interaction between slow $Kr^{8+}$ multicharged ions and the FeTPPCl molecule. By coincidence detection of recoil ions, ejected electrons and final charge state of the scattered projectile, the fragmentation schemes of $FeTPPCl^{+}$, $FeTPPCl^{2+}$ and $FeTPPCl^{3+}$ are measured precisely. For FeTPPCl at single charge state, the branching ratios have been measured for the emission of neutral and anion chloride. For $FeTPPCl^{2+}$ and $FeTPPCl^{3+}$ ions, the emission of neutral phenyl and cation phenyl molecules are also measured.

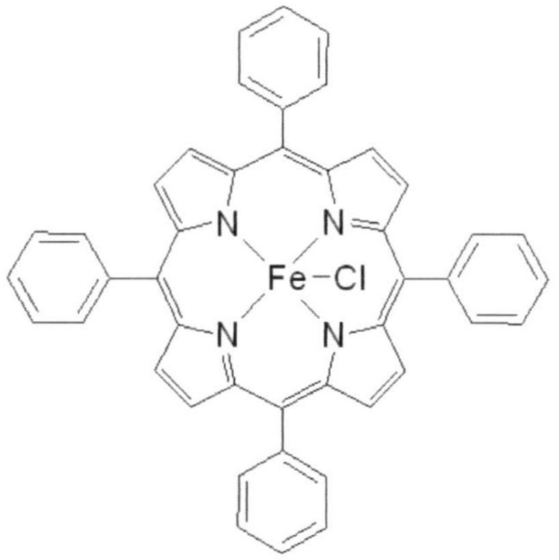


**Figure 1.** Structure of the tetraphenyl iron porphyrin chloride molecule.

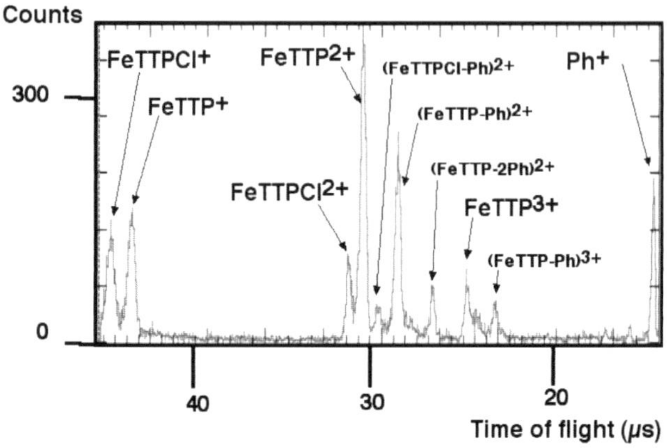


**Figure 2.** TOF spectra obtained in coincidence with the $Kr^{7+}$ outgoing projectile (s=1).

## 2. Experimental set-up

The experimental arrangement has been described in detail in ref. [7,8]. A pure FeTPPCl powder was evaporated from an oven heated at 350°C. The FeTPPCl jet crossed the Krypton ion beam delivered by an ECR ion source at the ARIBE Facility (GANIL, Caen, France) at 80 keV energy. Downstream from the interaction region, the scattered projectile ions were analyzed by a cylindrical electrostatic analyzer (R=200 mm) and detected with a channeltron. The charged FeTPPCl or fragments were extracted from the collision region and then analyzed with a Time of Flight (TOF) spectrometer equipped with a multi-anode. The electrons ejected during a collision event were accelerated to 20 keV and detected by a PIPS (silicon surface barrier detector) in coincidence with the other products in an event-by-event mode. Recoil ion spectra were recorded in coincidence with the detection of a $Kr^{(8-s)+}$ outgoing projectile (s is the number of electrons stabilized on the projectile) and the ejected electrons. Here, all spectra were obtained with the selection of $Kr^{7+}$ scattered projectiles (s=1) corresponding to collisions at large impact parameters. Under these conditions, a few electrons (typically 1 to 3) were involved in the charge exchange process associated with low excitation energy deposition on the target. The electron collection and detection efficiencies have been optimized and measured by the standard procedure [9,10]. Using an Argon gas target, the one-electron-detection efficiency has been found to be 96% in the following reaction: $Kr^{8+} + Ar \dashrightarrow Kr^{7+} + Ar^{2+} + 1e^{-}$.

## 3. Results and discussions

A typical recoil ion spectrum in coincidence with the detection of $Kr^{7+}$ (s=1) outgoing projectiles is presented in figure 2. Intact molecules are observed for singly and doubly charged states, $FeTPPCl^{+}$ and $FeTPPCl^{2+}$. Heavy charged fragments such as $FeTPP^{q+}$, $(FeTPP\text{-}Ph)^{q+}$ and $(FeTPP\text{-}2Ph)^{q+}$ are also identified, where Ph stands for a phenyl group. Apart from the Phenyl peak, no smaller fragments resulting from multifragmentation of the parent FeTPPCl molecules are observed on the spectrum except peaks corresponding to collisions with the background gas (not shown on the spectrum). This observation shows that under the present experimental conditions, via long distance electron capture process, multicharged molecules with relatively low excitation energy are prepared allowing only the emission of light fragment. In coherence with other experiments [4,5,6], the dominant low energy-cost channel for the three charge states of FeTPPCl ($FeTPPCl^{+}$, $FeTPPCl^{2+}$ and $FeTPPCl^{3+}$) is the emission of the Chlorine atom leading to the intense peaks $FeTPP^{+}$, $FeTPP^{2+}$ and $FeTPP^{3+}$. The small peak around 29.5 μs on the TOF spectrum has never been observed in previous experiments. It is attributed without ambiguity to $(FeTPPCl\text{-}Ph)^{2+}$ corresponding to the loss of a phenyl group directly from the FeTPPCl parent.

From the fragment ions $(FeTPP)^{2,3+}$, we observed mainly the emission of one or two Phenyl groups. The emission of a Benzene molecule $C_6H_6$ from $(FeTPP)^{+}$ fragment ions has been differentiated from the loss of a $C_6H_5$ fragment due to a simple C-$C_6H_5$ bond cleavage in the paper [4]. In our work, the mass resolution is not good enough to assign exactly the fragment ions with a precision of one unit of atomic mass. The peak corresponding to the emission of a phenyl group from singly charged $(FeTPP)^{+}$ ions is merely observable, while that from doubly charged $(FeTPP)^{2+}$ ions is the most prominent one of the spectrum. This indicates that the excitation energy of singly charged $(FeTPP)^{+}$ is so weak that any further dissociation is not possible in our experimental time window after the loss of the Chlorine atom from $FeTPPCl^{+}$. This is expected in the interaction between multicharged ions and molecules where one-electron capture process occurs at long distance and the energy deposited on the target is very low. On the other hand, doubly or triply charged FeTPPCl molecules are produced in two, three or a larger number of electron capture processes at shorter distance. A larger amount of energy is deposited on the parent molecular ions $FeTPPCl^{2+}$ and $FeTPPCl^{3+}$ leading to intense peaks of lower mass fragments, $(FeTPP\text{-}Ph)^{2+}$, $(FeTPP\text{-}2Ph)^{2+}$ and $(FeTPP\text{-}2Ph)^{3+}$. These peaks can be attributed to different fragmentation channels. For example, the peak $(FeTPP\text{-}Ph)^{2+}$ might be either due to the evaporation of a neutral Phenyl from doubly charged $(FeTPP)^{2+}$ or due to the asymmetrical fission by the loss of a monocharged $Ph^{+}$ from $(FeTPP)^{3+}$.

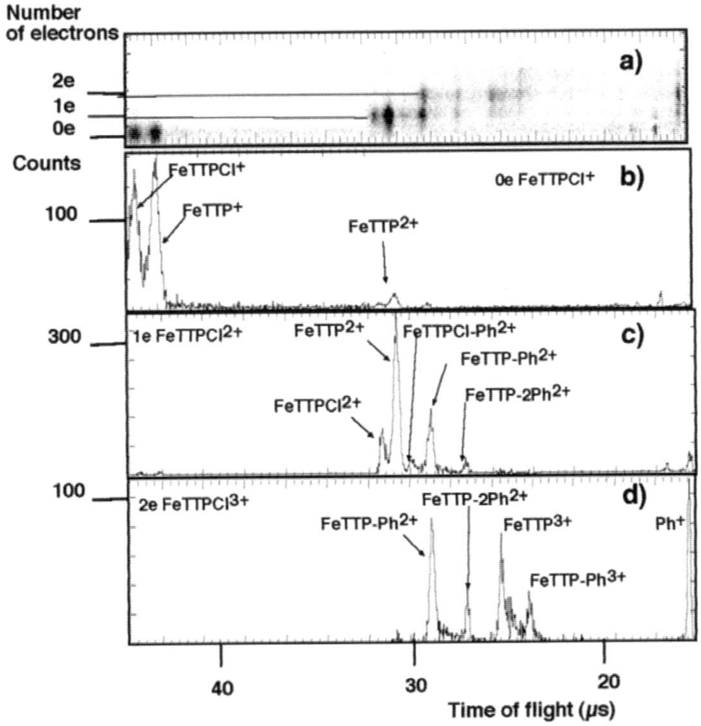


**Figure 3. a)** Two-dimensional spectrum: the TOF of the heaviest fragment of each event is plotted along the horizontal axis and the amplitude of the PIPS detector giving the number of detected electrons is plotted along the vertical axis. **b)** Partial TOF spectrum obtained by projection of the spectrum a) onto the horizontal axis with the selection of 0 electron. **c)** Idem as b) but with the selection of 1 electron. **d)** Idem as b) but with the selection of 2 electrons.

In order to measure precisely the branching ratios of different fragmentation channels related to a well-defined charge state of parent ions, we measured the number n of ejected electrons in coincidence with the recoil ions. Figure 3a displays a two-dimensional spectrum plotting the TOF of the heaviest fragment of each event (horizontal axis) and the amplitude of the PIPS detector signal (vertical axis). From this signal, we extracted information on the number of ejected electrons [8]. Figure 3b, 3c and 3d show the partial projection onto the horizontal axis of Figure 3a with the selection of n=0, 1 and 2 detected electrons. Based on the electron number conservation rule, r=n+s with s=1, the charge r of the parent molecules $FeTPPCl^{r+}$ related to the three fragmentation spectra is determined easily to be r=1, 2 and 3 for figure 3b, 3c and 3d respectively. For the $FeTPPCl^{+}$ parent ions (Figure 3a), the main decay channel is shown by the peak $(FeTPP)^{+}$ due to the emission of neutral chlorine. The small doubly charged peak, attributed to $FeTPP^{2+}$, could result from two channels. One involves a single electron transfer process followed by the emission of chlorine anion,

$Kr^{8+} + FeTPPCl \rightarrow Kr^{7+} + FeTPPCl^{+} \rightarrow Kr^{7+} + FeTPP^{2+} + Cl^{-}$.

The other is related to double electron transfer process, where one electron is lost by the projectile via autoionisation and it is missed from the detection due to the limited electron collection and detection efficiency,

$Kr^{8+} + FeTPPCl \rightarrow Kr^{6+} + FeTPPCl^{2+} \rightarrow Kr^{7+} + 1e^{-} + FeTPP^{2+} + Cl$.

After corrections of these spectra, taking into account of the electron collection and detection efficiency, as discussed in a previous paper [8], the number of counts of all observed dissociation channels related to parent $FeTPPCl^{r+}$ ions with r=1, 2 and 3 are given in the table 1. For singly charged parent molecule, $FeTPPCl^{+}$, the branching ratio between the emission of anion and neutral chlorine has been estimated to be 3%±1%. This surprisingly high negative ion ejection rate from a positively charged molecule can be explained qualitatively by the high electronegativity of the chlorine and by the particular ionic bond between iron and chlorine compared to the other valence bonds. The mobility of charge between the bonds like C-C or C-N is probably very different from that on the ionic bond. On the Fe-Cl ionic bond, the negative charge distribution should be situated near the chlorine atom. These arguments can explain why the chlorine cation was not observed in our experiment (s=1) whatever the initial charge state of parent ions (q=1-3).

In figure 3c, the peaks correspond mainly to the evaporation of neutral chlorine and Phenyl group from $FeTPPCl^{2+}$ parent ions. The number of count of singly charged fragment peaks $(FeTPP\text{-}Ph)^{+}$, $(FeTPP\text{-}2Ph)^{+}$ resulting from the emission of a charged phenyl are so small that the two peaks can hardly get out of the noise background. However, these channels are identified without ambiguity in coincidence spectrum between charged fragments $(FeTPP\text{-}Ph)^{+}$ and $Ph^{+}$ or $(FeTPP\text{-}2Ph)^{+}$ and $Ph^{+}$ (not shown here). The count numbers for these weak channels are also reported in table 1. From these measurements, the branching ratio for the emission of a charged fragment from $(FeTPP)^{2+}$ is estimated to 2.0%±0.5%. The evaporation channel, measured to be 98%, is the dominant fragmentation channel for the $FeTPPCl^{2+}$.

**Table 1.** Number of counts for stable and fragmentation channels from $FeTPPCl^{+}$, $FeTPPCl^{2+}$ and $FeTPPCl^{3+}$ parent ions after corrections taking into account of the electron collection and detection efficiency.

| Parent ions | stable | -Cl° | -Cl⁻ | -Ph° | -Cl°-Ph° | -Cl°-2Ph° | -Cl°-Ph⁺ | -Cl°-Ph⁺-Ph° |
|---|---|---|---|---|---|---|---|---|
| $FeTPPCl^{+}$ | 2750±50 | 3240±60 | 100±10 | | | | | |
| $FeTPPCl^{2+}$ | 1420±40 | 5720±80 | | 540±20 | 1960±40 | 240±15 | 34±6 | 17±4 |
| $FeTPPCl^{3+}$ | <10 | 840±30 | | | 330±20 | | 1300±40 | 300±20 |

On the figure 3d, the spectrum shows peaks corresponding to the fragmentation of $FeTPPCl^{3+}$ parent ions. The peak $(FeTPPCl\text{-}Ph)^{2+}$ noticed earlier in figure2, corresponding to the emission of a phenyl directly from the parent FeTPPCl molecule is observed only in Figure3c, but is absent in figure3d. This comparison confirms that the fragment ion $(FeTPPCl\text{-}Ph)^{2+}$ is the result of the loss of a neutral phenyl from a doubly charged FeTPPCl. The loss of a charged phenyl, $Ph^{+}$, from $FeTPPCl^{3+}$ is a negligible channel. In figure3d, intact $FeTPPCl^{3+}$ is not observed. $(FeTPP)^{3+}$ and $(FeTPP\text{-}Ph)^{3+}$ peaks correspond to the loss of Cl° and the loss of Cl° followed by the loss of a Ph° respectively. $(FeTPP\text{-}Ph)^{2+}$ and $(FeTPP\text{-}2Ph)^{2+}$ peaks correspond to the emission of a $Ph^{+}$ and a $Ph^{+}$ followed by the emission of a neutral Ph° from $(FeTPP)^{3+}$ respectively. From the table 1, the branching ratio for the asymmetrical fission of $(FeTPP)^{3+}$ can be estimated to 83%±4%. This high asymmetrical fission probability can be interpreted qualitatively by making an assumption within the charge distribution on the intermediate parent molecule $(FeTPP)^{3+}$. If we consider a good charge mobility on the molecule, the three positive charges should be distributed rather on the ring of FeTPP and more precisely on the four phenyl-groups due to the coulomb repulsion. To distribute statistically 3 charges on 4 Phenyl-groups, the probability that the group of phenyl is ejected with a charge is about 75%. This rough estimation is not too far from the experimental value of 83%.

## 4. Conclusion

As a conclusion, by analyzing the fragmentation TOF spectra of $FeTPPCl^{+}$, $FeTPPCl^{2+}$ and $FeTPPCl^{3+}$ parent molecules obtained with the $Kr^{8+}$ projectile, we have shown that the main low-energy-cost fragmentation channel is the emission of neutral chlorine independent of the charge state of the FeTPPCl molecule. By measuring the number of ejected electrons during the collision in coincidence with the recoil ions, the emission of chlorine anion from the monocharged FeTPPCl was identified and the branching ratio was measured to be 3 %. In future experiments, we plan to bring more direct evidence of this channel by analyzing the negative chlorine ion using a TOF tube in coincidence with the analysis of the corresponding positive ion $(FeTPPCl\text{-}Cl)^{2+}$. For $FeTPPCl^{2+}$ and $FeTPPCl^{3+}$ parent ions, the evaporation and asymmetrical fission processes have been measured showing that the emission of light charged fragment is always the phenyl group and never the monocharged chlorine cation. These observations were explained by different types of bounds for this molecule.

**Acknowledgments**
These experiments have been performed at the ARIBE-Facility, a part of the ITSLEIF Infrastructure, funded under the European Project n°026015

## References

[1] Battersby A R 1993 *Pure Appl. Chem.* **65** 1113
[2] Doiron D R and Gomer C J (Eds) 1984 *Porphyrin Localisation and treatment of tumor* (New York: A. R. Liss)
[3] Kadish K M, Smith K M and Guillard R (Eds) 2003 *The porphyrins Handbook* vols 1-20, (Oxford: Elsevier Science)
[4] Feil S, Winkler M, Sulzer F, Ptasinska S, Denifl S, Zappa F, Kräutler B, Märk T D and Scheier P 2006 *Int. J. Mass Spect.* **255-256** 232
[5] Dale M J, Costello K F, Jones A C and Langridge-Smith P R R 1996 *J. of Mass Spect.* **31** 590
[6] Bernigaud V, Manil B, Maunoury L, Rangama J and Huber B A 2008 *Eur. Phys. J.* D (in print)
[7] Martin S, Chen L, Bernard J, Brédy R, Buchet-Poulizac M C, Ma X and Wei B 2006 *Europhys. Lett.* **74** 985
[8] Martin S, Chen L, Denis A, Brédy R, Bernard J and Désesquelles J 2000 *Phys. Rev. A* **62** 022707
[9] Aumayr F, Lakits G and Winter H 1991 *App. Surf. Sci.* **47** 139
[10] Martin S, Chen L, Denis A and Désesquelles J 1999 *Phys. Rev. A* **59** R1734

# Irradiation of Water Ices by 2 keV Carbon Ions

**C. A. Hunniford[1], A. Dawes[2], D. Fulvio[3], B. Sivaraman[2], T.L. Merrigan[1], R. W. McCullough[1], N. J. Mason[2] and M.E. Palumbo[4]**

[1] Centre for Plasma Physics, School of Mathematics and Physics, Queen's University of Belfast, BT7 1NN, Northern Ireland

[2] Department of Physics and Astronomy, The Open University, Milton Keynes, MK7 6AA, United Kingdom

[3] Universita' degli Studi di Catania, Italy

[4] INAF - Osservatorio Astrofisico di Catania,Via Santa Sofia 78, I-95123 Catania, Italy

c.a.hunniford@qub.ac.uk

**Abstract.** In this paper, we report on the results of experimental studies of the irradiation of pure water ices using 2 keV $^{13}C^{+}$ and $^{13}C^{2+}$ ions. Studies have been carried out at two temperatures (30 and 90 K) and the influence of the different morphologies at these temperatures has also been studied. With singly charged ions, the formation of $^{13}CO_2$ was observed to be a strongly dependent upon morphology and showed a weaker dependence upon temperature. With doubly charged ions, the dependence upon temperature was significantly stronger. This is explained by enhanced production of reactive species resulting from the additional potential energy contained within the doubly charged ions. Increased temperature provides mobility to these species which may then yield additional $^{13}CO_2$.

## 1. Introduction

Ion processing plays an important role in the chemical and physical modification of ice surfaces in astrophysical environments. Magnetospheric ions surrounding the Gas Giants in the outer Solar System impinge upon and modify the icy satellite surfaces creating new chemical species [1], incorporating elements not originally present in the local ice composition [2]. These may be formed via ion induced dissociative or reactive processes.

Carbon is the fourth most abundant element in the interstellar medium (ISM) and plays an important role in many chemical networks, both in the ISM and in our own Solar System. Most of the carbon ions present in planetary magnetospheres, in our solar system, are thought to originate from the slow component of the solar wind [2] with charge states ranging from $4^{+}$ to $6^{+}$ [3] and energies in the low keV range [4]. However, surface sputtering and ionization of the atmospheres of local satellites can give rise to carbon ions with different charge state contributions. For example, the extraction of carbon bearing species (e.g. $CH_4$ from the atmosphere of Titan), may lead to the enrichment of the Saturnian magnetospheric plasma with carbon ions and carbon containing molecular ions. Similarly, the detection of $CO_2$ on the surface of the Galilean moons (Io, Europa, Ganymede and Callisto) and the surface of the Uranian satellite Ariel, suggests the presence of $CO_2^+$, $CO^+$ and $C^+$ ions in the local planetary magnetospheres [5]. Results from Cassini's Cosmic Dust analyser, on the Jovian and Saturnian stream particles, have revealed $C^+$ as a dominant component in the mass spectra [6]. The Cassini instruments will be the first to measure the charge states of low energy ions (2 – 30 keV/q) and fill the gap in the ion energy spectrum between 4 and 50 keV [7].

The majority of experiments have been carried out using ions with energies greater than 30 keV where electronic stopping (due to inelastic collisions) dominates the energy loss process. However, on a satellite surface where ice crusts are thick, an energetic ion will progressively lose energy via electronic processes until it reaches energies where nuclear stopping (due to elastic collisions) dominates and dissociative [8] and chemical processes may occur. Above about 30 keV, electronic stopping becomes increasingly larger than nuclear stopping. Thus it is expected that secondary processes involving secondary electrons and excited species will dominate reaction schemes.

In the present work we have investigated the interaction of low energy (2 keV) singly and doubly charged carbon ions with amorphous ice films at 30 K and 90 K. We also explore the role of the potential energy carried by doubly charged projectiles in the regime where nuclear stopping dominates.

## 2. Experimental Approach

Detailed descriptions of the irradiation chamber and ion beam arrangement have been given in previous publications [9,10] so that only a brief summary will be given here. Water ice samples were prepared in situ in an ultra-high vacuum chamber by directed vapour deposition onto a 20 mm diameter transmitting zinc selenide (ZnSe) substrate which was in thermal contact with a continuous flow open cycle cryostat. Beams of $^{13}C^{q+}$ (q = 1 or 2) ions were extracted from a 9.0 -10.5 GHz Electron Cyclotron Resonance (ECR) ion source.

Analysis of the irradiated water ice was carried out using in situ infrared spectroscopy derived from a purged Thermo-Nicolet Nexus 670 FTIR spectrometer. Changes in the infrared absorption spectrum of the ice were monitored as a function of ion fluence (ions cm$^{-2}$).

## 3. Results

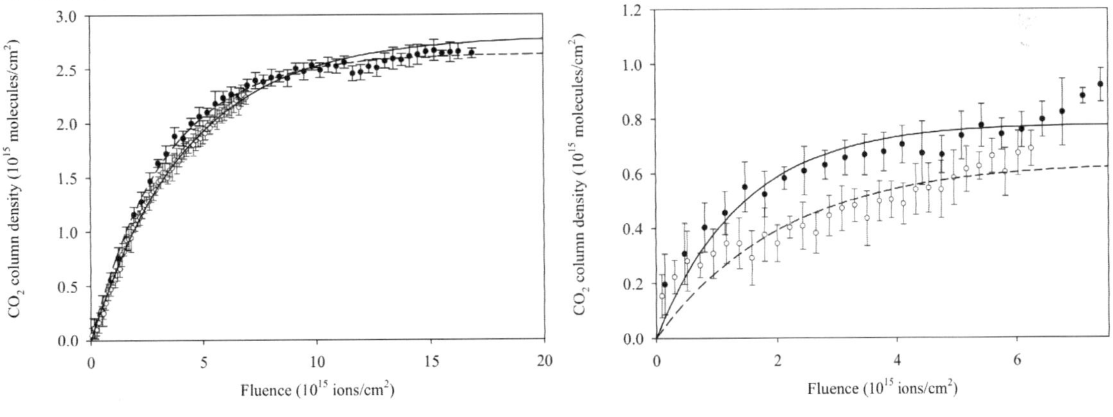


Figure 1 $CO_2$ yield, induced by 2 keV • $C^+$ (fit shown by solid line) and ○ $C^{2+}$ (fit shown by dashed line), in $H_2O$ ice (a) deposited and irradiated at 30 K ($30_{30}$) and (b) deposited at 90 K and irradiated at 30 K ($90_{30}$).

Figure 1(a) shows the yield of $^{13}CO_2$ in a water ice formed and irradiated at 30 K (denoted $30_{30}$) as a function of the total ion fluence (for both $C^+$ and $C^{2+}$). At low fluences, the $CO_2$ yield increased rapidly before beginning to saturate at fluences above several $10^{15}$ ions cm$^{-2}$. Beyond a fluence of ~$10^{16}$ ions cm$^{-2}$, little or no increase was observed in the $CO_2$ yield. This effect had been observed previously with both $CO_2$ [10,11] and $H_2O_2$ [12].

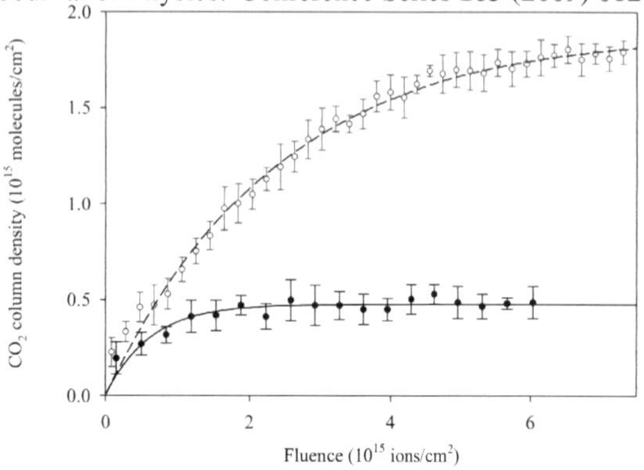


Figure 2 $CO_2$ yield in 90K water ice ($90_{90}$) induced by 2 keV • $C^+$ (fit shown by solid line) and ○ $C^{2+}$ (fit shown by dashed line).

Figure 1(b) shows the $^{13}CO_2$ yields in an ice formed at 90 K and irradiated at 30 K ($90_{30}$) – the same irradiation temperature as Figur but with the less porous morphology obtained when depositing at 90 K [10]. The $CO_2$ yields with $C^+$ were marginally higher than with $C^{2+}$.

Figure 2 shows the yield of $CO_2$ in an ice formed and irradiated at 90 K ($90_{90}$) – resulting in a low porosity ice. In this plot, a significant difference between $C^+$ and $C^{2+}$ irradiations was observed with the $CO_2$ yield higher for $C^{2+}$ irradiation.

## 4. Discussion

### *4.1. Morphological effects*

Comparing the $C^+$ results in figures 1(a) and 1(b), it can be seen that $CO_2$ yield was significantly greater with $30_{30}$ (30 K morphology) than $90_{30}$ (90 K morphology). The enhanced yield with the 30 K morphology was a result of the increased porosity of this ice and thus larger surface area on which reactions may take place– a result which agrees with previous experiments [10,12].

### *4.2. Temperature and Potential Energy Effects*

Ions approaching and entering the icy surface will capture electrons and deposit their potential energy close to the ice surface beyond which the projectile is likely to be neutralised. As a result of the initial kinetic energy of the projectile, it will continue to travel through the ice after it has been neutralised, progressively losing energy until it comes to rest. As a result of this, two spatially separated regions exist within the ice. In the absence of migration between these regions, the yield of $CO_2$ would be independent of the ion charge state since $CO_2$ would be formed in the chemical region whilst any effects due to the potential energy of the initial ion would be isolated in the capture region.

Considering the experiments carried out on the $30_{30}$ ice, it is clear that the $CO_2$ yield is independent of the ion charge state (figure 1(a)). Experiments performed on the $90_{30}$ ice (figure 1(b) - irradiation temperature again 30 K) demonstrate a similar trend with little difference observed between $C^+$ and $C^{2+}$ irradiations.

In contrast to irradiations carried out at 30K, when the irradiation was carried out at 90 K, ($90_{90}$ ice – 2), a notable difference was observed in the $CO_2$ yield for $C^+$ and $C^{2+}$ irradiation. Since little difference was observed between the different ions when this ice morphology was irradiated at 30 K, it is likely that the result at 90 K was a consequence of increased mobility of reactive species at the higher irradiation temperature. This additionally implies an enhanced yield of reactive species in the capture region when the ice is irradiated with a $C^{2+}$ ion. This is as would be expected when considering the potential energy released by the $C^+$ and $C^{2+}$ ions upon single electron capture. Such events would release 11.26 eV and 24.38 eV for $C^+$ and $C^{2+}$ respectively. Considering the energetics of ice phase water [13], it is clear that $C^{2+}$ deposits sufficient energy into the $H_2O$ molecules to exothermically access a dissociative ionization channel whereas the energy deposited by $C^+$ is insufficient to do so. It is therefore expected that $C^{2+}$ irradiation will result in more dissociation and fragmentation within the 'capture' region than will be the case with $C^+$ [14]. At 30 K, these products are isolated from the 'chemical region' and thus do not modify the $CO_2$ yield. At 90 K, the products of the capture processes (and products formed in the chemical region) are able to migrate and subsequently react yielding additional $CO_2$.

## 5. Conclusion

The results presented in this article have demonstrated several important effects which determine molecular formation within water ices.

- Higher charge state ions result in more $CO_2$ formation at irradiation temperatures of 90 K due to more fragmentation in the 'capture region'.
- At 90 K, the mobility of species which may subsequently yield $CO_2$ is higher than it is at 30 K.
- The more porous structure of 30 K ices results in a larger $CO_2$ yield which is thought to result from the enhancement of surface reactions.

However several important questions remain to be answered. The chemical reactions leading to the formation (or destruction) of $CO_2$ were not able to be confirmed. This was further complicated by the absence of species such as CO - expected either as precursors in the formation, or products from the destruction. Whilst the work demonstrated that some species, present or formed in the ice, are immobile at the lowest temperatures (30 K), not all species will be confined and thus may escape the ice as rapidly as they are formed or they may simply not be visible to the IR spectroscopic technique.

## Acknowledgements

This experiment has been performed at the QU-LEIF facility, part of the distributed LEIF infrastructure. The support received by the European Project ITS LEIF (RII3/026015) is gratefully acknowledged.

## References

[1] R. A. Baragiola, *Planetary and Space Sience*, 2003, **51**, 953.
[2] M. L. Delitsky and A. L. Lane, *Journal of Geophysical Research (Planets)*, 2002, **107**, 3-1.
[3] R. von Steiger, N. A. Schwadron, L. A. Fisk, J. Geiss, G. Gloeckler, S. Hefti, B. Wilken, R. F. Wimmer-Schweingruber and T. H. Zurbuchen, *Journal of Geophysical Research*, 2000, **105**, 27217-27238.
[4] J. L. Phillips, S. J. Bame, A. Barnes, B. L. Barraclough, W. C. Feldman, B. E. Goldstein, J. T. Gosling, G. W. Hoogeveen, D. McComas, J. M. Neugebauer and S. T. Suess, *Geophys. Res. Lett*, 1995, **268**.
[5] E. C. Sittler, R. E. Johnson, S. Jurac, J. D. Richardson, M. McGrath, F. Crary, D. T. Young and J. E. Nordholt, *Journal of Geophysical Research (Space Physics)*, 2004, **109**, 1214.
[6] S. Kempf, R. Srama, F. Postberg, M. Burton, S. F. Green, S. Helfert, J. K. Hillier, N. McBride, J. A. M. McDonnell, G. Moragas-Klostermeyer, M. Roy and E. Grün, *Science*, 2005, **307**, 1274-1276.
[7] N. Krupp, *Space Science Reviews*, 2005, **116**, 345-369.
[8] B. Seredyuk, R. W. McCullough, H. Tawara, H. B. Gilbody, D. Bodewits, R. Hoekstra, A. G. G. M. Tielens, P. Sobocinski, D. Pesic, R. Hellhammer, B. Sulik, N. Stolterfoht, O. Abu-Haija and E. Y. Kamber, *Physical Review A*, 2005, **71**, 22705.
[9] N. J. Mason, A. Dawes, P. D. Holtom, R. J. Mukerji, M. P. Davis, B. Sivaraman, R. I. Kaiser, S. V. Hoffmann and D. A. Shaw, *Faraday Discussions*, 2006, **133**, 311-329.
[10] Anita Dawes, Adam Hunniford, Philip D Holtom, Robin J Mukerji, Robert W McCullough and Nigel J Mason, *Phys. Chem. Chem. Phys,* 2007, **9** 2886-2893.
[11] Michael Davis, Ph.D. Thesis, 2006, The Open University
[12] M. J. Loeffler, U. Raut, R. A. Vidal, R. A. Baragiola and R. W. Carlson, *Icarus*, 2006, **180**, 265-273.
[13] N. Mårtensson, P-Å. Malmquist, S. Svensson, E. Basilier, J.J. Pireaux, U. Gelius and K.Siegbahn *Nouveau Journal de Chimie,* 1977, **1** 191
[14] Adam Hunniford, Ph.D. Thesis, 2006, Queen's University Belfast

Journal of Physics: Conference Series **163** (2009) 012079 doi:10.1088/1742-6596/163/1/012079

# Optical emission spectroscopy of excited atoms sputtered on a Ti surface under irradiation with multicharged Ar ions

**K Motohashi[1], Y Saitoh[2] and S Kitazawa[3]**

[1] Department of Applied Physics, Tokyo University of Agriculture and Technology, 2-24-16 Koganei-shi, Tokyo 184-8588, Japan

[2] Department of Advanced Radiation Technology, Takasaki Advanced Radiation Research Institute, Japan Atomic Energy Agency (JAEA), Takasaki, Gunma 370-1292, Japan

[3] Division of ITER Project, Fusion Research Development, Japan Atomic Energy Agency (JAEA), Naka, Ibaraki 311-0193, Japan

motohasi@cc.tuat.ac.jp

**Abstract**. Optical emission spectroscopy of excited atoms was carried out in order to investigate the sputtering processes on solid surfaces under irradiation of slow, multicharged ions. Many atomic lines of Ti I (neutral) and Ti II (single-charged ions) were observed in wavelengths from 250 to 750 nm with irradiation by $Ar^{3+}$ (30 keV) on a Ti surface which was placed in a low pressure $O_2$ atmosphere. The emission intensity of Ti I (520 nm) decreased monotonically with an increase of $O_2$ partial pressure, whereas that of Ti I / II (670 nm, a 2nd order wavelength of 335 nm) slightly increased. From a semi-logarithmic plot of emission intensity for the 670 nm spectrum as a function of distance from the surface, the mean velocity of the excited Ti atoms and ions in a normal direction parallel to the surface, or $\langle v_{\perp} \rangle$, is estimated.

## 1. Introduction

Electron capture by multicharged ions often induces desorption [1], sputtering [2], and nano-structure formation [3]. It is generally recognized that nano-structure formation is triggered by the electronic excitation of target materials followed by explosive secondary particle emissions. Positive secondary ions produced by multi-electron capture have been considered to play important roles in the initial stage of the Coulomb explosion [4]. Numerous papers have reported in the last decade that secondary ion yields increase with an increase of the charge state of multicharged ions [5]. However, few studies report the velocity, momentum, or kinetic energy of secondary particles, though the importance of the recoil momentum of secondary ions is widely admitted. Neutral atoms, as a majority of secondary particles especially might play key roles in nano-structure formation processes. Therefore, we focused in this study on the velocity measurements of neutral particles produced by low energy collisions of multicharged ions. In order to measure the mean velocity of sputtered atoms on solid surfaces, we performed optical emission spectroscopy. In our previous study [6], no noticeable change of mean velocity parallel to the surface, or $\langle v_{\parallel} \rangle$, among different charge states was observed in the $Al^*$ atoms which were emitted from Al and $Al_2O_3$ surfaces interacting with $Ar^{q+}$ ($q$ = 1–9). Results were obtained by analyzing the Doppler broadening in the emission spectrum.

## 2. Experimental details

A schematic illustration of the experimental setup is shown in figure 1. The multicharged Ar ions were extracted from a 10 GHz electron cyclotron resonance ion source [7], and their mass to charge ratios ($m/q$) were analysed by 90° sector magnets after being focused with an electrostatic lens. The ion beam irradiated a polycrystalline Ti surface at a normal angle after passing through a hole with a 10 mm inner diameter. One can adjust the surface linearly along the beam axis with a motion feedthrough. Light emitted from the surface was focused by an optical lens of magnification 1. Light was dispersed with a visible monochromator and detected with a photo-multiplier tube.

The hole was negatively biased (~ -100V) to repel secondary electrons emitted from the surface. The current of the ion beam (~ 2 μA) was measured with a current-voltage conversion amp (1 V/μA). The background pressure was $1\times10^{-6}$ Pa, whereas the partial pressure of oxygen molecules in the collision chamber was $1\times10^{-5}$–$1\times10^{-4}$ Pa. The vacuum was monitored with a cold cathode ion gauge. To eliminate any pre-existing oxidized layer and other impurities, the surface was cleaned by irradiation of an $Ar^{+}$ (10 keV) beam for an hour.

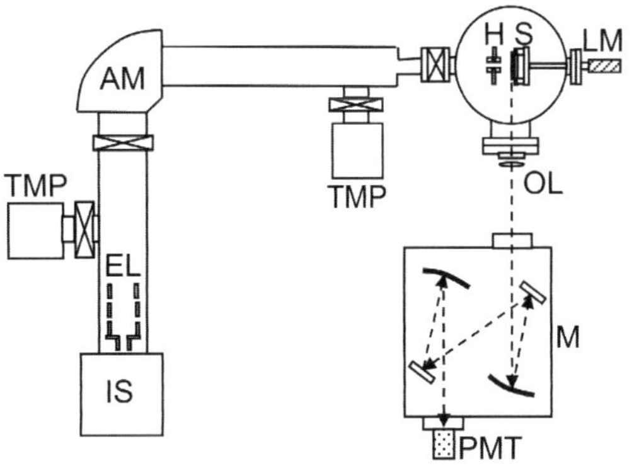


**Figure 1.** Experimental setup of the present study
IS, ECR ion source; EL, Electrostatic lenses; AM, Analyzing magnet; H, hole; S, Surfaces; LM, Linear motion feedthrough; OL, Optical lens; M, Monochromator; PMT, Photomultiplier tube; TMP, Turbo molecular pump

## 3. Results and discussion

Figure 2 shows the visible emission spectra measured on the Ti surface interacting with $Ar^{3+}$ (30 keV). The vertical axes of the figures indicate the number of photons normalized with the ion beam current. A number of atomic lines were clearly observed. The assigned spectra are summarized in Table 1. The lines observed in the present study correspond well with the results of pulsed laser [9] and $O^{+}$ [10] irradiation with $TiO_2$ surfaces except for some oxygen lines.

It is interesting that the spectra in the figure divide into two types, with type A including the spectra whose intensity increased with an increase in oxygen pressure, and type B including those whose intensity decreased with an increase in oxygen pressure. Most of the lines in the wavelength range of 300–380 nm belong to the former case, whereas those in 390–530 nm belong to the latter case. The 648 and 670 nm lines are the spectra of 2nd order wavelengths of 324 and 335 nm lines, respectively. From supplementary obtained results of high resolution wavelength scanning of the monochromator with a narrower slit width, we concluded that the spectra of Ti II belong to type A whereas those of Ti I belong to type B, though some Ti II spectra overlap with Ti I in figure 2.

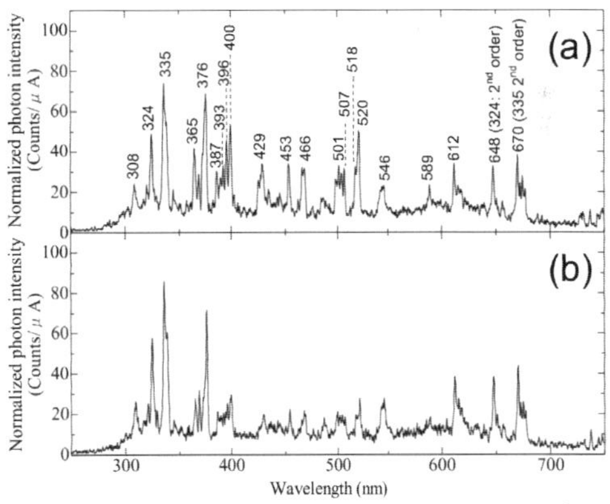


**Figure 2.** Optical emission spectra of $Ti^{*}$ and $Ti^{+*}$ sputtered from a polycrystalline Ti surface interacting with $Ar^{3+}$ (30 keV) in the $O_2$ partial pressure of (a) $<1\times10^{-8}$ Pa and (b) $1\times10^{-6}$ Pa

The photon intensities of 520 and 670 nm spectra as a function of partial oxygen pressure are shown in figure 3. A mutually opposite pressure dependence was clearly observed between 520 and 670 nm spectra. The divergent pressure dependence of emission intensity between neutral atoms and ions was also reported in an Mg surface irradiated with $Kr^+$ (6 keV) in an oxygen atmosphere [11]. In low $O_2$ pressure, the intensity of Mg I lines increased with an increase in $O_2$ pressure, whereas that of Mg II lines decreased slightly with an increase in $O_2$ pressure. One possible conclusion is that the different pressure dependence is due to two different mechanisms which produce $Mg^*$ and $Mg^{+*}$. One is the bond-breaking model [12] in Mg I and the other is a kinetic process leading to electron capture [13] in Mg II. Note that the pressure dependencies of both Mg I and II also reversed themselves between the two limited cases of low and high oxygen pressure. This suggests that the reverse was due to a change in the chemisorption of oxygen atoms related to the oxidation of the surface. We can not conclude the excitation mechanism of $Ti^*$ and $Ti^{+*}$ at the present stage. However, band gap formed in the partially oxidized Ti surfaces may have some relationship to the different pressure dependence between Ti I and II. The relative position of the energy levels between the band gap of the surface and the 4p electrons of $Ti^*$ or $Ti^{+*}$ is very important in both the excitation and de-excitaion processes.

Figure 4 shows the normalized intensity of 670 nm (2nd order wavelength of 335 nm) line as a function of distance between the observation point and the surface. The distance $Z$ was changed by linearly adjusting the surface with the motion feedthrough as shown in figure 1. The $I_0$ point in figure 4 represents the intensity at $Z = 0$ mm (surface). The experimental curve in $Z>0$ deconvoluted with the slit function (Gaussian curve) which was determined from the curve in $Z<0$ corresponds to the decay curve of emission intensity given by

$$I = \sum_k I_{0k} \exp\left( -\frac{Z}{\tau_k \cdot \langle v_\perp \rangle} \right) , \qquad (1)$$

**Table 1.** The assignments of the emission spectra observed in this study

| $\lambda_{exp}$ (nm) | $\lambda_{vac}$ (nm)† | Species | Transition |
|---|---|---|---|
| 308 | 308.8 | Ti II | $3d^24p \rightarrow 3d^24s$ |
| 324 | 323.45-324.19 | Ti II | $3d^24p \rightarrow 3d^24s$ |
| 335 | 334.19-337.76 | Ti I | $3d^24s4p \rightarrow 3d^24s^2$ |
| | 334.67 | Ti II | $3d^24p \rightarrow 3d^3$ |
| 365 | 364.27-365.35 | Ti I | $3d^24s4p \rightarrow 3d^24s^2$ |
| 376 | 375.29 | Ti I | $3d^24s4p \rightarrow 3d^24s^2$ |
| | 375.93-376.13 | Ti II | $3d^24p \rightarrow 3d^3$ |
| 387 | | unknown | |
| 393 | 392.45 | Ti I | $3d^24s4p \rightarrow 3d^24s^2$ |
| 396 | 395.63-395.82 | Ti I | $3d^24s4p \rightarrow 3d^24s^2$ |
| 400 | 399.86 | Ti I | $3d^24s4p \rightarrow 3d^24s^2$ |
| 429 | 428.74-431.48 | Ti I | $3d^34p \rightarrow 3d^34s$ |
| 453 | 453.32 | Ti I | $3d^34p \rightarrow 3d^34s$ |
| 466 | 461.73-468.19 | Ti I | $3d^24s4p \rightarrow 3d^24s^2$ |
| 501 | 501.41 | Ti I | $3d^24s4p \rightarrow 3d^24s^2$ |
| 507 | 506.47 | Ti I | $3d^24s4p \rightarrow 3d^24s^2$ |
| 517 | 517.37 | Ti I | $3d^24s4p \rightarrow 3d^24s^2$ |
| 520 | 519.3 | Ti I | $3d^24s4p \rightarrow 3d^24s^2$ |
| 546 | | unknown | |
| 589 | | unknown | |
| 612 | | unknown | |
| 648 | | 2nd order of 324 nm | |
| 670 | | 2nd order of 335 nm | |

† NIST database [8]

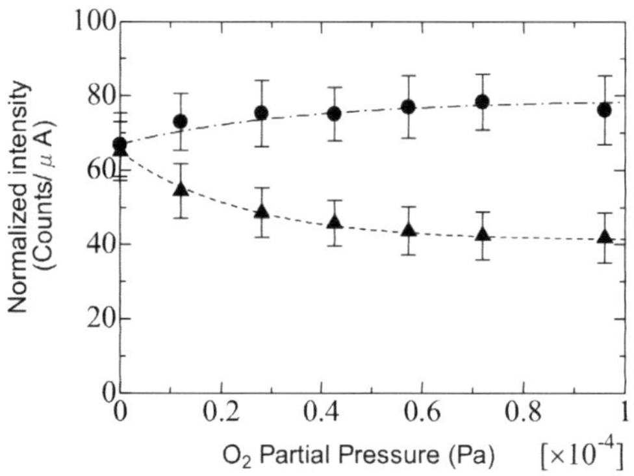


**Figure 3.** The emission intensities of Ti I 520 nm (▲) and Ti I / II 670 nm (●) as a function of partial oxygen pressure

where $\tau_k$ and $I_{0k}$ denote the life time and the spontaneous emission intensity at $Z = 0$, respectively, of the $k$-th excited state of the atom [14]. The terms in which $k>1$ define the cascade components. One can obtain the $\langle v_\perp \rangle$ value by fitting the eq. (1) with known life times into the decay curve in figure 4. The slow decay curve shown with the dashed-dotted curve in figure 4(b) is considered to be a cascade component with a long life time (> 10 ns). The dashed-double-dotted curve in figure 4(b) was drawn by assuming $\tau_k$ of $Ti^*$ ($3d^24s4p\ x^3G$) to be 10.6 ns [15]. As a result, $\langle v_\perp \rangle = (7 \pm 3) \times 10^4$ m/s was obtained for the $Ti^*$ atoms. However, the spectrum might include a significant contribution from Ti II as mentioned above. If the mean life time of $Ti^{+*}$ ($3d^24p\ z\ ^4G_J$ or $^4F_J$) ($J = 3/2$–$11/2$) is assumed to be

5.0 ns [16], the mean velocity becomes two times larger. Therefore we may conclude that the $<v_\perp>$ obtained in the present study was significantly larger than the $<v_\parallel>$ of $Al^*$ atoms which were emitted from Al and $Al_2O_3$ surfaces interacting with $Ar^{q+}$ ($q$ = 1–9) [6]. Generally speaking, a $<v_\perp>$ value depends on not only actual velocity of the sputtered atoms but also the escape probability of the corresponding electrons.

## 4. Conclusions

Optical emission spectroscopy of excited atoms emitted from a polycrystalline Ti surface interacting with $Ar^{3+}$ (30 keV) ions was conducted. Numerous line spectra assigned to Ti I and II were observed. The emission intensity of Ti I and II showed mutually divergent $O_2$-pressure dependencies. This fact suggests different emission mechanisms of $Ti^*$ and $Ti^{+*}$. The emission intensity of the 670 nm spectrum as a function of distance from the surface revealed a double exponential decay. The mean velocity $<v_\perp>$ of $Ti^*$ atoms was estimated to be larger than the $<v_\parallel>$ of $Al^*$ emitted from Al and $Al_2O_3$.

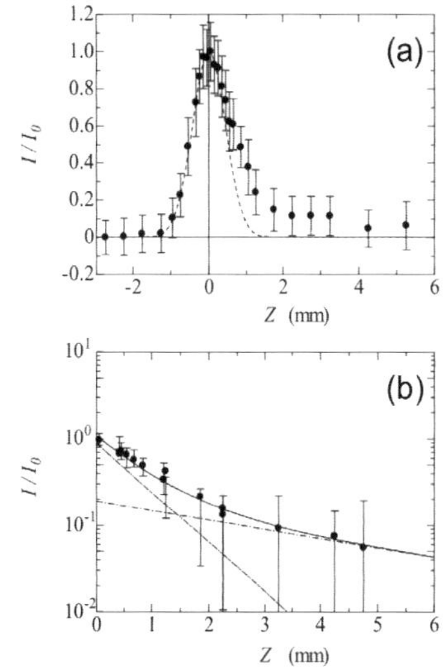


**Figure 4.** (a) Linear plot of the normalized emission intensity of the 670 nm spectrum as a function of distance $Z$ from the surface and (b) its semi-logarithmic plot at $Z \geq 0$
The broken curve, slit function; the solid, dashed-dotted, and dashed-double-dotted curves, fitted curves

## 5. Acknowledgements

The present study is supported by the Reimei Research Promotion project (Japan Atomic Energy Agency). Authors would like to thank Mr. Tod Tollefson for proofreading and English correction.

## References

[1] Burgdörfer J and Yamazaki Y 1996 Phys. Rev. A **54** 4140
[2] Schenkel T, Hamza A V, Barnes A V, Scneider D H, Banks J C, Doyle B L **1998** Phys. Rev Lett. 81 2590
[3] Tona M, Watanabe H, Takahashi S, Nakamura N, Yoshiyasu N, Sakurai M, Terui T, Mashiko S, Yamada C and Ohtani S 2007 Surf. Sci. **601** 723
[4] Cheng H P and Gillaspy J D 1997 Phys. Rev. B **55** 2628
[5] Tona M, Fujita Y, Yamada C and Ohtani S 2008 Phys. Rev. B **77** 155427
[6] Motohashi K, Tsurubuchi S, Shimizu H, Mochiji K, Tanuma H and Kobayashi N 1997 Phys Scr. **T73** 329
[7] Saitoh Y and Yokota W 1996 Rev. Sci. Instrum. **67** 1174
[8] Ralchenko Y, Kramida A E, Reader J and NIST ASD Team (2008). *NIST Atomic Spectra Database* (version 3.1.5), [Online]. Available: http://physics.nist.gov/asd3 [2008, April 23]. National Institute of Standards and Technology, Gaithersburg, MD
[9] Kitazawa S, Choi Y and Yamamoto S 2004 Vacuum **74** 637
[10] Kitazawa S, Yamamoto S, Asano M, Saitoh Y and Ishiyama S 2007 Nucl. Instrum. Meth. Phys. Res. B **256** 233
[11] Karmakar P, Agarwal P and Ghose D 2001 Appl. Surf. Sci. **178** 83
[12] Braise G 1976 Surf. Sci. **60** 65
[13] Joyes P 1973 Radiat. Effec. **19** 235
[14] Bashkin S 1976 *Beam-Foil Spectroscopy* (Berlin: Springer)
[15] Salih S and Lawler J E 1990 Astron. Astrophys. **239** 407
[16] Kwiatkowski M, Werner K and Zimmermann P 1985 Phys. Rev A **31** 2695

Journal of Physics: Conference Series **163** (2009) 012080 doi:10.1088/1742-6596/163/1/012080

# Secondary-ion emission from GaN(0001) and dodecanethiol/Au(111) surfaces irradiated with $Ar^{q+}$ ( $q$ = 4–8 )

**K Motohashi[1], M Flores[2] and Y Kanai[2]**

[1] Department of Applied Physics, Tokyo University of Agriculture and Technology, 2-24-16 Koganei-shi, Tokyo 184-8588, Japan
[2] Atomic Physics Laboratory, RIKEN, Wako-shi, Saitama 351-0198, Japan

motohasi@cc.tuat.ac.jp

**Abstract**. Secondary-ion mass spectroscopic (SIMS) studies have been performed to investigate desorption or sputtering processes of various solid surfaces interacting with highly charged ions (HCIs). We have recently developed an HCI-SIMS apparatus capable of detecting secondary ions, secondary electrons and scattered atoms/ions simultaneously. This apparatus allows us to analyze solid surfaces in four different operation modes: 1) low-energy ion scattering spectroscopy (LEIS), 2) SIMS in coincidence with LEIS, 3) SIMS triggered by scattered neutral atoms and 4) SIMS triggered by secondary electrons. The coincidence experiments with two different operation modes 2) and 4) were conducted with a GaN(0001) and a dodecanethiol/Au(111) self-assembled monolayer surface, respectively, in collisions with $Ar^{q+}$ ($q$ = 4–8) (2.5×$q$ keV). It was successful to measure two different SIMS spectra which were taken by triggering with the scattered $Ar^+$ ions and the secondary electrons in $Ar^{6+}$ collisions.

## 1. Introduction

The electron capture cross section increases almost linearly with an increase in the charge state of the incident ion [1]. In the classical over-the-barrier (COB) model, the critical distance between ion and target material, where the electron capture occurs, increases linearly with the square root of the charge state [2, 3]. The probability of electron capture is distributed within a limited collision energy range around the most probable populated level [4]. The width of the reaction window is proportional to the collision velocity under the assumption of uncertainty between the interaction time and the barrier height formed by HCI and the target matter. Therefore, the electron capture processes by highly charged ions have an important characteristic of state-selective reactions with large cross sections.

Fragmentation of molecules and desorption of surface atoms are often induced after electron capture, because the valence electrons corresponding to chemical bonds are preferentially captured by HCI [5, 6]. The electronic transitions may play key roles in the fragmentation or desorption processes. However, the details of the processes has not been cleared yet especially in desorption or sputtering on the surfaces. To clear the mechanisms of their processes, qualitative or spectroscopic studies as well as quantitative studies are necessary. The velocity, momentum and kinetic energy as well as the emission yields of product particles are very important. We have developed a novel surface analyzer which can detect scattered atoms/ions, secondary ions, and secondary electrons [7] to investigate the details of fragmentation and desorption processes induced by slow highly charged ions. The main objective of this study is to evaluate performance of this apparatus by two prepared surfaces of GaN and dodecanethiol/Au(111) self-assembled monolayer surfaces.

Journal of Physics: Conference Series **163** (2009) 012080 doi:10.1088/1742-6596/163/1/012080

## 2. Experimental

The schematic illustration of our experimental setup is shown in figure 1. The mass-to-charge ($m/q$) ratio of $Ar^{q+}$ ($q$ = 4–8) ions extracted from a compact electron beam ion source (EBIS) [8] was selected with an analyzer magnet. After adjusting their focus and trajectory with an electrostatic lens ($L_2$) and four parallel plates ($P_{1\perp}$, $P_{\parallel}$, $P_{2\perp}$ and $P_{3\perp}$), the ions collide with the target surface at an angle, $\theta \approx$ 10–15° tangential to the surface. The velocity and charge state of the scattered ions or atoms of $Ar^{(q-q')+}$ ($1 \leq q' \leq q$) that capture $q'$ electrons from the surface were analyzed with the Wien filter (WF) and the charge separator (CS), respectively [7]. When each scattered ion or atom hits the detector, three signal waveforms from the position-sensitive anode ($D_3$) were recorded with a digital storage oscilloscope (DSO). One can analyse the position ($x$, $y$) (see the coordinates in figure 1) by a simple arithmetic operation on the pulse heights of the waveforms [5, 6]. The Wien filter works as a velocity-dispersion filter; i.e. the spacing of exit slits extends along the $y$ direction [8]. Therefore, displacement along the $y$-axis of the sensed position of a scattered ion with respect to that of a neutral atom reflects its charge state and velocity. The charge state and velocity of the scattered ion can be analyzed from its displacement.

Secondary ions and electrons emitted from the surface were detected with two different multichannel plates (MCP) $D_1$ and $D_2$. The secondary ions were extracted with an electrostatic lens biased −30 V with respect to the surface potential, whereas the secondary electrons were extracted with a pair of rings biased +30 and +80 V. Since the surface of the MCP detector $D_2$ is biased about +390 V, the electrons with kinetic energy larger than 2 eV arrive at the detector after passing through two rings. The $m/q$ ratios of the secondary ions were analysed with a time-of-flight mass spectrometer

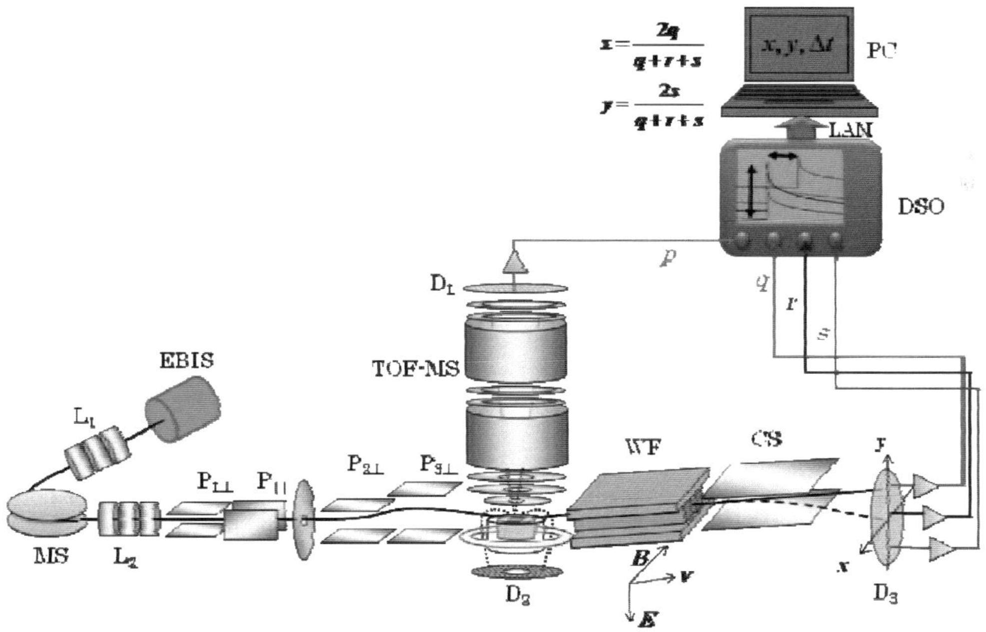


**Figure 1.** The schematic illustration of experimental setup
EBIS, Electron Beam Ion Source; $L_1$ and $L_2$, Einzel lenses; MS, $m/q$ selector; $P_{1\perp}$, $P_{\parallel}$, $P_{2\perp}$ and $P_{3\perp}$; Parallel plate deflectors; $D_1$ and $D_2$, MCP detectors of secondary ions and electrons; TOF-MS, Time-of-flight mass spectrometer; WF, Wien filter; CS, Charge separator; $D_3$, Position-sensitive MCP detector of the scattered ions and atoms; DSO, Digital storage oscilloscope; LAN, Local area network; PC, Personal computer

(TOF-MS).

This apparatus allows us to analyze solid surfaces in four different operation modes: 1) low-energy ion-scattering spectroscopy (LEIS), 2) SIMS in coincidence with LEIS, 3) SIMS triggered by scattered neutral atoms and 4) SIMS triggered by secondary electrons. The operation mode 1) needs only $D_3$ signals of scattered ions. The mode 2) needs the coincidence detection of secondary ions and scattered ions. The surface can be analyzed by mode 4) if TOF-MS is measured with the start signals of secondary electrons. The results of surface analyses by modes 2) and 4) will be presented in this paper.

The background pressure of the collision chamber was maintained below $6 \times 10^{-7}$ Pa. The kinetic energy of $Ar^{q+}$ was $2.5 \times q$ keV. The scattering angle was about 20° with respect to the direction of incident $Ar^{q+}$ ions. The flight length and acceleration voltage of secondary ions were 728 mm and 120 V, respectively, in the TOF-MS.

## 3. Results and discussion

Figure 2 shows the 2D-distributions of scattered Ar atoms and $Ar^+$ ions detected with $D_3$ in collisions of $Ar^{6+}$ ions with a GaN(0001) surface. The $y$ positions of Ar atoms stayed constant at 23 mm as shown with a white broken line, whereas those of $Ar^+$ ions decreased linearly with an increase in the electric field of the charge separator $E_{CS}$, as shown by a red solid line. These figures clearly demonstrate that the scattered atoms and ions were completely separated by the parallel plate CS.

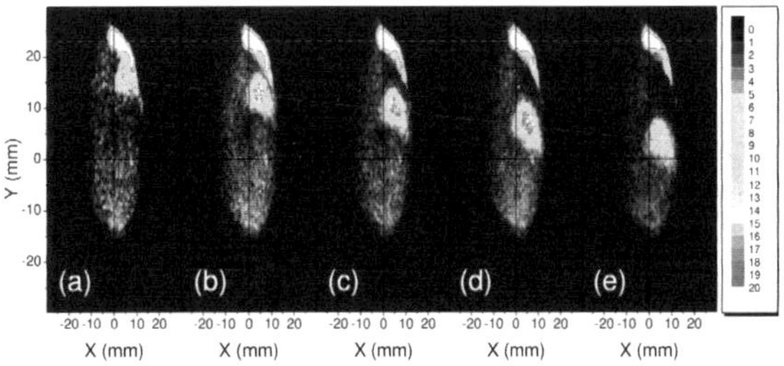


**Figure 2.** 2D-distributions of scattered Ar atoms and $Ar^+$ ions detected with the position-sensitive MCP detector $D_3$ with $E_{cs}$ = (a) 67, (b) 133, (c) 200, (d) 267 and (e) 333 V/cm

Figure 3 shows the correlation map of the time difference between $T_1$ (flight time of secondary ion in the TOF-MS) and $T_2$ (flight time of scattered ion from surface to $D_3$), and the displacement $y$ of the scattered $Ar^+$ ion at the $D_3$ detector (data taken under $E_{CS}$ = 267 V/cm). In the case of $Ar^{6+}$ incidence, $T_1$ (≥4.5 μs) was much longer than $T_2$ (≈ 1.9 μs). An intense proton spectrum was clearly observed in figure 3(a) and in the TOF spectrum, as shown in figure 3(b). The velocity dispersed pattern corresponding to LEIS spectrum is shown in figure 3(c), although we have not yet converted the displacement $y$ into the kinetic energy. The pass energy of the Wien filter was estimated to be 15.0 ± 0.6 keV. Three components of the velocity peaks, indicated by arrows, are seen in the figure. The three peaks were mainly due to the components of the proton spectrum. This suggests that proton emission from the GaN(0001) surface correlates with three different energy losses of scattered $Ar^+$ ions.

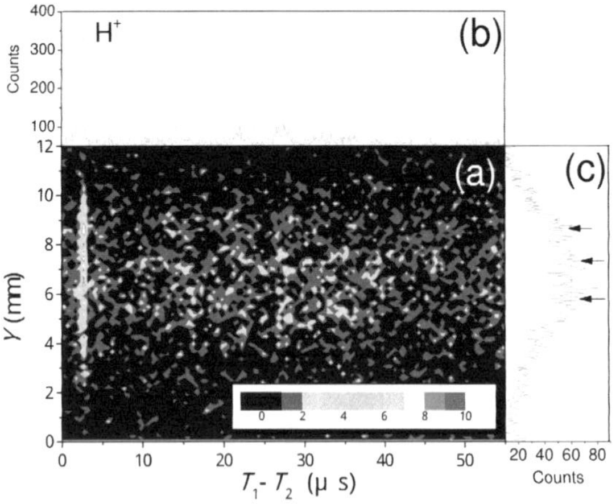


**Figure 3.** The correlation map of time difference $T_1$−$T_2$ and the displacement $y$ (a), and its histograms accumulated along each axis (b and c)

Figure 4 shows the TOF spectra of secondary ions emitted from a dodecanethiol/Au(111) surface [9] interacting with $Ar^{q+}$ ($q$ = 4, 6 and 8). These data were taken with the start signals of secondary electrons (operation mode 4). The other experimental conditions were the same as those in the case of GaN(0001). The surface is a self-assembled monolayer (SAM) comprising the ligand group (sulphur), the alkane chain ($(CH_2)_{11}$) and the terminal functional group ($CH_3$). Protons as well as molecular ions, such as $H_2^+$, $C_2H_n^+$ ($n$ = 2–6), and $C_3H_m^+$ ($m$ = 4–8) coming from the chain and the terminal group, were observed in the spectra. These secondary ions were also observed in collisions between $Ar^{q+}$ ($q$ = 5–9) and alkanethiol-SAMs at $\theta$ = 60° [9]. The relative yield of a specific ion with respect to the total number of detected ions is charge state dependent. For example, the relative yield of protons increased with an increase of $q$, whereas relative yields of molecular ions decreased. This tendency, i.e. the different $q$-dependence between protons and other ions with heavier mass, was also observed in GaN and SiC surfaces partially covered with hydrocarbon molecules [10].

The SIMS in coincidence with scattered ions which were dispersed with their kinetic energies was successful in the present study. However their absolute energies have not been determined yet. We are preparing to achieve the LEIS measurements by calibrating the Wien filter. Triple coincidence detection among the scattered ions, secondary electrons, and the secondary ions is expected in the next stage.

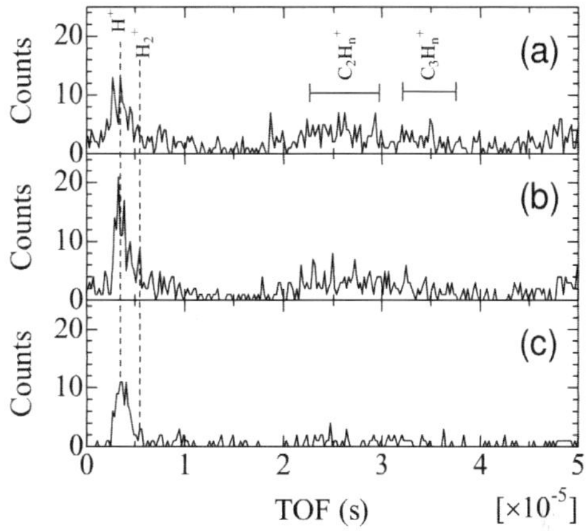


**Figure 4.** TOF spectra of secondary ions from a dodecanethiol/Au(111) surface interacting with (a) $Ar^{4+}$ (10 keV), (b) $Ar^{6+}$ (15 keV) and (c) $Ar^{8+}$ (20 keV) at glancing angle ($\theta \sim$ 12°).

## 4. Conclusions

A novel SIMS apparatus that allows us to perform coincidence detection between secondary ions and scattered atoms/ions or secondary electrons has been developed. A SIMS measurement by coincidence detection with scattered ions was performed in collisions between $Ar^{6+}$ ions and a GaN(0001) surface at $\theta \sim$ 12°. We recognized three velocity components of $Ar^+$ ions detected in coincidence with secondary ions. The secondary ions, such as protons and molecular fragments were observed in the collisions of $Ar^{q+}$ ($q$ = 4–8) with a dodecanethiol/Au(111) surface.

## 5. Acknowledgements

This work was partly supported by a Grant-in-Aid for Scientific Research (B) (19360016).

## References

[1] Kimura M, Nakamura N, Watanabe H, Yamada I, Danjo A, Hosaka K, Matsumoto A, Ohtani S, Sakaue H A, Sakurai M, Tawara H and Yoshino M 1995 J. Phys. B: At. Mol. Opt. Phys. **28** L643
[2] Niehaus A 1986 J. Phys. B: At. Mol. Phys. **19** 2925
[3] Burgdörfer J, Lerner P and Meyer F W 1991 Phys. Rev. A **44** 5674
[4] Taulbjerg K 1986 J. Phys. B: At. Mol. Phys. **19** L367
[5] Motohashi K and Tsurubuchi S 2003 J. Phys. B: At. Mol. Opt. Phys. **36** 1811
[6] Tsurubuchi S and Motohashi K 2003 J. Phys. B: At. Mol. Opt. Phys. **36** 3847
[7] Motohashi K, e-J. Surf. Sci. Nanotech. (submitted)
[8] Motohashi K, Moriya A, Yamada H and Tsurubuchi S 2000 Rev. Sci. Instrum. **71** 890
[9] Flores M, O'Rourke B, Esaulov V and Yamazaki Y, Phys. Rev. A, in press
[10] Motohashi K, Tsurubuchi S and Koukitu A 2005 Nucl. Instrum. Meth. Phys. Res. B **232** 254

# Energy dependence of ion guiding through nanocapillaries

**K Schiessl[1], C Lemell[1], K Tőkési[2] and J Burgdörfer[1]**

[1]Institute for Theoretical Physics, Vienna University of Technology, Wiedner Hauptstraße 8 – 10, A-1040 Vienna, Austria, EU

[2]Institute of Nuclear Research of the Hungarian Academy of Sciences (ATOMKI), H-4001 Debrecen, Hungary, EU

E-mail: klaus@concord.itp.tuwien.ac.at

**Abstract.** We present first simulation results for the energy dependence of the effective guiding of ions through nanocapillaries in insulating materials based on classical transport theory. We also investigate the dependence of guiding on the effective dielectric shielding near the insulating surface. We find agreement with experimental data over a range of energies without any freely adjustable parameter.

## 1. Introduction

The surprising discovery of guided transmission of highly charged ions (HCI) in their initial charge state through insulating nanocapillaries [1] has stimulated a large number of investigations employing various geometries and materials (e.g. [1-8]). Recently a guiding effect was found even for electrons with an energy of several hundred eV [9, 10]. While the mechanism for the latter is still unclear, for ion guiding the following scenario has emerged: during the charge-up phase, projectiles with charge state $q+$ entering the capillary under an angle of incidence $\theta_{in}$ larger than the geometric opening angle hit the capillary wall and deposit their charge. These charges are transported along the surface and into the bulk of the target material. Projectiles entering the capillary subsequently will be deflected due to the Coulomb field of the deposited charges and may eventually be transmitted (see Fig. 1). Guiding (stable transmission conditions) sets in as soon as a dynamical equilibrium between projectiles hitting the capillary wall (charging) and transport into the bulk or to the capillary exits (discharging) has been established.

A theoretical description and *abinitio* simulation of this process poses a considerable challenge in view of the widely disparate times scales simultaneously present in this problem [11]. We have therefore developed a mean-field classical transport theory [11] based on a microscopic classical-trajectory Monte Carlo simulation for the ion transported, self-consistently coupled to the charge-up of and charge transport at the internal capillary walls. The underlying idea is to relate the microscopic ion-surface scattering and ion transport to known macroscopic properties of the materials involved such as surface and bulk conductivity and the dielectric constant. This allows to be predictive and for a description of the ion guiding without resorting to freely adjustable parameters or additional ad-hoc assumptions. In the present communication we focus on the energy dependence of guiding, for which experimental data have recently become available [2]. We model transmission of $Ne^{7+}$ ions with kinetic energies ranging from 2 to 9 keV through

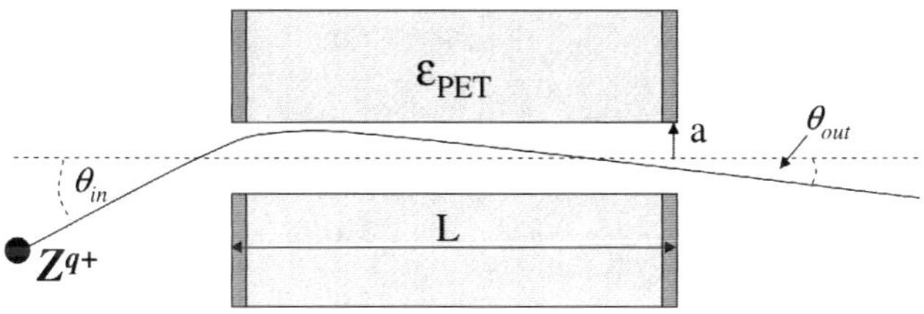


**Figure 1.** Schematic view of ion guiding through insulating capillaries. Parameters used in our simulation are projectile charge state $Ne^{7+}$, capillary radius $a = 100$ nm, length $L = 10\,\mu$m, entrance angle $\theta_{in}$, and dielectric constant $\varepsilon_{PET} = 3.3$.

PET nanocapillaries with a diameter of 200 nm (see schematic view in Fig. 1). The incident current was kept constant at a typical experimental value of 1 nA [1, 2]. Furthermore, we illustrate the importance of a proper description of dielectric shielding of the charges deposited on the surface.

## 2. Mean-field classical transport theory

We first briefly review the mean-field classical transport theory [11] employed in the present simulation. A projectile of charge state $q+$ enters a capillary under an angle of incidence $\theta_{in}$ and with kinetic energy $E$. It travels within the capillary under the influence of the Coulomb field $V_C$ of partially screened charges on the capillary wall deposited earlier until it either leaves the capillary, i.e. to contribute to transmission, or collides with the interior wall of the capillary (i.e. contributes to beam loss). In the latter case, it neutralizes on a (sub-)fs time scale and deposits $q$ at charges at the surface. Charges deposited may diffuse along the surface or into the bulk. These processes are governed by their diffusion constants $D_s$ and $D_b$ which can be derived from material properties [12]. Motion of surface charges is modeled by an random walk which may eventually end with transport to the metal cover of the capillary target. When diffusing into the bulk, decay of charge carriers is characterized by a time-constant $\tau_b$ which is inversely proportional to $D_b$ [11]. Both bulk and surface diffusion are assumed to be independent of the charge density at the surface, i.e. charge transport is treated within linear response. Non-linear charge transport as recently suggested [3] is neither invoked nor necessary.

Within the framework of linear response, the bare charge $q$ deposited by the ion is screened by the dielectric medium. As the deposition occurs at the interface to vacuum, proper treatment of dielectric screening is required [13]. Denoting the bare Coulomb interaction between the guided ion and the charge deposited by $V_C$ the effective screened interaction, denoted by $V_{SCR}$, depends on the location of the deposited charge $q$. In the limiting case that the charge has already diffused into the bulk, bulk screening would apply, i.e. $V_{SCR} = V_C/\varepsilon$. We use here and in the following the static limit $\varepsilon = \varepsilon(q \to 0, \omega \to 0)$ of the dielectric function since the characteristic time for build-up of screening (time interval $\Delta t$ between two subsequent ions entering the capillary $\approx 10^{-10}$ sec) is very large compared to the characteristic time for optical excitations, $\Delta t \gg 2\pi/\omega_{opt}$). The dielectric screening of surface charges, is determined by the induced polarization, "image charge" [13] $\chi = (\varepsilon - 1)/(\varepsilon + 1)$. The effective residual charge and hence the effective potential is given by $V_{SCR} = V_C\,(1 - \chi) = 2V_C/(\varepsilon + 1)$. Our simulation employs the latter expression if not otherwise stated. To illustrate the significance of dielectric screening we will also show results using bulk screening instead as well as for bare interactions ($V_{SCR} = V_C$). Trajectories of guided projectiles are, in addition to the charge-up within individual capillaries, also affected by the mesoscopic fields of nearby capillaries. The latter predicts a dependence of the angular distribution on the mean-distance between capillaries or, equivalently the capillary density, an example of which will be shown below.

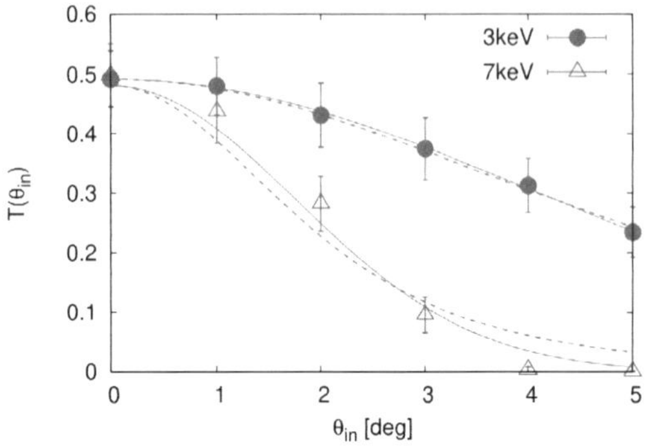


**Figure 2.** Transmission function $T(\theta_{in})$ as obtained for a capillary described in Fig. 1 for different energies. ●: $E = 3$ keV, △: $E = 7$ keV. Solid lines: Gaussian fits to the simulation [14] with $\theta_g = 5.8° \pm 0.1°$ (3 keV), and $\theta_g = 2.5° \pm 0.1°$ (7 keV). Dashed lines offer alternative fits to the data given by $T(\theta_{in}) = T(0°)/[1 + (\theta_{in}/\theta_l)^2]^2$ with $\theta_l = 7.7° \pm 0.1°$ (3 keV) and $\theta_l = 3.0° \pm 0.3°$ (7 keV).

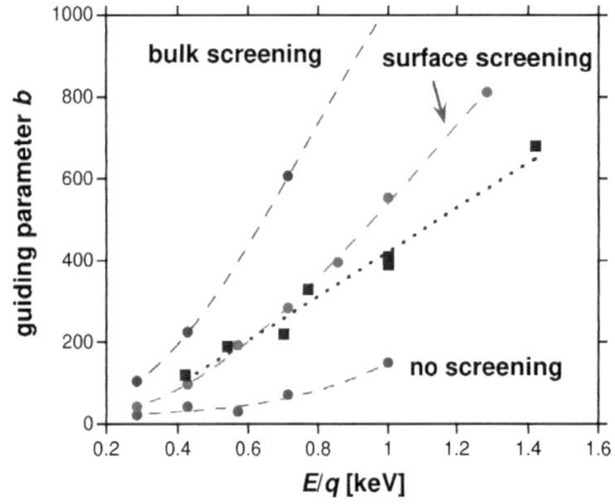


**Figure 3.** Guiding parameter $b$ as a function of $E/q$ of HCI. ■: Experimental data [2, 14]. Simulation (● ) of three models of dielectric shielding are shown (c.f. text), dashed lines shall guide the eye. The dotted line shows a linear fit to the experimental data.

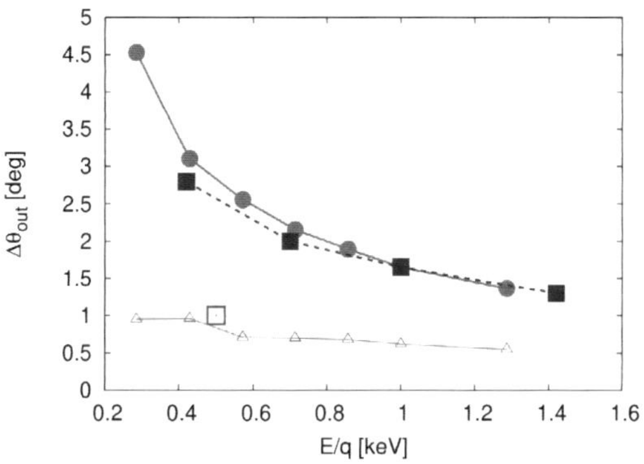


**Figure 4.** Angular width $\Delta\theta_{out}$ of trans–mitted HCI and deflection near exit (c.f. text), $\theta_{in} = 3°$. ■: Experiment [2] with $N = 4 \times 10^8$ cm$^2$. △: simulation of a single capillary, ● : including a collective field. □: Experiment [3] with $N = 4 \times 10^6$ cm$^2$ only.

## 3. Results

Figure 2 presents total transmission in equilibrium as a function of $\theta_{in}$ for a projectile energy of $E = 3$ keV and $E = 7$ keV. Following Hellhammer *et al.* [14] total transmission of HCI through nanocapillaries is fitted by a Gaussian $T(\theta_{in}) = T(0°)\exp(-\sin^2\theta/\sin^2\theta_g)$. Hence, transmission can be given in terms of the guiding parameter ("guiding power") $b = \sin^{-2}\theta_g$, or the effective guiding angle $\theta_g$. At the guiding angle the transmission function has decreased to $T(0°)/e$. The original motivation for the Gaussian ansatz was a Boltzmann type "thermalization" argument for the transverse energy. Note, however, that also other choices of fitting curves to $T(\theta_{in})$ perform similarly well. An example is shown for a square-Lorentzian in Fig. 2. By assuming guiding to be a statistical process, Refs. [2, 14] predict linear behavior for $b$ as a function of $E/q$. Fig. 3 shows the guiding parameter $b$ as a function of the scaled kinetic energy $E/q$. For comparison we show also simulation results assuming bulk screening as well as no screening at all ($V_{SCR} = V_C$). As expected, simulations invoking surface screening perform superior in reproducing the experimental data, the agreement is satisfactory. It should be noted that the approximately linear increase of $b$ indicated by Fig. 3 cannot continue for higher $E$. Instead, for large energies, $b$ approaches a constant as soon as guiding breaks down and $\theta_g$ becomes equal to the geometric opening angle. This saturation regime is not yet reached for $E/q \approx 1.5$ keV.

Another quantity of interest is the angular width $\Delta\theta_{out}$ (FWHM) of the guided beam upon exit. As transmission through a capillary with a geometric opening angle of $\theta_0 \simeq 1°$ should lead to a very narrow angular distribution, large values for $\Delta\theta_{out}$ observed in experiments (up to $5°$) have been surprising and have created considerable interest as to its underlying mechanism. Trajectory simulations for a *single* capillary do not feature such broad distributions but show a width close to $\theta_0$. In previous work we have suggested the collective field created by charge patches of the ensemble of all capillaries illuminated by the primary beam to be a possible source of enlarged angular width [11]. It acts as an einzellens in the exit region of the capillary deflecting projectiles close to the exit plane. Adding the influence of the resulting macroscopic field in our calculation leads to an energy dependence similar to the one observed in experiment (Fig. 4). The field strength is determined from the principal charge patch in the entrance region; the capillary density $N$ is assumed to be $N = 4 \times 10^8$ cm$^2$. The dependence of $\Delta\theta_{out}$ on the scaled energy $E/q$ can be well fitted by $\sin^2(\Delta\theta_{out}) \propto (E/q)^{-1}$ in agreement with experimental data [2]. Such a collective effect immediately implies that $\Delta\theta_{out}$ should depend on the density of the capillaries, or equivalently, on the mean spacing between them. In particular, drastic reduction in capillary density should reduce $\Delta\theta_{out}$ to close to the geometric opening angle. Indeed, samples with significantly lower capillary density ($N = 4 \times 10^6$ cm$^2$ ) show angular widths close to simulation results for a single capillary (see Fig. 4).

## 4. Conclusion

We have simulated ion transmission through insulating nanocapillaries using a mean-field classical-transport theory. Ion trajectories are propagated in the combined fields of charges deposited on the capillary wall, their polarization charges, the projectile image charge, and the macroscopic field from neighboring capillaries. The simulation avoids any freely adjustable parameters in order to be predictive and to provide qualitative insights into underlying mechanisms. We have varied projectile energy and angle of incidence and have tested different models of dielectric shielding. Best agreement with experimental data is found for dielectrically screened surfaces charges. Reasonable agreement with data could be found employing only macroscopic material parameters of PET like dielectric constant, surface and bulk conductivity.

## Acknowledgments

This work was supported by the Austrian Science Fund under grant no. FWF-SFB016 "ADLIS" and the TeT Grant No. AT-7/2007. K.T. was partially supported by the grant "Bolyai" from the Hungarian Academy of Sciences and the Hungarian National Office for Research and Technology. K.S. also acknowledges support by the IMPRS-APS program of the MPQ (Germany).

## References

[1] Stolterfoht N *et al.* 2002 *Phys. Rev. Lett.* **88** 133201; Stolterfoht N *et al.* 2007 *Phys. Rev.* A **76** 022712
[2] Stolterfoht N, Hellhammer R, Bundesmann J and Fink D 2008 *Phys. Rev.* A **77** 032905
[3] Stolterfoht N *et al.* 2007 *Phys. Rev.* A **76** 022712
[4] Vikor Gy, Rajendra Kumar R, Pešić Z, Stolterfoht N and Schuch R 2005 *Nucl. Instr. and Meth.* B **233** 218
[5] Sahana M B, Skog P, Vikor Gy, Rajendra Kumar R T and Schuch R 2006 *Phys. Rev.* A **73** 040901(R)
[6] Mátéfi-Tempfli S *et al.* 2005 *Nanotechnology* **17** 3915
[7] Skog P, Soroka I L, Johansson A and Schuch R 2007 *Nucl. Instr. and Meth.* B **258** 145
[8] Ikeda T *et al.* 2006 *Appl. Phys. Lett.* **89** 163502
[9] Milosavljević A R *et al.* L 2006 *Phys. Rev.* A **75** 030901(R)
[10] Das S, Dassanayake B, Winkworth M, Baran J, Stolterfoht N and Tanis J A 2007 *Phys. Rev.* A **76** 042716
[11] Schiessl K *et al.* 2005 *Phys. Rev.* A **62** 042902; Schiessl K *et al.* 2007 *Nucl. Instr. and Meth.* B **258** 150
[12] Data sheets of Mylar (http://www.dupontteijinfilms.com).
[13] Hägg L, Reinhold C O and Burgdörfer J 1997 *Phys. Rev.* A **55** 2097
[14] Hellhammer R, Bundesmann J, Fink D and Stolterfoht N 2007 *Nucl. Instr. and Meth.* B **258** 159

Journal of Physics: Conference Series **163** (2009) 012082 doi:10.1088/1742-6596/163/1/012082

# Surface nanostructure formation by the interaction of slow xenon ions on HOPG surfaces

**Y Y Wang[1, 2], G Q Xiao[1], Y T Zhao[1], D H Li[1, 2], D Zhao[2, 3], Z F Xu[3] and F L Li[3]**
[1] Institute of Modern Physics, Chinese Academy of Sciences, Lanzhou 730000, China
[2] Graduate University of the Chinese Academy of Sciences, Beijing 100049, China
[3] Department of Applied Physics, Xi'an Jiao Tong University, Xi'an 710049, China

E-mail: wangyuyu@impcas.ac.cn

**Abstract.** Experiments on nanostructure formation on highly oriented pyrolytic graphite (HOPG) surfaces irradiated by slow highly charged $Xe^{q+}$ ($q$=23-29) ions have been carried out at a new experimental terminal for surface physics on the 320 kV ECR platform at the Institute of Modern Physics (IMP). In the tapping mode atomic force microscope (AFM) image, the nano-sized hillocks protruding from the surfaces are probed. The height and diameter of the nanostructures increase with the projectile charge state. The present results reveal a similarity between the nanostructure formation produced by slow highly charged ions and the track formation induced by swift heavy ions.

## 1. Introduction

Ion-surface collisions have been studied intensively during recent years. Ions deposit energy either as kinetic energy dominated relevant to the interaction of swift heavy ions (SHI) in matter or as potential energy carried by slow (eV to keV) highly charged ions irradiated on surfaces. In comparison with SHI, highly charged ions (HCI) carry several tens of keV of potential energy which is delivered into only a few atomic layers of the surface, resulting in many different phenomena that are significantly dependent on the potential energy deposition, e.g., potential electron emission [1, 2], potential sputtering [3], X-ray emission [4] and surface defects [5]. Slow HCI can be used as a unique tool for surface etching, ultra-thin film growth and nanostructure fabrication [6].

Surface defects observed by a scanning probe microscope (SPM) in both scanning tunneling microscope (STM) and atomic force microscope (AFM) modes have been studied in some famous laboratories recently[5-10]. As is known, STM observation gives information on the electronic density of states, while AFM observation directly gives topographic surface structure. AFM has three modes: tapping, contact and non-contact. It is noted that a contact-AFM observation is often not very reliable in discussing topological structures while the tapping mode AFM was successfully applied to observe protrusions induced by 1 keV $Ar^+$ ions [11]. For $Xe^{46+}$ ions, the protrusion-like dots were visible not only in the STM image but also in the non-contact mode AFM image taken at the same impact site [12]. Up to now, a number of observations have been performed for ion-induced defects on surfaces like HOPG, mica, $CaF_2$, $Al_2O_3$, Si, $SiO_2$ and other materials [6-12]. HOPG is a common and useful material for surface etching and the nuclear fusion industry. Its lamellar structure gives rise to high thermal conductivity and semimetallic electrical conductivity parallel to the layers and poor conductivity normal to the planes.

## 2. Experimental Method

The experiments were carried out with the 14.5 GHz ECR ion source at the 320 kV experimental platform in the national laboratory of the Heavy Ion Research Facility in Lanzhou (HIRFL), China. The ion source used was the high charge state all-permanent magnet ECR ion source no. 2, where the extraction voltage of ions is 5-320 kV [13]. The ion beam was extracted from the ion source, selected by two 90° analyzing magnets and focused by two quadrupole lenses, then deflected by a 60° magnet and finally transported into the experimental chamber. By such procedure we can obtain a well-aligned ion beam with high purification. The beam was collimated to a divergence of about 0.7° by a pair of four-jaw slits and then impact onto the target holder where were mounted on a goniometer in the center of an ultra-high vacuum (UHV) chamber (base pressure is about $5\times10^{-10}$ mbar, $8\times10^{-9}$ mbar during sample irradiation). The projectile current was measured by a Faraday Cup with a pico-ampere meter (Model 6458, Keithley, USA). A cage which has an aperture of 5 mm with a pico-ampere meter was isolated in front of the target holder. The current at the target holder was also measured with a pico-ampere meter. By adjusting a pair of four-jaw slits, the incident ion beam was able to reach the target holder when the current of the cage was zero and the current of the target holder reached the maximum value. After focusing the ion beam, the HOPG samples which were freshly cleaved in the air with adhesive tape were transported to the target holder. During the experiment, we adjust the acceleration voltage for each charge state of xenon ions in order to remain the same projectile energy of 383 keV. The ion current density was kept below 5nA/cm$^2$. Total ion fluxes of about $1\times10^{10}$ ions/cm$^2$ were applied under normal incidence.

After irradiation, the HOPG samples were exposed to air, transported and set on a sample holder of a SPM (Nano III, USA). In the present study, all the measurements have been done in the tapping AFM mode in order to observe the real topographic surface structures.

## 3. Results and Discussion

Figure 1 shows a typical AFM image of the HOPG surfaces after xenon ion irradiation. The $1000\times1000$ nm$^2$ images show distinct hillock-like nanostructures on the initially flat surface for xenon ion with charge state 29 and 27. For $Xe^{23+}$ and $Xe^{25+}$, small dots are observed in the $300\times300$ nm$^2$ images. In contrast, we have also tested a freshly cleaved HOPG surface with AFM. It is almost flat and smooth before the irradiations.

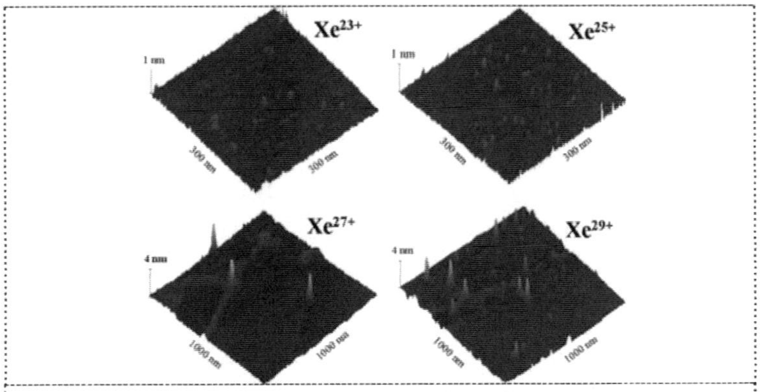


**Figure 1.** AFM images of nanostructures formed by irradiation of slow $Xe^{q+}$ ($q$=23, 25, 27, 29) ions on HOPG surfaces. Scan areas are $300\times300$ nm$^2$ and $1000\times1000$ nm$^2$, respectively.

Figure 2 shows the height and diameter of the dots as a function of charge state. More than 30 dots were probed with the AFM for ion beams in each charge state. The error bars in the figure represent the standard deviations. The height and diameter of the dots shows a strong charge state dependence.

When the charge state is equal to or higher than 27 there is a significant increase in the height and size of dots on the surfaces.

According to previous results obtained from the STM images, it was shown that there is no significant kinetic or potential energy dependence of the nanodot height and width irradiated by $Ar^{q+}$ (q=1, 8, 9, 11) ions on HOPG at the hundreds of eV kinetic energies range [14], in contrast with the conclusions of a strong charge state dependence and no kinetic energy dependence of the defect size from HOPG by xenon and argon ions at the large incident kinetic energy [15]. However, in our observation for $Xe^{27+, 29+}$ ions with kinetic energy of 383 keV impact on HOPG surfaces, nanodots were observed in the AFM tapping mode, and shown size increases as increasing charge state. For $Xe^{23+}$, a height of about 0.5 nm and a diameter of several nm have been observed in this measurement and in Nakamura's group with a STM [12, 15]. Our results clearly gave the evidence that the nanostructures correspond to the topographical change on the surface depend on the potential energy of incident ions. The dot size discrepancy between our and the previous results could be due to the differences in the kinetic energy of the incident ions or the SPM modes.

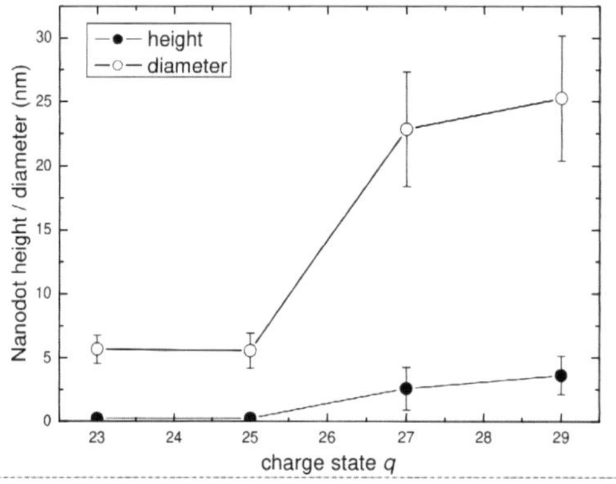


**Figure 2.** The incident ion charge state dependence of the height and diameter of nanodots. The error bars represent the standard deviations.

Since the kinetic energy of the incident ions was rather large, we could not specify which is more effective for the surface deformation between the kinetic and potential energies. Thus, in this study, we have studied the incident ion charge dependences of the nanodot size and height at the same kinetic energy range (383 keV, $v$=7.5×10$^5$ m/s). In the interaction of xenon ions on graphite surface, the nuclear stopping power $S_n$ is 2.3 keV/nm and the electronic stopping energy $S_e$ is 0.95 keV/nm (calculation using SRIM-2003) [16]. It was reported that for incident ions with $S_e \leq 7.3$ keV/nm, the damage process on HOPG surface is dominated by nuclear stopping [17]. The energy transferred to the lattice will eventually lead to heating and melting of the crystal. Until the relatively slow processes of melting and recrystallization occur, cooling process sets in, leading to the formation of nanostructures.

The average height of nanodots we have found for $Xe^{29+}$ ions irradiated is 3.6±1.5 nm. They are of hillock shape with a diameter of 25.3±4.9 nm and are similar to the nanodots and chains on $SrTiO_3$, $TiO_2$ and $Al_2O_3$ surfaces created by xenon ions with energies of several tens of MeV with under grazing angles [18]. The thermal spike model [17, 18] has been developed to describe the formation of cylindrical tracks or recrystallized materials in the solid irradiated by SHI, where electronic stopping dominates over nuclear stopping. As the model has described, swift ions deposit their kinetic energy mainly by inelastic collisions with lattice atoms. When this energy is enough to increase the lattice temperature locally above the melting temperature, a rapid quenching is followed, thereby causing excitation and ionization processes. How this electronic excitation finally leads to the modification of the material (craters, hillocks) is still not exactly known. Accordingly, a modified inelastic thermal spike model has been developed to consider the process of electronic excitation by the potential energy of HCI and the subsequent phase transitions (heating, melting and restructuring) induced by HCI impacting on the surface[10, 19]. Although the mechanism of energy deposition between HCI and SHI is fundamentally different in the beginning, strong electronic excitation on the surface within a nanometer scale region by HCI is very similar to the ion trajectory in the solid by SHI during the restructurization process. This may explain why nanostructures on the surface produced by slow HCI have a similarity of the tracks induced by SHI in the solid.

For further understanding, systematic studies are still ongoing for the energy and material dependences of nanostructures' height and size not only using STM and AFM, but also applying other complementary techniques e.g. Raman spectroscopy, low energy electron microscopy and small angle X-ray scattering, etc. In addition, simulations and calculations must be set up to clarify the surface modification mechanism.

## Acknowledgments

This work was supported by the National Natural Science Foundation of China through Grant No. 10405025 and Found of president's excellent scholarship of Chinese Academy of Sciences. We would like to thank J. Liu, C. Trautmann and Y. Yamazaki for many fruitful discussions, and J. Y. Li and P. Z. Wang of the 320 kV ECR platform for help during the experiments.

## References

[1] Meissl W, Winklehner D, and Aumayr F 2008 *e-J. Surf. Sci. Nanotech.* **6** 54
[2] Wang Y, Zhao Y, Qayyum A and Xiao G 2007 *Nucl. Instrum. Methods* B **265** 474
[3] Aumayr F and Winter H P 2004 *Phil. Trans. Roy. Soc.* (London) **362** 77
[4] Briand J P, Thuriez S, Giardino G, Borsoni G, Froment M, Eddrief M and Sebenne C 1996 *Phys. Rev. Lett.* **77** 1452
[5] El-Said A S, Heller R, Meissl W, Ritter R, Facsko S, Lemell C, Solleder B, Gebeshuber I C, Betz G, Toulemonde M, Moeller W, Burgdoerfer J and Aumayr 2008 *Phys. Rev. Lett.* **100** 237601
[6] Aumayr F and Winter H P 2003 *e-J. Surf. Sci. Nanotech.* **1** 171
[7] Mochiji K, Yamamoto S, Shimizu H, Ohtani S, Seguchi T and Kobayashi N 1997 *J. Appl. Phys.* **82** 6037
[8] Tona M, Watanabe H, Takahashi S, Fujita Y, Abe T, Jian S, Nakamura N, Yoshiyasu N, Yamada C, Sakurai M and Ohtani S 2007 *J. Phys: Conf. Ser.* **58** 331
[9] Minniti R, Ratliff L P and Gillaspy J D 2001 *Phys. Scr. T* **92** 22
[10] Aumayr F, El-Said A S and Meissl W 2008 *Nucl. Instr. and Meth.* B **266** 2729
[11] An B, Fukuyama, Yokogawa K and Yoshimura M 2002 *J. Appl. Phys.* **92** 2317
[12] Terada M, Nakamura N, Nakai Y, Kanai Y, Ohtani S, Komaki K and Yamazaki Y 2005 *Nucl. Instr. and Meth.* B **235** 452
[13] Sun L T, Zhao H W, Li J Y, Wang H, Ma B H, Zhang Z M, Zhang X Z, Guo X H, Shang Y, Li X X, Feng Y C, Zhu Y H, Wang P Z, Liu H P, Song M T, Ma X W and Zhan W L 2007 *Nucl. Instr. and Meth.* B **263** 503
[14] Mitsuda Y, O'Rourke, Nakamura N, Kainai Y, Ohtani S and Yamazaki Y 2007 *J. Phys: Conf. Ser.* **58** 351
[15] Nakamura N, Terada M, Nakai Y, Kanai Y, Ohtani S, Komaki K and Yamazaki Y 2005 *Nucl. Instr. and Meth.* B **232** 261
[16] Ziegler J F, Biersack J P and Littmark U *http://www.srim.org.*
[17] Liu J, Neumann R, Trautmann C and Mueller C 2001 *Phys. Rev. B* **64** 184115
[18] Akcoeltekin E, Peters T, Meyer R, Duvenbeck A, Klusmann M, Monnet I, Lebius H and Schleberger S 2007 *Nature Nanotechnology* **2** 290
[19] Lemell C, El-Said A S, Meissl W, Gebeshuber I C, Trautmann C, Toulemonde M, Burgdoerfer J and Aumayr F 2007 *Solid-State Electron.* **51** 1398

Journal of Physics: Conference Series **163** (2009) 012083 doi:10.1088/1742-6596/163/1/012083

# Absolute sputtering yields from solid Ne by low energy $He^+$ and $Ar^{q+}$ ($1 \leq q \leq 6$) impact

**Shinya Fujita**[1], **Takayuki Tachibana**[2], **Tetsuo Koizumi**[1,2] **and Takato Hirayama**[1,2]

[1]Department of Physics, Rikkyo University, Nishi-Ikebukuro, Toshimaku, Tokyo 171-8501 JAPAN

[2]Research Center for Measurements in Advanced Science (RCMAS), Rikkyo University, Nishi-Ikebukuro, Toshimaku, Tokyo 171-8501 JAPAN

E-mail: `hirayama@rikkyo.ac.jp`

**Abstract.** Absolute sputtering yields from the surface of solid Ne by low energy $He^+$ and $Ar^{q+}$ ($1 \leq q \leq 6$) impact are measured. Very large sputtering yields (300 atoms/ion for 1 keV $He^+$ impact, and 3000 atoms/ion for 1keV $Ar^+$ impact) have been observed. A significant dependence of the sputtering yields on the chage state, i.e. the potential energy, of the incident ion for $Ar^{q+}$ has not been observed because it is estimated to be much smaller than that of the kinetic sputtering, which suggests that the mechanism of potential sputtering is similar to those known for the electron- and photon-stimulated desorption processes.

## 1. Introduction

Interaction of rare gas solids (RGS) with low energy photons and electrons were well studied in these two decades [1, 2]. In these studies, creation of excitons and ions is found to play an important role in the desorption processes. Sputtering from RGS by singly charged keV-energy ion impact has been investigated by some groups [3–8], and kinetically induced sputtering mechanism has been discussed in detail. However little report can be found for multiply charged ion impact. Here we report the results for the absolute sputtering yields from solid Ne by singly and multiply charged ion impact.

## 2. Experimental

Singly and multiply charged ions are produced in an electron cyclotron resonance ion source (ECRIS, NANOGAN). The ion beam extracted from the ion source is mass-analyzed and focused onto the sample surface at the center of a main chamber. The incident angle of the ion beam to the sample is 10 deg. from the normal direction of the sample surface. Schematic of the main chamber is shown in fig.1. The main chamber is evacuated by a series of turbo molecular pumps and a Ti-getter pump, and the ultimate pressure is about $8 \times 10^{-9}$ Pa.

The sample film was prepared on a polycrystaline-Cu disk of 8 mm diameter. This was fixed to a mechanical cryostat and cooled down to about 4.5 K. The sample film was condensed on the Cu disk by filling the chamber with a gaseous Ne to a pressure of $10^{-6} \sim 10^{-4}$ Pa. The film thickness was estimated from the exposure assuming the condensation coefficient to be unity.

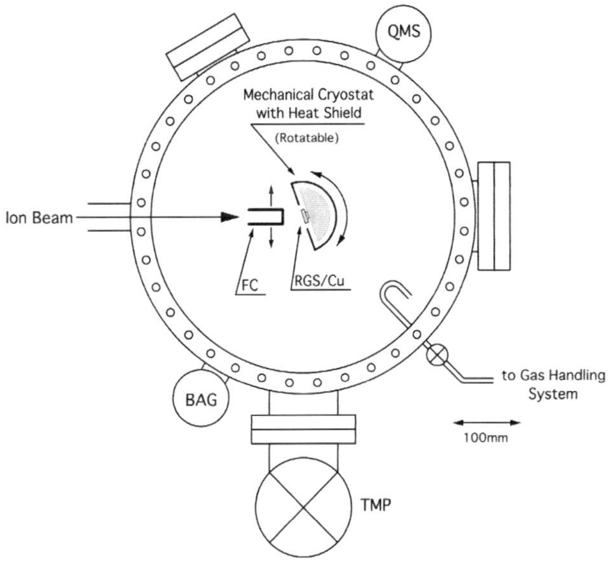


**Figure 1.** Schematic of the main chamber (top view). BAG: Bayard-Alpert ionization gauge, QMS: quadrupole mass spectrometer, TMP: tundem turbo molecular pump, FC: movable Faraday cup.

Absolute yield $Y_{\rm Ne}$ is estimated from the number of sputtered Ne atoms per second $N_{\rm d}$ and the number of incident ions per second $N_{\rm ion}$ by using the following equation, [9]

$$Y_{\rm Ne} = \frac{N_{\rm d}}{N_{\rm ion}} = \frac{S_{\rm Ne} \times \Delta P_{\rm Ne}}{kT} \cdot \frac{qe}{I_{\rm ion}} \tag{1}$$

where $S_{\rm Ne}, \Delta P_{\rm Ne}, k, T, q$ and $I_{\rm ion}$ are the pumping speed of the pumping system for Ne gas, the partial pressure change in the main chamber during the irradiation of the ion beam, Boltzmann constant, the ambient temperature, the charge state of the incident ion beam, and the ion beam current, respectively.

The sputtering rate was calculated from the pumping speed for Ne gas and the rise of the partial pressure in the vacuum chamber during irradiation of the sample. The total pumping speed of the turbo molecular pump and cold surfaces was measured to be $0.14 \pm 0.01$ m$^3$/s for Ne, which was determined from the Ne pressure measured by an ionization gauge installed in the main chamber and from the flow rate calibrated volumetrically using a reference volume and a Baratron pressure gauge as reference. The small rise of the partial pressure during irradiation was detected by a quadrupole mass spectrometer which was calibrated against the ionization gauge for each run of the experiment. It should be noted that the uncertainty of the relative sensitivity of the ionization gauge was cancelled in the present method for determining the sputtering rate.

Incident ion beam current is monitored by a movable Faraday cup just in front of the sample surface. Typical ion beam current is 0.1 - 1 nA and the beam is 3 mm in diameter.

The over-all uncertainty is estimated to be $\pm$ 40%. The largest source of the uncertainty is the absolute pressure measured by the BA-gauge.

## 3. Results and Discussion

Experimental results for absolute sputtering yields for $He^+$ and $Ar^{q+}$ ($q$ = 1, 4 and 6) impact are shown in fig.2 and fig.3, respectively. The thickness of the sample is about 500 ML. The total (electronic + nuclear) stopping power $S_{\rm Total}$ at the surface calculated by SRIM2008 [10] are also shown in the figures. The ranges of the incident ions at an energy of 1.5 keV for $He^+$

and 3 keV $Ar^+$ are estimated by SRIM2008 to be 31.6 nm and 9.0 nm, which correspond to about 110 ML and 30 ML of solid Ne, respectively.

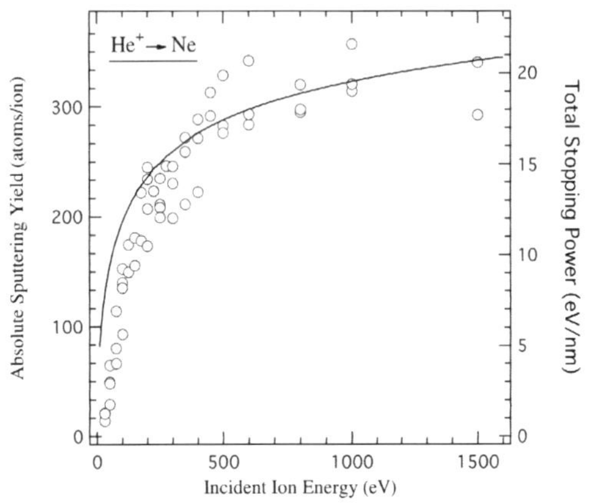


**Figure 2.** Absolute sputtering yields of solid Ne by $He^+$ impact as a function of the incident ion energy. The solid line is the total stopping power $S_{\mathrm{Total}}$ at the surface calculated by SRIM2008 [10]. Thickness of the sample film is about 500 atomic layers.

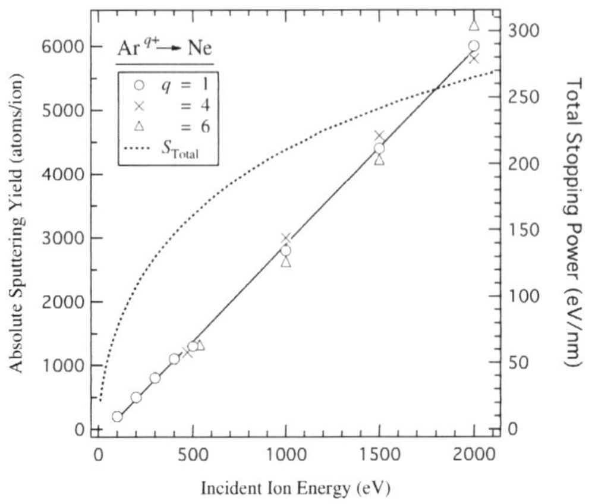


**Figure 3.** Absolute sputtering yields of solid Ne by $Ar^{q+}$ ($q = 1$ (○ ), 4 (×), 6 (△)) impact as a function of the incident ion energy. The dotted line shows the total stopping power $S_{\mathrm{Total}}$ at the surface calculated by SRIM2008 [10]. Thickness of the sample film is about 500 atomic layers. Some data points are slightly displaced horizontally for clarity of the presentation. The solid line is to guide the eye.

The sputtering yield for $He^+$ impact increases with the incident ion energy up to about 500 eV, and then saturates at about 300 (atoms/ion). This inicident energy dependence is similar to the total stopping power $S_{\mathrm{Total}}$ at the surface calculated by SRIM2008 [10] (solid line in fig.2), suggesting that the collision cascade mechanism is applicable to this system [5].

The results for $Ar^{q+}$ impact show surprisingly large sputtering yields even at relatively low incident energy, and that the yield is almost proportional to the incident ion energy. Charge state dependence of the sputtering yield is not observed within the experimental uncertainty. The fact that no dependence on the charge state, i.e. the potential energy of the incident ion, is observed in the present results can be explained as follows; an exciton or an ion created in solid Ne induces the desorption of 1 - 5 atoms [9]. If we assume that all of the potential energy of the incident ion can be used to create excitons ($E_{\mathrm{exciton}} = 17.2$ eV) or ions ($E_{\mathrm{g}} = 21.6$ eV for solid Ne), the "potential sputtering" [11, 12] (exciton- or ion-induced sputtering) yield will be at most 100 even by $Ar^{6+}$ (potential energy: 310 eV) impact, which is too small to be observed within the present experimental uncertainty. The present result suggests that the sputtering mechanism in the results of $Ar^{q+}$ impact is kinetic sputtering, and the mechanism of potential sputtering is similar to those known for the electron- and photon-stimulated desorption experiments [2].

All of the present results for the sputtering yields by singly-charged $He^+$ and $Ar^+$ impact are plotted in fig.4 together with the previously reported experimental data for $He^+$ [5, 13] and for $Ar^+$ [5]. Systematic differences with the data of Balaji et al. [5] by 50 to 100% were found, while the present results seem to be connected smoothly with the results by Ellegaard et al. [13].

In summary, we have measured the absolute sputtering yields from the surface of solid Ne by low energy $He^+$ and $Ar^{q+}$ ($q = 1 \sim 6$) ions. Surprisingly large sputtering yields for $Ar^{q+}$

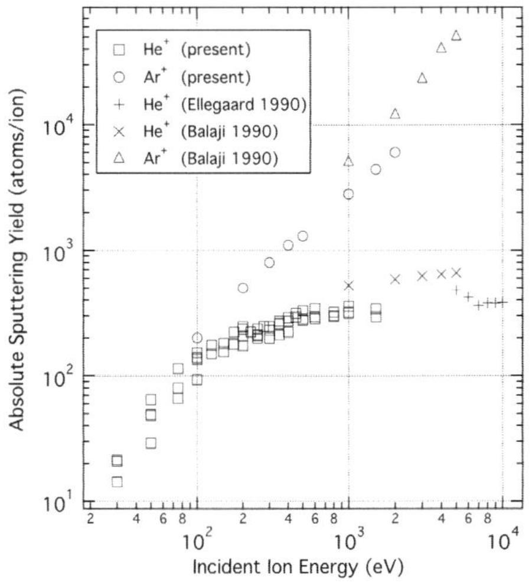


**Figure 4.** Absolute sputtering yields of solid Ne by $He^+$(□) and $Ar^+$(○) impact as a function of the incident ion energy. Also shown are experimental results for $He^+$ (+) [13] and (×) [5], and for $Ar^+$ (△) [5].

impact were found, where the kinetic sputtering process is dominant. Contributions from the potential sputtering have not been observed in the results of $Ar^{q+}$ impact because they are estimated to be much smaller than those of the kinetic sputtering. The presetn results suggest that the mechanism of potential sputtering is similar to those known for the electron- and photon-stimulated desorption processes.

**Acknowledgments**
The authors are grateful to Dr. Osamu Furuhashi for the significant contribution in the initial stage of this work. This work was partly supported by Grant-in-Aid for Scientific Research from the Ministry of Education, Science, Sports and Culture, Japan, and by Rikkyo University Special Fund for Research.

## References

[1] Zimmerer G 1994 *Nucl. Instrum. Meth. Phys. Res. B* **91** 601
[2] Hirayama T and Arakawa I 2006 *J. Phys. Cond. Matt.* **18** S1563
[3] Orth R G, Jonkman H T, Powell D H and Michl J 1981 *J. Am. Chem. Soc.* **103** 60269
[4] David D E, Magnera T F, Tian R, Stulik D and Michl J 1986 *Nucl. Instrum. Meth. Phys. Res. B* **14** 378
[5] Balaji V, David D, Magnera T and Michl J 1990 *Nucl. Instrum. Meth. Phys. Res. B* **46** 435
[6] Schou J, Ellegaard O, Sorensen H and Pedrys R 1988 *Nucl. Instrum. Meth. Phys. Res. B* **33** 808
[7] Ellegaard O, Schou J and Sørensen H 1986 *Nucl. Instrum. Meth. Phys. Res. B* **13** 567
[8] Pedrys R, Warczak B, Leskiewicz P, Schou J and Ellegaard O 1999 *Nucl. Instrum. Meth. Phys. Res. B* **157** 121
[9] Arakawa I, Adachi T, Hirayama T and Sakurai M 2000 *Surf. Sci.* **451** 136–142
[10] Ziegler J F 2008 Computer codes SRIM-2008, http://www.srim.org/
[11] Aumayr F, Varga P and Winter H 1999 *Int. J. Mass Spectrom.* **192** 415
[12] Aumayr F and Winter H 2004 *Phil. Trans. Roy. Soc. Lond. A* **362** 77
[13] Ellegaard O, Schou J and SØrensen H 1990 *Europhysics Lett.* **12** 459

# Secondary ion emission from a KCl(001) surface by grazing-angle incidence of swift ions

**K Nakajima, M Suzuki and K Kimura**

Department of Micro Engineering, Kyoto University,
Yoshida-honmachi, Sakyo-ku, Kyoto 606-8501, Japan

E-mail: kaoru@kues.kyoto-u.ac.jp

**Abstract**. The yield and the mass distribution of secondary ions emitted from a KCl(001) surface were measured, when various MeV ions (0.8 MeV $He^+$, 1.34 MeV $Li^+$, 2.1 MeV $B^{2+}$, 3 MeV $O^{2+}$, 1.5 MeV $Si^+$, 6 MeV $Si^{3+}$) were incident on the surface at grazing angles $\theta_i$ ranging from 1 to 6 mrad. Positive secondary ions such as $Cl^+$, $K^+$, $K_2^+$ and small clusters, $K(KCl)_n^+$ (n = 1–4) were detected coincidentally with specularly reflected projectile ions. Negative secondary ions, $Cl^-$, $(KCl)Cl^-$, $(KCl)_2Cl^-$, were much fewer by ~100. The secondary-ion production rate $P(x)$ as a function of the distance $x$ between the projectile and the surface atomic plane was derived from the $\theta_i$-dependence of the yield of the positive secondary ions. $P(x)$ showed the overlinear dependence on the electronic stopping power $S(x)$ for specularly reflected projectile ions. The power index varies from ~1 to ~3 depending on $S(x)$.

## 1. Introduction

When a solid target is bombarded with high-energy ions, the constituent atoms, molecules, clusters and their ions of the target are emitted from the surface in many cases. This phenomenon is well-known as sputtering. Electronic sputtering, which results from electronic excitation caused by ion bombardment, has been studied for a number of pairs of high-energy ions and targets (oxides, alkali halides, metals) [1-4].

When high-energy ions are incident on an atomically flat surface of a crystal at a grazing angle, ions are scattered at the angle for specular reflection with neither close collisions with the target atoms nor penetration through the surface. In the specular reflection the ions travel through the target electrons outside the surface atomic plane and lose the part of their energy by electron excitations on the surface. These features of the specular reflection of high-energy ions are favorable for the study of electronic sputtering, because only the surface electrons are excited and the contribution of nuclear sputtering is ideally ruled out even with bombardment by no very high-energy ions. Additionally, the trajectories of the specularly reflected ions are well-defined simply with the ion-surface repulsive potential, and so the impact parameter and the density of the consequent interactions (energy deposition, secondary electron emission, etc.) can be controlled by the variation of the angle of incidence of the ions.

In the present study, the yield and the mass distribution of secondary ions emitted from a KCl(001) surface was measured at grazing-angle incidence of various MeV ions. The secondary-ion production rate as a function of the distance between the projectile ion and the surface was estimated for each kind of projectile ions, and the relation between the production rate and the stopping power was investigated.

## 2. Experimental

Beams of 0.8 MeV $He^+$, 1.34 MeV $Li^+$, 2.1 MeV $B^{2+}$, 3 MeV $O^{2+}$, 1.5 MeV $Si^{2+}$, 6 MeV $Si^{3+}$ from a tandem-type accelerator at QSEC in Kyoto University were collimated to less than $0.1\times0.1$ mm$^2$ and to a divergence angle less than 0.1 mrad by a series of 4-jaw slits. The typical beam current was less than 1 pA. The collimated ion beams were incident on a KCl(001) surface at a grazing angle of 1–6 mrad from the surface plane in a ultra-high-vacuum (UHV) scattering chamber (base pressure $4\times10^{-8}$ Pa). The surface of KCl(001) was prepared by cleavage in air and cleaned by heating at 250°C for 1 hour in the UHV scattering chamber ($\sim1\times10^{-7}$ Pa). The KCl(001) target was at 180°C during the bombardment to keep the surface clean and to avoid macroscopic charging.

The ions reflected specularly from the surface of KCl(001) were selected by a aperture (Ø = 1 mm) placed 411 mm downstream from the target and their energies were analyzed using a 90° sector magnetic spectrometer and a 1-dimensional position-sensitive detector (1D-PSD). The energy analysis of reflected ions only with most probable charge state [5] was performed. For example, only $O^{4+}$ ions were energy analyzed for the incidence of 3 MeV $O^{2+}$. This can be excused by the fact that specularly reflected ions already reach their equilibrium charge distribution after many charge exchanges [6].

Positive secondary ions emitted from the surface were collected with electric field between the surface and a negatively biased grid (−1.5 kV) right in front of the surface. The secondary ions passing the grid travel through a drift tube (~0.4 m) and are detected by a micro-channel plate (MCP, effective diameter Ø = 20 mm). Negative secondary ions were also detected only in the case of 3 MeV $O^{2+}$ incidence. The mass to charge ratio ($m/z$) of the secondary ions is analyzed with the time-of-flight technique, in which the signals from the 1D-PSD (specularly reflected ions) and the MCP (secondary ions) are used as the start and stop signal, respectively.

## 3. Results and discussion

Figure 1 shows an example of the energy spectrum of projectile ions specularly reflected from the KCl(001) surface. The angle of incidence $\theta_i$ of $O^{2+}$ ions was 3.0 mrad from the surface plane. The energy spectrum of $O^{4+}$ ions through charge exchange with without reflection on the surface is also shown as the primary energy of incident ions. The energy of reflected ions forms a relatively narrow peak at ~2.93 MeV. This peak corresponds to the ions reflected from the surface without surface penetration or a close collision with the target atoms. Thus, this confirms that the secondary ions observed coincidentally with the detection of specularly reflected ions are generated only by electronic sputtering.

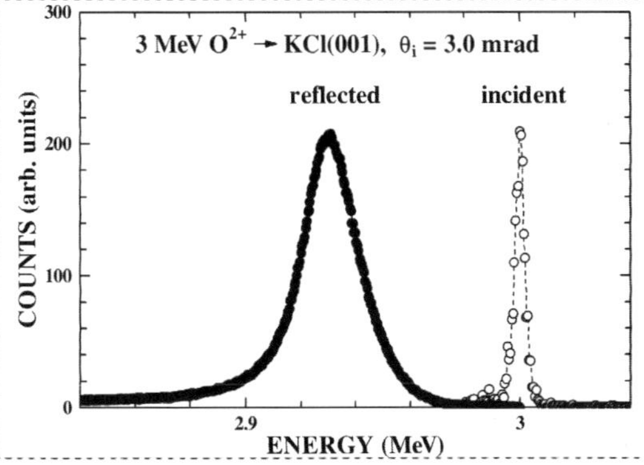


**Figure 1.** Energy spectrum of $O^{4+}$ ions reflected from the KCl(001) surface for a grazing-angle incidence of 3 MeV $O^{2+}$ ions.

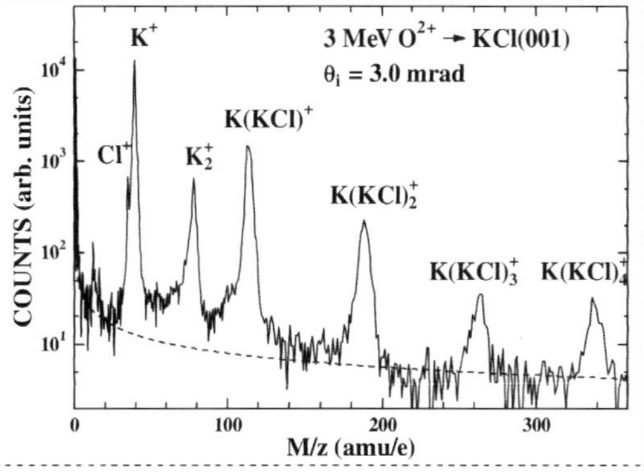


**Figure 2.** Mass spectrum of positive secondary ions from the KCl(001) surface bombarded with $O^{2+}$ ions at $\theta_i = 3.0$ mrad.

Figure 2 shows a typical mass spectrum of positive secondary ions from the KCl(001) surface for the incidence of $O^{2+}$ ions at 3.0 mrad. The abscissa is $m/z$, where $m$ is the mass of the secondary ion in the atomic mass unit (amu) and $z$ is the magnitude of electric charge of the secondary ion in the unit of elementary charge (e). The spectrum has several peaks assigned to $Cl^+$, $K^+$, $K_2^+$ and $K(KCl)_n^+$ (n = 1−4) ions. Secondary ions with higher charge states or with the larger mass ($m/z$ > 400 amu/e) were below the detection limit.

Emission of negative secondary ions was also observed for 3.0 MeV $O^{2+}$ incidence. $Cl^-$, $(KCl)Cl^-$ and $(KCl)_2Cl^-$ ions were detected, but the total yield of negative secondary ions was only about one-hundredth of that of positive secondary ions. This dominance of positive secondary ions indicates that the emission of secondary ions should be initiated with local charging due to the emission of hundreds of secondary electrons per projectile [7].

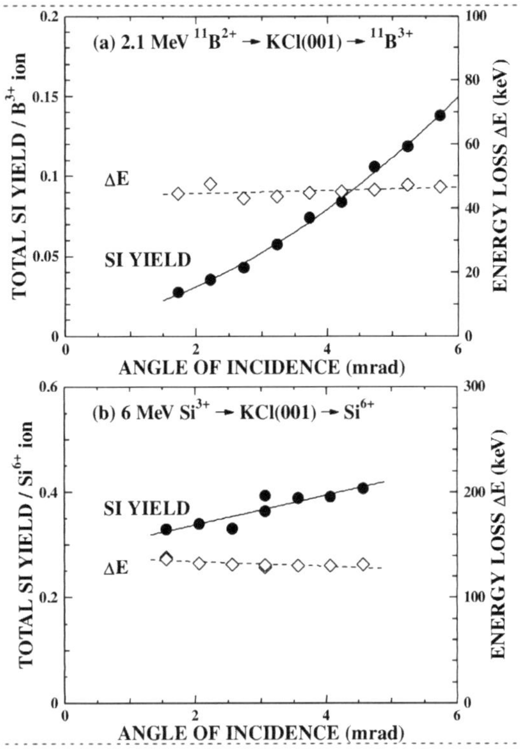


**Figure 3.** Total yield of positive secondary ions (SIs) and the most probable energy loss of reflected ions as a function of angle of incidence: (a) 2.1 MeV $B^{2+}$, (b) 6 MeV $Si^{3+}$.

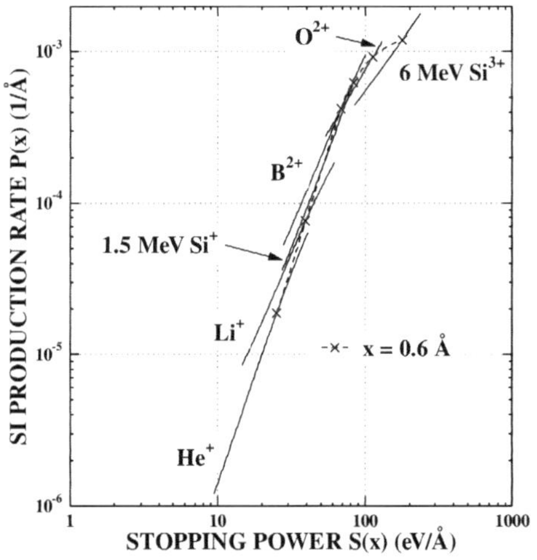


**Figure 4.** Secondary-ion (SI) production rate $P(x)$ versus stopping power $S(x)$ for various projectile ions. The cross marks (dashed curve) show the values at $x$ = 0.6 Å

Figure 3 shows the total yield of positive secondary ions as a function of the angle of incidence $\theta_i$ of (a) 2.1 MeV $B^{2+}$ and (b) 6 MeV $Si^{3+}$. The observed most probable energy losses of the reflected $B^{3+}$ and $Si^{6+}$ ions are also shown in Figure 3(a) and 3(b). The yields of secondary ions increase in yield with $\theta_i$, while the energy losses hardly depend on $\theta_i$.

The observed $\theta_i$-dependence of the secondary-ion yield $Y(\theta_i)$ gives the secondary-ion production rate per unit path length of a projectile as a function of the distance between the projectile and the surface, because the trajectory of a specularly reflected projectile ion is well-defined and the distance of closest approach of the projectile to the surface depends on the angle of incidence $\theta_i$. The measured secondary ion yield $Y(\theta_i)$ as a function of $\theta_i$ is represented as

$$Y(\theta_i) = Q\int_{traj.} P(x)dz \qquad (1)$$

where $P(x)$ is production rate of secondary ions per unit path length of a projectile traveling at the distance $x$ from the surface atomic plane of KCl(001) and the coordinate $z$ is along the trajectory of the projectile. $Q$ is the absolute detection efficiency of MCP. $Q$ is now assumed to be 0.5. To substitute the equation for trajectory of the projectile to equation (1) gives an Abel-type integral equation, which can be solved if $V_p(x) < V_p(x_{min})$ at any $x > x_{min}$ [8] as

$$P(x) = -\frac{1}{2\pi QE}\cdot\frac{dV_p(x)}{dx}\left\{Y(0)\sqrt{\frac{E}{V_p(x)}} + \int_0^{\pi/2}\left.\frac{dY(\theta_i)}{d\theta_i}\right|_{\sqrt{\frac{V_p(x)}{E}}\sin u} du\right\} \qquad (2)$$

where $E$ is the kinetic energy of the projectile, $x_{min}$ the distance of the closest approach of the projectile to the surface and $V_p(x)$ the surface continuum potential acting on the projectile. Molière potential is used as the ion-atom interaction potential for $V_p(x)$ and the polynomial expression of the fitting curve (shown by solid lines in Figure 3) is adopted as $Y(\theta_i)$, to obtain $P(x)$ with equation (2). In the same way, the position-dependent stopping power $S(x)$ for specularly reflected projectile ions were derived from the observed $\theta_i$-dependence of the energy loss $\Delta E(\theta_i)$ [9].

The obtained $P(x)$ and $S(x)$ decrease almost exponentially with the distance $x$, but the rate of decrease of $P(x)$ is larger than that of $S(x)$. Figure 4 shows $P(x)$ as a function of $S(x)$ for each kind of projectile ions. The distance $x$ as a parameter is limited to the range from ~0.4 to ~1.4 Å, where the closest distance between the projectiles and the surface atomic plane ranges for various angles of incidence. The production rate $P(x)$ seems to be proportional to power of $S(x)$ for each. The cross marks connected with dashed curve show the values at a constant distance $x$ of 0.6 Å. Along this curve, the power index decreases from ~3 to ~1 as the stopping power $S(x)$ increases from 25 to 180 eV/Å.

## 4. Conclusion

Secondary-ion emission from a KCl(001) surface was observed when various MeV ions are incident on the surface at grazing angles of 1–6 mrad. Positive secondary ions, such as $Cl^+$, $K^+$, $K_2^+$ and $K(KCl)_n^+$ (n = 1–4), are dominant in yield over negative secondary ions, such as $Cl^-$, $(KCl)Cl^-$ and $(KCl)_2Cl^-$, which indicates that secondary-ion emission should be initiated with local charging due to secondary-electron emission. The yield of positive secondary ions increases with $\theta_i$, while energy loss of the reflected ions is almost independent of $\theta_i$. Position-dependent production rate $P(x)$ for positive secondary ions decreases almost exponentially with the distance $x$ between the projectile and the surface. $P(x)$ shows the overlinear dependence on $S(x)$. The power index for the dependence of $P(x)$ on $S(x)$ decreases from ~3 to ~1 as $S(x)$ increases.

## References

[1] Sekioka T, Terasawa M, Sataka M, Kitazawa S and Niibe M 2002 *Nucl. Instr. and Meth.* B **193** 751
[2] Arnoldbik W M, Tomozeiu N and Habraken F H P M 2003 *Nucl. Instr. and Meth.* B **203** 151
[3] Neugebauer R, Jalowy T, Pereira J A M, de Silveira E F, Rothard H, Toulemonde M and Groeneveld K O 2003 *Nucl. Instr. and Meth.* B **209** 62
[4] Matsunami N, Sakata M, Iwase A and Okayasu S 2003 *Nucl. Instr. and Meth.* B **209** 288
[5] Shima K, Kuno N, Yamanouchi M and Tawara H 1992 *Atomic Data and Nuclear Data Tables* **51** 173
[6] Fritz M, Kimura K, Kuroda H and Mannami M 1996 *Phys. Rev.* A **54** 3139
[7] Kimura K, Usui S, Maeda K and Nakajima K 2002 *Nucl. Instr. and Meth.* B **193** 661
[8] Kimura K, Ooki S, Andou G, Nakajima K and Mannami M 1998 *Phys. Rev.* A **58** 1282
[9] Fujii Y, Narumi K, Kishine K, Kimura K and Mannami M 1992 *Nucl. Instr. and Meth.* B **67** 82

# Density matrix description of resonant coherent excitation of swift highly charged ions in oriented crystals

**V V Balashov[1], I V Bodrenko[1], V K Dolinov[1], A A Sokolik[2] and A V Stysin[1]**

[1]Skobeltsyn Institute of Nuclear Physics, Moscow State University, Moscow, Russia
[2]Institute of Spectroscopy, Russian Academy of Sciences, 142190 Troitsk, Russia

E-mail: `balvse@anna19.sinp.msu.ru` (V V Balashov)

**Abstract.** We report selected results of our latest resonant coherent excitation studies obtained within density matrix approach and present suggestions for new RCE experiments.

## 1. Introduction

Swift ion in matter is an open quantum system involved into both coherent and incoherent interactions with its surroundings. Starting from this understanding, we (contrary to theoretical approaches of other groups [1–5]) take advantages of density matrix formalism to describe resonant coherent excitation (RCE) of multiply charged ions in crystals and apply it to analysis of a number of experiments. Beginning our studies from the RCE of light ions at rather low energy [6, 7], the density matrix approach allowed us to achieve a unified description of both characteristics of the RCE process known from the pioneering papers of Okorokov [8] and early RCE experiments [9,10]. These are the charge state distribution at the exit of the target (survival fraction) and yield and angular distribution of the characteristic X-ray radiation produced by excited ions. Current experiments with relativistic highly charged ions reveal some new aspects of RCE process. We extend our approach in accordance with these new trends.

## 2. Density matrix analysis of recent RCE experimental data

### *2.1. Alignment of channeled RCE ion and angular anisotropy of its X-ray radiation*

Polarization of the electric field of the crystal lattice leads to alignment of the angular momentum of resonant coherently excited ions. General procedure of our calculations [6, 7] begins with solving the generalized Master equation to obtain time evolution of the density matrix for individual ion trajectories. Then, averaged over the whole ensemble of ions taking part in the RCE process, it is projected on the subspace of ionic states emitting photons whose yield, angular distribution and polarization are then calculated by using standard density matrix (statistical tensors) technique [11]. A strong evidence of the alignment effect was obtained in 2006 by the Tokyo RCE collaboration [12] by the measurements of angular asymmetry of photon yield from 423 MeV/u $Fe^{24+}$ ions in $(2\bar{2}0)$ planar channel of 21 $\mu$m-thick Si crystal. Our calculation showed close agreement with these data [13]. To go further to the origin of this effect, we undertook detailed investigation [14] of how certain geometrical properties of the in-crystal electric field manifest themselves in these X-ray observables.

### 2.2. *RCE resolved by ion trajectories*

Resolution of survival fraction RCE data by ion trajectory was done in [18] by using a thin ($\sim 1\,\mu$m) silicon crystal of thickness corresponding to one-quarter of the period of $Ar^{17+}$ ion transverse oscillations in the channel. Being motivated by this work, we performed calculations [14] of trajectory resolved anisotropy of X-rays from 423 MeV/u $Fe^{24+}$ ions in $(2\bar{2}0)$ channel of $0.9\,\mu$m Si crystal, coherently excited at the same $(k,l) = (2,-1)$ resonance as in work [12] on the angular anisotropy of X-rays from $Fe^{24+}$ ions in thick target. Calculation results show the origin of the horizontal (in the plane) to vertical (perpendicular to the plane) photon yield asymmetry ratio $H/V \approx 2$ observed in the experiment.

Experimental verification of these results demands coincident detection of the characteristic X-ray photons with emerging ions resolved by the deflection angle [10]. Inspiring results of recent photon-ion coincidence experiments [15] in near-lying domains of physics of highly charged ions support the idea of similar coincidence measurements in RCE studies as well.

### 2.3. *RCE without channeling (3D RCE)*

The recent experiment [16] proved the possibility of observation of RCE without channeling. When ions are sent through the crystal in non-channeling directions $(\theta_t, \phi_t)$, tuning to any specific $(k,l,m)$ harmonic may be achieved according to a general Okorokov resonance condition

$$\Delta E = \hbar\gamma(\mathbf{G}_{klm}\cdot\mathbf{v}) = \frac{2\pi\hbar}{a}\gamma v\left(\sqrt{2}(k\cos\phi_t + m\sin\phi_t)\cos\theta_t + l\sin\theta_t\right). \quad (1)$$

We performed calculations for 391 MeV/u $Ar^{17+}$ ions passing through 1 $\mu$m Si crystal. In Fig.1 we compare our results with the data [16] for the survival fraction. Considerable angular anisotropy is expected for corresponding in-target and out-of-target characteristic X-ray radiation.

## 3. Theoretical arguments for new RCE measurements

### 3.1. *Auger electrons from double RCE*

The experiment [19] on "ladder type" double RCE of autoionizing states of helium-like $Ar^{16+}$ ions passing through the $(2\bar{2}0)$ planar channel of a $27\,\mu$m-thick Si crystal connects physics of autoionization phenomena and channeling. Our calculation [14] performed for 387.90 MeV/u $Ar^{16+}$ ions, planar channeled in Si crystal and resonantly excited under simultaneous action of

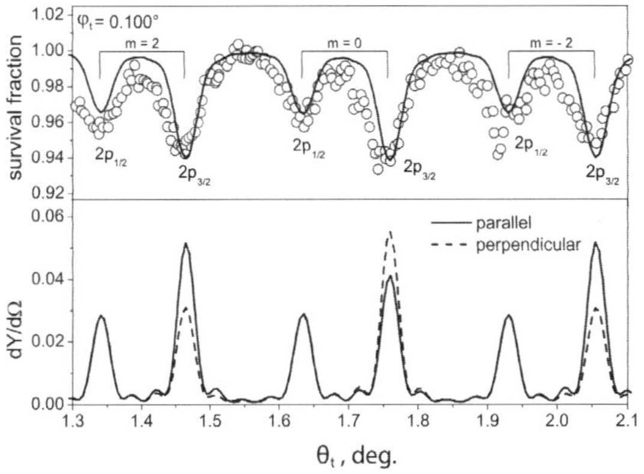


**Figure 1.** Calculated survival fraction (normalized) and differential X-ray yield under 3D RCE $(k,l) = (1,1)$ conditions for 391 MeV/u $Ar^{17+}$ in $1\,\mu$m thick Si crystal. Differential X-ray yield (summed contributions from excited ions decay inside and outside the target) is presented in two directions: in (220) plane (solid line) and perpendicular to this plane (dashed line). To fit the experimental data [16], ionization cross-sections [17] for n=2 states were multiplied by the factor $\sim 0.45$.

the $(k,l) = (1,-2)$ and $(k,l) = (1,1)$ harmonics, has revealed considerable Stark mixing between closely lying $2p^2 : {}^1D_2$ and $2s2p : {}^1P_1$ autoionizing states due to the Lindhard potential. As estimated in paper [14], the total yield of Auger electrons emitted in this process is about 40% of the total probability of the electron loss by the ion at the resonance maximum. Calculations also show that doubly excited states produced in the RCE process are aligned and, as a result, angular distribution of the Auger electrons is strongly anisotropic.

### *3.2. Metastable ion production*

Resonant monopole E0 transitions, such as $1s \to 2s$, are strongly suppressed in photoexcitation of free ions but turn out to be open in crystals due to their Stark mixing with corresponding optically allowed RCE transitions. Under channeling, the Lindhard continuous potential is responsible for this mixing; contribution of the metastable state to the survival fraction of the initial charge state is expected to be of the order of 10% (Fig. 2).

Without channeling such mixing was observed by the Tokyo collaboration [21] using double resonance technique in 3D resonant coherent excitation of $Ar^{16+}$. Ionic excited states were coupled by resonant strong oscillating electric field resulting in *Autler-Townes effect.* The density matrix calculation for the yield of the metastable $1s2s : {}^1S_1$ fraction in this experiment at the finest tuning on the resonance $1s2s : {}^1S_0 \to 1s2p : {}^1P_1$ is presented in Fig. 3.

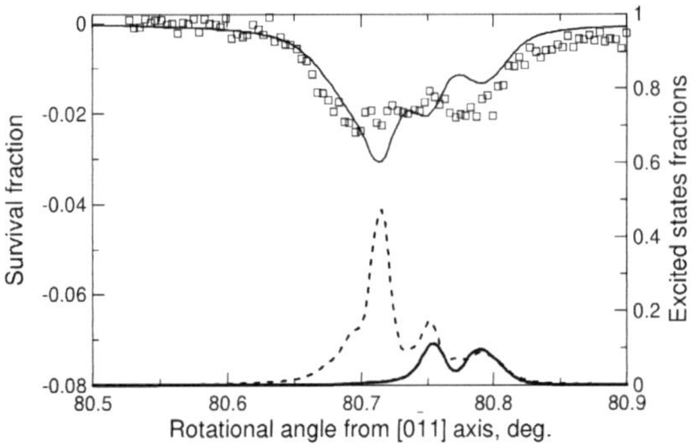


**Figure 2.** Calculated survival fraction (thin solid line) and excited states fractions (dashed lines — 2p, thick solid line — metastable 2s) for 390 MeV/u $Ar^{17+}$ passing through 0.74 $\mu$m thick silicon crystal under planar channeling conditions. The experimental data (squares) are taken from [20].

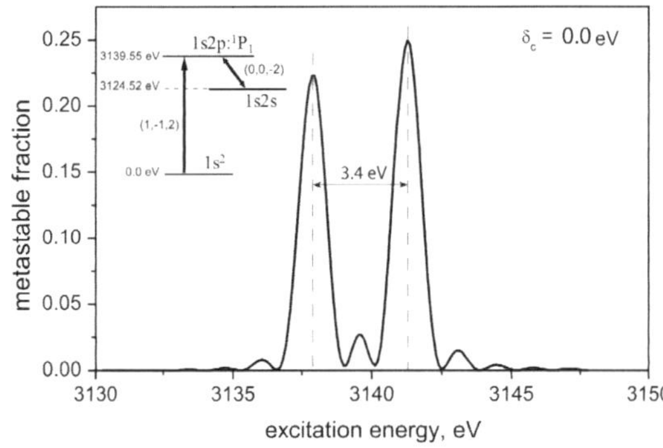


**Figure 3.** Calculated fraction of metastable $1s2s : {}^1S_0$ state at the exit of the $1\mu$m silicon crystal for 416 MeV/u $Ar^{16+}$ under double resonance in non-channeling conditions corresponding to the setup of the recent experiment [21].

### *3.3. Stokes parameters of X-ray radiation from RCE devices*

We consider observation of strong alignment of the angular momentum of relativistic RCE ions [12] as a signal to special experimental and theoretical investigations of this process as a candidate for a tunable source of polarized X-ray radiation. In Fig. 4 we demonstrate our calculation for the laboratory frame Stokes parameters of radiation from relativistic $Fe^{24+}$ ions under planar channeling in two cases of thick $21\,\mu$m and thin $0.9\,\mu$m Si crystal. The former illustrates high level of possible linear polarization of photons from $1s2p : {}^1P_1 \to 1s^2 : {}^1S_0$ deexcitation transition and possibility to vary its linear polarization degree in the interval $0 \le P_L \le 1$ on the $\theta \approx 40°$ cone around the direction of the incoming ion beam.

In Fig. 4b we suggest a special organization of RCE X-ray experiment to produce vector polarized RCE ions and, hence, to be able to observe its circular polarized radiation. Contrary to usual RCE X-ray measurements, photon detection should be done in coincidence with the exit ions leaving the channel asymmetrically relative to the channel plane. The example shown corresponds to RCE of 423 MeV/u $Fe^{24+}$ ions in thin 0.9 $\mu$m-thick Si crystal.

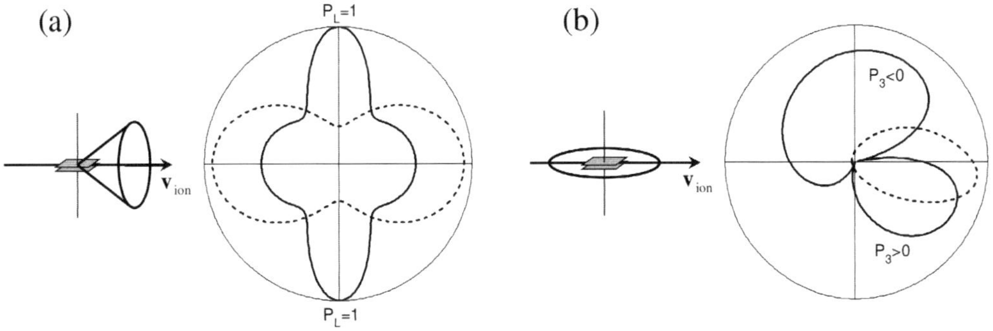


**Figure 4.** Angular distribution (dashed lines) and polarization of photons in the laboratory frame from 423 MeV/u $Fe^{24+}$ ions: a) linear polarization $P_L$ (solid line) of photons on the 41° cone, b) circular polarization $P_3$ (solid line) in the channel plane in coincidence with ions leaving the channel on one side from the channel plane.

## 4. Conclusion

We have reported selected results from our latest RCE studies and presented suggestions for new RCE experiments. Other aspects of performed calculations will be reported elsewhere.

## Acknowledgments

We would like to thank Prof. T. Azuma and his group at Tokyo Metropolitan University for useful discussions. The work was supported by RFBR (grants 06-02-17367, 08-02-08690).

## References

[1] Crawford O H and Ritchie R H 1979 *Phys. Rev.* A **20** 1848
[2] Datz S, Moak C D, Crawford O H et al 1980 *Nucl. Instr. Meth.* **170** 15
[3] Yamashita Y, Ohtsuki Y H 1980 *Phys. Rev.* B **22** 1183
[4] Salin A , Arnau A, Echenique P M 1998 *Phys. Rev.* A **57** 2772
[5] Komaki K, Azuma T, Ito T et al 1998 *Nucl. Instr. Meth.* B **146** 19
[6] Balashov V V, Bodrenko I V 2001 *Moscow University Physics Bulletin* **56** N1 33
[7] Balashov V V, Bodrenko I V 2006 *Phys. Lett.* A **352** 129; *Nucl. Instr. Meth.* B **245** 52
[8] Okorokov V V 1965 *JETP Lett.* **2** 111; *Yad. Phys.* **2** 1009 [*Sov. J. Nucl. Phys.* **2** 719].
[9] Datz S et al 1978 *Phys. Rev. Lett.* **40** 843
[10] Iwata Y, Komaki K, Yamazaki Y et al 1990 *Nucl. Instr. Meth.* B **48** 163
[11] Balashov V V, Grum-Grzhimailo A N, Kabachnik N M 2000 *Polarization and Correlation Phenomena in Atomic Collisions* Kluwer Academic (New York)
[12] Azuma T, Takabayashi Y, Kondo C et al 2006 *Phys. Rev. Lett.* **97** 145502
[13] Balashov V V, Sokolik A A, Stysin A V 2008 *JETP* **107** 133
[14] Balashov V V, Sokolik A A, Stysin A V 2008 *Nucl. Instr. Meth.* B (to be published)
[15] Dauvergne D et al 2003 *Nucl. Instr. Meth.* B **205** 773
[16] Kondo C, Masugi S, Nakano Y et al 2006 *Phys. Rev. Lett.* **97** 135503
[17] Rozet J P, Stephan C and Vernhet D 1996 *Nucl. Instr. Meth.* B **107** 67
[18] Kondo C, Masugi S, Muranaka T et al 2007 *Nucl. Instr. Meth.* B **256** 157
[19] Nakano Y, Masugi S, Muranaka T et al 2007 *J. Phys. Conf. Ser.* **58** 395
[20] Azuma T et al 2003 *Nucl. Instr. Meth.* **205** 779
[21] Nakai Y et al 2008 *Phys. Rev. Lett.* **101** 113201

Journal of Physics: Conference Series **163** (2009) 012088 doi:10.1088/1742-6596/163/1/012088

# Kinetic energy release distributions for $C_2^+$ emission from multiply charged $C_{60}$ and $C_{70}$ fullerenes

**H Cederquist[1], N Haag[1], Z Berényi[1], P Reinhed[1], D Fischer[1], M Gudmundsson[1], H A B Johansson[1], H T Schmidt[1] and H Zettergren[2]**

[1] Department of Physics, Stockholm University, AlbaNova University Center, S-10691 Stockholm, Sweden

[2] Department of Physics and Astronomy, University of Aarhus, DK-8000 Aarhus C, Denmark

E-mail: cederquist@physto.se

**Abstract.** We present a systematic study of experimental kinetic energy release distributions for the asymmetric fission processes $C_{60}^{q+} \rightarrow C_{58}^{(q-1)+} + C_2^+$ and $C_{70}^{q+} \rightarrow C_{68}^{(q-1)+} + C_2^+$ for mother ions in charge states $q = 4$–$8$ produced in collisions with slow highly charged ions. Somewhat to our surprise, we find that the KERD for asymmetric fission from $C_{60}^{q+}$ are considerably wider and have larger most likely values than the $C_{70}^{q+}$ distributions in the corresponding charge states when $q > 4$.

## 1. Introduction

Collision experiments on the extraordinary stable and symmetric fullerenes are particularly useful for the identification of the main mechanisms behind electron transfer processes, energy deposition and fragmentation. Indeed a very large number of such studies has been carried out since the 1990s (see e.g. [1] and references therein). When fullerenes collide with slow highly charged atomic ions several electrons may be transferred and multiply ionized intact fullerenes may be produced. Depending on their internal excitation energy, fullerene ions may also fragment within the experimental time frame of a few microseconds. The kinetic energy release distributions (KERD) in these unimolecular fragmentation processes provide valuable information on the energetics and dynamics of the individual reactions, with positions and shapes governed by the potential energy surfaces describing the interactions of the separating fragments in the exit channel. Unlike neutral $C_2$ evaporation,

$$C_n^{q+} \rightarrow C_{n-2}^{q+} + C_2, \quad (n = 60, 70) \tag{1}$$

for which kinetic energy release distributions have been measured earlier [2–4], much less is known about the $C_2^+$ emission process (asymmetric fission),

$$C_n^{q+} \rightarrow C_{n-2}^{(q-1)+} + C_2^+ \quad (n = 60, 70). \tag{2}$$

In this work, we present the first systematic study of experimental KERD for asymmetric fission (2) of $C_{60}^{q+}$ and $C_{70}^{q+}$ with $q = 4$–$8$. A single experimental KERD has been reported by Senn *et al.* [5] in 1998, but the more recent results on kinetic energy releases in asymmetric fission of multiply charged fullerenes [6, 7] are given as single (typical) values only. There are also no theoretical studies that could suggest a functional form for KERD of (2).

## 2. Experiment

For the present study, we have designed a new linear recoil-ion-momentum spectrometer which has been optimized for measurements of KERD for fragmenting complex molecules. The spectrometer, which is described in more detail in [8], consists of an acceleration region with 19 ring electrodes in a grounded housing with a small aperture for the *collinear* target jet, a field-free drift region, and a position sensitive detector with two microchannelplates (MCP, 40 mm in diameter) and a resistive anode. The dimensions of the spectrometer are chosen such that first order time focusing for different trajectory starting points is achieved. Collimated fullerene target jets effusing from a sublimation oven are crossed with a pulsed (2 kHz, 5 $\mu$s pulse length) beam of 57 keV $Xe^{19+}$ ions ($v$ = 0.4 a.u.) from a 14.5 GHz Electron Cyclotron Resonance (ECR) ion source.

Intact ionized fullerenes and charged fragments are extracted towards the detector directly after the passage of the ion beam by homogeneous acceleration fields of 6.0 or 9.0 V/mm. The ion flight times, as deduced from the time differences between the extraction pulses and signals from the MCP, and the corresponding four anode corner signals, yielding the position on the detector, are stored on an event-by-event basis.

## 3. Data analysis

Two-dimensional detector images for multiply charged $C_{58}$ and $C_{68}$ fragments from fission (2) and evaporation (1) (see figure 1, left, for an example) are converted to radial intensity distributions (cf. figure 1, right), which are then used to extract distributions of kinetic energy releases, $P(\epsilon)$, by means of a simulation and fitting procedure. SIMION 7.0 simulations of the radial distributions are performed taking into account the actual initial conditions and assumed

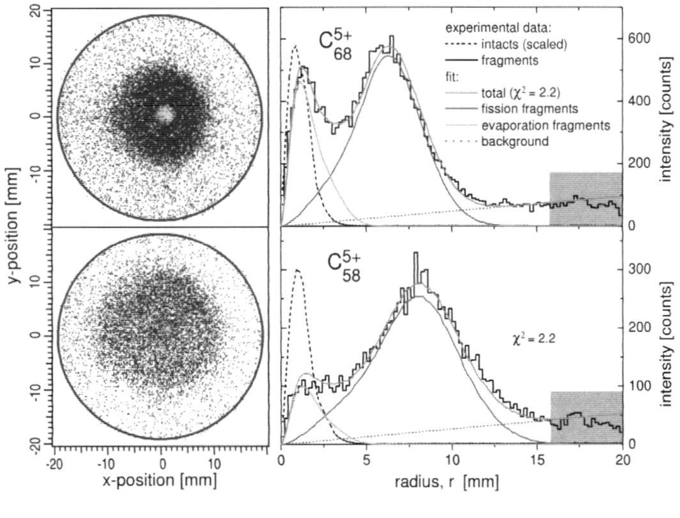


**Figure 1.** (Color online) Left: Two-dimensional detector images for $C_{68}^{5+}$ (upper panel) and $C_{58}^{5+}$ (lower panel) due to fragmentation of $C_{70}$ and $C_{60}$ mother ions, respectively. Right: Corresponding experimental radial distributions with fitted/simulated distributions. Contributions from fission and evaporation are indicated. The background is assumed to be uniform (intensity linear in $r$) up to $r$ = 16 mm. Events with $r$ > 16 mm are disregarded in the fit. For comparison, the radial distributions of intact $C_{70}^{5+}$ and $C_{60}^{5+}$ ions are indicated by dashed lines (not to scale).

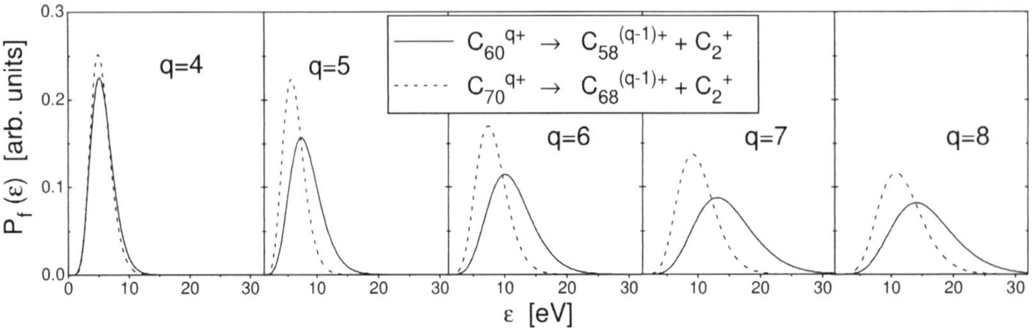


**Figure 2.** Kinetic energy release distributions in asymmetric fission reactions of multiply charged $C_{60}$ (——) and $C_{70}$ (- - - -), as functions of charge state, $q$. The distributions shown are optimized to reproduce the experimental data in the described simulation and fitting procedure.

probability distributions of the kinetic energy releases for evaporation, $P_e(\epsilon)$, and fission, $P_f(\epsilon)$. For different assumed analytical forms of the KERD for fission, the simulated total radial distributions (including contributions from evaporation, fission and background) are fitted to those obtained from the experiment by non-linear regression. The $(P_e(\epsilon),\ P_f(\epsilon))$-combinations which give the best fits are taken as the present results.

## 4. Results and discussion

The fission KERD, $P_f(\epsilon)$, yielding the best fit results are presented in figure 2. For both $C_{60}^{q+}$ and $C_{70}^{q+}$ the width of the distributions increases with $q$ (probably linked to the increase of the reverse activation barriers). We find that the KERD for $C_{60}^{4+} \rightarrow C_{58}^{3+} + C_2^+$ and $C_{70}^{4+} \rightarrow C_{68}^{3+} + C_2^+$ are very similar (cf. left panel of figure 2). For higher charge states $q$ of the mother ions, however, the distributions for $C_{70}$ are narrower and shifted towards lower energies.

These KERD have been obtained using the 'model-free' approach by Klots [9],

$$P(\epsilon) = a\epsilon^l exp\left(-l\frac{\epsilon}{\tilde{\epsilon}}\right), \tag{3}$$

where $a$ is a normalization factor, $\tilde{\epsilon}$ is the position of the maximum of the distribution, i.e. the most probable KER value, and $l$ is a parameter related to the interaction potential between the separating fragments ($0 \leq l \leq 1$). Following Głuch *et al.* [3] and Climen *et al.* [4], we take $l = 0.5$ for the evaporation KERD, corresponding to a Langevin-like long range interaction between $C_2$ and the heavier, charged fragment. The overall best fits to our experimental data are obtained by tentatively using equation 3 also for fission and taking $l = 8.5$ ($C_{60}$) and $l = 10.0$ ($C_{70}$). For the moment we refrain from an attempt to interpret these values and only regard the resulting functions $P_f(\epsilon)$ as efficient ways to parameterize the experimental data.

In figure 3, the most likely kinetic energy release values for *fission* of $C_{60}^{q+}$ and $C_{70}^{q+}$, $\tilde{\epsilon}_f$, are shown as functions of $q$. In the case of $C_{60}$, there is almost perfect agreement with theoretical values for the reverse activation barriers obtained by high level Density Functional Theory (DFT) transition state calculations [10] (cf. figure 3a). A comparison with earlier experimental kinetic energy release measurements is shown in part b of figure 3. The wide experimental distributions which we obtain show that kinetic energy releases may also be several eV smaller or larger than the reverse activation barrier. The lower energies may be explained as due to couplings of the reaction coordinate with other internal degrees of freedom leading to a situation in which the reverse barrier may partially be transformed to internal energy. The larger kinetic energy release

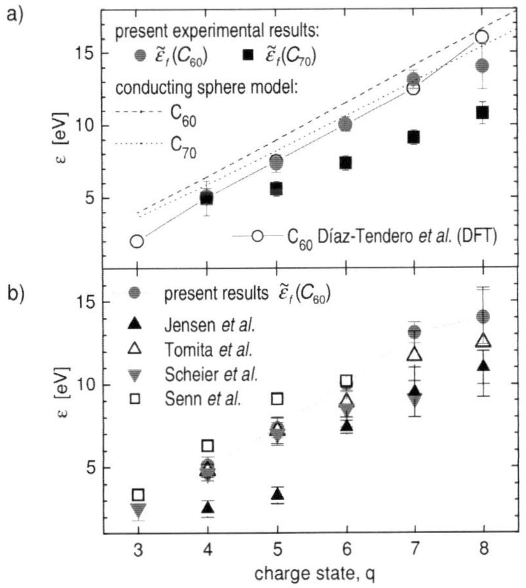


**Figure 3.** Comparison of the present most probable KER values for asymmetric fission of $C_{60}^{q+}$ and $C_{70}^{q+}$ with, (a), results from DFT calculations [10] (with lines to guide the eye) and an electrostatic model [6, 11], (b), other experiments on $C_{60}$ yielding single typical values by Scheier *et al.* [12], Tomita *et al.* [7], Senn *et al.* [5] and Jensen *et al.* [6].

values may possibly relate to remaining electronic excitations. At the moment, the mechanism behind this excitations is not completely clear and needs to be studied in further detail.

We believe that the observed differences between $C_{60}$ and $C_{70}$ are, at least partly, related to a larger polarizability due to the larger overall size of $C_{70}$. This effect, however, which is taken into account in the electrostatic model calculations [6, 11] plotted in figure 3a, cannot fully explain the observed differences, which are unexpectedly large. Additionally, the probably lower $C_{70}$ fission barriers [8], the larger number of vibrational degrees of freedom for $C_{70}$ and $C_{68}$ in comparison to $C_{60}$ and $C_{58}$, the inhomogeneous charge distribution on $C_{70}$ ions [13] and the non-spherical shape may play a role. Also the fact that there is a larger number of $C_{68}$ isomers and that several transformations between them are needed to reach the energetically most stable one, may lead to a situation where less excess energy is released as relative translational energy of the fragments.

## Acknowledgments

The support by the European Project ITS LEIF (No. RII3/026015) is gratefully acknowledged.

## References

[1] Campbell E E B and Rohmund F 2000 *Rep. Prog. Phys.* **63** 1061
[2] Matt S *et al* 1999 *Chem. Phys. Lett.* **303** 379
[3] Gluch K *et al* 2004 *Chem. Phys. Lett.* **385** 449
[4] Climen B *et al* 2007 *Chem. Phys. Lett.* **437** 17
[5] Senn G, Märk T D and Scheier P 1998 *J. Chem. Phys.* **108** 990
[6] Jensen J *et al* 2004 *Phys. Rev.* A **69** 053203
[7] Tomita S *et al* 2003 *Phys. Rev.* A **67** 063204
[8] Haag N *et al* 2008 *Phys. Rev.* A (accepted)
[9] Klots C E 1991 *Z. Phys.* D **21** 335
[10] Díaz-Tendero S, Alcamí M and Martín F 2005 *Phys. Rev. Lett.* **95** 013401
[11] Zettergren H *et al* 2002 *Phys. Rev.* A **66** 032710
[12] Scheier P, Dünser B and Märk T D 1995 *Phys. Rev. Lett.* **74** 3368
[13] Zettergren H, Alcamí M and Martín F 2008 *ChemPhysChem* **9** 861

# A newly built setup for small bio-molecule fragmentation study in Lanzhou

**B Li[1,2,3], X Ma[1], L Chen[3], R Brédy[3], X L Zhu[1], H P Liu[1], S F Zhang[1,2], W T Feng[1,2], D B Qian[1], D C Zhang[1], S Martin[3]**

1 Institute of Modern Physics, Chinese Academy of Sciences, Lanzhou 730000 China
2 Graduate University of Chinese Academy of Sciences, Beijing 100049 China
3 LASIM (UMR CNRS 5579), University Lyon 1, France

E-mail: x.ma@impcas.ac.cn

**Abstract**. A new setup for studying ion induced small bio-molecule fragmentation processes has been built in Lanzhou, China. A preliminary collision experiment between a 30 keV $He^{2+}$ ion beam and gas phase adenine molecules was performed. Partial TOF spectra associated to well-defined scattered projectile final charge states, $He^{+}$ or $He^{0}$ were recorded. Coincidence spectra between fragments were also obtained.

## 1. Introduction

In radiation therapy, the damage of DNA molecules is not only induced by direct impact of the primary high-energy particles, but also by low energy secondary species such as radicals, cations, anions, and secondary electrons, etc., generated in the track of the primary ionizing radiation. Studies involving low energy electron impact on DNA films and gas-phase bio-molecules showed that strand breaks in DNA or dissociation of the bio-molecules can occur at energies well below the ionization energy thresholds, and it was found that the formation of transient anions played an important role in the fragmentation processes [1, 2]. Fragmentation of bio-molecules induced by ion beams at low energy has also attracted much attention during the last years. Using multi-coincidence techniques, detailed information concerning fragmentation dynamics of molecule dissociation under ion impact has been revealed [3-8].

In this work, we report a setup built in Lanzhou aiming at performing collision experiments between ion and small bio-molecules. The specificity of this setup is to be able to detect scattered projectiles at all final charge states, including neutrals and negative ions, in coincidence with the detection of charged fragments. The first running experimental results using 30 keV $He^{2+}$ beam and Adenine ($C_5H_5N_5$) molecules are presented in this paper.

## 2. Experimental setup

A schematic view of the experimental setup is shown in Fig. 1. $He^{2+}$ projectiles at 30 keV were delivered by the ECR ion source of the Institute of Modern Physics in Lanzhou. The ion beam was guided into the experimental line after a 45° magnet. It was collimated by two sets of adjustable slits separated by a distance of 800 mm (not shown in the sketch) and then cleaned by a 30° coaxial electrostatic deflector (Beam cleaner). In the collision chamber, the beam crossed a gas-phase adenine jet evaporated from an oven heated at 135°C. A liquid nitrogen cooled trap was mounted above the collision region to collect the molecular jet. Two sets of electrostatic plates in horizontal and vertical planes were set before the interaction region in order to optimize the crossing between the projectile beam and the molecular jet. After the collision region, the outgoing projectiles were deflected according to the charge states by a parallel-plate analyzer. The intense primary beam was collected into a Faraday Cup (FC) beside the detector. Scattered particles that have undergone charge exchange were detected by a two-dimensional position sensitive detector (PSD). The charge of a scattered particle was mapped by the positions on the PSD. The hitting time of a scattered projectile on the PSD provided the time reference for each collision event. Charged recoil products were extracted by a homogeneous electrostatic field of 85V/cm perpendicular to both the ion beam and the target jet. After acceleration and passing through a field free drift Time Of Flight (TOF) tube, the recoil products were detected by a MCP detector (MCPs). The recoil signals were sent to a LeCroy 4208 TDC, which was operated at single channel multi-hit (8) mode, for analyzing the arrival times with respect to the reference signal provided by the PSD. The length of the acceleration region and that of the TOF tube were designed to fulfill the first-order focusing condition of Wiley-McLaren.

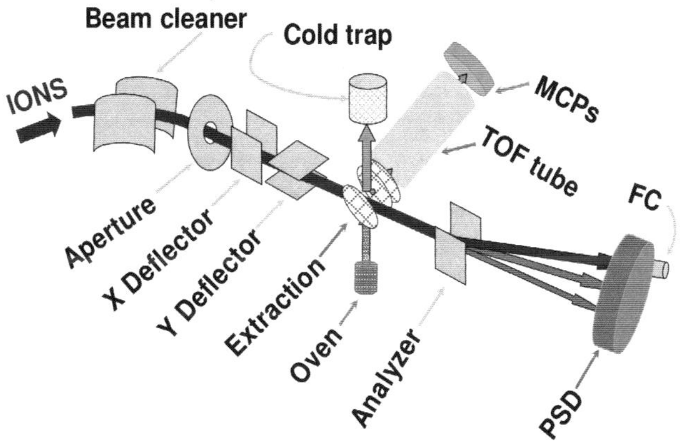


**Figure 1.** Schematic view of the experimental setup.

With the equipped Analyzer and PSD, the final charge states of the outgoing projectiles can be determined, and all projectiles undergoing charge exchange can be measured under the same experimental conditions. Partial TOF spectrum corresponding to a given scattered projectile charge state can be extracted from the total TOF spectrum by adding a filter on the charge of the scattered projectiles.

## 3. Results and discussion

In collisions between $He^{2+}$ ion beam and adenine molecules, one or two electrons can be kept by the projectile. These two channels were differentiated by the detection of two spots mapped on the PSD assigned respectively to mono-charged and neutral scattered projectiles, $He^{+}$ and $He^{0}$. By selecting the main spot associated to the scattered $He^{+}$ ions, partial TOF spectrum corresponding to single electron capture process was extracted, as shown in Fig. 2 a). In this spectrum, mono-charged stable parent ion $C_5H_5N_5^+$ is observed with high intensity. The small peak located next to the parent ion is assigned unambiguously to the heavy fragment $C_4H_4N_4^+$ resulted from the loss of a HCN fragment. Smaller fragments, such as $C_2H_mN_2^+$, $C_2H_mN^+$, $CH_mN^+$, $H^+$, etc (Here *m* stands for the number of hydrogen atom), are also present in the spectrum with lower intensities. Unfortunately, this part of the spectrum is dominated by $O_2^+$, $N_2^+$, and $H_2O^+$ peaks due to a leakage of our vacuum chamber.

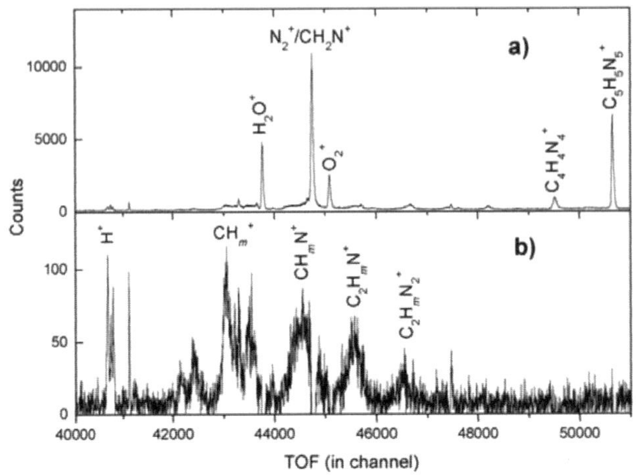


**Figure 2.** Partial TOF spectra for a) one electron stabilized on scattered the projectile; b) two electrons stabilized on scattered the projectile.

Fig. 2 b) presents the partial TOF spectrum obtained by selecting the spot associated to the neutral scattered projectiles $He^0$. It corresponds to processes where two electrons are kept by the scattered projectile. In this spectrum, only light fragments resulted from multi-fragmentation of the molecules are observed. This is in consistence with the formation mechanism of neutrals. Let's consider first two electrons primarily captured at large distance via two-electron-transfer process. Duo to the very different potential energy wells on the two doubly charged centers (79 eV for $He^{++}$ and 22 eV for adenine$^{++}$), these electrons occupy a doubly excited state on the projectile that decays mainly by autoionisation. Such events are classified in the partial TOF spectrum extracted in coincidence with the detection of $He^+$ (Fig.2.a). Therefore, in collisions between a highly charged ion and a target with limited dimension, the neutral formation channel is attributed to direct multi-electron-capture at short distance in frontal collisions [9]. In such cases, the energy deposited in the target during the collision leads to complete breakup of the target after the collision.

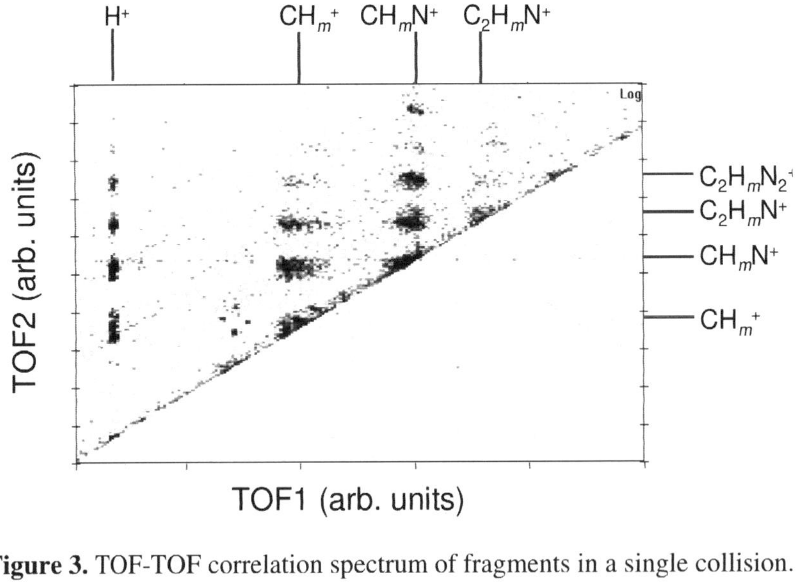


**Figure 3.** TOF-TOF correlation spectrum of fragments in a single collision.

When multifragmentation occurs, two or more charged particles are produced in one collision event. Fig. 3 is a correlation spectrum between TOF of the first two fragments in a single collision. Horizontal axis corresponds to the TOF of the faster/lighter one, and vertical axis to the TOF of the slower/heavier one. This spectrum was built without differentiation of the projectile final charge state. It shows that most of the coincidences occur between small fragments for both $He^+$ and $He^0$ channels. To get a better fragment ion collection efficiency, a stronger extraction field of 254V/cm was applied in this case.

## 4. Summary

With the experimental setup developed in Lanzhou, we are able to provide partial TOF spectra according to the final charge state of the scattered projectiles. The TOF spectrum obtained in coincidence with the detection of neutrals is a typical multifragmentation mass spectrum. It is attributed to short distance frontal collisions. TOF-TOF coincidence spectrum between fragments emerged in a single collision provides detailed information on the molecule dissociation dynamics. With this setup, we have also performed collision experiments using $F^{2+}$ projectiles. Partial TOF spectra have been obtained not only for neutral but also for negative scattered projectiles. These results will be published in a forthcoming paper. Based on the experiences from these preliminary experiments, an improved experimental setup is under development in Lanzhou.

**Acknowledgments**

This work was supported by the international collaboration convention CNRS in France (N°21657) and the National nature Science Foundation of China (No. 10434100). L. Chen gratefully acknowledges the support of K.C.Wong Education Foundation, Hong Kong.

**References**

[1] BoudaÏffa B, Cloutier P, Hunting D, Huels M A and Sanche L 2000 *Science* **287** 1658
[2] Sanche L 2005 *Eur. Phys. J. D* **35** 367
[3] deVries J, Hoekstra R, Morgenstern R and Schlathölter T 2003 *Phys. Rev. Lett.* **91** 053401
[4] Deng Z, Bald I, Illenberger E and Huels M A 2005 *Phys. Rev. Lett.* **95** 153201
[5] Alvarado F, Bari S, Hoekstra R and Schlathölter T 2007 *J. Chem. Phys.* **127** 034301
[6] Martin S, Brédy R, Allouche A R, Bernard J, Salmoun A, Li B and Chen L 2008 *Phys. Rev. A* **77** 062513
[7] Alvarado F, Bernard J, Li B, Brédy R, Chen L, Hoekstra R, Martin S and Schlathölter T 2008 *ChemPhysChem* **9** 1254
[8] Bernigaud V, Manil B, Maunoury L, Rangama J and Huber B A 2008 *Eur. Phys. J.* D (in print)
[9] Martin S, Chen L, Bernard J, Brédy R, Buchet-Poulizac M C, Ma X and Wei B 2006 *Europhys. Lett.* **74** 985

# Guiding of Argon ions through PET nano capillary foils

**M.Kreller[1], G. Zschornack[1] and U. Kentsch[2]**
[1]Technische Universität Dresden, Institute of Applied Physics, Helmholtzstr. 10, 01062 Dresden, Germany
[2]DREEBIT GmbH, Zur Wetterwarte 50, 01109 Dresden, Germany

E-mail: m.kreller@fzd.de

**Abstract.** The transmission of charge state separated slow ions through nano capillaries with an aspect ratio of 1/100 in insulating polyethylene terephthalate (PET) polymers was investigated on argon ions with different charge states from $Ar^{3+}$ up to $Ar^{12+}$. The experiments were performed with a beam line featuring an ion deceleration system. With it the ion guiding of projectiles of different kinetic energies ranging from 600eV up to 9.6 keV has been studied. The guiding power as well as the divergence angle of the transmitted ion beam have been investigated in dependence on the initial ion charge state and the ion energy. A scaling of the profile width $\tan \sigma_\phi \sim \sqrt{q}$ at a constant ion energy was found. Furthermore, the guiding angle was measured as independent on the ion charge state at an ion energy scaling with the ion charge state.

## 1. Introduction

The interaction of highly charged ions (HCI) with solid surfaces is described by the classical over the barrier model (COB) [1]. Using metallic microcapillaries as a target for a beam of slow HCI is a possibility to observe the decay of the hollow atoms that passed the capillaries [2]. Motivated by the extension of the COB to insulating surfaces [3] first measurements of the transmission of HCI through microcapillary targets in insulating materials were done [4].

In metallic materials the HCI can pass through the capillaries only when their axes are parallel to the incident beam direction. For insulating microcapillaries a transmission of HCI was also measured when the capillary axes are tilted in respect to the direction of the incident ion beam. The creation of charge patches on the inner walls of the capillaries and the deflection of the HCI enables the transmission of ions at tilt angles larger than the capillary aspect ratio. Since the first measurements in 2002 by N. Stolterfoht *et al.* several investigations have been done using different ion energies[5], ion charge states[6] target materials [7, 8] and different types of ions [9] but also electrons [10, 11] as projectiles. The aim of the miscellaneous experiments is to improve the theoretical models [12, 13] of the guiding effect. To investigate the influence of the charge state of the HCI we studied the guiding power as defined in [14] and the transmission profile of argon ions with different charge states from $Ar^{3+}$ up to $Ar^{12+}$ and with different kinetic energies ranging from 600 eV up to 9.6 keV.

## 2. Experimental techniques

The present experiments were performed at the micro beam facility of the TU Dresden at the Forschungszentrum Dresden-Rossendorf [15]. The ions are produced with an electron beam ion source[16]. The capillary foils are mounted on a target holder (Fig. 1) in the target chamber at a base pressure of $10^{-9}$ mbar. The target holder can be tilted with high precision relative to the direction of the incident ion beam to realise different tilt angles ($\Psi$) up to $\pm 10°$. The ion beam is focused and collimated to a diameter of 1 mm with a current of typically 30 – 800 pA. The divergence of the beam is better than 0.5° full width at half maximum (FWHM). The transmitted ion beam is collimated and deflected by a condenser perpendicular to the tilt direction so that different charged ions are detected at different positions in x-direction. The collimator can be removed to measure the transmission profile in the x-direction.

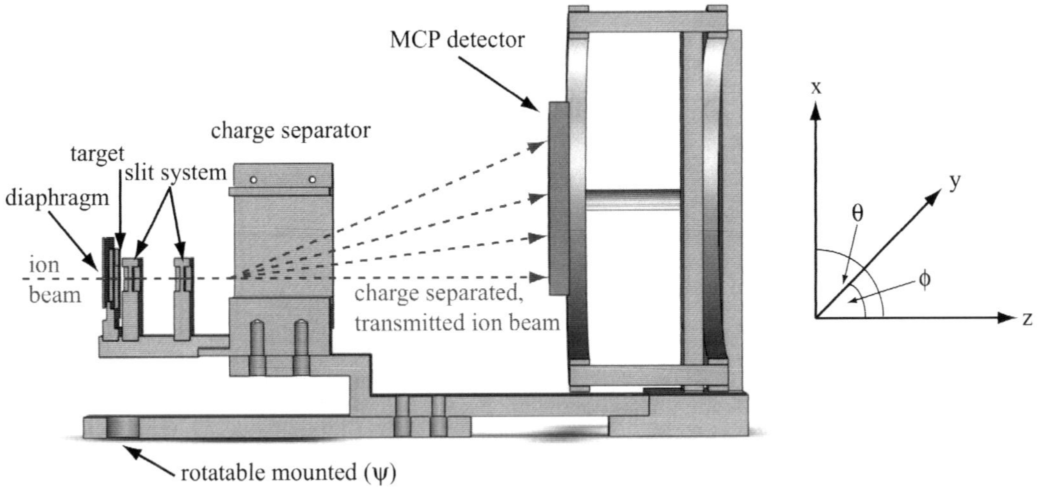


**Figure 1.** The Experimental setup is mounted a rotatable stage behind a deceleration lens in a ultrahigh vacuum chamber. Different ion energies in the range from several eV up to 20 keV times $q$ can be realised. A multi-channel-plate with a delay-line-anode is used as a position sensitive ion detector.

The PET capillary samples used for the experiments were produced by the irradiation of the foils with fast ions and etching the ion tracks. The capillaries have a diameter of about $130 \pm 20$ nm at a length of 10 $\mu$m. The front and the back of the PET foil is evaporated with Au forming a film of 10 nm thickness. The used capillary density of $10^7$ mm$^{-2}$ implies a geometric opening of about 13%.

## 3. Experimental results

### *3.1. Guiding power*

The first investigations were concentrated on the dependence of the guiding power[14] on the charge state of the incident ion beam. The guiding power, as a capability of the capillaries to guide ions, can be quantized by the tilt angle $\Psi_c$ for which the normalized transmission fraction drops as $f(\Psi_c)/f(0) = 1/e$. $\Psi_c$ is called the guiding angle.

First, we determined the fraction of transmitted ions by integration of the transmission profiles in $\theta$-direction. This intensity profile (Fig. 3, left graph) in dependence on $\phi$ is fitted by the Gaussian-like function

$$f(\phi) = f_0 + a \cdot \exp\left[-\left(\frac{\phi - \phi_0}{\sigma_\phi}\right)^2\right] \tag{1}$$

with the additional fitting parameter $f_0$. For calculating the guiding angle Eq. 1 is integrated neglecting the parameter $f_0$, that represents the background of the measurements. The transmission intensity calculated for different tilt angles is plotted in dependence on the tilt angle and fitted by

$$f(\Psi) = f(0) \cdot \exp\left[-\frac{\sin^2 \Psi}{\sin^2 \Psi_c}\right]. \tag{2}$$

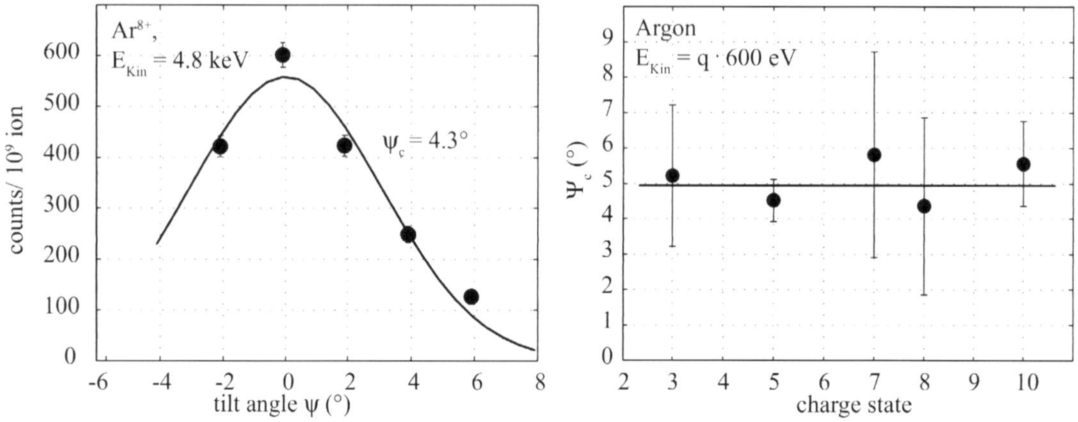


**Figure 2.** Guiding power of the transmission of Argon ions. The graph at the left shows the transmission of $Ar^{8+}$ in dependence on the tilt angle $\psi$ (dots) with the fit (black line) to calculate the guiding angle $\psi_c$. The graph at the right shows the guiding angle in dependence on the charge state of the incident ion beam. The energy of the incident ions is $600\,\text{eV} \cdot q$.

In Fig. 2 the measurement of the guiding angle for $Ar^{8+}$ and the dependence of the guiding angle on the ion charge state at an ion energy scaling linearly with the ion charge state are shown.

### *3.2. Transmission profile*

The relationship between the guiding power and the profile of the transmitted ion beam was measured. Hence, the dependence of the transmission profile on the charge state is analyzed in the following. The opening, defined as the width $\sigma_\phi$ in the $\phi$-direction of the transmitted ion beam is determined by the ion velocity in y- and z-direction:

$$\tan \sigma_\phi \sim \frac{v_y}{v_z} \tag{3}$$

The velocity in z-direction, depending on the ion energy, is chosen as a constant. The ion velocity in y-direction depends on the ion charge state $q$ and on the deflecting potential $E_y$ of the capillaries ($v_y \sim \sqrt{q \cdot E_y}$). Hence, if $E_y$ is independent on the ion charge that it is created by, Eq. 3 can be written as:

$$\tan \sigma_\phi = K(E_y) \cdot \sqrt{q}. \tag{4}$$

In Fig.3 the measurement of the transmission profile and the dependence of the transmission profile and the parameter $K(E_y)$ on the ion charge state at a constant kinetic energy are plotted. The width $\sigma_\phi$ is averaged over different tilt angles because no significant dependence of the width on the tilt angle was measured.

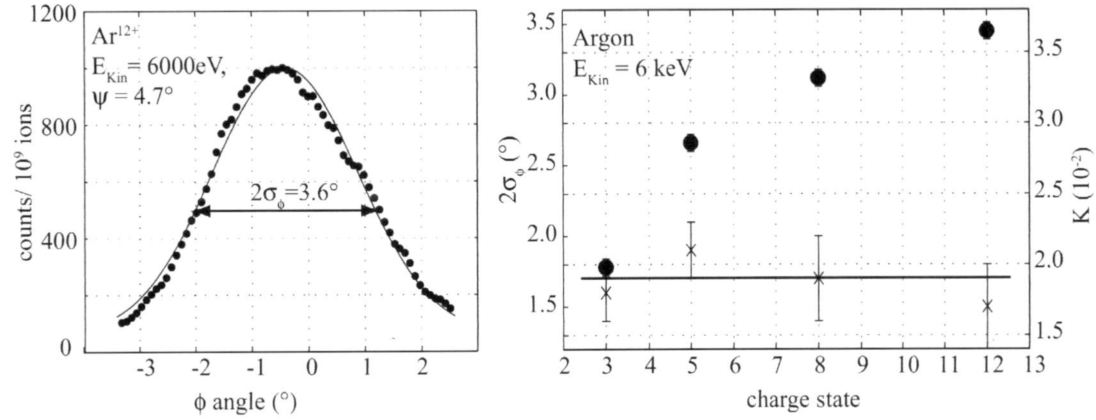


**Figure 3.** The graph on the left side shows the transmission profile of $Ar^{12+}$ ions. The dots represent the measurement and the solid line the Gaussian fit. In the graph on the right side the width of the transmission profile $\sigma_\phi$ (dots) and the parameter $K(E_y)$ (crosses) in dependence on the ion charge state are plotted. The energy of the incident ions is 6.0 keV.

## 4. Conclusion

The scaling of the transmission profile $\tan\sigma_\phi \sim \sqrt{q}$ and the independence of the parameter $K(E_y)$ on the charge state at a constant ion energy give evidences, that the deflecting potential of the capillaries depends only on the target properties such as surface and bulk conductivity. Furthermore, the independence of the guiding angle on the ion charge state at an ion energy scaling with the ion charge state shows, that the deflecting fields on the capillary walls are independent on the charge state of the initial ion beam.

## Acknowledgments

This work is supported by the Deutsche Forschungsgemeinschaft (project ZS 14/9-1) and by the EFRE fund of the EU and by the Freistaat Sachsen (Projects 12321/2000 and 12184/2000).

## References

[1] Burgdörfer J, Lerner P, Meyer FW, 1991 *Physical Review A* **44** 5674-5685
[2] Iwai Y, Murakoshi D, Kanai Y, Oyama H, Ando K, Masuda H, Nishio K, Nakao M, Tamamura T, Komaki K, Yamazaki Y, 2002 *Nucl. Instrum. Methods Phys. Res. B* **193** 504-507
[3] Ducree JJ, Casali F, Thumm U, 1998 *Physical Review A* **57** 338-350
[4] Stolterfoht N, Bremer JH, Hoffmann V, Hellhammer R, Fink D, Petrov A, Sulik B 2002 *Phys. Rev. Lett.* **88**
[5] Hellhammer R, Fink D, Stolterfoht N, 2007 *Nucl. Instrum. Methods Phys. Res. B* **261** 149-152
[6] Allen FI, Persaud A, Park SJ, Minor A, Sakurai M, Schneider DH, Schenkel T 2006 *Nucl. Instrum. Methods Phys. Res. B* **244** 323-326
[7] Schuch R, Johansson A, Kumar RTR, Sahana MB, Skog P, Soroka IL, Vikor G, Zhang HG 2008 *Canadian Journal of Physics* **86** 327-330
[8] Sahana MB, Skog P, Vikor G, Kumar RTR, Schuch R 2006 *Physical Review A* **73**
[9] Stolterfoht N, Hellhammer R, Bundesmann J, Fink D 2008 *Physical Review A* **77**
[10] Milosavljevic AR, Vikor G, Pesic ZD, Kolarz P, Sevic D, Marinkovic BP, Matefi-Tempfli S, Matefi-Tempfli M, Piraux L 2007 *Physical Review A* **75**
[11] Das S, Dassanayake BS, Winkworth M, Baran JL, Stolterfoht N, Tanis JA 2007 *Physical Review A* **76**
[12] Schiessl K, Palfinger W, Tokesi K, Nowotny H, Lemell C, Burgdorfer J 2005 *Physical Review A* **72**
[13] Lemell C, Schiessl K, Nowotny H, Burgdorfer J 2007 *Nucl. Instrum. Methods Phys. Res. B* **256** 66-70
[14] Hellhammer R, Bundesmann J, Fink D, Stolterfoht N 2007 *Nucl. Instrum. Methods Phys. Res. B* **258** 159-162
[15] Grossmann F, Heller R, Kreller M, Kentsch U, Landgraf S, Ovsyannikov VP, Schmidt M, Ullmann F, Zschornack G 2007 *Nucl. Instrum. Methods Phys. Res. B* **256** 565-570
[16] http://www.dreebit.com

# Discharging dynamics of insulator surfaces irradiated by highly charged ions

**N. Bundaleski[a], H. Khemliche[a], P. Rousseau[b], A. Cassimi[b], L. Maunoury[b], and P. Roncin[a]**

[a] Laboratoire des Collisions Atomiques et Moléculaires,
UMR 8625, CNRS -Université Paris-Sud, Orsay, F-91405, France

[b]CIMAP, CEA/CNRS/ENSICaen/Université de Caen Basse-Normandie,
BP 5133, F-14070 Caen Cedex 5, France

E-mail: nenad.bundaleski@u-psud.fr

**Abstract**. Guiding and focusing of keV to MeV ions by insulator micro-capillaries offer exciting perspectives for the production of low divergence micro-sized beams and the spatial control over the irradiated zone. These effects result from the local charging of the capillary inner wall and depend strongly on the charging/discharging dynamics of the insulating material. This dynamics has been studied on various glass insulator surfaces (borosilicate, fused silica and quartz) by grazing incidence highly charged ion beams. We propose simple experimental methods to derive the relevant time constant and study the influence of temperature and different material properties to the charging process.

## 1. Introduction

Bombardment of insulator surfaces by highly charged ions can be used for the production of nano-structures, opening perspectives for nano-fabrication [1,2]. Besides improving our understanding of the processes leading to material modification, it is equally important to be able to guide the highly charged ions at the desired spot. In this context, recent results on guiding and focusing of highly charged ions by taper-glass micro-capillaries [3,4] appear to be a very promising solution. The guiding, which was previously observed in PET nanocapillaries, was explained by the self-organizing process of the inner wall charge up [5]. Since the ion trajectories are nearly parallel to the inner walls, ions are focused in the tapered glass capillaries due to its geometry. However, a more detailed physical description is needed to better understand the guiding and focusing effects, and choose the best candidate material.

The tubular geometry is not necessarily optimum to access the relevant parameters such as surface and bulk diffusion of charges, secondary electron emission etc., governing space and time distribution of charges in a capillary. This task is best achieved by grazing incidence bombardment of a planar insulator surface by highly charged ions, since the primary mechanisms should be common. Such a geometry also offers a possibility to directly measure the above-mentioned material parameters and better understand the behavior of appropriate micro-capillaries.

We describe two simple methods for measuring discharge dynamics of insulator surfaces at different temperatures charged up by highly charged ions. The influence of the sample history (target annealing or change of its properties due to the ion bombardment) is also observed [6]. The results give valuable information to guide future experiments.

## 2. Measurement procedures

The insulator samples were prepared by prolonged heating at temperatures up to 600 K (phase separation, crystallization or a phase transition, in the case of quartz, might occur at higher temperatures) in UHV conditions for several hours. Sputter-cleaning was not employed to avoid preferential sputtering effects. The details of the experimental setup are given elsewhere [6] and its scheme is given in figure 1. The primary ion beam is extracted from an ECR ion source, mass analyzed and collimated before impinging the insulator surface with incident angle $\psi$ between 0.5° and 1.5°. The sample is progressively inserted into the beam so that part of the beam hits the surface while the rest flies over the surface without touching it and directly hits a position sensitive detector (PSD) located downstream. The quantity of ions impinging the surface is proportional to the intensity of scattered (mostly neutral) particles. The projectiles hitting the surface contribute to its charging by formation of electronic holes. Consequently, the surface charges repel the direct beam (thus reducing the rate of ions impinging the surface $\Phi$), and its deflection is monitored on the PSD (figure 1). This way, the ion beam is used for surface charging and monitoring of the charging process at the same time. Diffusion of holes into the bulk also takes place during the charging up, leading to surface discharging. Equilibrium deflection of the direct beam is achieved after some time. It is defined by the equality between the charging rate (proportional to $\Phi$) and the discharging rate. Once this steady state is reached, the beam is turned OFF to allow self-discharging of the surface. The discharging dynamics can be then determined by turning the beam ON for few seconds at regular time intervals to track the time evolution of the average direct beam deflection (here denoted as the beam deflection). The beam deflection decreases in time, which corresponds to surface discharging. This procedure can be used to determine discharging dynamics as a function of temperature.

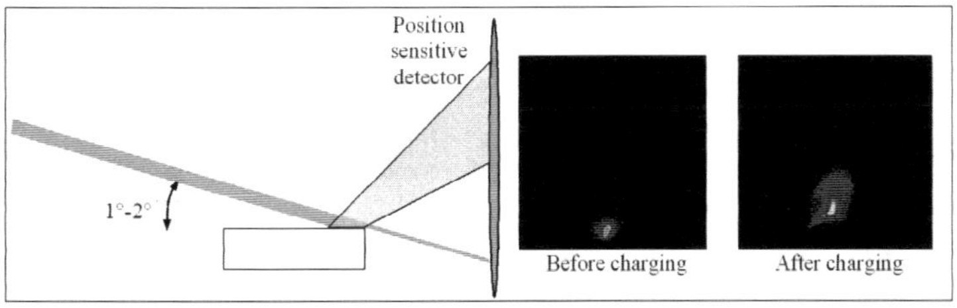


**Figure 1.** Experimental scheme and typical PSD images at the beginning and at the end of the charging process.

If the surface discharging is too fast to be monitored by this method, an alternative approach has to be developed, providing information on temperature dependence of discharging dynamics. As it is shown in the next section, the observed discharging time decay is exponential, so we approximate the discharge rate by $Q/\tau$, where $Q$ represents the overall charge in the subsurface region and $\tau$ is discharge time constant. Hence, the time evolution of $Q$ can be described by the following equation:

$$\frac{dQ}{dt} = \Phi \cdot (q + \gamma) - \frac{Q}{\tau} \tag{1}$$

where $q$ is the ion charge and $\gamma$ is the secondary electron yield per single ion. The proposed model is just an approximation since the surface distribution of charges is not considered. Nevertheless, it gives access to the dependence of $\tau$ on the sample temperature $T$, under the assumption that the direct beam deflection is directly proportional to $Q$. The equilibrium beam rate $\Phi_{eq}$, for which a constant deflection (i.e. approximately constant $Q$) is achieved, can be determined as a function of $T$. Then, since $dQ/dt = 0$ in the steady state, the following relation stands: $\tau(T) \sim 1/\Phi_{eq}(T)$. Here we assume that $\gamma$, which strongly depends on surface charge (which reduces electron emission), does not depend significantly on $T$ in the range of interest (typically 300 – 400 K). The explained approach is denoted here as dynamic equilibrium approach.

## 3. Results and discussion

The proposed methods for measuring the discharge dynamics are illustrated by examples of three different insulators of interest in tapered micro-capillary production: fused silica, quartz and borosilicate glass. The following ions were used in the experiments: $H^+$, $Ar^+$, $Ar^{4+}$, $Kr^{9+}$, $Xe^{12+}$, $He^+$, and $He^{2+}$. The primary ion energies were in the range of 1-2.5 keV per charge.

The time dependences of the $Ar^{4+}$ beam deflection from fused silica for two different temperatures are presented in figure 2. The origin of time corresponds to the beginning of the surface charging. Clearly, a temperature difference of 100 K produces a change of the discharge time constant by two orders of magnitude. The possible reason is the increased hole mobility – the residence time in the hole trap is decreased due to thermal activation. Besides, two slopes are visible in the case of higher temperature. As observed in [6], the time constants significantly depend on the sample history: they can be increased by as much as one order of magnitude if the measurement is done on a previously irradiated part of the surface. This fact, together with existence of two slopes in some cases, hinders the determination of $\tau(T)$. Decrease of $\tau$ with the incident angle in the investigated range (0.5°-1.5°) is also observed, which possibly implies influence of the charge flux. On the other hand, influence of the ion charge state is not observed [6].

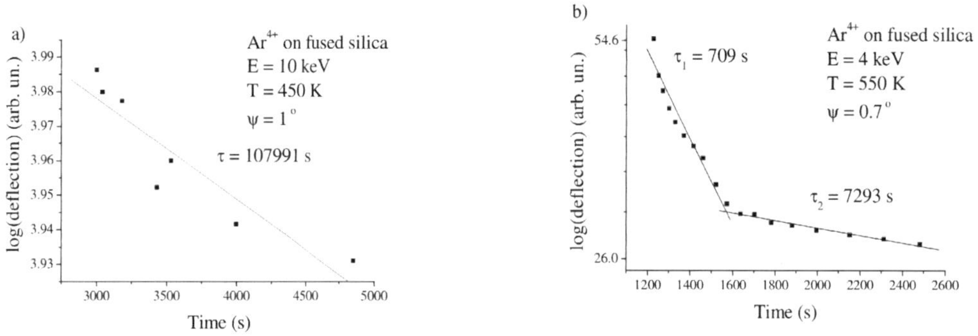


**Figure 2.** Time dependence of the $Ar^{4+}$ beam deflection ions from fused silica at temperatures: a) $T = 450$ K and b) $T = 550$ K.

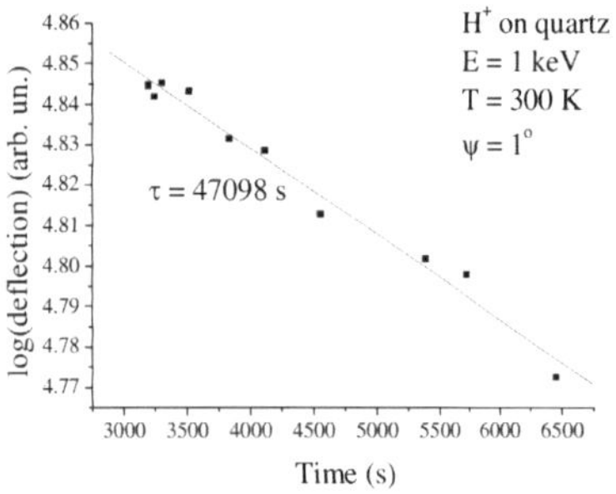


**Figure 3.** Time dependence of the beam deflection of $H^+$ ions from quartz surface. The charging experiment has been performed after 20 h annealing of the sample at $T = 780$ K.

The single crystal quartz sample has the same composition but quite different structure compared to amorphous fused silica. Its behavior in terms of discharging is completely different. In the first experiments, it was not possible to charge up as–received quartz. However, after a long annealing (20 h at 780 K, which is still below the phase transition temperature at 870 K) it was charged up with $H^+$ beam at room temperature. The discharge was very slow as can be seen from figure 2, which shows the discharge dynamics. A possible explanation of the observed behavior can be related to the fact that the hole traps in fused silica are defects [7], which should have low concentration at the quartz surface. Hence, while the majority of holes quickly migrates into the bulk, the small fraction is slowed down by filling the traps (this is coherent with the different activation energies for hole mobility – 0.16 eV for quartz and 0.6 eV for fused silica). Then, after annealing, the surface concentration of defects could be significantly increased by their segregation from bulk [6], so the discharging is significantly slowed down.

In contrast to fused silica and quartz, the former measurement procedure for the discharge dynamics could not be applied to the borosilicate sample ($SiO_{2(80.9\ \%)}$, $B_2O_{3(12.7\ \%)}$ $Al_2O_{3(2.3\ \%)}$ $Na_2O/K_2O_{(4.1\ \%)}$) since the discharging was too fast. However, the electrical properties of alkali doped borosilicates are mainly governed by the alkali ion mobility [8]. The dynamic equilibrium method, described in section 2, was applied to determine the temperature dependence of the discharge time constant. Assuming that $\tau = \tau_0 \cdot \exp(E_a/kT)$, where $E_a$ is activation energy for mobility of charges and $k$ is Boltzmann constant, $E_a$ can be determined from the slope of $\log(1/\Phi_{eq})$ vs. $1/T$ (figure 4). In this experiment 22.5 keV $Kr^{9+}$ ion beam was used. The obtained value of 0.48 ±0.04 eV is close to the activation energy for migration of $Na^+$ ions bound to non-bridging oxygen atoms in borosilicate glass (0.55 eV) [9]. In other words, it seems that surface discharging is not related to migration of holes, but of sodium ions [6].

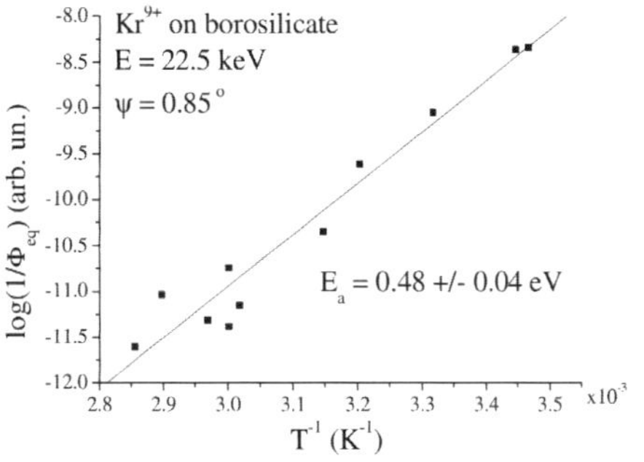


**Figure 4.** Logarithm of $1/\Phi_{eq}$ vs. $T^{-1}$ in the case of borosilicate charged by $Kr^{9+}$ ions. The obtained value for the activation energy is very close to that of the migration of sodium ions trapped in non-bridging oxygen defects in borosilicate [9].

## 4. Conclusion

Two simple experimental methods are proposed to measure the discharging dynamics of insulator surfaces previously charged by grazing incidence impact of highly charged ions. They can be used to determine characteristic time constants and the activation energy of the discharging process. To our knowledge, these magnitudes, providing macroscopic picture of the discharging process, has not been measured till now. On the other hand, measurement of surface charging properties on the nano scale is known for some time [10]. The presented experimental methods propose a link between the nano-scale approach, which provides fundamental information about the charging/discharging processes, and a specific technological problem such as ion guiding through insulator capillaries. Therefore, direct comparison of these two approaches would be valuable for better tailoring of taper-glass micro-capillaries and their future applications for highly charged ion guiding and focusing.

## Acknowledgments

We acknowledge support from the EEC FP6 network project 026015 "ITS LEIF".

## References

[1] Schenkel T, Hamza A V, Barnes A V, and Schneider D H 1999 *Progr. Surf. Sci* **61** 23
[2] El-Said A S *et al.* 2008 *Phys. Rev. Lett.* **100** 237601
[3] Ikeda T *et al.* 2006 *Appl. Phys. Lett.* **89** 163502
[4] Cassimi A *et al.* 2008 *Int. J. Nanotechnology* **5** 809
[5] Stolterfoht N *et al.* 2002 *Phys. Rev. Lett.* **88** 133201
[6] Bundaleski N *et al.* 2008 *J.Phys.: Conf. Ser.* **133** 012016
[7] Hughes R C 1977 *Phys. Rev. B* **15** 2012
[8] Wang Z, Hu Y, Lu H, Yu F 2008 *J. Non-Crystal. Solids* **354** 1128
[9] Grandjean A, Malki M, and Simonnet C 2006 *J. Non-Crystal. Solids* **352** 2731
[10] Schaadt D M *et al.* 1999 *Appl. Phys. Lett.* **74** 472

Journal of Physics: Conference Series **163** (2009) 012092 doi:10.1088/1742-6596/163/1/012092

# Guiding of slow highly charged ions through insulating nano-capillaries

**Zhang HQ, Skog P, and Schuch R**

Atomic physics, Fysikum, AlbaNova, Stockholm University, S-106 91 Stockholm, Sweden

E-mail: schuch@physto.se

**Abstract.** The guiding of highly charged ions through $SiO_2$ nano-capillaries has been investigated by our group, using 7 keV $Ne^{7+}$-ions. We studied in particular the transmission of ions incident at angles greater than the angle given by the capillary aspect ratio as a function of charge incident on the capillary membrane. In this report we show the re-arrangement of charge patches inside the capillary by observing the evolution of the two-dimensional angular distributions of the transmitted ions.

## 1. Introduction

The phenomenon of the creation of hollow atoms has attracted considerable attention [1-10]. There is a unique possibility for studying hollow atoms of the first generation, using metallic micro-capillaries as target for a beam of slow HCIs [1]. After highly charged ions have passed through the capillaries, one can observe the decay of hollow atoms in the vacuum by x-ray- or electron- spectroscopy [2]. The ions that have passed through the capillaries show an angular distribution that is very narrow (with a width given by the capillary aspect ratio).

When using membranes of nano-capillaries in insulating materials there is transmission of ions even when the capillary membrane is tilted at angles much larger than the angle given by the aspect ratio. This astounding feature of the interaction of slow HCIs with insulating nano-capillaries was first reported by Stolterfoht *et al.* for nano-capillaries in *polyethylene terephthalate* (PET) [3]. Due to a process which charges up the internal capillary walls, transmission is possible for ions with incident angles much larger than the angle corresponding to the capillary aspect ratio ($\approx 0.6°$). The first experiments with 3 keV $Ne^{7+}$-ions transmitted through capillaries in the insulating material PET reported transmission of ions for tilt angles up to 25°, with a full width at half maximum (FWHM) of 5° [3]. Our group performed experiments with nano-capillaries in PET of 100 nm in diameter and 10 $\mu$m in length with a beam of 7 keV $Ne^{7+}$-ions [8]. At this higher kinetic energy, the angular width (FWHM) of the transmitted ions is somewhat narrower; approximately 3.2° and ion transmission can be discerned for capillary tilt angles up to 4°.

The guiding of highly charged ions through insulating nano-capillaries has been attributed to the forming of charge patches on the inner walls of the capillaries, the main patch being located close to the entrance. For a fully discharged capillary membrane, oriented in such a way that the capillary axes are tilted some angle, with respect to the direction of the ion beam, there is no transmission of ions at the beginning. When ions collide with the capillary walls they deposit positive charge, which diffuse into the bulk and on the surface of the walls, through hopping of charges. Charge patches are formed at locations where most of the ion-wall collisions occur, which then deflect ions entering the capillaries at a later instant. Full transmission is achieved when the equilibrium point between deposition of charges on the walls - through ion-wall collisions - and diffusion of charges - through diffusion into the bulk and the hopping of charges on the surface - is reached [3, 7]. The guiding effect comes from the charging of the capillary inner wall, and depends on the charging dynamics of the insulating material. In this work, the re-arrangement of charge patches was observed from the evolution of the two-dimensional angular distributions of transmitted ions at tilt angles larger than the angle given by the capillary aspect ratio.

## 2. Experiments

The experiments were performed using a beam of 7 keV $Ne^{7+}$ ions from the 14.5 GHz ECR ion source at the Manne Siegbahn Laboratory (MSL) in Stockholm. The beam was analyzed with a bending magnet, focussed by two electrostatic quadrupole doublets and an einzel lens and then collimated by a pair of four-jaw slits set 1.55 m apart. The size of the beam was $2{\times}2$ mm$^2$ and the intensity was on the order of 0.1 nA/mm$^2$ at the capillary membrane. The angular divergence of the primary beam was measured to be less than 0.4° for all experiments. The primary beam charge state distribution shows a $\leq 0.03\%$ content of other charge states than $Ne^{7+}$. The base pressure in the beam line was approximately $1.5{\cdot}10^{-8}$ mbar and the experimental chamber was evacuated down to $10^{-9}$ mbar.

The $SiO_2$ nano-capillaries used in the experiments were fabricated by thermally oxidizing pores

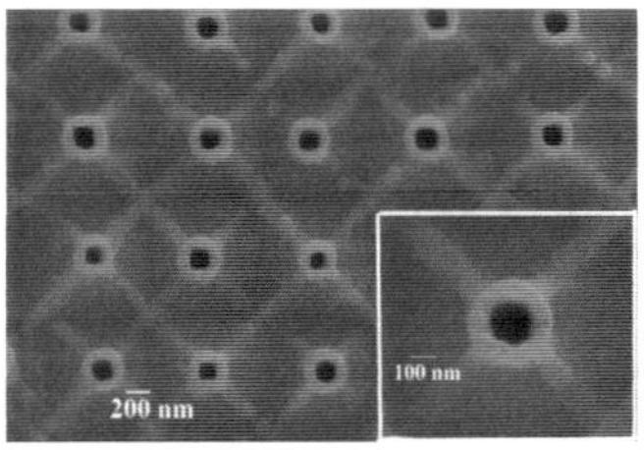


**Figure 1.** Top view of $SiO_2$ nano-capillaries.

in n-doped Si-100 made by photo-assisted electrochemical etching [11]. The silicon membrane contains regularly distributed, 1.4 $\mu$m apart, nano-capillaries of 100 nm in diameter and 25 $\mu$m in length (see figure 1). To get insulating electric properties, the capillary walls are covered by 100 nm thick $SiO_2$. Both front and back surfaces of the membrane were plated with 30 nm gold to prevent the surfaces from charging-up during ion-beam exposure. The nano-capillaries are, besides being very highly ordered, also highly parallel as the electro-chemical etching proceeds along the crystal axis. But the thin membrane is not perfectly flat; it has a slight overall curvature due to the strain induced by the differing lattice constants of the $SiO_2$ capillary walls and the Si bulk. However, the curvature is estimated to be much less than 0.1° over the area exposed to the ion beam in the experiments.

The capillary membrane was mounted in a holder on a goniometer, which allows independent adjustment of the target position in three axial dimensions and around two rotational axes, by this, at the beginning of each experiment the capillaries were aligned to the incoming beam by two angles - the azimuthal angle and the elevation angle. The azimuthal angle between the capillary axes and the incident beam is referred to as the tilt angle. A two-dimensional multi-channel plate with a resistive anode was used to detect the transmitted ions. The data acquisition system works in event mode.

## 3. Results and Discussions

We studied the angular distribution of 7 keV $Ne^{7+}$ transmitted through $SiO_2$ nano-capillaries as a function of charge incident on the capillary membrane when the tilt angle was changed from +2° to +1°. At first the capillaries were set at the tilt angle of +2° and charged up until equilibrium transmission was reached. Then the tilt angle of the capillary membrane was changed to +1° for re-distributing the charge patches in the so called re-charging process. Transmitted distributions measured at the tilt angle +1°, plotted as a function of incident charge on the capillary membrane, during the re-charge process are shown in figure 2. Corresponding projections of the transmitted distributions in the azimuthal angle plane (the plane of incidence) are shown in figure 3.

A double peak structure was observed in the transmitted distributions in the beginning of

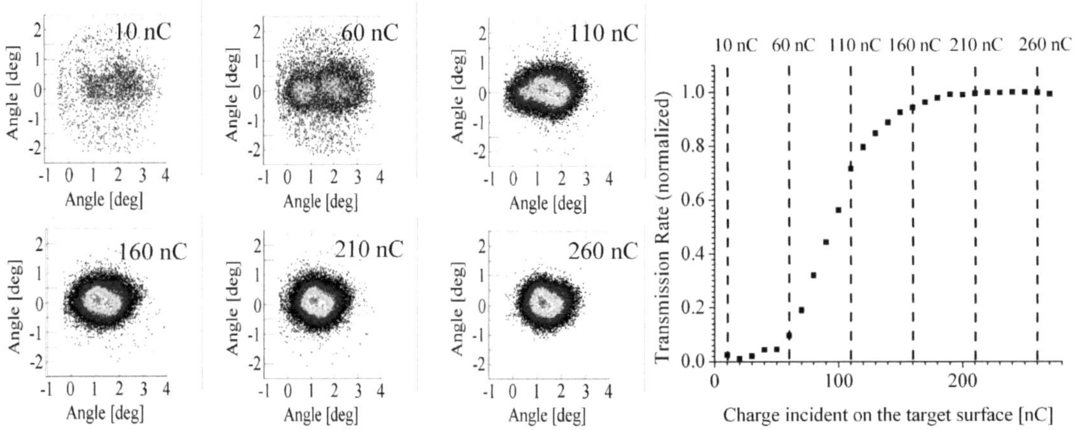


**Figure 2.** Two dimensional images of transmitted distributions (left side) during re-charging at tilt angle +1°, after rotating from +2°, sampled at various stages of the re-charge process indicated in the graph to the right, in which transmission rate vs. the charge incident on the target surface is shown. The dashed lines in the right graph indicate the stages at which the transmission images were sampled.

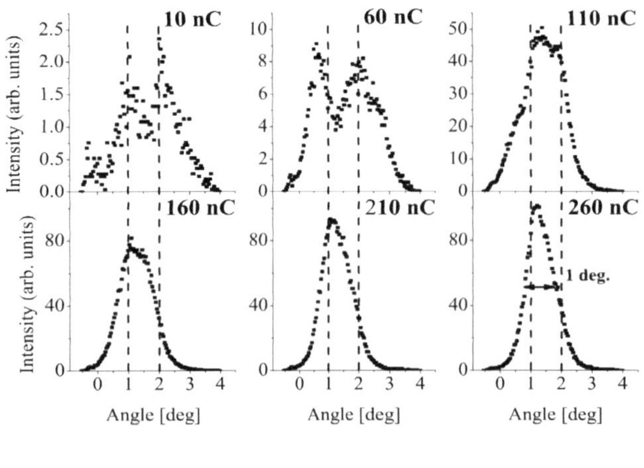


**Figure 3.** Distributions of transmitted ions projected onto the azimuthal angle plane at various stages of the re-charge process at tilt angle +1°, after rotating from +2°. The dashed lines indicate +1° and +2°, respectively.

re-charging at tilt angle $+1^\circ$ after rotating from tilt angle $+2^\circ$ ; one peak is centred around $+2^\circ$, while the other peak is centred around $+1^\circ$. The peak that was initially centred around the changed capillary axis ($+1^\circ$), shifts towards the beam direction and in the later stage of re-charge back to the capillary axis. The other peak that is initially centred around $+2^\circ$ shifts towards the changed orientation of the capillary axes and merges with the peak initially centred around $+1^\circ$ as the transmission again reaches equilibrium.

The angular width (FWHM) is $\approx 1^\circ$ for the tilt angle $+1^\circ$ at equilibrium. This can not be explained by the capillary aspect ratio and beam divergence. If there is only one entrance charge patch, the width of the transmitted distribution would be close to the angle given by the capillary aspect ratio ($0.23^\circ$). The broadening of the distribution was explained as the exit effect of a mesoscopic electric field from charging an ensemble of capillaries exposed to the beam [7, 12]. We calculated the influence of such an exit field effect on exiting ions and found the contribution to the broadening negligibly small in our case. Our study of the time-dependence of the transmitted distributions shed light on the issue (see below and reference [13]).

The development of transmitted distributions can be from the re-arrangement of the charge patches inside the capillary. The peak that initially is centred around $+2^\circ$ may be attributed to the old charge patches from the "memory" of the previous tilt angle ($+2^\circ$), while the other peak, which is initially centred around $+1^\circ$, is from the forming charge patches at the changed tilt angle $+1^\circ$. During the formation of the new charge patches, the peak centred around $+1^\circ$ from the newly formed charge patches oscillates back and forth around the capillary axis at the change tilt angle $+1^\circ$, while, due to the collapsing of the old charge patches, the peak centred around the previous tilt angle $+2^\circ$ shifts towards the new orientation of the capillaries. As seen from the shift of the transmitted distributions, the broadening is from the development of the charge patches inside the capillaries. The formation of the charge patches inside the capillary results in a reduction of collimation due to the capillary wall.

## Acknowledgments

Acknowledgments We would like to thank the staff of the Manne Siegbahn Laboratory (MSL) for providing the beam, the European network ITS-LEIF and the Swedish Research Council (VR) for financial support.

## References

[1] Ninomiya S, Yamazaki Y, Koike F, Masuda H, Azuma T, Komaki K, Kuroki K and Sekiguchi M 1997 *Phys. Rev. Lett.* **78** 4557
[2] Burgdörfer J, Lerner P and Meyer F W 1991 *Phys. Rev.* A **44** 5674
[3] Stolterfoht N, Bremer J H, Hoffmann V, Hellhammer R, Fink D, Petrov A and Sulik B 2002 *Phys. Rev. Lett.* **88** 133201
[4] Tokesi K, Wirtz L, Lemell C and Burgdörfer J 2001 *Phys. Rev.* A **64** 042902
[5] Tokesi K, Wirtz L, Lemell C and Burgdörfer J 2000 *Phys. Rev.* A **61** 020901
[6] Hellhammer R, Pesic Z D, Sobocinski P, Fink D, Bundesmann J and Stolterfoht N 2005 *Nucl. Instr. Meth. Res.* B **233** 213
[7] Schiessl K, Palfinger W, Tokesi K, Nowotny H, Lemell C and Burgdörfer J 2005 *Phys. Rev.* A **72** 062902
[8] Víkor Gy, Kumar R T R, Pešič Z D, Stolterfoht N and Schuch R 2005 *Nucl. Instr. Meth. Res.* B **233** 218
[9] Sahana M B, Skog P, Vikor G, Kumar R T R and Schuch R 2006 *Phys. Rev.* A **73** 040901(R)
[10] Mátéfi-Tempfli S *et al* 2006 *Nanotechnology* B **17** 3915
[11] Kumar R T R, Badel X, Vikor G, Linnros J and Schuch R 2005 *Nanotechnology* B **16** (9) 1697
[12] Schiessl K, Palfinger W, Tokesi K, Nowotny H, Lemell C and Burgdörfer J 2007 *Nucl. Instr. Meth. Res.* B **258** 150
[13] Skog P, Zhang HQ and Schuch R, to be published in PRL.

# Transmission of low-energy electrons through $SiO_2$ tube

**Wei Wang[1, 2], Dejun Qi [1], Deyang Yu [1]*, Mingwu Zhang [1, 2], Fangfang Ruan [1, 2], Jing Chen [1, 2] and Xiaohong Cai [1]***

[1] Institute of Modern Physics, Chinese Academy of Sciences, Lanzhou 730000, China
[2] Graduate School of Chinese Academy of Sciences, Beijing 100039, China
E-mail: d.yu@impcas.ac.cn, caixh@impcas.ac.cn

**Abstract** The transmission of low-energy (1100eV-1500eV) electrons through $SiO_2$ tube was investigated, the distribution of transmitted beam current was observed. Results indicate that the intensity of transmitted electrons is proportional to the relative incident beam intensity and the angular divergence of the transmitted electrons is reduced when incident energy increased.

## 1. Introduction

The interaction of charged particles with insulating capillaries has attracted lots of attention during last few years[1], especially after the guiding phenomena of highly charged ions (HCI) transmitted through the capillaries of polyethylene terephthalate (PET) was reported in 2002 [2]. Since then, a series of experimental and theoretical researches has been performed to investigate the HCI guiding effect in PET [3, 4, 5], $SiO_2$ [6], and $Al_2O_3$ [7]. Furthermore, particular attention has been paid to the production of HCI micro-beam with a single tapered glass capillary [8, 9, 10] and its application to cell biology [11]. Experiment has been done for the transmission of low-energy electrons through highly ordered $Al_2O_3$ [12] and polymer nanocapillaries [13].All the results clearly show the existence of the guiding effect as found for HCI.

To the best of our knowledge, the investigation of the guiding effect has been focused on the transmission of charged particles through micro-sized capillaries, and these studies have revealed much important information about the properties of the capillaries. Nevertheless, a full understanding of mechanism involved in the guiding phenomena has not been achieved yet. Particularly, the characteristics of charged particles travelling through single bended macro-sized tube are still unknown at present, and people wonder if the guiding effects would show any differences in case of the negatively charged particles such as electrons comparing with the results of HCI. The study of the guiding effects of electrons thus drew great attention in past two years.

In this works, we investigated the guiding effect of low energy (1100eV, 1300eV and 1500eV) electrons with beam current of several hundred nA through a bent macro-sized $SiO_2$ tube. It is interesting to compare the macro-sized transmission with the process in micro-sized capillaries because:1) Without the influence of neighboring tubes which is inevitable in the multi-capillaries, the electrons and holes (positive ions) may show different behavior in the tube ,and due to the low *q/m* ratio, electrons are more sensitive to electric field than HCI, that will lead to different processes in self-organizing charge up and discharge; 2) A higher transmission fraction in macro-sized tube would be useful in transmitting electron beam without electrostatic and magnetic lens.

## 2. Experimental setup

The experiment was performed in a high-vacuum chamber with a pressure of some $10^{-5}$ Pa, the electron beam was extracted from a $LaB_6$ electron gun. Experimental setup is shown seen in Fig.1.

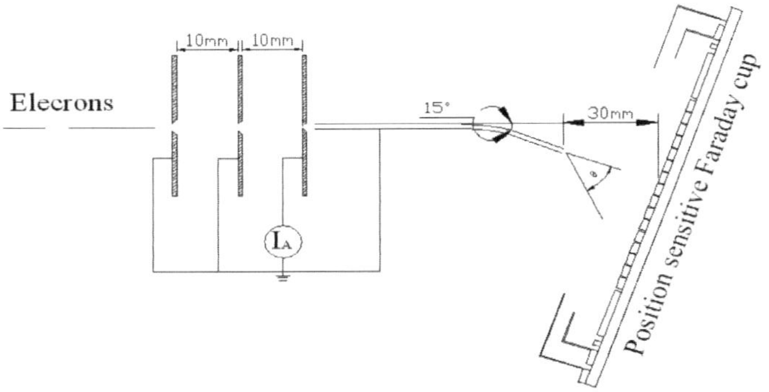


**Fig. 1** The schematic view of experimental setup

The electrons were collimated by three apertures with different diameter and then injected into the tube with the inlet and outlet angle of 15°. The outlet is located 30mm before the detector. This bent tube was prepared by heating a straight tube and then bending one end slowly. The total length of the tube is 50mm and inner and outer diameters are 2.3mm and 3.1 mm respectively. Its outer wall was carefully covered by aluminum foil and grounded.

To overcome the limitation that the normal PSD cannot work under high beam intensity, a position sensitive Faraday cup (PSFC) with 16 channels was developed to detect the transmitted electrons [14]. A 64-channel PSFC with better position resolution has been designed for further experiments.

## 3. Results and discussions

In this experiment we use the current on the last apertures $I_A$ to define the relative incident current intensity. The time dependence of transmitted electrons is shown in Fig. 2. It is found that the guiding effect was detected almost immediately after the beam was turned on ($I_A$=4.1μA, E=1100eV) which is similar to Milosavljevic's work [12] and is much faster than that observed in HCI experiments (about 10 min for PET) [2]. This is due to the high incident current intensity [12].

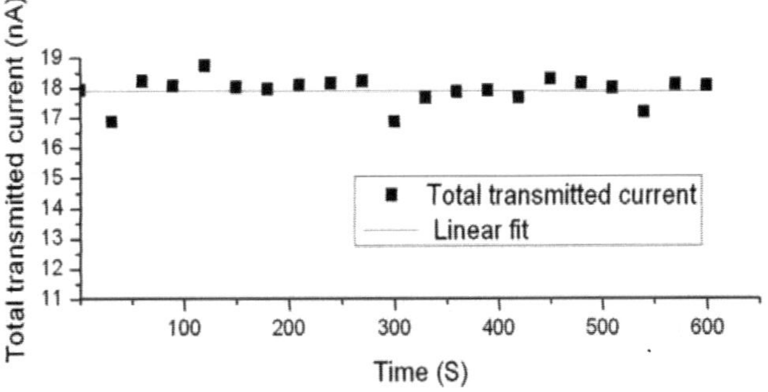


**Fig.2** Time dependence of transmitted electrons. Relative incident current is 4.1μA, and electron energy is 1100eV.

The distribution of transmitted electrons through bent tube is shown in Fig.3 (a). The relative intensity of incident current is $I_1>I_2>I_3>I_4$, and the incident energy are 1100eV, 1300eV, 1500eV from the top of the figure to the bottom. From Fig.3 (a) it is found that a remarkable fraction of electrons can transmit through the tube even though it was bent as large as 15° and the transmitted numbers of electrons is proportional to the relative incident current intensity.

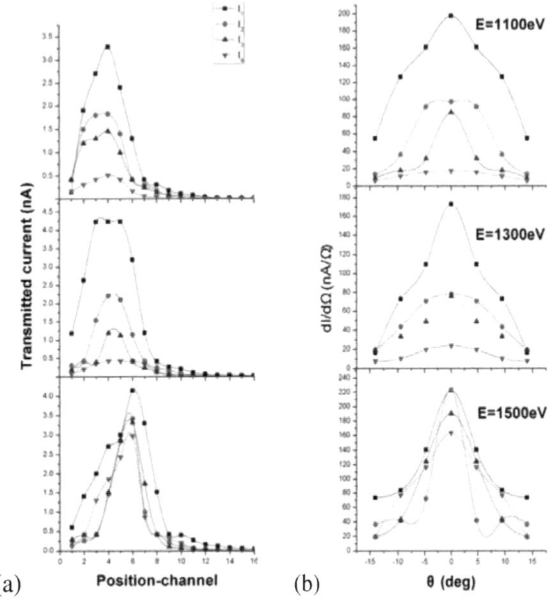


(a) (b)

**Fig.3** (a) Incident energy and current intensity dependence in bent tube. (b) Normalized current intensity to unit solid angel. The relative intensity of incident current is $I_1 > I_2 > I_3 > I_4$.

Fig.3 (b) shows the angular distribution of the current intensity normalized to unit solid angel. It is found that the angular distribution of the transmitted electrons is symmetric and nearly centralized within ±15°, and the divergence is reduced when incident energy increased. Considering the PSFC is 30mm away from the outlet of the tube, it suggests that this part of electrons transmit through the tube without any collision, which means these electrons undergo the "guiding effect".

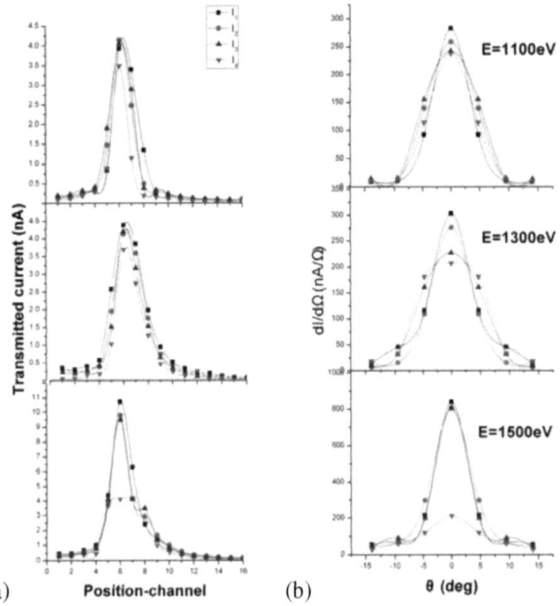


(a) (b)

**Fig.4** (a) Incident energy and current intensity dependence in straight tube. (b) Normalized current intensity with solid angel. The relative intensity of incident current is $I_1 > I_2 > I_3 > I_4$.

In Das's work [13], the transmitted electrons undergo significant energy losses processes that increase with the capillary tilt angle and even these "in-elastically scattered" electrons are still guided through the capillaries. This result is quite different from the earlier works for HCIs [2-7] and the result for 200-350 eV electrons in $Al_2O_3$ [12], for which no obvious energy lose was measured. In this work we did not measure the energy distribution of the transmitted electrons, and the present results are the sum of the elastically scattered and the possibly in-elastically scattered electrons. Further investigation should be done on order to clarify the scattering mechanisms.

We also measured the incident energy and current intensity dependence in a straight tube. As shown in Fig.4 (a), the curves have a narrower FWHM and more symmetric than in bent tube because majority of electrons transmit along the tube-axis without any collision in straight tube.

Above results are similar to the guiding effects of HCI and useful to clarify the guiding mechanism of charged ions including HCI and electron. Nevertheless, they are only primary results and the further experiments are necessary with improved accuracy and more information. To understand the experiment phenomena theoretical effort is also required.

### 4. Conclusion

The incident electrons cannot directly pass through the $SiO_2$ tube that is bent as large as 15°, but surely we observed the steady transmission of electrons with high intensity and low angular divergence. The present results indicate that transmitted current intensity in unit solid angel is proportional to the relative incident current intensity, and increased with incident energy, which requires further theoretical explanations.

## Acknowledgment

This work was supported by National Natural Science Foundation of China *NO.*10775160.

### References

[1] Yamazaki Y 2007 *Nucl. Instr. and Meth.* B **258** 139

[2] Stolterfoht N, Bremer J -H , Hoffmann V, Hellhammer R, Fink D, Petrov A and Sulik B 2002 *Phys. Rev. Lett.* **88** 133201

[3] Stolterfoht N, Hellhammer R, Bundesmann J, Fink D, Kanai Y, Hoshino M, Kambara T, Ikeda T and Yamazaki Y 2007 *Phys. Rev. A.* **76** 022712

[4] Schiessl K, Palfinger W, Tokesi K, Nowotny H, Lemell C and Burgdorfer J 2005 *Phys. Rev. A.* **72** 062902

[5] Schiessl K, Palfinger W, Tokesi K, Nowotny H, Lemell C and Burgdorfer J 2007 *Nucl. Instr. and Meth.* B **258** 150

[6] Sahana M-B, Skog P, Vikor G, Rajendra R-T and Schuch R 2006 *Phys. Rev. A.* **73** 040901

[7] Matefi S, Matefi M, Piraux L, Juhasz Z, Biri S, Fekete E, Ivan I, Gall, Sulik B, Vikor G, Palinkas J, and Stolterfoht N 2006 *Nanotechnology* **17** 3915

[8] Nebiki T, Yamamoto T, Narusawa T, Breese M-B, Teo E-J, and Watt F 2003 *J. Vac. Sci. Technol.* A **21(5)** 1671

[9] Ikeda T, Kanai Y, Kojima T-M, Iwai Y, Kambara T, Yamazaki Y, Hoshino M, Nebiki T and Narusawa T 2006 *J. Appl. Phys.* **89** 163502

[10] Ikeda T, Kojima T-M, Iwai Y, Kanai1 Y, Kambara1 T, Nebiki T, Narusawa T and Yamazaki Y 2007 *J. Phys: Conference Series.* **58** 68

[11] Iwai Y, Ikeda T, Kojima T-M, Yamazaki Y, Maeshima K, Imamoto N, Kobayashi T, Nebiki T, Narusawa T and Pokhil G-P 2008 *J. Appl. Phys.* **92** 023509

[12] Milosavljevic A-R, Vikor G, Pesic Z-D, Kolarz P, Sevic D and Marinkovic B-P 2007 *Phys. Rev. A.* **75** 030901

[13] Das S, Dassanayake B-S, Winkworth M, Baran J-L, Stolterfoht N, Tanis J-A 2007 *Phys. Rev. A.* **76** 042716

[14] Qi D, Yu D, Wang W, Zhang M and Cai X *Nucl .Phys. Rev.* (In publishing)

# Resonant Coherent Excitation of Li-like $Ar^{15+}$ Ions in a Thin Si Crystal

**Y. Nakano[1], T. Inoue[1], T. Azuma[1], A. Hatakeyama[2], Y. Nakai[3], K. Komaki[4,5], Y. Yamazaki[4,5], E. Takada[6] and T. Murakami[6]**

[1] Department of Physics, Tokyo Metropolitan University, Hachioji, Tokyo 192-0397, Japan
[2] Department of Applied Physics, Tokyo University of Agriculture and Technology, Koganei, Tokyo 184-8588, Japan
[3] Radioactive Isotope Physics Laboratory, RIKEN Nishina Center, Wako, Saitama 351-0198, Japan
[4] Graduate School of Arts and Sciences, University of Tokyo, Komaba, Meguro, Tokyo 153-8902, Japan
[5] Atomic Physics Laboratory, RIKEN, Wako, Saitama 351-0198, Japan
[6] National Institute of Radiological Sciences, Inage, Chiba 263-8555, Japan

E-mail: nakano-yuji@ed.tmu.ac.jp

**Abstract.** We investigated resonant coherent excitation (RCE) of 445 MeV/u Li-like $Ar^{15+}$ under the planar channeling and non-channeling conditions. The K-shell and L-shell excitations are observed as decreases in the survival fraction of the $Ar^{15+}$ ions after passing through a thin Si crystal. RCE under the planar channeling condition provides the opportunity to study a strong static Stark effect on the Li-like system with K-shell or L-shell vacancy. Under the non-channeling conditions, the observed excitation energies and the reported theoretical calculations were in reasonable agreement.

## 1. Introduction

When energetic ions are in flight through a crystal, they experience a temporally-oscillating field originating from the periodic crystalline structure. This oscillating field induces resonant coherent excitation (RCE) of the ions if the frequency matches their internal energy difference. It was predicted by Okorokov in 1965 [1] and extensively studied with H-like or He-like systems due to their simple atomic structures [2, 3, 4, 5, 6]. RCE of the L-shell electron in Li-like $Fe^{23+}$ was observed under the planar channeling condition, where the ions travel parallel to an array of atomic planes [7]. The channeling conditions have been generally preferred for the observation of RCE taking advantage of reducing the non-resonant collisional processes with the target atoms [8, 9]. On the other hand, study of RCE under the non-channeling conditions has greatly progressed in recent years with use of a relativistic ion beam and a thin crystal [10, 11]. RCE under the planar channeling and non-channeling conditions are referred to as two-dimensional and three-dimensional RCEs (2D- and 3D-RCEs), respectively. Here we report on 2D- and 3D-RCE of L-shell and K-shell electrons in Li-like $Ar^{15+}$ ion (Fig. 1).

The resonance condition of 3D-RCE is written as

$$\Delta E = \frac{h\gamma v}{a}\left\{\sqrt{2}\,(k\cos\phi + m\sin\phi)\cos\theta + l\sin\theta\right\}, \tag{1}$$

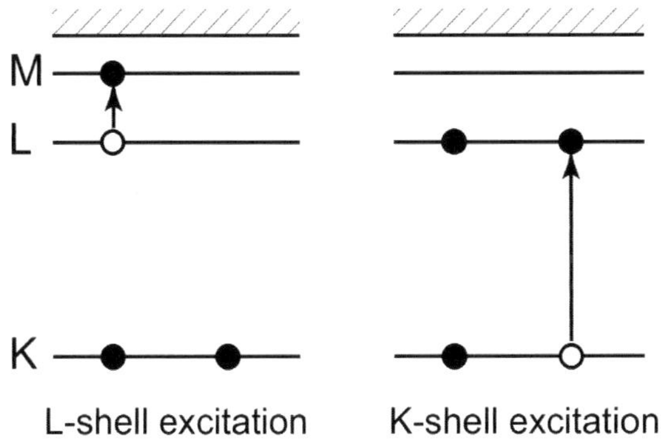


**Figure 1.** Schematic representation of L-shell excitation (left hand side) and K-shell excitation (right hand side) of Li-like ion.

where $h$ is Planck's constant, $v$ is the ion velocity, $\gamma$ is the Lorentz factor, $(k, l, m)$ is the Miller index, $\theta$ and $\phi$ are the rotation and tilt angles used to describe the orientation of the crystal with respect to the ion beam [10]. When $\phi = 0°$, the (220) planar channeling condition is satisfied and the index $m$ vanishes, leading to the 2D-RCE condition. In this condition, the ions experience not only the oscillating fields but also the static field due to the (220) atomic planes, which induce the static Stark effect [10]. In contrast, since the (220) planar potential is experienced as an oscillating field by the non-channeling ions, 3D-RCE is free from the static Stark effect. Note that ac Stark effect by the oscillating field remains in 3D-RCE [11].

## 2. Experiment

The experiments were performed at the heavy ion medical accelerator in Chiba (HIMAC). The 445.43 MeV/u $Ar^{15+}$ ion beam was charge-analyzed after passing the thin Si crystal mounted on a high precision goniometer. The path length in the crystal was 1.0 $\mu$m. We monitored the survival fraction (fraction of $Ar^{15+}$ in the emerged ion beam) as a function of the incident angle $\theta$ while fixing $\phi$. Since the cross-section of the collisional ionization in the crystal is larger for the excited states, RCE is observed as a decrease in the survival fraction. In the K-shell excitation, the KLL Auger process also contributes to the decrease of the survival fraction. Note that the electron capture cross-section is negligibly small in the present high-energy region. The beam velocity was determined by the well-known transition of $Ar^{16+}$ $1s^2$-$1s2p(^1P)$ (3139.55 eV) by 3D-RCE, which was produced by the collisional ionization of $Ar^{15+}$ in the crystal.

## 3. Result and Discussion

### *3.1. L-shell excitation*

We observed 2D- and 3D-RCEs of the $2s$ electron to the $n = 3$ states using the frequency components of the oscillating field specified by $(k, l) = (0, 1)$ and $(k, l, m) = (0, 1, -1)$, respectively. Figure 2(a) shows the survival fractions of $Ar^{15+}$ under the (220) planar channeling, where the tilt angle $\theta$ was converted to the transition energy using Eq. (1) with $(k, l) = (0, 1)$ and $\phi = 0$. The dip positions of 2D-RCE did not agree with the theoretical transition energies from $1s^2 2s$ to $1s^2 3s(^2S)$, $1s^2 3p(^2P)$ and $1s^2 3d(^2D)$ in vacuum [12], which are indicated by arrows in the figure. This is mainly due to the static Stark mixing between the $n = 3$ states induced by the (220) planar potential.

By contrast, in 3D-RCE observed at $\phi = -0.17°$ (Fig. 2(b)), the dip positions reasonably agreed with the theoretical transition energies, since 3D-RCE is free from the static Stark effect. Note that the $2s$-$3s$ and $2s$-$3d$ transitions were observed, although the electric dipole transition is dominant in 3D-RCE [13]. These forbidden transitions can be explained as a result of the ac Stark mixing between the $2s$ and $2p$ states induced by other harmonics of the oscillating fields near the $2s$-$2p$ resonance, that is, $(k, l, m) = (0, 0, 2)$ and $(0, 0, 4)$. In the frame work of

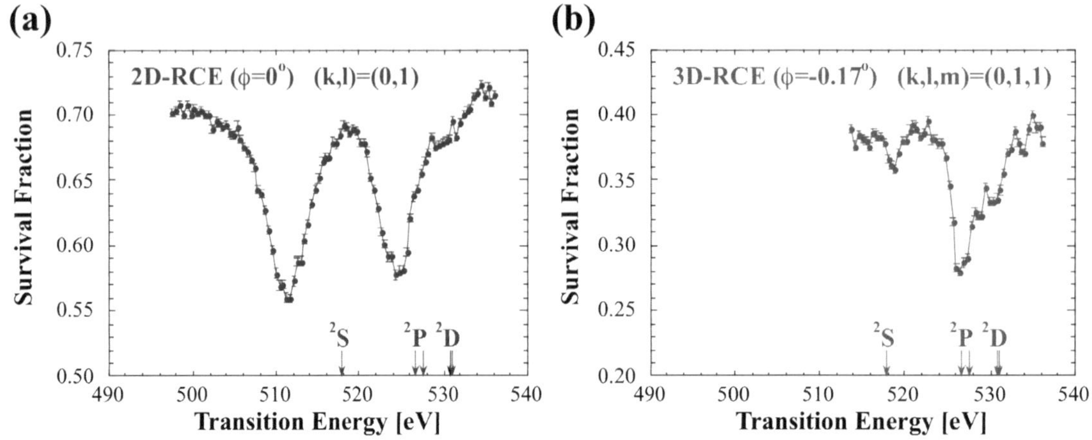


**Figure 2.** Survived fractions of $Ar^{15+}$ under the (a) planar channeling and (b) non-channeling conditions as a function of the transition energy. The arrows indicate the transition energies to $^2S_{1/2}$, $^2P_{1/2}$, $^2P_{3/2}$, $^2D_{3/2}$ and $^2D_{5/2}$ of the $1s^23l$ states in ascending order.

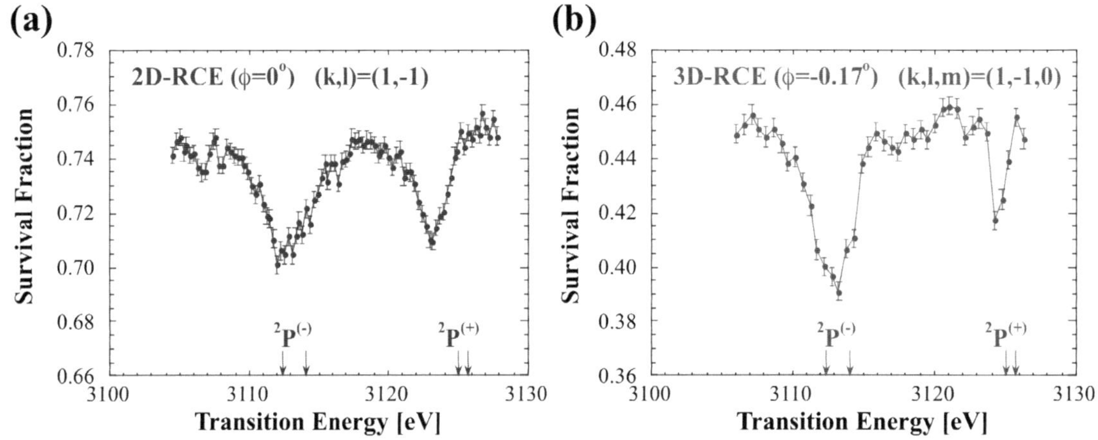


**Figure 3.** Survived fractions of $Ar^{15+}$ under the (a) planar channeling and (b) non-channeling conditions as a function of the transition energy. The arrows indicate the transition energies to $^2P^{(-)}_{1/2}$, $^2P^{(-)}_{3/2}$, $^2P^{(+)}_{1/2}$ and $^2P^{(+)}_{3/2}$ of the $1s2s2p$ states in ascending order.

the perturbative treatment, the maximum mixing is evaluated to be 5% and the corresponding energy shift is less than 0.2 eV.

### *3.2. K-shell excitation*

Excitation of the $1s$ electron to the $2p$ states was also observed. Figures 3(a) and (b) show the 2D- and 3D-RCE spectra induced by the frequency components $(k,l) = (1,-1)$ and $(k,l,m) = (1,-1,0)$, respectively. The theoretical transition energies of the $1s(^2S)2s2p(^1P)$ and $1s(^2S)2s2p(^3P)$, i.e., $^2P^{(+)}$ and $^2P^{(-)}$, calculated by Chen et.al. [14] are indicated in the figures. Since the energy differences between the mixable states are large in the $n = 2$ states compared to the $n = 3$ states, the positions of resonance dips in 2D-RCE are close to those in 3D-RCE. However, the intensity ratio of the two dips is apparently different between 2D- and 3D-RCEs, which sensitively reflect the static Stark mixing in 2D-RCE.

In 3D-RCE, as well as the L-shell excitation, the observed excitation energies reasonably agreed with the theoretical transition energies. It is noted that the ac Stark mixing between the $2s$ and $2p$ states discussed above does not affect the excitation energies of 3D-RCE by $(k,l,m) = (1,1,0)$ due to the fact that the polarization direction of the oscillating field specified by $(k,l,m) = (1,1,0)$ and $(0,0,m)$ are orthogonal. Therefore, the ions are not excited to the magnetic substates of the $1s2s2p$ states mixed with other states by the ac Stark effect.

## 4. Summary

We observed 2D-RCE and 3D-RCE of Li-like $Ar^{15+}$ in a thin Si crystal. In the L-shell excitation under the planar channeling condition, we observed strong static Stark effect in the 2D-RCE spectrum, which was not observed under the non-channeling condition. Furthermore, we succeeded in observing RCE of the K-shell electron in Li-like system for the first time. The intensity ratio of K-shell excitation to ${}^2P^{(+)}$ and ${}^2P^{(-)}$ was affected by the static Stark effect in 2D-RCE. In both cases, the transition energies observed by 3D-RCE well agreed with theoretical calculations. This work points the way toward the production of triply excited Li-like system in combination with the double resonance technique [11, 15]. The experiment for successive excitations of the $1s$ electrons in Li-like system are now in progress.

## Acknowledgment

The authors would like to thank Dr. Y. Kanai for helpful discussions. This work was supported in part by Grants-in-Aid for Scientific Research (No. 13440126, 16204030 and 19104010) from Japan Society for the Promotion of Science (JSPS). One of the authors (Y. Nakano) acknowledges support from JSPS (No. 20-10964) and another (A.H.) acknowledges support from the "Improvement of Research Environment for Young Researchers" program of The Ministry of Education, Culture, Science and Technology. This experiment is one of research projects with heavy ions at NIRS-HIMAC.

## References

[1] V. V. Okorokov, *JETP Lett.* **2**, 111 (1965).
[2] S. Datz, C. D. Moak, O. H. Crawford, H. F. Krause, P. F. Dittner, J. Gomez del Campo, J. A. Biggerstaff, P. D. Miller, P. Hvelpumd, and H. Knudsen, *Phys. Rev. Lett.* **40**, 843 (1978).
[3] S. Datz, C. D. Moak, O. H. Crawford, H. F. Krause, P. D. Miller, P. F. Dittner, J. Gomez del Campo, J. A. Biggerstaff, H. Knudsen, and P. Hvelplund, *Nucl. Instr. and Meth.* **170**, 15 (1980).
[4] F. J. García de Abajo and V. H. Ponce, *Adv. Quant. Chem* **46**, 65 (2004).
[5] T. Azuma, T. Ito, K. Komaki, Y. Yamazaki, M. Sano, M. Torikoshi, A. Kitagawa, E. Takada, and T. Murakami, *Phys. Rev. Lett.* **83**, 528 (1999).
[6] T. Azuma, Y. Takabayashi, C. Kondo, T. Muranaka, K. Komaki, Y. Yamazaki, E. Takada, and T. Murakami, *Phys. Rev. Lett.* **97**, 145502 (2006).
[7] Yoichi Nakai, T. Ikeda, Y. Kanai, T. Kambara, N. Fukunishi, K. Komaki, C. Kondo, T. Azuma, and Y. Yamazaki, *Nucl. Inst. and Meth. B* **230**, 90 (2005).
[8] D. S. Gemmell, *Rev. Mod. Phys.* **46**, 129 (1974).
[9] C. Cohen and D. Dauvergne, *Nucl. Inst. and Meth. B* **225**, 40 (2004).
[10] C. Kondo, S. Masugi, Y. Nakano, A. Hatakeyama, T. Azuma, K. Komaki, Y. Yamazaki, T. Murakami, and E. Takada, *Phys. Rev. Lett.* **97**, 135503 (2006).
[11] Y. Nakai, Y. Nakano, T. Azuma, A. Hatakeyama, C. Kondo, K. Komaki, Y. Yamazaki, E. Takada, and T. Murakami, *Phys. Rev. Lett.* **101**, 113201 (2008).
[12] K. McKeown, K. M. Aggarwal, F. P. Keenan, and S. J. Rose, *Phys. Scrp.* **70**, 295 (2004).
[13] Y. Nakano, C. Kondo, A. Hatakeyama, Y. Nakai, T. Azuma, K. Komaki, Y. Yamazaki, E. Takada, and T. Murakami, submitted.
[14] Mau Hsiung Chen, Bernd Crasemann, and Kh Rezaul Karim, *Phys. Rev. A* **24**, 1852 (1981).
[15] Y. Nakano, S. Masugi, T. Muranaka, T. Azuma, C. Kondo, A. Hatakeyama, K. Komaki, Y. Yamazaki, E. Takada, and T. Murakami, *J. Phys. Conf. Ser.* **58**, 359 (2007).

# Effect of Debye plasmas on the resonance states of two-electron highly stripped ions using the stabilization method

**Y. K. Ho**[1] **and Sabyasachi Kar**[1, 2]
[1]Institute of Atomic and Molecular Sciences, Academia Sinica, Taipei, Taiwan 106

[2]Center for Theoretical Atomic and Molecular Physics, The Academy of Fundamental and Interdisciplinary Sciences, Harbin Institute of Technology, Harbin 150080, China

E-mail: ykho@pub.iams.sinica.edu.tw

**Abstract**
Effect of Debye plasmas on the bound states and resonance states of two-electron highly stripped ions has been investigated using the stabilization method. Highly correlated exponential basis functions are used to extract resonance parameters. The doubly-excited $^1S^e$ resonance energies and widths of two-electron ions having nuclear charge Z are presented as functions of the screening parameters 1/D and 1/Z, with D being the Debye lengths. New results of ground-state energies, resonance energies and widths for the systems with nuclear charge Z=12 and 14, are reported.

## 1. Introduction

Plasma screening effects on the two-electron atomic systems has gained considerable attentions in recent years ([1-3], references therein). Plasma screening effect has taken care of by using the Debye screening concept of plasma modeling which is a good approximation for hot-dense plasmas and low-density warm plasmas. Recently, we have initiated resonance states calculations on the plasma-embedded He, $H^-$, $Ps^-$, $Li^+$, $C^{4+}$, $O^{6+}$, and $Ne^{8+}$ [3] in the framework of the Debye shielding approach for weakly coupled plasmas. In this work, our aim is to present the overall behavior of the resonance parameters (energy and width) of two-electron ions immersed in plasmas, as these ions occurs in astrophysical and laboratory tokamak plasmas. We also calculate the ground-state energies, the resonance energies and widths for the systems $Mg^{10+}$, and $Si^{12+}$ in plasmas for different screening parameters. The density of resonance states are calculated using stabilization method [4, 5]. Highly correlated basis functions are used to take care of the correlation effects between the charged particles.

## 2. Calculations

The non-relativistic Hamiltonian describing the two-electron system, with nuclear charge Z and embedded in Debye plasmas characterized by a parameter called Debye length D, is given by

$$H = -\tfrac{1}{2}\nabla_1^2 - \tfrac{1}{2}\nabla_2^2 - Z\left[\exp(-r_1/D)/r_1 + \exp(-r_2/D)/r_2\right] + \exp(-r_{12}/D)/r_{12}\ , \qquad (1)$$

where $r_1$ and $r_2$ are the radial coordinates of the two electrons and $r_{12}$ is their relative distance. A particular value of the Debye length $D$ (in $a_0$) (the Debye screening parameter $\mu=1/D$), corresponds to the range of plasma conditions, as $D$ is proportional to $(T/n)^{1/2}$, where $n$ being the plasma density and $T$ its temperature. Depending on plasma condition, the Debye model will work for astrophysical and laboratory tokamak plasmas. For the $^1S$ states of two-electron ions, we employ the wave functions:

$$\Psi = (1+\hat{O}_{12})\sum_{i=1}^{N} C_i \exp[(-\alpha_i r_1 - \beta_i r_2 - \gamma_i r_{12})\omega], \tag{2}$$

where the operator $\hat{O}_{12}$ is the permutation of two-identical particles. The scaling factor ω is set equal to 1 for bound states calculations and it is varied for resonance states calculations. The non-linear variation parameters $\alpha_i, \beta_i, \gamma_i$ are chosen from a widely used quasi-random process ([3, 6], references therein).

## 3. Results

To obtain resonance parameters using the stabilization method, one needs to obtain the energy levels E(ω) by diagonalizing the Hamiltonian (1) and with the basis functions (2). With the optimum choice of non-linear variational parameters in Eq. (2), one can obtain the best ground state energy that is upper bound to the true ground state energy of the system. The energy levels E (ω), obtained when ω is varied, are then plotted to obtain the stabilization diagram (see Ref. [3]). The stabilized plateau at the particular energy region indicates a resonance position. Details of the stabilization method can be obtained from the works of Ho and co-workers ([3, 5], references therein). The method will also work for high-Z ions like $Fe^{24+}$-ion within non-relativistic limit. To extract resonance parameters ($E_r$, $\Gamma$) for a particular resonance state, we calculate the density of the resonance states in the stabilized plateau of the energy level using the formula [3, 5 ]:

$$\rho_n(E) = \left|[E_n(\omega_{i+1}) - E_n(\omega_{i-1})]/(\omega_{i+1} - \omega_{i-1})\right|^{-1}_{E=E_n(\omega_i)}, \tag{3}$$

where the index $i$ is the $i$-th varied $\omega$ value, i.e, $\omega_i$ and the index n is for the n-th resonance. Also in Eq. (3), $\omega_{i-1}$ and $\omega_{i+1}$ are respectively the (i-1)$^{\text{th}}$ and (i+1)$^{\text{th}}$ varied $\omega$ values next to $\omega_i$. After calculating the density of resonance states $\rho_n(E)$ using formula (3), we fit it to the following Lorentzian form that yields resonance energy $E_r$ and a total width $\Gamma$, with

$$\rho_n(E) = y_0 + (A/\pi)\{(\Gamma/2)/[(E-E_r)^2 + (\Gamma/2)^2]\}, \tag{4}$$

where $y_0$ is the baseline offset, A is the total area under the curve, $E_r$ is the center of the peak, and $\Gamma$ denotes the full width of the peak of the curve at half height. Following the above procedure, we have obtained resonance energies and widths for several resonances for a given Debye length and for each atomic system. The results are presented in the figures 1-2 and table 1. In table 1 and figure 1, we present our new results for the plasma-embedded $Mg^{10+}$ and $Si^{12+}$ ions. The 2S threshold energies of the respective one-body subsystems are calculated using standard Slater-type orbitals and the results are well comparable with the reported results [7]. Our ground-state energies are lower than the results reported in the literature [1, 6]. The resonance parameters are reported for the first time for $Mg^{10+}$ and $Si^{12+}$ in plasmas. It is observed that the resonance energies are increased with increasing plasma strength, and ultimately close to the 2S threshold of the respective one-body subsystems. The resonance widths are decreasing for the '+'states and increasing for'–'states when the screening effect is increased. Details about the '+'and '–' states in plasmas are available in our earlier works [3]. In figure 2, we present the lowest resonance energy and width as a function of 1/Z and 1/D. The results for the system Z=2 to 10 are taken from our earlier works [3]. With the recent advancements in laser plasmas [8] and in astrophysical plasmas, we hope our finding will provide useful information for further studies in plasma physics and atomic physics.

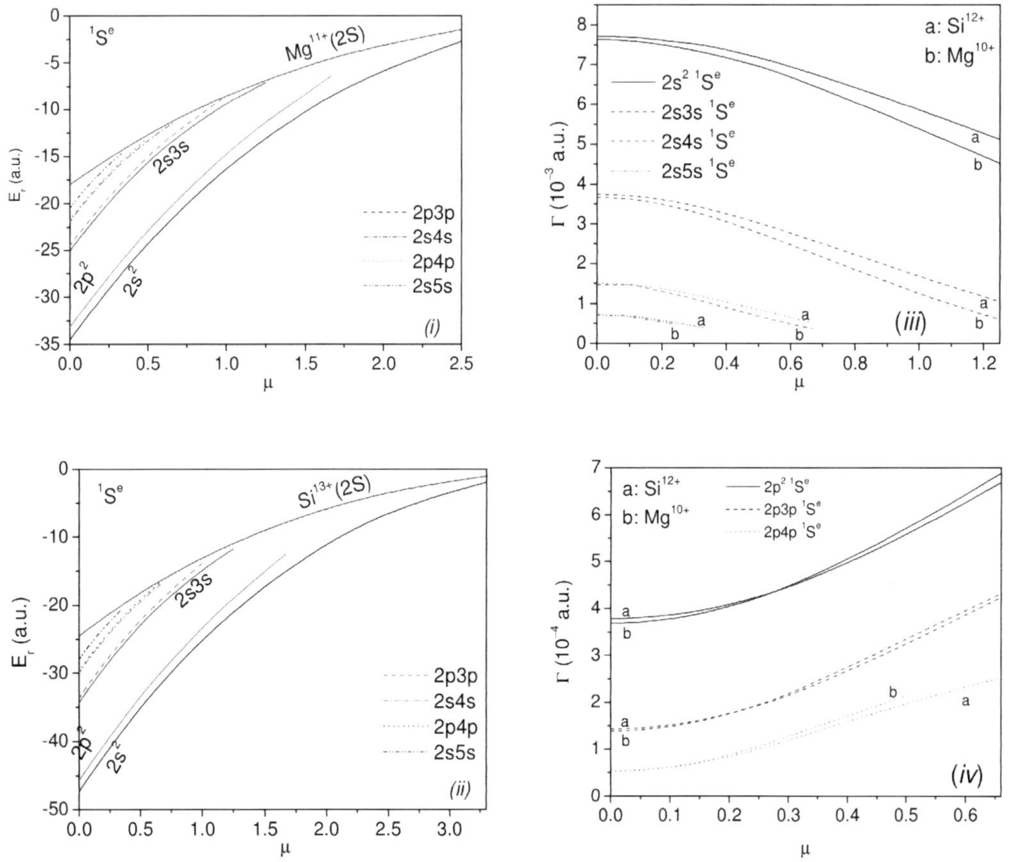


**Figure 1:** Resonance energy [(i)-(ii)] and resonance width [(iii)-(iv)] as functions of μ (in $a_0^{-1}$).

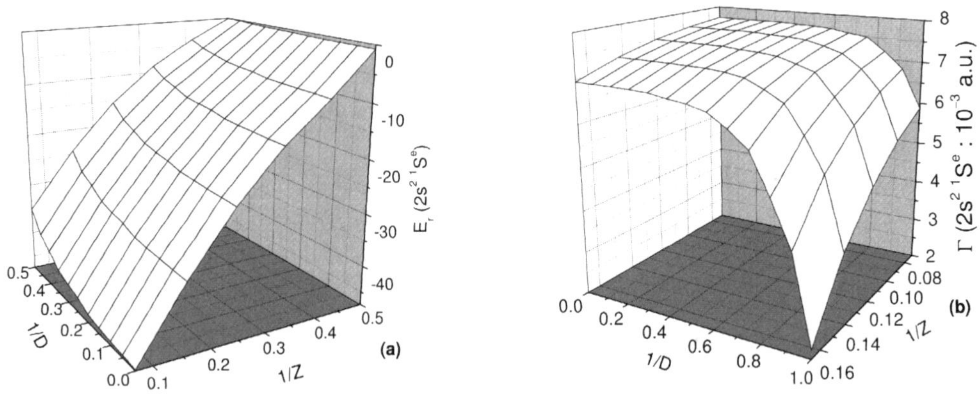


**Figure 2.** The $2s^2\ ^1S^e$ resonance energy $E_r$ (part a) and width Γ (part b) of two-electron ions as a function of 1/Z and 1/D.

**Table 1**. Ground-state energy (E), ${}^{1}S^{e}$ resonance energy ($E_r$) and width (Γ) of $Mg^{10+}$, $Si^{12+}$, the 2S threshold energies [E(2S)] of the respective subsystems for different Debye length *D* (in $a_0$). In the entry for the widths, the results shown as A-B denotes A x $10^{-B}$ . [a]Ref..[6]; [b]Ref. [1].

| | | | | | $Mg^{10+}$ | | | | | |
|---|---|---|---|---|---|---|---|---|---|---|
| *D* | $-E\,(1s^{2}\,{}^{1}S)$ | | $2s^2$ | $2p^2$ | 2s3s | 2p3p | 2s4s | 2p5p | 2s5s | −E(2S) |
| ∞ | 136.6569483 | $-E_r$ | 34.55124 | 33.18738 | 25.09158 | 24.51451 | 21.94427 | 21.69993 | 20.50474 | 18.00000 |
| | 136.65694[a] | Γ | 7.631-3 | 3.689-4 | 3.6662-3 | 1.389-4 | 1.4777-3 | 0.534-4 | 0.722-3 | |
| 50 | 136.1975291 | $-E_r$ | 34.09342 | 32.72946 | 24.63519 | 24.05815 | 21.48987 | 21.24566 | 20.05288 | 17.76120 |
| | | Γ | 7.629-3 | 3.692-4 | 3.6643-3 | 1.392-4 | 1.4756-3 | 0.538-4 | 0.719-3 | |
| 10 | 134.3714046 | $-E_r$ | 32.30506 | 30.93864 | 22.87952 | 22.30305 | 19.77789 | 19.53605 | 18.39404 | 16.82944 |
| | 134.37118[b] | Γ | 7.598-3 | 3.777-4 | 3.6212-3 | 1.479-4 | 1.4730-3 | 0.618-4 | 0.666-3 | |
| 3 | 129.1488914 | $-E_r$ | 27.46108 | 26.06830 | 18.34092 | 17.76290 | 15.61183 | 15.37608 | 14.637 | 14.31395 |
| | | Γ | 7.302-3 | 4.639-4 | 3.2365-3 | 2.335-4 | 1.048-3 | 1.417-4 | | |
| 2 | 125.5107175 | $-E_r$ | 24.31882 | 22.88728 | 15.57398 | 14.98249 | 13.26788 | 13.01832 | | 12.68762 |
| | | Γ | 6.986-3 | 5.669-4 | 2.7872-3 | 3.353-4 | 0.664-3 | 2.139-4 | | |
| 1 | 115.037212 | $-E_r$ | 16.32532 | 14.6608 | 9.27757 | 8.5724 | | | | 8.55877 |
| | 115.03709[b] | Γ | 5.390-3 | 9.467-4 | 1.226-3 | 7.31-4 | | | | |
| | | | | | $Si^{12+}$ | | | | | |
| *D* | $-E\,(1s^{2}\,{}^{1}S)$ | | $2s^2$ | $2p^2$ | 2s3s | 2p3p | 2s4s | 2p5p | 2s5s | −E(2S) |
| ∞ | 187.4070500 | $-E_r$ | 47.30540 | 45.69916 | 34.32488 | 33.63877 | 29.97406 | 29.68424 | 27.98032 | 24.50000 |
| | 187.40687[b] | Γ | 7.715-3 | 3.790-4 | 3.745-3 | 1.432-4 | 1.5102-3 | 0.540-4 | 0.709-3 | |
| 50 | 186.8676337 | $-E_r$ | 46.76760 | 45.16124 | 33.78852 | 33.10241 | 29.43970 | 29.14996 | 27.44853 | 24.22120 |
| | | Γ | 7.713-3 | 3.793-4 | 3.743-3 | 1.435-4 | 1.5086-3 | 0.543-4 | 0.707-3 | |
| 10 | 184.7215862 | $-E_r$ | 44.65978 | 43.05057 | 31.71381 | 31.02752 | 27.40954 | 27.12127 | 25.47292 | 23.12951 |
| | 184.72114[b] | Γ | 7.689-3 | 3.860-4 | 3.710-3 | 1.512-4 | 1.4731-3 | 0.614-4 | 0.692-3 | |
| 3 | 178.5668175 | $-E_r$ | 38.89069 | 37.25119 | 26.25843 | 25.5646 | 22.3433 | 22.05534 | 20.84132 | 20.14982 |
| | | Γ | 7.501-3 | 4.591-4 | 3.405-3 | 2.280-4 | 1.1660-3 | 1.319-4 | 0.397-3 | |
| 2 | 174.2638240 | $-E_r$ | 35.09559 | 33.41268 | 22.85384 | 22.14003 | 19.38966 | 19.08218 | | 18.19546 |
| | | Γ | 7.188-3 | 5.538-4 | 3.039-3 | 3.232-4 | 0.8418-3 | 1.985-4 | | |
| 1 | 161.8035595 | $-E_r$ | 25.19822 | 23.25546 | 14.75543 | 13.87773 | | | | 13.10924 |
| | 161.80347[b] | Γ | 5.875-3 | 9.376-4 | 1.668-3 | 5.462-4 | | | | |

**Acknowledgement**: Financial support from the National Science Council of Taiwan, R.O.C. is gratefully acknowledged.

**References**

[1] Saha B, Mukherjee T K, Mukherjee P K and Diercksen G H F 2002 Theor. Chem. Acc., **108**, 305
[2] A. N. Sil, Mukherjee P K, Int. J. Quantum Chem. **102**, 1061 (2005).
[3] Kar S and Ho Y K 2004 Phys. Rev. E**70**, 066411; 2005 Chem. Phys. Lett. **402**, 544; 2005 New J. Phys. **7**, 141; 2005 Phys. Rev. A **71**, 052503; 2005 Phys. Rev. A **72**, 010703 (R); 2005 J. Phys. B **38**, 3299; 2006 Phys. Rev. A **73**, 032502; 2006 J. Phys B **39**, 2455; 2006 Few-Body Systems **40**, 13; 2007 Euro. Phys. J. D **44**, 1; 2007 Phys. Scr. **75**, 13; J. Phys. B **40**, 1403; J. Phys. B: Special Issue on Resonances
[4] Mandelshtam V A, Ravuri T R, and Taylor H S 1993 Phys. Rev. Letts **70**, 1932
[5] Tan S S and Ho Y K 1997 Chin. J. Phys. **35**, 701; Kar S and Ho Y K 2004 J. Phys. B. **37**, 3177
[6] Thakkar A J and Smith V H, Jr. 1977 Phys. Rev. A **15**, 1,
[7] Rogers F J, Graboske H C, Jr., and Harwood D J 1970 Phys. Rev. A **1**, 1577
[8] NaKai S and Mima K 2004 Rep. Prog. Phys. **67**, 321

Journal of Physics: Conference Series **163** (2009) 012096 doi:10.1088/1742-6596/163/1/012096

# Two-photon transitions in the He$^+$ ion embedded in strongly-coupled plasmas

**H F Lai, Y C Lin and Y K Ho**

Institute of Atomic and Molecular Sciences, Academia Sinica, Taipei, Taiwan 106

E-mail: ykho@pub.iams.sinica.edu.tw

**Abstract**. In this work, we carry out an investigation of 1s-2s and 1s-3s two-photon transitions in the He$^+$ ions embedded in strongly-coupled plasmas. Using the ion-sphere (IS) model and with B-spline basis to represent the wave functions, absorption coefficients for different sizes of the sphere are calculated.

## 1. Introduction

There have been considerable interests in the investigations of atomic processes for atoms/ions embedded in weakly-coupled plasmas (Ref. [1] and references therein). Based on the values of the coupling constant ($\Gamma$) [2, 3], plasmas can be classified into two categories, the weakly-coupled ($\Gamma < 1$) and strongly-coupled ($\Gamma > 1$) plasmas (see Ref. [4] for example). The weakly-coupled plasmas can be modeled by the Debye-Hückel approach [5], and the strongly-coupled plasmas by the ion-sphere (IS) model [2]. Recently, Sil *et al.* [3] applied the ion-sphere (IS) model to study the effects of strongly-coupled plasmas on the energy levels and oscillator strengths for hydrogenic ions. In the present work, we employ the IS model to investigate the plasmas effects on the 1s-2s and 1s-3s two-photon transitions in the He$^+$ ion embedded in strongly-coupled plasmas.

## 2. Theory and calculation

The Hamiltonian of the system, with energy expressed in Rydberg units and Z=2, is

$$[-\nabla^2 + V(r)] \quad \Psi(\mathbf{r}) = E_0\Psi(\mathbf{r}), \text{ with } V(r) = -\frac{2Z}{r} + \frac{(Z-1)}{R}[3-(\frac{r}{R})^2] \text{ and } R = [\frac{Z-1}{4\pi n_e/3}]^{1/3}. \quad (1)$$

Here, $R$ is the radius of the IS sphere in units of Bohr radii $a_0$, and $n_e$ is the number density of the electrons in the plasmas. We use B-spline basis [6-8] to represent the wave functions. Following the notation of de Boor [6, 7], our wave functions are

$$\phi(r) = \sum_{j=1}^{n} c_j B_j(r) = r\chi(r), \text{ with } \int_0^R \chi^*(r)\chi(r)r^2 \mathrm{d}r = 1, \ \Psi(\mathbf{r}) = \chi(r)Y_{lm} \text{ and } \chi(r) = 0 \ \ at \ \ r = R. \quad (2)$$

The dimensionless transition amplitude, calculated using the pseudo-state method [9], and following the dipole selection rule in the length form, is given by

$$D = \frac{1}{2}\sum_{np}[\frac{1}{E_{np}-E_i-\omega} + \frac{1}{E_{np}+E_f+\omega}]R_{1s}^{np}R_{2s}^{np}, \quad \text{with } R_{ks}^{np} = \int_0^R \mathrm{d}r r^3 \chi_{np}\chi_{ks}, \quad (3)$$

where $\chi_{ks}$ are the bound *S*-states and $\chi_{np}$ are the intermediate *P*-wave states; $\omega$ is the energy in Rydbergs for one of the two photons. $E_i$, $E_f$, and $E_{np}$ are the energies for the initial *S*-state, final *S*-

states, and the intermediate $P$-states, respectively. The energy of the other photon is then fixed for a given set of $E_i$, $E_f$ and $\omega$.

## 3. Results and discussions

3.1. 1s-2s transition.
Following an earlier application using B-spline basis [10], here we use an order k = 13 with 40 terms B-spline basis to represent the wave functions. In Figure 1, we show the absorption coefficients $|D|^2$ *vs* the photon energy for four different $R$ values. Table 1 shows numerical results. For the entry under 'Absence of Plasmas' the results are obtained when the $He^+$ ion is confined by an impenetrable sphere (the wave functions vanish outside $R$) but without the plasma effects. Results under 'Presence of Plasmas' are obtained with the IS model of Eq. (1). In Table 1, we also show the changes of the resonance enhancement frequency as the sphere radius changes. The appearance of a resonance enhancement frequency is when the photon energy lies near an intermediate $P$-state. It is seen that in the absence of plasmas, the resonance enhance frequency is shifted to higher values when $R$ is decreased from the free-ion case. In the presence of plasmas, the enhancement frequency is first shifted down to lower energy, but it is then shifted to higher energy when $R$ is decreased further.

3.2. 1s-3s transition
Figure 2 shows the absorption coefficients $|D|^2$ *vs* the photon energy for four different $R$ values. Table 2 shows the transparency frequency (the absorption coefficient goes to zero) for different $R$ values. It is seen that the transparency frequency is shifted to lower values when $R$ is reduced from the free-ion case. When $R$ is reduced to about $4a_0$, the transparency frequency disappears. In Table 3, we show the changes of the resonance enhancement frequency for different $R$ values. In this energy region, there exist two resonance enhancement frequencies. It is observed that in the absence of plasmas, both frequencies are shifted to higher values when $R$ is decreased. In the presence of plasmas, both enhancement frequencies are first shifted down to lower energies, and then shifted to higher energies when the $R$ value is further reduced from the free-ion case.

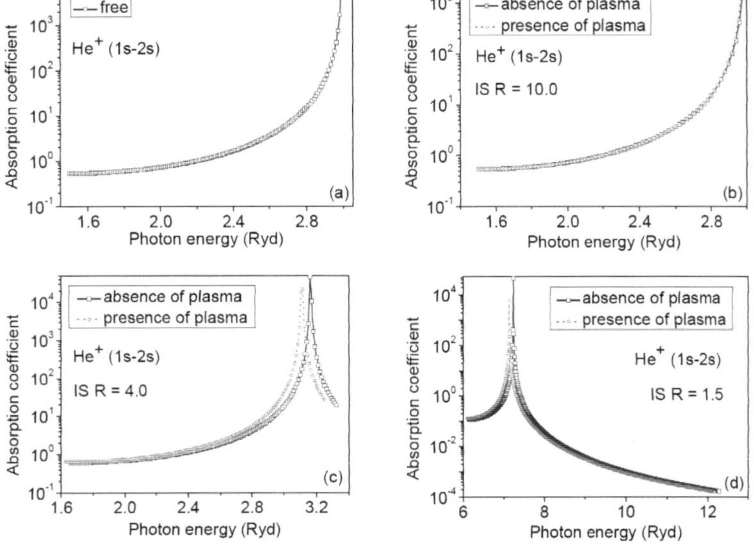


**Figure 1.** Absorption coefficient $|D|^2$ *vs* photon energy in the absence (solid blue line) and presence (red dot line) of plasma for $He^+$ in 1s-2s transition for four $R$ (in $a_0$) values.

**Table 1.** Resonance enhancement frequency (Ryd) *vs* $R$ ($a_0$) for $He^+$ in 1s-2s transition.

| $R$ | Resonance Enhancement Frequency | |
|---|---|---|
| | Absence of Plasma | Presence of Plasma |
| ∞ | 3.00000 | |
| 20 | 3.00000 | 2.99916 |
| 10 | 3.00004 | 2.99324 |
| 8 | 3.00089 | 2.98780 |
| 6 | 3.01396 | 2.98703 |
| 4 | 3.16420 | 3.11089 |
| 3 | 3.54977 | 3.47827 |
| 2 | 5.01434 | 4.92292 |
| 1.5 | 7.24174 | 7.13653 |

**Table 2.** Transparence frequency (Ryd) vs $R$ ($a_0$) for $He^+$ in the 1s-3s transition.

| $R$ | Transparence Frequency (Absence of Plasma) | Transparence Frequency (Presence of Plasma) |
|---|---|---|
| ∞ | 2.7748 | 2.7748 |
| 20 | 2.7748 | 2.7746 |
| 10 | 2.7703 | 2.7687 |
| 9 | 2.7664 | 2.7630 |
| 8 | 2.7581 | 2.7510 |
| 7 | 2.7417 | 2.7298 |
| 6 | 2.7100 | 2.6894 |
| 5 | 2.6007 | 2.5696 |

**Table 3.** Resonance enhancement frequency (Ryd) vs $R$ ($a_0$) for $He^+$ in the 1s-3s transition.

| $R$ | Frequency (1) | | Frequency (2) | |
|---|---|---|---|---|
| | Absence | Presence | Absence | Presence |
| ∞ | 3.00000 | | 3.55556 | |
| 20 | 3.00000 | 2.99916 | 3.55556 | 3.54996 |
| 10 | 3.00004 | 2.99323 | 3.58711 | 3.55500 |
| 9 | 3.00020 | 2.99088 | 3.61737 | 3.57967 |
| 8 | 3.00089 | 2.98780 | 3.67440 | 3.63057 |
| 7 | 3.00368 | 2.98494 | 3.78185 | 3.73137 |
| 6 | 3.01396 | 2.98703 | 3.98712 | 3.92928 |
| 5 | 3.04912 | 3.01081 | 4.39352 | 4.32707 |
| 4 | 3.16420 | 3.11089 | 5.25875 | 5.18161 |

## 4. Summary and conclusion

We have used the ion-sphere model and employed the B-spline basis to investigate the effect of strongly-coupled plasmas on atomic structures and transitions. By using the pseudo-state method, the two-photon absorption coefficients for the 1s-2s and 1s-3s transitions for plasma-embedded $He^+$ ions are calculated. It is observed that the resonance enhancement frequencies and the transparency frequency (for 1s-3s transition) are affected by the size of the sphere in the IS model. Owing to the recent developments in laboratory and astrophysics plasmas, it is hoped that our findings would play a useful role in plasma physics.

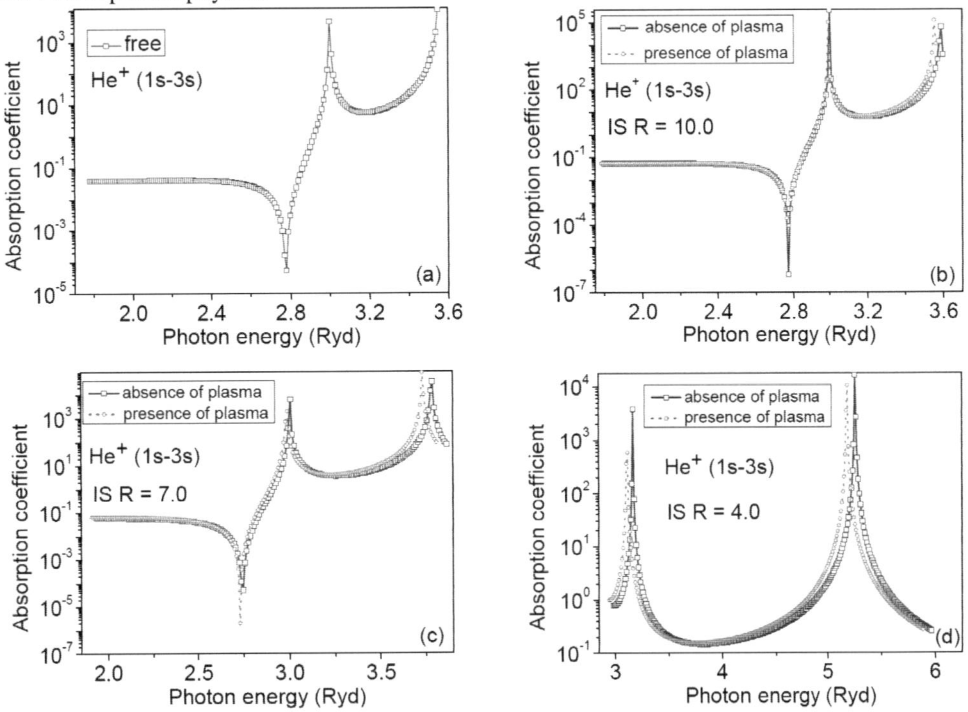


**Figure 2.** Absorption coefficient $|D|^2$ *vs* photon energy (in Ryd) in the absence (solid blue line) and presence (red dot line) of plasmas for $He^+$ in 1s-3s transition for four $R$ (in $a_0$) values.

**Acknowledgement**
Financial support from the National Science Council of Taiwan, R.O.C. is gratefully acknowledged.

**References**

[1] Zhao L B and Ho Y K 2004 *Phys.Plasmas* **11** 1695
Yu A and Ho Y K 2005 *Phys. Plasmas* **12** 043302
Sahoo S and Ho Y K 2006 *Phys. Plasmas* **13** 063301
Kar S and Ho Y K 2008 *Phys. Plasmas* **15** 013301
Paul S and Ho Y K 2008 *Phys. Plasmas* **15** 073301
[2] Ichimaru S 1982 *Rev. Mod. Phys.* **54** 1017
[3] Sil A N, Bhattacharyya S and Mukherjee P K 2007 *Int. J. Quantum. Chem.* **107** 2708
[4] Gurnett D A and Bhattacharjee A 2005 *Introduction to Plasma Physics*(Cambridge University Press) p.7
[5] Debye P and Hückel E 1923 *Z. Phys.* **24** 185
[6] de Boor C 1978 *A Practical Guide to Splines* (Springer: NY)
[7] Johnson W R, Blundell S A and Sapirstein J 1988 *Phys. Rev.* A **37** 307
[8] Chang T N 1994 Photoionization Dominated by Double-excitation in Two-electron and Divalent Atoms *Many-body Atomic Physics* ed J J Boyle and M S Pindzola (Cambridge University Press) chapter 5 pp 129-154
[9] Drachman R J, Bhatia A K and Shabazz A A 1990 *Phys. Rev.* A **42** 6333
[10] Fang T K, Ho Y K and Lin Y C 2002 B-spline-based Complex-rotation Method for Two-electron and Divalent Atoms, *Proc. of International Seminar on Photoionization in Atoms* (Kyoto University Press).

# Atomic processes for the damage on bio-molecules irradiated by XFEL

**K Moribayashi* and T Kai**

*Japan Atomic Energy Agency, 8-1, Umemidai, Kizugawa-city,, 619-0215, Japan*

* E-mail: moribayashi.kengo@jaea.go.jp

**Abstract.** We have calculated the change of the electronic states of C atoms irradiated by an XFEL as a function of time in order to study the damage of bio-molecules. We have treated various parameters such as pulse length, wavelength, flux of an XFEL, and bio-molecule size. This study gives valuable information for experiments of the diffraction patterns of single bio-molecules, which are employed to study their three dimensional structure.

## 1. Introduction

The analysis of three-dimensional structures of single bio-molecules such as proteins has lately attracted considerable attention as an application of an x-ray free electron lasers (XFEL) [1, 2]. This analysis will come from the diffraction patterns, which are produced from the irradiation of an XFEL onto bio-molecules. As the x-ray flux becomes larger, the better diffraction pattern is produced, however, the bio-molecules become more damaged, that is, the atoms in the bio-molecules become more ionized and more highly charged ions are produced. The highly charged ions lead to destroy the bio-molecules due to a Coulomb explosion. The damage and destroy, which appear as noise for the analysis of the three dimensional structure, depend on the experimental parameters such as flux, pulse length, wavelength of an XFEL, and bio-molecule size. The damage and destroy are studied in Refs. [1, 2] where the changes of the electronic states of atoms or ions as a function of time have been shown by using only one parameter of XFEL for a specific bio-molecule.

In this paper, we have treated various parameters in order to give significant information to the experiments of the diffraction patterns. Further, we employ more accurate atomic data for Auger, radiative transitions and electron impact ionization processes [3-5] which will be mentioned in Sec.2.

## 2. Models of the calculations

We treat C atoms with a spherical shape at solid density in this paper by the following reasons. (i) for the shapes of bio-molecules, we choose close to spherically shaped bio-molecules because of the lack of data concerning their three dimensional structures. (ii) as for carbon only choosing, the main elements in bio-molecules are carbon, nitrogen, and oxygen atoms. However, we can not see a big difference among their atomic data of cross sections and rates in terms of processes [3, 6]. Among these atoms, the number of carbon atoms is the largest in bio-molecules. (iii) as for density, the density of bio-molecules are almost the same as solid density.

For the calculation of the changes of the electronic states of atoms or ions in a bio-molecule as a function of time, we treat atomic processes such as photo absorption ionization (e.g., $C + h\nu \rightarrow C^+ + e^-$), Compton scattering (e.g., $C + h\nu \rightarrow C^+ + e^- + h\nu'$), Auger (e.g., $C^{+*} \rightarrow C^{2+} + e^-$), electron impact ionization (e.g., $C + e^- \rightarrow C^+ + 2e^-$), and radiative transitions (e.g., $C^{+*} \rightarrow C^+ + h\nu$), where $h\nu$ and $C^{+*}$ are an x-ray energy and an inner-shell excited state of the C ion. The atomic data of Auger and radiarive transitions are employed given in Ref. [3] where all the electronic states such as $1s^2\,2s^2\,2p^2$,

$1s\ 2s^2\ 2p^2$ listed in Table I and II of Ref. [3] are employed. The photo-ionization cross sections are calculated by Cowan's code [7]. The electron impact ionization cross sections and the energy distribution of the secondary, third order, fourth order electrons produced by electron impact ionization processes are employed in Ref. [6] and estimated by BED theory [4, 5], respectively. It should be noted that in this paper, the 2s and 2p states are treated separately and energy distributions of the secondary and highly order electrons produced by electron impact ionization processes are considered, but have not been in Refs.[1, 2].

Populations for the electronic states as a function of time are calculated by using the rate equations,

$$\frac{dN_0}{dt} = -\beta_0 N_0,$$

$$\frac{dN_1}{dt} = \alpha_{0,1} N_0 - \beta_1 N_1,$$

$$\bullet$$

$$\bullet$$

$$\frac{dN_n}{dt} = \alpha_{n-1,n} N_{n-1} - \beta_n N_n \ , \qquad (1)$$

where $N_0$, $N_1$,..., $N_n$ are the populations of the ground state of the atoms and ions, singly inner-shell ionized states and hollow atoms of ions, respectively, and $\alpha_{m,k}$ and $\beta_k$, which are produced by using the atomic data as mentioned in Refs. [3, 8-11], are the transition rate from $m$'th to the $k$'th state and the decay rate in the $k$'th state, respectively.

In our model, bio-molecules are treated as a spherical shape with a radius of $r$ and the solid density ( $\sim 3 \times 10^{22}/\text{cm}^3$ ). Electrons, in particular, photo-electrons produced due to photo absorption ionization processes give a significant contribution to the effect of bio-molecule size. Just after an XFEL starts to irradiate onto a bio-molecule, the photo-electrons, which have almost the same energy as that of an XFEL, can escape from the bio-molecule. We employ $r/v$ for the collision time between the photo-electron and the atoms or the ions in the bio-molecules where $v$ is a velocity of the photo-electron. The secondary and highly order electrons stay in the bio-molecules because their energies are too small to escape from it.

When one electron escapes from a bio-molecule, the charge of the bio-molecule increases. The Coulomb force due to this charge reduces the electron energy ($E_e$), as a result, the electron cannot escape from the bio-molecule when $E_e$ becomes 0 in the bio-molecule. We assume that the charge is concentrated at the center of the bio-molecule because of the spherical shape assumed. When the energy of a photo electron produced at the center of the bio-molecule becomes 0 in the bio-molecule, we assume that it cannot escape. For example, when $\lambda = 0.1$ nm, $r = 10$ nm, and $Q \geq 4 \times 10^{10}$, the photo-electrons cannot escape and stay in the bio-molecule, where $\lambda$ and $Q$ are the wavelength of the XFEL and the total charge in a bio-molecule, respectively.

## 3. Results and discussions

We have been studying the role of atomic processes such as photo-ionization, Auger, radiative transition processes [3], and electron impact ionization [5] for the damage of bio-molecules irradiated by an XFEL. By considering these roles, we have constructed the models mentioned in Sec.2. Furthermore, we have also proposed measuring the x-ray flux irradiated onto the bio-molecules by using hollow atoms [3, 9, 11]. For the measurement, the following property is used. The ratio of the number of fluorescent x-ray photons from the hollow atoms to that from the singly inner-shell ionized atoms is in proportion to the x-ray flux irradiating these atoms. In this paper, we have focused on the relationship of XFEL parameters to the damage. This will benefit future experiments using XFELs.

We have calculated the changes of the electronic states of atoms or ions in a bio-molecule as a function of time. However, the number of the electronic states is too large. Then, for clearer figures, we show the changes of charge determined from the electronic states.

Figures 1 (a - c) show the populations of the charge of C as a function of time for the pulse length ($\tau$) and wavelength ($\lambda$) of an XFEL of 100 fs and 0.1 nm, 10 fs and 0.1 nm and, 10 fs and 0.06 nm, respectively. The x-ray flux of the XFEL and the radius of bio-molecules are $10^{22}$/pulse/mm$^2$ and 10 nm, respectively. A gauss type time function is employed for the intensity of the XFEL (see upper sides of figures 1) and the time of 0 is set when the peak intensity of x-rays is located in the bio-molecule. We have found that a shorter value of $\tau$ produces smaller damage (see figures 1(a) and (b)). This is due to the fact that time scale of Auger processes is about 10 fs. Namely, photo-absorption ionization and Auger processes occur only once in the case of $\tau$ = 10 fs and several times in the case of 100 fs, respectively. Auger-electrons also give more significant contribution to the damage in the case of $\tau$ = 100 fs. We have also found that a shorter value of $\lambda$ produces smaller damage (see figure 1(c) and (b)). This is caused by the inner-shell photo-ionization cross sections. The photo absorption ionization cross sections for $\lambda$ = 0.1 nm is about 10 times larger than that for $\lambda$ = 0.06 nm [12].

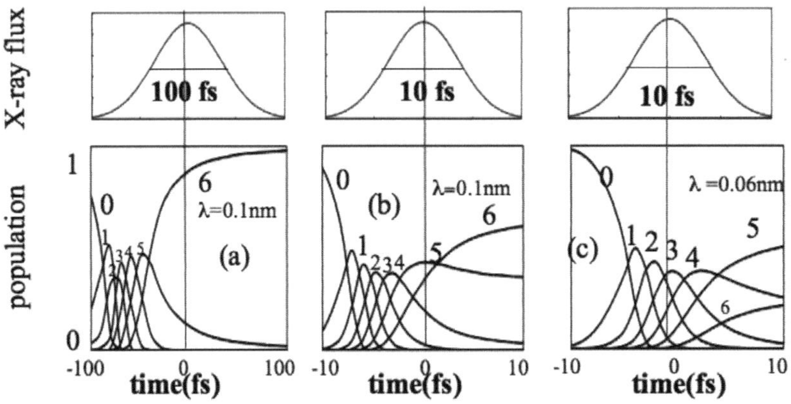


**Figure 1.** Population of the changes of the charge of C atom as a function of time for the x-ray flux of an XFEL of $10^{22}$/pulse/mm$^2$, the radius of bio-molecules of 10 nm, and (a) $\tau$ = 100 fs, $\lambda$ = 0.1 nm, (b) $\tau$ = 10 fs, $\lambda$ = 0.1 nm, (c) $\tau$ = 10 fs, $\lambda$ = 0.06 nm, respectively. The numbers written in the figures express charge (Lower figures). Upper figures: the x-ray intensity as a function of time. The time of 0 is set when the peak intensity of x-rays passes through the bio-molecule.

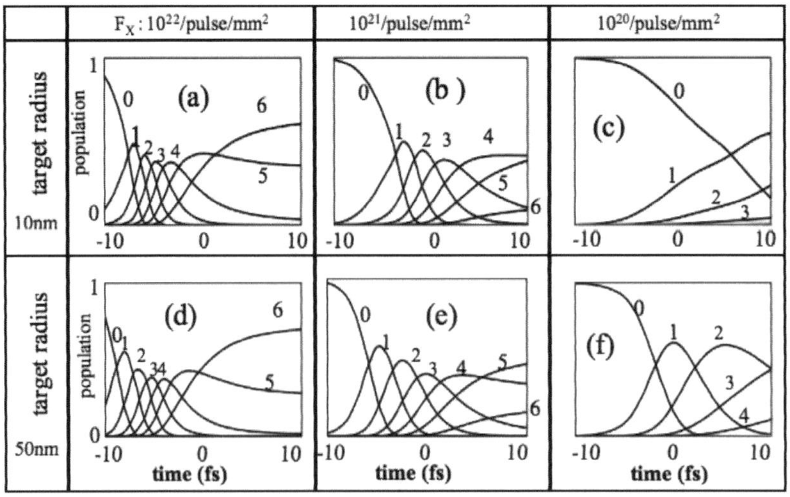


**Figure 2.** The same as figure 1(b) for the x-ray flux and the radius of a bio-molecule of (a) $10^{22}$/pulse/mm$^2$ and 10 nm, (b) $10^{21}$/pulse/mm$^2$ and 10 nm, (c) $10^{20}$/pulse/mm$^2$ and 10 nm, (d)$10^{22}$/pulse/mm$^2$ and 50 nm, (e)$10^{21}$/pulse/mm$^2$ and 50 nm, and (f)$10^{20}$/pulse/mm$^2$ and 50 nm, respectively.

Figures 2 ( a - f ) show the population of C with different charge states as a function of time for x-ray fluxes and radiuses of bio-molecule of $10^{22}$/pulse/mm$^2$ and 10 nm, $10^{21}$/pulse/mm$^2$ and 10 nm, $10^{20}$/pulse/mm$^2$ and 10 nm, $10^{22}$/pulse/mm$^2$ and 50 nm, $10^{21}$/pulse/mm$^2$ and 50 nm, and

$10^{20}$/pulse/mm$^2$ and 50 nm, respectively. X-ray fluxes corresponds to the time scale of the photo absorption ionization processes because the rates of photo absorption ionization processes ($R_{ap}$) are in proportion to the x-ray flux. On the other hand, the size dependence comes from electron impact ionization processes because the interaction time of the photo-electron and atoms or ions in a bio-molecule increases according to the the bio-molecule size as mentioned before. The electron impact ionization cross sections decrease as charge numbers become larger [6]. Namely, size effects appear clearly when a lot of neutral atoms exist in the bio-molecule and as the charge number becomes large, the size effect seems to reduce.

Smaller x-ray fluxes and smaller bio-molecule size produce smaller damage. However, the resolution power or the intensities for the diffraction pattern correspond to x-ray fluxes and bio-molecule size. Intensities for the diffraction patterns ($I_O$) are given by

$$I_o = I_{XFEL} \frac{r_c^{\,2} d^2}{R^2} \,|\, F(\vec{k}) \,|^2, \tag{2}$$

with

$$F(\vec{k}) = \int \rho(\vec{r}) e^{i\vec{k}\bullet\vec{r}} \, d\vec{r}, \tag{3}$$

where $I_{XFEL}$, $r_c$, $d^2$, $R$, and $\rho(r)$ are the intensity of XFEL, the classical radius for electrons, the distance between the target and detector, the area of the detector, the density of electrons, respectively and $\vec{k}$ is a scattering vector. As a bio-molecule size becomes larger, $F(\vec{k})$ increases because the number of electrons becomes larger. Therefore, we need to study the intensities of the diffraction patterns including the damage in order to propose the best parameter for the experiments in future.

## 4. Summary

We have studied the damage of a bio-molecule irradiated by an XFEL from the calculation of the changes of the charge of C atoms as a function of time for various parameters of XFELs. We have shown the relationship of the damage with the parameters of pulse length, wavelength, flux of an XFEL and bio-molecule size. We believe that these results become important for the experiment of the three-dimensional structure of a single bio-molecule. In the future, we will study the intensities of the diffraction patterns including the damage.

## Acknowledgements

We wish to thank Prof. N. Go, Drs. Y. Fukuda, H. Kono, A. Matsumoto, A. Tokushisa, J. Koga and M. Yamagiwa for their useful discussions. This study has been supported by the 'X-ray Free Electron Laser Utilization Research Project' of the Ministry of Education, Culture, Sports, Science and Technology of Japan (MEXT).

## References

[1] Neutze R, Wouts R, Spoel D, Weckert E, and Hajdu J 2000 *Nature* **406** 752
[2] Hau-Riege S P, London R , Szoke A 2004 *Phys.Rev.*E **69** 051906
[3] Moribayashi K 2008 *J. Phys. B* **41** 085602
[4] Kim Y K, Santos J P, and Parente F, 2000 *Phys.Rev.A* **62** 052710
[5] Kai T and Moribayashi K 2008 *J.Phys: Conference series* submitted
[6] Neeb M, Kempgens B, Kivimäki A, Köppe H M, Maier K, Hergenhahn U. Piancastelli M N, Rüdel A, and Bradshaw A M 1991 *J. Electron Spectr. Relat. Phenom.* **89-91** 19
[7] Cowan R D 1968 *J. Opt. Soc. Am.* **58** 808.
[8] Moribayashi K, Sasaki A, and Tajima T 1998 *Phys. Rev. A* **58** 2007
[9] Moribayashi K, Kagawa T, and Kim D E 2004 *J.Phys. B* **37** 4119
[10] Moribayashi K, Kagawa T, and Kim D E 2005 *J.Phys. B* **38** 2187
[11] Moribayashi K 2007 *Phys.Rev.A* **76** 042705
[12] Henke B L, Gullikson E M, and Davis J C 1993 *Atom. Data Nucl. Data Tables* **54** 181

# Progress of Opacity Experiment on the Shengguang II Laser Facility

**Jiamin Yang[1], Jiyan Zhang[1], Yan Xu[2], Jianmin Yuan[3], Jun Yan[2], Yaonan Ding[1], Yang Zhao[1], Fengtao Jin[1], Zhimin Hu[1], Guohong Yang[1] and Baohan Zhang[1]**

[1] Research Center of Laser Fusion, China Academy of Engineering Physics, P.O.Box 919-986, Mianyang, Sichuan 621900, P.R.China
[2] Beijing Institute of Applied Physics and Computational Mathematics, Beijing 10088, P.R.China
[3] Department of Applied Physics, National University of Defense Technology, Changsha 410073, P.R.China

E-mail: yjm70018@my-public.sc.cninfo.net

**Abstract.** In recent years, a series of opacity diagnostics have been developed at the Research Center of Laser Fusion in China. Two types of cavities (a conical cavity called a type I target, and a cylindrical cavity with foam baffles called a type II target ) were designed to convert the eight-beam laser into x-ray radiation to efficiently heat the sample and to prevent the sample from irradiation of the reflected laser and plasmas on the Shengguang II laser facility. Typical opacity experiments have been carried out using the two target designs, respectively. The results show that a sample temperature of about 95 *eV* has been reached using the type II target which is the highest obtained on the high power laser facility.

## 1. Introduction

The x-ray opacities of hot, dense plasmas have long been of interest due to their important and urgent need in studies of inertial confinement fusion (ICF), x-ray lasers and astrophysics. Theoretical calculations of opacities are quite complex and usually include numerous approximations. Therefore, experimental measurement of opacity is of importance to verify the theoretical models. In the past two decades, many opacity experiments have been done using high power lasers with the advent of high power lasers [1-7]. However, most of the opacity experiments reached sample temperatures of lower than 60*eV*. Experimental opacities with much higher sample temperatures are still scarce and challenging on laser facilities [8]. In recent years, a series of opacity diagnostics have been developed at the Research Center of Laser Fusion in China. A ninth laser beam has also been established on the Shengguang II eight-beam high power ns laser facility, which has duration of about 100ps and can be used as a short pulse backlighting beam in the opacity experiment. Two types of cavities have been designed to conduct opacity experiments on the Shengguang II laser facility, one of which has reached a sample temperature of about 95 *eV* and extended the sample state region of opacity studies.

## 2. Development of diagnostics for x-ray absorption spectrum

In the opacity experiment one of the crucial diagnostics is the x-ray absorption spectrum measurement. In the soft x-ray and super extra ultraviolet spectral regimes, two sets of flat-field grating spectrometers have been developed with 1200*l/mm* and 2400*l/mm* gratings, respectively. The 1200*l/mm* flat-field grating spectrometer covers the spectral regime of 5*nm* to 30*nm* with spectral resolution of about 0.02*nm*. The 2400*l/mm* grating spectrometer can measure the x-ray spectrum from 1*nm* to 5*nm* with higher spectral resolution of 0.01*nm*. Both of the grating spectrometers can be coupled to a gated micro-channel plate (MCP) detector with temporal resolution of about 100ps, as shown in figure 1. As an application of the flat-field grating spectrometer, the 2400*l/mm* grating spectrometer has been used to measure the x-ray absorption lines of low-Z mixture (CHO) [9]. The typical measured results are presented in figure 2. Shown in figure 2(a) are the recorded images of the x-ray source spectrum and the attenuated x-ray spectrum by the radiation-heated CHO sample, in which very clear absorption lines have been observed. The obtained x-ray absorption spectrum is compared with the unresolved-transition array (UTA) simulation [10]. As shown in figure 2(b), the simulated spectrum is in good agreement with the experimental one in general.

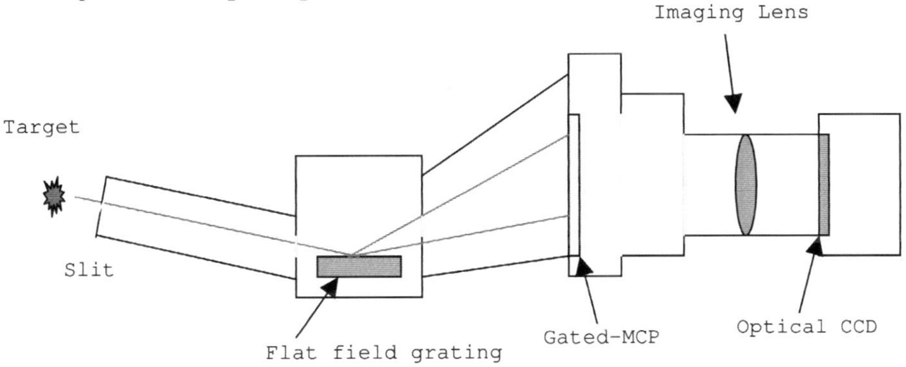


Figure 1. Schematic of time-resolving flat-field spectrometer

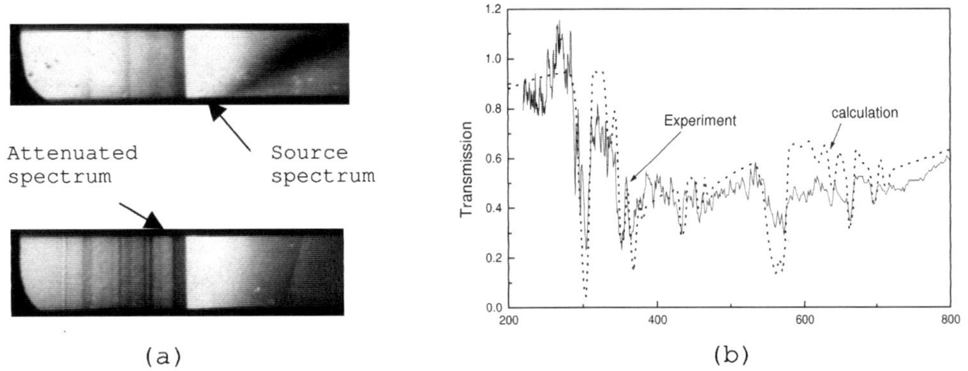


Figure 2. X-ray absorption result for low-Z mixture (CHO) measured with the 2400*l/mm* flat-field grating spectrometer. (a) raw x-ray spectrum image; (b) comparison of the measured absorption spectrum with the unresolved-transition array (UTA) calculation.

In the *keV* x-ray spectral regime, a series of crystal spectrometers have been established which cover the x-ray photon energies from 800 *eV* to 5000 *eV*. The crystal-dispersed spectrum can also be recorded with the gated MCP. One of the measured absorption spectra for an Al plasma with the crystal spectrometer is shown in figure 3 and compared with the calculation using the detailed-term accounting (DTA) model [11,12]. The absorption lines for *Li*- to *F*-like Al ions have been observed and reproduced with the simulation [11].

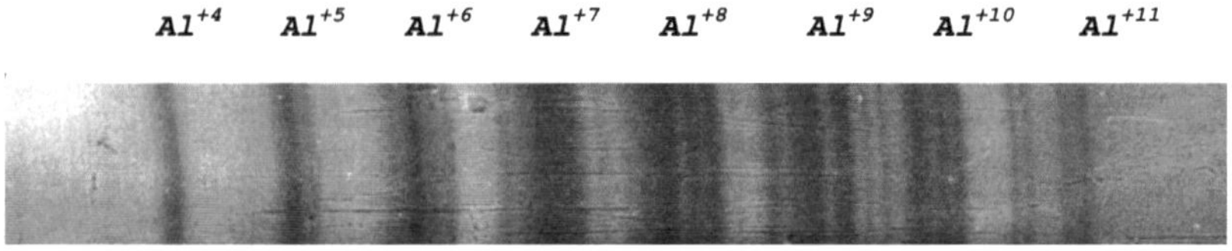


Figure 3. X-ray absorption spectrum image for Al plasmas with the crystal spectrometer.

## 3. Opacity experiment on the Shengguang II laser facility

The Shengguang II laser facility is a Neodymium glass laser facility, which has eight main laser beams and a 9th backlight beam. The main laser can deliver 250J per beam in 1ns with a laser wavelength of 351nm. The 9$^{th}$ laser beam can be operated in pulses of about 100ps or 2000ps. In the opacity experiment, the eight main laser beams are injected into the cavities and converted into intense x-ray radiation to heat the CH-tampered sample. The 9$^{th}$ beam has 2ω laser energy of 100J with pulse duration of 100ps and is delayed by 1ns with respect to the main laser beams. The x rays emitted from the 9$^{th}$ beam-irradiated gold dot are used as a backlighter. The experimental arrangement is shown in figure 4 and the x-ray absorption spectra have been measured with a crystal spectrometer.

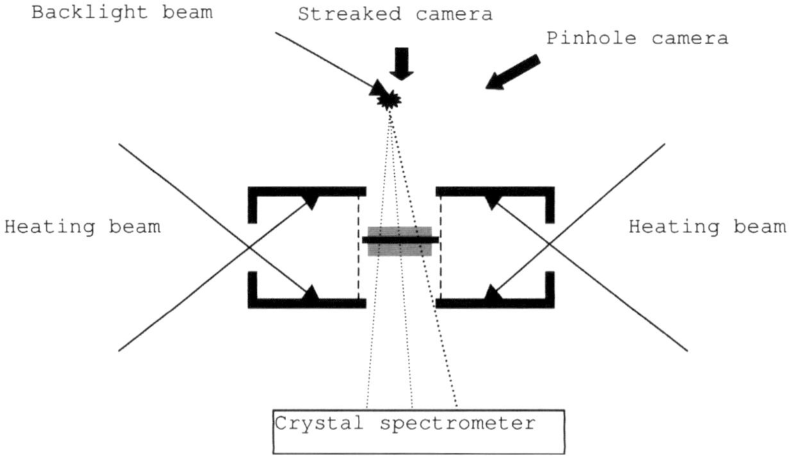


Figure 4. Schematic of the opacity experiment arrangement

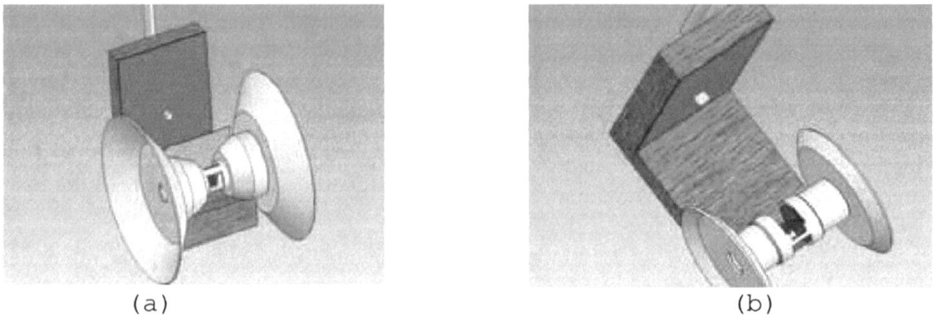


Figure 5. Schematic of the two target designs. (a) Conical Target; (b) Cylindrical target

As shown in figure 5, two types of cavities were designed to convert the eight-beam laser into x-ray radiation to efficiently heat the sample. To prevent the sample from irradiation of the reflected laser and plasmas, a conical cavity, called a type I target, has a conical shape at both ends of the cavity, and a cylindrical cavity, called a type II target, has foam baffles between the sample and the laser-x-ray conversion cavity. As the low density ($40mg/cm^3$) foam baffle can effectively transport the x-ray radiation, the sample between the two $200\mu m$ thick foam baffles can be heated to much higher temperature. The x-ray absorption spectra for Al plasmas using the two target designs are shown in figure 6 (a) and (b), respectively. The experimental results are simulated with a code based on the method given by Abdallah [13]. The results show that a sample temperature of about 95 eV has been reached using the type II target which is much higher than that of 54eV using the type I target [14,15].

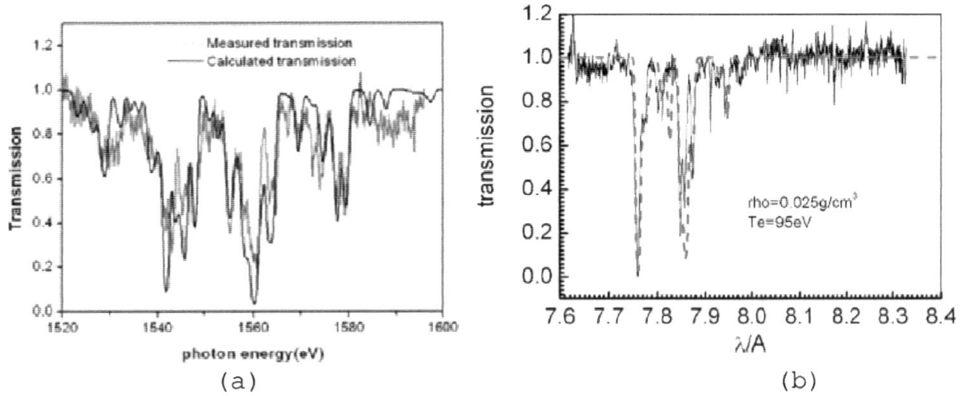


Figure 6. Comparison of experimental absorption x-ray spectra with the simulated ones. (a) using the conical target design; (b) using the cylindrical target design(experiment: solid line, calculation: dotted line).

## 4. Summary

In recent years, a series of opacity diagnostics have been developed at the Research Center of Laser Fusion. Two types of cavities were designed to convert the eight-beam laser into x-ray radiation to efficiently heat the sample on the Shengguang II laser facility. It was demonstrated that a sample temperature of about 95 *eV* has been reached using the type II target which is the highest obtained on the high power laser facility. The type II target can be further used in the opacity experiment for high temperature plasmas in the future.

[1] Perry T S, Davidson S J, Serduke F J *et al* 1991 *Phys. Rev. Letts*, 67(27), 3784
[2] Perry T S, Budil K S, Cauble R *et al* 1995 *J. Quant. Spectrosc. Radiat. Transfer*, 54(1/2), 317
[3] Iglesias C A, Chen M H *et al* 1995 *J. Quant. Spectrosc. Radiat. Transfer*, 54(1/2), 185
[4] Back C A, Perry T S, Bach D R *et al* 1997 *J. Quant. Spectrosc. Radiat. Transfer*, 58, 415
[5] Wang P, Macfarlane J J and Orzechowski T J 1997 *Rev.Sci. Instrum.* 68(1),1107
[6] Chenais-Popovics C, Fajardo M, Gilleron F *et al* 2001 *Phys. Rev. E* (65),016413
[7] Yang J, Zhang J, Ding Y *et al* 2003 *Phys. Plasmas* 10,4881
[8] Bailey J E, Rochau G A, Iglesias C A *et al* 2007 *Phys. Rev. Letts* 99, 265002
[9] Yang J, Ding Y, Yan J *et al* 2002 *Phys. Plasmas* 9(2), 678
[10] Yan J, Qu Y and Li J 1999 *High Power Laser and Particle Beams* 11,65
[11] Yang J, Zhang J, Ding Y, Yang G, et al 2007 *High Power Laser and Particle Beams* 19(9),1483
[12] Jin F, Zeng J, Yuan J 2004 *Phys. Plasmas* 11(9), 4318
[13] Josephy Abdallah, Jr. and Robert E.H.Clark 1991 *J. Appl. Phys.* 69(1), 23
[14] Zhang J, Yang J, Xu Y *et al* 2008 *ACTA PHYSICA SINICA*, 57(2), 985
[15] Xu Y, Zhang J, Yang J *et al* 2007 *Phys. Plasmas* 14, 052701

Journal of Physics: Conference Series **163** (2009) 012099 doi:10.1088/1742-6596/163/1/012099

# Production of multiply charged ions from the 4d photoionization of $Cs^+$

**Tetsuo Koizumi , Takao M. Kojima*, Mutsumi Sano**, Naoki Watanabe#**

Department of Physics, Rikkyo University, Tokyo 171-8501, Japan
(*)The Institute of Physical and Chemical Research (RIKEN), Saitama 351-0198, Japan
(**)Japan Synchrotron Radiation Research Institute, Hyogo 679-5198, Japan
(#)Institute of Low Temperature Science, Hokkaido University, Hokkaido 060-0819, Japan

E-mail:koizumit@rikkyo.ac.jp

**Abstract.** Using an ion-photon merged-beam apparatus, photoion yield spectra of $Cs^{2+}$ and $Cs^{3+}$ from the 4d photoionization of $Cs^+$ have been measured as a function of photon energy in the range 80-140 eV. Above 90 eV, the 4d giant resonance peak is observed clearly for $Cs^{3+}$ spectrum. Below 90 eV, some prominent discrete peaks are observed for both $Cs^{2+}$ and $Cs^{3+}$. These discrete peaks can be attributed to the 4d–np and 4d-nf transitions.

## 1. Introduction

The photoabsoption spectra of heavy atoms, such as Xe, Cs, Ba and the following lanthanides, show broad peaks called giant resonance in the 4d photoionization energy region. There has been great interest in systematic studies of 4d photoabsorption of atoms with increasing ionization along with isoelectronic sequence because these phenomena are good examples of the appearance of many-electron effects in atoms. However, very little is known experimentally on photoabsorption for ionic species. We have reported the photoion spectra of $Xe^{q+}$(q=1-3)[1-4], $Ba^+$[5], and $Eu^+$[6,7] in the 4d photoabsorption region. In this paper, we present the relative photoion yield spectra of $Cs^{2+}$ and $Cs^{3+}$ from a $Cs^+$ target within the 4d giant resonance region with an ion-photon merged-beam technique. The 4d photoabsorption behavior of $Cs^+$ should be intermediate between Xe and $Ba^{2+}$. As mentioned above, only few experimental data have been published for $Cs^+$. The 4d photoabsorption spectrum of $Cs^+$ has been measured by Cummings and O'Sullivan [8] with a dual-laser-produced plasma (DLP) technique. However, because of absorption saturation the shape of the 4d giant resonance could not be deduced in their experiment.

## 2. Experimental

The ion-photon merged-beam apparatus using in this experiment has been presented by Koizumi et al. [5]. Briefly, the measurements were done using synchrotron radiation from the 2.5 GeV storage ring of the Photon Factory at the High Energy Accelerator Research Organization (KEK) at an undulator beam line (BL-16B). Photons were monochromatized by a 24m spherical grating monochromator.

Photon intensity was measured as electric current by a photodiode located at the end of the chamber. Target ions, $Cs^+$, were produced by a surface ionization ion source, and extracted with 2 kV acceleration. The ions were deflected 90° by an electrostatic quadrupole and merged into the photon beam. After passing a 15 cm reaction region, the ions were demerged from the photon beam and charge analyzed by an electrostatic parallel plate analyzer. Primary ion current was monitored by a Faraday cup installed in the analyzer. The multiply-charged ions produced in the reaction region were detected by channel-type secondary electron multipliers combined with ion-to-electron converter plate in ion count mode. The vacuum pressure of the experimental chamber was kept lower than $4x10^{-8}$ Pa during the measurement. Photon energy of the monochromator has been calibrated by some photoabsorption lines. The accuracy of the energy scale in the present measurement was estimated to be better than 0.05 eV.

Relative photoion yield spectra of $Cs^{2+}$ and $Cs^{3+}$ from the 4d photoionization of $Cs^+$ are presented in this paper as a function of photon energy in the range 80-140 eV.

## 2. Results and Discussion

Figure 1 shows the relative photoion yield spectra of $Cs^{2+}$ and $Cs^{3+}$ from the 4d photoionization of $Cs^+$ as a function of photon energy. The energy resolution, $E/\Delta E$, is about 400 in this measurement. The "total ion" in the figure 1. a) means the sum of $Cs^{2+}$ and $Cs^{3+}$ yields. Because the ionization after photoabsorption is dominant process, and the triple-ionization process is negligibly small in this energy region, the total ion spectrum obtained in this experiment corresponds to the photo-absorption spectrum.

The 4d giant resonance peak, which is attributed to the 4d→εf transition, is observed clearly for $Cs^{3+}$ spectrum. It is worth noting that some window type resonances appear around at 100eV. Below 90 eV, some prominent discrete peaks are observed. There has been great interest in systematic studies of 4d photo-absorption of atoms along with isoelectronic sequence, Xe, $Cs^+$, and $Ba^{2+}$. For neutral Xe target, discrete peaks are not so prominent below 4d threshold [9]. On the contrary, for $Ba^{2+}$, Lucatorto et al. [10] showed that drastic change happens, and many very prominent discrete peaks appear in lower energy region of the 4d giant resonance peak. The $Cs^+$ case is intermediate between Xe and $Ba^{2+}$.

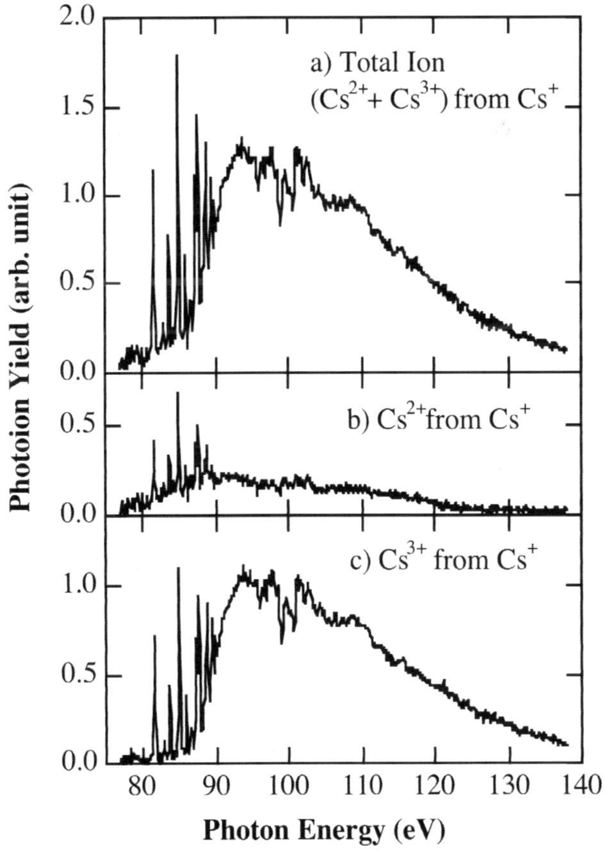


Fig. 1. The relative photoion yield spectra for $Cs^+$ target. a) Sum of $Cs^{2+}$ and $Cs^{3+}$ yield, b) $Cs^{2+}$ form $Cs^+$, c) $Cs^{3+}$ from $Cs^+$.

Sato et al. [11] calculated the giant resonances of Xe-like isoelectronic sequence by the time dependent local spin density approximation (TDLSDA). The agreement between their calculation spectrum and the present total ion one is fairly good.

Figure 2 shows photoion yield spectra of $Cs^{2+}$ and $Cs^{3+}$ with higher energy resolution ($E/\Delta E \approx 800$). Cummings and O'Sullivan [8] observed these discrete lines. The agreement between their result and present one is very good. They showed that these transitions are attributed to the 4d–np and 4d-nf transitions by Multiconfiguration Hartree-Fock (MCHF) calculations. Vertical lines in the figure 2. a) indicate their calculated transition energies. $d_{3/2}$ and $d_{5/2}$ represent the 4d hole states. MCHF calculations reproduce our experimental data fairly well.

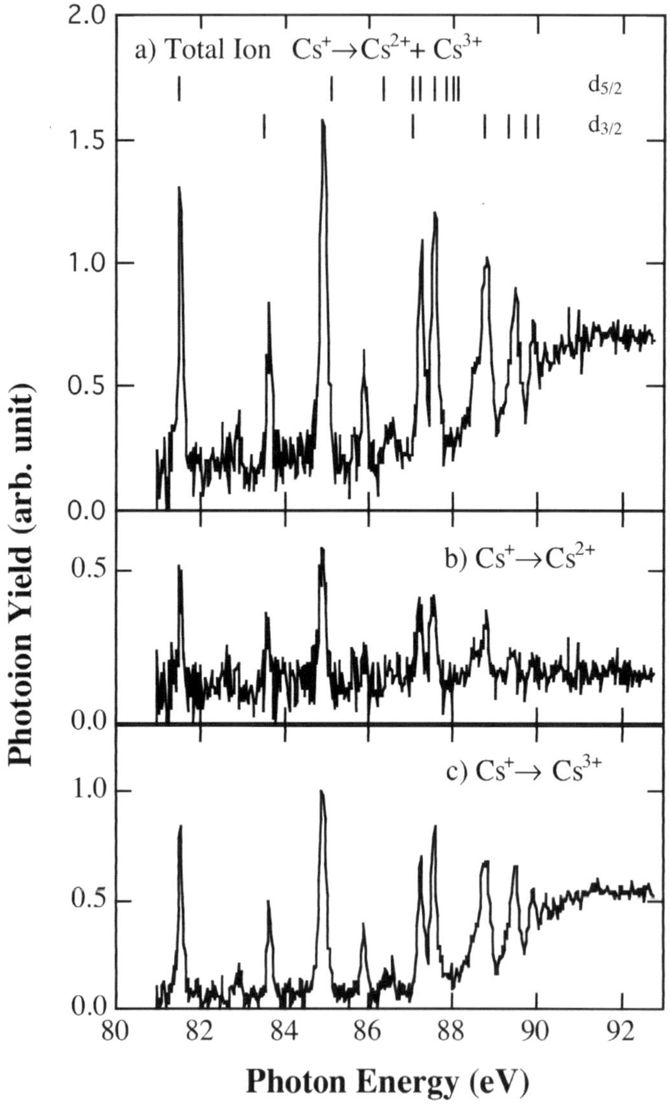


Fig. 2. Photoion yield spectra from $Cs^+$ ions with higher energy resolution of $E/\Delta E \approx 800$ as a function of photon energy. a) Sum of $Cs^{2+}$ and $Cs^{3+}$ yield, vertical lines indicate the calculated 4d–np and 4d-nf transition energies [8], b) $Cs^{2+}$ yield, c) $Cs^{3+}$ yield.

**Acknowledgement**
This experiment was done under approval of the Photon Factory of High Energy Accelerator Research Organization.

**References**

[1] M. Sano, Y. Itoh, T. Koizumi, T. M. Kojima, S. D. Kravis, M. Oura, T. Sekioka, N. Watanabe, Y. Awaya, and F. Koike; J. Phys B:At. Mol. Opt. Phys. **29** (1996) 5305-5313.
[2] T. Koizumi, Y. Awaya, A. Fujino, Y. Itoh, M. Kitajima, T. M. Kojima, M. Oura, M. Sano, T.

Sekioka, N. Watanabe, and F. Koike; Physica Scripta. T**73** (1997) 131-132.

[3] N. Watanabe, Y. Awaya, A. Fujino, Y. Itoh, M. Kitajima, T.M. Kojima, M. Oura, R. Okuma, M. Sano, T. Sekioka, and T. Koizumi; J. Phys B:At. Mol. Opt. Phys. **31** (1998) 4137-4141.

[4] Y. Itoh, A. Ito, M. Kitajima, T. Koizumi, T. M. Kojima, H. Sakai, M. Sano, and N. Watanabe; J. Phys B:At. Mol. Opt. Phys. **34** (2001) 3493-3499.

[5] T. Koizumi, Y. Itoh, M. Sano, M. Kimura, T. M. Kojima, S. Kravis, A. Matsumoto, M. Oura, T. Sekioka, and Y. Awaya; J. Phys B:At. Mol. Opt. Phys. **28** (1995) 609-616.

[6] T. M. Kojima, M. Oura, Y. Itoh, T. Koizumi, M. Sano, T. Sekioka, N. Watanabe, H. Yamaoka, and Y. Awaya; J. Phys B:At. Mol. Opt. Phys. **31** (1998) 1463-1468.

[7] T. M. Kojima, F. Chen, M. Kitajima, T. Koizumi, Y. Nakai, H. Yamaoka, and N. Watanabe; J. Electron Spectrosc. Relat. Phenom. **144-147** (2005) 71-74.

[8] A. Cummings and G. O'Sullivan, J. Phys. B**30**, (1997)5599-5607.

[9] See for example, J. B. West and J. Morton; At. Data Nucl. Data Tables **22** (1978) 103-107.

[10] T. B. Lucatorto, T. J. McIlrath, J. Suger, and S. M. Younger; Phys. Rev. Lett. **47** (1981) 1124-1128.

[11] M. Sano, X. M. Tong, and T. Watanabe; Physica Scripta. T**92** (2001) 388-390. and privert commucation.

# X-ray absorption by highly charged ions in plasmas: toward photo-pumping x-ray laser

**T Kawachi**[1] **and Y Kato**[2]
1. Quantum Beam Science Directorate, Japan Atomic Energy Agency (JAEA) 8-1, Umemidai, Kizugawa, Kyoto, 619-0216, Japan
2. The Graduate School for the Creation of New Photonics Industries, 1955-1, Gomatsu, Hamamatsu, Shizuoka, 431-1202, Japan

E-mail:kawachi.tetsuya@jaea.go.jp

**Abstract.** We have investigated new photo-pumping scheme, in which intense K$\alpha$ line is used as "emitter" and highly charged ions in plasma are used as "absorber". We focus the precise wavelength matching bewteen the aluminum K$\alpha$ line (0.833816 nm) and the resonance line $2p^6$-$2p^5 4d$ $(2p_{1/2}, 4d_{3/2})_1$ of the neon-like zinc ions (0.83400 nm) and calculate the excited level populations of the neonlike zinc ions by use of a Monte-Carlo simulation code. The calculated result shows that substantial amplification gain in the transition of $2p^5 3p$ - $2p^5 4d$ line at a wavelength of 3.5 nm can be expected in this scheme. We also propose the experimental set-up for this scheme, which implies this scheme is feasible under the present technology of ultra-short pulse laser and optics.

## 1. Introduction

In the photo-pumping x-ray laser scheme, spectral line emission from particular ions is absorbed by different element ions to create the population inversion in the latter. The success of this scheme as an x-ray laser depends upon exact spectral matching between the emission line and the absorption line. The widths of the spectral lines of ions in plasma, that are due mainly to the Doppler broadening and the Stark broadening, are typically ~ $\Delta\lambda/\lambda < 0.01\%$. Therefore high resolution spectroscopic studies are required to find an appropriate pair of the "emitter" and "absorber" ions. Indeed, in the early 1990s, the atomic physics group at the Lawrence Livermore National Laboratory (LLNL) has determined the accurate wavelengths of many spectral lines of highly charged ions by use of the electron beam ion trap (EBIT) [1, 2]. The data obtained by the EBIT group, in collaboration with the x-ray laser specialists at LLNL, was used to find the candidate pairs of ions for photo-pumping x-ray lasers.

However, the scheme using "emitter" ions and "absorber" ions has a technical difficulty: the emitter and the absorber ions should be located as close as possible so that the pumping emission reaches to the absorber ions efficiently. At the same time, the electron temperature should be high for the emitter to increase the emissivity of the ions, whereas the lower temperature is preferred for the absorber ions to avoid the "thermal" population in the lower lasing level, which reduces the amplification gain. This implies that the use of "emitter" ions and "absorber" ions is not practical under usual laser irradiation geometry [3]. In this presentation, we propose the use of Kα line from solid target as the emitter coupled with the laser-produced plasma as the absorber. We take aluminum Kα line (0.833816 nm) and resonance line $2p^6$-$2p^5 4d$ $(2p_{1/2}, 4d_{3/2})_1$ of the neon-like zinc ions (0.83400 nm) as an example and calculate the excited level populations of the neonlike zinc ions by use of a Monte-Carlo simulation code.

## 2. Monte-Carlo simulation and calculated result

Figure 1 shows the simplified energy level diagram of the neon-like zinc ions. In our model, the ground state $2p^6$, the $(2p_{1/2}, 3p_{1/2})_0$ level, $(2p^5 3p)$, the $(2p_{1/2}, 4d_{3/2})_1$ level, $(2p^5 4d)$, and the neighbouring levels of the $2p^5 4d$ level are included, and collisional-radiative processes among these levels are taken into account. The Kα line is absorbed by the transition of $2p^6$-$2p^5 4d$ resonance line, and the following collisional-radiative processes transfer the populations into other levels. The size of the zinc plasma is defined as 20 $\mu$m × 20 $\mu$m × 5 mm in the $x$-, $y$- and $z$-directions, respectively. We consider a sliced area in the plasma with the size of 20 $\mu$m × 20 $\mu$m in the $x$-$y$ plane and the thickness of 100 nm. This sliced area is divided into $(x, y)$ mesh with the size of 100 nm × 100 nm. The plasma parameters, *i.e.*, the electron density, $n_e$, the electron temperature, $T_e$, the neon-like ion density, $n_{\text{ne-like}}$, and the ion temperature, $T_i$, are assumed to be 100 eV, $2 \times 10^{21}$ cm$^{-3}$, 50 eV, $10^{20}$ cm$^{-3}$, respectively. The Kα photons come into the plasma from the $y$-direction. The linewidth $(\Delta\lambda/\lambda)$ of the Kα line is assumed to be ~ $10^{-3}$, and that of the $2p^6$-$2p^5 4d$ resonance line is determined by the Doppler broadening.

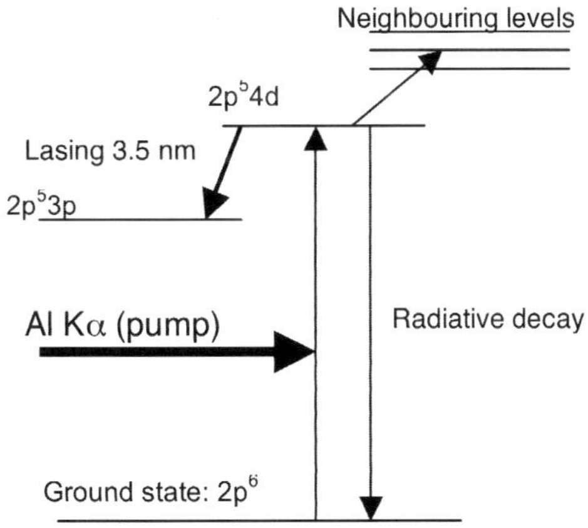


**Figure 1.** Energy level diagram of the Ne-like Zn ions. The aluminum Kα line can be absorbed by the transition of the $2p^6$-$2p^5 4d$ $(2p_{1/2}, 4d_{3/2})_1$.

The calculation procedure is as follows: Firstly the frequency of the Kα photon and the $x$-position are determined, randomly. Secondly the absorption length of the Kα photon in the plasma is calculated to determine the $y$-position where the Kα photon is absorbed. This routine is repeated $N$ times, where $N$ is the number of Kα photons coming into the calculation area in the time step. Thirdly, for each $(x, y)$ position, we determine the absorbed photon is lost by collisional-radiative processes to other excited levels or not. If the collisional-radiative processes take place, we count the increase of the other level populations. If not, it is determined that the resonance emission event occurs or not. In the case of the resonance emission process, the frequency of the resonance emission is determined randomly within the spectral linewidth, and the direction of the emission and the absorption length are calculated. From these information, we judge the photon is absorbed in new $(x, y)$-position or escapes from the plasma. The procedure 1~3 is iterated with 1 fs time step, and the temporal evolution of the excited level populations are calculated.

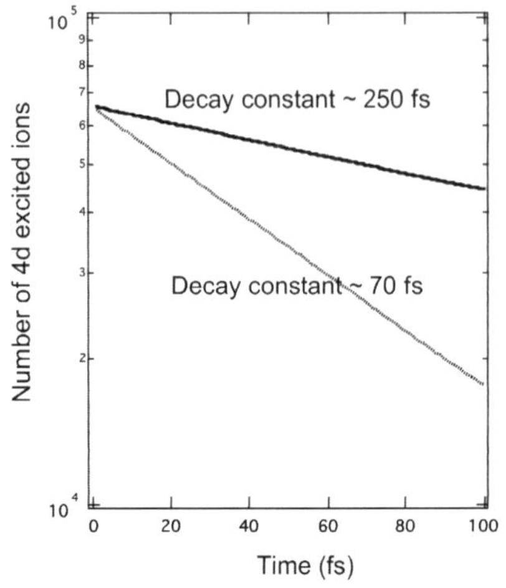


**Figure 2.** The temporal evolution of the $2p^54d$ level population under the condition that $3\times10^5$ photons come into one sliced area in 1 fs. Thick solid line is the case with the radiation trapping effect, and the thin solid line is the case without radiation trapping. Due to the radiation trapping effect, the decay constant becomes around 4 times longer.

Figure 2 shows a calculated result under the condition that the $3\times10^5$ photons comes into one sliced area in 1 fs. The lifetime of the $2p^54d$ level is presented. From the number of the $2p^54d$ excited ions in the early time region ($t \sim 0$ fs), it is found that around 20% of the Kα photons can be absorbed in the plasma. Under the assumption of the absence of the radiation trapping effect, the lifetime of the $2p^54d$ level is ~ 70 fs, which is in accordance with the spontaneous transition probability of this level, whereas with the radiation trapping effect, the effective lifetime extends up to 250 fs. It should be noted that under the present plasma parameters, the absorption length of the resonance line is around 500 nm. This implies that once the Kα photons are absorbed in the plasma, due to the small absorption length of the $2p^6$-$2p^54d$ resonance line, the photons can not escape from the plasma unless other collisional-radiative processes or the x-ray lasing process take place. This is one of the advantages of this scheme as the x-ray laser gain medium.

Figure 3 shows the temporal evolution of the $2p^53p$ and $2p^54d$ level population divided by their statistical weights (left-hand side ordinate) under the Kα pump with the duration of 500 fs (FWHM). The amplification gain of the $2p^53p$-$2p^54d$ line at a wavelength of 3.5 nm is also attached (right-hand side ordinate), where the line shape of the lasing line is assumed to be Gaussian due to Doppler broadening.

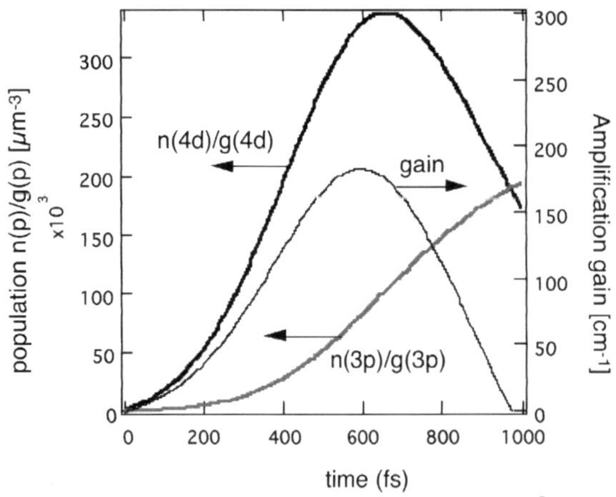


**Figure 3.** The temporal evolution of the $2p^54d$ and $2p^53p$ excited level populations divided by their statistical weights. The amplification gain coefficient of the $2p^53p$-$2p^54d$ line is also attached (See arrows).

The $n(4d)/g(4d)$ increases as time proceeds and decreases for $t > 600$ fs. This is due to the decrease in the pumping rate together with the collisional depopulation to the neighbouring levels whose time constant is ~ 300 fs. Substantial amplification gain is obtained in the time region of $t$ = 300 - 800 fs. This gain duration is much longer than that of other schemes for the shorter wavelength x-ray lasers using inner-shell ionized atoms [4, 5], in which the generation of the gain competes with Auger decay process, therefore the expected gain duration is less than 10 fs. This quite short duration induces the practical difficulties in the pumping method of shorter wavelength x-ray lasers so far. The long duration of the gain (~ 500 fs) in the present result indicates that the present scheme is feasible without any complicated pumping technique.

The $n(3p)/g(3p)$ increases gradually and at around $t$ = 1000 fs, the $n(3p)/g(3p)$ becomes larger than $n(4d)/g(4d)$. This result may be slightly overestimation. Because in the present calculation the collisional de-population process from the $2p^53p$ to the neighbouring levels, *e.g.*, $2p^53s$, $2p^53d$ are not taken into account. Including this effect, the $n(3p)/g(3p)$ may become smaller, which leads to the enhancement of the amplification gain.

## 3. Target design and pumping geometry

In Fig. 4, we propose the experimental set-up for this photo-pumping scheme. Target is an aluminum foil with the thickness of 4 $\mu$m, and 0.2 $\mu$m-thick zinc is deposited on it. The zinc side is irradiated by double pulses of pumping laser with both the intensities of $10^{13}$ W/cm$^2$, separated by 100 ps to create neon-like zinc ions in plasma. After a certain time delay, the opposite side (the aluminum side) is irradiated by an ultra-short pulse with 300fs duration and $10^{15}$-$10^{16}$ W/cm$^2$ intensity to generate intense aluminum K$\alpha$ line. The energy conversion efficiency from the pump laser to K$\alpha$ line is at least around $10^{-5}$/sr [6]. Since the distance from the K$\alpha$ source to the absorber is around 10 $\mu$m, the effective solid angle of the K$\alpha$ source to neon-like zinc plasma is 5 sr, therefore we can deliver 250 $\mu$J of K$\alpha$ photons into the zinc plasma, which is much larger than the assumption in the calculation in section 2. The transmittance of the 4 $\mu$m-thick aluminum foil for the K$\alpha$ photons is more than 70%.

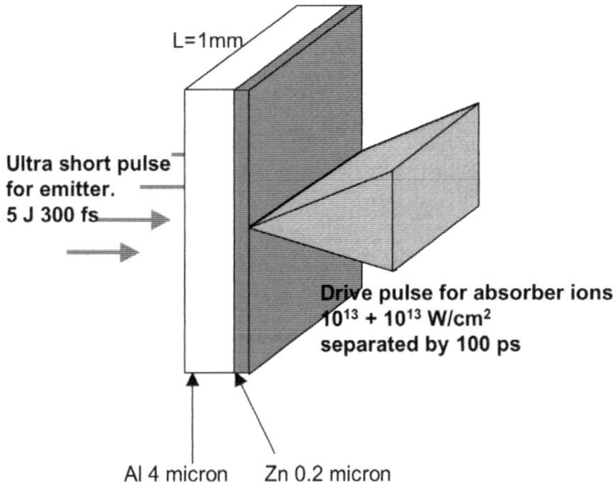


**Figure 4.** Experimental set-up for the photo-pumping scheme. Target is 4 $\mu$m-thick aluminum foil, and 0.2 $\mu$m-thick zinc is deposited on it. Zinc side is irradiated with 100 ps-duration double pulses and the aluminum side is irradiated by 300 fs-duration single pulse.

It is noted that under the intensity of $10^{15}$-$10^{16}$ W/cm$^2$, suprathermal electrons with the energy of several KeV are generated on the surface of the aluminum side, however virtually all the energy of the suprathermal electrons is transferred into bremsstrahlung due to 4 $\mu$m-thick aluminum foil. Since the aluminum foil works as a spectral window for the continuum emission, the heating effect of the plasma due to the continuum radiation becomes small.

## 4. Summary

We have investigated new photo-pumping scheme by use of the aluminum K$\alpha$ line and the $2p^6$-$2p^5 4d$ resonance line of the neonlike zinc ions. The population kinetics code shows that substantial amplification gain in the transition of $2p^5 3p$-$2p^5 4d$ line at a wavelength of 3.5 nm can be expected in this scheme. This scheme is one of the feasible "water-window" x-ray lasers using a laboratory size pumping source.

## 5. References

[1] Nilsen J, Biersdorfer P, et al., 1994 *Phys. Rev. A* **50** 2143
[2] Elliott S R, Biersdorfer P, MacGowan B J, and Nilsen J, 1995 *Phys. Rev. A* **52**: 2689
[3] Kato Y and Kawachi T, 2008, *Progress in Ultrafast Intense Laser Science IV*, (Berlin: Springer)
[4] Kim D, Son S H, Kim J H, Toth C and Barty C P J, 2001, *Phys. Rev. A* **63**, 023806
[5] Moon S J and Eder D C, 1998, *Phys. Rev A* **57**, 1391
[6] Rousse A *et al.*, 1994, *Phys. Rev E* **50**, 2200

Journal of Physics: Conference Series **163** (2009) 012101 doi:10.1088/1742-6596/163/1/012101

# Behavior of carbon impurity ions in radiation collapse in the Large Helical Device

**T Kato[1], H Funaba[1], K Sato[1], M Goto[1], B Peterson[1], I Murakami[1], D Kato[1] and N Yamamoto[2]**

[1]National Institute for Fusion Science, Toki, Gifu 509-5292, Japan
[2]Institute of Laser Engineering, Osaka University, Suita, Osaka 565-0871, Japan
E-mail: kato.takako@nifs.ac.jp

**Abstract.** We measured time-dependent EUV spectra to make a quantitative study of impurity spectral lines of C III, C IV and C V during radiative collapse in the Large Helical Device (LHD) at the National Institute for Fusion Science. We calculated the intensity ratios of spectral lines of C V and C III using our collisional-radiative model and compared to observed time-dependent intensity ratios. We find that intensity ratios of C V are affected by recombination at the end of plasma for an experiment showing radiation collapse; the population density of 2s3d $^3$P is increased by recombination.

## 1. Introduction

Plasma impurities are important from the viewpoint of energy balance and particle transport in plasma simulation or modeling of many kinds of plasmas, including fusion plasmas, astrophysical plasmas and industrial plasmas. Impurities play a key role for plasma evolution. We would like to know the absolute values of ion and electron densities, electron temperature, radiation power, etc. Spectroscopic measurements give a useful, quantitative way to study both plasma and impurity behavior. For that reason spectroscopy is a central technique for plasma diagnostics. In this paper, we present EUV spectra from fusion plasma experiments in the Large Helical Device (LHD), and study the behavior of impurities in representative cases showing radiation collapse.

Radiation collapse in a fusion machine is related to the problems of the density limit and the threshold temperature [1]. "Radiation collapse" occurs when the plasma temperature decays rapidly together with strong emission of radiation. It is related to the power balance between input power and emitted radiation power. Radiation collapse is thought to occur when the electron density increases and the electron temperature decreases, leading to a significant increase in radiation loss. Today this phenomenon is not understood quantitatively. The effect of the plasma electric field on radiation collapse was discussed recently [2]. In order to investigate how and why plasmas collapse, we measure time dependent impurity emission spectra and study the intensity ratios of carbon ions. In previous studies [3,4], we have measured the intensity ratios of carbon ions as plasma diagnostics to estimate the plasma temperature or density. In this paper we show some quantitative features of radiation collapse by comparing temporal behaviour of the intensity ratios of carbon ions with other measurement of plasma parameters.

## 2. Observed EUV spectra

We measured EUV spectra from LHD plasma with a grazing incident spectrometer SOXMOS [3] using a 133 $mm^{-1}$ grating. Two different wavelength regions, 200 – 346A and 953 – 1232A were measured together every 20 ms as shown in Fig.1. These spectral ranges are chosen to measure the emission lines from carbon ions. The line intensity ratios can give useful information for plasma diagnostics. In this paper we use the line intensity ratios to investigate the plasma state. The main observed line pairs used to obtain the intensity ratios are as follows; C V 227.18A(1s2s $^3S$ – 1s3p $^3P$) and 248.6A(2s2p $^3P$ – 1s3d $^3D$), C IV 312.4A(2s – 3p ) and 289.22A (2p – 4d), C III 977.02A (2s2 $^1S$ – 2s2p $^1P$) and 1175.5A ( 2s2p $^3P$ – $2p^2$ $^3P$), H I 1024A(1s – 2p) and 1215.7A(1s – 3p). Time-history comparisons suggest that the increase of radiation emission at the beginning is mainly from C IV lines and later from C V lines. [4]

We studied time dependent line intensity ratios of CIII and CV for a plasma with radiation collapse which happened in the experiment at NIFS-LHD shot number 55644. The time dependent spatial distribution of the electron temperature and density was measured by Thomson scattering [5] and the radiation power measured by imaging bolometers [1] in our experiments. The plasma was heated by Neutral Beam Injection (NBI) and a hydrogen gas puff was started at 0.5 s from the beginning of the plasma and continued until 1.1 s. The time history of the intensity ratios I(1175A)/I(977A) of CIII and I(248)/I(227) of CV is shown in Fig.2. Intensity ratios are almost constant during the heating until 0.94 s when the electron density begins to increase due to the gas injection. The intensity ratios increase from 0.94 s until 1.08 s. The electron temperature begins to decrease at 1.0 s and plasma decays before 1.2 s [4]. The lines from CV ions disappear after 1.18 s although lines from CIII are observed for a long time because the NBI heating continues and supports a low temperature plasma even after the radiation collapse. The low counting rates of CV lines before 0.3 s and after 1.2 s in Fig.2 (a) are probably noise due to low temperature.

Theoretical calculations for the line intensities are performed with the use of our collisional-radiative model for carbon ions. The intensity C V 227A is stronger than C V 248A in ionizing plasma because excitation from lower states is stronger to the level 1s3p than 1s3d. On the contrary the line CV 248.6A is stronger than C V 227A because recombination to 1s3d levels is larger than to 1s3p levels. The theoretical intensity ratios for a CV line pair I(248)/I(227) are shown in Fig.3 for ionizing, equilibrium and recombining plasmas as a function of electron temperature. The intensity ratios increase when recombination process increases.

The values of the observed line intensity ratios are around 0.7 during the heating. These small values indicate that the emission is produced by excitation in an ionizing plasma, and the electron temperature is estimated to be about 400 eV. After 0.94 s the intensity ratio increases rapidly and this indicates a contribution from recombination. We also estimated the temperature from another line pair (the intercombination line / resonance line pair of C V) for other experiments with Xe puffing and the derived temperature was about 200 eV. The radial position for 400 eV of our shot is $\rho = 0.95$ from the temperature distribution measured by Thomson scattering where $\rho$ is the scaled radius ($r = 1$ is the last closed magnetic surface). The spatial distribution of the radiation power measured by bolometer has a peak at $\rho = 0.94$. Therefore we assumed the electron temperature of C V region is the same as the peak region of the radiation power measured by the bolometer. In Fig. 3 we plot the observed intensity ratios against the temperature derived from the bolometric measurement. This figure shows that the intensity ratios are larger at lower temperature. The intensity I(248) increases more rapidly than I(227) at lower temperature. This indicates that the recombination effect on the population density of the upper level 1s3d $^3P$ of the line 248A is stronger than that of of 1s3p $^3P$ of the line 227A. In order to explain the increase of the ratios quantitatively it is necessary that the ion ratio C VI/ C V be about 0.95 – 0.98 which means the amount of H-like C is almost the same as He-like C. We made a time dependent theoretical calculation for carbon ions. It is found that the total radiation loss is mainly produced by excitation although the line CV 248A is also affected by recombination and the ion abundances are changing due to the recombination process.

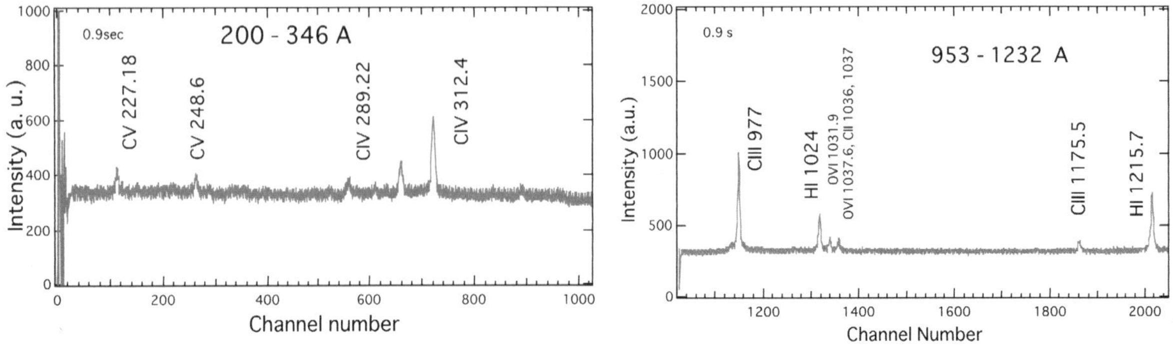


Fig.1 Observed spectra of carbon ions during the NBI heating (t = 0.9 s).

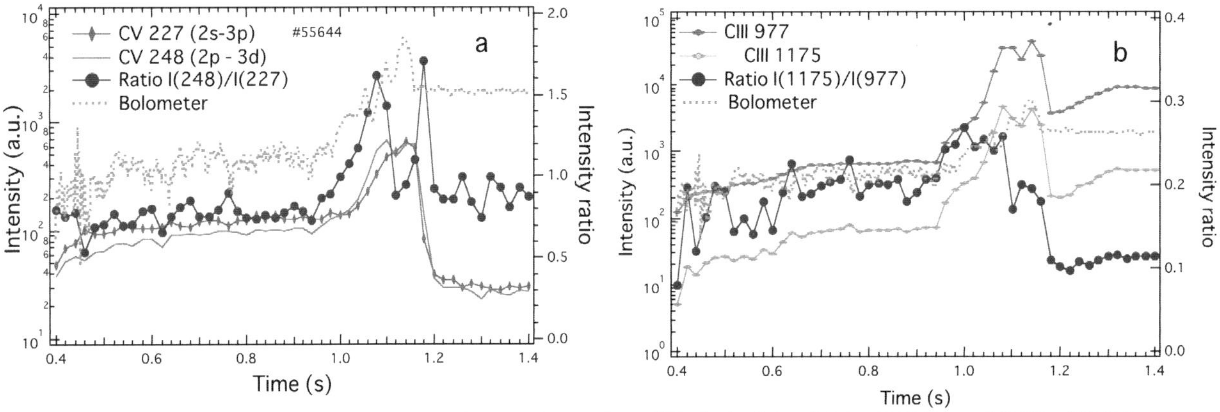


Fig.2 The time history of the observed intensity ratios (filled circle) of (a) C V lines I(248) /I(227) and (b) C III lines I(1175)/I(977). The time evolution of two lines is also plotted. Radiation power measured by the bolometer is shown by a dotted line.

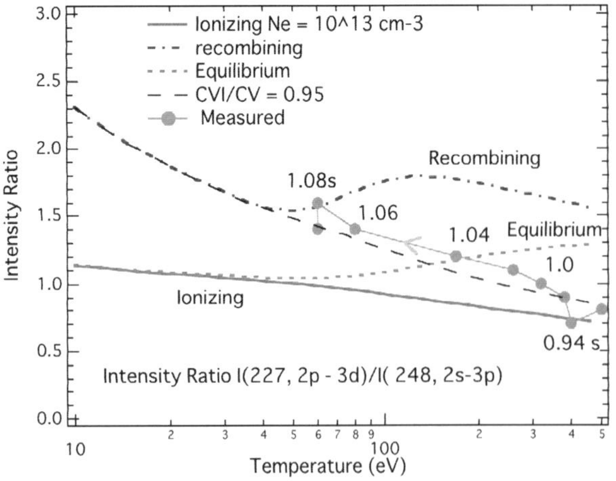


Fig. 3 The theoretical intensity ratios of CV I(1s2p – 1s3d)/I(1s2s – 1s3p) as a function of electron temperature (eV). The solid line indicates the case for ionizing plasma, the dot-dashed line for recombining plasma, the dotted line for equilibrium plasma conditions. The dashed line for the case the ion density ratio n(CVI)/n(CV) = 0.95.
The observed values are shown by filled circles with the measured time.

## 3. Energy balance at radiation collapse

When we compare the radiation power measured by the bolometer and the energy input from NBI heating, the NBI deposition power is much larger. In our case (#55644) the radiation power is less than 10% of the NBI deposition power. The radiation begins to increase at 0.94 s when the electron density begins to increase by gas puffing. However the electron temperature begins to decrease at 1 s. The radiation power has a peak at $\rho = 0.94$ which is the edge of the plasma, far from the center. How does the plasma cool? We consider the energy is lost by radiation at the edge of the plasma and the temperature at the center decreases only gradually. The stored energy is measured by a coil called a diamagnetic loop. The stored energy of the plasma reached the maximum value (about 700 KJ) at 0.96 s and decays within 0.2 s. The stored energy begins to decrease at 0.964 s and the loss rate of stored energy increases for a time. At around t = 1 s, the stored energy is decreasing by about 500 kW which is almost the same as radiation power measured by the bolometer.

The peak emission observed by the bolometer is $\rho = 0.9$ at 0.96 s and the emission rate Prad ~ 50 kW/ $m^3$. At this position the electron temperature is 500 eV and the electron density is 6 x $10^{13}$ $cm^{-3}$. The cooling time for this plasma $\tau = 3/2\ n_e\ kT_e$ / Prad ~ 0.1 s which is the same order of the decay time of the radiation collapse. We can consider the radiation loss at the edge is the origin of the collapse. However we do not yet know how the electron temperature at the center decreases.

## 4. Comparison with non-radiation collapse.

We also measured the spectra for a shot (#55642) in which the plasma decays without radiation collapse after the NBI heating terminates. The temperature and stored energy decay gradually over 0.4s. We derived the time history of the observed intensity ratios of carbon lines as in Sec.2. It is found that the behaviors of the intensity ratios are quite different from the case of radiation collapse. The intensity ratios of CV decrease rapidly although they increased in the case of radiation collapse. This means that the temperature where CV ions exist is increasing from our calculation shown in Fig.3. After the heating turns off, the plasma becomes smaller and the electron density increases according to the density measurement by Thomson scattering. Therefore it appears that CV ions move towards the center in the decay phase. However this phenomenon needs more careful analysis.

## 5. Summary and Discussion

We measured time-dependent spectra from carbon ions for shots with radiation collapse. The main part of the radiation loss is probably CIV and CV line emission from the time history of line intensities. Intensity ratios of CV and CIII indicate an increase of the recombination component after 0.92 s. We could explain the increase of CV lines by recombination processes from CVI to CV. However the main radiation is produced by excitation, not by recombination. We compared the time history of the intensity ratios for experiments with radiation collapse and non-radiative collapse. The intensity ratios for non-radiation collapse indicate an ionizing plasma, quite different from radiation collapse. We will study this difference further. We plan to extend our time-dependent model for carbon ions. We will also study radiation collapse caused by radiation from other elements such as neon, iron and xenon. We also would like to know why the central electron temperature decreases because of the radiation after 1 s.

Acknowledgement

This work is supported by NIFS/NINS under the project of Formation of International Network for Scientific Collaborations.

**References**

[1] B.J. Peterson et al 2006 Plasma Fusion Research, **1**, 45
[2] K. Ida et al, 2005 Nuc. Fusion **45**, 391
[3] T. Kato et al, 2002 Current Developments in atomic, molecular and chemical physics with applications, Kluwer Academic/Plenum Publishers, New York, 265 - 272
[4] T. Kato et al, 2007 Transactions of fusion science and technology, 126, 51
[5] K. Narihara, I. Yamada, H. Hayashi and K. Yamauchi, 2001 Rev. Sci. Instrum. , 72, 1122

# A tandem linear Paul trap as an ion source

**K. Izawa , H. Higaki, H. Yamate, H. Hitomi, K. Ito, M. Kuriki, and H. Okamoto**

Graduate School of Advanced Sciences of Matter, Hiroshima University
1-3-1 Kagamiyama, Higashi-Hiroshima, Hiroshima, Japan 739-8530

E-mail: kizawa-038@hiroshima-u.ac.jp

**Abstract.** Argon ions were produced by electron impact ionization in a linear Paul trap and confined there directly. The ions with a specific charge state ($Ar^+$ and $Ar^{2+}$ in this report) were successfully extracted as a pulsed beam by using another linear Paul trap as a mass filter. In principle, this technique is applicable for various highly charged ions generated and trapped in an ion source. The tandem linear Paul trap described here can also be employed, with the help of laser cooling, to produce nano-ion beams.

## 1. Introduction

Linear Paul traps have been widely employed nowadays for various purposes. For example, they work as a compact mass spectrometer. Applying the laser cooling technique, we can study frequency standards [1] and quantum computations [2] with ultra-cold ion ensembles in a Paul trap. It is also possible to use those confined one-component plasmas for the experimental study of charged-particle beam dynamics because the beam motion is physically almost equivalent to the plasma motion [3, 4].

Some techniques were previously reported of confining highly charged ions in a Kingdon trap [5] and a Penning trap [6]. It was also demonstrated that $Ne^{q+}(3 \leq q \leq 8)$ ions could be confined in a Paul trap [7]. Among these ion traps, a linear Paul trap is relatively advantageous to extract a particular ion species with a particular charge state. In comparison with other types of ion sources, such as electron cyclotron resonance ion sources (ECRIS) and electron beam ion sources (EBIS, EBIT), the present tandem linear Paul trap (TLPT) is quite compact and inexpensive. Neither intense magnetic fields nor microwaves are necessary to produce ions in TLPT, therefore it can be useful in many practical applications where a high ion current is not required.

In the following, we report on the application of TLPT as a highly charged ion source. Ar ions were employed to explore the basic properties of TLPT experimentally. The obtained data are compared with a rough estimate from the rate equation. $Ca^+$ ions were also confined in another linear Paul trap to make Coulomb crystals for nano-beam generation.

## 2. Experimental setup

A schematic diagram of TLPT is shown in Fig.1(a). This apparatus is basically the same as the one used in the previous study of resonant plasma instability [8, 9]. As illustrated in Fig.1, each electrode rod is axially divided into three pieces. The three sets of quadrupoles, which we call Ion Source (IS), Gate, and Experiment Region (ER), are electrically isolated, so that we can manipulate the plasma with various AC and DC voltages. By biasing the central Gate

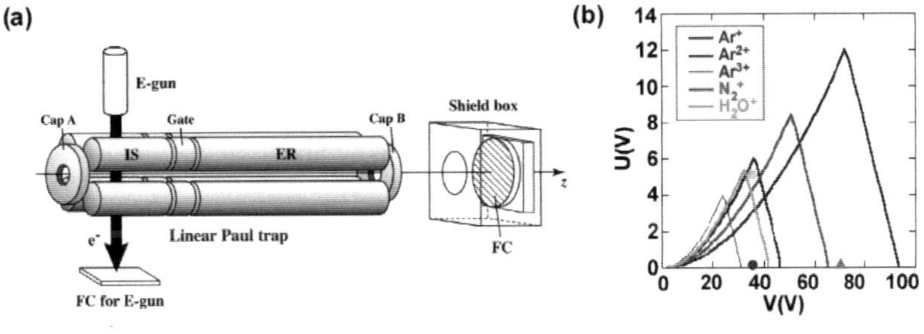


**Figure 1.** (a) A schematic diagram of a tandem linear Paul trap. (b) Stability boundaries for $Ar^{q+}(1 \leq q \leq 3)$ and ionized residual gas($N_2^+$:red line and $H_2O^+$:orange line) at the radio frequency $\Omega/2\pi$ of 1 MHz. The solid circle(blue) indicates the operating point of IS where both $Ar^+$ and $Ar^{2+}$ can be confined. The solid triangle(red) and square(green) correspond to the operating points of ER to extract $Ar^+$ or $Ar^{2+}$.

electrodes, it is possible to separate IS from ER. All electrodes have the diameter of 11.5 mm and are placed 5 mm apart from the trap axis. The axial lengths of IS, Gate, and ER are $L_{IS}$ = 5 cm, $L_G$ = 1 cm, and $L_{ER}$ = 10 cm, respectively. The typical vacuum pressure is $1.5 \times 10^{-7}$ Pa. When we introduce an Ar gas into the chamber, the vacuum is worsened to $5 \times 10^{-5}$ Pa corresponding to the Ar density of $n_0 \sim 1.3 \times 10^{10}$ cm$^{-3}$. Then, an electron beam is injected into IS to ionize Ar atoms. The typical energy of the electrons is 135 eV (i.e., average velocity $v_e \sim 6.8 \times 10^8$ cm/s), which is capable of producing $Ar^{6+}$. The electron beam current, measured by a Faraday cup placed under IS, was kept at about 6 $\mu$A.

Radial confinement of ions in the trap is accomplished by sinusoidal RF voltages of the form $\pm(U + V\cos\Omega t)$, where $U$ and $V$ represent the constant bias and the RF amplitude, respectively, and $\Omega$ is the RF frequency. Not only $V$ but also the bias $U$ are necessary to select specific ion species [10]. When we choose $\Omega/2\pi$ to be 1 MHz, the stability regions of various ion species are changed as shown in Fig.1(b). The RF amplitudes of the IS and Gate were fixed at $V_{IS} = V_G$ = 35.6 V. The corresponding operating point is indicated by the solid circle in Fig.1(b), where we can confine both $Ar^+$ and $Ar^{2+}$. The end plate" Cap A" and Gate electrodes were biased at $\phi_{CapA} = \phi_G$ = 6 V in order to achieve axial confinement of ions in IS. In addition, we biased the IS electrodes to $\phi_{IS} \sim$ 2.5 V for axially accelerating ions toward the Faraday cup. After a short accumulation of ions in IS, the bias voltage $\phi_G$ on the Gate electrodes is switched off to release the ions. They are then filtered in ER, depending on their charge states, and finally detected by a Faraday cup. The RF amplitude and bias voltage of ER were set at $V_{ER}$ = 70 V and $U_{ER}$ = 0 V for $Ar^+$ extraction while $V_{ER}$= 35.6 V and $U_{ER}$= 5 V to obtain $Ar^{2+}$ ions. These operating points are also shown in Fig.1(b) with the solid triangle and square, respectively.

## 3. Results and discussions

After a certain period of ionization time($t_i$), we shut down the electron beam and confine produced ions for a certain period($t_c$) in IS. Then, the ions are released toward the Faraday cup by switching off the bias $\phi_G$. Fig.2(a) shows the number of measured $Ar^{q+}$ (q = 1,2) ions plotted as a function of $t_i$. We see that the number of $Ar^+$ ions ($N_{Ar^+}$) and that of $Ar^{2+}$ ions ($N_{Ar^{2+}}$) have reached about $6 \times 10^6$ and $6 \times 10^4$, respectively, when the ionization time is sufficiently long. In both cases, the number of ions rapidly increases with $t_i$ and comes to a plateau after $t_i > 2$ s.

The lifetime $\tau_q$ of a confined plasma can be measured by changing $t_c$. The number of Ar ions that survive after a specific confinement period is plotted in Fig.2(b) where the ionization time

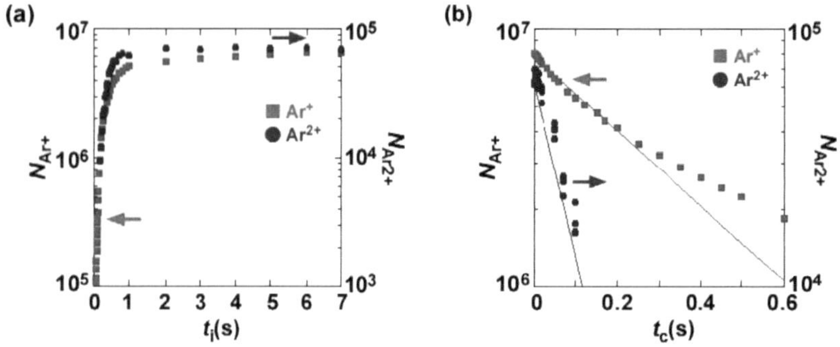


**Figure 2.** (a) Numbers of Ar ions $N_{Ar^+}$ and $N_{Ar^{2+}}$ as a function of ionization time $t_i$. The confinement times has been fixed at $t_c = 3$ ms. (b)Number of Ar ions extracted after different confinement periods. Exponential fitting of the data in the region $t_c < 300$ ms gives the life time of 300 ms for $Ar^+$ and 65 ms for $Ar^{2+}$

.

has now been fixed at $t_i \sim 6$ sec. The solid squares and the circles correspond to measurement data of $Ar^+$ and $Ar^{2+}$, respectively. We recognize that in the case of $Ar^+$, the ion loss somewhat slows down in the region $t_c > 300$ ms where the plasma confinement is limited mainly by collisions with residual gas atoms and RF heating [11]. The cause of the fast ion loss is suspected that it is due to the loss of trapped higher temperature ions which are created in the peripheral region around the electrodes. According to exponential fittings of these results, i.e. the solid lines in the figure, the lifetimes of the Ar plasmas under the present experimental conditions are about 300 ms for $Ar^+$ and 65 ms for $Ar^{2+}$. The lifetime of the $Ar^{2+}$ plasma is considerably shorter than that of $Ar^+$, which is probably due to charge exchange collisions with the Ar gas.

If we can assume that the neutral gas was ionized only by collisions with electrons, the density $n_q$ of $Ar^{q+}$ ions in IS can be estimated from the rate equations

$$\begin{cases} \dfrac{dn_1}{dt} = \dfrac{j}{e}\{n_0\sigma_0 - n_1\sigma_1\} - \dfrac{n_1}{\tau_1} \\ \dfrac{dn_2}{dt} = \dfrac{j}{e}\{n_0\sigma_0^{'} - n_2\sigma_2\} - \dfrac{n_2}{\tau_2} \end{cases} \tag{1}$$

where $j$ is the current density of electrons, $e$ is the elementary charge, $\sigma_q$ represents the cross section to generate $Ar^{(q+1)+}$ by electron impact on $Ar^{q+}$, and $\sigma_0^{'}$ is the cross section of multiple ionization that produces an $Ar^{2+}$ ion from an Ar atom. These cross sections are about $\sigma_0 \sim 2.7 \times 10^{-16}$, $\sigma_1 \sim 6.5 \times 10^{-17}$, $\sigma_2 \sim 3.6 \times 10^{-17}$, and $\sigma_0' \sim 1.8 \times 10^{-17}$ cm$^2$ according to Refs.[12, 13]. In the equilibrium state after $t_i \sim 2$ sec, the left hand side of the rate equations becomes zero. The ion densities are estimated to be $n_1 = 3 \times 10^7$ and $n_2 = 4 \times 10^5$ cm$^{-3}$. According to past experimental results [14], the radial extent of an Ar plasma is typically about 2 mm in diameter. Assuming that the ion distribution is roughly uniform, the calculated number of $Ar^+$ and $Ar^{2+}$ ions agrees fairly well with the observed number.

Assuming that the lifetime of $Ar^{3+}$ ions is equal to that of $Ar^{2+}$ in a proper operating point, $N_{Ar^{3+}}$ in IS is estimated to be 1.5 $\times 10^3$. However, $Ar^{3+}$ ions were not detected with a simple Faraday cup and an amplifier at the moment. Therefore, a multi-channel plate will be installed to confirm production of Ar ions in higher charge states.

The TLPT system can also be a good candidate for an ultra-low emittance ion beam source because it is straightforward to reduce the temperature of an ion plasma to near the absolute zero by employing the laser cooling technique. As is well-known, laser cooled ions eventually

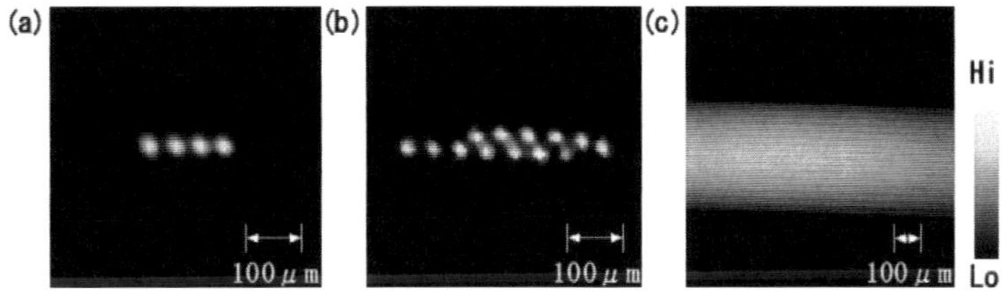


**Figure 3.** Coulomb crystals observed in a linear Paul trap. (a) 1D crystal consisting of 4 $Ca^+$ ions. (b) 2D crystal of 13 ions. (c) 3D multi-shell crystal of a few thousand ions.

form an ordered configuration referred to as a Coulomb crystal. Laser-induced fluorescence signals from Coulomb crystals generated in one of our Paul traps are depicted in Fig.3 where each bright dot represents a $^{40}Ca^+$ ion. Various crystalline structures can be formed depending on the line densities of ion plasmas. As pointed out previously [15], it is theoretically possible to extract a linear Coulomb crystal of nano-meter spot size. Additionally, the crystal can be accelerated without major heating. Note also that we can sympathetically cool various highly charged ions with an ultra-cold $Ca^+$ plasma in TLPT. The precise measurement of cold highly charged ions will be made easier with the use of a tandem linear Paul trap.

## 4. Conclusion

$6 \times 10^6$ $Ar^+$ ions and $6 \times 10^4$ $Ar^{2+}$ ions were successfully produced within a few seconds in TLPT and extracted as a pulsed beam of particular charge state. The obtained results are well explained by the rate equations. In addition to the Ar ions, other ion species with higher charge states will probably be obtainable although the possible current is limited. Therefore the TLPT system is useful as HCI source of table-top-size. The system can also be a nano-ion beam source with the use of the laser cooling technique.

## Acknowledgment

This work is partly supported by a Grant for Scientific Research by The Foundation for the Promotion Ion Engineering.

## References

[1] D. J. Berkeland, J. D. Miller, J. C. Bergquist, W. M. Itano, and D. J. Wineland, Phys. Rev. Lett **80**, 2089 (1998).
[2] J. I. Cirac and P. Zoller, Phys. Rev. Lett. **74**, 4091 (1995).
[3] H. Okamoto and H. Tanaka, Nucl. Instrum. Methods Phys. Res. Sect. A **437**, 178 (1999).
[4] H. Okamoto, Y. Wada, and R. Takai, Nucl. Instrum. Methods Phys. Res. Sect. A **485**, 244 (2002).
[5] L. Yang, and D. A. Church, Phys. Rev. Lett **70**, 3860 (1993).
[6] L. Gruber, J. P. Holder, B. R. Beck, J. Steiger, J. W. McDonald, J. Glassman, H. DeWitt, D. A. Church, and D. Schneider, Hyperfine Interactions **127**, 215 (2000)
[7] D. A. Church, H. M. Holzscheiter, and R. A. Kenefick, Appl. Phys. Lett **47**, 330 (1985)
[8] R. Takai, K. Nakayama, W. Saiki, K. Ito, and H. Okamoto, J. Phys. Soc. Jpn. **76**, 014802 (2007).
[9] H. Higaki, K. Ito, R. Takai, K. Nakayama, W. Saiki, K. Izawa, and H .Okamoto, Hyperfine Interactions, **174**, 77 (2007)
[10] P. K. Ghosh, "Ion Traps", Clarendon Press, Oxford(1995)
[11] M. N. Gaboriaud, M. Desaintfuscien and F. G. Major, Int. J. Mass Spectrom. Ion Phys. **41**, 109 (1981)
[12] A. Müller, E. Salzborn, R. Frodl, R. Becker, H. Klein, and H. Winter, J. Phys. B:At. Mol. Phys. **13**, 1877 (1980)
[13] H. C. Straub, P. Renault, B. G. Lindsay, K.A. Amith, and R. F. Stebbings, Phys. Rev. A **52**, 1115 (1995)
[14] K. Ito, K. Nakayama, S. Ohtsubo, H. Higaki, and H. Okamoto, Jpn. J. Appl. Phys. (2008) in press.
[15] M. Kano, S. Masuda, A. Ogata, K. Okabe, and H. Okamoto, J. Phys. Soc. Jpn. **73**, 760 (2004).

# The SPARC EBIT at GSI; Commissioning and Future Plans at the HITRAP Beamline

**B E O'Rourke[1,2], S Geyer[1,3], A Silze[4], A Solokov[1], G Vorobjev[1], D F A Winters[1,5], O Kester[1] and Th Stöhlker[1,5]**

[1] Gesellschaft für Schwerionenforschung (GSI), Planckstrasse 1, D-64291 Darmstadt, Germany
[2] Department of Physics and Astronomy, Queens University Belfast, Belfast, BT7 1NN, N. Ireland
[3] J. W. Goethe-Universität Frankfurt, Max-von-Laue Strasse 1, D-60438 Frankfurt, Germany
[4] Institut für Ionenstrahlphysik und Materialforschung, Forschungszentrum Rossendorf, Postfach 51 01 19, D-01314 Dresden, Germany
[5] Physikalisches Institut, University of Heidelberg, Philosophenweg 12, D-69120 Heidelberg, Germany

E-mail: `b.orourke@gsi.de`

**Abstract.** This contribution describes the current set-up and future plans for a small permanent magnet electron beam ion trap (EBIT) which has been installed at GSI. The EBIT is planned to serve as an off-line test ion source for experiments on the HITRAP project and also as a test-bed for instrumentation under development for the SPARC collaboration which is part of the new FAIR facility to be built at GSI. In order to increase the range of highly charged ion species which are produced by the source a charge breeding program has been initiated in which singly charged ions are externally injected into the EBIT with high efficiency. We describe recent results from the initial conditioning of the EBIT along with preliminary results of charge breeding tests.

## 1. Introduction

A small permanent magnet EBIT (Dresden EBIT [1, 2]) has been installed at GSI to serve as an off-line test ion source for the HITRAP project [3] and for use as a test setup for charge breeding explorations. The EBIT is named 'SPARC EBIT' as it will also be used as a test facility for instrumentation under development for the SPARC collaboration [4]. We describe some results of the initial conditioning of this source.

The characterisation of the EBIT is divided into two main areas; 1. x-ray measurements of the photons emitted from the trap, giving information on the charge state balance in the trap (section 3.1) and 2. measuring the composition of charge states in the ion beam extracted from the EBIT (section 3.2).

Since it is planned to use the EBIT as test source for HITRAP experiments charge breeding and efficient extraction from the EBIT is being investigated. Charge breeding involves the injection of atoms or low charge states into the EBIT and then extracting the HCI for use in another experiment.

## 2. Current Setup

The current status of the SPARC EBIT beamline is shown in figure 1. The EBIT is in the right hand side of the figure with ions extracted to the left and directed towards a quadrupole bender. The Dresden EBIT device has been described in detail elsewhere. Essentially it consists of a highly emissive Ir-alloy cathode which is biased up to -3 kV delivering a focussed electron beam of up to 50 mA. This beam is compressed by the magnetic field generated from a pair of SmCO permanent magnet rings placed inside the vacuum chamber. These magnets surround a cylindrically trap region and provide a maximum field of 250 mT at the centre of the trap. The trap region consists of 3 electrodes, by biasing the outer electrodes positively with respect to the center electrode, ions may be trapped in the central region. The compressed electron beam passes through the center of the trap interacting with the trapped ion plasma to create highly charged ions by successive ionization.

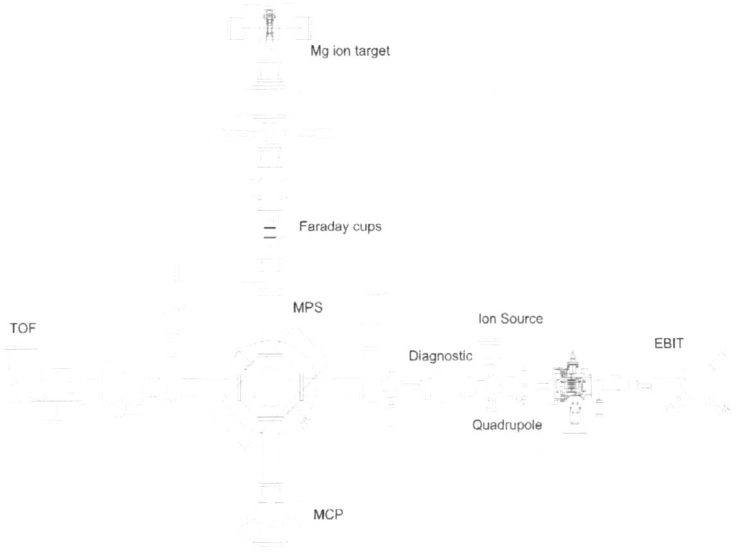


**Figure 1.** The current EBIT beamline

The trap electrodes may be biased positively up to a potential of 12 kV providing a total maximum electron beam energy of 15 keV. The power supply connected to the third trap electrode is controlled by a fast switching device allowing this supply to switch between two pre-determined voltages very rapidly. This set-up is used to switch on and off the axial trapping by switching the potential of this electrode from a value above the centre electrode, i.e. trap closed, to one equal to or slightly below the centre electrode, i.e. trap open. When the trap is opened the ion cloud is directed towards the extraction region and can be extracted into the connecting beamline.

The first part of the beam line consists of several focussing and deflection elements and a diagnostic chamber which is described in more detail later. A large multi-pass spectrometer (MPS) magnet is positioned in the central part of the beamline allowing the beam to be directed either 90° to the left, straight ahead or 90° to the right. The beamline to the left of the magnet contains a MCP detector which allows the charge balance of the beam to be determined by varying the current in the magnet and scanning the beam across the detector. When the ions are undeflected they are directed towards a time of flight (TOF) analysis chamber. By deflecting the beam 90° to the right the ions are directed towards a magnesium target. This target consists of a Mg oven which produces a thin beam of Mg atoms which traverses the ion beam perpendicularly from above. It is planned in the future to perform ion-atom collision experiments using this apparatus.

In the current setup singly charged alkali ions (e.g. K) are externally injected from a thermionic aluminosilicate ion source supplied by HeatWave Labs Inc. These ions are then deflected through 90° by a quadrupole bender and subsequently trapped in the EBIT and ionized. Simulations of ion injection into and extraction from the trap region have been performed using SIMION [5] which suggest that charge breeding is feasible. Expected yields of HCI can also be estimated using simulation codes which model the charge state evolution in the trap [6].

## 3. Results

### *3.1. X-Ray Spectra*

The Dresden EBIT is equipped with two windows providing line of sight to the trap region at 90° to the electron beam. We have used one of these windows to observe the x-rays emitted from interactions between HCI and electrons inside the trap region. An Amptek XR-100CR Si-pin diode detector was used for these measurements. This detector has a surface area of 13 $mm^2$ and provides a resolution of 260 eV at 10 keV.

A typical x-ray spectrum with Ar injection is shown in figure 2 (a) with the relevant peaks labeled. Figure 2 (b) shows two spectra plotted for different ion confinement times. In both figures the electron beam had an energy of 8.0 kV and a current of 22.4 mA. It can be clearly seen that increasing confinement times leads to increased production of the highest Ar ion charge states. The characteristic K-alpha and K-beta peaks of argon can be clearly seen along with the peaks due to recombination (RR) into the argon L and K shells. K-shell RR is only possible into fully stripped or hydrogen-like ions so existence of this peak is evidence that these high charge states are produced in the trap.

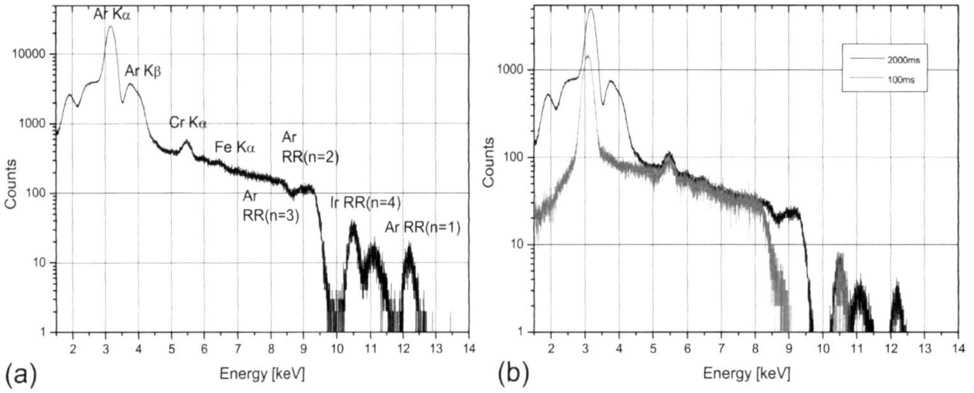


**Figure 2.** (a) Typical x-ray spectra of argon and (b) Argon spectra for two different confinement times

### *3.2. Ion Extraction*

It is planned to use the SPARC EBIT as an off-line source providing beams of HCI for the various HITRAP experiments. To measure the charge state distribution of extracted ions two separate techniques have been used; a) magnetic separation and b) time of flight (TOF). For magnetic separation the ion beam is deflected by a H shaped dipole magnet. By varying the magnetic field ions with a specific mass to charge ratio $(m/q)$ are deflected through 90° and are detected on a Farady cup. Figure 3 (a) shows a series of magnetic scans taken for a range of trap confinement times with argon gas injected into the trap region. As the confinement time increases from 50 ms up to 1000 ms it is clearly seen that the charge state balance shifts significantly towards higher charge states. However, the total ion current decreases due to increasing period between extractions.

TOF measurements were performed in a reflection geometry. Figure 3 (b) shows a series of TOF spectra taken as a function of trapping potential. All measurements were performed with argon gas injection at an electron beam current of 21 mA and energies of 8 keV. As the trapping potential increases it is observed that the trap capacitance increases leading to increased ion temperatures and hence increased line broadening in the spectra.

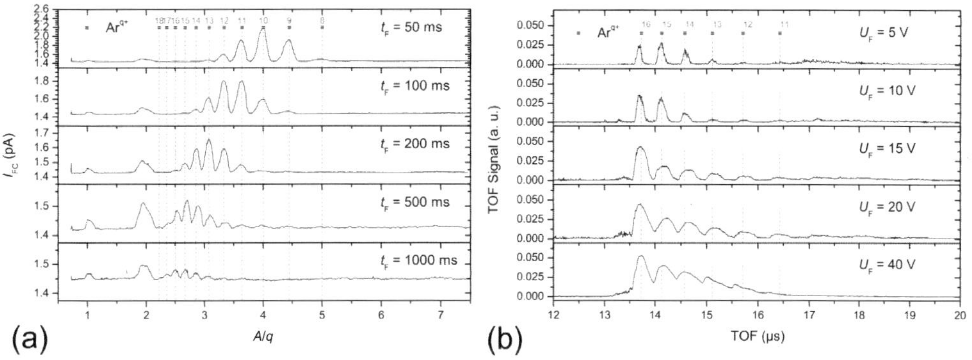


**Figure 3.** (a) Magnetic analysis scan of extracted ion beam from the SPARC EBIT with Ar injection. (b) TOF spectra of extracted ions from the SPARC EBIT with Ar injection.

## 4. Conclusion and Outlook

X-ray spectra of trapped ions and charge state analysis of extracted ion beams show that HCI up to $Ar^{18+}$, $Kr^{22+}$ and $Xe^{31+}$ can be produced and extracted from the SPARC EBIT. Simulations suggest efficient injection of ions from an external ion source for charge breeding is possible and experiments in this direction are continuing. A fast multi-parameter data acquisition system has recently been installed and future x-ray measurements will combine this system with a dedicated Si(Li) detector. Using this setup we plan a program of time-resolved x-ray measurements vital for characterization of the charge breeding process. In the near future the current test beamline will be transferred to the HITRAP experimental platform and the EBIT will eventually be used a an off-line source for various precision low energy HCI experiments.

## References

[1] Zschomack G, Heller R, Kreller M, Landgraf S, Grossmann F, Kentsch U, Ovsyannikov V P, Schmidt M and Ullmann F 2006 *Rev. Sci. Instrum.* **77** 03A904
[2] http://www.dreebit.com
[3] Beier T, Dahl L, Kluge H J, Kozhuharov C and Quint W 2003 Hitrap technical design report http://www.gsi.de/documents/DOC-2003-Dec-69-2.pdf
[4] Stöhlker T, Beyer H, Bräuning H, Bräuning-Demian A, Brandau C, Hagmann S, Kozhuharov C, Kluge H, Kühl T, Liesen D, Mann R, Nörtershäuser W, Quint W, Schramm U and Schuch R 2007 *Nucl. Instrum. Methods B* **261** 234
[5] http://www.simion.com
[6] Becker R, Kester O and Stöhlker T 2007 *Journal of Physics; Conference Series* **58** 443

# A Progress Report of 320 kV Multi-Discipline Research Platform For Highly Charged Ions

**X Ma[1], H. P. Liu[1], L. T. Sun[1], M. T. Song[1], X. L. Zhu[1], S. Sha[1], W.T. Feng[1,2], D. C. Zhang[1], S. F. Zhang[1,2], B. Li[1,2], J. Y. Li[1], D. B. Qian[1], S.Y. Xu[1,2], D. Q. Gao[1], P. Z. Wang[1], L. Z. Ma[1], K. D. Man[1], G. Q. Xiao[1], H. W. Zhao[1], W. L. Zhan[1]**

[1] Institute of Modern Physics, Chinese Academy of Sciences, Lanzhou, 730000, China

[2] The Graduate University of the Chinese Academy of Sciences, Beijing, 100049, China

E-mail: x.ma@impcas.ac.cn

**Abstract.** A dedicated platform for multi-disciplinary research with highly charged ions has been constructed, and an all-permanent magnet ECR ion source was built and installed in the beamline. Five experimental terminals are established for interdisciplinary Research. The high voltage supplied to the platform has reached 320 kV. The commissioning of the platform is successful, different ion beams have been provided for experimental studies, and the current status will be reported.

## 1. Introduction

The interactions between highly charged ions and materials (atoms, molecules, clusters, surfaces, and bio-molecules, etc) have attracted more and more interest not only in fundamental research but also in many application fields [1]. Low energy highly charged projectiles are usually provided by ECRIS, where their extraction voltage is below 50 kV. EBIS can provide even higher charge state ions but relatively low beam intensities. Several laboratories established very low energy setups or beamlines for surface and/or ion-atom/molecule studies, such as low energy beam line (LEBL) at CIRIL France, [2, 3], or the slow highly charged ion facility at RIKEN. Most big-scale laboratories provide projectile beam energies from several MeV/u up to GeV/u. At Frankfurt university the 14GHz ECRIS was complemented by a beamline for post accelerating ions [4]. As it is well known the stopping power of ions reaches a maximum (the Bragg peak) when the projectile energy is reduced down about from below hundreds keV/u to 1 MeV/u for different target materials. Due to the high density of ionization near the Bragg peak, it is of great importance for radiation therapy, surface modification, and radiation defects, etc. This energy range is therefore of practical interest for interdisciplinary research.

A dedicated platform for multi-disciplinary research with highly charged ions has been constructed at the Institute of Modern Physics (IMP) in Lanzhou, where the projectile energy range was extended to 320q keV (where q is the projectile charge state). The dynamical range of the projectile velocity will be extended from 0.1 a.u. to 2 a.u. covering electron capture dominant channels to ionisation dominant channels. Five experimental beamlines are equipped and the commissioning of the platform has been successful. The recent progress will be reported in the following.

## 2. The design and the performance of the platform

The new facility includes a 320kV high voltage platform, transport beamline, and five experimental terminals. The platform itself sits in an individual room. The new platform facility was planned to be installed in an existing building, where the space is very limited. According to this circumstance the layout of the facility was designed as shown in figure 1. The platform has the dimensions 3000×3100×1600mm. An ECR source, a Glaser lens, an analyzing magnet, electrostatic lens, correction magnets, microwave machine, power supplies, and some diagnostic devices are installed on the platform. All the devices on the platform are powered by an isolating transformer (400kV maximum, 80kW). The maximum voltage applied to the platform is designed to be 320kV.

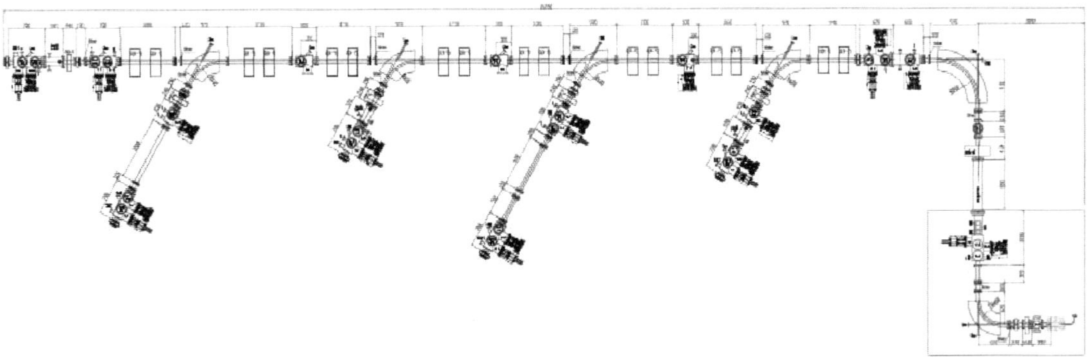

**Figure 1.** Layout of the high voltage platform and equipped experimental terminals.

To reduce the power load to the platform, an all permanent magnet ECR ion source was designed, which has a dimension of magnetic body of Ø650 mm × 562 mm. Its total weight is 950kg. The overall magnetic field has been measured and compared to the theoretical results; good agreement was obtained. The injection magnetic field of the source is 1.28 T and the extraction magnetic field is 1.07 T. The details of the magnetic design can be found in reference [5]. The ion source is implemented with an oven which can evaporate metallic elements. The ions extracted from the ECR source are first charge-state analyzed and then accelerated when leaving the platform. The extraction voltage of the ECR source can be varied from 7 kV to 30 kV. The ions can also be delivered to experimental area directly without post-acceleration. The beam envelopes for ion transport are shown in figure 2.

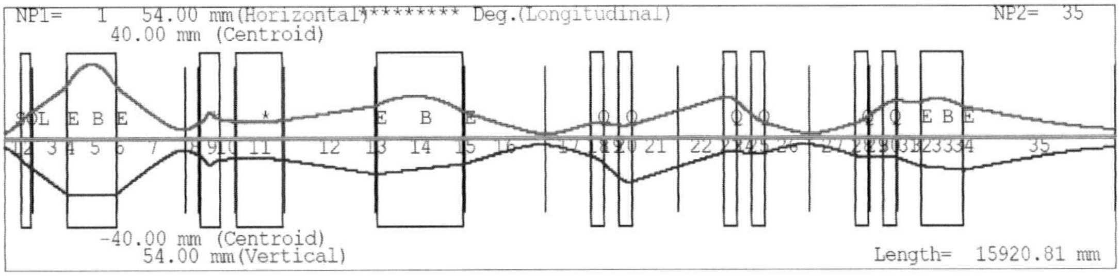


**Figure 2.** The beam envelopes of the ions from the ECR ion source to experimental terminal No.3.

The static vacuum of the transport beam line is better than $5\times10^{-10}$mbar. The vacuum is maintained at the $10^{-9}$mbar level during operation. The base vacuum condition is $2.0\times10^{-6}$ mbar at the injection side and $9.5\times10^{-7}$ mbar at the extraction side during the run of the facility. The control of the beamline and of the platform are done by a host computer in the control room. All the devices on the high voltage platform are controlled by a local computer which is accessed by the host computer at control room via wireless network. The controlling system functions very well for high voltages up to 320kV.

The commissioning of the beamline has been carried out successfully. During the preliminary commissioning, a 14.5 GHz microwave generator has been used to feed rf power directly into the plasma chamber through a WR62 rectangular wave guide. Highly charged ions such as $Ar^{13+}$ and $Xe^{30+}$ have been provided for radiation damage studies and surface experiments.The typical ion currents measured on the platform are listed in table 1. The transport efficiencies are from 60% to higher than 90% for post-accelerated ion beams. Figure 3 shows a typical charge state distribution spectrum optimized for highly charged Xe ions (microwave at 350W, extraction field at 22 kV). Several ion beams were accelerated to 320q keV already (q is the charge state of the ions) and provided to experiments.

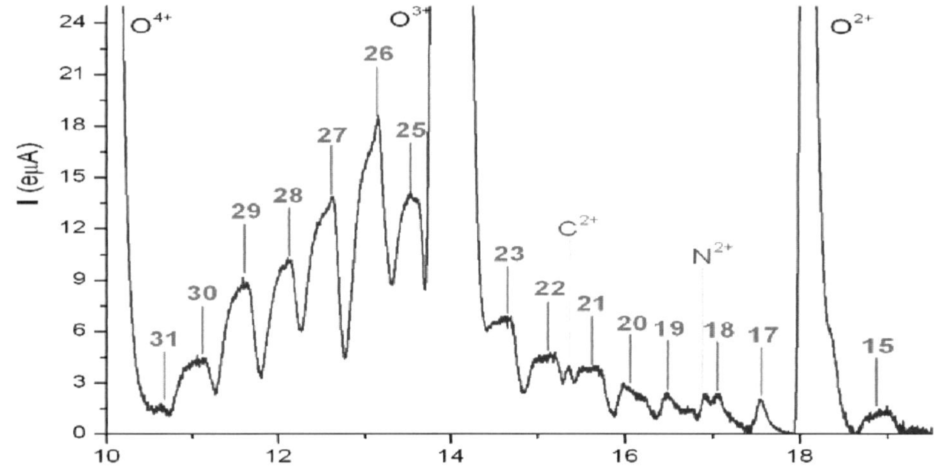


**Figure 3.** Typical charge state distribution spectrum for highly charged Xe ions.

Table 1. Typical beam intensities extracted from the all-permanent magnet ECR ion source

| Ion species | Charge states | Beam intensities (eμA ) | Extraction field（kV） | Microwave power（kW） |
|---|---|---|---|---|
| O | 6+ | 1000 | 20 | 1 |
| Ar | 8+ | 460 | 20 | 1 |
| | 11+ | 166 | | |
| | 12+ | 60 | | |
| | 14+ | 15.5 | | |
| | 16+ | 2 | | |
| Xe | 20+ | 83 | 23 | 0.4 |
| | 23+ | 75 | | |
| | 26+ | 46 | | |
| | 27+ | 27 | | |
| | 30+ | 7.5 | | |
| Ne | 7+ | 65 | 15 | 0.4 |
| | 8+ | 50 | | |
| C | 4+ | 155 | 15 | 0.35 |
| | 5+ | 72 | | |
| Bi | 19+ | 23 | 15 | 0.35 |
| | 21+ | 28 | | |
| | 31+ | 4.6 | | |

## 3. Experimental terminals

The beamlines are equipped with five experimental terminals as numbered in figure 1 [6]: (1) ultra-high vacuum chamber for surface studies, (2) reaction microscope for atomic physics collision dynamics as well as spectroscopy with atomic or molecular target beams, (3) multi-purpose experimental terminal for material science. (the design is flexible and allows the external users to bring their own set-ups for experiments), (4) atomic physics terminal for atomic collision and micro-capillary experiments, and (5) biophysics research, respectively. A photo of the five terminals is shown in figure 4. Experiments using the ions listed in table 1 have been carried out and part of the results are presented in these conference proceedings.

**Figure 4.** A photo of the transportation beam-line and the experimental terminals.

## 4. Outlook

The successful commissioning of the 320 kV multi-disciplinary research platform allow new opportunities for atomic physics, surface studies, and materials research. Various highly charged ions of metallic elements will be provided to experiments. A chopper will be implemented into the beamline soon to provide pulsed ion beams. This facility is open to outside and international users.

This work was supported by the key projects of Pilot Project of the Knowledge Innovation Program launched by the Chinese Academy of Sciences and partly by the National Nature Science Foundation of China under grant No. 10434100

## References

[1] H. Schmidt-boeking, et al., in AIP conference proceedings,Vol.604 (2002), pp120
[2] L. Maunoury et al, Rev. Sci. Instrum. 73, 561 (2002)
[3] H. Lebius, A. Brenac, B. A. Huber, et al., Rev. Sci. Instr. 74, 2276 (2003)
[4] O. Hohn, S. Runkel, A. Schempp, et al., Physica Scripta T80 (1999) 525
[5] L. T. Sun, H. W. Zhao, et al., Rev. Sci. Instr. 77 (2006) 03A319
[6] X. Ma, X. L. Zhu, et al, Sciences in China (Series G) 51 (2008)755

Journal of Physics: Conference Series **163** (2009) 012105 doi:10.1088/1742-6596/163/1/012105

# Evolution of X-ray Calorimeter Spectrometers at the Lawrence Livermore Electron Beam Ion Trap

**F S Porter[1], P Beiersdorfer[2], G V Brown[2], M F Gu[2], R L Kelley[1], S. Kahn[3], C A Kilbourne[1], D B Thorn[2]**

[1]Lawrence Livermore National Laboratory, Livermore, CA 94550, USA
[2]NASA Goddard Space Flight Center, Greenbelt, MD 20770, USA
[3]Department of Physics, Stanford University, Stanford, CA 94305, USA

E-mail: Frederick.S.Porter@nasa.gov

**Abstract.** High-resolution broadband, non-dispersive x-ray calorimeter spectrometers have been under development for spaceflight since 1984. As an offshoot of the significant NASA investment in this technology, we have developed a series of calorimeter instruments for laboratory use and installed them at the Electron Beam Ion Trap (EBIT) facility at the Lawrence Livermore National Laboratory. The calorimeter instruments at EBIT have significantly enhanced the capabilities of our laboratory astrophysics program including broad-band measurements of emission from charge exchange recombination and absolute cross sections for collisional excitation. The first Goddard Space Flight Center (GSFC) calorimeter instrument was installed at the EBIT facility in July of 2000 and has seen two major upgrades. The performance of the instrument has significantly improved from the initial instrument that had a resolving power of ~500 at 6 keV, and essentially no quantum efficiency at energies above 20 keV, to the current instrument that has a resolving power of 1350 and 95% quantum efficiency at 6 keV, and a resolving power of 1800 and 32% quantum efficiency at 60 keV.

## 1. Introduction

X-ray Calorimeter detectors have been in development since 1984 for use in non-dispersive spectrometers for astrophysics. Spaceflight calorimeter instruments use pixelated detectors that are broad-band, high resolution, and have nearly 100% quantum efficiency. The detector arrays are quite small with the current state-of-the-art still less than 1000 pixels. However, since each detector pixel is an independent spectrometer, the detectors are true spatial-spectral instruments. For astrophysical sources this is very important. Dispersive instruments for x-ray astrophysics are generally much less efficient and are typically slit-less spectrometers, relying on the source to limit the divergence of the beam onto the dispersing elements. Thus the dispersive spectrometers degrade rapidly when the source extends more than a few arcminutes. This limits the value of dispersive instruments for extended astrophysical sources such as supernova remnants, galaxies, clusters of galaxies, and the interstellar medium, all objects with rich spectral structure that would benefit substantially from high-resolution measurements. An x-ray calorimeter instrument is a self-contained spectrometer with no external optical components and can be placed directly at the focus of an x-ray telescope. Its performance is not limited by the spatial properties of the source. These characteristics make the calorimeter instrument highly attractive for x-ray astrophysics, and even though the instruments are large, require relatively large amounts of electrical power, and are complex, they are likely to be included on every major x-ray observatory for the foreseeable future including *Astro-H, Spectrum-Roentgen-Gamma, and the International X-ray Observatory*. During the last 8 years we have leveraged this extensive investment in high performance non-dispersive x-ray spectrometers to aid our laboratory astrophysics program that is centered around the Electron Beam Ion Trap (EBIT) at the Lawrence Livermore National Laboratory (LLNL).

**Table 1.** Calorimeter spectrometers at the LLNL EBIT. In each column the first values are for the low/midband pixels and the second values are for the high-energy pixels in the detector array.

| | XRS/EBIT v1 | XRS/EBIT v2 | ECS | TEMS |
|---|---|---|---|---|
| Operation (year) | 2000-2003 | 2003-2007 | 2007-- | 2010 |
| Array size | 32 pixels<br>none | 28 pixels<br>4 pixels | 16 pixels<br>14 pixels | 256 pixels<br>64 pixels |
| Pixel size | 0.64x0.64 mm | 0.64x0.64 mm | 0.64x0.64 mm | 0.3 x 0.3 mm<br>1.0 x 1.0 mm |
| Spectral resolution (FWHM) | 11.5 eV @ 6 keV | 6 eV @ 6 keV<br>70-150eV @ 60keV | 4.5 eV at 6 keV<br>32 eV @ 60keV | 0.8 eV @ 1 keV<br>2.0 eV @ 6 keV<br>30 eV @ 60 keV |
| Quantum efficiency | 95% @ 6 keV | 95% @ 6 keV<br>10% @ 60 keV | 95% @ 6 keV<br>32% @ 60 keV | 95% @ 6 keV<br>60% @ 60 keV |
| Operating temp. | 60 mK | 60 mK | 50 mK | 50 mK |
| Band pass | 0.1-12 keV | 0.1-12 keV<br>0.5-60 keV | 0.1-12 keV<br>0.5-100 keV | 0.1-12 keV<br>0.5-200 keV |
| Detector time constant | 9 ms | 3 ms<br>10 ms | 3 ms<br>40 ms | 0.3 ms<br>< 5 ms |

## 2. The EBIT calorimeter instruments

An x-ray calorimeter operates as a single photon thermal spectrometer [1]. The device absorbs each x-ray in a high quantum efficiency absorbing material and then measuring the temperature rise after the photoelectron has thermalized in the material using a very sensitive thermometer. Using low heat capacity materials, very low temperatures (< 100 mK), and sensitive thermometers, spectrometers with resolving powers of over 2000 with a broad dynamic range have been constructed.

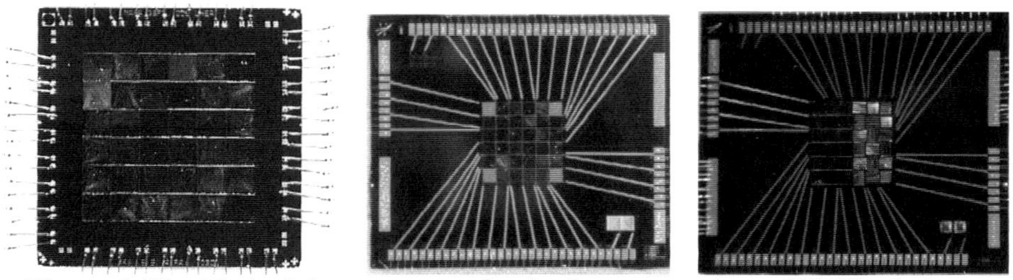

**Figure 1.** The 32 pixel detector arrays for the three generations of EBIT calorimeters. From left to right: the XRS/EBIT v1, The XRS/EBIT v2, and the ECS. The four corners of the middle array and the left side of the right array are high-energy detector pixels.

We have sequentially deployed three calorimeter instruments at the EBIT facility at Lawrence Livermore National Laboratory (LLNL) primarily for our laboratory astrophysics program (see [2] and references therein). The calorimeter instrument complements higher spectral resolution dispersive spectrometers at the EBIT facility by providing broad-band contextual information to the narrow band dispersive instruments. In addition, the calorimeter instruments are extremely efficient, allowing experiments to be performed that would be prohibitive using the dispersive instruments in the laboratory alone. Our laboratory astrophysics program concentrates on benchmarking the atomic codes used in the spectral synthesis models that form the basis for interpreting astrophysical observations. This includes, for example, measurements of line ratios, cross sections, and line identification in collisional excitation and charge exchange recombination.

The calorimeter spectrometers at EBIT are summarized in Table 1 and the detector arrays for each instrument are shown in Figure 1. The first generation spectrometer (XRS/EBIT v1) used a spare detector array and an engineering model focal plane assembly from the Astro-E observatory [3]. The array is composed of 32 pixels in a square 6 x 6 geometry (the corner pixels are not read-out) with a 0.64 mm pitch. The absorber material is 8 μm of HgTe, a material that is a compromise between fast thermalization and low heat capacity. The thermometers are formed *in-situ* on the silicon substrate of the array by ion implantation directly into each pixel. The pixels and their thermal isolation structures are then defined using wet chemical etching. The v1 instrument had a Gaussian instrumental response with a spectral resolution of 11.5 eV full width at half maximum (FWHM) and a quantum efficiency of 95% at 6 keV, and a simultaneous bandpass from 0.1 to ~12 keV. The instrument was first deployed in July of 2000 and ran almost continuously until it was upgraded in 2003 as described below.

The second generation instrument (XRS EBIT v2) is based on an improved version of the ion-implanted detectors from the Astro-E instrument described above [4]. The improvements were part of the development program for the Astro-E2 (Suzaku) observatory that was launched in 2005. The basic improvements were in the noise performance of the implanted thermometers and a switch from wet etching to reactive ion etching in defining the individual pixels and their thermal and mechanical support structures in the array to improve the reproducibility and the mechanical robustness of the detector. The performance of the v2 instrument is significantly enhanced. The spectral resolution of the detector improved by a factor of two to 6 eV FWHM at 6 keV. In addition, we implemented four pixels with increased quantum efficiency at high energies by using 30 μm of Bi in place of the 8 μm of HgTe in the x-ray absorber. This improved the QE at 60 keV from 3% to about 10%. Unfortunately the spectral resolution of these early high-energy detectors was between 70 and 150 eV at 60 keV, limiting their utility.

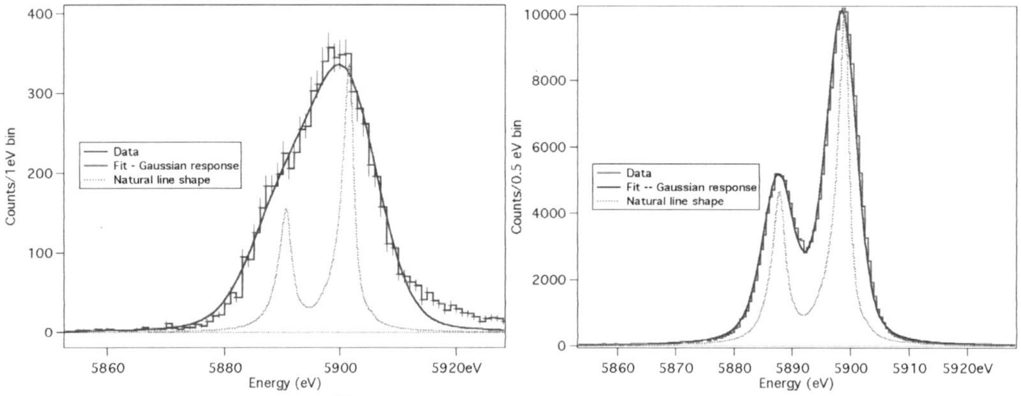


**Figure 2.** Mn $K_{\alpha}$ emission from a $^{55}$Fe radioactive source as measured by the XRS/EBIT v1 (left) with a resolution of 11.5 eV FWHM and with the ECS (right) with a resolution of 4.5 eV FWHM.

Both the v1 and the v2 XRS/EBIT instruments used the same cryogenic assembly consisting of a 14 liter liquid helium tank pumped to 1.5 K, a liquid nitrogen thermal shield, and a single stage adiabatic demagnetization refrigerator (ADR) to cool the detector array to its operating temperature of 60 mK. The focal plane assembly itself was, in both cases, an engineering model unit from the Astro-E flight program. The ADR and dewar were assembled quickly from spare parts in order to field the instrument as cheaply and quickly as possible. The infrastructure, including the analog and digital readout electronics, was composed largely of engineering model and ground support equipment from the Astro-E flight program. This, unfortunately, made the instrument somewhat cumbersome to use. The pumped helium system and the small ADR limited the experimental time to about 12 hours per cycle, and required cryogen fills and pumpdowns at least every other day. Even with their deficiencies the XRS/EBIT instruments were in almost continuous use for over seven years. In contrast, the third generation instrument is a completely new instrument that was designed from the ground up to be a low maintenance permanent facility at the EBIT laboratory.

The EBIT Calorimeter Spectrometer (ECS) is the third generation calorimeter instrument at the EBIT facility [5]. It was completed and installed in November 2007 and has been running continuously for almost a year. The cryogenic system is completely new based on an atmospheric pressure (unpumped) liquid helium cryostat at 4.2K, a closed cycle $^{3}$He/$^{4}$He getter pumped refrigerator [6], and a single stage ADR. The ECS uses the $^{3}$He/$^{4}$He system, which has a base temperature of 340 mK, to cool the focal plane outer housing and to pre-cool the ADR. The ECS runs continuously for 65 hours at 50 mK before requiring a 3 hour recharge of the ADR and $^{3}$He/$^{4}$He system. Cryogens are filled every 2 weeks (liquid helium) and 3 days (liquid Nitrogen) and filling doesn't interfere with the operation of the spectrometer. Small improvements to the focal plane assembly improve the spectrometer performance to 4.5 eV FWHM at 6 keV for the 16 mid-band detectors as shown in Figure 2. The ECS detector array also includes 14 high-energy detector pixels that use 100 μm HgTe absorbers to increase the quantum efficiency at 60 keV to 32%. The spectral resolution of the high-energy detectors is also substantially improved to 32 eV at 60 keV. The ECS dramatically increases the efficiency of operating the calorimeter instrument at the EBIT facility, adds a substantial capability at high energies, and improves the gain stability and spectral resolution of the mid-band detectors. The fourth generation instrument, currently under construction, uses newer detector technology to, again, substantially improve the spectral performance across the entire band from 0.1 to 200 keV.

## 3. The Next generation EBIT calorimeter instrument

We are currently designing and constructing the next generation EBIT calorimeter instrument based on significant improvements in calorimeter technology developed for the Constellation-X (now the International X-ray Observatory) program. These detectors use superconducting transition edge thermometers and lithographically patterned x-ray absorbers to substantially improve the performance of the detector system. In addition, these detectors are readout using superconducting quantum interference device (SQUID) amplifiers that can be multiplexed to dramatically increase the pixel count of the instrument. The fourth generation instrument, termed the Transition Edge Microcalorimeter Spectrometer (TEMS), will use the same cryogenic package as the ECS described above but in a cryogen free pulse-tube dewar that will require no servicing. The prototype of this system is currently operating in our laboratory and runs for over 100 hours at 50 mK before an automatic 2.5 hour recycle is required. As shown in Table 1, the array scale, detector time constant, and spectral resolution are substantially improved over the ECS. The large simultaneous bandpass, efficiency, and spectral performance of the TEMS instrument will substantially increase the capability of our laboratory astrophysics program.

## Acknowledgments

The authors gratefully acknowledge support from NASA's ROSES program. In addition, part of this work was performed under the auspices of the U. S. Department of Energy by Lawrence Livermore National Laboratory under Contract DE-AC52-07NA27344.

## References

[1] McCammon D 2005 *Cryogenic Particle Detection*, ed C Enss (Heidelberg: Springer) p 1
[2] Porter F S, Gygax J, Kelley R L, Kilbourne C A, King J, Beiersdorfer P, Brown G V, Thorn D B and Kahn S M 2008 *Can. J. Phys.* **86** 231
[3] Stahle C K, Audley M D, Boyce K R, Brekosky R P, Fujimoto R, Gendreau K C, Gygax J D, Ishisaki Y, Kelley R L, McClanahan *et al.* 1995 *Proc. SPIE* **3765** 128
[4] Stahle C K, Allen C A, Boyce K R, Brekosky R P, Brown G V, Cottam J, Figueroa-Feliciano E, Galeazzi M, Gygax J D, Jacobson M B *et al.* 2004 *Nucl. Inst. and Meth. A.* **520** 466
[5] Porter F S, Gygax J, Kelley R L, Kilbourne C A, King J, Beiersdorfer P, Brown G V, Thorn D B and Kahn S M 2008 *Rev. Sci. Inst.* **79** *in press*
[6] Chase Research Cryogenics Ltd., Uplands, 140 Manchester Road, Sheffield S105DL, UK

# Enhancement of soft X-ray emission from fs laser plasma by using mixture of molecule and atomic gases as cluster jet targets and its application for nanostructure imaging

**T A Pikuz[1,2], A Ya Faenov[1,2], I Yu Skobelev[2], S V Gasilov[2], A S Boldarev[3], V A Gasilov[3], Y Fukuda[1], M Kando[1], H Kotaki[1], I Daito[1], T Homma[1], K Kawase[1], T Kameshima[1], T Kawachi[1], P Bolton[1], H Daido[1], T Kimura[1], T Tajima[1], Y Kato[1], S V Bulanov[1]**

[1] Kansai Photon Science Institute (KPSI), Japan Atomic Energy Agency (JAEA), Kizugawa-city, Kyoto 619-0215, Japan
[2] Joint Institute for High Temperatures, Russian Academy of Sciences, Moscow, Russia
[3] Institute for Mathematical Modelling, Russian Academy of Sciences, Moscow, Russia

E-mail: faenov.anatoly@jaea.go.jp

**Abstract**. Submicron scale clusters, produced by mixture of molecular ($CO_2$) and atomic (He) gases expanding from the specifically designed super-sonic nozzle, have been irradiated by femtosecond laser pulses in order to build up a bright debris-free soft X-ray source. Spectroscopy measurements showed that X-ray flux in the strongest oxygen spectral lines ($He_\beta$ at 665.7 eV and $Ly_\alpha$ at 653.7 eV) produced by the plasma from gas mixture clusters is 2-8 times larger than in the case of using pure $CO_2$ or $N_2O$ clusters as targets and reaches values about $2.8 \times 10^{10}$ ph/(sr · pulse). Lines intensity and spatial dimension were measured from two observation point – along and perpendicular to the laser beam propagation directions. Images of 100 nm thick Mo foils in a wide field of view ($cm^2$ scale) with high spatial resolution (700 nm) were obtained using LiF crystals as soft X-ray imaging detectors. Agreement between calculated and measured intensity of the image fragment which illustrates overlapping of Mo foil layers confirms accuracy of the plasma parameters measurements.

## 1. Generation of soft X-ray emission by fs laser irradiation of atomic and molecular gas mixture cluster targets

Different types of nanostructure-based targets which can significantly increase the efficiency of fs laser pulses coupling into the target have been proposed recently [1-3]. Such targets allow to increase considerably the fs laser radiation coupling with targets and consequently to enhance generation of X-ray emission. However a drawback of this approach is that the production of nanostructured targets and their manipulation are complicated processes. Expansion of different gases in supersonic nozzles is rather simple way to produce targets with nanoscale dimensions [4]. Cluster targets' using allows one to reach a very efficient fs laser pulses coupling with plasma. Such approach gives possibility to create very bright point plasma X-ray sources which generate small amount of hot electrons and do not produce debris.

Unfortunately, only 10-30 % of expanded in supersonic nozzle gas could be aggregated in clusters [5]. It means that except clusters, which absorb laser energy well there is the residual gas with enough high pressure in the plasma generation volume, which absorbs very efficiently soft X-rays (SXR) and VUV radiation of plasma. Additional decreasing of the soft x-rays flux also takes place in surrounding area consisting of cold gas clusters (figure 1a,b). This absorption could be reduced by means of using light atomic gas helium which has a very small absorption even when the density of this gas is high (figure 1c). However, pure He gas practically do not produce clusters during expansion in supersonic nozzle and it is impossible to produce intensive soft X-ray radiation using such gas. On the other hand, the presence of He gas helps clusterization of molecule gases. Obviously, it is very attractive to use the mixture of atomic He gas with some molecule gases for production of clusters targets. Indeed, such gases combination will have the advantage of the effective cluster production due to the expansion of molecule gas in mixture with He gas [4]. The other benefit is the small absorption of the useful soft X-ray radiation in residual gas (figure 1b).

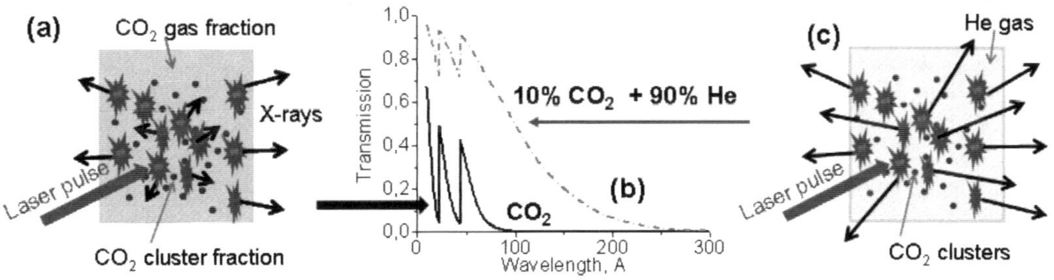


**Figure 1.** Influence of the residual gas on the soft X-ray flux. (a) Pure $CO_2$ clusters; (b) transmission function of the $CO_2$ and the mixture of the $CO_2$ and He gases proposed in this work; (c) Schematic draw of the $CO_2$ clusters in the residual He gas which is transparent for the SXR.

Especially designed [6] supersonic gas jet nozzle with input and output diameters of 0.5 and 2.0 mm, respectively, and a length of 75 mm was used in our experiments to produce clusters. Figure 2 presents results of the numerical simulations which have been made to optimize the size of produced clusters, which allowed to increase SXR output from the cluster targets [7,8]. According to the modeling the best X-ray flux for the mixture of atomic (90 % He) and molecular (10% $CO_2$) could be reached at 60 bars backing pressure.

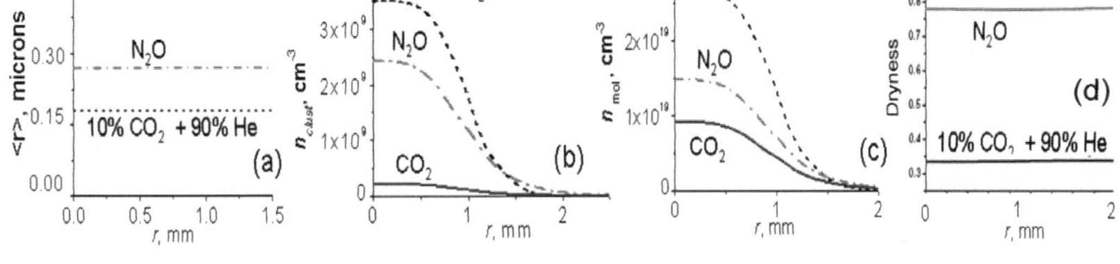


**Figure 2** Dependencies of the cluster parameters versus distance from the vertical nozzle axis at 1.5 mm from the nozzle outlet: (a) cluster radius; (b) cluster concentration; (c) molecular concentration; (d) - Dryness: $CO_2$ -20 bars, $N_2O$ - 30 bars, mixture - 60 bars. Such gas pressures values correspond to the largest size of clusters and the highest observed SXR intensity.

Experimental setup scheme is shown in the figure 3. Cluster target was irradiated by the 36 fs Ti:Sa laser pulses with energy of 160 mJ focused to the 50 μm focal spot. The spatially resolved X-ray spectra were measured form side and front observation points using two Focusing Spectrometers with Spatial Resolution (FSSR). Spectrographs were adjusted to measure spectra of $Ly_\alpha$ (energy 653.7 eV) and $He_\beta$ (energy 665.7 eV) lines of oxygen with both high spectral ($\delta\lambda/\lambda \sim 3000$) and spatial (~ 40 μm) resolutions. Modeling of obtained spectra allowed us to measure the plasma parameters: $T_e$ = 110 eV, $N_e = 10^{20}$ cm$^{-3}$ and to calculate other spectral line profiles. Calculated emission spectra of the plasma components are shown in the figure 4. Table 1 contains main SXR parameters measured for the used laser plasma sources. It shows that X-ray flux produced by the mixture target increases in 2-8 times compare with the case of using pure $CO_2$ or $N_2O$ clusters and reach values about $2.8\text{x}10^{10}$ ph/(sr · pulse). Considering 1 ps X-ray pulse duration we estimated peak spectral brightness of the developed source. For the observed lines it reaches $1.3\text{x}10^{23}$ ph/s/mm$^2$/mrad$^2$/0.1%BW.

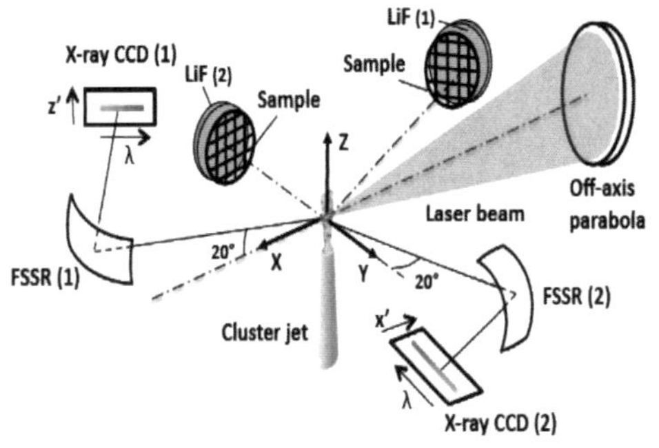


**Figure 3.** Experimental setup.

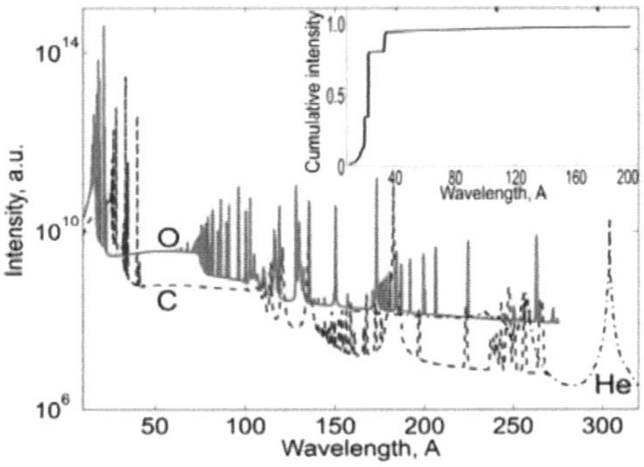


**Figure 4.** Plasma emission spectrum.

**Table 1**. Plasma size and X-ray flux measured by the front and side spectrographs

| Gas | $P_{gas}$, bar | Parameter | $He_\beta$ ($\lambda$=18.267 Å) | | $Ly_\alpha$ ($\lambda$=18.970 Å) | |
|---|---|---|---|---|---|---|
| | | | Front | Side | Front | Side |
| $N_2O$ | 30 | Photon intensity ph/(sr · pulse) | $3.3\text{x}10^9$ | $2.7\text{x}10^9$ | $4.4\text{x}10^9$ | $2.6\text{x}10^9$ |
| | | Source size, μm | 80×80 | 80×460 | 90×90 | 90×400 |
| | | Brightness, ph/s/mm$^2$/mrad$^2$/0.1%BW | $3.1\text{x}10^{22}$ | $3.7\text{x}10^{21}$ | $3.0\text{x}10^{22}$ | $4.2\text{x}10^{21}$ |
| $CO_2$ | 20 | Photon intensity ph/(sr · pulse) | $4.1\text{x}10^9$ | $1.9\text{x}10^9$ | $6.6\text{x}10^9$ | $3.1\text{x}10^9$ |
| | | Source size, μm | 120×120 | 120×570 | 100×100 | 100×530 |
| | | Brightness, ph/s/mm$^2$/mrad$^2$/0.1%BW | $1.7\text{x}10^{22}$ | $1.6\text{x}10^{21}$ | $3.2\text{x}10^{22}$ | $2.8\text{x}10^{21}$ |
| $CO_2$ (10%) + He (90%) | 60 | Photon intensity, ph/(sr · pulse) | $2.8\text{x}10^{10}$ | $1.5\text{x}10^{10}$ | $3.7\text{x}10^{10}$ | $2.1\text{x}10^{10}$ |
| | | Source size, μm | 120×120 | 120×700 | 100×100 | 100×620 |
| | | Brightness, ph/s/mm$^2$/mrad$^2$/0.1%BW | $1.3\text{x}10^{23}$ | $9.2\text{x}10^{21}$ | $1.6\text{x}10^{23}$ | $7.9\text{x}10^{21}$ |

## 2. Application of source for the radiography of the nm-thick foils

Soft X-ray absorption images (figure 6) of 100 nm thick Mo foils in a wide field of view (mm$^2$ scale) with high spatial resolution (700 nm – see figure 5) have been obtained using the LiF crystals [9] as soft X-ray imaging detectors. High dynamic range and sensitivity of used detector for SXR allow clear resolving up to 4 layers of 100 nm Mo foils. It is possible to reproduce the intensity levels by integrating the product between the plasma spectrum and the transmission function of Mo foils. Layer intensities, normalized at the maximum value of the gray levels, are reported as red dashed lines over the trace in figure 6d. As can be seen from figure 6d, the calculated intensity values are in good agreement with the experimental results. That also proves accuracy of the plasma parameters estimation.

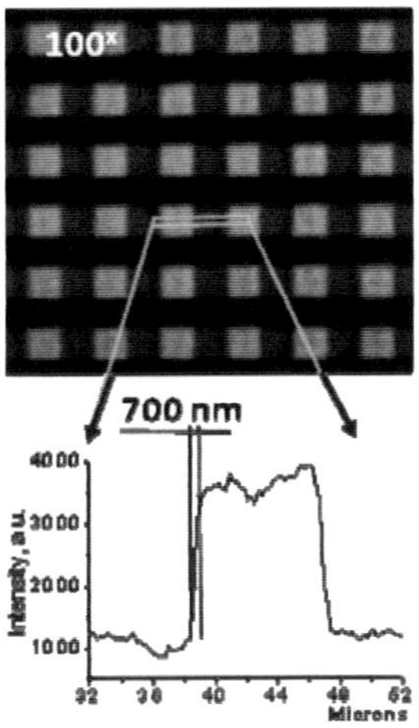


**Figure 5.** The image of the 1500 lpi grid demonstrates a spatial resolution of 700 nm.

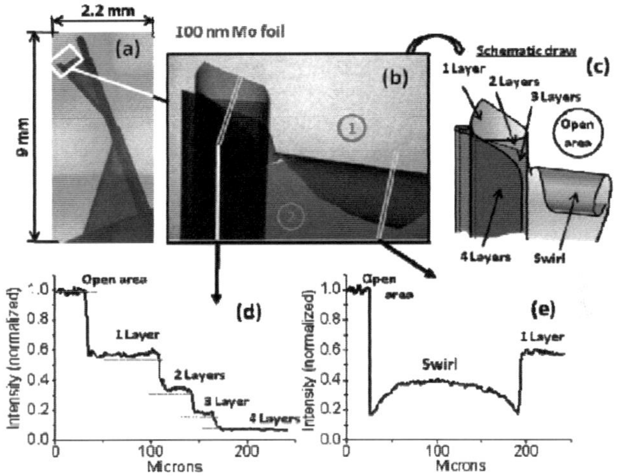


**Figure 6.** SXCI of 100 nm thick damaged Mo foil. (a) – Large scale image; (b) – Magnified fragment ($40^x$); (c) Schematic draw of the curl; (d) – Intensity profile, illustrating overlapping foil layers; (e) – Intensity profile showing curled Mo layer.

## 3. Conclusion

By optimizing gas target parameters we have developed compact and debris free tabletop soft x-ray source operating at 1-10 Hz which has the following parameters: 10-50 A polychromatic spectrum, 90% of radiation contains in the spectral lines of carbon and oxygen, X-ray flux in the individual lines exceeds $3\times10^{10}$ photon/sr/pulse while total x-ray flux is at least $10^{12}$ photons/sr/pulse, peak brightness $1.6\times10^{23}$ ph/s/mm$^2$/mrad$^2$/0.1%BW, source size is about 100 μm x 100 μm. To achieve such parameters we used mixture of the atomic (He) and molecular ($CO_2$) gases. Atomic gas enhances clusterization of the $CO_2$ gas and He is transparent for the soft x-ray radiation. The possible application of this source is the metrology and qualitative analysis of thin foils and biological objects with nanoscale thickness.

This research was partially supported by the Japan Ministry of Education, Science, Sports and Culture, Grant - in - Aid for Kiban A No 20244065, Kiban B No. 21360364, by the Russian Fund of Basic Research ( Project No. 09 - 02 - 92482 - MNKS_a, 09 - 02 - 12226 -ofi-m), and by the RAS Presidium Program of basic researches No. 27

[1] Nishikawa T, Suzuki S, Watanabe Y, Zhou O and Nakano H 2004 *Appl. Phys. B.* **78,** 885
[2] Sumeruk H.A, Kneip S, Symes D.R et al. 2007 *Phys. Rev. Letters.* **98**, 045001
[3] Palchan T, Pecker S, Henis Z, Eisenmann S and Zigler A 2007 *Appl. Phys. Lett.* **90,** 041501
[4] Smirnov B.M. Clusters and Small Particles (Springer- Verlag, New York, 2000).
[5] Boldarev A.S, Gasilov V.A, Faenov A.Ya et.al., 2006 Rev. Rev. Sci. Instr. **77**, 083112
[6] Boldarev A.S, Gasilov V.A and Faenov A.Ya 2004 Technical Physics **49**, 388
[7] Abdallah J, Jr, Csanak G, Fukuda Y et al., 2003 Phys. Rev.A, **68**, 063201
[8] Fukuda Y, Akahane Y, Aoyama M, et al. 2004 Laser and Particle Beam **22**, 215
[9] Baldacchini G, Bollanti S, Bonfigli F et al., 2005 Rev. Sci. Instrum. **76**, 113104

# The atomic model of the Sn plasmas for the EUV source

**Akira Sasaki, Atsushi Sunahara[1], Katsunobu Nishihara[2], Takeshi Nishikawa[3], Fumihiro Koike[4], Hajime Tanuma[5]**

Quantum Beam Science Directorate, Japan Atomic Energy Reseach Agency, 8-1 Umemidai, Kizugawa-shi, Kyoto, Japan 619-0215

[1]Institute for Laser Technology, 1-8-4, Utsubohonmachi, Nishi-ku, Osaka, 550-0004

[2]Institute for Laser Engineering, Osaka University, 2-6 Yamadaoka, Suita, Osaka, 565-0871

[3]Okayama University, 1-1-1 Tsushimanaka, Okayama, 700-8530

[4]Kitasato University, 1-15-1 Kitasato, Sagamihara, Kanagawa, 228-8555

[5]Tokyo Metropolitan University, 1-1 Minamiosawa, Hachioji, Tokyo, 192-0397

E-mail: sasaki.akira@jaea.go.jp

**Abstract**. An atomic model of Sn to study the level population and radiative properties of near 10 times ionized states in laser produced plasmas is presented. The extreme ultra-violet (EUV) spectrum is investigated in conjunction with the effect of the detailed structure of resonance lines as well as satellite lines from multiply- and inner-shell excited states. The coefficients of radiative transfer are calculated with present model, which used in the radiation hydrodynamics simulation for the optimization of the EUV light source.

## 1. Introduction

Sn plasmas are being intensively studied as a candidate of the extreme ultra-violet (EUV) light source for the next generation microlithography. More than 180W of EUV output power at a wavelength of $\lambda$=13.5nm within the 2% bandwidth is expected using laser produced plasmas (LPP) or discharge pumped plasma (DPP) sources [1].

Laser produced plasmas have been interested for their capacity to produce incoherent and coherent emission up to x-ray wavelength region. However, those sources required extreme conditions to work, therefore industrial application has not been demonstrated. Recently, it is found that plasmas of mid-z ions (z≈50) excited by laser irradiation with a moderate intensity (≈$10^{10}$W/cm$^2$) or discharge can be applicable to high power EUV sources.

The highly charged ions of mid- to high-z elements have characteristic emission spectrum in several different UV to x-ray bands depending on the plasma temperature and corresponding ion charge. In contrast to the wavelength of $\Delta n\neq 0$ transitions, which shifts according to the ion charge, the wavelength of $\Delta n=0$ transitions are nearly constant over several charge states, which is useful for application to light sources, as shown in Figure 1.

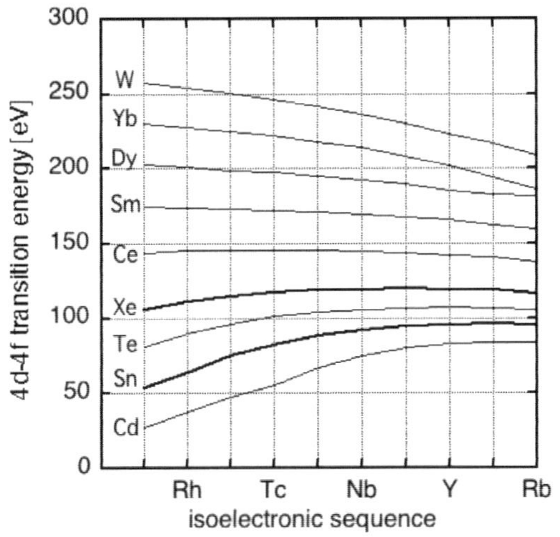


**Figure 1.** Energies of 4d-4f transitions of Pd-($4d^{10}$ isoelectronic sequence) to Rb-($4p^64d$) like ions of Cd to W.

Sn plasma, heated to electron temperatures of 20-50eV, is ionized up to a state with a mean charge of ≈10, which have a ground configuration of $4d^i nl$, where i is the occupation number. In the case of these 4d ions, strong configuration mixing takes place between $4d^{i-1}4f$ and $4p^54d^{i+1}$ configurations, so that predominant emission is obtained at λ=13.5nm, because a large number of 4d-4f and 4p-4d fine structure transitions overwrap each other in a same wavelength region [2]. Ions beyond Kr-like state still have intense emission near λ=13.5nm through 4p-4d transitions, therefore Sn is an efficient emitter over a wide range of plasma temperature.

In addition to the 4d-4f resonance lines, satellite lines from multiply- and inner-shell excited states contribute to the EUV emission. As the optical depth of the plasma increases, the emission intensity from resonance lines at λ=13.5nm saturates, whereas relative intensity of satellite lines in the longer wavelength increases [3]. Then the spectral width increases resulting in more than 50% of pumping energy is converted to the total EUV emission. However, the conversion efficiency to the 2% bandwidth at λ=13.5nm should rather increase by decreasing the optical depth of the plasma, by reducing the spectral width and excess emission outside the required emission band, therefore the pumping condition requires optimization [4].

We develop an atomic model of Sn ions and calculate the coefficients of radiative transfer of Sn plasmas, in order to investigate the emission spectrum and conversion efficiency of the EUV sources through the radiation hydrodynamics simulation.

## 2. The atomic model

We develop a collisional radiative model of Sn ions, and the level population is calculated assuming collisional radiative equilibrium (CRE) as a function of electron temperature and ion density. In order to investigate the emission over a wide temperature range, the neutral to Ar-like Sn ions are taking into account.

The atomic states are defined in terms of their non-relativestic electron configuration. A set of rate equations corresponding to each state is solved including ionization and excitation processes in plasmas through radiation and electron collisions. The energy of each state, radiative transition probability and autoionization rates are calculated using the HULLAC code [5]. Rates of collisional excitation and ionization are calculated using empirical formulas as a function of the transition energy, as well as the oscillator strength in the case of excitation [6]. Rates of radiative recombination are also calculated using an empirical formula. Rates of inverse processes such as collisional deexcitation and three-body recombination are determined from the detailed balance.

The ion abundance and level population usually depend on the set of atomic states included in the model. In particular, atomic states with low excitation energies should be included, because these states have a large population. However, Sn ions have a really large number of atomic states. We propose a method to choose an appropriate set of states.

We define a set of excited states of each ion, from low excited states of the next charge state. For example, from the ground and excited states of Mo-like Sn such as $4d^6$, $4d^55s$, $4d^55p$, $4d^54f$, $4p^54d^7$, according to the order of excitation energy, groups of excited states of Tc-like Sn such as $4d^6nl$, $4d^55snl$, $4d^55pnl$, $4d^54fnl$, $4p^54d^7nl$ can be generated. The atomic model for all charge states can be determined recursively. Then we perform iterative calculations of level population increasing the number of groups of excited states of each ion, until the ion abundance and radiation loss converge. Typically, we take 5 groups of states into account, for example in the cases of Pd-like to Rb-like ions, we typically include $4d^{i-1}nl$, $4d^{i-2}5snl$, $4d^{i-2}5pnl$, $4d^{i-2}4fnl$, and $4p^54d^inl$ configurations with maximum principal and orbital quantum number of $n \leq 8$ and $l \leq 3$. Figure 2 shows the temperature dependence of the mean charge, which converges including more than 5 groups of states [7].

Using the calculated level population, we calculated the emissivity and opacity of the plasma. It is shown that each emission line consists of a large number of fine structure transitions called unresolved transition array (UTA) [8]. Although, as the spectral width of 4d-4f transition arrays are 0.3-0.5nm, the accurate central wavelength and structure should be determined, in order to apply calculated radiative coefficients to optimization of the efficiency.

We calculate the wavelength and detailed spectral structure of 4d-4f and 4p-4d transitions using the Hullac code considering the effect of configuration interaction (CI). We see energy levels converge after including sufficient CI, and spectral profile agrees well with those observed in the charge exchange spectroscopy (CXS) [9]. Therefore, we used calculated level energies and radiative rates in the collisional radiative model, however transition wavelengths are corrected according to the observed shift between Hullac code and CXS.

Calculated spectrum also depends on how many satellite channels are included in the model, as well as their spectral structure. We calculated the detailed structure of satellite lines, which have considerable contribution to the spectrum. We also corrected the wavelength of 4d-4f, and 4p-4d satellite lines so as their wavelength converge to those of the resonance line of the next charge state, at the limit of the principal quantum number of the spectator electron of $n=\infty$. The same iterative method, which is used to determine level population, is applied to find dominant satellite channels. As shown in Figure 3, the emissivity in the EUV wavelength region increases as the number of groups of levels included in the model increases. Although, the peak emissivity at $\lambda=13.5$nm is almost the same, the spectral width of the emissivity increases considerably by increases number of atomic states in the model. The emissivity also converges if the number of groups of states is increased above 5, where we include $4d^{i-1}nl$, $4d^{i-2}5snl$, $4d^{i-2}5pnl$, $4d^{i-2}4fnl$, and $4p^54d^inl$ configurations to include resonance and satellite lines of 4d-4f, 4d-5p, and 4p-4d transitions with one spectator electron in an outer orbital.

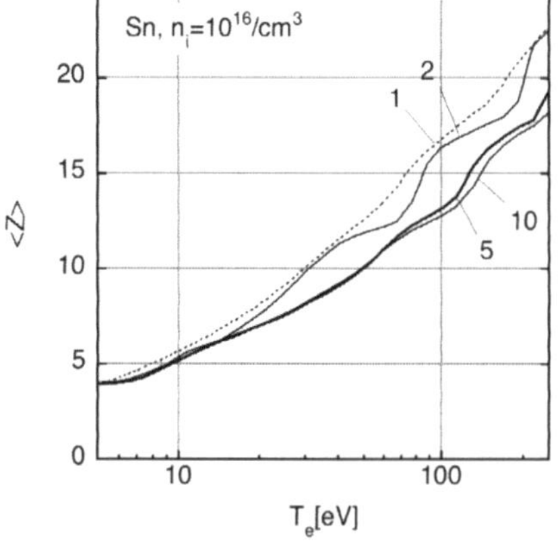


**Figure 2.** Electron temperature dependence of the mean charge of Sn plasmas and its atomic model dependence.

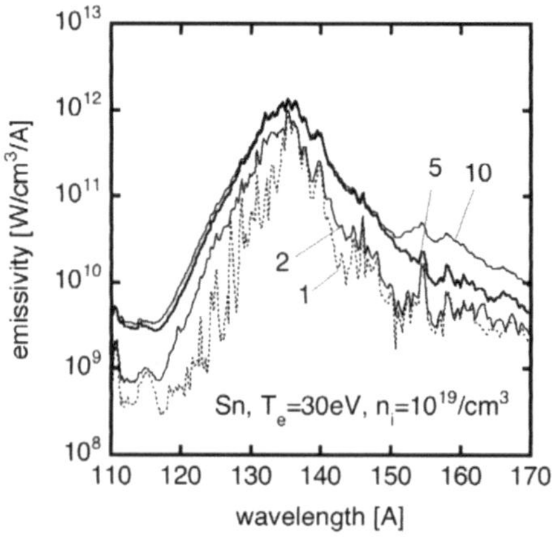


**Figure 3.** Calculated emissivity of Sn plasmas in the EUV wavelength region.

## 3. Result and discussion

Figure 4 shows calculated specific spectral emissivity and its temperature dependence for the ion density ranging from $10^{17}$ to $10^{21}/cm^3$. It is shown that the spectral width increases as the ion density increases, due to increased contribution of the satellite lines.

It is found that the spectral efficiency, which is defined by the ratio of inband emissivity to the total, has a maximum value where the mean charge of the plasma becomes ≈10, which increase as density decreases up to 40% at $n_i=10^{17}/cm^3$.

Peak wavelength decreases as temperature increases until it reaches the lowest value of 13nm, and then increases up to 16nm at the electron temperature of 100eV. This arises from the z-dependence of the wavelength of the 4d-4f transitions.

Finally, we carry out radiation hydrodynamics simulation with present atomic data. We compared calculated and measured spectrum as well as conversion efficiency to see reasonable agreement [4]. We also find improvement of conversion efficiency using low density plasmas excited by $CO_2$ lasers, which is implied in the dependence of the spectral efficiency on ion density shown in Figure 4.

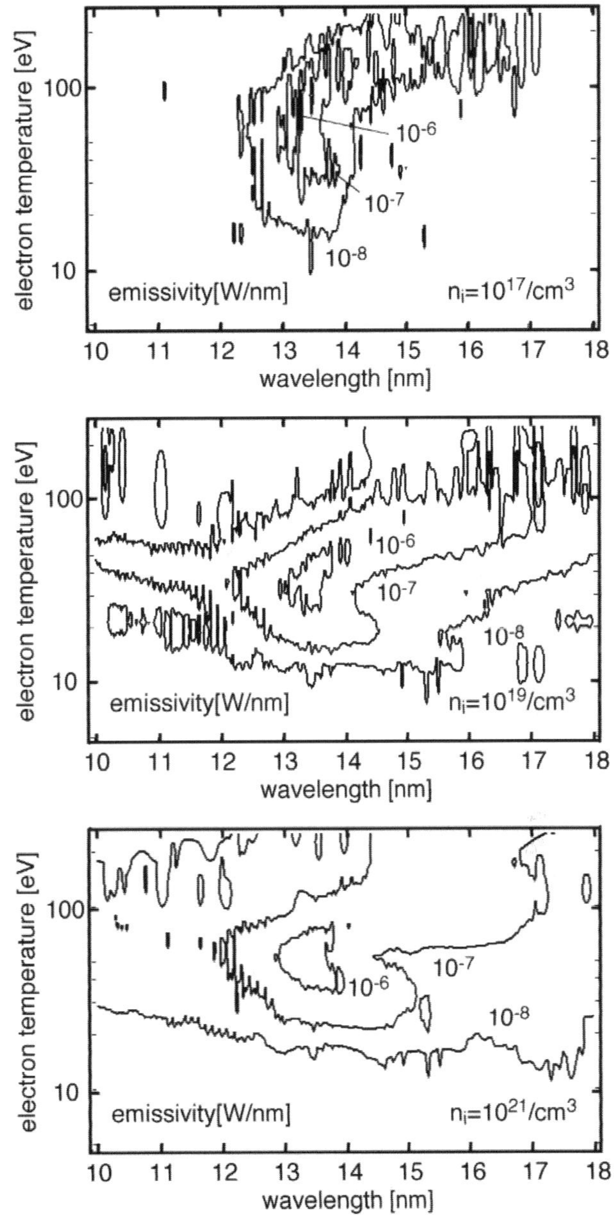


**Figure 4.** Calculated specific spectral emissivity (per ion) as a function of plasma electron temperature, for ion density of $10^{17}$, $10^{19}$, and $10^{21}/cm^3$.

**Acknowledgement**

This work was performed under the auspices of a Leading Project promoted by MEXT Japanese Ministry.

## References

[1] EUV Sources for Lithography, edited by V. Bakshi, 2005 SPIE press.
[2] W. Svendsen and G. O'Sullivan Phys. Rev. **A50**, 3710 (1994)
[3] A. Sasaki, et al., Appl. Phys. Lett., **85**, 5857 (2004).
[4] K. Nishihara, et al., Phys. Plasmas, **15**, 056708 (2008).
[5] A. Bar-Shalom, M. Klapisch, and J. Oreg., J. Quant. Spectrosc. Radiat. Transfer **71**, 169 (2001).
[6] W. A. Lokke and W. H. Grasberger, Lawrence Livermore Laboratory Report UCRL-52276 (1977).
[7] A. Sasaki, et al., HEDP, **3**, 250 (2007).
[8] C. Bauche-Arnoult, J. Bauche, and M. Klapisch, Phys. Rev. **A 31**, 2248 (1985).
[9] H. Tanuma, et al., J. Phys. Conf. Ser. **58**, 231 (2007).

# Creation of highly-charged calcium ions for the *g*-factor determination of the bound electron

**B Schabinger[1] [1], S Sturm[1], K Blaum[2,3], W Quint[3,4], A Wagner[1] and G Werth[1]**

[1] Institut für Physik, Johannes Gutenberg-Universität Mainz, 55099 Mainz, Germany
[2] MPI für Kernphysik, 69029 Heidelberg, Germany
[3] Ruprecht-Karls-Universität Heidelberg, 69120 Heidelberg, Germany
[4] GSI, 64291 Darmstadt, Germany

E-mail: schabing@uni-mainz.de

**Abstract.** High-precision measurements of the magnetic moment of the electron bound in hydrogen- and lithium-like ions can be used to test bound-state quantum electrodynamical calculations. In the past measurements with relative experimental uncertainties as low as $2\times10^{-9}$ were performed on hydrogen-like carbon and oxygen ions. In the current experiment we plan to measure the *g*-factor of hydrogen-like and lithium-like calcium ions. A relative uncertainty $\delta g/g$ in the order of $10^{-9}$ is aspired. Here, we will give the motivation for the experiment, present the experimental techniques and first results.

## 1. Introduction

The *g*-factor of an electron is a proportionality constant between its angular momentum $\overrightarrow{J}$ and its magnetic moment $\overrightarrow{\mu}$

$$\overrightarrow{\mu} = -g\frac{e}{2m_e}\overrightarrow{J}. \tag{1}$$

For a Dirac-point-like free electron the *g*-factor is predicted to be exactly 2 [1]. Because of quantum electrodynamic [2] and relativistic [3] contributions the *g*-factor is shifted from the predicted value. In case the electron is bound to an ion, the electron is affected by the strong electromagnetic fields of the nucleons at its position. The strength of the e.m. fields ranges from $10^9$ V/cm for helium ($Z$=2) to $10^{15}$ V/cm for uranium ($Z$=92). Measurements on hydrogen-like carbon and oxygen ions were performed with relative uncertainties in the level of $10^{-9}$ [4, 5] to test highly accurate calculations [6] for the *g*-factor of a single electron bound to a spinless nucleus. In combination both led to the most precise determinations of the electron mass in atomic units [7, 8].

The effects of the bound-state quantum electrodynamics (BS-QED) increase with the nuclear charge $Z$, therefore heavier elements like calcium in the current experiment [9] are particularly interesting. The element calcium has two spinless isotopes (masses 40 u and 48 u), where the total spin of the system is the electron spin. The binding energy of the second last electron in the 1s-state of hydrogen-like calcium $Ca^{19+}$ is ~5 keV, which is within reach of the experimental setup. Not only the *g*-factor of hydrogen-like calcium $Ca^{19+}$ will be measured, it is also planned to investigate lithium-like calcium $Ca^{17+}$. Here the calculations take into account the electron-electron interactions and claim an accuracy of $9{\cdot}10^{-8}$ [10]. For studying nuclear structure effect one can compare *g*-factor measurements of the doubly-magic isotopes $^{40}Ca$ and $^{48}Ca$, which is also planned.

[1] The results which are presented here are part of the PhD thesis of Birgit Schabinger and Sven Sturm

## 2. *g*-factor measurement

The *g*-factor determination of a single stored ion is based on the "continuous Stern-Gerlach-effect" [11]. The Larmor frequency

$$\omega_L = g\frac{e}{2m_e}B \tag{2}$$

is the spin precession frequency in a magnetic field $B$ and can be determined by measuring the spin flip probability for different excitation frequencies. Therefore the trapped ion is irradiated with microwaves at a frequency close to the Larmor frequency and the spin state before and after irradiation is observed. At the Larmor frequency the spin-flip probability [12] reaches a maximum for ions with negligible temperature. The value of the magnetic field at the ion's position can be determined from its cyclotron frequency, given by

$$\omega_c = \frac{q}{m_{ion}}B, \tag{3}$$

$q$ being the charge of the ion and $m_{ion}$ its mass. The $g$-factor is thus given by

$$g = 2\frac{\omega_L}{\omega_c}\frac{q}{e}\frac{m_e}{m_{ion}}, \tag{4}$$

where the masses of the ion and the electron are known from other experiments [8, 13].

### 2.1. *Triple-Penning trap setup*

The experimental setup is based on a set of three cylindrical Penning traps [14, 15], creation, analysis and precision trap, for more details see [16, 17]. The axial confinement of the ions is achieved with an electrostatic potential, superimposed to a constant magnetic field in the direction of the trap axis for radial confinement. The ion motion in a Penning trap is a composition of three independent oscillations. One is in the axial direction with the frequency

$$\omega_z = (qU_0/(m_{ion}d^2))^{1/2}, \tag{5}$$

which depends on the trap potential $U_0$ and the characteristic trap size parameter $d$. In the radial direction the motion can be decomposed into two circular motions, with the reduced cyclotron frequency $\omega_+$ and the magnetron drift frequency $\omega_-$, given by

$$\omega_\pm = \frac{\omega_c}{2} \pm \sqrt{\left(\frac{\omega_c}{2}\right)^2 - \frac{\omega_z^2}{2}}. \tag{6}$$

### 2.2. *Ion creation*

The charge breeding of the ions is performed inside the trap with an electron beam ion source (EBIS) in combination with a Penning trap located inside a sealed vacuum chamber [16, 17] to achieve extreme vacuum conditions ($p \leq 10^{-16}$ mbar) at liquid helium temperature. Electrons are emitted from a single field emission point (FEP) by applying a voltage difference between an acceleration electrode and the FEP. The electron beam is guided by the magnetic field through the creation trap and is reflected by the strong negative potential at the reflector electrode. The electrons oscillate several times between the reflector and the FEP until the Coulomb-repulsion causes the beam to expand radially and hit the target made out of graphite with a sputtered layer of mainly $^{40,44,48}$Ca and some contamination of silicon. The atoms are ablated and ionized by the electrons. The ions are trapped by the negative potential of the creation trap where charge breeding takes place by consecutive electron-impact ionization.

### *2.3. Mass spectrum in the Penning trap*

The charge breeding process creates several charge states of different elements, for example $^{12}C^{6+}$, $^{16}O^{7+}$ and $^{28}Si^{11+}$ (see Fig. 1 a) . The axial frequency of an ion in a Penning trap is determined by measuring the frequency of the image current induced by the ion across an electronic resonance circuit attached to the trap electrodes. To identify the different ions one can take advantage of the dependence of the axial motion on the trapping potential $U_0$. By scanning the potential several species with different $q/m_{ion}$ come into resonance with the axial tank circuit. To create a mass spectrum as shown in figure (Fig. 1 a) a special amplifier is used. The ion signal is already amplified by a cryogenic preamplifier and again amplified and filtered by a second amplifier at room temperature. After a narrow bandpass filter, the signal is rectified and integrated. With the knowledge of one $q/m_{ion}$ all other signals can be related to the various ion species.

### *2.4. Measurement of the axial frequency of a single ion*

If the isolated ion is in thermal equilibrium with the tank circuit a minimum in the Johnson noise density of the resonator appears exactly at the ion's oscillating frequency (Fig. 1 b) . Because the axial frequency depends on the trapping potential $U_0$ it is possible to bring ions with different charge-to-mass ratios $q/m_{ion}$ into resonance with the tank circuit at 754 kHz in case of the precision Penning trap.

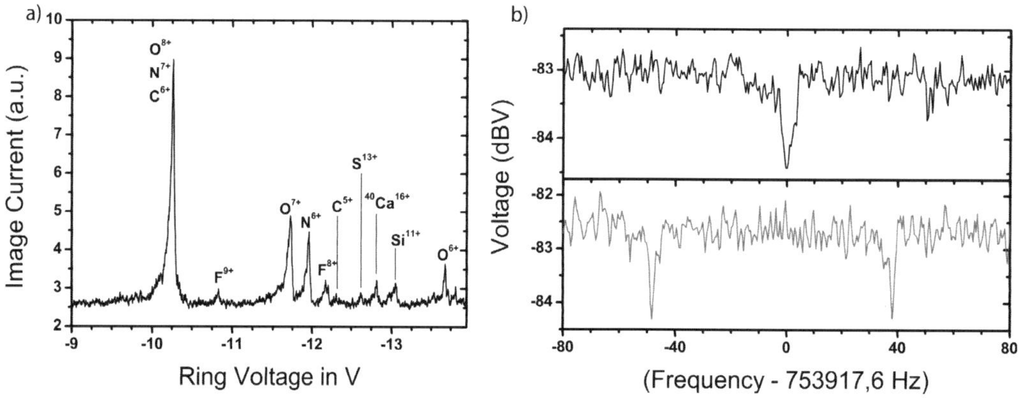


**Figure 1.** a) A mass spectrum is shown, which was taken in the precision trap. To get a better signal-to-noise level several spectra were summed. At the time the spectra were taken, a target consisting of $CaF_2$ was used. b) In the upper graph the Johnson noise density of the tank circuit including a minimum at the axial frequency of some contaminant $^{28}Si^{12+}$ ion is shown. By coupling the axial and the radial motion via irradiation of a rf signal near the sideband frequency $\omega_+ - \omega_z$ the resonance splits into two components separated by the Rabi-frequency, shown in the lower graph. For more details see text.

### *2.5. Determination of the free cyclotron frequency*

The free cyclotron frequency is not a real eigenmotion of the ion in the Penning trap, therefore it cannot be measured directly. The relation between the three eigenmotions and the free cyclotron frequency is expressed in the invariance theorem [18]:

$$\omega_c = \sqrt{\omega_+^2 + \omega_-^2 + \omega_z^2}. \tag{7}$$

### *2.6. Sideband coupling*

The reduced cyclotron and the magnetron frequency can be determined by resonant coupling of two ion motions [12]. Therefore a suited electric radio-frequency field (rf) is irradiated, causing an amplitude modulation of the corresponding motions. In this experiment the "double-dip method" is used, where a rf signal at the sideband frequency $\omega_+ - \omega_z$ causing the axial resonance to split into two. The components $\omega_{r,l}$ are given by

$$\omega_{r,l} = \omega_z - \frac{\delta}{2} \pm \sqrt{\delta^2 + |A|^2} \tag{8}$$

where $|A|$ is a proportion for the measure of the coupling in units of frequency and $\delta$ is a measure for the detuning of the coupling frequency to the sideband frequency $\omega_+ - \omega_z$. Finally, the reduced cyclotron frequency can be expressed as

$$\omega_+ = (\omega_+ - \omega_z + \delta) + \omega_z + 2\left(\frac{\omega_l + \omega_r}{2} - \omega_z\right) \tag{9}$$

where "$\omega_+ - \omega_z + \delta$" is the known irradiated coupling frequency and $\omega_z$ and $\omega_{r,l}$ respectively are observables. For the magnetron motion $\omega_-$ an analogous analysis can be used.
By using the "double-dip method" the reduced cyclotron frequency $\omega_+$ was determined for several highly charged ions, e.g. $^{28}Si^{12+}$, with an accuracy of $10^{-6}$. This allowed the determination of the field strength of the superconducting magnet to $B = 3.7642\,\mathrm{T}$.

## 3. Current status

Highly charged ions have been created in a cryogenic EBIS and transported to the precision trap where single species have been prepared. The axial and reduced cyclotron frequency was measured for several species. Presently, the charge state breeding process is being investigated in more as required to create lithium-like calcium ions. The next steps will be the detection of single highly-charged ion and finally the determination of the $g$-factor. In the future, we plan to extend our $g$-factor measurements to heavy highly-charged ions up to uranium $^{238}U^{91+}$ at the HITRAP facility [19, 20].

## Acknowledgments

We acknowledge the financial support from the EU (HPRI-CT-2001-50036), the BMBF (06Mz180), the Helmholtz Association of National Research Centers (VH-NG-037) and the DFG (Vo-980-1).

[1] Dirac P A M 1928 *Proc. R. Soc.* **A117** 610
[2] Gabrielse G, Hanneke D, Kinoshita T, Nio M and Odom B 2006 *Phys. Rev. Lett.* **97** 030802
[3] Breit G 1928 *Nature* **122** 649
[4] Häffner H, Beier T, Hermanspahn K, Kluge H-J, Quint W, Stahl S, Verdú J and Werth G 2000 *Phys. Rev. Lett.* **85** 5308
[5] Verdú J, Beier T, Djekic S, Kluge H-J, Quint W, Stahl S, Valenzuela T, Vogel M and Werth G 2004 *Phys. Rev. Lett.* **92** 093002
[6] Pachucki K, Czarnecki A, Jentschura U D and Yerokhin V A 2005 *Phys. Rev.* A **72** 022108
[7] Beier T, Häffner H, Karshenboim S G, Kluge H-J, Quint W, Stahl S, Verdú J and Werth G 2002 *Phys. Rev. Lett.* **88** 011603
[8] Mohr P J and Taylor B N 2005 *Rev. Mod. Phys.* **77** 1
[9] Vogel M, Alonso J, Djekic S, Kluge H-J, Quint W, Stahl S, Verdú J and Werth G 2005 *Nucl. Inst. Meth.* B **235** 7
[10] Glazov D A, Shabaev V M, Tupitsyn I I, Volotka A V, Yerokhin V A, Plunien G and Soff G 2004 *Phys. Rev.* A **70** 062104
[11] Dehmelt H G 1986 *Proc. Natl. Acad. Sci. USA* **53** 2291
[12] Verdú J, Djekic S, Kluge H-J, Quint W, Stahl S, Valenzuela T, Vogel M and Werth G 2004 *Phys. Scr.* **T112** 68
[13] Nagy S, Fritioff T, Solders A, Schuch R, Björkhage and Bergström I 2006 *Eur. Phys. J.* D **39** 1
[14] Brown L S and Gabrielse G 1986 *Rev. Mod. Phys.* **58** 233
[15] Blaum K 2006 *Phys. Rep.* **425** 1
[16] Alonso J, Blaum K, Djekic S, Kluge H-J, Quint W, Schabinger B, Stahl S, Verdú J, Vogel M and Werth G 2006 *Rev. Sci. Inst.* **77** 03A901
[17] Schabinger B, Alonso J, Blaum K, Kluge J, Quint W, Stahl S, Vogel M and Werth G 2007 *J. Phys. Conf. Ser.* **58** 121
[18] Brown L S and Gabrielse G 1982 *Phys. Rev. A* **25** 2423
[19] Herfurth F, Beier T, Dahl L, Eliseev S, Heinz S, Kester O, Kluge H-J, Kozhuharov C, Maero G and the Hitrap Collaboration 2005 *AIP Conf. Proc.* **793** 278
[20] Kluge H-J et al. 2008 *Adv. Quantum Chem.* **53** 83

# Re-trapping and cooling of highly-charged ions

**M Hobein, A Solders, M Suhonen, Y Liu, O Kamalou, R Schuch**

Atomic Physics, AlbaNova, Stockholm University, 10691 Stockholm, Sweden

E-mail: hobein@physto.se

**Abstract**. Presently, a trapping system for cooling highly-charged ions (HCI) is being set up at AlbaNova at Stockholm University. The experiment aims at production of low temperature (emittance) HCI at very low energy. HCI are extracted from the new Stockholm EBIT (S-EBIT) before evaporative cooling is applied in a Penning trap. In the future the cooled ions will be injected into the precision trap of the high-precision mass spectrometer SMILETRAP II. In first tests the emittance of trapped ions was measured and it was shown that highly and low-charged ions could be simultaneously stored.

## 1. Introduction

SMILETRAP II is a new Penning-trap mass-spectrometer at Stockholm University designed for high-precision mass measurements using the merits of highly-charged ions (HCI). The concept is based on the former experiment SMILETRAP [1] with which masses of a wide range of elements were determined reaching uncertainties of a few parts in $10^{-10}$ [2,3]. The precision of the mass measurements increases linearly with the cyclotron frequency $\omega_c = Bq/m$ of an ion in a Penning trap, which in turn depends linearly on the charge of the ion and the strength of the magnetic field $B$ in the trap. Therefore, SMILETRAP II was recently connected to the new Stockholm electron-beam ion trap (S-EBIT).

Presently, the uncertainty of the mass measurements is limited, among other things, by the homogeneity of the magnetic field in the precision trap: Approximately every 2 seconds, a pulse of several hundred HCIs arrives at the precision trap of which only the coldest ion is selected and stored. Then quadrupole excitation with frequencies $\nu$ scanned around $\omega_c$ is applied to determining the mass of the ion using the TOF technique. A spread in the position of the ion in the trap leads to a deviation of the cyclotron frequency contributing to the overall uncertainty. Therefore, a precise centering of each ion in the precision trap is necessary for confining all ions to a volume in which the homogeneity of the magnetic field $B$ is sufficiently high. This centering can be achieved by cooling the HCIs in a second trap, the so-called cooling trap, prior to the injection into the precision trap. As cooling technique evaporative cooling was chosen as it is more flexible in its application to ions of different $q/m$ ratio in contrast to, for instance, resistive cooling. Cooling results in a decrease of the emittance of the ions after extraction and can therefore be monitored by an emittance measurement.

Evaporative cooling and its effect on the emittance of ions will be discussed in the following section, before the setup of the experiment is presented in section 3. Finally, first results of re-trapping experiments are given in section 4.

## 2. Evaporative cooling and emittance

Evaporative cooling has been successfully applied to highly-charged ions in EBITs [4]. The principle is as follows: The trapping potential for an ion with charge $q$ in a Penning trap is $qV$ (figure 1), where $V$ is the trap voltage, i.e. the voltage between centre and end electrodes of the Penning trap. Ions with a higher energy will escape the trap. Mixing HCIs and low-q ions in the trap, temperature equilibrium will be achieved after equilibration time $\tau_{eq}$ given in seconds by [5]:

$$\tau_{eq} = 5.87 \frac{A_1 A_2}{n q_1^2 q_2^2 \ln \Lambda} \left( \frac{T_1}{A_1} + \frac{T_2}{A_2} \right)^{3/2} \tag{1}$$

where $A$ is the mass number of the ion, $n$ the particle density, $q$ the charge, $T$ the temperature in degrees K and $\Lambda$ is defined as $\Lambda = \langle h/p_0 \rangle$ with Debye shielding distance $h$ and initial impact parameter $p_0$. The equilibration time $\tau_{eq}$ is in the range of 100 ms for our typical values of ion density, temperature, charges and masses in the cooling trap.

By slightly lowering the trapping potential, preferentially the hottest low-q ions will leave the trap, while basically all HCIs will remain in the trap as they see a much higher potential well (see figure 1). After restoring the full trapping potential, the HCIs will once again equilibrate after $\tau'_{eq}$ with the now colder low-q ions and thus cooling of the HCIs is achieved.

The normalized emittance of ions with charge $q$ can be derived from the temperature of the ions in the trap $T_{ion}$. The emittance $\varepsilon$ is given by:

$$\varepsilon = \pi \cdot r_0 \sqrt{\frac{kT_{ion}}{qU}} \tag{2}$$

Where $r_0$ is the ion cloud radius in the trap, $k$ the Boltzmann constant, $T_{ion}$ the ion temperature and $U$ the extraction voltage or acceleration potential.

For example, to reach with SMILETRAP II uncertainties in the mass measurements in the order of $10^{-11}$, the emittance of the injected ions should be better than 1 π·mm·mrad.

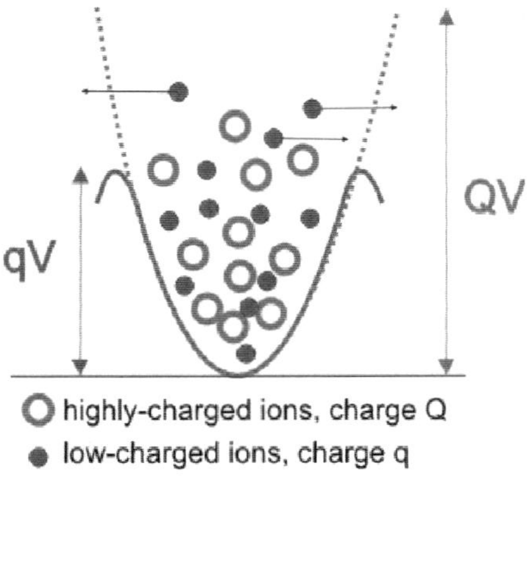


**Figure 1.** Evaporative cooling. For a trap with voltage V between centre and end electrodes, the trapping potential is qV: Higher charged ions are more strongly bound.

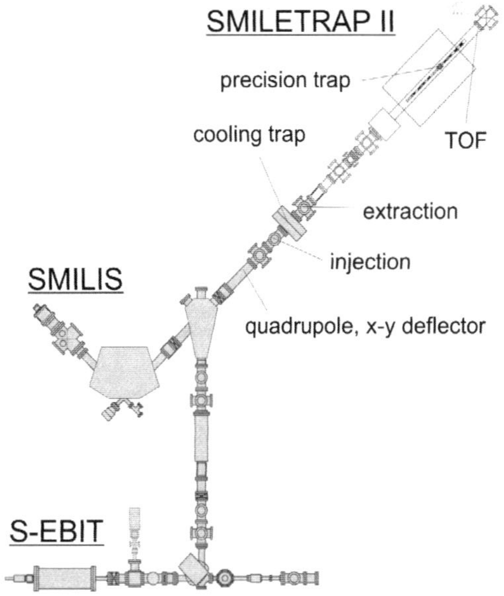


**Figure 2.** SMILETRAP II at AlbaNova.

## 3. Setup

In figure 2 the layout of SMILETRAP II at the facility for HCI at AlbaNova at Stockholm University is shown. Ions are created in two external ion sources, S-EBIT and SMILIS, at 3.6 kV, from where they are extracted to the beam line at ground potential and transported to SMILETRAP II. While low-q ions can be produced with the PIG ion source SMILIS, e.g. $p^+$, $He^+$ or $Ar^{5+}$, S-EBIT yields HCIs. Presently, S-EBIT is upgraded from a 30 keV to a 250 keV electron beam, enough to ionize any element completely.

At SMILETRAP II the incoming ions are retarded and trapped at 3.6 kV in the first trap, the so-called cooling trap which is a cylindrical Penning trap with a magnetic field of 1.1 Tesla. An injection system, which consists of 10 ring electrodes and a cylindrical cone at the entrance of the trap for decelerating the ions, facilitates the injection. After trapping the high-voltage of the trap is ramped down to ground potential and from here, the ions will be transported to the precision trap, a hyperbolical Penning trap in a 6 Tesla superconducting magnet, where the actual mass measurement is performed. In the cooling trap rf-excitation can be applied to the segmented center electrode. This feature will be used for "side-band cooling", i.e. centering the ions by converting the cooled cyclotron motion into the magnetron motion.

For measuring the emittance of trapped ions a pepper-pot emittance meter was placed at the exit of the cooling trap. This is a construction of a RoentDek DLD40 position sensitive delay-line detector and a 50 μm thin stainless steel foil in front. The foil has 10x10 holes with a diameter of 100 μm spaced by 2 mm. Ions of the extracted beam are transmitted through the holes and appear as spots on the detector. By using only geometrical parameters of the foil and the spots, i.e. hole position, mean position of the spots and r.m.s. size of the spots, the emittance can be determined by a formula given in [6].

## 4. First results

In first tests with HCIs from S-EBIT ($Si^{12+}$, $S^{14+}$, $Ar^{16+}$) around $10^3$ ions could successfully be captured in the cooling trap. Using $He^+$ from SMILIS more than 3000 ions per pulse could be trapped. For singly-charged helium storage times of up to 500 ms were achieved, while $Ar^{16+}$ could only be stored for up to 10 ms limited by the moderate vacuum of $10^{-8}$ mbar. Two preparative experiments were performed, namely the measurement of the emittance of trapped ions and the simultaneous storage of HCIs and low-q ions.

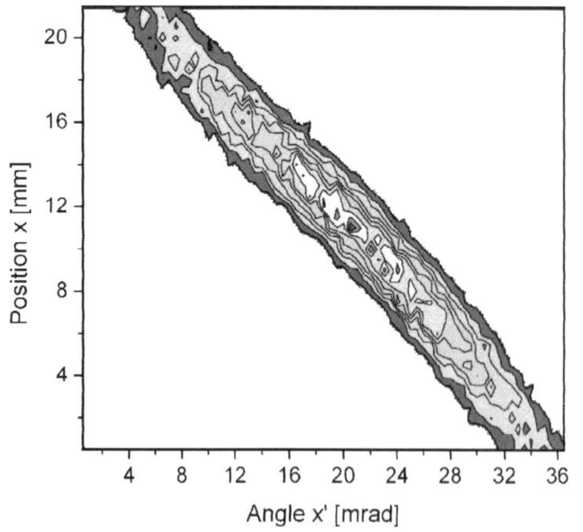


**Figure 3.** Vertical phase space of $Ar^{16+}$ from S-EBIT trapped for 100 us. $\varepsilon_{95\%}$ = 3.1 π·mm·mrad at 3.6 q keV.

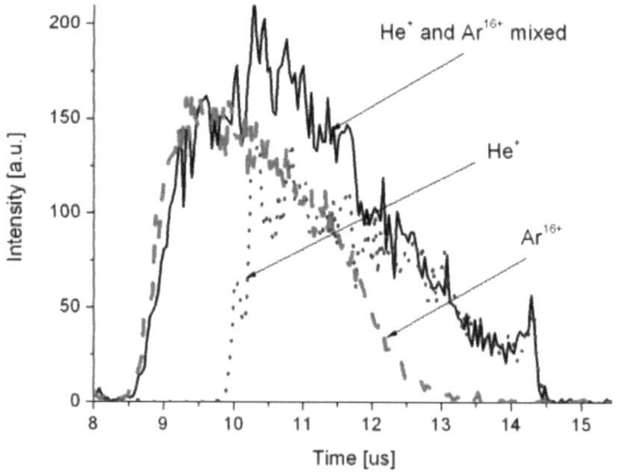


**Figure 4.** Simultaneous storage of $Ar^{16+}$ and $He^+$. The TOF spectrum of simultaneously stored $He^+$ and $Ar^{16+}$ is the sum of the individual spectra.

### 4.1. Emittance of trapped ions

The vertical and horizontal emittance of various ions from S-EBIT and SMILIS, trapped in the cooling trap and extracted with an energy of 3.6 keV per charge, was determined. The storage time was varied between 100 μs and 5.1 ms for HCIs and 20 ms for singly-charged ions, but no significant dependence on the emittance was found. Emittances of few π·mm·mrad were measured, for example, for $S^{14+}$ and $Ar^{16+}$ ions (figure 3) extracted from S-EBIT a r.m.s emittance of 3-4.8 π·mm·mrad was determined, while for $He^{+}$ ions from SMILIS an emittance of 6.2 π·mm·mrad was measured. These results are in very good agreement with values from equation 2. For example, for $Ar^{16+}$ ions an emittance of 4.17 π·mm·mrad is obtained from this equation assuming an ion cloud radius of 1 mm, an ion temperature of 1 eV in the cooling trap and an acceleration potential of 3.6 kV.

### 4.2. Simultaneous storage of $Ar^{16+}$ and $He^{+}$

Synchronizing the ion pulses from S-EBIT and SMILIS simultaneous storage of HCIs ($Ar^{16+}$) and low-q ions ($He^{+}$) in the cooling trap was achieved: First, $Ar^{16+}$ ions from S-EBIT were trapped in the 10 V deep trap and stored for few hundred microseconds. When the $He^{+}$ pulse from SMILIS arrived the trap voltage was lowered to 5 V whereby the singly-charged helium ions could enter the trap, while the highly-charged argon ions were still bound by the 16 times higher trapping potential due to their charge. Having both ions in the trap region the full trap potential was restored and thus simultaneous storage was achieved. The individual TOF spectra of $Ar^{16+}$ and $He^{+}$ can be clearly distinguished as presented in figure 4 showing a difference in the time of flight peak of about 1.5 μs, while the spectrum of simultaneously trapped $Ar^{16+}$ and $He^{+}$ is simply given the sum of the individual spectra. However, due to the too short storage times given by the vacuum compared to the estimated equilibration time (equation 1) no evaporative cooling could be applied.

## 5. Summary and outlook

In order to achieve uncertainties below $10^{-10}$ in the mass measurements with SMILETRAP II, cooling of ions is necessary. The ion temperature in the trap can be derived from the emittance of extracted ions. Emittances of about 5 π·mm·mrad were determined for various highly and low-charged trapped ions. Furthermore, it was shown that HCIs and low-q ions can be stored simultaneously in the cooling trap. This allows for the application of evaporative cooling in the future when longer storage times are reached by a better vacuum. Furthermore, an increase in the precision with SMILETRAP II will be achieved by using higher charge states with the latest S-EBIT upgrade.

## Acknowledgments

The present research has been sponsored by the European ITS-LEIF project and the Wallenberg foundation. We also thank Dr. Peter G. Thirolf and Prof. D. Habs for letting us use their emittance meter.

## References

[1] Bergström I, Carlberg C, Fritioff T, Douysset G, Schönfelder J and Schuch R 2002 *Nucl. Instr. And Meth. In Phys. Res.A* **487** 618-51
[2] Nagy S, Fritioff T, Solders A, Schuch R, Björkhage M and Bergström I 2006 *Europ. Phys. J. D* **39** 1-4
[3] Solders A, Bergström I, Nagy S, Suhonen M and Schuch R 2008 *Phys. Rev. A* **78** 012514
[4] Penetrante B M, Bardsley J N, Levine M A, Knapp D A and Marrs R E 1991 *Phys. Rev. A* **43** 4873-82
[5] Spitzer Jr. L 1956 *Physics of fully ionized gases* (New York: Interscience)
[6] Zhang M 1996 *Emittamce formula for slits and pepper-pot measurements* FERMILAB-TM-1988

# Trapping of highly charged ions with an electrostatic ion trap

**A Gumberidze[1], D Attia[1], C I Szabo[1], P Indelicato[1], A Vallette[1], S Carmo[2]**

[1] Laboratoire Kastler Brossel, CNRS, ENS, Université P. et M. Curie Paris-6, 75252 Paris CEDEX 05, France

[2] Department of Physics, Coimbra University, P-3000 Coimbra, Portugal

E-mail: a.gumberidze@gsi.de

**Abstract.** In the current contribution, we present our first results with an electrostatic ion trap which has been constructed at LKB (Laboratoire Kastler Brossel) in collaboration with the Weizmann institute. The electrostatic ion trap is composed of two coaxial electrostatic mirrors, working for charged particles in the same way as a Fabry-Perot interferometer works for light in the optical regime. The trap has been attached to the beamline of SIMPA (Source d'ions Multichargs de Paris) an Electron Cyclotron Resonance Ion Source (ECRIS) at LKB (Laboratoire Kastler Brossel) and INSP (Institut des NanoSciences de Paris). The trap is foreseen to be applied for various studies on highly charged ions; as measurements of the life times of metastable states, mass spectroscopy etc. First results for trapping highly-charged ions are presented and simulations in order to explain the observed behavior are discussed.

## 1. The experimental setup

The developments and applications of ion traps and heavy-ion storage rings has had a significant impact on many branches of physics. These devices enable storing of ions for a relatively long time, with meV to eV kinetic energies, in ion traps and to GeV kinetic energies, in heavy-ion storage rings. Therefore, they provide unique experimental conditions for a broad range of experimental investigations. Ion traps are able to trap charged particles for a long time, and they have been successfully applied in several areas of physics. Recently, a novel class of ion traps has been developed [1, 2], in which ions oscillate between a pair of electrostatic mirrors, much like photons in an optical resonator. The trapped ions have kinetic energies of a few keV in the central part of the trap while inside the mirrors they have only a few meV near their turning points, where they are also subject to radial focusing forces. Several experiments have already been performed with these devices, such as lifetime [3, 4] as well as studies of charge exchange [5].

Following a design from the Weizmann Institute [1], we have constructed an electrostatic trap and attached it to the beam line of "SIMPA" (Source d'ions Multichargs de Paris) an Electron Cyclotron Resonance (ECR) ion source at LKB (Laboratoire Kastler Brossel) and INSP (Institut des NanoSciences de Paris). The experimental setup including the ECR ion source, the beamline and the trap is presented in Fig. 1. A schematic drawing of the trap together with a photography is displayed in Fig. 2. The ion trap consists of two coaxial electrostatic mirrors, each composed of a stack of eight cylindrical electrodes. The configuration of the trap is characterized by the

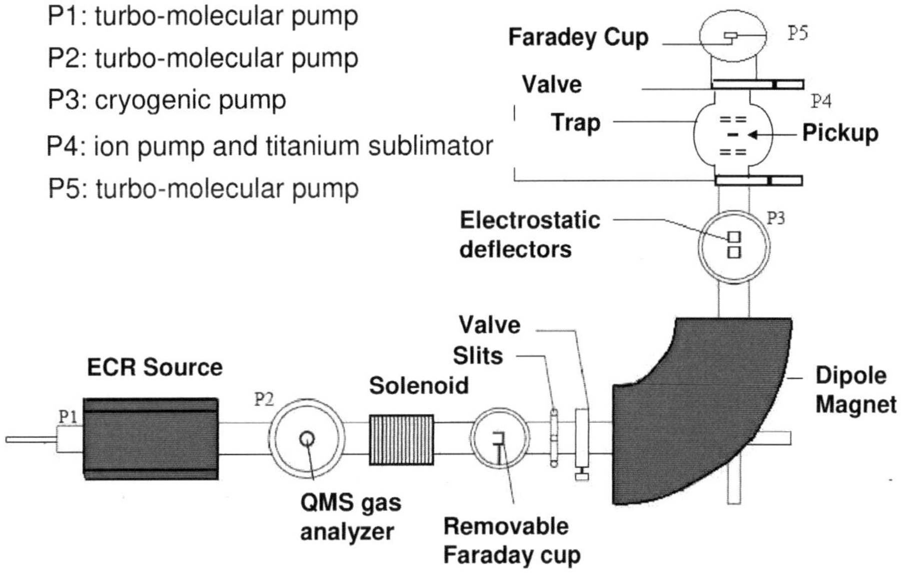


**Figure 1.** The SIMPA ECR ion source and the electrostatic trap at the Laboratoire Kastler Brossel.

potentials on five of these electrodes, V1, V2 ,V3 ,V4, Vz (see Fig. 2), the other three being grounded. Provided that the field generated by these potentials fulfills certain criteria, ions can be trapped oscillating between the two mirrors [1]. The vacuum system of the ion trap includes an ionic pump and a Ti getter pump that can reach low pressures down to $\sim 10^{-11}$ mbar. Injection of an ion bunch into the trap is performed by grounding all the entrance electrodes. The electrodes on the other side are kept at high potential so that the bunch is reflected toward the entrance. Before the ion bunch returns to the entrance mirror, the potentials of its electrodes are rapidly ($\sim$ 100 - 200 ns) raised and the ion bunch is thus confined between the mirrors.

The highly-charged ions are produced in the SIMPA ECR ion source and are extracted and accelerated by the extraction voltage of 4.2 kV. Then, the ions are focused and analyzed by a solenoid and a dipole magnet downstream to the ion source before injection into the trap (see Fig. 1). In addition, steering of the ions between the dipole magnet and the ion trap can be done with the help of two electrostatic deflectors. An electrostatic chopper located before the first magnet was used to create bunches of ions with a temporal extension of 0.05 - 1 $\mu s$. The evolution of ion bunches during storage was monitored with a cylindrical pickup electrode made with a length of 7 mm and inner diameter of 18 mm located at the center of the trap (see Fig. 2). The amplified pickup signal was recorded on a digital oscilloscope and a spectrum analyzer.

## 2. The experimental data and simulations

With the help of the pick-up electrode in the linear center of the trap we have proved trapping for tens of milliseconds in the case of a variety of high charge states of O, Ar, Kr and Xe ions (up to $O^{5+}$, $Ar^{13+}$, $Kr^{21+}$ and $Xe^{20+}$). Using a low ($< 20V$) radio frequency voltage with frequencies close to the oscillation frequency of the ions in the trap we were able to observe trapping times up to 50 ms. Running the ion trap in the so called "bunching" or synchronization mode [6] in the time domain of the pickup signal we observed a strange oscillation on the millisecond scale superposing the ion oscillation in the trap on the microsecond scale. This is shown in Fig. 3 for

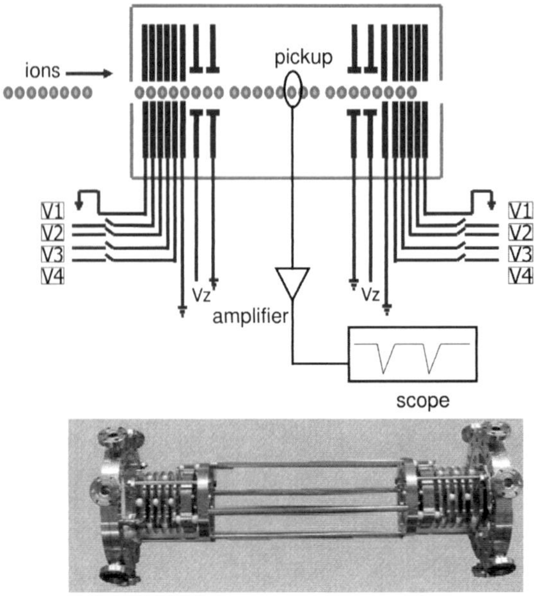


**Figure 2.** The schematic view and a photography of the electrostatic trap at the Laboratoire Kastler Brossel.

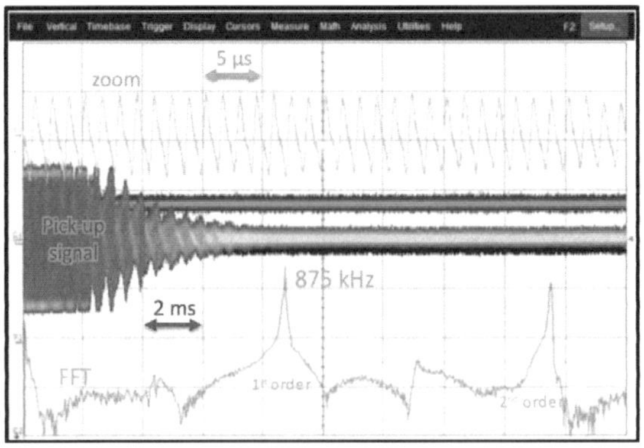


**Figure 3.** Data from the pick-up electrode, for $Ar^{8+}$ trapping, on the ms time scale (green), a zoom on the $\mu$s time scale (yellow) and FFT of the 20 ms range of data from pick up (pink).

the case of $Ar^{8+}$.

Our goal is to explain the observed oscillations with computer simulations. We mainly use two types of simulations for this purpose. The first one is based on the 1D model by Pedersen et al [7] where behavior of a test ion close to a bunch of ions in the trap is investigated. The estimate for the bunch size dynamics can be obtained from observing variation of the distance between the test ion and the ion cloud as function of time. This is shown in Fig. 4 where millisecond scale oscillation of the bunch size is clearly visible. This could in principle, be considered as a possible reason for the observed oscillations in the experiment because the change in the bunch size could affect the pickup signal. In addition to the 1D model, we are developing an N-particle 3D simulation code where we include the interaction between the ions fully and observe dynamics of all the particles in the potential of the trap. First results of this simulation are shown in fig 4. Here, we show that if there is a slight misalignment during the injection of the ion beam in the trap, the radial velocity given to the bunch leads to radial oscillations of the packet. These radial oscillations could perturb the signal on the pickup as well because it depends on the distance from the ions. Further checks for various effects and consistency are underway. After successfully trapping highly-charged ions, we plan, as a next step, to utilize the ion trap for measurement of the life times of metastable states in highly charged ions.

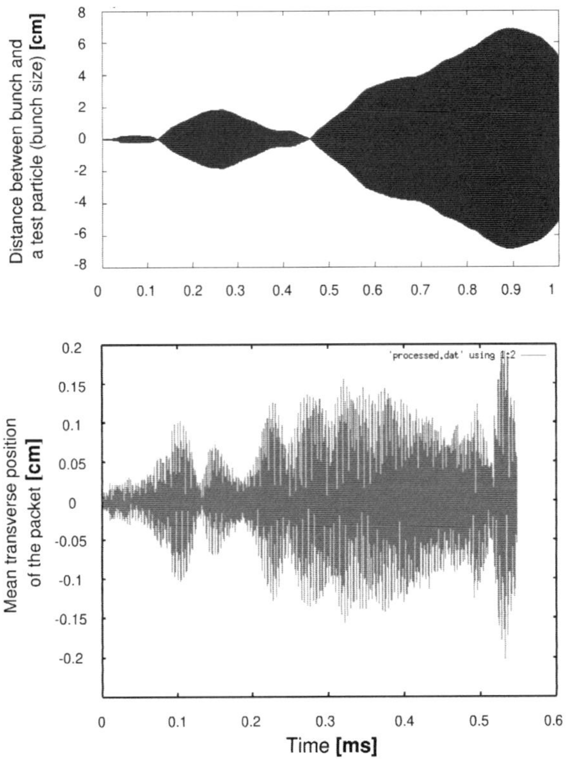


**Figure 4.** Top: 1D simulation showing variation of the distance between the test ion and the ion cloud as function of time (bunch-size dynamics); Bottom: N-particle 3D simulation of the bunch radial position as function of time.

### *2.1. Acknowledgments*

Authors wish to acknowledge assistance and encouragement from colleagues; M. Trassinelli, O Heber, D. Zajfmann and special work by the technical staff; J. Mérot, J. M. Isac, J. P. Okpisz, B. Delamour, P. Travers.

## References

[1] Zajfman D, Heber O, Vejby-Christensen L, Ben-Itzhak I, Rappaport M, Fishman R, and Dahan M 1997 *Phys. Rev.* A **55** R1577

[2] Dahan M, Fishman R, Heber O, Rappaport M, Altshtein N, van der Zande W J and Zajfman D, 1998 *Rev. Sci. Instrum.* **69** 76

[3] Wester R, Bhushan K G, Altstein N, Zajfman D, Heber O, and Rappaport M L 1999 *J. Chem. Phys.* **110** 11 830

[4] Knoll L, Bhushan K G, Altstein N, Zajfman D, Heber O, and Rappaport M L 1999 *Phys. Rev. A* **60** 1710

[5] Strasser D, Bhushan K G, Pedersen H B, Wester R, Heber O, Lafosse A, Rappaport M L, Altstein N, and Zajfman D 2000 *Phys. Rev. A* **61** 060705

[6] Heber O, Witte P D, Diner A, Bhushan K G, Strasser D, and Toker Y 2005 *Rev. Sci. Instrum.* **76** 013104

[7] Pedersen H B, Strasser D, Heber O, Rappaport M L, and Zajfman D 2002 *Phys. Rev.* A **65** 042704

# X-ray spectroscopy characterization of $Ar^{17+}$ produced by an ECRIS in the afterglow mode

**C. Prigent[1), E. Lamour[1), J. Mérot[1), B. Pascal[1), J.-P. Rozet[1), M. Trassinelli[1), D. Vernhet[1), J.-Y. Pacquet[2), L. Maunoury[3), F. Noury[3), J.-M. Ramillon[3)**

[1) Institut des NanoSciences de Paris, CNRS UMR 75-88, Université Pierre et Marie Curie, Paris 6, 75015 Paris, France

[2) GANIL, bd Henri Becquerel, BP 55027, F-14076 Caen cedex 05, France, EU

[3) CIMAP, 6 bd du Maréchal Juin, F-14050 Caen cedex 04, France, EU

Corresponding author's e-mail address: christophe.prigent@insp.jussieu.fr

**Abstract**. An electron–cyclotron–resonance ion source (ECRIS) operating in the afterglow mode allows to produce pulsed highly charged ion beams. We present a characterization of the first production of pulsed $Ar^{17+}$ ions from an ECRIS using X-ray spectroscopy techniques. An ion current increase of a factor 3-4 compared to the continuous operating mode with a short rise time (< 100 μs) and a full width at half maximum of 500 μs has been obtained.

## 1. Introduction

Electron – cyclotron – resonance ion sources (ECRIS) are devices designed to produce intense beams of highly charged ions (HCI). They are mirror machines where the plasma is heated by a radio frequency (RF) in the GHz range [1]. A resonant coupling between wave and electron occurs when electrons cross a particular magnetic surface – the resonant surface – where the Larmor frequency matches the wave frequency. It leads to an increase of the perpendicular velocity (relative to the magnetic flux lines) and the heated electrons are trapped in a magnetic bottle (minimum – B structure). When the RF power is turned off, the electrons cool down in a short time, the confinement gets worse and consequently the electron density is reduced. The potential barrier is therefore lost, and all ions can escape from the trap. The extracted ion current of heavy ions in high charge states can increase dramatically in a short time (around few ms), determined principally by the electron confinement lifetime. This scenario is well known as the afterglow pulse [2]. Effects occurring in the afterglow mode may be optimized to a few microamperes of pulsed particle current of high charge state and a few hundreds of μs long at a frequency of a few Hz. Several ways to increase the amplitude and/or to drop the pulse duration have been proposed and supported by experimental and theoretical studies, as adding a pulsed magnetic field to change the magnetic mirror gradient [3], or injecting a pulsed ion beam instead of neutral gas [4]. The afterglow mode allows good confinement to obtain the desired charge state, while continuous operating mode (cw) leads to a compromise between confinement and steady state extraction of ions.

At present, the afterglow operation mode in the ECRIS has been mostly developed in a few applications as beam-injector of high-energy (> MeV/A) accelerators. Acceleration of heavy ions demands very short pulses (<1 ms) in high charge states. In particular in the frame of the CERN program, lead ions in charge states up to 28+ have been obtained with a peak current around 100 eμA, a bunch duration of 400 μs and at a repetition frequency of 1 to 5 Hz [5, 6]. It is also of big interest for atomic physics experiments making direct use of H-like or even bare ions in the low energy regime (i.e. a few keV/q). However, control of pulse widths, optimization of the operation conditions and measurements of bare or H-like ions currents are not trivial. For most HCI of medium Z (like argon), a complete characterization of the ion beams delivered in the afterglow mode is needed, since the current is around a few enA or lower, and possible contamination by ions having close m/q ratio may occur. Classical measurement techniques, based on ion current measurements in a Faraday cup, lead to ambiguous results.

To overcome this problem, we make use of more sophisticated X-ray spectroscopy techniques. Having Ar ions of a few hundreds of keV impinging on a thin carbon foil, we have been able to characterize, for the first time, the production of $Ar^{17+}$ delivered in the afterglow mode. The ion-solid interaction leads to K-line emission coming from the deexcitation of projectile atomic states populated by charge transfer from the target (single or multiple capture processes occur). As the lifetime of these excited states is well below a ps, X–ray spectroscopy allows to perform measurement on ultra short time scale. Consequently, measurements of the X–ray yield over a few ms show a direct image of the ion bunch in terms of intensity and time distribution. Moreover, this emission provides a clear signature of both the atomic number **and** the charge state of the emitting projectile. Characteristic K lines in the 3-4 keV range have been observed, allowing for a precise identification and quantification of the $Ar^{17+}$ ion beam extracted from the plasma explosion after the RF power has been turned off.

In the next section, we will present the experimental set-up together with the main results. A conclusion with outlook will be given in the last part.

## 2. Experiment and results

The experimental set-up is schematically depicted in figure 1. The ECRIS used in these studies was the SUPERSHyPIE ion source [7] of GANIL (Grand Accélérateur National d'Ions Lourds) at the ARIBE facility (French acronym for "accelerator dedicated to research with low energy ions"). The RF power supply provides a microwave at 14.5 GHz frequency with a maximum power of 1 kW during the pulse. The repetition rate is fixed at 10 Hz for all the measurements reported here, with rise and fall time of the pulse generator (Tektronix AFG3021) of 18 ns. The time required to obtain the full RF power is 5 μs. The current set for the magnetic coils of the ECRIS is optimized in order to get the best possible plasma confinement (1.2 T at the injection side and 1 T at the extraction side). The extraction voltage is 15 kV and in order to keep it stable, a high voltage capacity of 1 μF has been added at the output of the High Voltage (HV) power supply. This capacity is mounted in parallel to the 30 MΩ resistance to the ground. For all experiments $O_2$ was used as support gas. The ion bunch current extracted from the source can be measured with a Faraday cup (F1 in fig. 1) located at the focal plane of the q/m analyzing dipole (D1 in fig. 1). The Faraday cup is connected to ground via a 100 kΩ resistance across which the voltage drop is read out with an oscilloscope (Tektronix TDS1012). From the first dipole to the collision chamber, the ion beam is transported along a beam line of 14 m equipped with several distributing dipoles (one used, D2 shown in fig. 1), 12 quadrupoles and 6 magnetic steerers. The vacuum after the first dipole is around $10^{-7}$ mbar while it reaches $10^{-9}$ in the last section of the beam line up to the collision chamber. The ion beam is directed in the collision chamber onto a thin (15 μg/cm$^2$) carbon foil. For beam current measurements, the target is removed and the beam enters a biased Faraday cup (F2). An estimation of the ion beam transmission may be extracted comparing the measured ion current in the first Faraday cup (F1) with the value recorded in the collision chamber (F2). We found 10 – 15 % in the cw mode as well as in the afterglow mode.

When the target is "in beam", the X rays characterizing the ion-solid interaction are recorded by a Silicon-Drift-Detector (SDD) (XFlash model detector from RONTEC) of well known efficiency [8]. The X-ray yield is measured also during operation in the continuous mode, thus allowing a well controlled relationship between instantaneous current and X-ray yield in the afterglow mode. A typical spectrum, showing np → 1s transitions in $Ar^{q+}$ ions resulting from multiple captures of target electrons in highly excited states of $Ar^{17+}$, is given in figure 2. No additional lines are recorded, proving without ambiguity the sole presence of $Ar^{17+}$ ions in the incident beam. Nevertheless, to avoid any spurious signal recording, an energy window on these transitions was set in the data acquisition system (DAS). Thus, only X-ray emission from 2.9 to 4 keV contributes to the recorded time spectra.

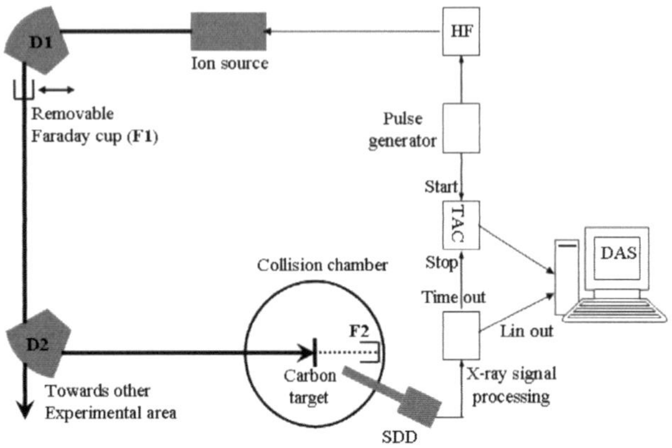


**Figure 1.** Scheme of the experimental set-up together with the associated electronics and data acquisition system (DAS).

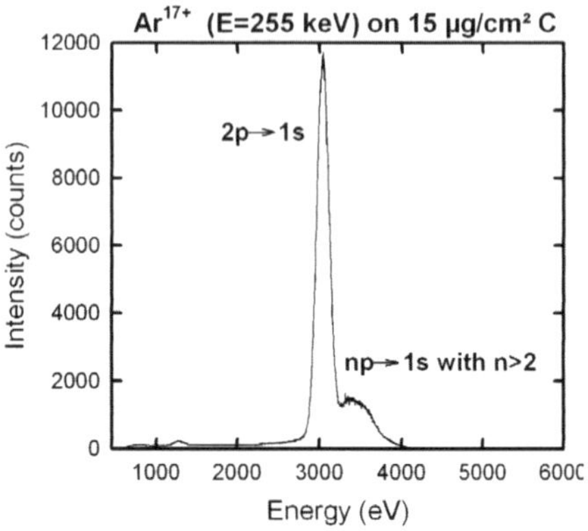


**Figure 2.** Typical spectrum showing the 2p→1s and np→1s with n > 2 transitions in $Ar^{q+}$ resulting from multiple capture of carbon target electrons in highly excited states of $Ar^{17+}$ ions at 255 keV.

In afterglow mode, X rays are recorded as a function of time delay after the end of RF pulse. A time to amplitude converter (TAC) of 2 ms dynamic range is started 300 µs before the end of the RF pulse and stopped by a time signal from the SDD with negligible delay. Time resolution can be as good as a few tens of ns, but was limited here to ~ 6 µs / channel. Great care was taken to insure that single stop condition over the dynamic range was achieved: a typical 0.1 probability of X-ray recording per shot was obtained by limiting the solid angle of the SDD with well defined slits.

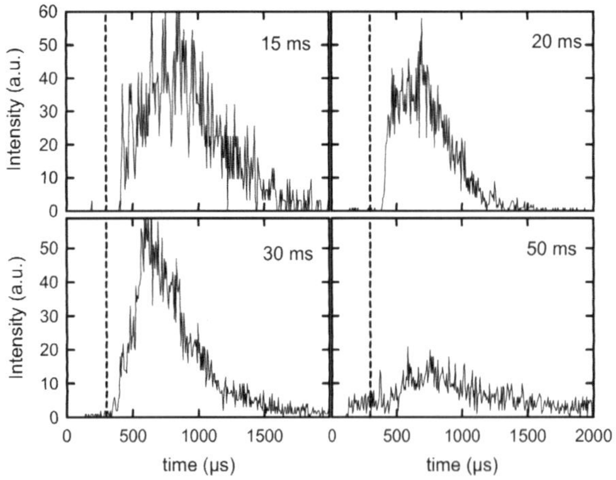


**Figure 3.** Time spectra for four HF pulse lengths (from 15 to 50 ms with a repetition rate of 10 Hz) exhibiting the X-ray time distribution after the turnoff of the RF power (indicated by vertical dashed lines).

| HF activation time (ms) | FWHM (µs) | Rising time (µs) | Decay time (µs) |
|---|---|---|---|
| 15 | 800 | 240 | 630 |
| 20 | 500 | 100 | 500 |
| 30 | 500 | 240 | 1000 |
| 50 | ~ CW | ~ CW | ~ CW |

**Table 1.** Full Width Half Maximum (in ms), rising and decay time (in µs) for different HF activation time.

Figure 3 shows spectra of the X-ray emission recorded for different RF pulse widths namely of 15, 20, 30 and 50 ms. These four spectra have been recorded for the same acquisition time (2000 s) and for constant values of the tuning parameters of the ECRIS. Measurements have been found highly reproducible and used as a control during two weeks for the investigation of the ion-cluster dynamics.

For RF pulse durations ≤ 30 ms, the afterglow current is found to be around 3-4 times higher than in cw mode, analogously to previous measurements performed with lower argon ion charge states [9]. Starting from 50 ms, the gain factor is not significant and the X-ray distribution tends towards a cw mode. The full width at half maximum of the afterglow bunch is shorter for 20 and 30 ms RF pulse duration (around 500 μs) than for 15 ms (800 μs). These results show evidence for an optimum pulse duration to efficiently and rapidly empty the ion reservoir in the confinement volume. Moreover, the rise time is found to be a bit faster when using a 20 ms RF pulse width than 15 or 30 ms (see Table 1). A short rise time may be interesting for measurements based on coincidence techniques as in ion – matter interaction studies.

We remark that the decrease of the ion current with the increase of the RF pulse length (heating time) is opposite to what has been previously measured [6]. The authors show a drop of a factor 5 in the ion current when decreasing the RF length from 50 to 20 ms with the same repetition rate. Contrary to these studies where medium ion charge states are considered (i.e. maximum in the ion charge state distribution as $Pb^{27+}$, $Xe^{20\text{-}22+}$ [10]), here, we explore, with $Ar^{17+}$ ions, the highly charged states. This opposite behaviour suggests different charge breeding mechanisms for these two kinds of charge state.

## 3. Conclusion

We have presented the first characterization of pulsed $Ar^{17+}$ ion beams from an ECRIS by x–ray spectroscopy techniques in the keV range. In our experimental conditions, an optimal RF width of 20 ms has been obtained leading to an ion current increase of a factor 3-4 with respect to the cw mode, a short rise time of less than 100 μs and a full width at half maximum of 500 μs.

Combined with theoretical developments, the use of x–ray spectroscopy in the afterglow mode could give additional information to climb a new step in understanding of plasma and highly charged ion production in the ECR ion sources. In particular, x–ray spectroscopy acts as a probe on short time scales of the plasma itself and sheds some light on the competition between the heating, the breeding and the de-confinement time.

## Acknowledgement

This work has been supported by the French national agency of research (ANR: Agence Nationale de Recherche) contract ANR 06 BLAN 0233.

[1] Geller R. et al., « ECRIS : the electron cyclotron ion source « , Annu. Rev. Part. Sci. **40**, 15 (1990).
[2] Briand P. et al., NIMA, **294**, 673 (1990).
[3] Hill. C.E. and Langbein K., Rev. Sci. Instrum., **67** (3), 1328 (1996). Mühle C. et al., Rev. Sci. Instrum, **67** (3), 1331 (1996).
[4] Shirkov G. et al. NIM B, **95**, 527 (1995). Shirkov G et al., Rev. Sci. Instrum., **67** (3), 1158 (1996).
[5] Sortais P., Rev. Sci. Instrum **63** (4), 2801 (1992).
[6] Hill C.E. et al., Rev. Sci. Instrum., **73** (2), 564 (2002).
[7] Maunoury L. et al., Rev. Sci. Instrum., **73** (2), 561 (2002).
[8] Lamour E. et al., submitted to Rev. Sci. Instrum.
[9] Müller L., et al, Rev. Sci. Instrum, **73** (3) 1140 (2002).
[10] Tinschert K., et al, Rev. Sci. Instrum, **75** (5) 1407 (2004).

# Highly Charged Ion Injector in the Terminal of Tandem Accelerator

**M. Matsuda, T. Asozu, T. Nakanoya, K. Kutsukake, S. Hanashima and S. Takeuchi**
Japan Atomic Energy Agency, Tokai Research and Development Center, Nuclear Science Research Institute, Tandem Accelerator Section
2-4 Shirakata-Shirane, Tokai, Naka, Ibaraki, Japan 319-1195

E-mail: matsuda.makoto@jaea.go.jp

**Abstract.** A highly charged heavy ion injector using an all permanent magnet type electron cyclotron resonance ion source (ECRIS) has been constructed in the high voltage terminal of the vertical and folded type 20UR Pelletron tandem accelerator at Japan Atomic Energy Agency at Tokai. The new in-terminal injector made it possible to accelerate highly charged heavy ions which have not been obtained from the tandem accelerator. Beam energy and beam intensity have been remarkably increased and the noble gas ion beams have become available.

## 1. Introduction

The tandem accelerator efficiently works on high energy heavy ion acceleration with an electron stripping carbon foil in the terminal which changes negative ions to multi-charged positive ions. However, carbon foils are short-lived under an intense heavy ion beam. It is difficult to greatly increase the beam intensity without loosing long time stability. On the other hand, charge states and intensities of the highly charged heavy ions from a high performance ECRIS are larger than those obtained after a charge exchange by carbon foil. In figure 1, charge states of ions with an intensity of several eμA expected for an ECRIS are compared with the most probable charge states obtained by electron stripping at a terminal voltage of 16 MV. For ions over a mass number of 100, the charge states higher than 20+ can be available from a 14.5 GHz ECRIS, compared to 13+ from the foil stripping. This fact motivated us to develop an in-terminal ECRIS (TECRIS) injector. Problems with stripper foil, such as short life time, energy straggling, emittance growth and beam intensity reduction could be cleared away also. We started with a 10 GHz compact ECRIS for this TECRIS project in 1998 [1], and succeeded in increasing the beam intensity and accelerating noble gas ions for many experiments. The TECRIS using a high performance 14.5 GHz ECRIS was newly developed as a second phase of this project.

## 2. Description of the TECRIS injector

The high voltage terminal is in a severe environment, i.e. it is filled with the pressurized insulation gas to 0.55 MPa(abs) and itself is held at a voltage of 20 MV at maximum. Therefore, the capability to withstand the high pressure and electrical discharges is indispensable to the in-terminal injector. We use a 14.5 GHz high performance all permanent magnet type ECRIS SuperNanogan [2], so that it can work within the limited electricity and installation space in the high voltage terminal.
We considered the following in the design of the TECRIS injector;

1) the system be simple, 2) the injector should coexist with the conventional tandem acceleration system, 3) the injector and related equipments be as compact as possible, 4) every thing be securely resistant against pressure and electrical discharge, 5) the system should work certainly, simply, stably and safely, 6) every thing should work for a long time without maintenance, at least for a period of the beam time of about four months.

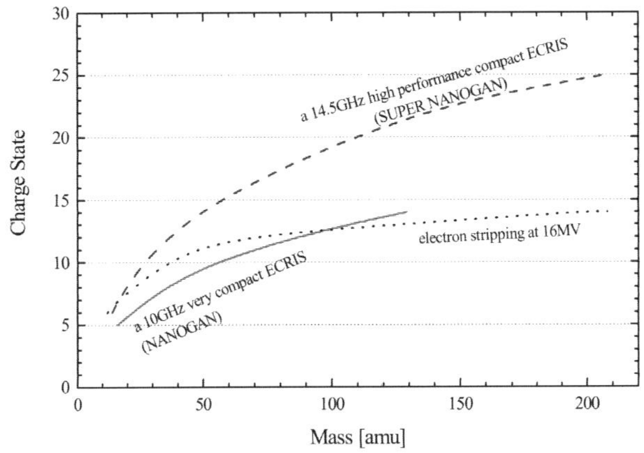


**Figure 1.** Charge states of ions with an intensity of several eμA expected for compact ECRISs are compared with the most probable charge states obtained at a terminal voltage of 16 MV.

*2.1. Layout and device control*

The layout of the TECRIS injector is illustrated in figure 2. In usual tandem acceleration, the negative ions enter from the low energy (LE) side acceleration tube, charge exchanged to a multi-charged positive ion by a foil stripper, charge-selected by the 180° bending magnet and re-accelerated by the high energy (HE) side acceleration tube. The injector was horizontally connected to the LE side beam line by the 90° injection magnet in order to coexist with the tandem acceleration mode. A schematic diagram of the TECRIS injector is shown in figure 3. There is an 80 kV deck for pre-acceleration. Power supplies for the 30kV extraction, einzel lens, steerer and 90° injection magnet, source gas control circuits and gas bottles were put in the 80 kV deck. All devices in the high voltage deck are controlled through an optically linked CAMAC communication system. The electric power is provided by means of an insulating transformer. The high voltage power supply for pre-acceleration, the RF amplifier, the cooling system, and the pump station are installed on the terminal potential.

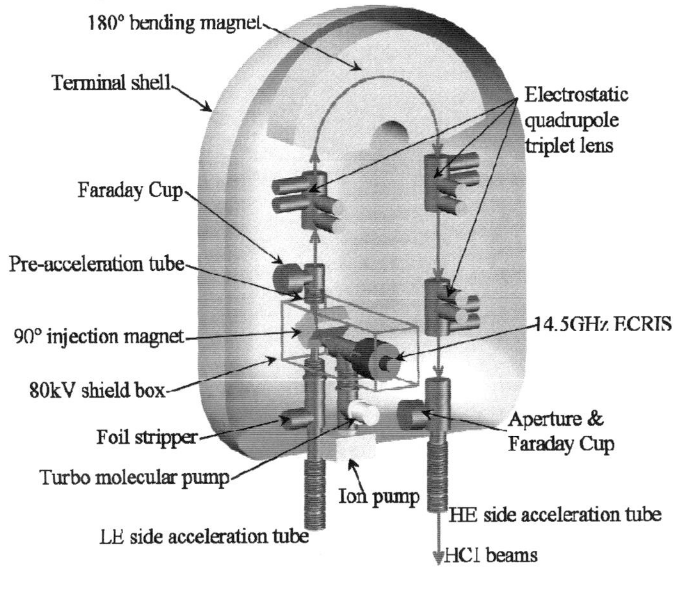


**Figure 2.** 14.5 GHz TECRIS injector developed for the tandem accelerator. The ion beam generated by the ECRIS is bent with 90° injection magnet and 180° bending magnet, and then accelerated through the main 20 MV acceleration tube toward the earth potential.

### *2.2. Ion source*

The operating conditions of the ECRIS were optimized prior to the installation in order to obtain a simple and reliable operation. The ion source is controlled only by adjusting the RF power and the voltage of the bias electrode and by selecting the source gas. A gas mixing method is used for the source gas. The gas bottles are filled by the mixed gas of the main gas (Ar, Kr, Xe or etc.) and the support gas ($N_2$ or $O_2$). A calibrated leak source with a shut-off solenoid valve was employed for simplification. Eight gas bottles with a calibrated leak valve were installed on the 80 kV deck, and each gas pressure was adjusted to the best flow rate through the leak valve. These gas flow rate were about $1{\sim}3\times10^{-5}$ Pa$\cdot$m$^3$/sec. Though the maximum performance was not always obtained, this fixed flow method was quite effective to make the ion source operation easy and certain. The beam current adjustment was mainly done by RF power control. The beam intensities were increased by a factor of ten using the bias method that applied a negative dc voltage to the center conductor of RF guide inserted into the plasma chamber. In addition to these gas bottles, one gas line has been installed with a thermo-mechanical-leak valve connected to motor driven sliding transformer for the gas flow control so that the condition is tuned to a high charge state more than 25+ for Xe etc.

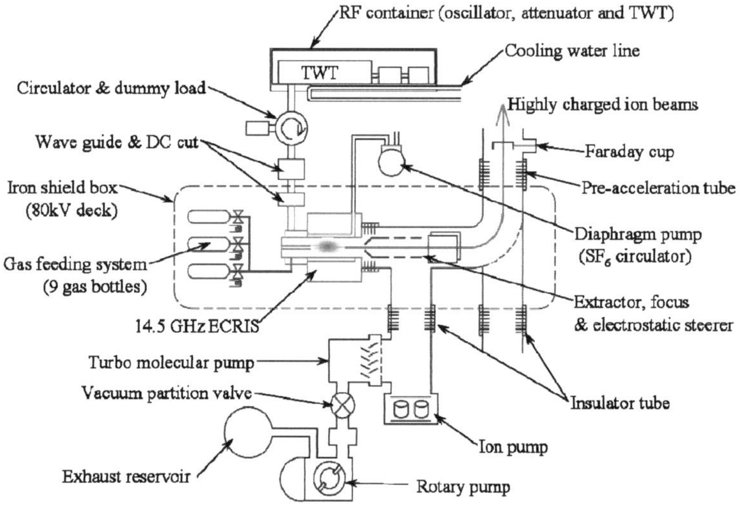


**Figure 3.** Schematic diagram of the TECRIS injector.

### *2.3. RF source*

The RF source is composed of a dielectric oscillator, a 200 W traveling wave tube amplifier and a variable attenuator. They were set in an airtight container to keep them an atmospheric pressure and prevent them from electrical discharges. The circulator to protect the amplifier from the reflected RF power was connected to the output port on the container. The RF amplifier of which power consumption is 1 kW is cooled by water. The plasma chamber of the ECRIS is cooled by $SF_6$ gas flow.

### *2.4. Pumping station*

Since the ion pumps do not work for noble gases, a turbo molecular pump (TMP) of a pumping speed of 450 l/sec and a rotary pump (RP) were employed for pumping out the gas from the source. These pumps have a protection against high pressure. The TMP was directly attached to the vacuum chamber without a gate valve. An electro-magnetic valve for the vacuum partition was located between the TMP and RP. The exhaust gas from the pumps was designed to accumulate in a closed vessel. The TMP driver was set in an airtight vessel to keep it atmospheric pressure and protect the complicated electric circuit from electric discharge. The system was arranged to work in fail-safe. Even when the pumping system fails, the high vacuum is held by the ion pump which is installed for holding the vacuum while the source is off.

### *2.5. Magnetic shielding for ECRIS*

For an ECRIS with of powerful permanent magnets, the influence of a leakage of magnetic flux on an ion beam can not be ignored. The magnetic field on vertical beam line was 2 mT at maximum. We made a shield panel which covered up the ion source and its extraction and focus chamber with 3.2 mm thick iron plates. With the iron shield, the leakage magnetic field was suppressed to about 1/10 and there was no beam steering problem in actual acceleration of negative ion beams.

## 3. Acceleration results and performance

Highly charged noble gas ions were accelerated from the TECRIS by the terminal voltage of 15 MV. The intensities of these ions are shown in figure 4. Ions of $Ne^{8+}$, $Ar^{12+}$, $Kr^{15+}$ and $Xe^{22+}$ were obtained with a beam intensity of 100 pnA or more, and these charge states agreed with the curve for the 14.5GHz ECRIS in figure 1. The ion source operation and beam handling of the new injector was easily carried out. The accelerated ion beams were stable and the beam transmission efficiency was nearly 100% for acceleration from the terminal to the exit of tandem accelerator. The intensities of highly charged heavy ions were increased by a factor of 10~100 compared with the 10 GHz ECRIS or the tandem acceleration mode. Xenon ion energy reached 420 MeV which was the highest energy this tandem accelerator has ever had, and it is possible to provide beams in a wide energy range of 50~500 MeV. The beam quality was greatly improved compared with the tandem acceleration mode because there was neither the energy spread nor the beam divergence due to the carbon foil.

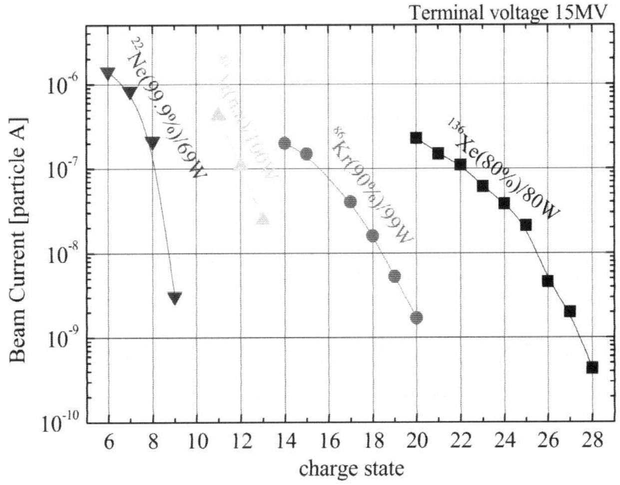


**Figure 4.** Charge state distributions of noble gas ions from 14.5 GHz TECRIS injector. Beam intensities were measured at the exit of tandem accelerator.

## 4. Conclusion

A 14.5GHz TECRIS injector was constructed at the high voltage terminal of the JAEA tandem accelerator. Highly charged heavy ions which were not obtained by the tandem acceleration mode in the past were successfully accelerated. We can use a high energy and intense beam without the problem with stripper foils. Highly charged and intense beams from TECRIS have become available for studying atomic and nuclear physics and radiation effects [3].

## References

[1] Matsuda M, Kobayashi C and Takeuchi S 1999 *Proc. of the 14th Int. Workshop on ECRIS,* CERN, Geneva 176-179
[2] PANTECHNIK S.A. www.pantechnik.fr
[3] Kawatsura K *et al* 2008 these proceedings

# The Commissioning Progress of the Cooler Storage Ring HIRFL-CSR in Lanzhou

**Xiaohong Cai[1], Jiawen Xia, Wenlong Zhan, Hushan Xu and the CSR Commissioning Group**

Institute of Modern Physics, Chinese Academy of Sciences, Lanzhou 730000, China

E-mail: caixh@impcas.ac.cn

**Abstract.** HIRFL-CSR is a multi-purpose cooler-storage-ring system constructed at the Institute of Modern Physics, Chinese Academy of Sciences in Lanzhou, China. Construction of the HIRFL-CSR storage ring complex has been finished, and passed the government check and acceptance, recently. The first stored beam in the CSRm was obtained in January 2006 using stripping injection. The commissioning got great progress in 2007. In early January 2007 electron cooling in CSRm was successfully done, and the multiple multi-turn injection was successively realized for $^{12}C^{6+}$, $^{36}Ar^{18+}$ and $^{129}Xe^{27+}$ beams, respectively. The $^{129}Xe^{27+}$ beams was extracted from the main ring by fast extraction. The 660MeV/u $^{12}C^{6+}$ was injected into the experimental ring and reached an intensity of 15mA there. The first two physics experiments were done in December 2007 including the mass measurement in isochronous mode in the experimental ring. The 300MeV/u $^{12}C^{4+}$ ions were successfully slow-extracted from the main ring in early 2008. This paper presents the main commissioning results.

## 1. Introduction

HIRFL-CSR is a multi-purpose cooler-storage-ring system in the National Laboratory of Heavy Ion Research (HIRFL) In Lanzhou, China [1,2] including a main ring (CSRm), an experimental ring (CSRe), and a radioactive beam line (RIBLL II) connecting the two rings. Heavy ion beams in an energy range of 8–30 MeV/u from the HIRFL will be accumulated, cooled and accelerated to the energy of 1 GeV/u ($^{12}C^{6+}$) and 500 MeV/u ($^{238}U^{72+}$) in the main ring, and then extracted to produce secondary beams (radioactive ion beam or highly charged heavy ions). The secondary beams will be accepted by and stored in the experimental ring for internal-target experiments or for high-precision spectroscopy with beam cooling. The experimental ring CSRe can accept highly charged ions with energies up to 750 MeV/u ($^{12}C^{6+}$) and 500 MeV/u ($^{238}U^{92+}$). The double-ring-system provides flexibility in the production of highly charged ions and of radioactive ion beams, thus offering opportunities for nuclear physics and atomic physics research as the facility at GSI in Germany [3]. Construction of the HIRFL-CSR storage ring complex has been finished, and passed the check and acceptance organized by the National Development and Reform Commission (NDRC) in July 30, 2008.

## 2. The commissioning of the HIRFL-CSR project

[1] To whom any correspondence should be addressed.

The commissioning of the HIRFL-CSR system began in 2005. The first stored beam in the CSRm was obtained on January 23, 2006 using 6.897 MeV/u $C^{4+}$ as the injected beam in combination with stripping injection mode. The commissioning got considerable progress in 2007 (Table 1). In early January 2007 electron cooling in CSRm was successfully done and 1GeV/u $^{12}C^{6+}$ ions up to 1.6mA (corresponding to the ion number of $8\times10^9$) have been accumulated by using cooling stacking. The multiple multi-turn injection (MMI) was then realized for $^{12}C^{6+}$, $^{36}Ar^{18+}$ and $^{129}Xe^{27+}$ (Fig.1) beams, respectively. In this injection mode the ions are injected into an unoccupied outer region of phase space with non-integer tune which ensures many turns before the injected beam re-occupies the same region [4].The first MMI was done for $^{12}C^{6+}$ in CSRm with e-cooling on April 1, and MMI for $^{36}Ar^{18+}$ in CSRm with e-cooling was tested on April 25 reaching a intensity of 180μA. For the MMI test for Xe ion, the 2.9MeV/u $^{129}Xe^{27+}$ ions provided by HIRFL-SFC was injected into CSRm, accumulated to 500μA ($1\times10^8$ ions) by using MMI and accelerated to 235MeV/u. The kicker for fast extraction with the parameters of $I_{max}$ = 2700A, $V_{max}$= 60 kV and rising time=150ns was tested on-line, and the 235MeV/u $^{129}Xe^{27+}$ beams was first extracted from CSRm by fast extraction in August 2007.

In October the 7MeV/u $^{12}C^{4+}$ provided by SFC was accumulated and accelerated to 660MeV/u in CSRm and then extracted fast to RIBLL II. The 660MeV/u $^{12}C^{6+}$ transferred through RIBLL II was injected into CSRe and reached an intensity of 15mA (corresponding to the ion number of $1.56\times10^{16}$pps) in CSRe (Fig.2). This was the first successful commissioning of the whole HIRFL-CSR system.

Table 1 Commissioning progresses of HIRFL-CSR project

| **Date** | **Commissioning conditions** | **Results** |
|---|---|---|
| January 23, 2006 | Stripping injection 6.897 MeV/u $C^{4+}$ | Stored beam in CSRm |
| January 2, 2007 | $^{12}C^{6+}$ | Cooling Stacking + Ramping, 1GeV/u , 1.6mA, CSRm |
| April 1, 2007 | $^{12}C^{6+}$ | multiple multi-turn injection (MMI), CSRm |
| April 25, 2007 | $^{36}Ar^{18+}$ | MMI, 1 GeV/u, 180μA, CSRm |
| June 25, 2007 | 2.9MeV/u $^{129}Xe^{27+}$ | MMI, 235MeV/u , 500μA, CSRm |
| August 4, 2007 | 235MeV/u $^{129}Xe^{27+}$ | Fast extraction from CSRm |
| October 6, 2007 | 660MeV/u $^{12}C^{6+}$ | Stored in CSRe,15 mA |
| December 12, 2007 | Radioactive ions produced, isochronous mode in CSRe | Mass measurements, $\Delta M/M\sim10^{-5}$ |

The first two physics experiments were done in December 2007. In one of the experiments a Be target of 5 mm was installed at the primary target position of RIBLL II, secondary beams such as $^{34}Cl$, $^{32}S$ and $^{30}P$ and so on were produced and injected into CSRe which was set in the isochronous mode. The masses of the stored secondary ions could be determined from the revolution time spectrum of the ions measured by using a time-of-flight (TOF) detector system. The stored ions went through a carbon foil of the TOF detector and emitted secondary electrons. The electrons were deflected by a perpendicular magnetic field and detected by a multi-channel plate giving the timing signal. The frequency resolution of the stored secondary ions reached $\Delta f/f \sim10^{-7}$, and the mass resolution of $10^{-5}$ was obtained (Fig.3).

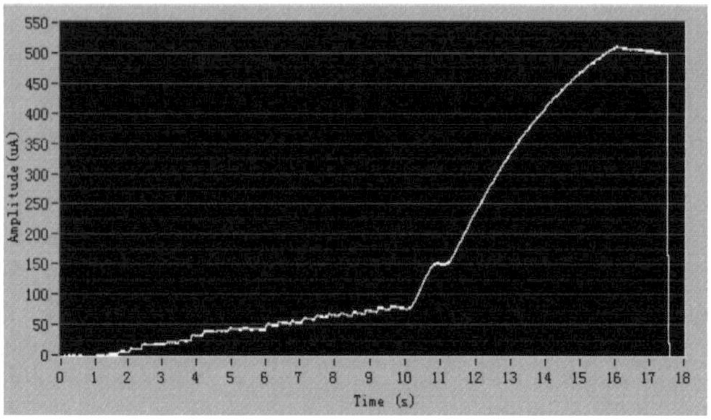


Fig. 1 Multiple multi-turn injection and acceleration in CSRm for 2.9 MeV/u $^{129}Xe^{27+}$

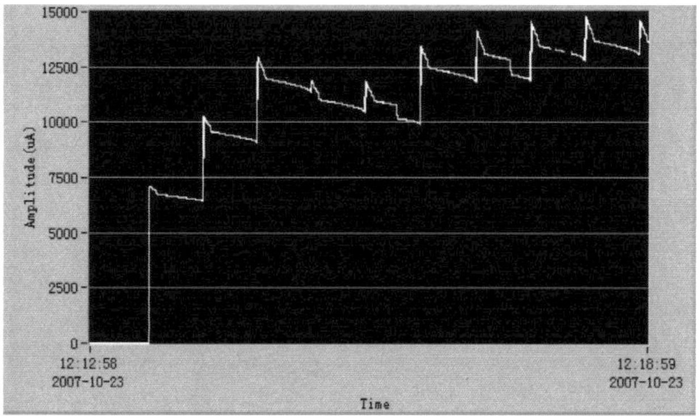


Fig. 2 Multi time injection in CSRe for 660 MeV/u $^{12}C^{6+}$

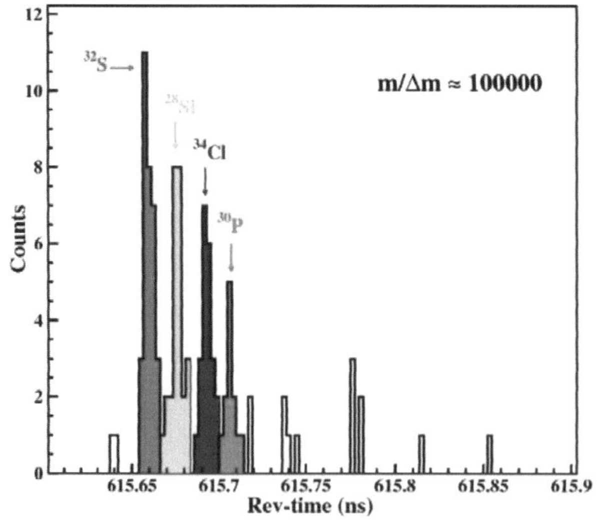


Fig. 3 Mass measurement in CSRe in isochronous mode

During the first mass measurement experiment, the HIRFL-CSR system was set in the full running mode shown in Fig. 4. In this full running mode the $^{36}Ar^{8+}$ ions produced by ECR source was first pre-accelerated to 2 MeV/u by HIRFL-SFC, and to 22MeV/u by HIRFL-SSC. The 22MeV/u $^{36}Ar^{8+}$ changed to $^{36}Ar^{18+}$ after passing through a stripper, and then injected into CSRm by MMI, accelerated to 368MeV/u by CSRm. The 368MeV/u $^{36}Ar^{18+}$ ions extracted from CSRm transferred through RIBLL II, and injected into CSRe for the mass measurement experiment.

In early 2008, the 300MeV/u $^{12}C^{4+}$ ions were successfully slow-extracted from CSRm. This is an important milestone of the HIRFL-CSR project which makes the external target experiments at RIBLL II and the cancer therapy at the therapy terminal possible.

The first phase experiments at the HIRFL-CSR include the internal target experiments in the CSRm, the external target experiments with RIBLL II and the internal target experiments in the CSRe. The experiments cover a wide range of hadron physics, radioactive ion beam physics, highly charged atomic physics, high energy density physics, biological medical physics and so on.

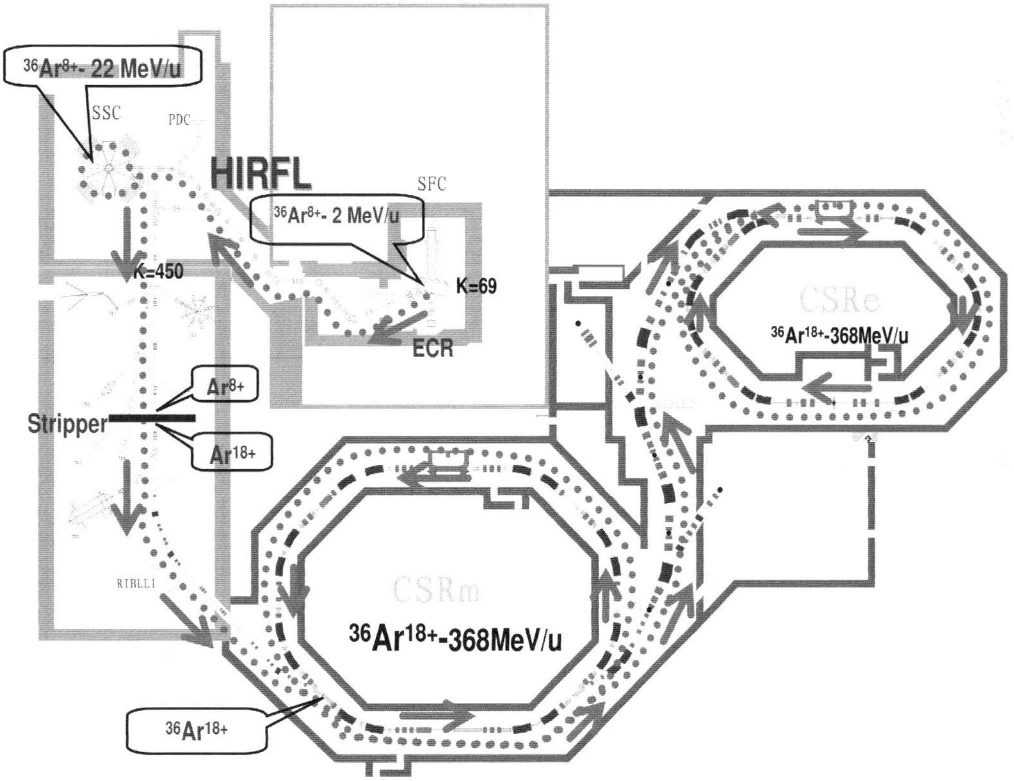


Fig. 4 Full running mode of the HIRFL-CSR system

## 3. References

[1] Xia J W, Zhan W L, Wei B W, Yuan Y J, Song M T, Zhang W Z, Yang X D, Yuan P, Gao D Q, Zhao H W, Yang X T, Xiao G Q, Man K D, Dang J R, Cai X, Wang Y F, Tang J Y, Qiao W M, Rao Y N, He Y, Ma L Z and Zhou Z Z, 2002 *Nucl. Instr. Meth.* A**488** 11
[2] Cai X, Lu R, Shao C J, Ruan F F, Yu D, Li M S, Torpokov D K, Shestakov Yu V, Sadykov R S, Zevakov S A, Xia J W and Zhan W L, 2007 *J. Phys. Conference Series*, **58** 55
[3] Kienle P, 1988 *Nuclear Physics* A**478** 84
[4] Yuan Y J, Xia J W, Zhang W Z, Song M T and Yang X D, 2001, *Nucl. Phys. Rev.* **18** 39

# Absolute detection efficiencies of an ion-counting system with a channel-electron multiplier

**Tetsuo Koizumi and Yuta Chihara**

Department of Physics, Rikkyo University, Tokyo 171-8501, Japan

E-mail:koizumit@rikkyo.ac.jp

**Abstract**. Absolute detection efficiencies of an ion-counting system have been examined for $Li^+$, $Rb^+$, $Cs^+$, $Ar^+$, and $Ar^{2+}$ ions as a function of the incident ion-impact energy. The ion-counting system consists of a converter plate (SUS), an electrostatic-lens system, and a channel-type secondary-electron-multiplier. For all ions examined in this study, the detection efficiencies are about unity if the ion energy is greater than 5 keV. The charge-state dependence is not as high for $Ar^+$ and $Ar^{2+}$.

## 1. Introduction

Ion detection in counting mode, using a channel-type secondary-electron-multiplier (CEM) or a microchannel plate (MCP), is commonly applied in ion–neutral-collision and photoabsorption experiments involving ions. In experiments where one requires the absolute number of ions produced, knowing the detector's absolute detection efficiency is essential. Because the detection efficiency depends on the kinetic energy of the ions as well as their nature, it is important to examine the energy dependence of the absolute detection efficiency for various ions.

Several experimental studies of the ion-detection characteristics of CEMs or MCPs in counting mode have been reported. Burrous et al. [1] and Ravon [2] studied the relative detection efficiencies of a channeltron for several ions. Absolute detection efficiencies were studied by Iglesias and McGarity [3] for $H^+$ of energy 0.2–10 keV, Crandall and Ray [4] for $H^+$ and $H^-$ of 0.3–5 keV, and Fricke et al. [5] for H, He, Ar, Kr, and Xe ions above 4 keV. Using MCPs, relative detection efficiencies have been measured by Luhmann et al. [6] for $Xe^{q+}$ ($q$ = 1, 2, 3) and by Yagi et al. [7] for Li, Ar, Ba, and Yb ions. Takagi et al. [8] reported the gain characteristics of a CEM and an MCP for multiply charged ions.

The purpose of the present study was to examine the absolute ion-detection efficiency in single-ion counting mode by an ion-counting system using a CEM (Channeltron 4139S). This counting system is the same as the ion detector used for the absolute photoionization cross-section measurements of $Xe^+$ [9]. The experiment was carried out with $Li^+$, $Rb^+$, $Cs^+$, $Ar^+$, and $Ar^{2+}$ ions in the energy range from 0.2 to 6 keV.

## 2. Experimental set-up

Figure 1 shows a schematic diagram of an ion-counting system and the detection-efficiency measurement set-up. The ion-counting system consists of a converter plate (SUS), an electrostatic-lens system, and a CEM. A surface-ionization-type ion source was used for singly charged alkali ions and

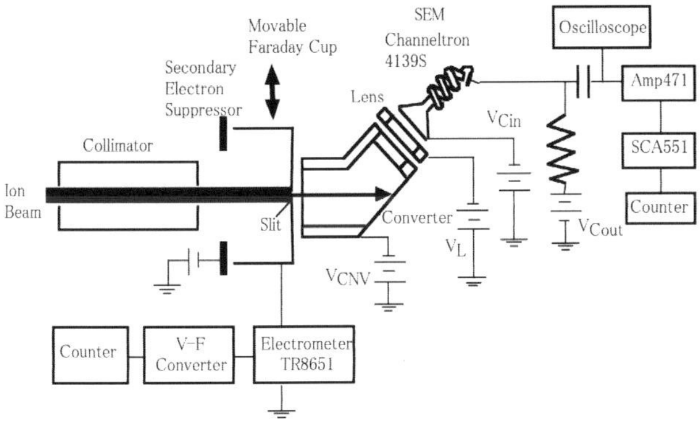


Fig.1. Schematic diagram of an ion-counting system and the detection-efficiency measurement set-up.

an Electron Cyclotron Resonance ion source (NANOGAN) for $Ar^+$ and $Ar^{2+}$. The incident ions passing through a narrow slit in the bottom of a movable Faraday cup hit the converter plate and the secondary electrons emitted from the converter are subsequently accelerated to the CEM. The output signals from the CEM are counted using conventional ion-counting electronics. The Faraday cup has an incident-beam aperture diameter of 16 mm and can be moved vertically along the direction normal to the ion beam with a linear-motion feedthrough.

The absolute detection efficiency, $D$, is defined as

$$D = \frac{qe}{I_f} \int n(x)dx \approx \frac{qe}{I_f} \sum_i \frac{n(x_i)}{w} \Delta x \quad , \qquad (1)$$

where $I_f$ is the ion current measured at the Faraday cup, $x$ the position of the slit, $n(x)$ the count rate from the CEM (counts $s^{-1}$; cps) at $x$, $x_i$ the $i$'th position of the slit, $\Delta x$ the stepping distance of the slit, $w$ the slit height, $q$ the charge state, and $e$ the elementary charge.

The total ion current, $I_f$, was measured by an electrometer (Advantest TR8651) calibrated with a standard power supply and a standard resistor. The typical total primary ion-beam current in this set-up is in the order of $10^{-14}$ to $10^{-13}$ A. The slit in the bottom of the Faraday cup is 16 mm wide and 0.030 mm high. The FWHM of the measured primary-beam profile is typically 1.5 mm. A very small fraction of primary ions passes through the slit and then hits the converter plate. The secondary electrons emitted from the converter are accelerated and focused on the CEM by the lens system. The acceleration voltage between the converter and the CEM, and the CEM bias voltage ($V_{\mathrm{Cin}}-V_{\mathrm{Cout}}$) are fixed at 1.9 and 2.6 kV, respectively. These voltages represent the optimum condition for ion counting. A typical ion-count rate at the beam's peak position is about $10^4$ cps. The measurements were made at several count rates, from 8000 to 15000 cps. Because we do not observe any count-rate dependence in this experiment, we can neglect ion-count loss for count rates of this order. The stepping distance of the slit is normally 0.10 mm. We have examined the effect of using a smaller stepping distance (0.04 mm), but no stepping-distance dependence was observed.

The main origin of the systematic experimental error is the determination of the slit height and beam fluctuations during the measurements. The slit height is determined experimentally as

$$w = \frac{1}{I_f} \sum_i I_C(x_i) \Delta x \quad , \qquad (2)$$

where $I_C(x)$ is the ion current of the converter plate at the slit position, $x$. In this measurement, $I_C$ can be measured in direct-current mode, because $I_f$ is in the order of $10^{-6}$ A. The slit height determined in this experiment is 0.0308±0.002 mm. Fluctuations in the ion beam are about 1%. The total systematic error in the detection efficiency is estimated at about ±2%.

## 3. Results and discussion

Figure 2 shows a typical pulse-height distribution (PHD) from a main amplifier (ORTEC471) for $Cs^+$ ions of different ion-impact energies. The bias voltage is fixed at 2.6 kV. All PHDs show peaked distributions. This is the result of saturation caused by the space-charge limitation of the secondary-electron yield in the channels. The peak positions of the PHDs do not depend on the ion-impact energy, except for the low-impact energy of 2.6 keV. The present result shows that the saturation distribution is reached at a bias voltage of 2.6 kV, except for the low-energy regime. It is necessary to set the discriminator level of a single-channel analyser (SCA:ORTEC 551) appropriately to obtain the correct detection efficiency. The results in Fig. 2 indicate that setting the discriminator level is not difficult for this system, and that any dark current or low-noise effect is negligible in this experiment.

Figures 3 and 4 show the absolute detection efficiencies, *D*, for alkali ions and for $Ar^+$ and $Ar^{2+}$, respectively, as a function of the ion-impact energy at the converter plate. Each point in the figures is the mean of several measurements. The estimated statistical uncertainties in this experiment were within ±2%.

*D* increases with increasing ion energy and approaches a constant value in the higher-energy regime. For all ions examined in this study, *D* is about unity if the ion energy is greater than 5 keV.

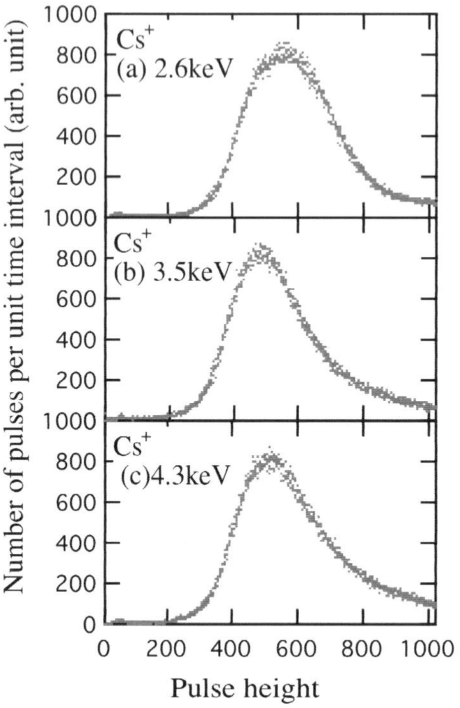


Fig. 2. Typical pulse-height distributions for $Cs^+$ ions with different ion impact energies: (a) 2.6 keV, (b) 3.5 keV, and (c) 4.3 keV.

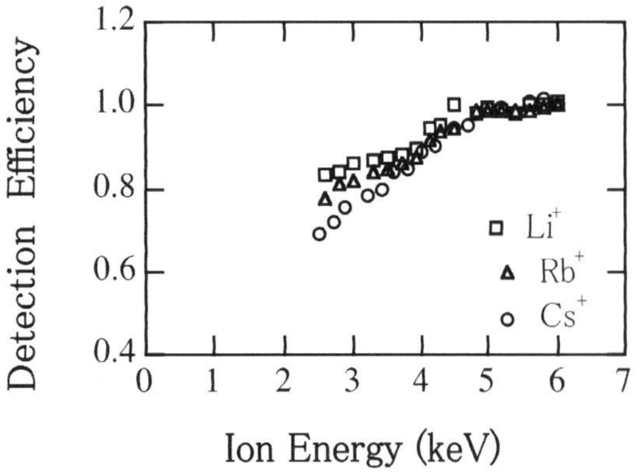


Fig. 3. Absolute detection efficiencies for $Li^+$, $Rb^+$, and $Cs^+$ as a function of the ion-impact energy.

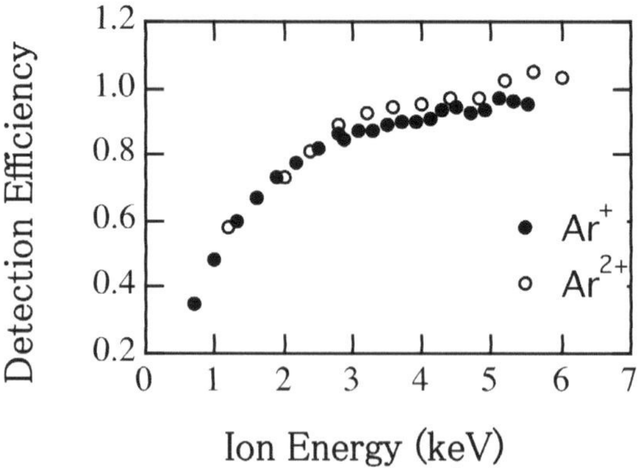


Fig. 4. Absolute detection efficiencies for $Ar^+$ and $Ar^{2+}$ as a function of the ion-impact energy.

According to the specifications of the channeltron used, its electron-detection efficiency has a peak value of 0.6–0.7 around 1 keV. To obtain $D$ = 1, more than one electron must be emitted from the converter plate by the ion impact. The average number of secondary electrons emitted from the converter depends on the ion-impact energy. This is the origin of the energy dependence of $D$.

The charge-state dependence is not as high for $Ar^+$ and $Ar^{2+}$. $D$ depends only on the ion energy for a given atomic species. This feature is in qualitative agreement with the results of Ravon [2], Luhmann et al. [6], and Yagi et al. [7]. This means that the kinetic-emission mechanism dominates secondary-electron emission in this energy range.

In the lower-energy regime, $D$ not only depends on the ion energy but also on the atomic mass. Lighter atoms have a higher $D$ for the same ion energy. Burrous et al. [1] proposed an empirical universal proportionality relation, as in

$$D \propto m^{\alpha} v \quad , \qquad (3)$$

where $v$ is the ion velocity, $m$ the atomic mass, and $\alpha = 0.4$. We obtained the same type of mass dependence, but with $\alpha \approx 0.48$, in this experiment.

## References

[1] C. N. Burrous, A. J. Lieber, and V. T. Zaviantseff; Rev. Sci. Instr. **38** (1967) 1477-1481.
[2] J. P. Ravon; Nucl. Instr. and Meth. **211** (1983) 7-15.
[3] G. E. Iglesias and J. O. McGarity; Rev. Sci. Instr. **42** (1971) 1728-1729.
[4] D. H. Crandall and J. A. Ray; Rev. Sci. Instr. **46** (1975) 562-564
[5] J. Fricke, A. Müller, and E. Salzborn; Nucl. Instr. and Meth. **175** (1980) 379-384.
[6] T. Luhmann, Ch. Gerth, M. Groen, M. Martins, B. Obst, M. Richter, and P. Zimmermann; Phys. Rev. A**57** (1998) 282-291.
[7] S. Yagi, T. Nagata, M. Koide, Y. Itoh, T. Koizumi, and Y. Azuma; Nucl. Instr. and Meth. B **183** (2001) 476-486.
[8] S. Takagi, T. Iwai, Y. Kaneko, M. Kimura, N. Kobayashi, A. Matsumoto, S. Ohtani, K. Okuno, H. Tawara, and S. Tsurubuchi; Nucl. Instr. and Meth. B **215** (1983) 207211.
[9] Y. Itoh, A. Ito, M. Kitajima, T. Koizumi, T. M. Kojima, H. Sakai, M. Sano, and N. Watanabe; J. Phys. B:At. Mol. Opt. Phys. **34** (2001) 3493-3499.

Journal of Physics: Conference Series **163** (2009) 012115 doi:10.1088/1742-6596/163/1/012115

# Development and applications of electron beam ion source for nanoprocesses

**M Sakurai[1], M. Tona[2], H. Watanabe[3], Nobuyuki Nakamura[3], Shunsuke Ohtani[3], T.Terui[4], S. Mashiko[4] and H. A. Sakaue[5]**

[1] Department of Physics, Kobe University, Kobe 657-8501, Japan

[2]Department of Chemistry, Kobe University, Kobe 657-8501, Japan

[3]Institute for Laser Science, University of Electro-Communications, Chofu 182-8585, Japan

[4]National Institute of Information and Communications Technology, Kobe 651-2492, Japan

[5]National Institute for Fusion Science, Oroshi-cho, Toki 509-5292, Japan

E-mail:msakurai@kobe-u.ac.jp

**Abstract.** An electron beam ion source (EBIS) was developed and applied to create nanostructures. The EBIS uses a commercial superconducting magnet (3T) with vacuum system independent from another vacuum system which comprises all of the ion source constituents except the magnet. The attained parameters of electron beam are emission current of 180mA and acceleration voltage of 20kV. The EBIS creates HCIs in nA region (e.g. 2nA for $Ar^{12+}$). Higher charge states are suppressed by residual gases. The HCIs were irradiated to HOPG and the structure of irradiated surface was observed by STM, and the image indicates the fluence as much as $10^{13}$ion/cm$^2$ is accessible with the present EBIS.

## 1. Introduction

The interaction of slow highly charged ions (HCIs) with solid surfaces is useful for 'nanoprocess'; the modification, activation, machining and analysis in nanometer scale. The most serious problem in the application of HCIs for nanoprocesses is the limited intensity of HCI beam per specific area in nanometer size. Since the emittance of EBIS is not sufficiently small to focus the beam into nanometer size, screening aperture has to be used to create nano-beam. Assuming that the typical beam size of EBIS is 0.1mm in diameter, the beam intensity must be as high as $10^8$ ion/s in order to inject an ion over 10x10 nm$^2$ area every second. An electron beam ion source, 'Kobe EBIS' has been developed to improve the beam intensity problem [1-3]. In the present paper, the final design after several remodeling and the present performance of the EBIS is described and some data of applications for irradiation of samples by HCIs are presented.

## 2. Kobe EBIS

In order to reduce the running cost and simplify the operational procedure, the Kobe EBIS uses a commercial superconducting magnet (3T) with a bore of 180 mm in diameter cooled by a

[1] To whom any correspondence should be addressed.

closed-cycle refrigerator. The schematic diagram of the EBIS after the remodeling is shown in figure 1. The detail of the original design including electron beam and ion beam simulation is described on the literature [1,2]. All the components except the magnet are made bakable up to 250 deg C in order to minimize the contribution of background pressure when the EBIS is in operation. Electron gun and collector, which are operated on high voltage, are cooled by circulating coolant during operation, while the coolant has to be removed (not stick to apparatus) during the bakeout procedure. So, we used Vertrel XF (DuPont-Mitsui Fluorochemicals) as coolant because of its high vapor pressure and high electric resistivity. Prior to the operation, discharge cleaning of the drift tubes is also necessary, and is performed applying high voltage (5kV) to drift tubes under the maximum magnetic field.

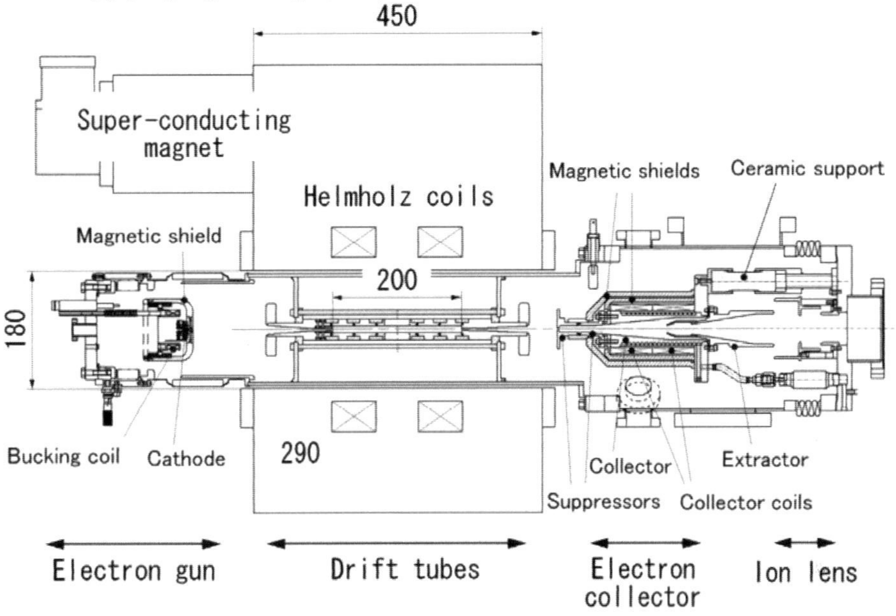


**Figure 1.** Schematic diagram of Kobe EBIS

After the initial operation, we found some errors in the calculation of the magnetic field distribution revealing that magnetic shields at the electron gun and collector were not efficient enough. The electron beam current was limited to less than 20 mA due to the return current of electrons (going back to the electron gun), which was not collected by the collector. To solve this problem, the electron gun has been moved away from the superconducting magnet by 9 cm and the magnetic shield at collector has been reconstructed with thicker shield structure. The pair of solenoid coils has been inserted inside the shield envelope and the ceramic supports have been reinforced [3]. The present device has been designed to perform degassing treatment thoroughly. Consequently, the maximum electron current has been increased up to 180mA. The acceleration voltage has been set to 20kV to insure stability of the whole system.

We have measured the charge state distribution of extracted ions introducing Ar gas at $5\times10^{-8}$ Pa over a base pressure of $\sim1\times10^{-7}$Pa, as shown figure 2. The base pressure of the ion source in the presence of electron beam is $5\text{x}10^{-8}$ Pa $\sim1\text{x}10^{-7}$ Pa. The electron beam current was 150mA and acceleration voltage was 18kV. The potential of the drift tubes was 3kV. The ion current was measured by a Faraday cup located just after the exit slit of an analyzing magnet as a function of the current of the analyzing magnet. In order to maximize the ion currents, the slits were fully open. This leads to a low mass resolution in the spectrum of figure 2 where peaks of $Ar^{q+}$ ($q\leq12$) can be identified. The peak intensities of $Ar^{12+}$, $Ar^{11+}$ and $Ar^{9+}$ ions are 2.0, 3.9 and 9.4nA, respectively. An adjustment of the magnetic field axis and the electron beam direction within an error of 0.1 mm is mandatory to reach such current intensities of the extracted ions. The peaks of ions which are ascribed to residual gas elements exist, and get dominant at lower m/q region. By the measurement with higher resolution (narrower slit width), the peak of $Ar^{16+}$ was observable. It is essential to improve the vacuum for producing higher charge states abundantly. The charge state distribution strongly depends on the vacuum inside the drift tubes, and the most effective measure to improve the vacuum is to cool the drift tubes down to cryogenic temperature. The main residual gases under operation are $H_2$, CO and $CO_2$. Especially the partial pressure of CO and $CO_2$ increases during the presence of electron beam. So it is desirable to cool the drift tubes lower than 20K where the vapor pressure of CO is sufficiently low.

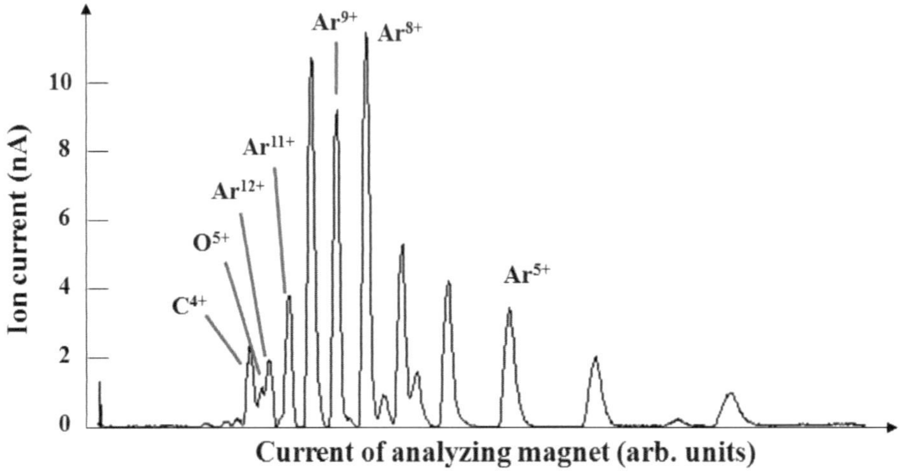


**Figure 2.** Charge state distribution of extracted Ar HCIs. For detailed description of the running conditions of the EBIS, see text.

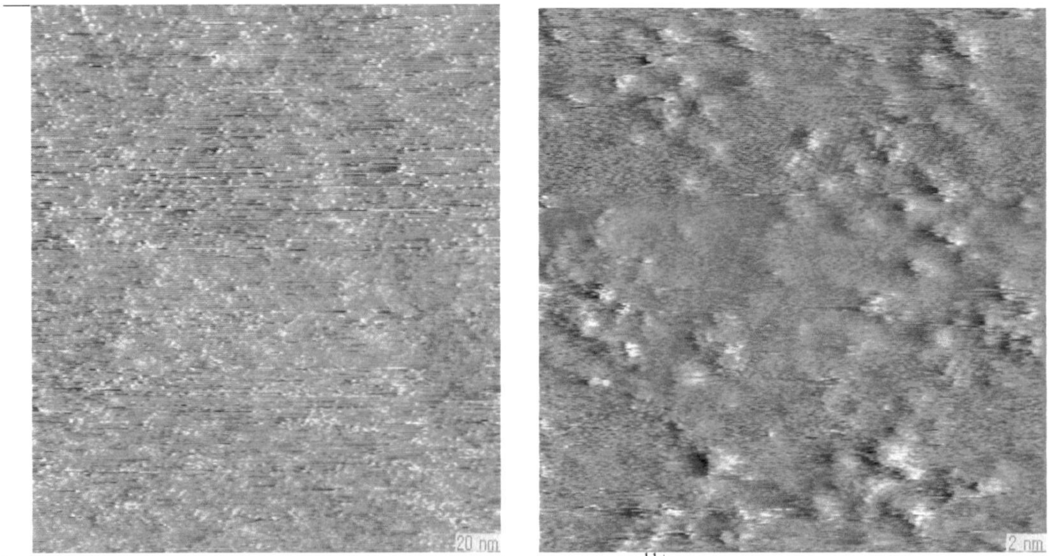


**Figure 3.** STM image of the HOPG surface irradiated with $Ar^{11+}$ ions. The fluence of irradiated HCIs is estimated at $10^{13}/cm^2$.

## 3. Irradiation experiments

The irradiation experiments were performed using $Ar^{11+}$ ions. Concerning the experimental setup, mass separated HCIs are focused through an electrostatic lens system and introduced to an experimental chamber equipped with a precise sample holder [4]. The irradiated sample was transferred to a STM system keeping UHV environment. Target of highly oriented pyrolytic graphite (HOPG) was used and the surface images after irradiation were measured by a scanning tunneling microscope (STM). The STM images of HOPG surface irradiated with $Ar^{11+}$ ions for about 1 hour are shown in figure 3. The ion current from the sample was 0.3 nA. The images exhibit many protruded structures which are typical as irradiation traces on HOPG [5]. The fluence of irradiated HCIs is estimated at $10^{13}/cm^2$ which corresponds to an irradiation density of 10 traces per $100nm^2$ area. For future applications of nanoprocesses induced by HCIs, the flux of the beam must be sufficiently high in order to irradiate nm size regions within a realistic time. There should be some application with the present status, however, further improvement of the beam flux at the sample by 1~2 orders magnitude is necessary for the general use of HCI for nanoprocess applications.

## References

[1] Sakurai M, Nakajima F, Fukumoto T, Nakamura N, Ohtani S and Mashiko S 2004 *J. Phys. Conf. Ser.* **2** 52-56

[2] Sakurai M, Nakajima F, Fukumoto T, Nakamura N, Ohtani S, Mashiko S and Sakaue H 2005 *Nucl. Instr. Meth.* B**235** 519

[3] Sakurai M, Kawase Y, Mitsumori K, Sakaue HA, Watanabe H, Nakamura N, Ohtani S, Terui T, Mashiko S 2007 *Shinku* **50** 390 [in Japanese]

[4] Sakurai M, Tona M, Takahashi S, Watanabe H, Nakamura N, Yoshiyasu N, Yamada C, Ohtani S, Sakaue HA, Kawase Y, Mitsumori K, Terui T and Mashiko S 2007 *Nucl. Instr. Meth.* B**257** 297

[5] Mochiji K, Yamamoto S, Shimizu H, Ohtani S, Seguchi T and Kobayashi N 1997 *J. Appl. Phys.* **82** 6037

Journal of Physics: Conference Series **163** (2009) 012116 doi:10.1088/1742-6596/163/1/012116

# Investigation of ions emitted from a tin fuelled laser produced plasma source

**A O'Connor, P Dunne, O Morris, F O'Reilly, G O'Sullivan, and E Sokell**

School of Physics, University College Dublin, Ireland

E-mail: aodh.oconnor@ucd.ie

**Abstract.** In this work ions emitted from a laser produced plasma created from a solid tin target have been analyzed using a spherical sector energy analyzer (ESA). Employing time of flight analysis, the ESA allows determination of ion energy and charge stage as a function of ion emission angle. Ions of charge Sn $^{+1}$ - Sn$^{+9}$ with energy to charge ratios of 0.22 - 3.2keV have been investigated over a range of emission angles from 20 - 80 degrees, relative to the target normal, for three distinct laser pulse energies.

## 1. Introduction

With the commercial drive to create ever smaller microprocessor features, in accordance with Moore's law, the physical limits of current photolithographic methods are fast approaching. Extreme Ultra-Violet (EUV) lithography at 13.5 nm is favoured as the successor to current techniques for printing circuits at the 32 nm node and beyond [1]. As EUV radiation is absorbed strongly by all materials, Bragg reflectors must be employed as collector optics. Mo/Si multilayered Bragg reflectors are the driving factor in the choice of EUV wavelength, as they possess a high reflectivity at 13.5 nm within a 2% bandpass. Tin and xenon are candidate targets for laser produced plasma based EUV sources as they emit strongly in the 13.5 nm region [2, 3]. Laser produced plasmas (LPP's) that are favourable for EUV emission emit energetic ions [4, 5, 6]. Such energetic ions would sputter EUV collector optics, greatly reducing their reflectivity [7]. For laser produced EUV sources to be viable for high volume manufacture a source cleanliness of $\geq$30,000 hours is required [1, 7]. In order to preserve collector optic lifetime, a number of debris mitigation techniques have been proposed [8]. To inform the optimization of such debris mitigation techniques, the energy and charge of ions ejected from LPP's that emit strongly at 13.5 nm has been characterized using a spherical sector energy analyzer for ion emission angles spanning $20 - 80$ degrees to the target normal. Previous studies on solid tin targets have demonstrated a strong dependence of the conversion efficiency, into the 13.5 nm band, on laser power density [3], thus the ion characterization measurements were performed over a range of laser power densities.

## 2. Experiment

The experimental arrangement is similar to one described previously [9]. A Comstock Model AC-902B Electrostatic Spherical sector energy Analyzer (ESA) was employed to characterize ions ejected from tin fueled laser produced plasmas according to their energy and charge state.

The ESA consists of two concentric hemi-spherical plates between which a potential difference is applied, so that only ions of a particular energy to charge ratio ($\frac{E}{q}$) reach the detector. The detector is a pair of micro-channel plates (Comstock Model CP-602B) and the resulting signals are recorded on a digital oscilloscope. Time of flight analysis is used to differentiate between ions of different charge states with the same value of $\frac{E}{q}$.

The plasmas were produced using a Nd:YAG laser that delivered 60 - 325 mJ per pulse, in a pulse duration of 7 ns (full-width at half-maximum intensity) at 1064 nm and over a spot size with a diameter of approximately 200 $\mu$m . As previously described, a rotating stage is used to mount a solid planar tin target [9]. The axis of rotation of the stage is centered along the optical axis to ensure that, after three reflections, the incoming laser beam strikes normal to the target face irrespective of any rotation undergone by the stage. This arrangement allows the fixed in space ESA to be exposed to ions emitted at a range of angles to the target normal.

For each spectrum recorded the tin target was shot twice with the laser. The initial shot cleared the oxidation and surface defects from the target face and the second was used to record the ion spectrum. Five individual ion spectra were recorded for each ($\frac{E}{q}$) and emission angle setting. The ions were banded according to their charge and these signals integrated. Over the five shots the signal area for each charge state remained relatively consistent, the average being used as a measure of the ion signal for that ($\frac{E}{q}$) and angle. Ion signals were recorded for emission angles from 20 to 80 degree angles, relative to the target normal, over a energy/charge range of 0.22 - 3.2 keV. These measurements were recorded for three laser pulse energies (60, 105 and 325 mJ). Each laser energy corresponded to a distinct power density, of the order of 2.5, 4.5, and 14 $\times$ $10^{10}\frac{\text{W}}{\text{cm}^2}$ respectively, the spot size and pulse duration being kept constant throughout.

## 3. Results

### *3.1. 60mJ Laser Pulse Results*

In Fig.1 energy distributions for ions emitted from a tin plasma created using a 60mJ laser pulse are plotted for angles of emission from 20 to 80 degrees, with respect to the target normal. With increasing angle, the number of ions ejected decreases. However, the ion energy distribution remains similar from 20 - 50 degrees. Beyond 50 degrees the mean ion kinetic energy shifts to lower energies but the energy distribution profile is essentially the same shape for all emission angles. These results would indicate that ions are preferentially ejected close to target normal, parallel to the incoming laser, with the more energetic ions tending to be emitted in this direction.

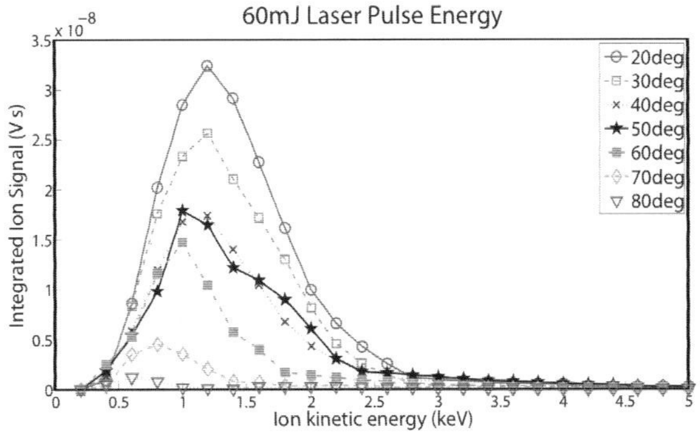


**Figure 1.** Energy distributions for ions emitted from plasmas created with a 60mJ laser pulse energy, data are shown for angles of emission from 20 to 80 degrees.

### *3.2. 105mJ Laser Pulse Results*

Increasing the laser energy from 60mJ to 105mJ, Fig.2, gives rise to an increase in the number of ions ejected as well as the mean kinetic energy of the ions. As with the results obtained for a 60mJ pulse energy, the ion energy distribution remains relatively similar in shape over all angles, but again the data show preferential ion emission close to target normal. The average ion kinetic energy is once again shifted to lower energies with increasing angle, although this effect is apparent over all angles and not just above 50 degrees as observed for 60mJ pulses (Fig.1).

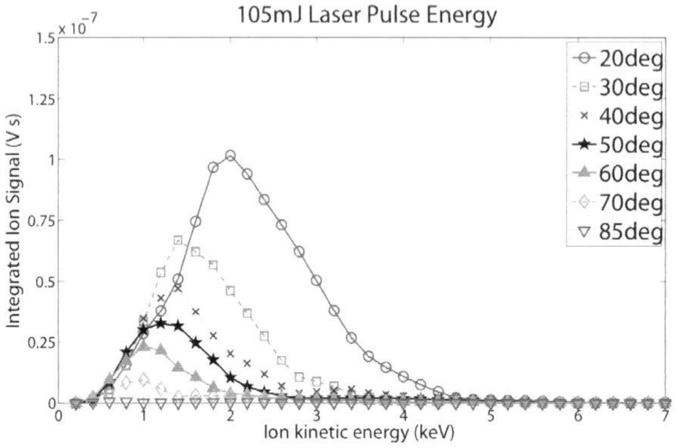


**Figure 2.** Energy distributions for ions emitted from plasmas created with a 105mJ laser pulse energy, data are shown for angles of emission from 20 to 85 degrees.

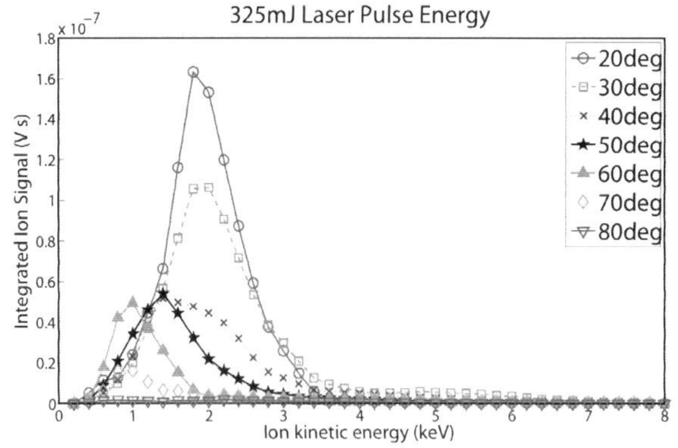


**Figure 3.** Ion energy distributions for ions emitted from tin plasmas created using 325mJ laser pulses.

### *3.3. 325mJ Laser Pulse Results*

The highest pulse energy used was 325mJ, Fig.3 shows that further increasing the laser energy does not result in an increased mean ion kinetic energy at 20 degrees. However, some of the additional energy appears to be distributed over ions ejected in the 30 to 50 degree range, shifting their mean kinetic energy to larger values in comparison with the data shown in Fig.2. The number of ions ejected has once again increased and the preferential ejection of energetic ions close to the target normal continues. At this increased pulse energy, the peak number of ions ejected at 20 and 30 degrees is increased relative to those emitted at larger angles.

Increasing the pulse energy from 105mJ to 325mJ resulted in no significant difference in the maximum ion kinetic energy, which was found to be somewhat short of 10keV. The maximum energy recorded for the 60mJ pulse energy being lower at approximately 5keV. This is in agreement with previous work, performed on tin plasmas with a Nd:YAG laser power density of $7 \times 10^{10} \frac{\mathrm{W}}{\mathrm{cm}^2}$, which indicated a maximum kinetic energy of 7keV [4].

## 4. Conclusion

The energy and angular distributions of ions emitted from tin plasmas created using Nd:YAG laser pulses have been recorded. The number of ions as well as their mean kinetic energy was found to increase with increased laser power density. The energy distribution profiles for each angle studied were all relatively similar in shape, although the mean ion kinetic energy showed dependence on both laser pulse energy and emission angle. In all cases studied, there was evidence for preferential ion emission close to target normal (parallel to the incoming laser beam), along with a decrease in ion energy with increased angle from target normal. The maximum kinetic energy of the ejected ions appears to plateau at laser pulse energies between 105mJ and 325mJ. These measurements provide information on ejected ion energies and angular distributions that should inform debris mitigation strategies for efficient, next generation EUV lithography sources.

## 5. Acknowledgements

This work was supported by Science Foundation Ireland grant number 06/RFP/PHY017 and UCD seed funding. We are grateful to Dave Ruzik and members of his group at the University of Illinois Urbana-Champaign for supply of the ESA.

## References

[1] International Technology Roadmap for Semiconductors Webpage. Itrs reports lithography edition, 2007.

[2] N. Bowering, M. Martins, W.N. Partlo, and I. V. Fomenkov. 2004. Extreme ultraviolet emission spectra of highly ionized xenon and their comparison with model calculations. *J. Appl. Phys.*, **95** 51–52.

[3] P. Hayden, A. Cummings, N. Murphy, G. O'Sullivan, P. Sheridan, J. White, and P. Dunne. 2006. 13.5 nm extreme ultraviolet emission from tin based laser produced plasma sources. *J. Appl. Phys.*, **99** 093302.

[4] T. Higashiguchi, C. Rajyaguru, N. Dojyo, Y. Taniguchi, K. Sakita, W. Sasaki, and S. Kubodera. 2005. 13.5 nm extreme ultraviolet emission from tin based laser produced plasma sources. *Rev. Sci. Instruments*, **76** 126102.

[5] E. Antonsen, K C. Thompson, M. R. Hendricks, D. A. Alman, B. E. Jurczyk, and D. N. Ruzic. 2006. Ion debris characterization from a z-pinch extreme ultraviolet light source. *J. Appl. Phys.*, **99** 063301.

[6] K. Takenoshita, C. Koay, S. Teerawattansook, and M. Richardson. 2005. Mitigation of fast ions from laser produced sn plasma for extreme ultraviolet source. *Emerging Lithographic Technologies IX*, **Proceedings of SPIE Vol. 5751**.

[7] V. Bakshi. 2006. *EUV sources for lithography*. SPIE Press.

[8] S. S. Harilal, B O'Shay, and M. Tillack. 2005. Debris mitigation in laser-produced tin plume using magnetic field. *J. Appl. Phys.*, **98** 036102.

[9] O. Morris, P. Hayden, P. Dunne, F. O'Reilly, G. O'Sullivan, E. Sokell, E. L. Antonsen, S. N. Srivastava, K. C. Thompson, and D. N. Ruzic. 2007. Determination of charge state, energy and angular distributions of tin ions emitted from laser produced plasma based euv sources. *J. Phys.: Conf. Ser*, **58** 391–294.

# AUTHOR INDEX

# AUTHOR INDEX

# AUTHOR INDEX

# AUTHOR INDEX

# AUTHOR INDEX